EVOLUTION OF
PRIMARY PRODUCERS
IN THE SEA

EVOLUTION OF PRIMARY PRODUCERS IN THE SEA

Edited by

PAUL G. FALKOWSKI

Institute of Marine and Coastal Sciences and
Department of Geological Sciences
Rutgers University
New Brunswick, New Jersey

ANDREW H. KNOLL

Department of Organismic and Evolutionary Biology
Harvard University
Cambridge, Massachusetts

ELSEVIER

AMSTERDAM • BOSTON • HEIDELBERG • LONDON
NEW YORK • OXFORD • PARIS • SAN DIEGO
SAN FRANCISCO • SINGAPORE • SYDNEY • TOKYO
Academic Press is an imprint of Elsevier

Elsevier Academic Press
30 Corporate Drive, Suite 400, Burlington, MA 01803, USA
525 B Street, Suite 1900, San Diego, California 92101-4495, USA
84 Theobald's Road, London WC1X 8RR, UK

This book is printed on acid-free paper. ∞

Library of Congress Cataloging-in-Publication Data
Evolution of primary producers in the sea / editors, Paul G. Falkowski, Andrew H.
    Knoll. p. cm.
  Includes bibliographical references and index.
  ISBN-13: 978-0-12-370518-1
  1. Marine productivity. 2. Marine plants–Evolution. 3. Bacteria, Autotrophic–Evolution.
I. Falkowski, Paul G. II. Knoll, Andrew H.
  QH91.8.M34E97 2007
  577.7′15–dc22   2007025654

British Library Cataloguing in Publication Data
A catalogue record for this book is available from the British Library.

ISBN: 978-0-12-370518-1

For all information on all Elsevier Academic Press publications
visit our Web site at www.books.elsevier.com

Printed and bound by CPI Group (UK) Ltd, Croydon, CR0 4YY
Transferred to digital print 2012

Contents

Contributors

Numbers in parentheses indicate the pages on which author's contributions begin.

Rolf S. Arvidson (377), Department of Earth Science, Rice University, Houston, Texas

Marie-Pierre Aubry (251), Department of Geological Sciences, Rutgers University, Piscataway, New Jersey

Sandra L. Baldauf (75), Department of Biology, University of York, York, United Kingdom

Debashish Bhattacharya (109), Department of Biological Sciences, Roy J. Carver Center for Comparative Genomics, University of Iowa, Iowa City, Iowa

Robert E. Blankenship (21), Departments of Biology and Chemistry, Washington University, St. Louis, Missouri

Nicholas J. Butterfield (109), Department of Earth Sciences, University of Cambridge, Cambridge, United Kingdom

Charles F. Delwiche (191), Cell Biology and Molecular Genetics, University of Maryland—College Park, College Park, Maryland

Colomban de Vargas (251), Station Biologique de Roscoff, Equipe "Evolution du Plancton et Paléoceans," CNRS, France

Paul G. Falkowski (1, 405), Institute of Marine and Coastal Sciences and Department of Geological Sciences, Rutgers University, New Brunswick, New Jersey

Johanna Fehling (75), Department of Biology, University of York, York, United Kingdom

Katja Fennel (405), Department of Oceanography, Dalhousie University, Halifax, Nova Scotia, Canada

Z. V. Finkel (333), Environmental Science Program, Mount Allison University, Sackville, New Brunswick, Canada

Rainer Gersonde (207), Alfred Wegener Institute for Polar and Marine Research, Columbusstrasse, Bremerhaven, Germany

Beverley R. Green (37), Botany Department, University of British Columbia, Vancouver, British Columbia, Canada

Michael W. Guidry (377), Department of Oceanography, School of Ocean and Earth Science and Technology University of Hawaii, Honolulu, Hawaii

Jeremiah D. Hackett (109), Woods Hole Oceanographic Institution, Woods Hole, Massachusetts

Christian Hamm (311), Alfred Wegener Institute for Polar and Marine Research, Am Handelshafen, Bremerhaven, Germany

Miriam E. Katz (405), Department of Geological Sciences, Rutgers University, Piscataway, New Jersey

Andrew H. Knoll (1, 133), Department of Organismic and Evolutionary Biology, Harvard University, Cambridge, Massachusetts

Wiebe H.C.F. Kooistra (207), Stazione Zoologica "Anton Dohrn," Villa Comunale, Naples, Italy

Elena Litchman (351), W.K. Kellogg Biological Station, Michigan State University, Hickory Corners, Michigan

Fred T. MacKenzie (377), Department of Oceanography, School of Ocean and Earth Science and Technology, University of Hawaii, Honolulu, Hawaii

David G. Mann (207), Royal Botanic Garden, Edinburgh Scotland, United Kingdom

William Martin (55), Evolutionary Biology Program, University of Düsseldorf, Düsseldorf, Germany

David Mauzerall (7), Rockefeller University, New York, New York

Linda K. Medlin (207), Alfred Wegener Institute for Polar and Marine Research, Am Handelshafen, Bremerhaven, Germany

Charles J. O'Kelly (287), Bigelow Laboratory for Ocean Sciences, West Boothbay Harbor, Maine

Jonathan L. Payne (165), Department of Geological and Environmental Sciences, Stanford University, Stanford, California

Ian Probert (251), Station Biologique de Roscoff, Equipe "Evolution du Plancton et Paléoceans," CNRS, France

Jason Raymond (21), Microbial Systems Division, Biosciences Directorate Lawrence Livermore National Laboratory, Livermore, California

Sumedha Sadekar (21), Program in Computational Biosciences, Arizona State University, Tempe, Arizona

Michael J. Sanderson (109), Section of Evolution and Ecology, University of California—Davis, Davis, California

Victor Smetacek (311), Alfred Wegener Institute for Polar and Marine Research, Am Handelshafen, Bremerhaven, Germany

Diane Stoecker (75), Horn Point Laboratory, University of Maryland Center for Environmental Science, Cambridge, Maryland

Roger E. Summons (133), Department of Earth, Atmospheric, and Planetary Sciences, Massachusetts Institute of Technology, Cambridge, Massachusetts

Bas van de Schootbrugge (165), Institute of Geology and Paleontology, Johann Wolfgang Goethe University Frankfurt, Frankfurt am Main, Germany

Jacob R. Waldbauer (133), Joint Program in Chemical Oceanography, Massachusetts Institute of Technology and Woods Hole Oceanographic Institution, Cambridge, Massachusetts

Hwan Su Yoon (109), Department of Biological Sciences, Roy J. Carver Center for Comparative Genomics, University of Iowa, Iowa City, Iowa

Jeremy Young (251), Paleontology Department, The Natural History Museum of London, London, United Kingdom

John E. Zumberge (133), GeoMark Research, Houston, Texas

Preface

In most contemporary texts on biological oceanography, the basic evolutionary history of the organisms is ignored. Hence, while most students in the field are well aware of the existence of, for example, diatoms and dinoflagellates, they have little understanding of when the various taxa arose, how they are related, and their impact on the biogeochemical cycles on Earth over geological time.

The idea to begin a systematic, integrated study of the evolution of primary producers in the ocean originated at a meeting held in 1999 on the shore of Lake Balaton, Hungary. It subsequently evolved into a formal collaborative research program, supported largely by the National Science Foundation. Not all the authors for these chapters participated in that program, but all were consulted and invited to workshops over the years. The volume is structured to follow the evolutionary history of marine photosynthetic organisms from the Archean to the present. While each chapter is self-contained, they are interrelated and reflect the complexity of the subject, spanning organic and isotopic geochemistry, molecular phylogeny, micropaleontology, cell physiology, and paleoecology. We believe the volume summarizes our knowledge of the overall topic of the evolution of marine photoautotrophs at this time; clearly more detailed understanding will emerge as whole genome analyses become more widely available, and as progress is made in geochemistry and structural biology.

We are deeply grateful to the National Science Foundation and the Agouron Institute for their support, and to Rutgers University and Academic Press for helping to sponsor the symposium, held in January 2006, from which this volume is derived. We thank Yana Zeltser and Marge Piechota for their extremely valuable administrative help.

Paul G. Falkowski and Andrew H. Knoll

1

An Introduction to Primary Producers in the Sea: Who They Are, What They Do, and When They Evolved

PAUL G. FALKOWSKI AND ANDREW H. KNOLL

Earth is approximately 4.6 billion years old, and for the past 4.3 billion years or so there has been a persistent film of liquid water on its surface (Watson and Harrison 2005). The original source of the water is not known with certainty (Robert 2001; Drake and Righter 2002); however, it is one of the most important features that distinguishes this planet from all others in our solar system. A second distinguishing feature is the abundance of molecular oxygen in the atmosphere. Based on the isotopic fractionation of sulfur, it would appear that oxygen began to accumulate in the atmosphere and surface ocean between 2.4 and 2.3 billion years ago (Ga) (Farquhar *et al.* 2000; Bekker *et al.* 2004), and in this case, the source *is* known: the oxidation, or "splitting," of liquid water in oceans and/or lakes by a group of organisms that evolved to utilize the energy of the sun to catalyze the reaction (Knoll *et al.*, Chapter 8, this volume). All complex life ultimately came to be dependent on oxygenic photosynthesis (Falkowski 2006; Raymond and Segre 2006). How and when this metabolic capacity evolved remains one of the great unsolved scientific questions (Blankenship *et al.*, Chapter 3, this volume). Once it did evolve, however, the genetic imprint spread via horizontal gene transfer and a series of symbiotic associations to form a diverse photosynthetic biota that would prove resilient to planetary catastrophes including global glaciations, meteorite bombardments, and massive volcanic eruptions while profoundly and irreversibly altering Earth's chemistry.

In this book, we examine both the molecular biological issue of how water came to be oxidized and the ecological and evolutionary issues of this process, once evolved, which were appropriated and perpetuated by a wide variety of organisms living in the oceans. As in all complex stories, there remain many unanswered questions.

I. WHAT IS PRIMARY PRODUCTION?

On Earth, six major elements, H, C, O, N, P, and S overwhelmingly comprise the ingredients of life (Schlesinger 1997). With the single exception of P, the elements that form the major biopolymers, including proteins, lipids, polysaccharides, and nucleic acids, are incorporated primarily in reduced form; that is, they have received electrons and/or protons from some source. Indeed, electron or hydrogen transfer (redox) reactions form the backbone of biological chemistry (Mauzerall, Chapter 2, this volume).

Earth's early atmosphere is thought to have been mildly reducing, containing CO_2, N_2, H_2O, and possibly CO in significant amounts but probably not much CH_4, H_2S, or NH_3 and almost certainly very little if any O_2 (Kasting 1993). The addition of H_2 to inorganic carbon (i.e., CO_2) to form organic matter (e.g., sugars, $[CH_2O]_n$) is endothermic, requiring an input of energy. Hence, this reaction does not occur spontaneously on Earth's surface at temperatures and pressures compatible with the co-occurrence of liquid water. Conversely, the oxidation of organic carbon compounds produces energy that can be used by organisms to make biopolymers. The ability to reduce inorganic carbon to organic matter is restricted to a relatively small subset of metabolic pathways. Some bacteria and archaea exploit nonphotochemical reactions to reduce inorganic carbon, but by far, photosynthesis is the most efficient, familiar, and widespread means of accomplishing this end (Falkowski and Raven 2007). Because the organisms capable of this metabolic feat provide organic matter for all other organisms in the ecosystem, they

are called "primary producers." Although not all primary producers are photosynthetic, all photosynthetic organisms are primary producers. The rate of production of organic matter by the ensemble of primary producers determines the rate of energy flow, and hence production, of all other trophic levels in nearly every ecosystem (Lindeman 1942).

Photosynthesis uses the energy of the sun to catalyze a redox reaction. The process requires an electron donor/acceptor pair. The electron donor is coupled to a photoreceptor, such that upon absorption of a single quantum at the appropriate wavelength, a single electron is transferred to the acceptor. This process takes approximately 1 picosecond. The primary acceptor, in turn, rapidly donates the electron in a stepwise fashion to other, lower energy acceptors, thereby both preventing a direct backreaction with the donor (which would lead to a useless, Sisyphean electron cycle) and allowing the electron transfers to slow down to millisecond time scales, thereby accommodating the kinetics of biochemical reactions (Blankenship 2002). Ultimately the electron, accompanied by a proton, is used to reduce CO_2 to the equivalent of a carboxyl group, COOH. Further electron transfers yield increasingly reduced forms of organic carbon: the carboxyl group is reduced to an aldehyde and/or ketone (intermediate metabolites), then to an alcohol (found in sugars and polysaccharides), and ultimately to an alkane (C-H, found in lipids). The donor is rereduced by an electron ultimately extracted from a substrate external to the cell, such as H_2S, CH_2O, Fe^{2+}, or H_2O. Of these potential substrates, H_2O is the most abundant on Earth's surface, but it also requires the most energy to oxidize.

The machinery that evolved to use water as a source of reductant is the most complex energy transduction system in nature. In all oxygenic photosynthetic organisms, there are two photochemical reactions connected by a cytochrome. Molecular structural analyses clearly indicate that the two photochemical reaction centers (each of which contains the primary donor and acceptors covalently bound to specific amino residues in a protein complex) have surprisingly similar structural topologies (Blankenship *et al.*, Chapter 3, this volume). Both types of reaction centers are composed of two different polypeptides (i.e., heterodimers) embedded within and spanning a nonphospholipid bilayer membrane. The amino acid sequences of the proteins in the two reaction centers are very different, however. Both structural and amino acid sequence homologies strongly suggest that one of the reaction center dimers, designated photosystem II, is derived from purple bacteria, a group of anaerobic photosynthetic organisms incapable of oxidizing water. However, unlike in the purple sulfur bacteria, photosystem II contains a quartet of Mn atoms and a Ca atom bound to amino acids in the protein heterodimer on one side of the membrane (Ferreira *et al.* 2004). This metal center forms the heart of the water oxidizing machine; it has no known analogue elsewhere in nature. The second photosystem (photosystem I) is derived from green sulfur bacteria and uses a set of iron sulfur clusters as primary electron acceptors. The primary role of this photosystem is to use the energy of light to drive electrons extracted from water by photosystem II to lower (more electrically negative) potentials, where ultimately the electron is used to reduce ferridoxin. Although both photosystems (like their anoxygenic, bacterial counterparts) can operate in a cycle to generate transmembrane electrical fields that can be coupled to adenosine triphosphate (ATP) formation (Blankenship 1992), the efficiency of cyclic electron transport around photosystem I is extremely high. Indeed, in the biological reduction of N_2 in some species of cyanobacteria, a special, differentiated cell, the heterocyst, loses all photosystem II activity (and hence no longer generates oxygen) but retains cyclic photochemically driven electron flow around photosystem I to provide energy (Wolk *et al.* 1994).

II. HOW IS PHOTOSYNTHESIS DISTRIBUTED IN THE OCEANS?

This question has two answers, one ecological and one phylogenetic. How and when the two photosystems of anoxygenic photobacteria fused into one remains poorly understood (see Martin, Chapter 5, this volume). What we do know is that it happened exactly once—in the common ancestor of extant cyanobacteria, the only prokaryotes capable of oxygenic photosynthesis. Today, cyanobacteria form a moderately diverse clade, with species attaining ecological importance in eutrophic fresh waters, in the peritidal benthos where salinity or migrating sands limit algal competitors and animal grazers (see Hamm and Smetacek, Chapter 14, this volume), and in mid-gyre phytoplankton. The group has attained far greater distribution, however, as the plastids of photosynthetic eukaryotes (Bhattacharya and Medlin 1998). Oxygenic photosynthesis in eukaryotic cells originated via an endosymbiotic event in which cyanobacteria were incorporated as symbionts and subsequently reduced to metabolic slaves within their host cells. The progeny of this fusion not only diversified to become the hundreds of thousands of glaucophyte, red algal, green algal, and land plant species found today but also provided the autotrophic partner for six or more new rounds of endosymbiosis that spread photosynthesis widely throughout the eukaryotic domain (see Hackett *et al.*, Chapter 7; Fehling *et al.*, Chapter 6, this volume).

On land, photosynthesis is dominated by a single clade derived from the charophyte green algae, the embryophytic land plants (see O'Kelly, Chapter 13, this volume). A few vascular plants have secondarily recolonized coastal marine waters, but photoautotrophy in the oceans springs from much more diverse phylogenetic sources. Green algae play a role, especially small flagellates, common in coastal blooms and in open ocean picoplankton. Secondary endosymbioses, involving green algae as the autotrophic partner, have resulted in three further groups of algae: the chlorarachniophytes, the photosynthetic euglenids, and certain dinoflagellates. None are ecologically prominent in the oceans.

Red algae are diverse and ecologically important as seaweeds, but they do not at present play a significant role in the phytoplankton. In contrast, photosynthetic clades containing plastids that originated as red algal symbionts dominate primary production in many parts of the oceans. The heterokonts, surely one of evolution's great success stories, include both abundant and diverse seaweeds (e.g., the kelps) and the ubiquitous diatoms found on land and in the sea as microbenthos and phytoplankton (see Kooistra *et al.*, Chapter 11, this volume). Another ecologically important group in the marine phytoplankton is the haptophyte algae, especially the calcite-precipitating coccolithophorids (see de Vargas *et al.*, Chapter 12, this volume). Like photosynthetic heterokonts, haptophytes capable of photosynthesis have red-algal–derived plastids.

About half of known dinoflagellate species are photosynthetic, and most of these also contain plastids derived from the red algal line (see Delwiche, Chapter 10, this volume). Dinoflagellates are photosynthetically promiscuous, however. In addition to the red and green plastids already mentioned, they include species with plastids derived from a *tertiary* endosymbiosis that incorporated a haptophyte alga.

The phylogenetic diversity of marine photoautotrophs correlates with observed ecological heterogeneity of primary producers, with green, red, and brown seaweeds along coasts (and the remarkable floating brown alga *Sargassum* proliferating far from shore); diatoms, dinoflagellates, and coccolithophorids dominating shelf phytoplankton; and cyanobacterial and green picoplankton in oligotrophic mid-ocean environments.

What biological or environmental conditions drove the spread of photosynthesis through the Eucarya? And why did green algae come to cover the land, whereas algae with "red" plastids dominate many parts of the oceans? The factors that selected for primarily red secondary endosymbiotic algae in the modern ocean are not well understood

(Falkowski *et al.* 2003); however, it appears that the trace element composition of these organisms differs significantly from that of green algae (Quigg *et al.* 2003), potentially reflecting the redox conditions of the oceans, especially from the end-Permian extinction to present. It has been hypothesized that red plastids originated once, early in the history of the so-called chromalveolate clade, and spread during the radiation of these diverse protists (a group that includes dinoflagellates, heterokonts, and haptophytes, all of which contain heterotrophic lineages in their basal branches) (Cavalier-Smith 2002). This "Chromalveolate hypothesis" implies that the extant heterotrophic species in this group had plastids but somehow lost them for unknown reasons. The hypothesis has some support from molecular phylogeny (see Hackett *et al.*, Chapter 7; Delwiche, Chapter 10; de Vargas *et al.*, Chapter 12, this volume) but remains controversial (Grzebyk *et al.* 2003). Regardless of whether red secondary plastids were incorporated into host cells once or multiple times, organisms possessing this type of plastid have generally extremely large absorption cross sections for light, and red plastids appear to be well suited for photosynthesis at very low photon fluxes (see Green, Chapter 4, this volume).

III. WHAT IS THE EVOLUTIONARY HISTORY OF PRIMARY PRODUCTION IN THE OCEANS?

The distribution of photosynthesis on the tree of life implies a complex history of photosynthesis in the oceans, and the geological record confirms that this is the case. The present structure—both ecological and phylogenetic—of autotrophy in marine ecosystems originated only about 200 million years ago (Ma) (see Katz *et al.*, Chapter 18, this volume). What governed the successive Mesozoic radiations of dinoflagellates, coccolithophorids, and diatoms, and what did earlier oceans look like? It appears that green algae were more abundant in the Paleozoic oceans (see Payne and van de Schootbrugge, Chapter 9; Knoll *et al.*, Chapter 8, this volume).

Neither do the environmental influences of the algae stop with the carbon cycle (see Guidry *et al.*, Chapter 17, this volume). The calcitic coccoliths precipitated by coccolithoporids constitute a major sink for calcium carbonate, for the first time providing a significant flux of carbonate to the deep seafloor (Litchman, Chapter 16, this volume). Diatoms, in turn, have come to dominate the oceans' silica cycle, influencing the evolutionary trajectories of heterotrophs such as radiolarians and siliceous sponges. The fluxes of these "hard parts," which probably evolved as a protective mechanism against grazing (see Hamm and Smetacek, Chapter 14; Finkel, Chapter 15, this volume), were also accompanied by the flux of organic matter into the ocean interior. The degradation and oxidation of this organic matter has "imprinted" the ocean interior with a nitrogen–phosphate ratio that is unique to that ecosystem and is an example of an "emergent" property of the co-evolution of the chemistry of the sea and the organisms that live in it (Falkowski 2001).

IV. CONCLUDING COMMENTS

The study of the evolution of marine photoautotrophs has a long and venerable history but, perhaps surprisingly, has been largely overlooked by both students and scholars of biological oceanography. We hope that this book will give such students access to a new understanding of the organisms responsible for half of the primary production on the planet. Our understanding of evolution of marine photoautotrophs has been greatly enriched in recent years as information in the fields of molecular biology, algal physiology and biophysics, paleontology, genomics, and Earth systems history have been more integrated. The chapters in this volume are an attempt at such an integration. It is our hope that the readers of this book will not only find the information useful but also, that students especially will be inspired to identify and address new questions.

References

Bekker, A., Holland, H.D., Wang, P.-L., Rumble III, D., Stein, H.J., Hannah, J.L., Coetzee, L., and Beukes, N.J. (2004). Dating the rise of atmospheric oxygen. *Nature* **427**: 117–120.

Bhattacharya, D., and Medlin, L. (1998). Algal phylogeny and the origin of land plants. *Plant Physiol.* **116**: 9–15.

Blankenship, R.E. (1992). Origin and early evolution of photosynthesis. *Photosyn. Res.* **33**: 91–111.

Blankenship, R.E. (2002). *Molecular Mechanisms of Photosynthesis*. Oxford, Blackwell Science.

Blankenship, R.E., Sadekar, S., and Raymond, J. (2007). The evolutionary transition from anoxygenic to oxygenic photosynthesis. *Evolution of Primary Producers in the Sea*. P.G. Falkowski and A.H. Knoll, eds. Boston, Elsevier, pp. 21–35.

Cavalier-Smith, T. (2002). Chloroplast evolution: secondary symbiogenesis and multiple losses. *Curr. Biol.* **12**(2): R62–R64.

Delwiche, C.F. (2007). The origin and evolution of dinoflagellates. *Evolution of Primary Producers in the Sea*. P.G. Falkowski and A.H. Knoll, eds. Boston, Elsevier, pp. 191–205.

de Vargas, C., Aubry, M.-P., Probert, I., and Young, J. (2007). Origin and evolution of coccolithophores: From coastal hunters to oceanic farmers. *Evolution of Primary Producers in the Sea*. P.G. Falkowski and A.H. Knoll, eds. Boston, Elsevier, pp. 251–285.

Drake, M.J., and Righter, K. (2002). Determining the composition of the Earth. *Nature* **416**(6876): 39–44.

Falkowski, P.G. (2001). Biogeochemical cycles. *Encyclopedia of Biodiversity* **1**: 437–453.

Falkowski, P.G. (2006). Tracing oxygen's imprint on Earth's metabolic evolution. *Science* **311**: 1724–1725.

Falkowski, P.G., Katz, M., van Schootenbrugge, B., Schofield, O., and Knoll, A.H. (2003). Why is the land green and the ocean red? *Coccolithophores—From Molecular Processes to Global Impact*. J. Young and H.R. Thierstein, eds. Berlin, Springer-Verlag.

Falkowksi, P.G., and Raven J.A. (2007). *Aquatic Photosynthesis*. Princeton, Princeton University Press.

Farquhar, J., Bao, H., and Thiemens, M. (2000). Atmospheric influence of Earth's earliest sulfur cycle. *Science* **289**(5480): 756–758.

Fehling, J., Stoecker, D., and Baldauf, S.L. (2007). Photosynthesis and the eukaryote tree of life. *Evolution of Primary Producers in the Sea*. P. G. Falkowski and A.H. Knoll, eds. Boston, Elsevier, pp. 75–107.

Ferreira, K.N., Iverson, T.M., Maghlaoui, K., Barber, J., and Iwata, S. (2004). Architecture of the photosynthetic oxygen-evolving center. *Science* **303**: 1831–1838.

Finkel, Z.V. (2007). Does phytoplankton cell size matter? The evolution of modern marine food webs. *Evolution of Primary Producers in the Sea*. P.G. Falkowski and A.H. Knoll, eds. Boston, Elsevier, pp. 333–350.

Green, B.R. (2007). Evolution of light-harvesting antennas in an oxygen world. *Evolution of Primary*

Producers in the Sea. P.G. Falkowski and A.H. Knoll, eds. Boston, Elsevier, pp. 37–53.

Grzebyk, D., Schofield, O., Vetriani, C., and Falkowski, P.G. (2003). The Mesozoic radiation of eukaryotic algae: The portable plastid hypothesis. *J. Phycol.* **39**: 259–267.

Guidry, M.W., Arvidson, R.S., and Mackenzie, F.T. (2007). Biological and geochemical forcings to phanerozoic change in seawater, atmosphere, and carbonate precipitate composition. *Evolution of Primary Producers in the Sea.* P.G. Falkowski and A.H. Knoll, eds. Boston, Elsevier, pp. 377–403.

Hackett, J.D., Yoon, H.S., Butterfield, N.J., Sanderson, M.J., and Bhattacharya, D. (2007). Plastid endosymbiosis: Sources and timing of the major events. *Evolution of Primary Producers in the Sea.* P.G. Falkowski and A.H. Knoll, eds. Boston, Elsevier, pp. 109–131.

Hamm, C., and Smetacek, V. (2007). Armor: Why, when, and how. *Evolution of Primary Producers in the Sea.* P. G. Falkowski and A. H. Knoll, eds. Boston, Elsevier, pp. 311–332.

Kasting, J.F. (1993). Earth's early atmosphere. *Science* **259**: 920–926.

Katz, M.E., Fennel, K., and Falkowski, P.G. (2007). Geochemical and biological consequences of phytoplankton. *Evolution of Primary Producers in the Sea.* P.G. Falkowski and A.H. Knoll, eds. Boston, Elsevier, pp. 405–429.

Knoll, A.H., Summons, R.E., Waldbauer, J.R., and Zumberge, J.E. (2007). The geological succession of primary producers in the oceans. *Evolution of Primary Producers in the Sea.* P.G. Falkowski and A.H. Knoll, eds. Boston, Elsevier, pp. 133–163.

Kooistra, W., Gersonde, R., Medlin, L.K., and Mann, D.G. (2007). The origin and evolution of diatoms: Their adaptation to a planktonic existence. *Evolution of Primary Producers in the Sea.* P.G. Falkowski and A.H. Knoll, eds. Boston, Elsevier, pp. 207–249.

Lindeman, R. (1942). The trophic-dynamic aspect of ecology. *Ecology* **23**: 399–418.

Litchman, E. (2007). Resource competition and the ecological success of phytoplankton. *Evolution of*

Primary Producers in the Sea. P.G. Falkowski and A.H. Knoll, eds. Boston, Elsevier, pp. 351–375.

Martin, W. (2007). Eukaryote and mitochondrial origins: Two sides of the same coin and too much ado about oxygen. *Evolution of Primary Producers in the Sea.* P.G. Falkowski and A.H. Knoll, eds. Boston, Elsevier, pp. 55–73.

Mauzerall, D. (2007). Oceanic photochemistry and evolution of elements and cofactors in the early stages of the evolution of life. *Evolution of Primary Producers in the Sea.* P.G. Falkowski and A.H. Knoll, eds. Boston, Elsevier, pp. 7–19.

O'Kelly, C.J. (2007). The origin and early evolution of green plants. *Evolution of Primary Producers in the Sea.* P.G. Falkowski and A.H. Knoll, eds. Boston, Elsevier, pp. 287–309.

Payne, J.L., and van de Schootbrugge, B. (2007). Life in Triassic oceans: Links between planktonic and benthic recovery and radiation. *Evolution of Primary Producers in the Sea.* P.G. Falkowski and A.H. Knoll, eds. Boston, Elsevier, pp. 165–189.

Quigg, A., Finkel, Z.V., Irwin, A.J., Rosenthal, Y., Ho, T.-Y., Reinfelder, J.R., Schofield, O., Morel, F., and Falkowski, P.G. (2003). The evolutionary inheritance of elemental stoichiometry in marine phytoplankton. *Nature* **425**: 291–294.

Raymond, J., and Segre, D. (2006). The effect of oxygen on biochemical networks and the evolution of complex life. *Science* **311**: 1764–1767.

Robert, F. (2001). The origin of water on Earth. *Science* **293**: 1056–1058.

Schlesinger, W.H. (1997). *Biogeochemistry: An Analysis of Global Change.* New York, Academic Press.

Watson, E.B., and Harrison, T.M. (2005). Zircon thermometer reveals minimum melting conditions on earliest Earth. *Science* **308**: 841–844.

Wolk, C.P., Ernst, A., and Elhai, J. (1994). Heterocyst metabolism and development. *The Molecular Biology of Cyanobacteria.* E.D.E. Bryant, ed. Dordecht, Kluwer Academic Publishers: 769–823.

2

Oceanic Photochemistry and Evolution of Elements and Cofactors in the Early Stages of the Evolution of Life

DAVID MAUZERALL

In this chapter I attempt to outline a dominant role of photochemistry in the evolution, and possibly the origin, of life and to explain the observed chemicals found useful in the process. The enormous energy requirement for biological evolution as we know it can only be supplied by solar energy. The inevitable oceanic photochemistry caused by the ultraviolet (UV) radiation passing through the anoxic atmosphere supplied the reducing solvated electrons and the oxidizing radicals to generate the various molecules needed for metabolism. These early reactions would occur largely by free radical mechanisms favored by the absence of oxygen and so be different from modern metabolism. The redox reactions are catalyzed by the variable valence transition metals, which were incorporated into the metabolic pathways. Iron was exceedingly useful because its redox potential can be varied over the entire biological range by simple substitution of ligands. The porphyrins are also unique as the pigments of life. They are components of all the electron transfer pathways of energy transduction. The simple biosynthetic pathway of the porphyrins is conserved throughout all of biology. These compounds support all photosynthetic systems. Other cofactors have received less attention, and their evolution is less clear. The transamination reactions catalyzed by pyridoxal may have been a valuable component of prebiotic metabolism. In most of these paths, evolution has been conservative, modifying used pathways rather than inventing new ones.

I. ENERGY REQUIREMENTS FOR LIFE

Given a biomass of ~6×10^{17} g carbon (C), the area of the earth, 5×10^{18} cm^2, and the minimal energy requirement of a resting cell, 10^{-3} W (gC)$^{-1}$, 10 times less than that of a quiescent yeast cell, an energy flux of 10^{-4} W cm^{-2} is required to support present life on Earth. Alternatively, given the energy needed to fix a mole of carbon 500 kJ (mole)$^{-1}$ and a long cellular reproduction time of 2 years, one again reaches the requirement of 10^{-4} W cm^{-2}. Based on isotopic ratios of carbon and sulfur, Schidlowski *et al.* (1983) argue that the biomass has been roughly constant for 3.5 billion years. Only solar energy at 0.1 W cm^{-2} can supply this source of energy. The next most prevalent source, lightning, fails by more than two orders of magnitude (Miller and Orgel 1974). Cells are little black holes for energy. Broda (1975) has pointed this out along with the fact that the sun generates less than one hundred thousandth of the energy consumed by living cells per gram of material. This is a reason, aside from nuclear physics, that it is so large. This solar energy requirement for evolution does not translate into an equal requirement for the origin of life. However, simple continuity argues that this is the case. It has often been argued that there were many origins of life. My point is that the form connected to photosynthesis was successful.

II. PREBIOTIC PHOTOCHEMISTRY—UV AND OCEANIC PHOTOCHEMISTRY

The lack of oxygen, and thus of ozone, in the prebiotic atmosphere allowed the full spectrum (>180 nm) of sunlight to penetrate to the oceans (Figure 1). Liquid water absorbs strongly at wavelengths less than 180 nm. Ions such as chloride, sulfate, and phosphate are thus protected, but anions such as Br$^-$ and I$^-$ would absorb this plentiful UV, forming reactive oxidizing radicals and solvated electrons (for details see Mauzerall 1992a, 1995, and Table 1). The quantum yields are >0.1 but less than unity because of the proximity effect. The electron is trapped in the water close to the radical and recombination occurs before it escapes by diffusion. The solvated electrons can react with CO_2 to form oxalate, formate, and other organics. The oxidizing species can halogenate aromatic and aliphatic hydrocarbons, which can hydrolyze to form phenols and alcohols, regenerating the halide ions. The total amount of primary products rivals that of modern photosynthesis, 3×10^{15} moles of fixed carbon per year. UV photochemistry must have contributed significantly to prebiotic reactions. This photochemistry vanished with the formation of the ozone layer, which absorbs all wavelengths <~300 nm. The present ocean is still a very large photoreactor, but, aside from photosynthesis, the reactions are now all photo-oxidations because of the presence of oxygen and organic materials, producing the prevalent "gelbstoffe."

Nitrite ion, formed via NO by lightning and far UV in the atmosphere, on excitation will form NO• and OH•, highly oxidizing species useful in the anaerobic world. Fe^{+2} at neutral pH forms plentiful H_2 (Mauzerall *et al.* 1993) and is discussed later. It is worth stressing that this oceanic photochemistry occurs without *any* assumptions aside from the given environment. The ocean is a natural, very large photochemical reactor.

These reactions are crucial because of the currently accepted view that the early environment was only mildly reducing, not highly so as in the Urey model, having mostly N_2 and CO_2 in place of NH_3 and CH_4. The yield of biomolecules by electric discharge, following the results of the Miller experiment (Miller 1953), decreases with

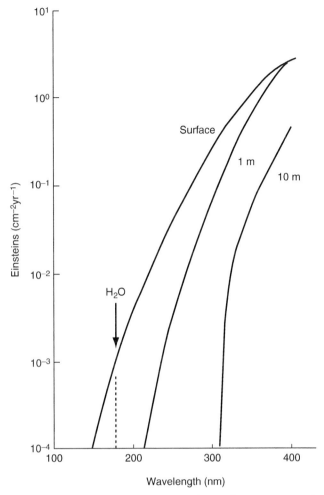

FIGURE 1.   The cumulative Einsteins (moles of solar photons) shorter than a given wavelength plotted on a logarithmic scale versus wavelength at the surface, 1- and 10-m depths of the primitive ocean. The cutoff at ~180 nm by the strong water absorption is indicated. The attenuation is based on data of Withrow and Withrow (1956). Taken with permission from Mauzerall, D. (1992a).

a decreasing H to C ratio (Stribling and Miller 1987). Thus, under the presently accepted conditions of the early Earth, biomolecules would not be formed.

Sources of reducing and oxidizing agents are required to establish a free energy gradient necessary to form the required biomolecules and sustain the organization of living things. The oceanic photoreactions could kick-start the chemical evolution that preceded the origin of self-duplicating entities. Once these entities were established

with the aid of photochemistry, they quickly exhausted the essential ingredients (Malthus is right on these time scales), and selection pressure ensured the development of photosynthesis, efficient cellular transport systems, and useful metabolic sequences. This required the development of the enzymatic reactions to carry out the processes, and the selection of metals and cofactors to accomplish these reactions.

Studies of protein sequence phylogenies suggest that chemoautotrophs were

TABLE 1.   Order of magnitude estimate of photochemical yields for various ions in the early ocean

Ion	λ, nm	ϕ	Conc, M	Moles/Earth	Photons, Einsteins/Earth year[c]	Turnover/year	Moles product/year[d]
I$^-$	254	0.2	10^{-6}	$5 \times 10^{11}/1\,m$	10^{17}	4×10^4	2×10^{16}
Br$^-$	229	0.5	10^{-3}	$5 \times 10^{14}/1\,m$	10^{16}	10	5×10^{15}
Fe(OH)$_2$	350	0.6[a]	$10^{-6,b}$	$5 \times 10^{12}/10\,m$	5×10^{17}	4×10^4	2×10^{17}

The primary products are halide radicals and solvated electrons for I$^-$ and Br$^-$ and H$_2$ for Fe(OH)$_2$. The data for this table can be found in Mauzerall (1992a, 1995).

[a]Per electron. The quantum yield of hydrogen is 0.3.

[b]The concentration of Fe^{+2} is estimated to be 10^{-4} in the Banded Iron Formation regions (Trendall 1972). The given concentration is an estimate for the entire ocean. The solubility of Fe(OH)$_2$ is ~10^{-4} M.

[c]The photon flux is the cumulative amount from the given wavelength to shorter wavelengths (see Figure 1). The irradiated region is limited to 1 m because of ultraviolet (UV) absorption by water except at 350 nm where it extends to >10 m. All photons are absorbed in this layer again except for 350 nm where about half are absorbed.

[d]The amount of primary product rivals the output of modern photosynthesis, ~3×10^{15} moles carbon per year, which requires 10–20 times that amount of primary product.

the first cellular forms. However, with the growing evidence for facile "horizontal" gene transfer and the acceptance of the artifactuality of rooted trees, the original cell becomes a matter of opinion. Although the Archaea are assumed to be primitive because of their mostly anaerobic metabolism, suited for conditions on the early Earth, the absence of large amounts of methane and hydrogen would have limited the energy available to reduce carbon dioxide and thus their evolution. It may be possible that they have de-evolved to the highly specific niches they now occupy.

III. EVOLUTION OF COFACTORS

Williams and da Silva (1996) have given a lengthy compilation of the natural selection of elements in both the early (inorganic) and later (organic) Earth. It is an excellent source for this chapter. I restrict this discussion to the elements found most useful by evolution. I am particularly concerned with the chemical reasons that these particular elements were so used.

A. Metals

1. Iron

By metals we mean the ions. Of all metals, Fe has a special place. Not only is it the most

prevalent metal following Al in the Earth's crust, but also it is the most useful for biological reactions. These reactions can be roughly divided into two categories: redox reactions and group transfer, including hydrolytic, reactions. The redox reactions are required to reduce the then prevalent CO$_2$ and N$_2$ to organics and ammonia, and to shuffle between the various oxidized and reduced organic compounds necessary for metabolism. In this category, the variable valence metals are supreme. Fe is particularly useful because its redox potential can be changed by about ±1 V by simply substituting ligands. Such an opportunity could not be overlooked by evolution. This was stressed by Granick (1957), and he has given a rationale of how the modern catalase could have arisen by natural selection from the catalytic properties of Fe(OH)$_3$, through organic complexes, to the powerful catalases and peroxidases now found in almost all cells. We believe the use of Fe may have begun with the prevalence of Fe^{+2} in the Archaen oceans. At neutral pH the Fe(OH)$_2$ thus formed can reduce water to H$_2$ both thermally and photochemically with the abundant near UV light (Borowska and Mauzerall 1987 and references therein; Mauzerall et al. 1993). The Fe^{+3} so formed is a useful oxidant, still used by some bacteria (Lovley et al. 2004).

2. Manganese

The highlight of the evolution of redox chemistry was the discovery of the oxygen evolving complex (OEC). This discovery freed biological evolution from its dependence on limited supplies of redox agents, such as sulfide or ferrous ions, and opened evolution to water as the near infinite supply of both reducing and oxidizing agents. The large free energy gradient from the reaction of oxygen with reducing agents allowed complex multicellular forms of life to develop. The gaseous form of oxygen equally ensured that this advantage would be available to all, following adaptation. Our present limited knowledge of the OEC (Kern *et al*. 2005; Loll *et al*. 2005) suggests that a very particular arrangement of four manganese ions and one calcium ion together with specific ligands allows a sequential four-stage oxidation of the manganese ions, culminating in the oxidation of water to oxygen and four protons. It may well have required 1–2 billion years of evolution to produce such a feat, which still cannot be duplicated by chemical models. Without this discovery, evolution would still be near the single cell level. It is likely that the discovery proceeded through stages, producing higher oxidation states of the manganese ions, which could be used for reactions such as C-H oxidations now carried out by the heme enzyme P_{450} and O_2. This evolution proceeded synergistically with that of the photosynthetic system (see Section III.B.1.a.).

The intermediate stages may also have produced peroxides, which could have been used to oxygenate compounds such as steroids, for which oxygen is now used. This hypothesis is consistent with the widespread occurrence of peroxidases and catalases in ancient lineages without requiring the prevalence of free oxygen. It is possible that only a few new genes were required to make the OEC and thus oxygen. The delicacy of the OEC process is underlined by the fact that these intermediate high potential oxidants can easily oxidize every one of the organic molecules that they are made of! Yet these intermediates must live long enough (seconds) to be further oxidized by the gentle rain of photons onto the photosynthetic system. In fact, the D1 component of the OEC has a rapid turnover in saturating light, most likely because of excess oxidation of the OEC. Photosynthetic systems generate food and oxygen far in excess of their own requirements. It is this altruism that allowed evolution as we know it. Thus, the systems can simply resynthesize a damaged component rather than attempt the difficult task of evolving a more resistant component. Maybe it is not surprising that our knowledge of this magnificent system is still incomplete.

3. Other Metals

The other transition metals, Cu, Ni, and Co, were rapidly utilized by evolution because of their prevalence and their variable valences with redox potentials within the useful range of biology, $\sim\pm1$ V. Cobalt has the added advantage of forming bonds to carbon and allowing complex synthetic reactions. Each of these metals has developed complexes with proteins and cofactors (see Section III.B.) that allow specific reactions to be catalyzed (see, e.g., Kehres and Maguire 2003).

Nickel may have been more prominent in early life. The anaerobic conditions would have favored the methane- and hydrogen-using (archae) bacteria, which use CO_2 as an oxidant. These organisms use nickel preuroporphyrin derivatives (Factor 430, see Section III.B.1.a.) as catalysts. As the biosphere became more oxidized, Fe was found to be more useful because of its variable redox properties. It is possible that some trace element requirements are residues of ancient useful metals. One suspects that biological evolution used its powerful selection method on whatever was available. Metals such as Ag and Pb may have been used for special reactions and then dropped. As organisms grew more complex the deleterious effects of these metals likely overwhelmed their usefulness and substitutions using the more common metals arose.

Divalent Ca and Mg were found very useful in neutralizing charge on the nucleic acids and arranging the structure of RNA into useful and even enzymatic (ribozyme) structures. Even the small differences between monovalent Na and K were used later in evolution to form potentials across the cell membrane, resulting in the specialized nerve cells, which, in aggregate, allow us to now analyze the evolutionary process itself.

Other metals such as Au and Pt are good catalysts as free metals but are too unreactive to be used to mediate biological redox reactions in evolution. Whereas biology makes great use of electron transfer reactions, electron conductors are only artifacts of our more recent evolution. Having said this, a recent article describes conductive pili in *Geobacter sulfurreducens*, which are presumably used to donate electrons to insoluble ferric oxide without the need of outer-membrane c-type cytochromes (Reguera *et al.* 2005)! It is difficult to make restrictive general statements in biology.

B. Cofactors

The term cofactor requires a definition. It is usually applied to small molecules that enable proteins to transfer electrons, hydrogen, and chemical groups. However, several of the active amino acids, for example, histidine, serine, tyrosine, glutamate, and aspartate have similar functions, including binding of metal ions. Is the heme in a cytochrome a cofactor? To simplify we consider all these compounds as cofactors.

Cofactors can be divided into three groups: (1) ligands to metal ions involved in electron transfer, (2) molecules involved in carbon–hydrogen transfer, and (3) molecules involved in chemical group transfer. Let us consider them sequentially but stress those involved in electron and hydrogen transfers.

1. *Electron Transfer*

a. Porphyrins

Foremost among these molecules are the porphyrins, the pigments of life (Battersby 1994). They are believed to have been formed very early because of their incredible (to chemists) formation from very simple precursors (Figure 2). Shemin and coworkers (Wittenberg and Shemin 1950; Shemin *et al.* 1955) showed that the fourfold symmetric porphyrin is an octamer of delta-aminolevulinate, itself a simple condensation product of glycine and succinate or a reduction and transamination product of glutamate (Beale 1994). All of the early reactions of the biosynthesis of porphyrins have been duplicated in the test tube, often with high yields (Mauzerall 1960a, b), with the one crucial exception of the self-condensation of delta-aminolevulinate to the pyrrole, porphobilinogen, where the yield is very low. It is the Achilles heel of the otherwise attractive hypothesis (Mauzerall 1976) of the early or even prebiotic formation of porphyrins. The biological synthesis first forms the hexahydroporphyrin or porphyrinogen, which cannot bind metals or absorb visible light. However, they are easily oxidized by UV light or mild oxidants (Mercer-Smith *et al.* 1985). The intermediate porphomethenes have been shown to disproportionate to porphyrins and the more reduced forms (Mauzerall 1962). Thus, the reactive and light-absorbing porphyrins would be readily available even in the absence of oxygen. The biosynthetic pathway to chlorophyll has been rationalized as a progressive change from ionic, water-soluble pigments suitable for early or even prebiotic photosynthesis, through less soluble protein-bound pigments, to the lipid-soluble chlorophylls localized in hydrophobic proteins in lipid membranes for modern photosynthesis, Figure 1 (Mauzerall 1992b). The change from photoreactions in solution or aggregates to closed lipid bilayers, liposomes, resulted in the formation of electrochemical gradients and the characteristic cellular potentials. The development of transmembrane potentials and ion transport was an inevitable consequence of photo-driven charge transfer across the membrane (Mauzerall and Sun 2003). The liposomes themselves served

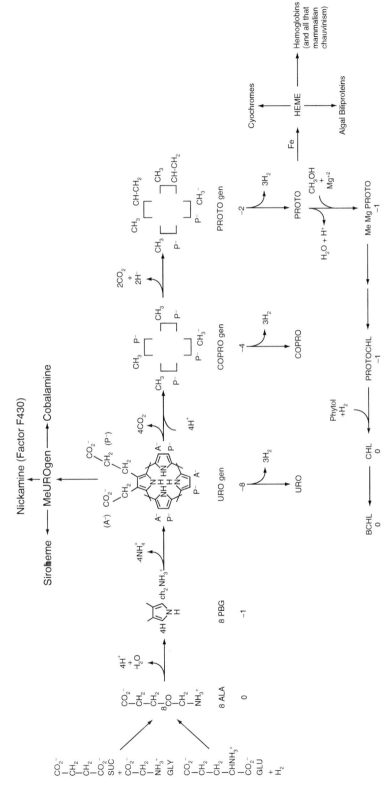

FIGURE 2. Schematic of the biosynthetic pathway of porphyrins. The numbers below the abbreviations are the ionic charges per molecule at neutral pH. SUC, succinate; ALA, 5-aminolevulinate; BCHL, bacteriochlorophyll; CHL, chlorophyll; COPRO, coproporphyrin; gen, porphyrinogen; hexahydro-porphyrin; MeMgPROTO, magnesium protoporphyrin monomethyl ester; PBG, porphobilinogen; PROTO, protoporphyrin; PROTOCHL, protochlorophylide; URO, uroporphyrin. Taken with permission from Mauzerall, D. (1992b).

to keep the important metabolic machinery concentrated inside and required importation of outside nutrients, defining the cell as we know it.

The porphyrin pathway suggests that enzymes originated by the time of the coproporphyrin to protoporphyrin step, where two specific propionic acid groups out of four are oxidatively decarboxylated to vinyl groups, following a series of chemically simple reactions.

My hypothesis is that early photosynthesis occurred using free base porphyrins that, in their excited state, oxidize the reduced compounds prevalent at that time and are themselves reduced to become more powerful donors. These products produced a free energy gradient, which allowed the formation of an early metabolism and of organized systems. As the donors on the early Earth diminished, the reducing power of the excited states of filled orbital metallo-porphyrins became useful and were selected. In the absence of oxygen, these photochemical reactions could occur with the triplet state of the porphyrins, which have a millionfold longer lifetime than the singlet excited states and would not require a highly organized reaction center. This allowed quantitative reaction with micromolar concentration of reactants, not the ~0.1 M concentrations or organized reaction centers required by singlet state reactions.

Insertion of ferrous ion into porphyrins produces the hemes of cytochromes (see Figure 1) ubiquitous in all electron transfer pathways of biology. Their usefulness resides in the wide variety of redox potentials available by changing the fifth and sixth ligands on binding to the protein. A modern variation allows the reversible binding of oxygen in the hemoglobins and the evolution of complex organisms. It is easy to see a progression in function of these remarkable molecules, beginning with simple unicellular organisms that do simple electron transfers, to the more complex organisms that carry the crucial oxygen to deep tissues and even act in control and signaling pathways (Uchida and Kitagawa 2005).

Insertion of Co into the cobalamines, which are modified preporphyrins formed early in the biosynthetic pathway (see Figure 1), allows useful carbon synthesis via the unusual but stable cobalt–carbon bond. Some of these enzymes use the adenosyl radical as the reactive species (Frey 1990, 2001), including a ribonucleotide reductase (see Section III.B.2.). Insertion of Ni into these porphyrin-like molecules led to enzymes capable of attacking unreactive C-H bonds, which allowed the methanogens to exist.

But it is the insertion of Mg into chlorophylls that changed the world. This occurred in spite of the fact that Mg is difficult to insert in the place of two protons inside the porphyrin macrocycle. Possibly its high concentration in the ocean contributed to this choice. Mg may also be the mini-max solution to the problem of the minimum redox potential versus stability in water (Fuhrhop and Mauzerall 1969), required for early photosynthesis. Interestingly, when the OEC was discovered, evolution did not change the metal ion, for example, to Zn, which although scarce is very readily bound and raises the redox potential by >0.1 V. Zinc is a filled shell ion and thus photochemically active as opposed to the transition metals with empty d-orbitals, which self-quench the excited state. Zinc has in fact been found in a photosynthetic bacterium living at very low pH (Hiraishi and Shimada 2001). As the OEC was being developed, evolution changed the protein-binding site in ways not yet understood. It did not change the biosynthesis of chlorophyll to reach the high potential, ~1.2 V, needed to form oxygen from water. It is easier for mutations to change amino acid sequences than biosynthetic pathways.

b. Iron–Sulfur Complexes

Another very widespread metal-ion binding group in biology is the Fe_nS_n complex. These cubane derivatives form spontaneously from ferrous and sulfide ions, supporting their presence in early metabolism. They are enzymatically assembled in vivo (Sutak *et al.* 2004). They come in a variety of stoichiometric ratios. Their redox potentials are all on the reducing side, often near the hydrogen potential at neutral pH. They are exemplified by the ferrodoxins, hydrogenases, and nitrogenases. All were very useful in the anaerobic environment of the early biosphere. The ferrodoxins have made the transition to aerobic organisms, possibly by developing rapid electron transfer reactions that can compete with the limited presence of molecular oxygen. Granick (1957, 1965) anticipated such an FeS cofactor in a model for an ur-photosynthetic unit. A crystal of magnetite was assumed to be covered with FeS on one face and MnO_x on the opposite face. Photo-dissociation of electrons and holes was hypothesized to cause reductions at the FeS face and oxidations at the MnO_x face. Unfortunately, the model has not yet been realized.

The ability of the nitrogenases to reduce free N_2 to NH_3 was crucial to early, as it still is for present, life (Falkowski 1997). If the escape of hydrogen from the early Earth was much slower than previously thought (Tian *et al.* 2005), the H_2 from volcanoes and the then prevalent ferrous hydroxide could accumulate and be used for reduction of N_2. Our attempts to reduce N_2 with ferrous hydroxide have so far been ambiguous (Mauzerall 2005). The nitrogenases, which use Fe and Mo, have been found to contain CO and $CN-$ as cofactors (Rees 2002). These may be necessary to achieve the very negative potential required to initiate the reduction of N_2. The biochemical reaction also requires at least a dozen ATP molecules in addition to the six electrons and six protons to reduce N_2 to two NH_3. *Both* the electrons and the ATP are supplied by photosynthesis. The chemical industrial process supplies the energy required to overcome the barrier to reduction of N_2 and to shift the equilibrium by high pressure and temperature in the presence of an iron catalyst.

2. H-Transfer

The metal complexes transfer one electron at a time, but a C-H bond is electron paired and, assuming a cationic carbon, requires a proton, H+, and two electrons; a hydrogen atom, H•, and one electron, or a hydride ion, H^-, to be formed. Simple one electron transfers are most rapid and thus involve chemically reactive free radicals. Evolution selected the pyridinium ion-dihydropyridine system (NAD-NADH) as an effective hydride acceptor-donor. It can reversibly reduce C=O to CHOH and vice versa. The hydride moiety is directly transferred between the carbon atoms (Westheimer *et al.* 1951). The biosynthesis of NAD is relatively straightforward. In *Escherichia coli* (Begley *et al.* 2001), aspartate is oxidized to iminosuccinate, which is condensed with dihydroxyacetone-phosphate to a quinolinate. Alternatively, 3-hydroxy anthranilate is ring opened to 2-amino carboxy muconic semi-aldehyde, which is cyclized to a quinolinate (Colabroy *et al.* 2005). The latter is condensed with phosphoribose, decarboxylated and ring opened to nicotinamide mono nucleotide (NMN). Condensation with adenosine yields NAD. A more likely prebiotic path is the Hantzsh reaction, which produces dihydropyridines by self condensation of an aldehyde, a β-dicarbonyl and an amine, all likely present in the early mix of organic compounds.

However, the reaction of NAD with one electron systems is a problem because the intermediate radical is highly reactive and of high energy. This difficulty was solved by the flavins, whose biosynthesis is complex (Bacher *et al.* 2001). The imidazole ring of GTP is opened to form a pyrimidinedione. Condensation of this with a butanone phosphate, obtained from ribulose 5-phosphate, forms a lumazine derivative. Dismutation of two molecules of the lumazine forms riboflavin and a molecule of the pyrimidinedione, which is recycled. The reactions are mechanistically complex and may well not be the primitive synthesis, but there does not seem to be much work done on a simpler chemical synthesis.

The flavins are ambidextrous in that they undergo both one- and two-electron reactions as do the other large class of electron transfer agents, the quinones. The reason for this ability is that the intermediate free radicals, one electron species, are relatively stable, as first shown by Michaelis (1931). A likely reason for the selection of flavins over quinones is that the redox potentials of most quinones are oxidizing, while that of the flavins is reducing, and a reducing potential was needed. The quinones, ubiquitous in electron transfer chains, added the critical contribution of protons, as required by the quinone–hydroquinone transformation. With lipid-soluble quinones, this led to proton gradients across the lipid bilayers and their use in energy transduction. However, none of these reduced radicals are stable to oxygen. Thus, in the modern biosphere they are restricted to anaerobic conditions, protection by impervious protein structures or must have short lifetimes. However, in the anaerobic early biosphere these free radicals could be used *directly* to carry out useful reactions. Free radicals can break or insert directly into C-H or C-C bonds. Thus, it is likely that early metabolism was chemically simpler than that of modern times. Flavin radicals could have been used to perform organic reactions that now require the NAD-NADH systems.

The rise of oxygen shifted the evolutionary selection pressure from radical enzymes to the NAD-NADH system. A strong hint that radicals were so used is that some have survived to this day (Frey 1990, 2001). The ribonucleotide reductase, which forms deoxyribotides, uses a tyrosinyl radical, but it presently requires oxygen for its formation. The anaerobic ribonucleotide reductase in *E. coli* uses a glycil radical, and the reducing agent is formate. As mentioned in Section III.A.1.a., some B_{12} enzymes use adenosyl radicals. The class of photolyases that repair UV-induced dimerization of thymine in DNA are also radical enzymes. Derivatives of flavins have been used to carry out the photo reversal of thymine dimerization in vitro (Niemz *et al.* 1999; Wiest *et al.* 2004).

Flavins are very photoreactive using near-UV or blue light, easily oxidizing organic compounds such as tertiary amines (Mauzerall 1960b). They could have been an intermediate stage between the UV photochemistry of the early prebiotic ocean and the visible light photochemistry of the porphyrins. Unfortunately, the lack of a simple biomimetic scheme for their synthesis weakens this hypothesis.

3. Chemical Group Transfer

Chemical group transfer is at the heart of all metabolism. The cofactors for group transfer are determined by the chemistry involved in the reaction. Sulfhydryl molecules were found very suitable for acetyl group transfer and phosphate molecules for ester group transfer. DeDuve (1991) has presented good arguments for considering thiol esters as the original "high energy" bond. What is required, aside from the catalytic enzyme, is that the reactive species be stable to water near neutral pH. This is what the phosphate group does very well, in spite of its low availability, resisting hydrolysis both by the particular bonding and by its negative charge, repelling attack by the hydroxide ion (Westheimer 1987).

One cannot omit the cofactor pyridoxal, as it is probably the most useful of all group transfer cofactors. The transamination reactions alone are very numerous and it is involved in the synthesis of amino acids, and even of porphyrins through δ-aminolevulinate. Surprisingly, the biosynthesis of pyridoxine has been determined only recently and involves the condensation of 1-deoxy-D-xylulose with 4-hydroxy-L-threonine (Spenser and Hill 1995). These are made from simple intermediates such as triose phosphate. The oxidation of the 4'alcohol group of pyridoxine yields pyridoxal. The extensive work of Snell (see, e.g., Ikawa and Snell 1954) has shown that the many reactions catalyzed by pyridoxal can be duplicated in vitro. As an example, pyridoxamine and glyoxal form pyridoxal and glycine. These reactions are catalyzed by multivalent ions such as Al^{+3} or Cu^{+2}. A review of these reactions is given by Martell (1989). They could have contributed greatly to early metabolism.

It is of interest that both the H and group transfer cofactors are attached to nucleotides: phospho-ribose-adenosine. The synthesis of δ-aminolevulinate, the ultimate precursor of porphyrins from glutamate, uses glutamyl t-RNA as the substrate. The possibility that these residues are the vestiges of the "RNA world" is reduced by the scarcity of ribozymes (Steitz and Moore 2003) that catalyze reactions such as H transfer or electron transfer. However, for the group transfer class this remains a possibility. A more mundane explanation for this occurrence is that these residues increase the water solubility and simplify specific binding of the cofactors. Only one type of binding module in the enzymes is necessary for all of the reactions. A more basic explanation is that evolution resists reinventing the wheel: given a biosynthetic pathway, it modifies or adds on to the path, instead of developing a new one. Evolution being a stepwise, quite random process, it is essentially conservative. It is the large number of intersections between these reactions and pathways that lead to the richness of biology.

A comparison of the biosynthetic pathways of the porphyrins with that of other cofactors brings out a striking difference. The porphyrins pathway is optimal in chemical terms and is highly conserved in biology. The other pathways have an accidental quality and are not conserved in any detail. This reflects both on the importance of the porphyrins and on the opportunistic, nondesigned paths of evolution.

IV. CONCLUSIONS

An argument is presented to support the thesis that only photosynthesis supplies sufficient energy to support biological evolution. There may have been several origins of life but only that tied to photosynthesis

evolved. The early anaerobic atmosphere favored UV reactions in that largest of all photochemical reactors, the ocean. It also favored free radical reactions, some of which survive to this day. Thus, early metabolism was probably quite different from that of modern times. The remarkably simple and universal biosynthesis of porphyrins supports the claim that they were very early involved in energy metabolism. In contrast, although the chemistry of many other cofactors justifies their selection by evolution, the paucity of information, particularly on biomimetic pathways, limits our ability to place them in the conservative, opportunistic paths of evolution.

Acknowledgments

I thank Paul Falkowski and Andrew Knoll for helpful suggestions. This work was supported by the Rockefeller University.

References

Bacher, A., Eberhardt, S., Eisenreich, W., Fischer, M., Herz, S., Illarionov, B., Kis, K., and Richter, G. (2001). Biosynthesis of riboflavin. *Vitam. Horm.* **61**: 1–49.

Battersby, A.R. (1994). How nature builds the pigments of life: the conquest of Vitamin B_{12}. *Science* **264**: 1551–1557.

Beale, S.I. (1994). Biosynthesis of cyanobacterial tetrapyrrole pigments: hemes, chlorophylls and phycobilins. *Molecular Biology of Cyanobacteria*. D.A. Bryant, ed. Dordrecht, Kluwer Academic Publishers, pp. 519–558.

Begley, T.P., Kinsland, C., Mehl, R.A., Osterman, A., and Dorrestein, P. (2001). The biosynthesis of nicotinamide adenine dinucleotide in bacteria. *Vitam. Horm.* **61**: 103–119.

Borowska, Z., and Mauzerall, D. (1987). Efficient near ultraviolet light induced formation of hydrogen by ferrous hydroxide. *Origins Life* **17**: 251–259.

Broda, E. (1975). *The Evolution of the Bioenergetic Process*. Oxford, Pergammon Press, pp. 40–41.

Colabroy, K.L., Li, T., Ge, Y., Zhang, Y., Liu, A., Ealick, S.E., McLafferty, F.W., and Begley, T.P. (2005). The mechanism of inactivation of 3-hydroxyanthranilate-3,4-dioxygenase by 4-chloro-3-hydroxyanthranilate. *Biochemistry* **44**: 7623–7831.

DeDuve, C. (1991). *Blueprint for a Cell: The Nature and Origin of Life*. Burlington, NC, Patterson Publisher.

Falkowski, P.G. (1997). Evolution of the nitrogen cycle and its influence on the biological sequestration of CO_2 in the ocean. *Nature* **387**: 272–275.

Frey, P.A. (1990). Importance of organic radicals in enzymatic cleavage of unactivated C-H bonds. *Chem. Rev.* **90:** 1343–1357.

Frey, P.A. (2001). Radical mechanisms of enzymatic catalysis. *Ann. Rev. Biochem.* **70:** 121–148.

Fuhrhop, J., and Mauzerall, D. (1969). The one-electron oxidation of metalloporphyrins. *J. Am. Chem. Soc.* **91:** 4174–4181.

Granick, S. (1957). Speculations on the origin and evolution of photosynthesis. *Ann. N Y Acad. Sci.* **69:** 292–308.

Granick, S. (1965). Evolution of heme and chlorophyll. *Evolving Genes and Proteins.* V. Bryson and H.J. Vogel, eds. New York, Academic Press, pp. 67–88.

Hiraishi, A., and Shimada, K. (2001). Aerobic anoxygenic photosynthetic bacteria with zinc-bacteriochlorophyll. *J. Gen. App. Microbiol.* **47:** 161–180.

Ikawa, M., and Snell, E.E. (1954). Benzene analogs of pyridoxal. The reactions of 4-nitrosalicylaldhyde with amino acids. *J. Am. Chem. Soc.* **74:** 653–660.

Kehres, D.G., and Maguire, M.E. (2003). Emerging themes in manganese transport, biochemistry and pathogenesis in bacteria. FEMS *Microbiol. Rev.* **27:** 263–290.

Kern, J., Loll, B., Zouni, A., Saenger, W., Irrgang, K.-D., and Biesiadka, J. (2005). Cyanobacterial Photosystem II at 3.2 A resolution-the plastoquinone binding pockets. *Photosyn. Res.* **84:** 133–159.

Loll, B., Kern, J., Saenger, W., Zouni, A., and Biesiadka, J. (2005). Towards complete cofactor arrangement in the 3.0 A resolution structure of photosystem II. *Nature* **438:** 1040–1044.

Lovley, D.R., Holmes, D.E., and Nevin, K.P. (2004). Dissimilatory Fe(III) and Mn(IV) reduction. *Adv. Microbial Physiol.* **49:** 219–286.

Martell, A.E. (1989). Vitamin B_6 catalysed reactions of α-amino acids and α-keto acids: model systems. *Accts. Chem. Res.* **22:** 115–124.

Mauzerall, D. (1960a). The photoreduction of porphyrins and the oxidation of amines by photo-excited dyes. *J. Am. Chem. Soc.* **82,** 1832–1833.

Mauzerall, D. (1960b). The condensation of porphobilinogen to uroporphyrinogen. *J. Am. Chem. Soc.* **82:** 2605–2609.

Mauzerall, D. (1962). The photoreduction of porphyrins: structure of the products. *J. Am. Chem. Soc.* **84:** 2437–2445.

Mauzerall, D. (1976). Chlorophyll and photosynthesis. *Phil. Trans. R. Soc. Lond. Series B.* **273:** 287–294.

Mauzerall, D. (1992a). Oceanic sunlight and the origin of life. *Encyclopedia of Earth System Science,* Vol. 3. New York, Academic Press, pp. 445–453.

Mauzerall, D. (1992b). Light, iron, Sam Granick and the origin of life. *Photosyn. Res.* **33:** 63–170.

Mauzerall, D. (1995). Ultraviolet light and the origin of life. *Ultra Violet Radiation and Coral Reefs.* D. Gulko and P.L. Jokiel, eds. Honolulu, HIMB Tech Report #41; UNIHI-CSEA-Grant-CR-95–03, pp. 1–13.

Mauzerall, D. (2005). The reduction of nitrogen to ammonia by ferrous hydroxide. Abstract O-1 The 14th International Conference on the Origin of Life. June 19–24, Beijing, China.

Mauzerall, D., Borowska, Z., and Zielinski, I. (1993). Photo and thermal reactions of ferrous hydroxide. *Origins Life* **23:** 105–114.

Mauzerall, D., and Sun, K. (2003). Photoinduced charge separation in lipid bilayers. *Planar Lipid Bilayers (BLMs) and Their Applications,* Vol. 7. H.T. Tien and A. Ottova, eds. Amsterdam, Elsevier, pp. 963–979.

Mercer-Smith, J.A., Raudino, A., and Mauzerall, D.C. (1985). A model for the origin of photosynthesis III. The ultraviolet photochemistry of uroporphyrinogen. *Photochem. Photobiol.* **42:** 239–244.

Michaelis, L. (1931). The formation of semiquinones as intermediary reduction products from pyocyanin and some other dyestuffs. *J. Biol. Chem.* **92:** 211–223.

Miller, S.L. (1953). A production of amino acids under possible primitive earth conditions. *Science* **117:** 528–529.

Miller, S.L., and Orgel, L.E. (1974). *The Origins of Life on the Earth.* Englewood Cliffs, NJ, Prentice Hall, p. 55.

Niemz, A., and Rotello, V.M. (1999). From enzyme to molecular device. Exploring the interdependence of redox and molecular recognition. *Accts. Chem. Res.* **32:** 44–52.

Rees, D.C. (2002). Great metalloclusters in enzymology. *Ann. Rev. Biochem.* **71:** 221–246.

Reguera, G., McCarthy, K.D., Mehta, T., Nicoll, J.S., and Derek, R. (2005). Extracellular electron transfer via microbial nanowires. *Nature* **435:** 1098–1101.

Schidlowski, M., Hayes, J.M., and Kaplan, I.R. (1983). Isotopic inferences of ancient biochemistries: carbon, sulfur, hydrogen and nitrogen. *Earth's Earliest Biosphere.* J.W. Schopf, ed. Princeton, Princeton University Press, pp. 149–186.

Shemin, D., Russell, C., and Abramsky, T. (1955). The succinate-glycine cycle. I. The mechanism of pyrrole synthesis. *J. Biol. Chem.* **215:** 613–626.

Spenser, I.D., and Hill, R.E. (1995). The biosynthesis of pyridoxine. *Natural Product Rep.* **12:** 555–565.

Steitz, T.A., and Moore, P.B. (2003). RNA, the first macromolecular catalyst: the ribosome is a ribozyme. *Trends Biochem. Sci.* **28:** 411–418.

Stribling, R., and Miller, S. L. (1987). Energy yields for hydrogen cyanide and formaldehyde syntheses: the HCN and amino acid concentrations in the primitive ocean. *Origins Life* **17:** 261–273.

Sutak, R., Dolezal, P., Fiumera, H.L., Hrdy, I., Dancis, A., Delgadillo-Correa, M., Johnson, P.J., Muller, M., and Tachezy, T. (2004). Mitochondrial-type assembly of FeS centers in the hydrogenosomes of the amitochondriate eukaryote *Trichomonas vaginalis. Proc. Natl. Acad. Sci. U S A* **101:** 10368–10373.

Tian, F., Toon, O.B., Pavlov, A.A., and De Sterck, H. (2005). A hydrogen-rich early earth atmosphere. *Science* **308:** 1014–1017.

Trendall, A.F. (1972). Revolution in earth history. *J. Geol. Soc. Australia* **19**: 287–311.

Uchida, T., and Kitagawa, T. (2005). Mechanism for transduction of the ligand-binding signal in heme-based gas sensory proteins revealed by resonance Raman spectroscopy. *Accts. Chem. Res.* **38**: 662–670.

Westheimer, F.H. (1987). Why nature chose phosphates. *Science* **235**: 1173–1178.

Westheimer, F.H., Fisher, H.F., Conn, E.E., and Vennesland, B. (1951). The enzymatic transfer of hydrogen from alcohol to DPN. *J. Am. Chem. Soc.* **73**: 2403.

Wiest, O., Harrison, C.B., Saettel, N.J., Sax, M., and Konig, B. (2004). Design, synthesis, and evaluation of a biomimetic artificial photolyase model. *J. Org. Chem.* **69**: 8183–8185.

Williams, R.J.P., and Frausto da Silva, J.J.R. (1996). *The Natural Selection of the Chemical Elements: The Environment and Life's Chemistry*. Oxford, Oxford University Press.

Withrow, R.B., and Withrow, A.P. (1956). Generation, control and measurement of visible and near-visible radiant energy. *Radiation Biology*, Vol. 3. A. Hollander, ed. New York, McGraw-Hill, p. 96.

Wittenberg, J., and Shemin, D. (1950). The location in protoporphyrin of the carbon atoms derived from the α-carbon atom of glycine. *J. Biol. Chem.* **185**: 103–116.

3

The Evolutionary Transition from Anoxygenic to Oxygenic Photosynthesis

ROBERT E. BLANKENSHIP, SUMEDHA SADEKAR,
AND JASON RAYMOND

Free molecular oxygen first appeared in the Earth's atmosphere in significant quantities about 2.4 billion years ago (Ga), a byproduct of oxygenic photosynthesis invented by primitive cyanobacteria. It has long been supposed that anoxygenic photosynthesis existed prior to the invention of oxygenic photosynthesis and that modern groups of anoxygenic phototrophs are the descendants of these ancient bacteria. However, the evidence for this is indirect and primarily consists of arguments based on the lesser complexity of the photosynthetic apparatus of these organisms. The evolutionary pathway that connects the anoxygenic and the oxygenic phototrophs is a complex one. Several lines of evidence suggest that horizontal gene transfer has been important in bacterial evolution, in general, and also in the evolution of photosynthesis.

The advent of oxygenic photosynthesis stands as one of the few truly transformative events in the history of life, one in which a new type of metabolism was invented that changed the rules of the game for all other organisms. Almost certainly many, if not most, organisms alive at the time this process began were killed by the poisonous oxygen that was produced as a waste product when water was oxidized, including, most importantly, the organisms that produced it.

Despite the overriding importance of this metabolic process, we have almost no information as to how it developed from the process carried out by mechanistically simpler, and therefore presumably more primitive, anoxygenic (non–oxygen-evolving) forms of photosynthesis, or even if oxygenic photosynthesis might have been the earliest form of light-energy capture, and the existing groups of anoxygenic phototrophs may have been derived by loss of function.

This chapter reviews briefly the early geologic evidence for photosynthesis and the structural and functional differences between anoxygenic and oxygenic photosynthesis and then goes on to discuss the evolutionary transitions that have led us to the current situation in which oxygenic photosynthetic organisms dominate the biosphere and anoxygenic organisms have largely retreated to specific environmental niches.

I. EARLIEST EVIDENCE FOR PHOTOSYNTHESIS AND THE NATURE OF THE EARLIEST PHOTOTROPHS

The origin and evolution of photosynthesis played a pivotal role in the development of life on Earth. Understanding this historical milestone provides unique insight not only into the evolutionary process, but also into how organism–environment interactions effect global-scale changes, ranging from carbon cycling and sequestration to composition of the atmosphere and oceans. Current evidence for the nature of early life and of photosynthesis comes by way of two methodologically distinct routes: direct observation of biological, chemical, or isotopic features that have been preserved in the geological record and reconstruction of ancestral organisms based on conserved traits of their modern descendants. Both of these lines of evidence are outlined briefly, although by no means comprehensively, in this section.

The earliest evidence for life itself comes from the study of isotope fractionation in graphitic carbon inclusions found in ancient rocks. The signature manifests as a carbon-12 enrichment in these inclusions in 3.7- to 3.85-billion-year-old Isua Supracrustal Belt (ISB) in Greenland (Schidlowski *et al.* 1979; Schidlowski 1988; Mojzsis *et al.* 1996). The degree of enrichment of "light" carbon 12 over "heavy" carbon 13 can be indicative of biological processes, including specific modes of autotrophic carbon fixation. In particular, the degree of enrichment found in many ISB rocks is characteristic of organisms fixing CO_2 via the Calvin cycle, common to all known oxygenic phototrophs and many proteobacteria (not all of which are photosynthetic). Mojzsis *et al.* (1996) caused considerable excitement with their discovery of such signatures in "Earth's oldest rocks," whose origin dates back to within a geological eyeblink of the late heavy bombardment that kept the Earth's surface largely molten and likely sterile. Their conclusions invoke the existence of relatively complex, possibly photosynthetic prokaryotes almost as soon as the planet became habitable. However, this evidence for earliest life has been seriously questioned. Several recent articles cast doubt on the age of these rocks and their degree of alteration (Fedo and Whitehouse 2002; van Zuilen *et al.* 2002), and even question the presence of carbon-bearing graphite (Lepland *et al.* 2005) in the formations originally studied (notably, much of this follow-up work has been done in the same lab as the original work). van Zuilen *et al.* (2002) conclude that most of the ISB carbonate-bearing rocks have been altered by fluid interactions, precluding their use in inferring biological activity with the notable exception of a site studied by Rosing (1999) on a less highly metamorphosed 3.7-billion-year-old section of the ISB. Although Rosing concludes his observed isotope signatures are consistent with photoautotrophy, Olson (2006) points out that such fractionations may result from nonphotosynthetic autotrophs using the reductive acetyl-CoA pathway. Additionally, the reductive tricarboxylic acid (TCA) cycle as well as numerous nonautotrophic C fixation pathways can feasibly result in such a fractionation (Kopp *et al.* 2005; Zerkle *et al.* 2005). Based on co-occurring enrichments of carbon 12 and uranium in the ISB, Rosing and Frei (2004) have invoked oxygenic photosynthesis at 3.7 Ga, although Kopp *et al.* (2005) and Tice and Lowe (2006) (see later) provide explanations for both enrichment observations that do not require the presence of oxygen. On the whole, there is currently no compelling evidence that photosynthesis was taking place in this most ancient environment.

The next milepost in the evolution of photosynthesis comes at roughly 3.5 Ga and has also recently been the subject of intense debate. Microfossils—microscopic structures morphologically and, in many cases, compositionally consistent with fossilized microorganisms—have been described across a wide swath of the geological record. However, they are quite rare in Precambrian (>540 million years old) rocks, and among the earliest such microfossils known are filamentous structures in 3.47-billion-year-old cherts that have been likened to trichome-forming cyanobacteria (Schopf 1993; see Knoll *et al.*, Chapter 8, this volume for further information on this subject). This claim has been questioned by microbiologists based on many examples of noncyanobacterial (and nonphotosynthetic) filamentous organisms and the tenuousness of linking microbial morphology to taxonomy or metabolic lifestyle (e.g., Kopp *et al.* 2005). Furthermore, the biogenicity of many of these structures has been called into question by Brasier *et al.* (2002), who suggest these structures are nonbiological and their associated carbon isotope signatures result from hydrothermally driven, Fischer–Tropsch reactions of volcanic CO gas. Although the dust has yet to settle in the ensuing debate on biogenicity, the original claim that these structures represent ancient cyanobacteria is uncertain and perhaps questionable.

Microbial mats and stromatolites, most famously associated with cyanobacteia

as exemplified by a few extant environments such as Shark Bay in Western Australia, have records extending roughly 3.5 billion years into the past. Although once argued as a hallmark of photosynthesis—for example, as evidence for oxygenic photosynthesis at 2.7 Ga (Buick 1992)—stromatolite-like structures can indeed result from both abiotic/sedimentary processes and formation by nonphotosynthetic biota (Grotzinger and Knoll 1999), typified by meters-thick microbial mats that have been found at methane seeps devoid of phototrophs (Michaelis *et al.* 2002). Without other evidence, the biogenicity of such structures can be difficult to interpret. Grotzinger and Knoll (1999) point out that probably less than 1% of stromatolites have associated fossilized microorganisms that might give some clue to their generation. However, matlike structures can be quite informative when information about their geochemical context is preserved. Recent analysis of microbial mats and redox speciation in the 3.4-billion-year-old Buck Reef Chert by Tice and Lowe (2004, 2006) allowed detailed inference of a putative anoxygenic photosynthetic community functioning in an anoxic, H_2-rich marine environment. The authors provide an anaerobic alternative to oxygen in the redox cycling of uranium, as proposed by Rosing and Frei (2004), and go on to argue that hydrogen might have been the prevalent, if not only, abundant electron donor for photosynthesis. Recent atmospheric models suggest that hydrogen was a more abundant component of the Archean atmosphere than previously assumed (Tian *et al.* 2005), and Tice and Lowe (2006) propose its inescapable decline and ensuing electron-donor limitation may have precipitated the evolution of oxygenic photosynthesis.

Chemical fossils or "molecular biomarkers" are unique among early evidence for life by way of the information that they convey, as their chemical complexity invokes a very specific series of reactions often specific to a known biochemical pathway. Although many biomolecules are widely distributed among modern organisms, some harbor chemical substitutions diagnostic for single clades of modern organisms (although such interpretations assume our knowledge of such diversity is complete) (e.g., Knoll 2003; Raymond and Blankenship 2004a). Of direct relevance to the early record of photosynthesis are biomarkers found in 2.7-billion-year-old shales (Brocks *et al.* 1999; Summons *et al.* 1999). These include hopanoid derivatives with specific methyl-group substitutions observed in extant methane oxidizing bacteria and cyanobacteria, as well as in steranes found in modern eukaryotes and in a few prokaryotes wherein biosynthesis is found to require molecular oxygen. Taken together, these biomarkers suggest cyanobacteria had evolved oxygenic photosynthesis by 2.7 Ga, although the oxygen they produced took another 400–500 million years to become abundant enough for wide scale geological changes to occur (e.g., Des Marais 2000). More recently, biomarkers have been described from a 1.64-billion-year-old marine basin (Brocks *et al.* 2005), including products of the carotenoids okenone, chlorobactene, and isorenieratene. The first one is found in modern purple sulfur proteobacteria, and the latter two in green sulfur bacteria. Brocks *et al.* (2005) discuss the presence of these organisms as a barometer for an ancient anoxic, sulfide-rich marine basin similar to that proposed to have persisted throughout much of the Proterozoic (Canfield 1998; Anbar and Knoll 2002).

The second type of evidence used to infer the characteristics of ancient organisms comes from the analysis of the genes, proteins, and genomes of their modern descendants. Although from the onset of molecular phylogeny there was hope that would entail "simply" reconstructing the speciation of extant organisms, it has become abundantly clear that, for prokaryotes in particular, the evolution of individual genes and pathways does not necessarily parallel the evolution of the species, because of nonvertical

evolution, including horizontal gene transfer and gene duplication, recombination, and gene loss. This problem is particularly evident when studying the evolution of photosynthesis. The five phyla among which photosynthetic organisms are found are not closely related based on their distribution on the tree of life (typically reconstructed using the 16S ribosomal RNA gene) (Blankenship 1992). Thus, explaining the evolution of photosynthesis in the context of the tree of life requires multiple instances of transfer and/or loss of the capability. Considering individual components of the photosynthetic machinery only confounds the problem; subsets of these five phyla share common reaction centers (RCs), light-harvesting structures, and pigment biosynthesis pathways.

A number of hypotheses have been proposed to explain these complex distributions, and accruing evidence has brought about consensus on a few key points. The first phototrophs were almost certainly anoxygenic (see later) and possessed a single, homodimeric RC, which gave rise to both extant types (I and II) through gene duplication (Schubert *et al.* 1998; Sadekar *et al.* 2006). The nature of the ancestral RC has been the subject of much conjecture, but determining its nature (i.e., Type I-like versus Type II-like) is currently beyond the resolving capability of molecular phylogenetics. Allen (2005) points out that the ancestral RC may not directly correspond to either of the two modern types of RCs, feasibly having characteristics of both types. Bacteriochlorophyll preceded chlorophyll as a primary pigment (Xiong *et al.* 2000), although these may have been preceded by other biosynthetically simpler porphyrins (not found among extant phototrophs) (Mauzerall 1992). Horizontal gene transfer has been a pivotal driving force in the evolution of photosynthetic organisms (Raymond *et al.* 2002), bolstering alternatives to theories based on differential gene loss from an imagined highly complex photosynthetic ancestor.

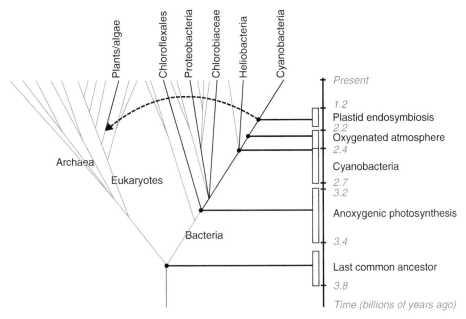

FIGURE 1. Schematic of major events in the evolution of photosynthesis through time, and their inferred distribution on the tree of life. Dark lines on the phylogeny indicate lineages with known photosynthetic members, with names given at the top of the tree. Dates given are approximations (uncertainties indicated with white boxes) based on geological evidence discussed in the text.

As summarized in Figure 1, evidence for the evolution of photosynthesis suggests anoxygenic phototrophs may have been active from at least 3.4 billion years onward, with earlier evidence presently circumstantial or inconclusive. These photosynthetic communities may have been using H_2 as an electron donor, although other donors such as reduced iron and sulfide are also plausible. The earliest direct evidence as to the identity of phototrophs occurs with putatively cyanobacterial-specific biomarkers at 2.7 billion years, and global scale signatures of oxidation of the atmosphere appear between 2.4–2.3 billion years. Biomarkers for photosynthetic green sulfur bacteria and proteobacteria occur beginning about 1.6 Ga, as does additional evidence for the widespread occurrence of cyanobacteria. As discussed in Section II, structural comparisons and molecular phylogeny supports that anoxygenic photosynthesis preceded the evolution of cyanobacteria and the ability to evolve O_2.

II. STRUCTURAL CONSERVATION OF THE CORE STRUCTURE OF PHOTOSYNTHETIC REACTION CENTERS DURING EVOLUTION

Proteins sharing a common evolutionary origin have high primary sequence identity and similar three-dimensional structures. This sequence identity decreases and the structural similarity diminishes with increase in evolutionary distance between proteins. Below about 25% sequence identity, the sequences are said to be in "twilight zone" of molecular evolution studies (Doolittle 1986; Rost 1999). In such situations, it is usually not possible to reliably infer common ancestry from sequence comparisons alone. However, similar structures can persist well into the twilight zone and structural comparisons can sometimes be used to infer common ancestry of more distantly related proteins.

Photosynthetic RCs are integral membrane proteins (Blankenship 2002) and are categorized into two classes—Type I and

Type II, based on the identity of the early electron acceptors. From biophysical and sequence analysis of RCs, it has been understood that the proteins within these two classes are structurally, as well as functionally, similar and probably are descended from a single common ancestor. However, it had not been realized until structural data became available that each of the two classes of RCs are probably themselves descended from a very distant common ancestor, as the residual sequence identity between the two classes is less than 10%, pushing them well into the twilight zone of sequence similarities.

Structural overlay of these RCs using a Combinatorial Extension Monte Carlo algorithm (Guda *et al.* 2004), shown in Figure 2, includes proteobacterial RCs and cyano-

bacterial photosystems I and II. All known RCs have a dimeric core of proteins, in most cases heterodimers that consist of two similar but distinct subunits (Schubert *et al.* 1998; Blankenship 2002). These heterodimers are the Type II RCs consisting of the L and M subunits of the proteobacterial RCs. The cyanobacterial photosystem II consists of the D1 and D2 proteins, whereas the Type I RCs are exemplified by the PsaA and PsaB subunits of the cyanobacterial photosystem I. Figure 2 portrays 10 subunits superimposed from five independent structures. Although the structures of these complexes are less conserved in the loops linking the transmembrane domains, they are remarkably well conserved in the five transmembrane helical domains. The two halves of

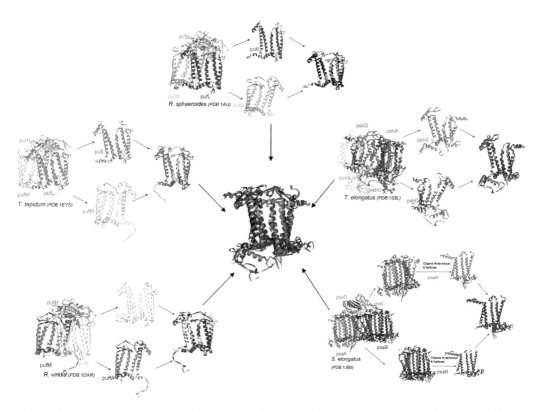

FIGURE 2. Structural overlay of reaction centers. The central figure portrays α-proteobacteria: *Rhodobacter sphaeroides* (1AIJ), L, M chains; *Rhodopseudomonas viridis* (1DXR), L, M chains; *Thermochromatium tepidum* (1EYS), L, M chains; Cyanobacteria: *Thermosynechococcus elongatus* (1S5L), D1, D2 chains of photosystem II; *Synechococcus elongatus* (1JB0), A1, A2 chains of photosystem I. The figures along the boundaries exhibit similar structure of the two heterodimers of each reaction center. (See color plate.)

the heterodimeric complexes in each case are very similar to each other, indicating that they have arisen from gene duplication and divergence events. The sequence identities range from 60–70% for the same subunits from different species of proteobacteria to less than 10% for comparisons between the Type I and Type II complexes (Sadekar *et al.* 2006).

An unrooted phylogenetic tree generated with sequence alignments derived from the structural alignments is shown in Figure 3. The tree is derived from a sequence alignment derived from structurally aligned proteins and is identical in topology to a sequence-based tree produced from a larger group of RC sequences representing all known classes of photosynthetic organisms. Three putative ancient gene duplication events indicated by stars in the figure represent gene duplications and subsequent divergences that led to a heterodimeric RC core structure. The ancestral state is inferred to

be a homodimeric structure. There are two groups of photosynthetic RCs for which structural information is not yet available, the green sulfur bacteria and the heliobacteria. These are both Type I RCs generally similar in biochemical and biophysical properties to photosystem I, although in both cases they are known to be homodimeric complexes (Büttner *et al.* 1992; Liebl *et al.* 1993).

The result that the structure-based tree has an identical topology to trees based on sequence alignments (although the details of the alignment are somewhat different) suggests that this tree represents the true evolutionary relationships of photosynthetic RCs, including three independent gene duplication events that gave rise to heterodimeric complexes. Whether these apparent multiple gene duplication events, especially with respect to the Type II RCs, represent a correct topology or distortion by other processes such as gene conversion has been discussed by various authors

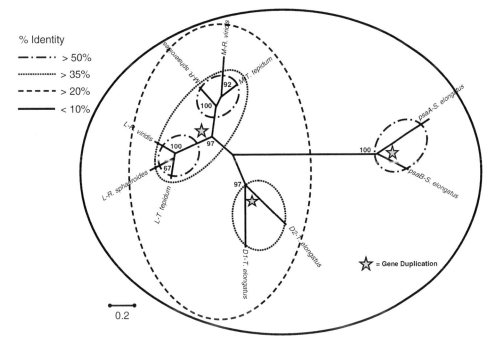

FIGURE 3.   Unrooted neighbor-joining phylogenetic tree of photosynthetic reaction centers based on a sequence alignment derived from the structural alignments shown in Figure 2 (central alignment). The stars represent inferred gene duplication events. The boundaries enclose proteins sharing a particular percentage of similarity.

(Blankenship 1994; Lockhart *et al.* 1996; Blankenship 2002).

In conclusion, we can say that despite extensive sequence divergence, the structures of the transmembrane portions of core RC complexes have remained remarkably conserved and provide insights into the nature of the earliest RCs.

III. THE STRUCTURAL AND MECHANISTIC DIFFERENCES BETWEEN THE ANOXYGENIC REACTION CENTERS OF TYPE II AND PHOTOSYSTEM II OF OXYGENIC ORGANISMS

The structural comparisons shown in the previous section establish clearly that all known classes of photosynthetic RCs derive from a single common ancestor. This unity of overall structure reflects the conserved overall function of light-induced electron transfer chemistry that all RCs carry out. However, there are still many differences among them that are crucial to the distinctive functions of each of the classes of RCs. Here we focus on the Type II RCs that are found in two classes of anoxygenic phototrophs: purple photosynthetic bacteria and green filamentous photosynthetic bacteria (RC II) as well as photosystem II of oxygenic photosynthetic organisms (PS II).

These Type II RCs have a core structure that consists of a protein heterodimer that forms the central scaffold for the electron transfer cofactors. All Type II RCs oxidize an electron donor and reduce a quinone acceptor to a quinol during two successive turnovers. The quinone acceptor side of the complex is remarkably similar in both the RC II and PS II complexes. The first photoinduced electron transfer reduces Q_A to Q_A^-, and the electron is subsequently transferred to Q_B, where the semiquinone Q_B^- is relatively stable. A second turnover reduces Q_A again, and this electron is also transferred to Q_B, coupled to the uptake of two protons to form the fully reduced neutral quinol

Q_BH_2. This species then dissociates from the RC and enters the quinone pool in the membrane, where it is oxidized by another integral membrane oxidoreductase complex. In purple photosynthetic bacteria this is the cytochrome bc_1 complex, whereas in cyanobacteria the related cytochrome b_6f complex oxidizes the quinol. The complex that oxidizes the quinol in green filamentous bacteria has not been conclusively identified, although it has been proposed to be a novel type of oxidoreductase that is not related to either the cytochrome bc_1 or b_6f complexes (Yanyushin *et al.* 2005).

The acceptor side of all Type II complexes is thus very similar in all known cases, with the differences among the various types of complexes being relatively minor, such as the precise quinone (ubiquinone or menaquinone in the RC II complexes found in anoxygenic organisms and plastoquinone in PS II). In contrast, the electron donor sides of the RC II and PS II complexes are dramatically different. RC II complexes almost always have a *c*-type cytochrome as an immediate electron donor, either a soluble monoheme cytochrome such as cytochrome c_2 or a membrane-bound tetraheme cytochrome. Ultimately, electrons either re-enter the complex from cyclic electron flow or from an electron donor, such as H_2S, that is oxidized by another enzyme complex. PS II has a very different electron donor system in which electrons are supplied by the oxidation of water, forming O_2 as a waste product. This is accomplished by an oxygen evolving complex (OEC) that consists of a unique protein-bound complex of four Mn ions as well as Ca and Cl⁻ cofactors (Ferreira *et al.* 2004).

The chemistry carried out by the OEC is the four-electron oxidation of two water molecules to form O_2 (McEvoy *et al.* 2005). This takes place with four successive photoinduced turnovers of the PS II complex, with the oxidizing equivalents stored as more and more oxidized states of the Mn complex (S states), until the four-electron oxidized state called S4 is achieved, which then rapidly reacts to produce O_2. The oxidation of water to form molecular oxygen is one of the most difficult thermodynamic steps in all of biology and requires an average oxidizing potential of at least 0.8 V versus normal hydrogen electrode (NHE), although a recent estimate of the redox potential generated by PS II is greater than 1.3 V (Rappaport *et al.* 2002). This extraordinarily highly oxidizing potential imposes very difficult structural and mechanistic constraints on the complex. For example, pigments such as carotenoids and even other chlorophyll molecules are not found near the OEC, as they would be very rapidly oxidized by the highly oxidizing conditions. The electron transfer cofactor that connects the OEC to the PS II photoactive pigment(s) P680 is a tyrosine amino acid side chain.

In addition to the presence of the OEC in PS II, which is primarily bound to the D1 protein of the core complex, there is a remarkably more complex protein complement in PS II compared to RC II (Raymond and Blankenship 2004b). All known PS II RCs contain upward of 25 protein subunits, compared to a total of three, or in some cases four, in RC II complexes in anoxygenic bacteria. The precise functional roles of all of these subunits are not known, although many of them are clearly involved in protecting PS II from the highly oxidizing species that it generates as part of its normal function. Even with all this protection, the complex is still extremely vulnerable to oxidative damage and the entire PS II complex is completely disassembled, the D1 core protein subunit replaced, and the complex reassembled approximately once every half hour during normal function. This stands in sharp contrast to the case in RC II, in which the two proteins that make up the core of the complex (L and M) are indefinitely stable, even under repeated turnovers.

In addition to the very different protein complements of the two Type II RCs, there is an important difference in the pigment composition. All RC II complexes found in anoxygenic bacteria contain either bacteriochlorophyll *a* or *b*. Almost all PS II complexes contain chlorophyll *a*, although a few are known that contain chlorophyll *d* (Chen *et al.* 2005). The structures of these different pigments lead to significantly different optical and redox properties of the pigments. The shorter wavelength photons absorbed by the chlorophyll *a* pigments in PS II have a significantly higher energy than the near infrared (IR) photons absorbed by the bacteriochlorophyll pigments in RC II complexes (680 nm versus 870 to 960 nm). These higher energy photons are needed to produce the much more highly oxidizing species generated by PS II, compared to RC II, assuming that the electron acceptors are similar in redox properties in the two types of complexes, which appears to be the case in all known examples.

IV. EVOLUTIONARY SCENARIOS FOR HOW THE TRANSITION FROM ANOXYGENIC TO OXYGENIC PHOTOSYNTHESIS MAY HAVE TAKEN PLACE

How did the remarkable molecular machine of PS II come about? Although it is clear that it shares a basic core structure with the presumably more ancient RC II complexes, there are major differences, including the Mn-containing OEC, the much more extensive protein complement in PS II, and a different pigment complement. It is a challenge to try to understand how this complex evolved because the differences are so significant and there are no known organisms that have intermediate complexes or transitional forms (Dismukes and Blankenship 2005). The evolutionary development of PS II is one of the most poorly understood aspects of the evolution of the entire process of photosynthesis, and most of the available information is informed speculation rather than established fact.

Each of these differences between RC II and PS II, the OEC, the protein subunit complement, and the pigment composition are discussed briefly and independently, although there were undoubtedly intermediate forms of all of them, and the precise order of how they developed is not known. The increased protein complexity probably was the last thing to develop and was a response to protect the system from the highly damaging species generated once the more strongly oxidizing pigments and other intermediates were already present.

A plausible evolutionary scenario focusing on the cofactor evolutionary development that begins with RC II and ends with PS II was proposed by Blankenship and Hartman (1998) and is shown schematically in Figure 4. It includes transitional electron donors such as H_2O_2 that are easier to oxidize than H_2O and a primitive OEC that carries out two-electron oxidation rather than four-electron oxidation chemistry. The available concentration of H_2O_2 was probably relatively low and it may have initially served as only a supplemental electron donor. The

primitive OEC was proposed to have been derived from Mn catalase enzymes, which contain a binuclear Mn complex with oxo bridges. This metal complex bears some structural similarity, albeit not a strong one, to the tetra-Mn complex of the OEC. Other possible intermediates that have been proposed include Mn bicarbonate complexes (Dismukes *et al.* 2001) and Mn-containing minerals (Sauer and Yachandra 2002; Russell and Arndt 2005). It must be stressed that all these ideas are informed suggestions but have no direct experimental support. The evolutionary origin of the OEC is the single most obscure aspect of the entire process of the origin and evolution of PS II.

The pigment development is somewhat easier to track because a considerable amount is known about the biosynthesis of chlorophyll-type pigments. This aspect of the problem has been discussed in detail (Blankenship 2002; Mauzerall, Chapter 2, this volume) and is only summarized

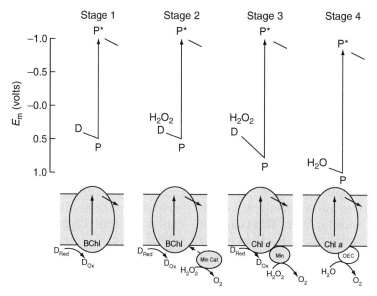

FIGURE 4. Evolutionary scenario for the transition from anoxygenic to oxygenic photosynthesis as proposed by Blankenship and Hartman (1998). The upper portion of this figure depicts the energetics of the oxidation reaction by way of a redox potential scale, with the photon absorbances represented by vertical arrows with length proportional to photon energy. The lower portion represents the cofactor composition, including the type of pigment and the Mn complex, beginning with an Mn catalase and ending with the oxygen evolving complex (OEC). Taken from Blankenship and Hartman (1998), reproduced by permission from Elsevier.

here. Chlorophylls are the product of a very long (17 or more steps) biosynthetic pathway that probably, at least in part, recapitulates their evolutionary development. This idea was first proposed by Sam Granick and is known as the Granick hypothesis (Mauzerall 1992). The first half of the biosynthesis of chlorophylls is identical to heme biosynthesis and presumably reflects the more ancient origin of cytochromes in nonphotosynthetic organisms (Beale 1999). The first committed step of chlorophyll biosynthesis is the insertion of Mg instead of Fe into the center of the porphyrin. This change produces a complex with a long-lived excited state that is much more suitable for light-induced energy storage by electron transfer than the hemes, which have ultrashort excited-state lifetimes dominated by internal conversion processes that dissipate the energy as heat. Mg^{2+} is a closed-shell ion and does not have the efficient internal conversion pathways, resulting in a much longer excited-state lifetime for the pigment. The excited state in the Mg porphyrins lasts long enough to react by electron transfer, leading to energy storage. The pigments that are later in the biosynthetic pathway have strong absorbance bands in the visible region of the spectrum. These properties derive from the structural modifications such as ring substituents and ring asymmetry induced by reduction of some double bonds. These modifications serve to tune the properties of the chlorophylls to increase photon absorption in the blue and red regions of the spectrum and are almost certainly the result of a series of evolutionary developments that gradually made the pigments more efficient at light absorption and photochemistry. The Granick hypothesis proposes that those steps that are later in the biosynthetic sequence were added last. This produces an apparent contradiction with the Granick hypothesis because the bacteriochlorophylls are later products than the chlorophylls, although they are always associated with much simpler photocom-

plexes and, by all analyses, more primitive photosynthetic organisms.

The most important of the later steps in the biosynthesis of chlorophylls are the ring reductions in pyrrole rings D and B carried out by the protochlorophyllide reductase and chlorophyllide-reductase enzymes, respectively. The former enzyme is found in all photosynthetic organisms, whereas the latter is found only in those organisms that make bacteriochlorophyll. The two enzyme complexes are clearly related to each other by divergence from an ancestral enzyme that is also distantly related to nitrogenase (Raymond *et al.* 2004). The important implication of this finding is the proposal that an ancient nonspecific enzyme reduced both rings B and D and produced bacteriochlorophyll in two successive turnovers, without accumulation of significant amounts of chlorophyll. The loss of the chlorophyllide-reductase enzyme, leaving only the protochlorophyllide-reductase enzyme that is active in chlorophyll biosynthesis, is the key evolutionary step that produced the chlorophyll-type pigments with their shorter wavelength absorbance that is essential for oxygenic photosynthesis. Thus, bacteriochlorophyll pigments probably preceded chlorophyll pigments in evolution, despite being later in the modern-day biosynthetic pathway, thereby resolving the apparent contradiction with the Granick hypothesis discussed previously. This issue is discussed at length by Blankenship (2002).

Another enigmatic aspect of the evolutionary development of oxygenic photosynthesis is how the two types of photosystems ended up in the same organism and became functionally linked. Although the two classes of RCs almost certainly have a distant common ancestor, as discussed previously, it is not clear what evolutionary path the complexes followed to arrive at cyanobacteria. Did the two types of RCs develop separately in the same organism and then later become functionally linked or, alternatively, did they develop separately in distinct organisms and then come together by a large-scale lateral gene transfer event? If the former scenario is true, what was

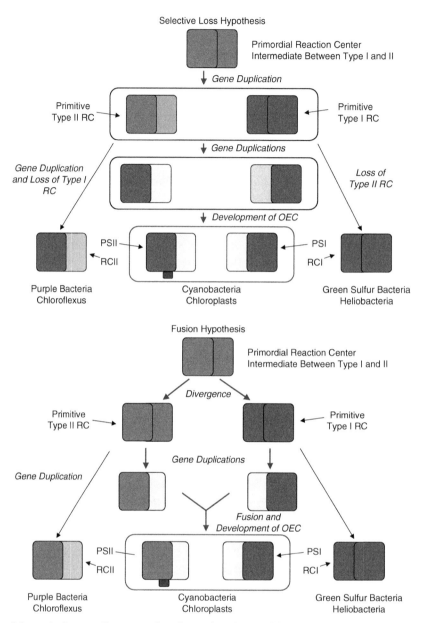

FIGURE 5. Schematic diagram illustrating the selective loss (top) and fusion (bottom) hypotheses for the evolutionary development of photosynthetic reaction centers. The core protein subunits of the various types of reaction centers are shown as colored figures. Homodimeric complexes have two identical subunits, whereas heterodimeric complexes have two similar yet distinct subunits, represented by slightly different colors. The gene duplication, divergence, and loss events that led to existing organisms are indicated. Cells containing two types of reaction centers are shown enclosed within an outline. Black outline indicates an anoxygenic organism and red outline indicates an oxygenic organism. The oxygen evolving complex (OEC) is shown as a red figure. Time flows from the top to the bottom of the diagram, with the primordial homodimeric reaction center at the top and the five existing groups of phototrophic prokaryotes as well as the eukaryotic chloroplast at the bottom. PS, photosystem; RC, reaction center. (See color plate.)

the evolutionary origin of the single-photosystem anoxygenic phototrophs? Were they derived from a primitive cyanobacterium by loss of either the Type I or Type II RC? These two scenarios for the evolutionary origin of existing phototrophs have been discussed extensively in the literature and are usually given the names fusion hypothesis and selective loss hypothesis (Olson and Pierson 1987; Mathis 1990; Blankenship 1992, 2002; Xiong *et al.* 2000; Olson and Blankenship 2004; Allen 2005). These two different hypotheses are diagrammed in Figure 5. There is not yet enough information available to choose between these two extremes or any number of intermediate scenarios. However, it may be possible to resolve this question when complete genome sequences are available for a significant number of phototrophs, as the two scenarios make quite different predictions about the evolutionary history of the organisms. This information should be contained in the genomes. However, it is a significant challenge to resolve this important question, and it may require the development of new molecular evolution methods.

V. CONCLUSIONS AND PROSPECTS FOR THE FUTURE

A number of lines of evidence agree that early RC complexes were found in anoxygenic phototrophs and were probably protein homodimers. The transition from these primitive anoxygenic complexes with an simple protein complement and bacteriochlorophyll pigments to the oxygenic PSII with chlorophyll pigments, an Mn-containing OEC, and a much more complex protein complement was a remarkably dramatic evolutionary development that is still very poorly understood.

Acknowledgments

Research into the evolution of photosynthesis is funded by a grant to REB from the NASA Exobiology Program. JRR is supported by a Lawrence Fellowship at Lawrence Livermore National Laboratory.

References

Allen, J.F. (2005). A redox switch hypothesis for the origin of two light reactions in photosynthesis. *FEBS Lett.* **579**: 963–968.

Anbar, A.D., and Knoll, A.H. (2002). Proterozoic ocean chemistry and evolution: a bioinorganic bridge? *Science* **297**: 1137–1142.

Beale, S. (1999). Enzymes of chlorophyll biosynthesis. *Photosynth. Res.* **60**: 43–73.

Blankenship, R.E. (1992). Origin and early evolution of photosynthesis. *Photosynth. Res.* **33**: 91–111.

Blankenship, R.E. (1994). Protein-structure, electron-transfer and evolution of prokaryotic photosynthetic reaction centers. *Antonie van Leeuwenhoek* **65**: 311–329.

Blankenship, R.E. (2002). *Molecular Mechanisms of Photosynthesis*. Oxford, Blackwell Science, Oxford.

Blankenship, R.E., and Hartman, H. (1998). The origin and evolution of oxygenic photosynthesis. *Trends Biochem. Sci.* **23**: 94–97.

Brasier, M.D., Green, O.R., Jephcoat, A.P., Kleppe, A.K., van Kranendonk, M.J., Lindsay, J.F., Steele, A., and Grassineau, N.V. (2002). Questioning the evidence for Earth's oldest fossils. *Nature* **416**: 76–81.

Brocks, J.J., Logan, G.A., Buick, R., and Summons, R.E. (1999). Archean molecular fossils and the early rise of eukaryotes. *Science* **285**: 1033–1036.

Brocks, J.J., Love, G.D., Summons, R.E., Knoll, A.H. Logan, G.A., and Bowden, S.A. (2005). Biomarker evidence for green and purple sulphur bacteria in a stratified Palaeoproterozoic sea. *Nature* **437**: 866–870.

Buick, R. (1992) The antiquity of oxygenic photosynthesis—evidence from stromatolites in sulfate-deficient Archean lakes. *Science* **255**: 74–77.

Büttner, M., Xie, D.L., Nelson, H., Pinther, W., Hauska, G., and Nelson, N. (1992). Photosynthetic reaction center genes in green sulfur bacteria and in photosystem 1 are related. *Proc. Natl. Acad. Sci. U S A* **89**: 8135–8139.

Canfield, D.E. (1998). A new model for Proterozoic ocean chemistry. *Nature* **396**: 450–453.

Chen, M., Telfer, A., Lin, S., Larkum, A.W.D., Barber, J., and Blankenship, R.E. (2005). The nature of the Photosystem II reaction centre in the chlorophyll *d* containing prokaryote, *Acaryochloris marina. Photochem. Photobiol. Sci.* **4**: 1060–1064.

Des Marais, D.J. (2000). Evolution. When did photosynthesis emerge on Earth? *Science* **289**: 1703–1705.

Dismukes, G.C., and Blankenship, R.E. (2005). The origin and evolution of photosynthetic oxygen production. *Photosystem II: The Water/ Plastoquinone Oxido-Reductase in Photosynthesis.* T. Wydryzynski and K. Satoh, eds. Dordrecht, Springer, pp. 683–695.

Dismukes, G.C., Klimov, V., Baranov, S.V., Kozlov, Y.N., DasGupta, J., and Tyryshkin, A. (2001). The origin of atmospheric oxygen on Earth: the innovation of oxygenic photosynthesis. *Proc. Natl. Acad. Sci. U S A* **98**: 2170–2175.

Doolittle, R.F. (1986). *Of Urfs and Orfs: A Primer on How to Analyze Derived Amino Acid Sequences*. Mill Valley, CA, University Science Books.

Fedo, C.M., and Whitehouse, M. J. (2002). Metasomatic origin of quartz-pyroxene rock, Akilia, Greenland, and implications for Earth's earliest life. *Science* **296**: 1448–1452.

Ferreira, K., Iverson, T., Maghlaoui, K., Barber, J., and Iwata, S. (2004). Architecture of the photosynthetic oxygen-evolving center. *Science* **303**: 1831–1838.

Grotzinger, J.P., and Knoll, A.H. (1999). Stromatolites in Precambrian carbonates: evolutionary mileposts or environmental dipsticks? *Annu. Rev. Earth Planetary Sci.* **27**: 313–358.

Guda, C., Lu, S., Scheeff, E.D., Bourne, P.E., and Shindyalov, I.N. (2004). CE-MC: a multiple protein structure alignment server. *Nucleic Acids Res.* **32**: W100–103.

Knoll, A.H. (2003). The geological consequences of evolution. *Geobiology* **1**: 3–14.

Knoll, A.H., Summons, R.E., Waldbauer, J.R., and Zumberge, J.E. (2007). The geological sucession of primary producers in the oceans. *Evolution of Primary Producers in the sea*. P.G. Falkowski and A.H. Knoll, eds. Boston, Elsevier, pp. 133–163.

Kopp, R.E., Kirschvink, J.L., Hilburn, I.A., and Nash, C.Z. (2005). The paleoproterozoic snowball Earth: a climate disaster triggered by the evolution of oxygenic photosynthesis. *Proc. Natl. Acad. Sci. U S A* **102**: 11131–11136.

Lepland, A., van Zuilen, M.A., Arrhenius, G., Whitehouse, M.J., and Fedo, C.M. (2005). Questioning the evidence for Earth's earliest life—Akilia revisited. *Geology* **33**: 77–79.

Liebl, U., Mockensturm-Wilson, M., Trost, J.T., Brune, D.C., Blankenship, R.E., and Vermaas, W. (1993). Single core polypeptide in the reaction center of the photosynthetic bacterium *Heliobacillus mobilis*: structural implications and relations to other photosystems. *Proc. Natl. Acad. Sci. U S A* **90**: 7124–7128.

Lockhart, P., Steel, M., and Larkum A. (1996). Gene duplication and the evolution of photosynthetic reaction center proteins. *FEBS Lett.* **385**: 193–196.

Mathis, P. (1990). Compared structure of plant and bacterial photosynthetic reaction centers—evolutionary implications. *Biochim. Biophys. Acta* **1018**: 163–167.

Mauzerall, D. (1992). The Granick hypothesis—porphyrin biosynthesis and the evolution of photosynthesis. *Photosynth. Res.* **33**: 163–170.

Mauzerall, D. (2007). Oceanic Photochemistry and evolution of elements and cofactors in the early stages of the evolution of like. *Evolution of Primary Producers in the Sea*. P.G. Falkowski and A.H. Knoll, eds. Boston, Elsevier, pp. 7–19.

McEvoy, J.P., Gascon, J.A., Batista, V.S., and Brudvig, G.W. (2005). The mechanism of photosynthetic water splitting. *Photochem. Photobiol. Sci.* **4**: 940–949.

Michaelis, W., Seifert, R., Nauhaus, K., Thiel, V., Blumenberg, M., Knittel, K., Gieseke, A., Peterknecht, K., Pape, T., Boetius, A., Amann, R., Jorgensen, B.B.,

Widdel, F., Peckmann, J.R., Pimenov, N.V., and Gulin, M.B. (2002). Microbial reefs in the Black Sea fueled by anaerobic oxidation of methane. *Science* **297**: 1013–1015.

Mojzsis, S.J., Arrhenius, G., McKeegan, K.D., Harrison, T.M., Nutman, A.P., and Friend, C.R.L. (1996). Evidence for life on Earth before 3,800 million years ago. *Nature* **384**: 55–59.

Olson, J.M. (2006). Photosynthesis in the Archean Era. *Photosynth. Res.* **88**: 109–117.

Olson, J.M., and Blankenship, R.E. (2004). Thinking about the evolution of photosynthesis. *Photosynth. Res.* **80**: 373–386.

Olson, J.M., and Pierson, B.K. (1987). Evolution of reaction centers in photosynthetic prokaryotes. *Int. Rev. Cytol.* **108**: 209–248.

Rappaport, F., Guergova-Kuras, M., Nixon, P.J., Diner, B.A., and Lavergne, J. (2002). Kinetics and pathways of charge recombination in photosystem II. *Biochemistry* **41**: 8518–8527.

Raymond, J., and Blankenship, R.E. (2004a). Biosynthetic pathways, gene replacement, and the antiquity of life. *Geobiology* **2**: 199–201.

Raymond, J., and Blankenship, R.E. (2004b). The evolutionary development of the protein complement of Photosystem 2. *Biochim. Biophys. Acta* **1655**: 133–139.

Raymond, J., Zhaxybayeva, O., Gogarten, J.P., Gerdes, S.Y., and Blankenship, R.E. (2002). Whole-genome analysis of photosynthetic prokaryotes. *Science* **298**: 1616–1620.

Raymond, J., Siefert, J.L., Staples, C.R., and Blankenship, R.E. (2004). The natural history of nitrogen fixation. *Mol. Biol. Evol.* **21**: 541–554.

Rosing, M.T. (1999). ^{13}C-Depleted carbon microparticles in >3700-Ma sea-floor sedimentary rocks from west Greenland. *Science* **283**: 674–676.

Rosing, M.T., and Frei, R. (2004). U-rich Archean seafloor sediments from Greenland—indications of >3700 Ma oxygenic photosynthesis. *Earth Planet. Sci. Lett.* **217**: 237–244.

Rost, B. (1999). Twilight zone of protein sequence alignments. *Protein Eng.* **12**: 85–94.

Russell, M.J., and Arndt, N.T. (2005). Geodynamic and metabolic cycles in the Hadean. *Biogeosciences* **2**: 97–111.

Sadekar, S., Raymond, J., and Blankenship, R.E. (2006). Conservation of distantly related membrane proteins: photosynthetic reaction centers share a common structural core. *Mol. Biol. Evol.* **23**: 2001–2007.

Sauer, K., and Yachandra, V. (2002). A possible evolutionary origin for the Mn-4 cluster of the photosynthetic water oxidation complex from natural MnO2 precipitates in the early ocean. *Proc. Natl. Acad. Sci. U S A* **99**: 8631–8636.

Schidlowski, M. (1988). A 3,800-million-year isotopic record of life from carbon in sedimentary-rocks. *Nature* **333**: 313–318.

Schidlowski, M., Appel, P.W.U., Eichmann, R., and Junge, C.E. (1979). Carbon isotope geochemistry of the 3.7 × 109-yr-old Isua sediments, west Greenland—implications for the Archaean carbon and oxygen cycles. *Geochim. Cosmochim. Acta* **43**: 189–199.

Schopf, J.W. (1993). Microfossils of the early Archean apex chert—new evidence of the antiquity of life. *Science* **260**: 640–646.

Schubert, W.D., Klukas, O., Saenger, W., Witt, H.T., Fromme, P., and Krauss, N. (1998). A common ancestor for oxygenic and anoxygenic photosynthetic systems: a comparison based on the structural model of photosystem I. *J. Mol. Biol.* **280**: 297–314.

Summons, R.E., Jahnke, L.L., Hope, J.M., and Logan, G.A. (1999). 2-Methylhopanoids as biomarkers for cyanobacterial oxygenic photosynthesis. *Nature* **400**: 554–557.

Tian, F., Toon, O.B., Pavlov, A.A., and De Sterck, H. (2005). A hydrogen-rich early Earth atmosphere. *Science* **308**: 1014–1017.

Tice, M.M., and Lowe, D.R. (2004). Photosynthetic microbial mats in the 3,416-Myr-old ocean. *Nature* **431**: 549–552.

Tice, M.M., and Lowe, D.R. (2006). Hydrogen-based carbon fixation in the earliest known photosynthetic organisms. *Geology* **34**: 37–40.

van Zuilen, M.A., Lepland, A., and Arrhenius, G. (2002). Reassessing the evidence for the earliest traces of life. *Nature* **418**: 627–630.

Xiong, J., Fischer, W.M., Inoue, K., Nakahara, M., and Bauer, C.E. (2000). Molecular evidence for the early evolution of photosynthesis. *Science* **289**: 1724–1730.

Yanyushin, M., del Rosario, M., Brune, D., and Blankenship, R.E. (2005). A new class of bacterial membrane oxidoreductases. *Biochemistry* **44**: 10037–10045.

Zerkle, A.L., House, C.H., and Brantley, S.L. (2005). Biogeochemical signatures through time as inferred from whole microbial genomes. *Am. J. Sci.* **305**: 467–502.

Evolution of Light-Harvesting Antennas in an Oxygen World

BEVERLEY R. GREEN

Light-harvesting antennas are pigment-protein complexes that absorb light energy and pass it along to the photosynthetic reaction centers, where the photochemical reactions occur. Some of the antenna pigments are part of the macromolecular core complexes of Photosystem I (PS I) and Photosystem II (PS II) (Green 2003; Blankenship *et al.*, Chapter 3, this volume). This chapter is concerned with the peripheral antennas and is restricted to those of cyanobacteria and chloroplasts. Major players in this story are the events of primary and secondary endosymbiosis that gave rise to the extant groups of photosynthetic eukaryotes. The cast of characters involves "all the usual suspects": gene duplications, gains and losses, relocations, recruitments, lateral gene transfer, and molecular opportunism at all levels.

The rise of atmospheric oxygen played an important role in shaping the evolution of light-harvesting antennas. This is because molecular oxygen can be activated by the transfer of energy from excited states of chlorophyll (Chl), or can be converted to highly reactive superoxide or peroxide by the redirection of electrons from the electron transport chain. Every photosynthetic organism must balance the efficient harvesting of light with measures to counter the damaging effects of excess light energy. Using light energy is a dangerous business, especially in the presence of molecular oxygen. As a result, the evolution of light-harvesting antennas is closely connected with the evolution of photoprotective mechanisms. Different groups of photosynthetic organisms have evolved different adaptations, and these are reflected in striking differences in their membrane structure.

I. HOW CYANOBACTERIA CHANGED THE WORLD

It was the cyanobacteria that evolved the ability to extract electrons from water with the help of two photosystems linked in series. The molecular oxygen produced by them changed the history of the Earth. As an electron source, water is "cheap," that is, very accessible and available in unlimited quantity. It would have given the cyanobacteria a big advantage over their neighbors with single photosystems, who relied on sulfide, other inorganic ions, and trace organics to donate electrons (Nisbet and Sleep 2001). It is hardly surprising that the cyanobacteria prospered.

For the first 2 billion years of its existence, Earth is thought to have had an almost completely anaerobic atmosphere (Anbar and Knoll 2002). As the O_2 levels rose, first in the atmosphere and then in shallow waters, every living organism had to adapt to cope with this "pollutant." Photosynthetic organisms were particularly susceptible to photo-oxidative damage because an excited Chl molecule that cannot transfer its energy rapidly to a reaction center may decay to the triplet state and react with O_2 to give singlet O_2, a reactive oxygen species. Furthermore, electrons from the electron transport chain can flow from PS I to O_2, producing superoxide (O_2^-), another reactive species that disproportionates to give hydrogen peroxide (H_2O_2) and water, the "Asada cycle" (Asada 1999). This is significant when the pool of terminal electron acceptors is limited, for example, by low CO_2 or NO_3^-, or when so much energy is being absorbed by the photosystems that all the electron acceptors are reduced. Reactive oxygen species are disposed of by a variety

of reactions catalyzed by such enzymes as superoxide dismutatase, ascorbate peroxidase, and glutathione reductase, which can now be found in almost all living organisms (reviewed in Noctor and Foyer 1998).

On the global scale, another serious effect of the rise of O_2 was the oxidation of much of the dissolved iron Fe(II) to Fe(III), which is much less soluble in water. This resulted in biologically accessible Fe(II) becoming a scarce commodity both in the sea and on land. As a result, we see that Fe levels have a significant regulatory effect on some aspects of light-harvesting, effects that can be compounded by, and are correlated with, oxidative stress (Michel and Pistorius 2004).

However, the rise of atmospheric oxygen provided a new and better terminal electron acceptor for respiratory chains. Oxidative respiration provides much more energy than fermentation or respiration using other electron acceptors. The availability of oxygen promoted the diversification of heterotrophic prokaryotes. At a later stage, it would have provided selective pressure promoting the endosymbiosis of one of these prokaryotes, which became the first mitochondrion.

The most important factor in preventing damage to the photosynthetic reaction centers is to get rid of excess excitation energy. Any energy not used to push electrons around must be dissipated as fluorescence or as heat, or it will have a destructive effect on the Chls and reaction centers themselves. This is a particular problem in PS II, which constantly undergoes repair. Each group of oxygenic photosynthesizers has its own methods for disposing of this excess energy. The mechanisms are not well understood but generally involve transfer of energy to carotenoids, which are part of the antenna complexes, and/or regulation of the pH gradient across the thylakoid membrane. The evolution of the light-harvesting antennas, which are responsible for the initial absorption of solar radiation, is therefore intimately connected with the evolution of mechanisms for getting rid of excess excitation energy.

II. LIGHT-HARVESTING ANTENNAS AND THE EVOLUTION OF THE ALGAE

The major light-harvesting antennas of cyanobacteria and chloroplasts involve members of just three protein families: the phycobiliproteins, the IsiA-Pcb Chl-proteins (part of the Core Complex family), and the light-harvesting complex (LHC) superfamily. Members of each protein family bind pigments that are variations of only three basic types: phycobilins, Chls, and carotenoids. Phycobilins are open-chain tetrapyrroles, each of which absorbs a specific band of wavelengths, determined by a variety of substituents at conserved positions on the chain and by pigment–protein interactions. Chl a plays the central role in photosynthesis, being involved in charge separation in the reaction centers as well as being the major Chl in light-harvesting antennas. The accessory Chls, found only in antennas and not in reaction centers, are Chl b in green algae and plants and Chl c in the large group of algae, mostly marine, that obtained their chloroplasts by secondary endosymbiosis. The Chls are always found in association with carotenoids, long linear isoprenoids with the ability to release excitation energy as vibrational energy (heat), as well as the ability to transfer energy to reaction centers, depending on the respective energy levels of the pigments. A graphic summary of the antenna types and where they are found is shown in Figure 1.

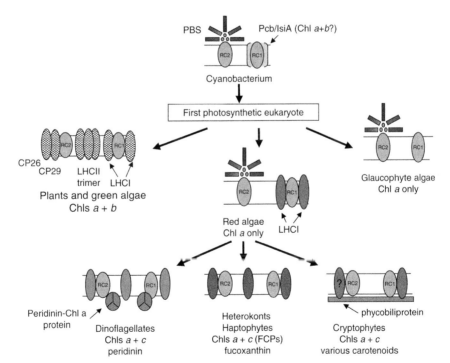

FIGURE 1. Light-harvesting antennas of cyanobacteria and chloroplasts. The cyanobacterial ancestor of the chloroplast had phycobilisomes (PBSs) and an antenna of the Pcb/IsiA family that may have been associated with PS I. It may have synthesized chlorophyll (Chl) b. Descendants of the first photosynthetic eukaryote diversified into the green line (plants and green algae) with the loss of PBSs, the red line (rhodophyte algae) with the loss of Chl b, and the glaucophyte algae. Green line and red line chloroplasts have antennas of the light-harvesting complex (LHC) superfamily, binding Chl $a+b$ (pale grey ovals) and Chl a (hatched ovals) respectively. One or more secondary endosymbioses involving a red algal endosymbiont gave rise to the four major algal groups with Chl a/c LHC antennas (patterned grey ovals). The peridinin-Chl a protein is found only in dinoflagellates (Macpherson and Hiller 2003). RC1, RC2 (black ovals); reaction center cores of Photosystem I and Photosystem II, respectively. (See color plate.)

The evolution of all photosynthetic eukaryotes is a history of endosymbiosis. A number of lines of evidence support the idea that a single primary endosymbiosis gave rise to the first photosynthetic eukaryote, and subsequent diversification led to three algal lineages with primary plastids: the green algae (chlorophytes), the red algae (rhodophytes), and the glaucophyte algae (Rodríguez-Ezpeleta *et al.* 2005). However, the four algal groups that use Chl *c* in their light-harvesting antennas (heterokonts, haptophytes, cryptophytes, and dinoflagellates) are the result of secondary endosymbiosis, in which nonphotosynthetic eukaryote(s) acquired a red algal endosymbiont (Gibbs 1981). Phylogenetic trees have not resolved the question of whether they were all the descendants of a single secondary endosymbiotic event; this subject is extensively reviewed elsewhere (Cavalier-Smith 2000; McFadden 2001; Green 2003, see chapters by Delwiche and Hackett *et al.*, Chapters 10 and 7, respectively, this volume). The endosymbiosis story does not stop there, however. Some dinoflagellates have dispensed with their original peridinin-type plastid and acquired new ones from cryptophytes, chlorophytes, and haptophytes by tertiary endosymbiosis (Ishida and Green 2002; Delwiche, Chapter 10, this volume).

III. PHYCOBILISOMES

The phycobilisomes (PBSs) are complex macromolecular structures attached to the cytoplasmic or stromal surfaces of thylakoid membranes (Figure 2). They are found in cyanobacteria and two of the three lineages of algae with primary chloroplasts, the red and glaucophyte algae (see Figure 1). The PBS has been completely lost in the third primary lineage, that of the green algae and plants. The similarities between cyanobacterial and algal PBSs are one of the strongest pieces of evidence for the cyanobacterial ancestry of the chloroplast.

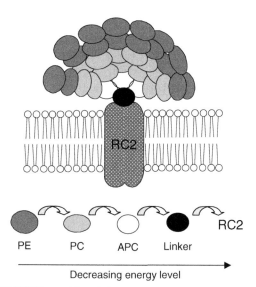

FIGURE 2. Phycobilisome (PBS) energy cascade. Simplified model of PBSs attached to Photosystem II core dimer (RC2) by linker (black). APC, allophycocyanin hexamers; PC, phycocyanin hexamers; PE, phycoerythrin hexamers. The exact absorption wavelengths depend on the species but increase from PE to RC2.

The PBS is a highly efficient energy funnel directing energy primarily to PS II (reviewed by Gantt *et al.* 2003; Green and Gantt 2005). It is made up of phycobiliprotein hexamers stacked into rods stabilized by linker proteins. Each hexamer is made up of 3 α and 3 β polypeptide chains, each binding one to three specific bilin pigments via thioether linkages to cysteine side chains. Phycoerythrin and phycocyanin hexamers are stacked into rods, organized such that the phycoerythrin chromophores on the outer part of the rod pass excitation energy to the lower energy level phycocyanin chromophores on the inner part of the rod (see Figure 2). They in turn pass it down to the chromophores of the basal allophycocyanin rods. The energy is then transferred, sometimes via a pigmented linker, to the PS II reaction center core complex. Energy transfer is very efficient because the close packing of hexamer units in the rods positions the chromophores close enough to each other to allow transfer by the Förster exchange mechanism (Mimuro and Kikuchi 2003).

The PBS is the only light-harvesting antenna that does not involve Chl and does not bind carotenoids. As in all oxygenic photosynthetic organisms, there are carotenoids associated with the core complexes of the photosystems, which can dissipate some energy. However, it has recently been shown that a soluble carotenoid protein that is associated with thylakoids and PBSs in *Synechocystis* PCC6803 is involved in energy dissipation (Wilson *et al.* 2006). The fact that the PBS is outside the membrane and that its energy transfer is so efficient probably also decreases the potential for photodamage. In addition, it appears that cyanobacterial PBSs are able to move along the plane of the membrane in response to high light, in such a way that some of their absorbed energy is transferred to PS I rather than PS II. This would tend to make the energy reaching each photosystem approximately equal, rather than biased in favor of PS II and would thus prevent overreduction of the quinone pool (Joshua and Mullineaux 2005). However, in spite of the fact that red algal PBSs resemble cyanobacterial PBSs in many ways, there is no evidence that red algal PBSs can move (Gantt *et al.* 2003). This may have something to do with the fact that red algae have a membrane intrinsic antenna of the LHC superfamily associated exclusively with PS I (see later).

Phylogenetic trees (Apt *et al.* 1995) suggest that a single gene for an ancestral biliprotein was duplicated to give the α and β polypeptides (Figure 3). Successive rounds of gene duplication and divergence gave rise to allophycocyanin, phycocyanin, and finally phycoerythrin. The biliprotein genes remain in the chloroplast genome in red and glaucophyte algae, although the genes for some linkers and most of the genes encoding enzymes for bilin pigment synthesis have been transferred to the nucleus. Structural variations of the bilin pigments, combined with their interactions with the protein environment, provide a diversity of absorption wavelengths that enable different species and populations to inhabit

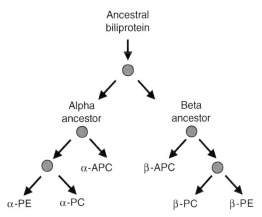

FIGURE 3. Evolutionary origin of the phycobiliprotein family by successive rounds of gene duplication (circles) and divergence. The ancestral pigment-binding protein gave rise to the α and β polypeptides, which then diversified into α and β polypeptides of allophycocyanin (APC), phycocyanin (PC), and phycoerythrin (PE). Based on the work of Apt *et al.* (1995).

a variety of environmental niches (Gantt *et al.* 2003). The amazing variety of pigment structures are illustrated in several reviews (Wedemeyer *et al.* 1996; Scheer 2003; Toole and Alnutt 2003).

There are several examples where the PBSs have degenerated but phycobiliprotein genes remain. In the cyanobacteria with Chl *a/b* antennas (see later), a few phycoerythrin genes are still expressed (Ting *et al.* 2002), although whether phycoerythrin contributes significantly to enhancing photosynthesis has not been resolved. In the cryptophyte algae, which acquired their chloroplasts by secondary endosymbiosis involving a red alga (see Figure 1), the β subunit gene is conserved and still located on the chloroplast genome, but the α subunit gene has been completely replaced by a totally different, nuclear-encoded gene of unknown origin (Wilk *et al.* 1999; Macpherson and Hiller 2003; Broughton *et al.* 2006). The antenna assembles in the thylakoid lumen in the form of 2α-2β tetramers with a unique three-dimensional structure, completely different from that of the PBS hexamer (Wilk *et al.* 1999). This evolutionary story thus involves recruitment of a protein to a new function, the generation of a novel structure from pre-existing parts, and retargeting to a different cellular compartment (Broughton *et al.* 2006).

IV. THE ISIA/PCB FAMILY

Most cyanobacteria placed under Fe-limited conditions downregulate PS I and stop making PBSs (Sandström *et al.* 2002). In their place, they synthesize a Chl *a*-binding membrane-intrinsic protein called IsiA, which forms rings of 18 subunits around each PS I trimer (Bibby *et al.* 2001; Boekema *et al.* 2001). With about 13 Chl *a* per IsiA, this more than doubles the light-harvesting antenna of PS I, because each PS I core complex binds about 90 Chls. Even if there are fewer PS I units, the units that are still present are able to function more effectively and balance electron flow through PS I and PS II, thus preventing damage to PS II (Sandström *et al.* 2002). Sequence analysis suggested that the *isiA* gene was closest to the *psbC* gene, which encodes one of the core complex antenna proteins of PS II (CP43), and that they had probably been derived by gene duplication from a common ancestor (Burnap *et al.* 1993).

The IsiA protein is also induced in response to oxidative stress (Singh *et al.* 2003, 2005; Michel and Pistorius 2004) and by exposure to high light intensity even in the presence of adequate levels of Fe (Havaux *et al.* 2005). In the latter study, it was shown that the high light stress induced the formation of reactive oxygen species, which may be the primary signal for induction of these proteins (Michel and Pistorius 2004). In one species of cyanobacteria it has been shown that severe Fe limitation results in an excess of IsiA units, which effectively shield the photosystems by absorbing energy and releasing it as heat via carotenoids (Yeremenko *et al.* 2004; Ihalainen *et al.* 2005). Here we see a macromolecular pigment-binding complex that can function in both light-harvesting and in photoprotection.

Several cyanobacteria have dispensed with PBSs and use a membrane-intrinsic Chl *a/b* antenna related to IsiA and CP43 (reviewed in Green 2003; Chen and Bibby 2005). These cyanobacteria are referred to as "prochlorophytes," even though the group consists of three unrelated genera: *Prochloron*, *Prochlorothrix*, and *Prochlorococcus*. Figure 4 shows a phylogenetic tree based on the amino acid sequences of this family. A group containing most of the prochlorophyte Chl *a/b* (Pcb) proteins is sister group to the IsiA proteins, suggesting both diverged from a common ancestor. However, several of the Chl *a/b* proteins, encoded by the *pcbC* genes, form a separate branch, which appears to have originated via an earlier duplication of a CP43-related gene (van der Staay *et al.* 1998; Green 2003). It is particularly striking that a strain of *Prochlorococcus* adapted to a high-light surface environment (Med in Figure 4) has only one *pcbA* gene, whereas a strain living in a very low-light environment (1375) has a large number of *pcbA*-like genes (Garczarek *et al.* 2000, 2001).

Acaryochloris, an unusual cyanobacterium containing mostly Chl *d*, is often grouped with the prochlorophytes because its Chl *d*-binding antenna proteins are related to the prochlorophyte Chl *a/b* proteins. It also lacks PBSs, although it has short rods made of four hexamer units (Hu *et al.* 1999). It has a PcbC protein closely related to *Prochlorothrix* PcbC and a PcbA protein related to the main group of Pcb proteins (Chen and Bibby 2005; Chen *et al.* 2005a). In both *Acaryochloris* and *Prochlorothrix*, expression of the *pcbC* gene is induced in response to Fe deprivation and the Chl-proteins make a ring of 18 units around the PS I trimers (Bumba *et al.* 2005; Chen *et al.* 2005b), just like the IsiA rings in cyanobacteria without Chl *b*.

In *Acaryochloris*, *Prochloron*, and *Prochlorothrix*, the *pcbC* gene is just downstream of the *pcbA* gene (Chen *et al.* 2005a). *Prochlorothrix* has an intervening *pcbB* gene, obviously derived by duplication from the *pcbA* gene, and all three are cotranscribed

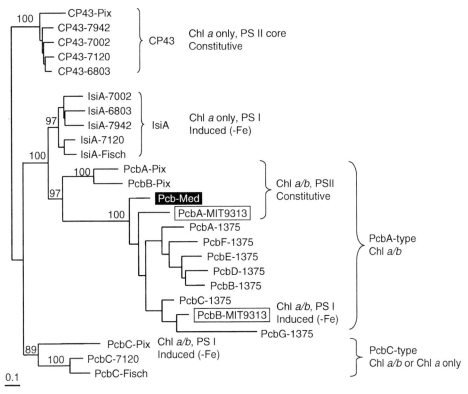

FIGURE 4. Phylogenetic tree of IsiA-Pcb-CP43 protein family, constructed from amino acid sequences using the neighbor-joining method with rate-corrected distances. Pix, *Prochlorothrix*; Fisch, *Fischerella*. Med, MIT9313, and 1375 are high-light, medium-light, and low-light ecotypes of *Prochlorococcus*. Other four-digit numbers are cyanobacterial strains: *Anabaena* PCC7120; *Anacystis nidulans* PCC7042; *Synechococcus* PCC7002; *Synechocystis* 6803. Bar represents 0.1 changes per amino acid position. Numbers on branches represent bootstrap support for that clade. Adapted from Green (2003).

(van der Staay *et al.* 1998). Interestingly, a similar gene order is found in *Fischerella*, one of two cyanobacteria that do not make Chl *b* at all but nevertheless have a *pcbC*-like gene (Geiss *et al.* 2001). In *Fischerella*, the *pcbC* gene is just downstream of the *isiA* gene, and both are induced under Fe deficiency, although they do not appear to be cotranscribed. A common gene order among unrelated organisms, which have a variety of different Chls, would usually be considered strong evidence of lateral gene transfer, as suggested by Chen *et al.* (2005a). However, the evolutionary model proposed by Chen and colleagues would not necessarily involve lateral gene transfer (Chen and Bibby 2005; Chen *et al.* 2005a).

Several recent articles have shown a novel arrangement of Pcb proteins flanking PS II dimers in *Prochloron* and *Prochlorococcus* (Bibby *et al.* 2003a, b). A similar arrangement has been found in *Acaryochloris* (Chen *et al.* 2005c). The PS II-associated proteins appear to be encoded by the *pcbA*-type genes, which are constitutively expressed and not induced in response to Fe limitation. Members of the IsiA/Pcb family are therefore involved in acclimation to light limitation as well as acclimation to Fe limitation by balancing delivery of energy to both photosystems and dissipating excess energy. This prevents direct damage due to light energy and secondary damage due to the production of active oxygen species.

Given the general usefulness of this antenna family, it is rather surprising that it was completely lost during the primary endosymbiosis that gave rise to the chloroplast. Absolutely no trace of the family can be found in the genomes of red or green algae or plants or in the EST collections from the glaucophyte *Cyanophora paradoxa* (D. G. Durnford, personal communication; D. Bhattacharya, personal communication). Its functions appear to have been taken over by members of the LHC superfamily (Section VI.). A possible hint may be found in the presence of a putative ancestor of the LHC superfamily in most cyanobacterial genomes (see later).

V. ABOUT CHLOROPHYLLS

The major groups of algae are distinguished by their Chls (see Figure 1). However, more and more evidence has accumulated to suggest that the evolution of Chl synthesis occurred independently of the evolution of the Chl-binding proteins

(Green 2003). In fact, the major divisions reflect relatively minor modifications of Chl structure (Figure 5) and the presence or absence of a very small number of enzymes. The antenna protein families appear to have adapted to bind whatever Chls their owners can synthesize (Green 2001, 2003). This concept is supported by in vitro studies where red algal and dinoflagellate antenna proteins have been reconstituted successfully with Chls their ancestors never encountered; the reconstituted pigment-proteins showed native spectroscopic properties (Grabowski *et al.* 2001; Miller *et al.* 2005).

Chl *b* differs from Chl *a* only in having a formyl group rather than a methyl group at C-7 (see Figure 5). Protochlorophyllide *b* is synthesized from protochlorophyllide *a* by protochlorophyllide *a* oxygenase (CaO), a mixed-function oxygenase (Tanaka *et al.* 1998; Oster *et al.* 2000). Both protochlorophyllides are converted to their respective Chls by the same set of enzymes. Chl *b* extends red light absorption slightly to the blue but probably makes very little difference except in an extremely light-limited

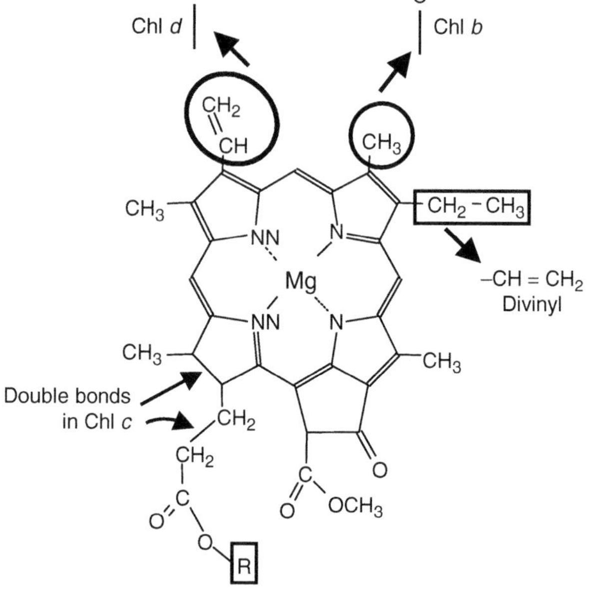

FIGURE 5. The chlorophylls (Chls). The Chl *a* structure is shown, with the modifications found in Chls *b*, *d*, and *c*. R = H in most Chls *c* and a phytyl group in the other Chls. Numbering according to Scheer (2003).

habitat. *Prochlorococcus* uses divinyl-Chl *a* and divinyl-Chl *b* simply because it lacks the divinyl reductase enzyme (Nagata *et al.* 2005). The CaO sequences of *Chlamydomonas*, *Arabidopsis*, *Prochloron*, and *Prochlorothrix* are all clearly related even though the two prochlorophytes are not related to each other or to the ancestor of the chloroplast (Tomitani *et al.* 1999).

Because most cyanobacteria do not make Chl *b* and do not have any *pcb*-like genes, this supports the idea that the *cao* gene was spread by lateral transfer among the three prochlorophyte genera and the cyanobacterial ancestor of the chloroplast (Green 2003). Those cyanobacteria that had the ability to make Chl *b* adapted members of the IsiA/Pcb family to bind it. Tomitani *et al.* (1999) suggested an alternative explanation: that all cyanobacteria originally had the *cao* gene but most cyanobacterial lines subsequently lost it. Regardless of which scenario might be more plausible, the close relatedness of the plant and cyanobacterial CaO implies that the first photosynthetic eukaryote had a *cao* gene and made Chl *b*. The gene would subsequently have been lost in the red and glaucophyte lineages (see Figure 1).

Chl *d* has a formyl group instead of a vinyl group at C-3 (see Figure 5). This does move the absorption maximum substantially to the red, giving *Acaryochloris* a distinct ecological niche in a low-light, red-enriched environment (Kühl *et al.* 2005). *Acaryochloris* makes very little Chl *a*, which means that the PS I core complex and reaction center use Chl *d* exclusively (Hu *et al.* 1998) and that most of the Chl in PS II is Chl *d*, although there is still some discussion about the precise role of Chl *a* in PS II (Chen and Bibby 2005). This means that Chl *a* is not unique in its photosynthetic role—that role can be filled by other actors. It also means that the proteins that bind this Chl *d* have adapted to do so quite effectively. As pointed out in the previous section, the closely related Pcb proteins bind Chl *a* only (*Fischerella*), Chl *d* only (*Acaryochloris*), and Chls *a* and *b* in various ratios (*Prochloron*, *Prochlorothrix*, and *Prochlorococcus*).

The Chls *c* show the greatest structural differences compared to Chl *a*: the C17-C18 double bond is not reduced by protochlorophyllide oxidoreductase, and there is a double bond in the C-17 side chain (see Figure 5). This modification has the effect of diminishing the red Chl peak and increasing the Soret band absorption strength (Scheer 2003). In addition, most Chls *c* do not have a long hydrocarbon tail esterified onto C-21, although there is one Chl c_2 that has a galactolipid tail (Garrido *et al.* 2000). Chl c_2 differs from Chl c_1 in retaining the divinyl group at C-8. The synthesis of Chl *c* requires that there be at least one novel enzymatic activity, the hypothetical 17^2 oxidase (Green 2003).

The Chl *c* family of pigments is found only in the four groups of algae that are the result of secondary endosymbiosis involving a red algal endosymbiont (see Figure 1). This implies that either the red algal endosymbiont(s) made Chl *c*, or that the four groups had a common ancestor that evolved the ability to do so.

VI. THE LHC SUPERFAMILY

A. The Light-Harvesting Antennas

The LHC superfamily includes a large group of Chl-proteins that have three transmembrane helices and bind Chl and carotenoids. Sequence relatedness of all these proteins, whether they bind Chl *a* and *b*, Chl *a* and *c*, or only Chl *a* shows that they had a common ancestor (Green and Pichersky 1994; Durnford *et al.* 1999; Durnford 2003; Green 2003). Based on their sequence similarity, they should have the same overall three-dimensional structure as pea LHCII, which has been determined by electron and x-ray crystallography (Kühlbrandt *et al.* 1994; Liu *et al.* 2004; Standfuss *et al.* 2005). The most conserved parts of the sequence are the first and third transmembrane helices, which are so similar to each other that they must have originated by gene duplication (Green and Pichersky 1994).

The family has diversified most extensively in the green lineage, with antennas specific for PS I (LHCI) and PS II (CP29, CP26, and CP24) (Figure 6). Most of the predominant Chl *a/b* protein, LHCII, is associated with PS II in the grana (stacked) thylakoids, but under higher light intensities a fraction of the population is able to migrate to the stroma (single) thylakoids to transfer excitation to PS I (reviewed in Green *et al.* 2003). Phylogenetic trees show that the different types of Chl *a/b* protein must have evolved after the green lineage and red

lineage separated, because they always form a separate branch from the red algal and Chl *a/c* proteins (Figure 6; Green and Durnford 1996; Durnford *et al.* 1999). This means that we should not expect to find functional analogs of specific Chl *a/b* complexes in the red lineage.

As more sequence data became available from the red lineage, it was found that most of the Chl *a/c* proteins are in a group well separated from the red algal Chl *a* proteins (see Figure 6). In the red algae, the LHC is only associated with PS I (LHCI) (Gantt *et al.*

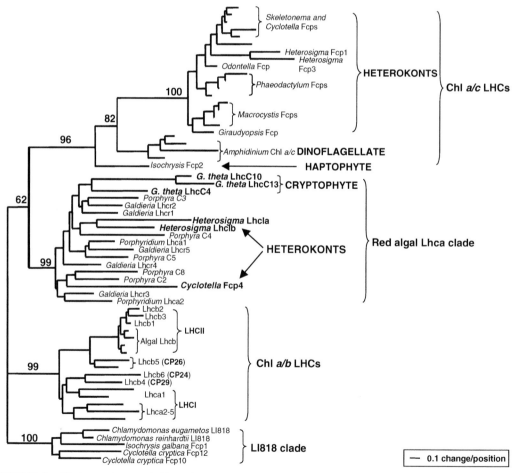

FIGURE 6. Phylogenetic tree of chlorophyll (Chl)-binding members of light-harvesting complex (LHC) superfamily, constructed from amino acid sequences using neighbor-joining. Bar represents 0.1 changes per amino acid position. Numbers on branches represent maximum likelihood bootstrap support for that clade. Adapted from Green (2003).

2003). However, several Chl *a/c* proteins can be seen in the red algal LHCI cluster. In the algae with Chl *c,* there is little evidence for the association of specific LHCs with any particular photosystem, with the possible exception of the cryptophyte algae where a Chl *a/c* complex may be associated primarily with PS I (Bathke *et al.* 1999). It was therefore not surprising to see the three cryptophyte sequences in the red cluster, but there were also sequences from two heterokonts: the diatom *Cyclotella* and the raphidophyte *Heterosigma.* Addition of sequences from the diatom *Thalassiosira pseudonana* showed that some of them also fall into this clade, with good bootstrap support (K. Ishida and B.R. Green, unpublished). We have not yet been able to determine if any of these LHCI-like proteins are preferentially associated with PS I (Bentley *et al.* 2005).

There is also a new clade that we have called the LI818 clade, because it contains sequences from a haptophyte and a diatom that are most closely related to the LI818 protein of *Chlamydomonas* (Richard *et al.* 2000). It is still not clear whether the LI818 protein binds Chl, but its expression pattern is different from that of other LHC family members (Savard *et al.* 1996; Oeltjen *et al.* 2002, 2004). It is the only clade that has members from both the red and green lineages.

B. The Stress–Response Connection

The green lineage has several additional members of the LHC superfamily that are involved in responding to high light stress (Figure 7) (reviewed in Green and Kühlbrandt 1995; Jansson 1999; Heddad and Adamska 2002; Teramoto *et al.* 2004). First to be discovered were the early light-induced proteins (Elips), which bind carotenoids but little or no Chl, even though they retain some of the conserved Chl-binding sites in the first and third helices (Green *et al.* 1991). Then it was discovered that a 22-kDa protein of PS II (now called PsbS) had four transmembrane helices and appeared to be the

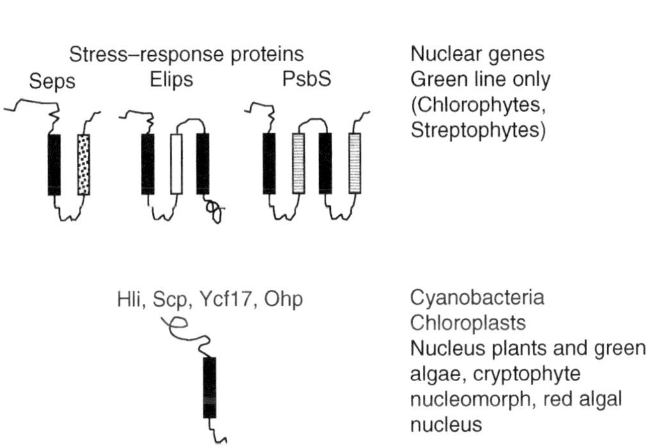

Type	Where found
Light-harvesting Chl-proteins Chl *a/b*, Chl *a/c*, Chl *a* LHCs	Nuclear genes All eukaryotes except glaucophytes

Stress–response proteins Seps Elips PsbS	Nuclear genes Green line only (Chlorophytes, Streptophytes)

Hli, Scp, Ycf17, Ohp	Cyanobacteria Chloroplasts Nucleus plants and green algae, cryptophyte nucleomorph, red algal nucleus

FIGURE 7. Members of the light-harvesting complex (LHC) superfamily. Black bars: transmembrane helices with significant sequence similarity and at least two conserved chlorophyll (Chl)-binding sites. Patterned bars: transmembrane helices that share similarity only with those of same pattern. N- and C-termini and nonmembrane segments have lengths that vary widely among members of a group. Elips, early light-induced proteins; Hli, high light-induced; LHCs, light-harvesting complexes; PsbS, 22-kDa protein of PS II; Scp, small Cab-like proteins; Seps, stress-expressed proteins; Ycf17, plastid-encoded homolog of Hli and Scp found in red line algae.

result of tandem duplication of a gene for a two-helix protein (Green and Pichersky 1994). A few years later the two-helix stress-expressed proteins (Seps) were discovered, which fit that model very well (Heddad and Adamska 2000). PsbS can bind Chl and carotenoids in vitro, although whether it does so in its native habitat is unclear.

In higher plants PsbS is essential for non-photochemical quenching, the process by which excitation energy transfer to PS II RC is downregulated (Li *et al.* 2000). However, it is not the only member of the family to contribute to nonphotochemical quenching: LHCII and the minor Chl *a/b* complexes of PS II are also involved (reviewed by Bassi and Caffarri 2000). Homologs of the plant Seps, Elips, and PsbS have been found in *Chlamydomonas*; many are induced by transfer to high light and/or oxidative stress (Ledford *et al.* 2004; Teramoto *et al.* 2004). However, genome sequencing revealed no sign of PsbS or Seps in the diatoms *Thalassiosira pseudonana* or *Phaeodactylum tricornutum* nor in the red algae *Porphyra purpurea* or *Gauldiaria* sp. (Armbrust *et al.* 2004; B.R. Green and K. Ishida, unpublished). This suggests that the two- and four-helix members of the LHC family evolved in the green line, after the chlorophyte–rhodophyte divergence. Given the important role of PsbS in plants, it also implies that the molecular mechanisms for dissipating excitation energy must be different in the green and red lineages.

C. Prokaryotic Ancestry of the LHC Superfamily

The light-harvesting members of the LHC family are clearly a eukaryotic innovation, because no genes for any two-helix, three-helix, or four-helix members of the LHC family have been found in any cyanobacterial genome. They probably originated after the glaucophyte line diverged from the common ancestor of the red and green lines, because no LHCs have been detected in glaucophytes (Steiner and Löffelhardt 2002).

However, a putative ancestor involved in resistance to high light stress is found in most cyanobacteria. The high light-induced (Hli) proteins are small and have one trans-membrane helix, which shares considerable sequence relatedness with the first and third helices of the light-harvesting proteins and the Elips (reviewed in Green and Kühlbrandt 1995; Heddad and Adamska 2002; Green 2003). Based on molecular replacement using the pea LHCII structure as a template, we proposed that two Hli proteins could be cross-linked by ionic interactions in the same way as the first and third helices of LHCII, making a homodimer with conserved Chl and carotenoid binding sites (Green and Kühlbrandt 1995). In *Synechocystis* 6803 and *Synechococcus* PCC7942, the *hli* genes are induced in response to high light stress or nutrient deprivation (Funk and Vermaas 1999; He *et al.* 2001), and there is some evidence that the Hli proteins bind carotenoids (Dolganov *et al.* 1994).

Genes encoding Hli-like proteins (annotated as *ycf17*) are found on the plastid genomes of glaucophytes, rhodophytes, and cryptophytes. In rhodophytes and cryptophytes, copies of the *ycf17* gene are also found in the nuclear and nucleomorph genomes, respectively, whereas the gene has disappeared from diatom plastid genomes and is found only in the diatom nucleus (Armbrust *et al.* 2004). This clearly shows successive stages in plastid-to-nucleus gene transfer. Genes encoding Hli-like proteins are also found in the *Arabidopsis* and *Chlamydomonas* nuclear genomes, and both are upregulated in response to high light stress (Jansson 1999; Andersson *et al.* 2003; Teramoto *et al.* 2004). Unfortunately, nothing is known about the function of any of the proteins encoded by plastid *ycf17* genes.

We have proposed that the original function of the LHC superfamily was acclimation to high light and other stresses and that the light-harvesting function was acquired later (Green and Pichersky 1994; Green and Kühlbrandt 1995; see also Montané and Kloppstech 2000). Our model for the ancestor of the three-helix LHCs was an Hli-like protein (Green and Kühlbrandt 1995). At some point after the origin of the chloroplast, a copy of an *hli* gene was transferred to the nucleus, where it was expressed and continued to play a photoprotective role. We originally proposed that an ancestral *hli* gene fused with another sequence to encode a two-helix protein, a tandem duplication produced the gene for a four-helix protein (like PsbS), and the fourth helix was lost to give the three-helix members of the family, which just retain a small amphipathic helix at the C-terminus (Green and Pichersky 1994; Green and Kühlbrandt 1995). The involvement of two- and four-helix intermediates now seems unlikely, because the extant two- and four-helix members of the family are found only in the green line. However, the striking sequence relatedness between Hlips and the first and third LHC helices supports the idea that the latter had its origin in the tandem duplication of an *hli*-like gene, with intervening sequence becoming the second transmembrane helix and allowing the first and third helices to be oriented so that they could form the ionic bridges. The first three-helix pigment-binding protein would have had a predominantly photoprotective role, but some of its descendants would have acquired the ability to bind more Chl and eventually become light-harvesting antennas with "built-in" photoprotection.

VII. OVERVIEW

The evolution of the light-harvesting antennas must be intimately connected with the evolution of mechanisms for getting rid of excess excitation energy, especially in an oxygen-rich environment. Carotenoids are well suited as energy dissipators, so it is no surprise that they are found along with Chls in most parts of the photosynthetic apparatus: reaction centers (mostly β-carotene), the internal antennas of PS II (CP47 and CP43), the IsiA/Pcb antennas of cyanobacteria,

and the LHC superfamily of photosynthetic eukaryotes. Because of their variety in structure, there are also some carotenoids that can act as antennas, particularly fucoxanthin and its derivatives, found in heterokonts and haptophytes, and the peridinin-Chl *a* protein of dinoflagellates (Macpherson and Hiller 2003). The PBSs are the only antenna complex that does not contain carotenoids, but it has recently been shown that a caroteno-protin is associated with cyanobacterial thylakoids and appears to play a photoprotective role (Wilson *et al.* 2006).

There are still many unanswered questions about the evolution of light-harvesting antennas. Why should PBSs have been lost from the green lineage, while they were retained in the glaucophytes and the red algae? We understand the genetic mechanisms involved in the proliferation and diversification of gene families, but the replacement of the IsiA-Pcb family by the LHC superfamily in photosynthetic eukaryotes continues to be a puzzle. If the cyanobacterial ancestor of the chloroplast made Chl *b*, it was most likely bound in a prochlorophyte-like Chl *a/b* antenna (Tomitani *et al.* 1999). Last but not least, there is the mystery of why all the terrestrial photosynthesizers have Chl *a/b* antennas, whereas the majority of photosynthetic eukaryotes in the marine environment have Chl *a/c* antennas. Because both types of antenna are efficient, the answer may turn out to have more to do with the fact that the latter acquired their chloroplasts via secondary endosymbiosis(es). The successful integration of one eukaryotic cell into another must have required a huge number of adaptations in all aspects of cellular structure and function, only some of which involved the photosynthetic apparatus.

Acknowledgments

Work in the author's laboratory was supported by the Natural Sciences and Engineering Research Council of Canada and the award of a Killam Research Fellowship from the Canada Council. I thank the members of my laboratory for proofreading the manuscript and providing helpful input and Dr. Ken-Ichiro Ishida for Figure 6.

An earlier version of this chapter was presented at a symposium honoring the award of the Japan Prize to Thomas Cavalier-Smith, Tokyo, December 2004.

References

Anbar, A.D., and Knoll, A.H. (2002). Proterozoic ocean chemistry and evolution: a bioinorganic bridge? *Science* **297**: 1137–1142.

Andersson, U., Heddad, M., and Adamska, I. (2003). Light stress-induced one-helix protein of the chlorophyll *a/b*-binding family associated with photosystem I. *Plant Physiol.* **132**: 811–821.

Apt, K.E., Collier, J.L., and Grossman, A.R. (1995). Evolution of the phycobiliproteins. *J. Mol. Biol.* **248**: 79–96.

Armbrust, E.V., Berges, J.A., Bowler, C., Green, B.R., Martinez, D., and 38 others. (2004). The genome of the diatom *Thalassiosira pseudonana*: ecology, evolution, and metabolism. *Science* **306**: 79–86.

Asada, K. (1999). The water-water cycle in chloroplasts: scavenging of active oxygens and dissipation of excess photons. *Annu. Rev. Plant Physiol. Plant Mol. Biol.* **50**: 601–639.

Bassi, R., and Caffarri, S. (2000). Lhc proteins and the regulation of photosynthetic light harvesting function by xanthophylls. *Photosynth. Res.* **64**: 243–256.

Bathke, L., Rhiel, E., Krumbein, W.E., and Marquardt, J. (1999). Biochemical and immunochemical investigations on the light-harvesting system of the cryptophyte *Rhodomonas* sp.: evidence for a photosystem I specific antenna. *Plant Biol.* **1**: 516–523.

Bentley, F., Harnett, J., Ishida, K., and Green, B.R. (2005). Investigation of PSI-associated light-harvesting proteins in a chromophyte alga. *Photosynthesis: Fundamental Aspects to Global Perspectives*. A. van der Est and D. Bruce, eds. Lawrence, KS, ACP Press, pp. 161–163.

Blankenship, R.E., Sadekal, S., and Raymond, J. (2007). The evolutionary transition from anoxygenic to oxygenic photosynthesis. *Evolution of Primary Producers in the Sea*. P.G. Falkowski and A.H. Knoll, eds. Boston, Elsevier, pp. 21–38.

Bibby, T.S., Mary, I., Nield, J., Partensky, F., and Barber, J. (2003a). Low-light-adapted *Prochlorococcus* species possess specific antennae for each photosystem. *Nature* **424**: 1051–1054.

Bibby, T.S., Nield, J., and Barber, J. (2001). Iron deficiency induces the formation of an antenna ring around trimeric photosystem I in cyanobacteria. *Nature* **412**: 743–745.

Bibby, T.S., Nield, J., Chen, M., Larkum, A.W.D., and Barber, J. (2003b). Structure of a photosystem II supercomplex isolated from *Prochloron didemni* retaining its chlorophyll *a/b* light-harvesting system. *Proc. Natl. Acad. Sci. U S A* **100**: 9050–9054.

Boekema, E.J., Hifney, A., Yakushevska, A.E., Piotrowski, M., Keegstra, W., Berry, S., Michel, K.-P., Pistorius, E.K., and Kruip, J. (2001). A giant chlorophyll-protein complex induced by iron deficiency in cyanobacteria. *Nature* **412**: 745–748.

Broughton, M.J., Howe, C.J., and Hiller, R.G. (2006). Distinctive organization of genes for light-harvesting proteins in the cryptophyte alga *Rhodomonas*. *Gene* **369**: 72–79.

Bumba, L., Prasil, O., and Vacha, F. (2005). Antenna ring around trimeric Photosystem I in chlorophyll *b* containing cyanobacterium *Prochlorothrix hollandica*. *Biochim. Biophys. Acta* **1704**: 1–5.

Burnap, R.L., Troyan, T., and Sherman, L.A. (1993). The highly abundant chlorophyll-protein complex of iron-deficient *Synechococcus* sp. PCC7942 (CP43) is encoded by the *isiA* gene. *Plant Physiol.* **103**: 893–902.

Cavalier-Smith, T. (2000). Membrane heredity and early chloroplast evolution. *Trends Plant Sci.* **5**: 174–182.

Chen, M., and Bibby, T.S. (2005). Photosynthetic apparatus of antenna-reaction centres supercomplexes in oxyphotobacteria: insight through significance of Pcb/IsiA proteins. *Photosyn. Res.* **86**: 165–173.

Chen, M., Bibby, T.S., Nield, J., Larkum, A., and Barber, J. (2005b). Iron deficiency induces a chlorophyll *d*-binding Pcb antenna system around Photosystem I in *Acaryochloris marina*. *Biochim. Biophys. Acta* **1708**: 367–374.

Chen, M., Bibby, T.S., Nield, J., Larkum, A.W.D., and Barber, J. (2005c). Structure of a large photosystem II supercomplex from *Acaryochloris marina*. *FEBS Lett.* **579**: 1306–1310.

Chen, M., Hiller, R.G., Howe, C.J., and Larkum, A.W.D. (2005a). Unique origin and lateral transfer of prokaryotic chlorophyll-*b* and chlorophyll-*d* light-harvesting systems. *Mol. Biol. Evol.* **22**: 21–28.

Delwiche, C.F. (2007). The origin and evolution of dinoflagellates. *Evolution of Primary Producers in the Sea*. P.G. Falkowski and A.H. Knoll, eds. Boston, Elsevier, pp. 191–205.

Dolganov, N.A.M., Bhaya, D., and Grossman, A.R. (1994). Cyanobacterial protein with similarity to the chlorophyll *a/b* binding proteins of higher plants: evolution and regulation. *Proc. Natl. Acad. Sci. U S A* **92**: 636–640.

Durnford, D.G. (2003). Structure and regulation of algal light-harvesting complex genes. *Photosynthesis in Algae*. A.W.D. Larkum, S.E. Douglas, and J.A. Raven, eds. Dordrecht, Kluwer Academic Publishers, pp. 63–82.

Durnford, D.G., Deane, J.A., Tan, S., McFadden, G.I., Gantt, E., and Green, B.R. (1999). A phylogenetic assessment of the eukaryotic light-harvesting antenna proteins, with implications for plastid evolution. *J. Mol. Evol.* **48**: 59–68.

Funk, C., and Vermaas, W. (1999). A cyanobacterial gene family coding for single-helix proteins resembling part of the light-harvesting proteins from higher plants. *Biochemistry* **38**: 9397–9404.

Gantt, E., Grabowski, B., and Cunningham, F.X. (2003). Antenna systems of red algae: Phycobilisomes with photosystem II and chlorophyll complexes with photosystem I. *Light-Harvesting Antennas in Photosynthesis*. B.R. Green and W.W. Parson, eds. Dordrecht, Kluwer Academic Publishers, pp. 307–322.

Garczarek, L., Hess, W.R., Holtzendorff, J., van der Staay, G.W.M., and Partensky, F. (2000). Multiplication of antenna genes as a major adaptation to low light in a marine prokaryote. *Proc. Natl. Acad. Sci. U S A* **97**: 4098–4101.

Garczarek, L., van der Staay, G.W.M., Hess, W.R., Le Gall, F., and Partensky, F. (2001). Expression and phylogeny of the multiple antenna genes of the low-light adapted strain *Prochlorococcus marinus* SS120 (Oxyphotobacteria). *Plant Mol. Biol.* **46**: 683–693.

Garrido, J.L., Otero, J., Maestro, M.A., and Zapata, M. (2000). The main nonpolar chlorophyll *c* from *Emiliana huxleyi* (Prymnesiophyceae) is a chlorophyll c_2-monogalactosyldiacylglyceride ester. A mass spectrometry study. *J. Phycol.* **36**: 497–505.

Geiss, U., Vinnemeier, J., Schoor, A., and Hagemann, M. (2001). The iron-regulated *isiA* gene of *Fischerella muscicola* strain PCC73103 is linked to a likewise regulated gene encoding a Pcb-like chlorophyll-binding protein. *FEMS Microbiol. Lett.* **197**: 123–129.

Gibbs, S.P. (1981). The chloroplasts of some algal groups may have evolved from endosymbiotic eukaryotic algae. *Ann. N Y Acad. Sci.* **361**: 193–208.

Grabowski, B., Cunningham, F.X. Jr., and Gantt, E. (2001). Chlorophyll and carotenoid binding in a simple red algal light-harvesting complex crosses phylogenetic lines. *Proc. Natl. Acad. Sci. U S A* **98**: 2911–2916.

Green, B.R. (2001). Was "molecular opportunism" a factor in the evolution of different photosynthetic light-harvesting pigment systems? *Proc. Natl. Acad. Sci. U S A* **98**: 2119–2121.

Green, B.R. (2003). The evolution of light-harvesting antennas. *Light-Harvesting Antennas in Photosynthesis*. B.R. Green and W.W. Parson, eds. Dordrecht, Kluwer Academic Publishers, pp. 129–168.

Green, B.R., and Durnford, D.G. (1996). The chlorophyll-carotenoid proteins of oxygenic photosynthesis. *Annu. Rev. Plant Physiol. Plant Mol. Biol.* **47**: 685–714.

Green, B.R., and Gantt, E. (2005). Distal Antennas of Photosystem II. *Photosystem II: The Water/Plastoquinone Oxido-Reductase In Photosynthesis*. T. Wydrzynski and K. Satoh, eds. Dordrecht, Kluwer Academic Press, pp. 23–44.

Green, B.R., and Kühlbrandt, W. (1995). Sequence conservation of light-harvesting and stress-response proteins in relation to the three-dimensional molecular structure of LHCII. *Photosyn. Res.* **44**: 139–148.

Green, B.R., and Pichersky, E. (1994). Hypothesis for the evolution of three-helix Chl a/b and Chl a/c light-harvesting antenna proteins from two-helix and four-helix ancestors. *Photosyn. Res.* **39**: 149–162.

Green, B.R., Pichersky, E., and Kloppstech, K. (1991). The chlorophyll *a/b*-binding light-harvesting antennas of green plants: the story of an extended gene family. *Trends Biochem. Sci.* **16**: 181–186.

Green, B.R., Anderson, J.M., and Parson, W.W. (2003). Photosynthetic membranes and their light-harvesting antennas. *Light-Harvesting Antennas in Photosynthesis*. B.R. Green and W.W. Parson, eds. Dordrecht, Kluwer Academic Publishers, pp. 1–28.

Hackett, J.D., Yoon, H.S., Butterfield, N.J., Sanderson, M.J., and Bhattacharya, B. (2007). Plastid endosymbiosis: Sources and timing of major events. *Evolution of Primary Producers in the Sea*. P.G. Falkowski and A.H. Knoll, eds. Boston, Elsevier, pp. 109–131.

Havaux, M., Guedeney, G., Hagemann, M., Yeremenko, N., Matthijs, H.C.P., and Jeanjean, R. (2005). The chlorophyll-binding protein IsiA is inducible by high light and protects the cyanobacterium *Synechocystis* PCC6803 from photooxidative stress. *FEBS Lett.* **579**: 2289–2293.

He, Q., Dolganov, N., Björkman, O., and Grossman, A.R. (2001). The high light-inducible polypeptides in *Synechocystis* PCC6803. Expression and function in high light. *J. Biol. Chem.* **276**: 306–314.

Heddad, M., and Adamska, I. (2000). Light stress-regulated two-helix proteins in *Arabidopsis thaliana* related to the chlorophyll *a/b*-binding gene family. *Proc. Natl. Acad. Sci. U S A* **97**: 3741–3746.

Heddad, M., and Adamska, I. (2002). The evolution of light stress proteins in photosynthetic organisms. *Comp. Funct. Genomics* **3**: 504–510.

Hu, Q., Marquardt, J., Iwasaki, I., Miyashita, H., Kurano, N., Mörschel, E., and Miyachi, S. (1999). Molecular structure, localization and function of biliproteins in the chlorophyll *a/d* containing oxygenic photosynthetic prokaryote *Acaryochloris marina*. *Biochim. Biophys. Acta* **1412**: 250–261.

Hu, Q., Miyashita, H., Miyashita, H., Iwasaki, I., Kurano, N., Miyachi, S., Iwaki, M., and Itoh, S. (1998). A photosystem I reaction center driven by chlorophyll *d* in oxygenic photosynthesis. *Proc. Natl. Acad. Sci. U S A* **95**: 13319–13323.

Ihalainen, J.A., D'Haene, S., Yeremenko, N., van Roon, H., Arteni, A.A., Boekema, E.J., van Grondelle, R., Matthijs, H.C.P., and Dekker, J.P. (2005). Aggregates of the chlorophyll-binding protein IsiA (CP43') dissipate energy in cyanobacteria. *Biochemistry* **44**: 10846–10853.

Ishida, K., and Green, B.R. (2002). Second- and third-hand chloroplasts in dinoflagellates: Phylogeny of oxygen-evolving enhancer 1 (PsbO) protein reveals replacement of a nuclear-encoded plastid gene by that of a haptophyte tertiary endosymbiont. *Proc. Natl. Acad. Sci. U S A* **99**: 9294–9299.

Jansson, S. (1999). A guide to the Lhc genes and their relatives in *Arabidopsis*. *Trends Plant Sci*. **4**: 236–240.

Joshua, S., and Mullineaux, C.W. (2005). Phycobilisome diffusion is required for light-state transitions in cyanobacteria. *Plant Physiol*. **135**: 2112–2119.

Kühl, M., Chen, M., Ralph, P.J., Schreiber, U., and Larkum, A.W.D. (2005). A niche for cyanobacteria containing chlorophyll *d*. *Nature* **433**: 820.

Kühlbrandt, W., Wang, D.N., and Fujiyoshi, Y. (1994). Atomic model of plant light-harvesting complex by electron crystallography. *Nature* **367**: 614–621.

Ledford, H.K., Baroli, I., Shin, J.W., Fischer, B.B., Eggen, R.I.L., and Niyogi, K. K. (2004). Comparative profiling of lipid-soluble antioxidants and transcripts reveals two phases of photo-oxidative stress in a xanthophyll-deficient mutant of *Chlamydomonas reinhardtii*. *Mol. Gen. Genomics* **272**: 470–479.

Li, X.P., Bjorkman, O., Shih, C., Grossman, A.R., Rosenquist, M., Jansson, S., and Niyogi, K.K. (2000). A pigment-binding protein essential for regulation of photosynthetic light harvesting. *Nature* **403**: 391–395.

Liu, Z., Yan, H., Wang, K., Kuang, T.Y., Zhang, J., Gul, L., An, X., and Chang, W.R. (2004). Crystal structure of spinach major light-harvesting complex at 2.72Å resolution. *Nature* **428**: 287–292.

Macpherson, A.N., and Hiller, R.G. (2003). Light-harvesting systems in chlorophyll *c*-containing algae. *Light-Harvesting Antennas in Photosynthesis*. B.R. Green and W.W. Parson, eds. Dordrecht, Kluwer Academic Publishers, pp. 323–352.

McFadden, G.I. (2001). Primary and secondary endosymbiosis and the origin of plastids. *J. Phycol*. **37**: 951–959.

Michel, K.-P., and Pistorius, E.K. (2004). Adaptation of the photosynthetic electron transport chain in cyanobacteria to iron deficiency: The function of IdiA and IsaA. *Physiol. Plant* **120**: 36–50.

Miller, D.J., Catmull, J., Puskeiler, R., Tweedale, H., Sharples, F.P., and Hiller, R.G. (2005). Reconstitution of the peridinin-chlorophyll a protein (PCP): evidence for functional flexibility in chlorophyll binding. *Photosyn. Res*. **86**: 229–240.

Mimuro, M., and Kikuchi, H. (2003). Antenna systems and energy transfer in cyanophyta and rhodophyta. *Light-Harvesting Antennas in Photosynthesis*. B.R. Green and W.W. Parson, eds. Dordrecht, Kluwer Academic Publishers, pp. 281–306.

Montané, M.-H., and Kloppstech, K. (2000). The family of light-harvesting-related proteins (LHCs, ELIPs, HLIPs): was the harvesting of light their primary function? *Gene* **258**: 1–8.

Nagata, N., Tanaka, R., Satoh, S., and Tanaka, A. (2005). Identification of a vinyl reductase gene for chlorophyll synthesis in *Arabidopsis thaliana* and implications for the evolution of *Prochlorococcus* species. *Plant Cell* **17**: 233–240.

Nisbet, E.G., and Sleep, N.H. (2001). The habitat and nature of early life. *Nature* **409**: 1083–1091.

Noctor, G., and Foyer, C.H. (1998). Ascorbate and glutathione: keeping active oxygen under control. *Annu. Rev. Plant Physiol. Plant Mol. Biol*. **49**: 249–279.

Oeltjen, A., Krumbein, W.E., and Rhiel, E. (2002). Investigations on transcript sizes, steady state mRNA concentrations and diurnal expression of genes encoding fucoxanthin chlorophyll *a/c* light harvesting polypeptides in the centric diatom *Cyclotella cryptica*. *Plant Biol*. **4**: 250–257.

Oeltjen, A., Marquardt, J., and Rhiel, E. (2004). Differential circadian expression of genes fcp2 and fcp6 in *Cyclotella cryptica*. *Int. Microbiol*. **7**: 127–131.

Oster, U., Tanaka, R., Tanaka, A., and Rüdiger, W. (2000). Cloning and functional expression of the gene encoding the key enzyme for chlorophyll *b* biosynthesis (CAO) from *Arabidopsis thaliana*. *Plant J*. **21**: 305–310.

Richard, C., Ouellet, H., and Guertin, M. (2000). Characterization of the LI818 polypeptide from the green unicellular alga *Chlamydomonas reinhardtii*. *Plant Mol. Biol*. **42**: 303–316.

Rodríguez-Ezpeleta, N., Brinkmann, H., Burey, S.C., Roure, B., Burger, G., Löffelhardt, W., Bohnert, H.J., Philippe, H., and Lang, B.F. (2005). Monophyly of primary photosynthetic eukaryotes: green plants, red algae, and glaucophytes. *Curr. Biol*. **15**: 1325–1330.

Sandström, S., Ivanov, A.G., Park, Y.-I., Öquist, G., and Gustafsson, P. (2002). Iron stress responses in the cyanobacterium *Synechococcus* sp. PCC 7942. *Physiol. Plant* **116**: 255–263.

Savard, F., Richard, C., and Guertin, M. (1996). The *Chlamydomonas reinhardtii* LI818 gene represents a distant relative of the cabI/II genes that is regulated during the cell cycle and in response to illumination. *Plant Mol. Biol*. **32**: 461–473.

Scheer, H. (2003). The Pigments. *Light-Harvesting Antennas in Photosynthesis*. B.R. Green and W.W. Parson, eds. Dordrecht, Kluwer Academic Publishers, pp. 29–81.

Singh, A.K., Li, H., Bono, L., and Sherman, L.A. (2005). Novel adaptive responses revealed by transcription profiling of a *Synechocystis* sp. PCC6803 ΔIsiA mutant in the presence and absence of hydrogen peroxide. *Photosyn. Res*. **84**: 229–240.

Singh, A.K., McIntyre, L.M., and Sherman, L.A. (2003). Microarray analysis of the genome-wide response to iron deficiency and iron reconstitution in the cyanobacterium *Synechocystis* sp. PCC 6803. *Plant Physiol*. **132**: 1825–1839.

Standfuss, J., van Scheltinga, A.C.T., Lamborghini, M., and Kühlbrandt, W. (2005). Mechanisms of photoprotection and nonphotochemical quenching in pea light-harvesting complex at 2.5 A° resolution. *EMBO J*. **24**: 919–928.

Steiner, J.M., and Löffelhardt, W. (2002). Protein import into cyanelles. *Trends Plant Sci*. **7**: 72–77.

Tanaka, A., Ito, H., Tanaka, R., Tanaka, N.K., and Yoshida, K. (1998). Chlorophyll *a* oxygenase (CAO) is involved in chlorophyll *b* formation from chlorophyll *a*. *Proc. Natl. Acad. Sci. U S A* **95:** 12719–12723.

Teramoto, H., Itoh, T., and Ono, T. (2004). High-intensity-light-dependent and transient expression of new genes encoding distant relatives of light-harvesting chlorophyll *a/b* proteins in *Chlamydomonas reinhardtii*. *Plant Cell Physiol.* **45:** 1221–1232.

Ting, C.S., Rocap, G., King, J., and Chisholm, S.W. (2002). Cyanobacterial photosynthesis in the oceans: the origins and significance of divergent light-harvesting strategies. *Trends Microbiol.* **10:** 134–142.

Tomitani, A., Okada, K., Miyashita, H., Matthijs, H.C., Ohno, T., and Tanaka, A. (1999). Chlorophyll *b* and phycobilins in the common ancestor of cyanobacteria and chloroplasts. *Nature* **400:** 159–162.

Toole, C.A., and Allnutt, F.C.T. (2003). Red, cryptomonad and glaucocystophyte algal phycobiliproteins. *Light-Harvesting Antennas in Photosynthesis*. B.R. Green and W.W. Parson, eds. Dordrecht, Kluwer Academic Publishers, pp. 305–334

van der Staay, G.W.M., Yurkova, N., and Green, B.R. (1998). The 38 kDa chlorophyll *a/b* protein of the prokaryote *Prochlorothrix hollandica* is encoded by a divergent *pcb* gene. *Plant Mol. Biol.* **36:** 709–716.

Wedemayer, G., Kidd, D., and Glazer, A.N. (1996). Cryptopomonad biliproteins, bilin types and location. *Photosyn. Res.* **48:** 163–170.

Wilk, K.E., Harrop, S.J., Jankova, L., Edler, D., Keenan, G., Sharples, F., Hiller, R.G., and Curmi, P.M.G. (1999). Evolution of a light-harvesting protein by addition of new subunits and rearrangement of conserved elements: crystal structure of a cryptophyte phycoerythrin at 1.63Å resolution. *Proc. Natl. Acad. Sci. U S A* **96:** 8901–8906.

Wilson, A., Ajlani, G., Verbavatz, J.-M., Vass, I., Kerfeld, C.A., and Kirilovsky, D. (2006). A soluble carotenoid protein involved in phycobilisome-related energy dissipation in cyanobacteria. *Plant Cell* **18:** 992–1007.

Yeremenko, N., Kouril, R., Ihalainen, J.A., D'Haene, S., van Oosterwijk, N., Andrizhiyevskaya, E.G., Keegstra, W., Dekker, H.L., Hagemann, M., Boekema, E.J., Matthijs, H.C.P., and Dekker, J.P. (2004). Supramolecular organization and dual function of the IsiA chlorophyll-binding protein in cyanobacteria. *Biochemistry* **43:** 10308–10313.

Eukaryote and Mitochondrial Origins: Two Sides of the Same Coin and Too Much Ado About Oxygen

WILLIAM MARTIN

I. CELL EVOLUTION WITH AND WITHOUT ENDOSYMBIOSIS

This book is mainly about photosynthesis, and many chapters deal with eukaryotic photosynthesis, which took root in the cyanobacterial origin of chloroplasts via endosymbiosis. The host that acquired the cyanobacterial ancestor of plastids was certainly a eukaryote and certainly possessed mitochondria (see Fehling *et al.*, Chapter 6; Hackett *et al.*, Chapter 7, this volume). This chapter deals with the question of whence that eukaryote and its mitochondria arose.

On the issue of eukaryote origins, biologists are divided into various camps. The camps express and defend fundamentally different views about how eukaryotes are linked to prokaryotes in both the phylogenetic sense and in the sense of evolutionary mechanisms. Several decades of literature on the topic are marked by controversy, opinion, and mutually incompatible, and sometimes even diametrically opposed, interpretations of the same observations. Such observations can entail morphological traits, phylogenies of individual genes, genes shared with eubacteria and archaebacteria, eukaryote-specific genes, the microfossil record, ultrastructure, and so forth. When it comes to the origin of eukaryotes as viewed from the standpoint of their core metabolism, that is, the pathways through which eukaryotes supply their cells with carbon and ATP, the literature is somewhat more narrow, often inexplicit, and often contains some weighty assumptions that are critically reinspected here. In this chapter, I examine hypotheses concerning the origin of mitochondria and hydrogenosomes (anaerobic forms of mitochondria), the kind of host that might have acquired their common ancestral endosymbiont, how "Canfield" oceans fit into the picture at mitochondrial origin, and how the origin of mitochondria might have precipitated the origin of the nucleus, had the host been a *bona fide* prokaryote.

There was a time about 100 years ago when biologists were excited about the idea that chloroplasts and mitochondria might be descendants of free-living bacteria via endosymbiosis. The literature introducing endosymbiotic theory in the modern sense (or symbiogenesis, as it was originally called) as an explanatory principle in cell evolution probably took root in a 1905 article by the Russian biologist Constantin Mereschkowsky (1905) about chloroplast origin; the article was published in German and has only recently been translated into English (Mereschkowsky 1905). The literature for symbiotic origins of mitochondria probably starts with the 1918 book by the French biologist Paul Portier, *Les Symbiotes*, which I unfortunately cannot read in the original but whose gist has been summarized by Sapp (1994). Into the 1920s, the idea of symbiogenesis for the origins of organelles was taken seriously, especially by botanists. However, the idea soon fell out of grace, possibly owing to Wilson's (1928) condemnation of endosymbiosis (excerpted in Martin and Kowallik 1999) as fanciful imagination in his influential textbook *The Cell in Development and Heredity*, which banned the notion from the realm of serious scientific endeavor or, as he put it, from "polite biological society" for several decades.

Accordingly, well into the 1970s, many mainstream views on the relatedness of prokaryotes and eukaryotes, terms championed by Stanier and van Niel (1962) (see Sapp 2005 for a historical summary), entailed the notion that eukaryotes descend via direct filiation from cyanobacteria (photoautotrophs) and that algae are hence the most primitive eukaryotes, from which the remaining eukaryotes arose through loss of autotrophy (and plastids) and a transition to the heterotrophic lifestyle. That view, whose possibly last hurrah can be found in Cavalier-Smith (1975), is no longer modern, but it represents what many of today's senior scientists were taught in college, as Doolittle (1980), for example, has explained. An alternative view concerning the origin of eukaryotes from prokaryotes and not involving endosymbiosis was put forth by de Duve (1969); it focused on microbodies and involved the origin of pinocytosis in a wall-less prokaryote and the subsequent origin of phagocytosis, not dissimilar to Stanier's (1970) suggestion but differing with regard to the origin of chloroplasts, which Stanier suggested to have preceded mitochondria.

Lynn Margulis (Sagan 1967) did a great service to everyone by rediscovering some of the older literature on endosymbiosis and by summarizing a good deal of supporting biochemical evidence in its favor (Margulis 1970). Her version of endosymbiotic theory, dubbed "serial endosymbiosis theory," or SET, by F.R.J. Taylor (1974), has, however, always entailed an additional prokaryotic partner at eukaryote origins above and beyond the mitochondrion and the host (and later, the chloroplast in plants). This additional partner was a spirochaete, which was seen as the ancestor of eukaryotic flagella. Few specialists have found the spirochaete part of Margulis' endosymbiotic theory either convincing or necessary, and no independent or reproducible data have ever come out in its support. Nonetheless, then (Margulis 1970) as today, it is a salient element of Margulis *et al.*'s (2000, 2005) version of SET, which had a broad influence on thinking about eukarayote origins, irrespective of its generally unpopular spirochaete element. Endosymbiotic theory predicted the sequences of genes encoded in organelle DNA to be more similar to their homologues among free-living prokaryotes than to nuclear-encoded counterparts, whereas competing alternatives did not. That strong line of evidence eventually brought virtually all resistance against endosymbiotic theory for mitochondria (and chloroplasts) to an end (Gray and Doolittle 1982), which can be seen as one of the first examples in which molecular evolution served to discriminate between two competing alternative theories about cell evolution.

II. THE STANDARD MODEL OF HOW AND WHY THE MITOCHONDRION BECOME ESTABLISHED

For about 40 years, traditional reasoning on the rationale behind mitochondrial origins has focused on oxygen, ATP, and improved energy yield from glucose breakdown through oxidative phosphorylation. However, in order for there to be a formulatable rationale behind mitochondrial origins, one has to formulate a null hypothesis about the nature of the host, which specifies the nature of host–symbiont interactions that lead to a stable symbiosis and hence, in turn, to specify the selective advantages for either partner during the transition from endosymbiont to organelle.

Margulis (named Sagan in 1967) suggested that the host that acquired the mitochondrial endosymbiont was an anaerobic, heterotrophic, fermenting, cell-wall–lacking (amoeboid, in the broad sense) prokaryote perhaps similar to modern *Mycoplasma* (Sagan 1967; Margulis 1970). This host corresponded, in terms of cell topology, to the nucleocytoplasmic component of eukaryotes. It is noteworthy that Mereschkowsky (1910) also envisaged an amoeboid and nucleus-lacking host as the ancestor of his nucleocytoplasmatic lineage. In his view,

however, this host acquired a bacterium that became the nucleus, whereby the origin of his host and the origin of his symbiont lineages were attributed to two independent origins of life, one early (bacteria, *das Mycoplasma*) and one later (the host lineage, *das Amoeboplasma*). There was no room for an endosymbiotic origin of mitochondria in Mereschkowsky's model of cell evolution.

Margulis clearly voiced the opinion that the selective advantage that the mitochondrion conferred upon its assumedly fermenting host was improved ATP yield from glucose breakdown by virtue of oxygen respiration and oxidative phosphorylation. For example, Sagan (1967, p. 229) wrote: "The anaerobic breakdown of glucose to pyruvate along the Embden-Meyerhof pathway occurred in the soluble cytoplasm under the direction of the host genome. [...] The greater amounts of energy available after the incorporation of the mitochondrion resulted in large cells with amoeboid and cyclotic movement." This tied in very nicely with another view that emerged at the time, namely that the origin of eukaryotes (and their mitochondria) corresponded temporally and causally to the global rise in atmospheric oxygen levels ~2 billion years ago (Ga). For example, she wrote (Sagan 1967, p. 225): "The subsequent evolution of aerobic metabolism in prokaryotes to form aerobic bacteria (protoflagella and protomitochondria) presumably occurred during the transition to the oxidizing atmosphere."

This specific assumption of Margulis' theory, which I call the "oxygen/ATP" argument, was accepted by many subsequent authors, including de Duve (1969, 2005), Stanier (1970), Cavalier-Smith (1987a, 2002), and Kurland (Andersson and Kurland, 1999). However, just because many scientists accept a certain assumption does not provide evidence that it is correct.

However, even those authors who followed Margulis' oxygen/ATP argument did not accept the spirochaete part of her theory. Probably as a consequence of that, a new view emerged in about 1980 that placed the oxygen/ATP advantage in the context of a modified form of Margulis' SET that (1) lacked the spirochaete endosymbiont and (2) assumed the newly discovered archae-bacteria to be relatives of the host lineage, flanked by the assumption that the host was a phagocytotic, anaerobic, fermenting eukaryote (possessing a nucleus and other salient eukaryotic features).

Accordingly, that 1980 view assumed that the prokaryote-to-eukaryote transition, leading to a eukaryote that initially lacked mitochondria, occurred via gradualist mechanisms such as point mutation and hence did not involve symbiosis at all (Doolittle 1980; van Valen and Maiorana 1980). This is what Doolittle (1998) has called the "standard model," which can be found under other names elsewhere in recent literature (de Duve 2005). In that view, mitochondria are interpreted as a small tack-on to, and mechanistically unrelated to, the process that made eukaryotic cells nucleated and complex (Cavalier-Smith 2002). Thus, in the standard model, mitochondria (and chloroplasts) were endosymbionts all right, but the meat-and-potatoes of the prokaryote-to-eukaryote transition (the origin of eukaryote-specific traits) was seen as having occurred independently from, and prior to, the origin of mitochondria.

The crisply argued article by van Valen and Maiorana (1980) is an excellent, explicit early formulation of the standard model in terms of its stated assumptions about the overall course of the evolution of energy metabolism and the relationship of eukaryotes to archeabacteria (the host was an amoeboid, anaerobic, fermenting cell related to archaebacteria, the advantage of the mitochondrial endosymbiont was to supply ATP). For some reason, however, it is very seldom cited. The seminal comparative-biochemical article by John and Whatley (1975) is equally explicit about the nature of the host but appeared before the archaebacteria had been discovered, and hence lacks that phylogenetic element of the standard model.

III. THERE ARE AT LEAST 12 SUBSTANTIAL PROBLEMS WITH THE STANDARD MODEL

The standard model has some problems that are gladly overlooked by its proponents, so the onus is upon those who challenge the standard model to articulate them. The following is a short list of problems with the standard model.

First, as Whatley et al. (1979) pointed out, the premise of SET that ATP was the initial benefit of the mitochondrial symbiosis can hardly be correct. The idea that the mitochondrial endosymbiont should be able to export ATP to its host from the beginning is quite unreasonable because it would require the mitochondrial endosymbiont to be able and willing to export ATP to its environment, something that no cells are known to do (Whatley et al. 1979). Hence Whatley et al. (1979) suggested that the symbiont might have initially oxidized products of the host's assumed fermentative metabolism (e.g., lactate) as a possible initial benefit of the symbiosis.

A second difficulty with the standard model is that in order for the host to actually realize the assumed benefit of ATP exported by the symbiont, the host must have had a form of physiology in which it was somehow unable to synthesize the amount of ATP that it needed for its lifestyle. What cells are known that cannot satisfy their own ATP needs, given the proper substrates and conditions? Anaerobic fermenters (both prokaryotic and eukaryotic) do just fine in nature. Furthermore, they have not been outcompeted during evolution by cells that can gain more ATP from glucose oxidations, hence efficiency in ATP yield itself does not seem to be a prime determinant of microbial success in the wild. A good example of this is found in *Escherichia coli*, which in the presence of oxygen, where ~38 moles of ATP per glucose could in principle be gained, expresses the non–proton-pumping version of complex I (NADH dehydrogenase, *ndh* operon), so that lower ATP yields are attained, while only the proton-pumping version of complex I (*nuo* operon) is expressed under anaerobic conditions (Tran *et al.* 1997).

Third, in the standard model, both the symbiont and the host are heterotrophs. It is not evident why two heterotrophs would undergo a symbiotic relationship to begin with: they would be competitors, not natural partners. Numerous and fascinating endosymbiotic relationships of prokaryotes living within eukaryotic cells are known, for example among the deep-sea tubeworms, which cannot eat for lack of a gut; their symbiotic bacteria are chemoautotrophs (not heterotrophs) that make carbohydrates for their hosts from CO_2 via H_2S-dependent chemosynthesis (Dubilier *et al.* 2001).

Fourth, there is the problem that the standard model's host is assumed to have been an anaerobe, whereas the symbiont was supposedly an O_2-dependent aerobe. Exceptions to that assumption are de Duve's (1969) and Cavalier-Smith's (1987b) early formulations, which entertain the possibility that oxygen-dependent metabolism in peroxisomes, with or without an endosymbiotic origin of these single-membrane–bounded organelles, preceded the origin of the double-membrane–bounded mitochondrion. However, as Blackstone (1995) has clearly pointed out, it is hard to imagine how a strict aerobe (the mitochondrial endosymbiont) and a strict anaerobe (the host) would come to form a stable symbiotic relationship, and I can think of no examples from the literature. There are, however, many examples of strict aerobes interacting with one another, for example *Rickettsia* pathogens in human cells (Andersson and Kurland 1999) or anaerobes interacting with one another (Fenchel and Finlay 1995).

Fifth, the anaerobe-meets-aerobe problem might not be so severe after all, because it is unclear what sort of microbial environment would allow them to enter into a stable symbiosis in the first place. Strict aerobes and strict anaerobes do not usually inhabit the same environments in nature.

Of course, there is the possibility that the symbiont might have been a facultative anaerobe that was able to survive either with or without oxygen (Fenchel and Bernard 1993; Martin and Müller 1998), but most formulations for the origin of mitochondria assume a specialized physiology for the endosymbiont.

Sixth, the view that the appearance of oxygen in the atmosphere at 2.4 Ga before present corresponded to some sort of oxygen catastrophe for the entire planet, which is a central assumption embraced by the standard model, is no longer current among geologists studying the history of oxygen on Earth. It has given way to a newer model designated as "intermediate oxidation state" or "Canfield ocean" (Canfield 1998; Canfield *et al.* 2000; Anbar and Knoll 2002; Shen *et al.* 2003; Arnold *et al.* 2004; Poulton *et al.* 2004; Brocks *et al.* 2005; see also Knoll *et al.*, Chapter 8, this volume). In a nutshell, this newer view suggests that during the time from the appearance of oxygen in the atmosphere at 2.4 Ga up until about ~0.6 Ga, marine sulfate reduction was globally widespread in the oceans, leading to sulfidic and anoxic water below the photic zone. Accordingly, there probably was no global oxygen crisis or oxygen catastrophe of the type that Margulis *et al.* (2000) or Anderson and Kurland (1999) have assumed to be crucial to mitochondrial origin. Furthermore, if one follows the argument that an oxygen catastrophe should have been the selective pressure that led to the origin of mitochondria via their ability to save their host through oxygen detoxification (Anderson and Kurland 1999), then the same oxygen catastrophe should have eliminated the prokaryotic anaerobes altogether for the same reasons, or alternatively, those prokaryotic anaerobes that survived the oxygen catastrophe should possess mitochondria for the same reasons as eukaryotes do. Hence there is cause to suspect that the role of oxygen in eukaryote origins might have been dramatically overrated in the past. Furthermore, in the newer model of ocean geochemistry, high sulfide and anaerobiosis would be the global ecological setting for eukaryote origins. Alternative models for the origins of mitochondria (and eukaryotes) that entail anaerobic ecological interactions between the host and its symbiont would fit much more comfortably into the context of Canfield oceans (Martin *et al*. 2003; Embley and Martin 2006). Of course, one could suppose that eukaryotes arose in the photic zone of Canfield oceans, or at the border of the photic zone and the anoxic layer, but no one has suggested that in the literature so far, and most recitations of the standard model see the host that acquired the mitochondrion as an amoeba that lacked a flagellum for swimming.

Seventh, the existence of eukaryotic anaerobes that possess mitochondria and whose mitochondria produce ATP under anaerobic conditions, of which there are many (Tielens *et al*. 2002; Müller 2003), is a big problem for the standard model. If the ancestral function of mitochondria was a specialized oxygen-dependent ATP-synthesis, as many would still like to believe (Andersson and Kurland 1999; Kurland and Andersson 2000; Margulis *et al*. 2005), then the anaerobic functions in mitochondria can only be readily explained by lateral gene transfer (LGT). This is noteworthy, because the most adamant opponents of the idea that LGT had anything to do with evolution (Kurland *et al*. 2003), still resort to LGT as the preferred explanation for anaerobic ATP synthesis in eukaryotes (Andersson and Kurland 1999), because it is required to uphold the standard model. Hydrogenosomes are, perhaps, the worst offenders of the standard model with regard to the role of oxygen at mitochondrial origin. Hydrogenosomes are anaerobic forms of mitochondria that were discovered over 30 years ago (Lindmark and Müller 1973, 1993; Müller 2003) and have hence been around in the literature as long as the SET (under that name) has and longer than the standard model. Up until the late 1990s hydrogenosomes were usually left unmentioned

in the mainstream literature concerning the origin of mitochondria (Gray 1992; Andersson *et al*. 1998). However, that has changed in recent years with the renewed interest in the possible evolutionary significance of eukaryotic anaerobes and the recognition that hydrogenosomes are anaerobic forms of mitochondria (Embley *et al*. 1997; Gray *et al*. 1999; Embley *et al*. 2003; Dyall *et al*. 2004; van der Giezen and Tovar 2005; van der Giezen *et al*. 2005; Embley and Martin 2006), as clearly demonstrated by the presence of a typical mitochondrial genome in the hydrogenosomes of a ciliate (Boxma *et al*. 2005) and by the homology of the protein-import machinery in mitochondria and hydrogenosomes (Dolezal *et al*. 2006). The ATP-producing biochemistry of hydrogenosomes has been summarized elsewhere (Müller 1993, 2003; Martin and Müller 1998; Hrdy *et al*. 2004; van der Giezen *et al*. 2005) but is shown again in Figure 1 for the example of *Trichomonas vaginalis* in comparison to a typical anaerobic eukaryote that has no organelles involved in ATP synthesis. Hydrogenosomes contain several enzymes that are central in energy metabolism among eukaryotes that possess neither mitochondria nor hydrogenosomes; these are shown in Figure 1B for the example of *Giardia intestinalis* (Müller, 1993, 2003; Martin and Müller 1998), which also produces hydrogen, albeit only under highly anoxic conditions and in lower amounts than eukaryotes that contain hydrogenosomes (Lloyd *et al*. 2002). How unusual and how diverse is energy metabolism among eukaryotes? A comparison of the end products of energy metabolism in a typical enteric proteobacterium—respiration in the presence of oxygen and mixed acid fermentations in the absence of oxygen (Fenchel and Finlay 1995)—is contrasted to the brunt of diversity found among virtually all eukaryotes in Figure 2; note that the latter is a group-specific subset of the former, and that none of the eukaryotic metabolic types correspond strictly with phylogeny, suggesting that the set of enzymes that make up the core of

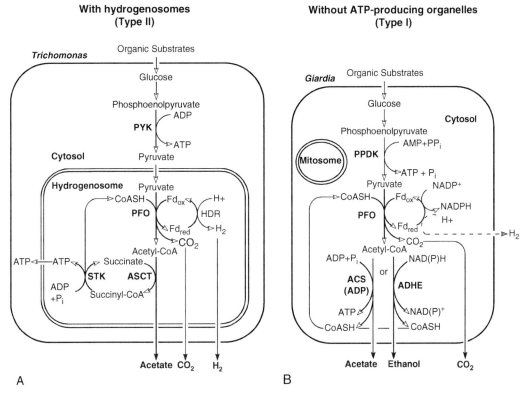

FIGURE 1. Simplified summary of energy metabolism in eukaryotes that lack cytochromes, as previously summarized by others (Müller 1993, 2003; Yarlett 1994; Martin and Müller 1998; Tielens *et al*. 2002; Hrdy *et al*. 2004; Atteia *et al*. 2006; Embley and Martin 2006). The designations Type I (**A**) and Type II (**B**) refer to the nomenclature used by Martin and Müller (1998) to designate energy metabolism in eukaryotes that lack typical mitochondria, but synthesize ATP in hydrogenosomes (Type II) or in the cytosol only (Type I). ACS, acetyl-CoA synthase (ADP-forming); ADHE, bifunctional aldehyde/alcohol dehydrogenase; ASCT, acetate:succinate CoA-transferase; HDR, iron-only hydrogenase; PFO, pyruvate:ferredoxin oxidoreductase; PPDK, pyruvate: phosphate dikinase; PYK, pyruvate kinase; STK, succinate thiokinase. The dotted arrow in (**B**) indicates that *Giardia* does produce some H_2 under certain conditions (Lloyd *et al*. 2002).

eukaryote energy metabolism were present in the eukaryote common ancestor (Martin and Müller 1998; Embley and Martin 2006; Atteia *et al*. 2006).

Eighth, anaerobic mitochondria pose yet another difficult problem to the standard model, because if the ancestral function that allowed mitochondria to become established in the eukaryotic cell in the first place was improved ATP-synthesis through oxidative phosphorylation (38 moles of ATP per mole of glucose) as opposed to fermentations (2 ATP per glucose through typical ethanol or lactate fermentations), how can one explain that numerous eukaryotic groups

with anaerobic mitochondria synthesize 2–5 moles of ATP per mole of glucose (Tielens *et al*. 2002)? In terms of the standard model, the ancestral eukaryote (the host) obtained a 20-fold increase in energy yield from glucose via the mitochondrion (38 versus 2 ATP per glucose), allowing it to outcompete anaerobes. However, multiple times in independent lineages, eukaryotes have returned to the fully satisfactory yield of 2 ATP per glucose (Martin and Müller 1998; Embley and Martin 2006). If the selective advantage of increased energy yield were so great, *why would any eukaryote give it up*, let alone in multiple independent lineages?

The typical end products of energy metabolism in enteric bacteria

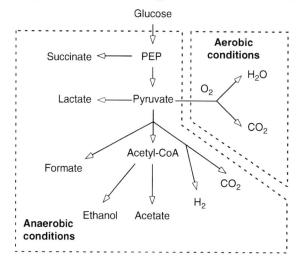

A

The typical end products of energy metabolism among eukaryotes

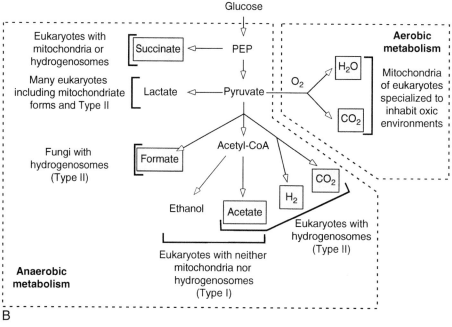

B

FIGURE 2. Comparison of energy metabolism of eukaryotes to a proteobacterium. **(A)** The end products of energy metabolism in a typical enteric bacterium (modified from Fenchel and Finlay 1995). **(B)** The end products of energy metabolism in eukaryotes that are specialized to different kinds of environments. (For details, see Müller 1993, 2003; Yarlett 1994; Martin and Müller 1998; Tielens *et al.* 2002; Hrdy *et al.* 2004; Atteia *et al.* 2006; Embley and Martin 2006.) The arginine dihydrolase pathway (Biagini *et al.* 1998), alanine production (Müller 2003), and propionate production (Biagini *et al.* 1998; Tielens *et al.* 2002) were neglected here for simplicity. The designations Type I and Type II refer to the nomenclature used by Martin and Müller (1998). The energy metabolic capacity of many α-proteobacteria exceeds that of the enterics (Imhoff and Trüper 1992), while containing the pathways shown in the figure. Gray boxes around end products indicate that they are formed in mitochondria or hydrogenosomes.

The answer is probably that increased energy yield was not the advantage conferred by mitochondria in the first place. In a theory that was built upon the role of oxygen and ATP in eukaryote evolution as a rationale for the acquisition of mitochondria (Sagan 1967; Margulis 1970), there was no room for anaerobic mitochondria or hydrogenosomes, because they do not mesh with the salient assumption of the theory. It is probably for this reason that staunch proponents for the decisive role of oxygen at mitochondrial origin still reject the evidence that mitochondria and hydrogenosomes share a common origin (Margulis et al. 2000, 2005; de Duve 2005).

Ninth, if eukaryotes are descendants of archaebacterial relatives, as the standard model prescribes, why is it that they have a eubacterial glycolytic pathway in their cytosol, when the standard model would clearly predict them to have an archaebacterial glycolytic pathway as evidence for their ancestry? Eukaryotes unquestionably possess eubacterial genes for the glycolytic pathway (Martin and Schnarrenberger 1997; Martin and Russell 2003; Esser et al. 2004), and the circumstance that proponents of the standard model deny this to be true (Canback et al. 2002) does not change the data or otherwise negate the observation.

Tenth, if eukaryotes are descendants of archaebacterial relatives, why is it that they have eubacterial-type lipids instead of the archaebacterial-type lipids, which should prevail if, as de Duve (2005) and Kurland et al. (2003) have argued, the rRNA tree is largely an accurate description of life's history. Cavalier-Smith (2002) has suggested that the easiest explanation for this would be the assumption that the common ancestor of all archaebacteria possessed eubacterial type lipids but reinvented its membrane lipid biosynthesis in the presence of a perfectly good solution to the problem of making membranes. In my opinion, that suggestion, seen from the standpoint of a microbe in the wild, is like trying to invent the bicycle while riding one.

Eleventh, if eukaryotes are descendants of archaebacterial relatives, as the standard model would have it, why is it that eukaryotes possess more eubacterially related genes than archaebacterially related genes, as recent genome comparisons have shown (Rivera et al. 1998; Rivera and Lake 2004; Esser et al. 2004)? This observation could, in principle, be accounted for by LGT one gene at a time from eubacteria to eukaryotes, but were that the case, then different eukaryotic groups would be predicted to possess fundamentally different collections of genes, which they do not (Embley and Martin 2006), except in the plants, which have acquired a hefty influx of about 18% of their protein coding genes from cyanobacteria (Martin et al. 2002).

Twelfth, if the prokaryote-to-eukaryote transition occurred solely through gradualist mechanisms (such as mutation and gene duplication, but without gene acquisitions from organelles), as the standard model prescribes (Cavalier-Smith 2002), why is it that the suspected primitively amitochondriate protists—those eukaryotes that presumably never possessed a mitochondrion, descendants of the host under the standard model—have turned out to possess mitochondria after all, anaerobic ones called hydrogenosomes (Müller 1993) or ones that are not involved in ATP synthesis at all called mitosomes (Tovar et al. 1999), as recently reviewed elsewhere (van der Giezen and Tovar 2005; Embley and Martin 2006)? Clearly, the standard model can offer mass microbial extinctions among host lineages as an explanation for this observation. However, then we must ask: for what reason did they become extinct? Energy metabolism cannot readily be invoked here because modern eukaryotic anaerobes get just as little out of glucose fermentations (2 ATP per glucose) as the host in any formulation of the standard model does, and hence should have become extinct for the same reasons.

To summarize, the standard model seems unconvincing upon closer inspection.

IV. THE SAME 12 ISSUES FROM THE STANDPOINT OF AN ALTERNATIVE THEORY

It could be that the standard model is wrong and best discarded in favor of alternatives. One alternative model for the origin of mitochondria and eukaryotes is called the hydrogen hypothesis (Martin and Müller 1998). It differs from the standard model in the following 12 points (Martin and Müller 1998).

First, the nature of the symbiosis between host and symbiont probably did not involve ATP and oxygen at all but hydrogen transfer (anaerobic syntrophy) instead, hydrogen being a waste product of the mitochondrial symbiont's ancestrally facultative anaerobic metabolism and the host being a hydrogen-dependent chemoautotroph.

Second, the advantage that the host realized was waste H_2 produced by the symbiont as an electron donor for its H_2-dependent carbon and energy metabolism.

Third, the symbiont was a facultative anaerobic heterotroph that produced H_2 under anaerobic conditions, and the host was an H_2-dependent chemoautotroph (with a carbon and energy metabolism perhaps similar to methanogens). Modern associations between H_2-producing heterotrophs and H_2-dependent chemoautotrophs are very common. Such physiological associations are called anaerobic syntrophy (Fenchel and Finlay 1995) and have been well studied (Schink 1997; Bernhard et al. 2000).

Fourth, the host in the hydrogen hypothesis is assumed to have been a strict anaerobe, whereas the symbiont was a facultative anaerobe and hence was able to respire but produced hydrogen in the anaerobic conditions that allowed the host to survive. To clarify this point, a facultative anaerobe is able to respire when oxygen is available but generates organic waste products and often H_2 when no oxygen is available (see Figure 1). In this sense, the mitochondrial endosymbiont would have brought along the genes and enzymes for both O_2-dependent

oxidative phosphorylation and anaerobic fermentations (Martin and Müller 1998; Embley and Martin 2006).

Fifth, the anaerobe-meets-aerobe problem germane to the standard model is not a problem for the hydrogen hypothesis because the principle of hydrogen transfer (anaerobic syntrophy) at the seat of the model is abundantly observable in nature today (Fenchel and Finlay 1995; Schink 1997).

Sixth, the anaerobic nature of the initial host–symbiont relationship in the hydrogen hypothesis is highly compatible with the concept of sulfidic and anoxic Canfield oceans during the time from 2.4 Ga up until about 0.6 Ga and is furthermore compatible with the appearance of the earliest microfossil eukaryotes (~1.45 Ga: Javaux et al. 2001) and plants (~1.2 Ga: Butterfield 2000) at that time. It hence requires no "oxygen catastrophe" (Sagan 1967; Margulis 1970; Woese 1977) in order to operate. Moreover, the hydrogen hypothesis predicts that starting from a common facultatively anaerobic ancestral state, eukaryotes would diversify and specialize to anaerobic, microaerophilic, and aerobic habitats as they arose over geological time, such that the anaerobic lifestyle would not be a trait depending upon phylogeny but a trait depending upon ecological specialization instead.

Seventh, the existence of eukaryotic anaerobes that possess mitochondria and whose mitochondria produce ATP under anaerobic conditions is directly accounted for via descent with modification of biochemical attributes possessed by the ancestral endosymbiont in the hydrogen hypothesis. In that sense, the hydrogen hypothesis hinged upon the common ancestry of mitochondria and hydrogenosomes and furthermore predicted eukaryotes such as *Giardia* and *Entamoeba* to have possessed a mitochondrion in their evolutionary past, a prediction that was borne out through the discovery of reduced mitochondria (mitosomes) in both organisms (Tovar et al. 1999, 2003). No version of the standard model ever made a similar prediction.

The origin of the nucleus.
Group II introns spread to many sites and undergo the transition
to spliceosomal introns. This causes problems because

i. Ribosomes
are fast, and

ii. Spliceosomes
are slow,

iii. Hence gene expression is severely impaired.

Solution:
Either   1) Invent ultra-fast spliceosomes (unlikely),
           2) Get rid of the introns (didn't happen), or
           3) Separate translation from splicing (below).

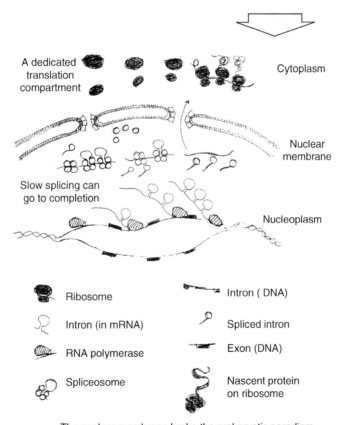

A dedicated
translation
compartment

Cytoplasm

Nuclear
membrane

Slow splicing can
go to completion

Nucleoplasm

Ribosome

Intron (DNA)

Intron (in mRNA)

Spliced intron

RNA polymerase

Exon (DNA)

Spliceosome

Nascent protein
on ribosome

The nuclear membrane broke the prokaryotic paradigm
of cotranscriptional translation, but the reason was to
separate splicing (not transcription) from translation.

FIGURE 3.   A schematic suggestion for the origin of the nucleus following the origin of mitochondria in a prokaryotic host (for details, see Martin and Koonin 2006). For an overview of other ideas regarding the origin of the nucleus, see Martin (2005) and Lopez-Garcia and Moreira (2006).

Eighth, the hydrogen hypothesis accounts directly for the anaerobic lifestyle of various eukaryotic groups via independent adaptations to aerobic and anaerobic habitats from a facultatively anaerobic ancestral state. Cavalier-Smith (2004) has also recently suggested that the ancestor of mitochondria was a facultative anaerobe but misattributed this concept to a 2002 article by himself in which the idea is neither developed nor inferred.

Ninth, the hydrogen hypothesis specifically addresses the circumstance that eukaryotes have a eubacterial glycolytic pathway in their cytosol; it is accounted for by endosymbiotic gene transfer, the process through which eukaryotes acquire genes from organelles (Timmis *et al.* 2004).

Tenth, the hydrogen hypothesis also directly accounts for the circumstance that eukaryotes have eubacterial type lipids instead of the archaebacterial type lipids via endosymbiotic gene transfer (Timmis *et al.* 2004) of the corresponding genes. Such replacements of functionally redundant biosynthetic pathways during endosymbiosis (Martin and Schnarrenberger 1997) are neither rare nor surprising. For example, eukaryotes do not possess two complete sets of pathways for amino acid biosyntheses, nucleotide biosyntheses, or cofactor biosyntheses. Instead, they possess only one each, even though both host and endosymbiont should have been competent for such biochemistry in order to be viable from the onset. It is noteworthy that amino acid biosynthesis, nucleotide biosynthesis, and cofactor biosyntheses are performed predominantly by eubacterial enzymes in eukaryotes (Esser *et al.* 2004), a circumstance for which no formulation of the standard model entailing an archaebacterially related host can account.

Eleventh, the hydrogen hypothesis directly accounts via endosymbiotic gene transfer (Timmis *et al.* 2004) for the circumstance that eukaryotes possess more eubacterially related genes than archaebacterially related genes (Martin and Koonin 2006).

Twelfth, the hydrogen hypothesis operates with a prokaryotic, archaebacterial host (Martin and Müller 1998; Martin and Koonin 2006); hence it predicts that primitively amitochondriate protists never existed. It therefore requires no corollary hypothesis of massive lineage extinctions to account for the lack thereof among modern eukaryotes (van der Giezen and Tovar 2005; Embley and Martin 2006).

V. CRITICISM AND DEFENSE OF THE HYDROGEN HYPOTHESIS

The hydrogen hypothesis has received some criticisms that are to be taken seriously and some that are not. The most serious critique has come from Cavalier-Smith (2002), whose main complaints have been twofold.

First, he has lamented the lack of a specific selective pressure or biological rationale behind the origin of the nucleus. This criticism has recently been addressed with the suggestion that one can account simply and directly for the origin of the nucleus itself (more specifically, for the origin of nucleus–cytosol compartmentation) as a consequence of group II introns entering the eukaryotic lineage via the mitochondrial symbiont through endosymbiotic gene transfer to the cytosolic chromosomes of a prokaryotic (archaebacterial) host (Koonin and Martin 2006). Group II introns, so introduced, apparently spread to many positions in the chromosomes of the host and underwent the transition from group II introns to spliceosomal introns at those positions. (For an explanation of what group II introns and spliceosomes are, the reader is referred to Lambowitz and Zimmerly [2004] and Collins and Penny [2005]). However, because splicing on spliceosomes is slow (about a minute per intron), whereas translation is fast (about 10 peptide bonds per second), the coexistence of intron-bearing transcripts and fast ribosomes in the same compartment (the host's cytosol) would have made it virtually impossible for such a cell to express its genes because ribosomes would translate unspliced transcripts, as both Doolittle (1991) and Cavalier-Smith (1991) once pointed out. Possible solutions to this fortuitously arisen problem (a consequence of the mitochondrial symbiosis in a prokaryotic host) could have entailed (1) the spontaneous origin of spliceosomes that are orders of magnitude faster than modern ones (unlikely), (2) the removal of introns from the DNA (which did not occur, because many introns in eukaryote genes are as old as eukaryotes themselves), or (3) the introduction of a means to separate splicing spatiotemporally (physically) from translation, in the most simple case by the introduction of a membrane that separates the nucleoplasm from the cytoplasm (Martin and Koonin 2006). This would allow the (slow) process of splicing to go to completion in the absence of (fast) active ribosomes (Figure 3). This view implies a chromosome-free cytosol, not the nucleus, as the genuinely novel compartment in eukaryotes. A cytosol that is specifically dedicated to translation and that is free of transcriptionally active chromatin is something that no prokaryote can boast. At the same time, the very general sorts of structures and processes that distinguish eukaryotes from prokaryotes (cytoskeleton, membrane trafficking, complex signal transduction pathways, etc.) are mainly attributes of the cytosol, a compartment where protein–protein interactions have evolved in the absence of interfering protein–DNA interactions.

Second, Cavalier-Smith (2002) has taken an adamantine stance on the issue that phagocytosis was an absolutely essential prerequisite (a *conditio sine qua non*) for the origin of mitochondria. This is most forcefully argued in his 2002 contribution that contains the term "phagotrophic origin" in the title. However, there are two problems with that argument. First, there are at least two cases reported in which prokaryotes exist as endosymbionts within the cytosol of other prokaryotes. In one case the host is a β-proteobacterium (von Dohlen *et al.* 2001);

in the other case the host is a cyanobacterium (Wujek 1978). Therefore, we can be certain that phagocytosis is not a *conditio sine qua non* for the establishment of intracellular endosymbiosis (the unpopular parasitic *Bdellovibrio* model of Margulis [1981] for the entry of the mitochondrial endosymbiont notwithstanding). Note that the prediction of the hydrogen hypothesis that symbiont entry into a prokaryotic host should be possible has been borne out, whereas the prediction of the standard model that primitively amitochondriate eukaryotes should exist has not.

There is a second problem with the phagotrophy argument. It is unquestionably true that phagocytosis promotes the establishment of intracellular endosymbiotic bacteria. Paul Buchner's (1953) seminal book on the issue contains over 700 pages describing examples of photosynthetic symbionts within animals, and a quick search with the keyword "endosymbiotic bacteria" will turn up several hundred more recent citations. Indeed, the prevalence of endosymbiotic bacteria among protists is one of the reasons that axenic protist cultures are very often difficult to obtain. There can be no question that phagocytosis promotes the establishment of intracellular symbioses. However, let us grant the existence of countless intracellular symbioses since the origin of phagocytosis over geological time, and contrast that to the striking observation that mitochondria arose from proteobacteria (and plastids arose from cyanobacteria) only once in all of Earth's history. No other organelles of eukaryotes are known that descend from prokaryotic symbionts. Hence *phagocytosis promotes the establishment of intracellular symbioses, but it has no influence whatsoever on the rate at which organelles arise from prokaryotic endosymbionts*. That observation, coupled with rare examples of prokaryotic hosts mentioned previously, reveals that phagocytosis is not strictly required for the origin of mitochondria; moreover it appears to be irrelevant to the issue altogether. Phagotrophy does not influence the rate at which organelles arise in evolution; it only influences the rate at which endosymbionts arise in evolution. The difference between organelles and endosymbionts is clear and simple: endosymbionts encode all of the proteins that they contain in their own genome, organelles import many proteins that they require for function from the cytosol. The difference between organelles and endosymbionts is the protein import apparatus.

Additional critique, but of less substantial nature, has come from Kurland, who asserted to have falsified the hydrogen hypothesis (Canback *et al.* 2002) by claiming (1) that glycolysis is a universal pathway, (2) that glycolytic enzymes reveal a classical three-domain tree of life that is indistinguishable from the RNA tree, and (3) that all genes that eukaryotes acquired from the mitochondrial endosymbiont must be expected to branch with α-proteobacterial homologues. Regarding the first two claims, scientists familiar with carbohydrate metabolism in archaebacteria know well that archaebacteria rarely possess the Embden-Meyerhof at all and that when they do, most of the enzymes involved are altogether unrelated to the eukaryote (and eubacterial) versions (Verhees *et al.* 2003; Siebers and Schönheit 2005). Archaebacteria generally do not possess homologues of the genes that specify the activities of the eukaryotic glycolytic pathway (Doolittle *et al.* 2003; Martin and Russell 2003). Regarding the third claim, α-proteobacteria have undergone a considerable amount of LGT during evolutionary time, as some authors of Canback *et al.* (2002) admit in a context where the origin of mitochondria is not an issue (Boussau *et al.* 2004). The ancient allele-sampling problem that LGT introduces into the origin of eukaryote genes that were acquired either from the mitochondrion or from the chloroplast is obvious, has been discussed elsewhere (Martin 1999; Embley and Martin 2006), and need not be reiterated here.

Other authors have found it preferable to reject the hydrogen hypothesis on the assumption that mitochondria and hydrogenosomes do not share a common ancestor (de Duve 2005; Margulis *et al.* 2005). However, the finding that some hydrogenosomes have retained a genome that unambiguously identifies the organelle as a mitochondrion in the evolutionary sense (Boxma *et al.* 2005), in addition to the common protein import machinery shared by mitochondria and hydrogenosomes (Dolezal *et al.* 2006), dispels such concerns.

VI. INTERMEZZO

The editors of this volume requested that I discuss in more detail the possibility that the host was a heterotroph, that the mitochondrial endosymbiont was an anaerobic photosynthesizer, and that the benefit of that symbiosis was carbohydrate production for the host. The most explicit version in the literature that would come close to that suggestion was formulated by Dennis Searcy (1992, 2002); it is based on sulfur cycling entailing a *Thermoplasma*-like (a prokaryotic, not a eukaryotic) host not on carbohydrate production. Woese (1977) offered the suggestion that the mitochondrion might have been an anaerobic phototrophic (eu)bacterium, while the host was a eukaryote that represented the ancestral organizational state of all life, such that the cell-wall–bearing state of archaebacteria and eubacteria evolved independently from their ancestrally eukaryotic organization. Woese suggested that the anaerobic phototrophic endosymbiont evolved the capacity to respire oxygen after becoming an endosymbiont and, following Margulis, surmised that, "Clearly the oxygen catastrophe is the key element in this scenario," because he assumed that "bacteria and anaerobic eucaryotes would encounter the increasing oxygen level simultaneously. Therefore no fully fashioned aerobic bacteria exist at the onset for the anaerobic eucaryote to engulf" (Woese 1977, p. 94). Woese's model thus did not entail the inheritance of respiration from respiring prokaryotes, at odds with both older (John and Whatley 1975) and more modern comparative data (Lang *et al.* 1999).

It also entailed no prokaryote-to-eukaryote transition at all but a eukaryote-to-prokaryote transition instead; as such it did not address the origin of eukaryotes (because they, we assumed, represent the ancestral state of cell organization). That "eukaryotes-early" view still has some supporters (Forterre and Philippe 1999; Poole *et al.* 1999) and draws upon the introns-early hypothesis (Doolittle 1980) as evidence in its favor, even though the architect of introns early (Doolittle) abandoned the theory years ago (Stoltzfus *et al.* 1994). Woese (1977) also did not suggest carbohydrate production in the context of the origin of mitochondria, but Fenchel and Bernard (1993) did, indirectly, by suggesting that facultative anaerobic photosynthesizers in the modern ciliate *Strombidium prupureum* might serve as a model for the origin of mitochondria, and by noting that excretion of photosynthate might benefit the host. However, the question of whence and how such a eukaryotic host might arise in evolution was an issue neither for Fenchel and Bernard (1993) nor for Fenchel and Finlay (1995). Implicit was the standard model regarding the origin of the host (Fenchel and Finlay 1995), but the assumption that the host was a eukaryote and that it was a heterotroph entails some problems, as discussed previously.

My own view on the physiology of the mitochondrial symbiont is that "From molecular phylogeny we can assume that it was a member of the α-proteobacteria, and that it therefore may have been photosynthetic or non-photosynthetic, autotrophic (able to satisfy its carbon needs from CO_2 alone) or heterotrophic, anaerobic or aerobic, or all of the above, as is the case for many contemporary representatives of the group, such as Rhodobacter sphaeroides" (Martin and Müller 1998, p. 38). The mitochondria of eukaryotes have quite a diversity of biochemical attributes with regard to ATP synthesis, even involving inorganic electron donors, as recently summarized by Tielens *et al.* (2002). However, the sum of all diversity in eukaryotic core physiology is surpassed by single α-proteobacterial species such as members

of the genus *Rhodospirillum* (Imhoff and Trüper 1992). No biochemical attributes of mitochondria provide any link whatsoever with photosynthesis. Accordingly, there is certainly no reason to exclude the possibility that the mitochondrial symbiont could have been photosynthetic (Martin and Müller 1998). However, because there is no recognizable trace of photosynthetic physiology among modern eukaryotes that never possessed a plastid, it would seem that photosynthesis, if possessed by the mitochondrial endosymbiont, played no role in the establishment of that symbiosis and hence that if the ancestor of mitochondria was photosynthetic that capacity was lost before the diversification of major eukaryotic groups.

The standard model focuses on oxygen for the origin of mitochondria and mostly neglects anaerobic energy metabolism in eukaryotes. Much attention is currently given to hopanoids, 2-methylhopanoids in particular, as possible biomarkers for cyanobacteria, hence, oxygen production (see Knoll *et al.*, Chapter 8, this volume). The hydrogen hypothesis focuses on anaerobic energy metabolism and addresses the circumstance that energy metabolism in eukaryotes is basically that of a facultatively anaerobic bacterium (see Figure 2) and that it consists of eubacterial enzymes (Embley and Martin 2006), even in the glycolytic pathway (Esser *et al.* 2004). The hydrogen hypothesis also assumes an important role for oxygen, namely as a utilitous waste product of energy metabolism in chloroplasts, which might have facilitated the origin of plastids (Martin and Müller 1998; see also Hackett *et al.*, Chapter 7, this volume).

VII. CONCLUSIONS

The standard model of eukaryote and mitochondrial origin posits that the host was a eukaryote related to archaebacteria, that the prime advantage of the mitochondrion at the onset of symbiosis was improved ATP yield from glucose by virtue of oxygen respiration, and that the nucleus arose before the mitochondrion. It predicts that primitively amitochondriate eukaryotes should exist, and it cannot directly account for anaerobic mitochondria or hydrogenosomes. The hydrogen hypothesis (Martin and Müller 1998) posits that the host was a hydrogen-dependent archaebacterium, that the prime advantage conferred by the mitochondrial endosymbiont was H_2-generation for anaerobic syntrophy, and that the nucleus arose in the wake of mitochondrial origin (Martin and Koonin 2006). It predicts that primitively amitochondriate eukaryotes never existed, it directly accounts for anaerobic mitochondria and hydrogenosomes (Embley and Martin 2006), and as it assumes a prokaryotic host, it directly accounts for the coexistence of spliceosomal introns, nucleus–cytosol compartmentation, and mitochondria among eukaryotes (Martin and Koonin 2006).

The standard model is couched in an older view of Earth history that entails an oxygen catastrophe, during which a radical transition from globally anoxic habitats to globally oxic habitats is assumed. The hydrogen hypothesis is couched in a geological setting that entails predominantly anaerobic habitats, with an increasing prevalence of oxic niches during the latter half of Earth's history and hence is highly compatible with the concept of Canfield oceans (see Knoll *et al.*, Chapter 8, this volume).

There are many suggestions in the literature for the origin of mitochondria (reviewed in Martin *et al.* 2001) besides the standard model and the hydrogen hypothesis, but all of them (except one) operate with a heterotrophic host for the origin of mitochondria. The one that operates with an autotrophic host is also the one that can directly account (1) for hydrogenosomes, (2) for the circumstance that eukaryotes have an archaebacterial genetic apparatus that survives through eubacterial energy metabolism, and (3) for the circumstance that all eukaryotes either possess or possessed a eubacterial

endosymbiont, the common ancestor of mitochondria and hydrogenosomes.

Acknowledgments

I thank the Deutsche Forschungsgemeinschaft for funding and many colleagues for discussions. My thanks to the critical referee who issued many questions about hydrosomes [sic].

References

Anbar, A.D., and Knoll, A.H. (2002). Proterozoic ocean chemistry and evolution: a bioinorganic bridge. *Science* **297**: 1137–1142.

Andersson, S.G.E., and Kurland, C.G. (1999). Origins of mitochondria and hydrogenosomes. *Curr. Opin. Microbiol.* **2**: 535–541.

Andersson, S.G.E., Zomorodipour, A., Andersson, J.O., Sicheritz-Ponten, T., Alsmark, U.C.M., Podowski, R.M., Naslund, A.K., Eriksson, A.S., Winkler, H.H., and Kurland, C.G. (1998). The genome sequence of *Rickettsia prowazekii* and the origin of mitochondria. *Nature* **396**: 133–140.

Arnold, G.L., Anbar, A.D., Barling, J., and Lyons, T.W. (2004). Molybdenum isotope evidence for widespread anoxia in mid-Proterozoic oceans. *Science* **304**: 87–90.

Atteia, A., van Lis, R., Gelius-Dietrich, G., Adrait, A., Garin, J., Joyard, J., Rolland, N., and Martin, W. (2006). Pyruvate:formate lyase and a novel route of eukaryotic ATP-synthesis in anaerobic *Chlamydomonas* mitochondria. *J. Biol. Chem.* **281**: 9909–9918.

Bernhard, J.M., Buck, K.R., Farmer, M.A., and Bowser, S.S. (2000). The Santa Barbara Basin is a symbiosis oasis. *Nature* **403**: 77–80.

Biagini, G.A., McIntyre, P.S., Finlay, B.J., and Lloyd, D. (1998). Carbohydrate and amino acid fermentation in the free-living primitive protozoon *Hexamita* sp. *Appl. Environ. Microbiol.* **64**: 203–207.

Blackstone, N. (1995). A units-of-evolution perspective on the endosymbiont theory of the origin of the mitochondrion. *Evolution* **49**: 785–796.

Boussau, B., Karlberg, E.O., Frank, A.C., Legault, B.A., and Andersson, S.G. (2004). Computational inference of scenarios for alpha-proteobacterial genome evolution. *Proc. Natl. Acad. Sci. U S A* **101**: 9722–9727.

Boxma, B., de Graaf, R.M., van der Staay, G.W., van Alen, T.A., Ricard, G., Gabaldon, T., van Hoek, A.H., Moon-van der Staay, S.Y., Koopman, W.J., van Hellemond, J.J., Tielens, A.G.M., Friedrich, T., Veenhuis, M., Huynen, M.A., and Hackstein, J.H.P. (2005). An anaerobic mitochondrion that produces hydrogen. *Nature* **434**: 74–79.

Brocks, J.J., Love, G.D., Summons, R.E., Knoll, A.H., Logan, G.A., and Bowden, S.A. (2005). Biomarker evidence for green and purple sulphur bacteria in a stratified Palaeoproterozoic sea. *Nature* **437**: 866–870.

Buchner, P. (1953). *Endosymbiose der Tiere mit pflanzlichen Mikroorganismen*. Basel, Birkhäuser.

Butterfield, N.J. (2000). *Bangiomorpha pubescens* n. gen., n. sp.: implications for the evolution of sex, multicellularity, and the Mesoproterozoic/Neoproterozoic radiation of eukaryotes. *Paleobiology* **263**: 386–404.

Canback, B., Andersson, S.G., and Kurland, C.G. (2002). The global phylogeny of glycolytic enzymes. *Proc. Natl. Acad. Sci. U S A* **99**: 6097–6102.

Canfield, D.E. (1998). A new model for Proterozoic ocean chemistry. *Nature* **396**: 450–453.

Canfield, D.E., Habicht, K.S., and Thamdrup, B. (2000). The Archean sulfur cycle and the early history of atmospheric oxygen. *Science* **288**: 658–661.

Cavalier-Smith, T. (1975). The origin of nuclei and of eukaryotic cells. *Nature* **256**: 463–468.

Cavalier-Smith, T. (1987a). The origin of eukaryote and archaebacterial cells. *Ann. N Y Acad. Sci.* **503**: 17–54.

Cavalier-Smith, T. (1987b). The simultaneous symbiotic origin of mitochondria, chloroplasts and microbodies. *Ann. N Y Acad. Sci.* **503**: 55–71.

Cavalier-Smith, T. (1991). Intron phylogeny: A new hypothesis. *Trends Genet.* **7**: 145–148.

Cavalier-Smith, T. (2002). The phagotrophic origin of eukaryotes and phylogenetic classification of Protozoa. *Int. J. Syst. Evol. Microbiol.* **52**: 297–354.

Cavalier-Smith, T. (2004). Only six kingdoms of life. *Proc. Roy. Soc. Lond.* B **271**: 1251–1262.

Collins, L., and Penny, D. (2005) Complex spliceosomal organization ancestral to extant eukaryotes. *Mol. Biol. Evol.* **22**, 1053–1066.

de Duve, C. (1969). Evolution of the peroxisome. *Ann. N Y Acad. Sci.* **168**: 369–381.

de Duve, C. (2005). *Singularities. Landmarks on the Pathways of Life*. Cambridge, Cambridge University Press.

Dolezal, P., Likic, V., Tachezy, J., and Lithgow, T. (2006). Evolution of the molecular machines for protein import into mitochondria. *Science* **313**: 314–318.

Doolittle, W.F. (1980). Revolutionary concepts in evolutionary biology. *Trends Biochem. Sci.* **5**: 146–149.

Doolittle, W.F. (1991). The origin of introns. *Curr. Biol.* **1**: 145–146.

Doolittle, W.F. (1998). A paradigm gets shifty. *Nature* **392**: 15 16.

Doolittle, W.F., Boucher, Y., Nesbo, C.L., Douady, C.J., Andersson, J.O., and Roger, A.J. (2003). How big is the iceberg of which organellar genes in nuclear genomes are but the tip? *Phil. Trans. R. Soc. Lond. B.* **358**: 39–58.

Dubilier, N., Mülders, C., Ferdelman, T., de Beer, D., Pernthaler, A., Klein, M., Wagner, M., Erséus, C., Thiermann, F., Krieger, J., Giere, O., and Amann, R. (2001). Endosymbiotic sulphate-reducing and sulphide-oxidizing bacteria in an oligochaete worm. *Nature* **411**: 298–302.

Dyall, S.D., Brown, M.T., and Johnson, P.J. (2004). Ancient invasions: From endosymbionts to organelles. *Science* **304**: 253–257.

Embley, T.M., Horner, D.S., and Hirt, R.P. (1997). Anaerobic eukaryote evolution: hydrogenosomes as biochemically modified mitochondria? *Trend. Ecol. Evol.* **12**: 437–441.

Embley, T.M., and Martin, W. (2006). Eukaryotic evolution, changes and challenges. *Nature* **440**: 623–630.

Embley, T.M., van der Giezen, M., Horner, D.S., Dyal, P.L., Bell, S., and Foster, P.G. (2003). Hydrogenosomes, mitochondria and early eukaryotic evolution. *IUBMB Life* **55**: 387–395

Esser, C., Ahmadinejad, N., Wiegand, C., Rotte, C., Sebastiani, F., Gelius-Dietrich, G., Henze, K., Kretschmann, E., Richly, E., Leister, D., Bryant, D., Steel, M.A., Lockhart, P.J., Penny, D., and Martin, W. (2004). A genome phylogeny for mitochondria among alpha-proteobacteria and a predominantly eubacterial ancestry of yeast nuclear genes. *Mol. Biol. Evol.* **21**: 1643–1660.

Fehling, J., Stoecker, D., and Baldauf, S.L. (2007). Photosynthesis and the eukaryote tree of life. *Evolution of Primary Producers in the Sea*. P.G. Falkowski and A.H. Knoll, eds. Boston, Elsevier, pp. 75–107.

Fenchel, T., and Bernard, C. (1993). A purple protist. *Nature* **362**: 300.

Fenchel, T., and Finlay, B.J. (1995). *Ecology and Evolution in Anoxic Worlds*. Oxford, Oxford University Press.

Forterre, P., and Philippe, H. (1999). Where is the root of the universal tree of life? *BioEssays* **21**: 871–879.

Gray, M.W. (1992). The endosymbiont hypothesis revisited. *Int. Rev. Cytol.* **141**: 223–357.

Gray, M.W., Burger, G., and Lang, B.F. (1999). Mitochondrial evolution. *Science* **283**: 1476–1481.

Gray, M.W., and Doolittle, W.F. (1982). Has the endosymbiont hypothesis been proven? *Microbol. Rev.* **46**: 1–42.

Hackett, J.D., Yoon, H.S., Butterfield, N.J., Sauderson, M.J., and Bhattacharya, (2007). Plastid endosymbiosis. Sources and timing of the major events. *Evolution of Primary Producers in the Sea*. P.G. Falkowski and A.H. Knoll, eds. Boston, Elsevier, pp. 109–131.

Hrdy, I., Hirt, R.P., Dolezal, P., Bardonova, L., Foster, P.G., Tachezy, J., and Embley, T.M. (2004). *Trichomonas* hydrogenosomes contain the NADH dehydrogenase module of mitochondrial complex I. *Nature* **432**: 618–622.

Imhoff, J.F., and Trüper, H.G. (1992). The genus *Rhodospirillum* and related genera. In: *The Prokaryotes*, 2nd ed, Vol III. A. Balows, H.G. Trüper, M. Dworkin, W. Harder, and K.-H. Schleifer, eds. New York, Springer-Verlag, pp. 2141–2159.

Javaux, E.J., Knoll, A.H., and Walter, M.R. (2001). Morphological and ecological complexity in early eukaryotic ecosystems. *Nature* **412**: 66–69.

John, P., and Whatley, F.R. (1975). *Paracoccus denitrificans* and the evolutionary origin of the mitochondrion. *Nature* **254**: 495–498.

Knoll, A.H., Summons, R.E., Waldbauer, J.B., and Zumberge, J.E. (2007). The geological succession of primary producers in the oceans. *Evolution of Primary Producers in the Sea*. P.G. Falkowski and A.H. Knoll, eds. Boston, Elsevier, pp. 133–163.

Kurland, C.G., and Andersson, S.G. (2000). Origin and evolution of the mitochondrial proteome. *Microbiol. Mol. Biol. Rev.* **64**: 786–820.

Kurland, C.G., Canback, B., and Berg, O.G. (2003). Horizontal gene transfer: a critical view. *Proc. Natl. Acad. Sci. U S A* **100**: 9658–9662.

Lambowitz, A.M., and Zimmerly, S. (2004). Mobile group II introns. *Annu. Rev. Genet.* **38**: 1–35.

Lang, B.F., Gray, M.W., and Burger, G. (1999). Mitochondrial genome evolution and the origin of eukaryotes. *Annu. Rev. Genet.* **33**: 351–397.

Lindmark, D.G., and Müller, M. (1973). Hydrogenosome, a cytoplasmic organelle of the anaerobic flagellate, *Tritrichomonas foetus*, and its role in pyruvate metabolism. *J. Biol. Chem.* **248**: 7724–7728.

Lloyd, D., Ralphs, J. R., and Harris, J. C. (2002). *Giardia intestinalis*, a eukaryote without hydrogenosomes, produces hydrogen. *Microbiology* **148**, 727–733.

Lopez-Garcia, P., and Moreira, D. (2006). Selective forces for the origin of the eukaryotic nucleus. *BioEssays* **28**: 525–533.

Margulis, L. (1970). *Origin of Eukaryotic Cells*. New Haven, Yale University Press.

Margulis, L. (1981). *Symbiosis in Cell Evolution*. San Francisco, Freeman.

Margulis, L., Dolan, M.F., and Guerrero, R. (2000). The chimeric eukaryote: Origin of the nucleus from the karyomastigont in amitochondriate protists. *Proc. Natl. Acad. Sci. U S A* **97**: 6954–6959.

Margulis, L., Dolan, M.F., and Whiteside, J.H. (2005). Imperfections and oddities in the origin of the nucleus. *Palaeobiology* **31**: 175–191.

Martin, W. (1999). Mosaic bacterial chromosomes—a challenge en route to a tree of genomes. *BioEssays* **21**: 99–104.

Martin, W. (2005). Archaebacteria (Archaea) and the origin of the eukaryotic nucleus. *Curr. Opin. Microbiol.* **8**: 630–637.

Martin, W., Hoffmeister, M., Rotte, C., and Henze, K. (2001). An overview of endosymbiotic models for the origins of eukaryotes, their ATP-producing organelles (mitochondria and hydrogenosomes), and their heterotrophic lifestyle. *Biol. Chem.* **382**: 1521–1539.

Martin, W., and Koonin, E.V. (2006). Introns and the origin of nucleus-cytosol compartmentalization. *Nature* **440**: 41–45.

Martin, W., and Kowallik, K.V. (1999). Annotated English translation of Mereschkowsky's 1905 paper "Über Natur und Ursprung der Chromatophoren im Pflanzenreiche." *Eur. J. Phycol.* **34**: 287–295.

Martin, W., and Müller, M. (1998). The hydrogen hypothesis for the first eukaryote. *Nature* **392**: 37–41.

Martin, W., Rotte, C., Hoffmeister, M., Theissen, U., Gelius-Dietrich, G., Ahr, S., and Henze, K. (2003). Early cell evolution, eukaryotes, anoxia, sulfide, oxygen, fungi first (?), and a tree of genomes revisited. *IUBMB Life* **55**: 193–204.

Martin, W., Rujan, T., Richly, E., Hansen, A., Cornelsen, S., Lins, T., Leister, D., Stoebe, B., Hasegawa, M., and Penny, D. (2002). Evolutionary analysis of *Arabidopsis*, cyanobacterial, and chloroplast genomes reveals plastid phylogeny and thousands of cyanobacterial genes in the nucleus. *Proc. Natl. Acad. Sci. U S A* **99**: 12246–12251.

Martin, W., and Russell, M.J. (2003). On the origins of cells: a hypothesis for the evolutionary transitions from abiotic geochemistry to chemoautotrophic prokaryotes, and from prokaryotes to nucleated cells. *Phil. Trans. Roy. Soc. Lond. B* **358**: 59–85.

Martin, W., and Schnarrenberger, C. (1997). The evolution of the Calvin cycle from prokaryotic to eukaryotic chromosomes: a case study of functional redundancy in ancient pathways through endosymbiosis. *Curr. Genet.* **32**: 1–18.

Mereschkowsky, C. (1905). Über Natur und Ursprung der Chromatophoren im Pflanzenreiche. *Biol. Centralbl.* **25**: 593–604.

Mereschkowsky, C. (1910). Theorie der zwei Plasmaarten als Grundlage der Symbiogenesis, einer neuen Lehre von der Entstehung der Organismen. *Biol. Centralbl.* **30**: 278–288, 289–303, 321–347, 353–367.

Müller, M. (2003). Energy metabolism. Part I: Anaerobic protozoa. *Molecular Medical Parasitology*. J Marr, ed. London, Academic Press, pp. 125–139.

Poole, A., Jeffares, D., and Penny, D. (1999). Prokaryotes, the new kids on the block. *BioEssays* **21**: 880–889.

Poulton, S.W., Fralick, P.W., and Canfield, D.E. (2004). The transition to a sulphidic ocean ~1.84 billion years ago. *Nature* **431**: 173–177.

Rivera, M.C., Jain, R., Moore, J.E., and Lake, J.A. (1998). Genomic evidence for two functionally distinct gene classes. *Proc. Natl. Acad. Sci. U S A* **95**: 6239–6244.

Rivera, M.C., and Lake, J.A. (2004). The ring of life provides evidence for a genome fusion origin of eukaryotes. *Nature* **431**: 152–155.

Sagan, L. (1967). On the origin of mitosing cells. *J. Theoret. Biol.* **14**: 225–274.

Sapp, J. (1994). *Evolution by Association: A History of Symbiosis*. New York, Oxford University Press.

Sapp, J. (2005). The bacterium's place in nature. *Microbial Phylogeny and Evolution: Concepts and Controversies*. J. Sapp, ed. Oxford, Oxford University Press, pp. 3–52.

Schink, B. (1997). Energetics of syntrophic cooperation in methanogenic degradation. *Microbiol. Mol. Biol. Rev.* **61**: 262–280.

Searcy, D.G. (1992). Origins of mitochondria and chloroplasts from sulphur-based symbioses. *The Origin and Evolution of the Cell*. H. Hartman and K. Matsuno, eds. Singapore, World Scientific, pp. 47–78.

Searcy, D.G. (2002). Syntrophic models for mitochondrial origin. *Symbiosis: Mechanisms and Model Systems*. J. Seckbach, ed. Doordrecht, Kluwer, pp. 163–183.

Shen, Y., Knoll, A.H., and Walter, M.R. (2003). Evidence for low sulphate and anoxia in a mid-Proterozoic marine basin. *Nature* 423: 632–635.

Siebers, B., and Schönheit, P. (2005). Unusual pathways and enzymes of central carbohydrate metabolism in Archaea. *Curr. Opin. Microbiol.* 8: 695–705.

Stanier, R.Y. (1970). Some aspects of the biology of cells and their possible evolutionary significance. *Symp. Soc. Gen. Microbiol.* 20: 1–38.

Stanier, R.Y., and van Niel, C.B. (1962). The concept of a bacterium. *Arch. Mikrobiol.* 42: 17–35.

Stoltzfus, A., Spencer, D. F., Zuker, M., Logsdon, J.M., and Doolittle, W.F. (1994). Testing the exon theory of genes: The evidence from protein structure. *Science* 265: 202–207.

Taylor, F.R.J. (1974). Implications and extensions of the serial endosymbiosis theory of the origin of eukaryotes. *Taxon* 23: 229–258.

Tielens, A.G., Rotte, C., van Hellemond, J.J., and Martin, W. (2002). Mitochondria as we don't know them. *Trends Biochem. Sci.* 27: 564–572.

Timmis, J.N., Ayliffe, M.A., Huang, C.Y., and Martin, W. (2004). Endosymbiotic gene transfer: organelle genomes forge eukaryotic chromosomes. *Nat. Rev. Genet.* 5: 123–135.

Tovar, J., Fischer, A., and Clark, C.G. (1999). The mitosome, a novel organelle related to mitochondria in the amitochondrial parasite *Entamoeba histolytica*. *Mol. Microbiol.* 32: 1013–1021.

Tovar, J., León-Avila, G., Sánchez, L.B., Sutak, R., Tachezy, J., van der Giezen, M., Hernández, M., Müller, M., and Lucocq, J.M. (2003). Mitochondrial remnant organelles of *Giardia* function in iron-sulphur protein maturation. *Nature* 426: 172–176.

Tran, Q.H., Bongaerts, J., Vlad, D., and Unden, G. (1997). Requirement for the proton-pumping NADH dehydrogenase I of *Escherichia coli* in respiration of NADH to fumarate and its bioenergetic implications. *Eur. J. Biochem.* 244: 155–160.

van der Giezen, M., and Tovar, J. (2005). Degenerate mitochondria. *EMBO Rep.* 6: 525–530.

van der Giezen, M., Tovar, J., and Clark, C.G. (2005). Mitochondrion-derived organelles in protists and fungi. *Int. Rev. Cytol.* 244: 175–225.

van Valen, L.M., and Maiorana, V.C. (1980). The archaebacteria and eukaryotic origins. *Nature* 287: 248–250.

Verhees, C.H., Kengen, S.W., Tuininga, J.E., Schut, G.J., Adams, M.W., De Vos, W.M., and Van der Oost, J. (2003). The unique features of glycolytic pathways in Archaea. *Biochem. J.* 375: 231–246.

von Dohlen, C.D., Kohler, S., Alsop, S.T., and McManus, W.R. (2001). Mealybug β-proteobacterial endosymbionts contain γ-proteobacterial symbionts. *Nature* 412: 433–436.

Whatley, J.M., John, P., and Whatley, F.R. (1979). From extracellular to intracellular: the establishment of mitochondria and chloroplasts. *Proc. R. Soc. Lond. B* 204: 165–187.

Wilson, E.B. (1928). *The Cell in Development and Heredity*, 3rd revised edition. New York, Macmillan. Reprinted in 1987 by Garland Publishing, New York.

Woese, C.R. (1977) Endosymbionts and mitochondrial origin. *J. Mol. Evol.* 10: 93–96.

Wujek, D.E. (1979). Intracellular bacteria in the blue-green-alga *Pleurocapsa minor*. *Trans. Am. Micros. Soc.* 98: 143–145.

Yarlett, N. (1994). Fermentation product formation. *Anaerobic Fungi: Biology, Ecology and Function*. D.O. Mountfort, C.G., Orpin, eds. New York, Marcel Dekker, pp. 129–146.

Photosynthesis and the Eukaryote Tree of Life

JOHANNA FEHLING, DIANE STOECKER, AND
SANDRA L. BALDAUF

Our current understanding of the eukaryote tree of life is a work in progress, based on a patchwork of data. Parts of the tree date as far back as the earliest morphology-based trees of Haeckel (see Hamm and Smetaceck, Chapter 14, this volume). These were repeatedly revised, often radically, based on data from improved microscopes and microscopic techniques (Whittaker and Margulis 1978), and this continues to the present day. Most recently, molecular phylogenetic data have radically revised parts of the tree again. Although many eukaryote groups have been defined by one type of data and confirmed by others, there are a surprising number that are supported by only a single line of evidence, either ultrastructural or morphological. Thus, there are still major groups of eukaryotes that lack either molecular phylogenetic or phenotypic justification, and it may be some time before all lines of evidence converge.

This is contrary to early expectations that molecular phylogeny could solve the whole tree in "one go." It was thought that it was only necessary to sequence the right gene from every organism, and then simple phylogenetic analysis could be used to reconstruct the full tree of life. It quickly emerged that the best candidate for this was the small ribosomal subunit RNA (SSU rRNA). SSU rRNA is a large molecule, with parts that are conserved across all life, and others that vary among even the most closely related species (Pace *et al.* 1986). Thus, it can potentially resolve relationships at multiple taxonomic levels. Of equal importance is the fact that it is universal, often multicopy, and, together with the large subunit RNA (LSU rRNA), the most abundant RNA molecule in the cell. This makes SSU rRNA sequences relatively easy to obtain from almost any organism and even with minute amounts of starting material.

SSU rRNA phylogeny revolutionized our understanding of the eukaryote tree and identified a number of important groups for the first time. However, it eventually emerged that there were serious flaws in these trees, particularly with regard to the deepest branches (Philippe and Adoutte 1998). This is because genes can evolve at very different rates and with different patterns in different lineages, sometimes even quite closely related ones. This can lead to the artifactual grouping of unrelated sequences, particularly fast or oddly evolving ones, a problem that has become known as long branch attraction (LBA; Felsenstein 1978; Philippe and Germot 2000). LBA is essentially the inability of phylogenetic methods to distinguish truly old sequences, which should have long branches, from fast evolving sequences, which also have long branches. Although phylogenetic methods continue to improve, this inability to resolve "old" from "odd" based on sequence analyses of single genes confounds all trees to some extent (Hillis *et al.* 1994).

The flaws in the SSU rRNA tree only became apparent with the accumulation of similar data from other genes. Unfortunately, these trees had their own flaws. For example, inspection of tubulin trees shows that the gene evolves slowly in organisms with flagellated stages, because tubulin is a major flagellar protein, but very fast in nonflagellate species (Keeling and Doolittle 1996). This makes it a potentially good gene for deep phylogeny among flagellates and for shallow phylogeny among nonflagellates. It now appears that all genes have some "phylogenetic" flaw(s), that is, clocklike evolution in some lineages but idiosyncratic evolution in others, such that they reconstruct some relationships well but misconstrue others (Baldauf *et al.* 2000).

This inability of single genes to consistently reconstruct deep eukaryote branches was interpreted by some to mean that the major groups had arisen in an explosive radiation, and the relationships among them could never be recovered (Philippe and Adoutte 1998). However, this is not consistent with the fact that many genes do reconstruct deep branches, although different genes resolve different branches and none resolve all of them (Baldauf *et al.* 2000). It is also inconsistent with the fact that, when genes are combined into multigene datasets, that is, concatenated end-to-end as if they were one long sequence, most deep eukaryote branches are recovered with strong and consistent support (Baldauf *et al.* 2000; Bapteste *et al.* 2002; Harper *et al.* 2005; Rodriguez-Ezpeleta *et al.* 2005; Li *et al.* 2006). Although the theoretical issues concerning the combining of data have been debated at length (Huelsenbeck *et al.* 1996), one can simply think of combined sequences as the whole genome with very large holes in it.

I. THE EUKARYOTES

Eukaryotes are clearly a monophyletic group. They possess numerous unique and complex synapomorphies. All eukaryotes have multiple linear chromosomes, which they sequester in one or more nuclei (Alberts *et al*. 2002). These nuclei are enclosed in a double membrane punctuated with pores, which are large multiprotein complexes that both conduct and regulate macromolecular transport (Schwartz 2005). The nuclear membrane is continuous with a larger endomembrane system including the endoplasmic reticulum and Golgi apparatus, the site of membrane construction, recycling, and protein modification (Alberts *et al*. 2002). All eukaryotes have or belong to lineages that once had mitochondria, an endosymbiotic organelle derived from an alpha proteobacterium. These organelles have been lost, reduced, or converted to fermentative, hydrogen-producing organelles (hydrogenosomes) multiple times (Dyall and Johnson 2000; Martin, Chapter 5, this volume).

Evidence for the endosymbiotic origin of plastids is actually much stronger than that for mitochondria. Plastids retain at least 10-fold more genes in their genomes than the average mitochondrion. Most of these genes are organized into operons, and the order of the genes in the operons and the sequences of the genes themselves are clearly cyanobacterial (Douglas 1998). Although long a matter of debate, the evidence is now strongly in favor of a single primary origin for eukaryotic plastids. This appears to have occurred early in the evolution of the group now known as Archaeplastida (see Section II.D.; O'Kelly, Chapter 13, this volume). From this starting point, however, photosynthesis spread widely among eukaryotes by a variety of means, as described in more detail at the end of this chapter.

Other salient features of eukaryotic cells include their size, which is generally around 10 to 100 μm in diameter, roughly an order of magnitude larger than most prokaryotic cells. However, recent culture-independent surveys using polymerase chain reaction technology (ciPCR surveys) have identified a vast array of eukaryotes in the 1–10 μm range (Moreira and López-García 2002). Many eukaryotes are motile or feed by means of flagella. These are large, multiprotein complexes, unlike bacterial flagella, which are constructed from a single protein (Alberts *et al*. 2002). There is no fundamental difference between eukaryotic cilia and flagella, and it has often been suggested that both should be referred to as cilia to distinguish them from the nonhomologous bacterial structure.

Approximately 14 amoeboid types of eukaryotes are recognized. These are scattered across the tree, which may suggest multiple independent origins from flagellate ancestors (Cavalier-Smith 1998, 2002; Patterson *et al*. 2000b) or the ancestral eukaryotic state. Traditionally, amoebae have been classified based on the morphology of their pseudopodia, and, where present, the structure and composition of extracellular scales or shells (tests). These now appear to be largely robust phylogenetic characters, based on the limited amount of molecular phylogenetic data that is beginning to emerge (Roger *et al*. 1996; Nikolaev *et al*. 2004; Smirnov *et al*. 2005). As a result, the majority of amoeboid types can now be assigned to three main eukaryote groups—Amoebozoa, Rhizaria, and the heterolobosean subdivision of discicristates (mitochondriate excavates) (Figure 1).

II. OVERVIEW OF THE TREE

Most eukaryotes can be assigned to one of eight major groups, six of which can be further grouped to make three supergroups (see Figure 1). Opisthokonts include animals, fungi, and several primitively single-celled lineages. Amoebozoa consists mostly of naked amoebae with lobose pseudopodia plus the social amoebas. Together, Amoebozoa and Opisthokonta form the supergroup "unikonts," probably the only major group of eukaryotes to be ancestrally uniflagellate (Richards and Cavalier-Smith 2005). Rhizaria are mostly testate (shelled) amoebas with fine, anastomosing pseudopodia. The Archaeplastida (formerly

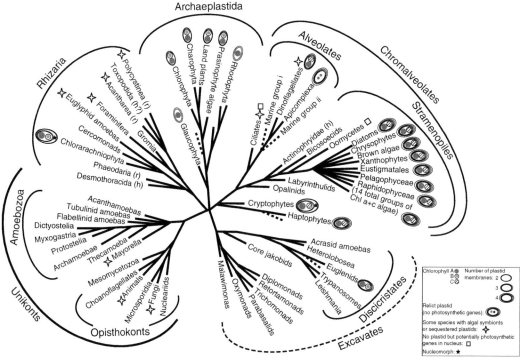

FIGURE 1. Consensus phylogeny of eukaryotes showing the distribution of photosynthesis among the major groups. The tree shown is a consensus phylogeny of eukaryotes based on a combination of molecular phylogenetic and ultrastructural data. The tree is further annotated to indicate the distribution of photosynthesis across the tree. Groups harboring endosymbiotic organelles are indicated by schematic plastids with variable numbers of membranes and chlorophyll composition, as indicated in the key to the lower right of the figure. Groups with members harboring transient algal endosymbionts are indicated by stars. Taxa classified as Radiolaria are indicated by an "r" following their names, and former "Heliozoa" are indicated by an "h." (See color plate.)

Plantae) are the group in which eukaryotic photosynthesis first arose. Chromalveolates (alveolates + stramenopiles + cryptophytes + haptophytes) include all major groups of marine algae. Finally, the excavates (discicristates + amitochondrial excavates) is a tenuous grouping of extremely diverse unicellular taxa. The main treatments of the higher level classification of eukaryotes are by Adl *et al.* (2005) and Cavalier-Smith (2004). Detailed organismal descriptions can be found in Lee *et al.* (2000) and Hausman and Hülsmann (1996). There are many recent reviews of higher-level classification (Baldauf 2003; Cavalier-Smith, 2004; Simpson and Roger 2004; Keeling *et al.* 2005). Especially useful websites include the Tree of Life site (http://tolweb.org/tree/), and David Patterson's Micro°scope, which

contains a wealth of protist images and much more (http://starcentral.mbl.edu/microscope).

A. Opisthokonts

The sisterhood of animals and fungi is supported by all large, broadly taxonomically sampled molecular datasets (Baldauf *et al.* 2000), and, most importantly, all large multigene trees (Baldauf *et al.* 2000; Moreira *et al.* 2000; Bapteste *et al.* 2002; Lang *et al.* 2002; Steenkamp *et al.* 2006). Nonetheless, the validity of the taxon continues to be debated (Loytynoja and Milinkovitch 2001; Philip *et al.* 2005). Animals and fungi together also possess the unique combination of flattened mitochondrial cristae, and, when flagellate, a single basal flagellum on reproductive cells (Cavalier-Smith 1998, 2002). These flagella also have unique similarities in their anchorage systems (Patterson 1999). However, none of these characters is universally present in all animals and fungi or uniquely absent in all other eukaryotes (Steenkamp and Baldauf 2004). The only universal synapomorphy for animals and fungi is a 9–17 amino acid insertion in protein synthesis elongation factor 1A (EF1A; Baldauf and Palmer 1993; Steenkamp et al. 2006), although some choanoflagellates appear to have lost this gene and the insertion may have evolved a second time independently (Atkinson and Baldauf unpublished; Keeling and Inagaki 2005).

One reason for the lack of phenotypic justification for opisthokonts is probably that animals and fungi are such highly morphologically derived lineages. They have independently invented very different types of multicellularity, as indicated by the fact that the earliest branches in both lineages are single-celled taxa (Steenkamp et al. 2006). Therefore, it is perhaps among these single-celled lineages that additional opisthokont synapomorphies may be found. Recent molecular phylogenetic data have identified several primitively unicellular lineages specifically allied with either animals or fungi (ministeriids, choanoflagellates, and mesomycetozoa; Steenkamp et al. 2006, and references therein).

Possibly the closest sister taxon to animals is *Ministeria vibrans,* the only currently cultured member of its genus. It is a stalked unicell, the body of which is surrounded by stiff radiating arms of unknown composition, and with flattened mitochondrial cristae (Cavalier-Smith and Chao 2003). However, support for *M. vibrans* as the sister taxon to animals is not unambiguous (Cavalier-Smith and Chao 2003; Steenkamp et al. 2006), and it may just be an unusual choanoflagellate with fast-evolving sequences (Steenkamp et al. 2006). On the other hand, there is now very strong support for choanoflagellates as the next closest relatives to animals (Lang et al. 2002; Medina et al. 2005; Steenkamp et

al. 2006). These aquatic uniflagellates with flattened mitochondrial cristae have long been noted for their resemblance to the collar cells of sponges. The next deepest branch leading to animals is the Mesomycetozoa (Lang et al. 2002; Mendozoa et al. 2002; Steenkamp et al. 2006). These are a grab bag of obligate intracellular parasites of aquatic animals, whose relationship to each other is only obvious in molecular trees (Ragan et al. 1996). The enigmatic taxon *Capsaspora prowozekei* may also be a mesomycetozoan, or it may, as some now believe, form an even deeper branch, making it the deepest branch of the entire holozoa (Lang et al. 2002; Ruiz-Trillo et al. 2004). It is an amoeba with filose pseudopodia, formerly classified as a nucleariid. This is particularly intriguing, as most of the rest of the nucleariids appear to be the sister group to the fungi (Medina et al. 2005; Steenkamp et al. 2006).

Within fungi, the earliest major branches are chytrids, which may be paraphyletic with respect to the rest of the fungi (Keeling et al., 2000; James et al. 2006). These unicells form pseudo-hyphae and have zoospores with a single basal flagellum. Thus, chytrids are the only known flagellated members of the fungi. The first branch of multicellular fungi is the Zygomycetes, followed by the Glomales (Schüßler et al. 2001). The latter are the arbuscular mycorrhizal fungi, which form symbioses with the vast majority of land plants. Their hyphae invade plant roots where they proliferate, exchanging mineral nutrients for host photosynthate. So far, no reproductive stage is known. The Ascomycetes and Basidiomycetes form the remaining two major groups of fungi. Both appear to be monophyletic and each other's sister taxon. It has long been suspected, and has now been confirmed by ciPCR survey that a vast diversity of fungi remain undescribed (Vandenkoornhuyse et al. 2002).

Deep relationships within the metazoa have proven much harder to resolve. This may be due to explosive radiation (Rokas et al. 2006) or to multiple genome duplications and differential gene loss such that orthologous genes are difficult to identify (Wolfe 2001). The limited amount of data currently available suggest that sponges are paraphyletic, with the glass sponges (Hexactinellida) branching off first, followed by the group of Calcarea + Demospongiae (Borchiellini et al. 2001). These appear to be followed by various branches of diploblasts, such as comb jellies (Ctenophora) and corals and jellyfish (Hydrozoa) (Halanych 2004). For the "higher" animals, current consensus is that they form three separate groups. The Deuterostomes include echinoderms and chordates, and their sister group is the Ecdysozoa + Lophotrochozoa. Ecdysozoa include the molting animals such as nematodes and arthropods, and Lophotrochozoa include most platyhelminthes, brachiopods, annelids, mollusks, and others (Wolf et al. 2004; Philippe and Telford 2006).

B. Amoebozoa

The Amoebozoa include several subdivisions of free-living amoebae, a group of amitochondriate amoeba/amoeboflagellates (Archamoebae), and the social amoebas (Mycetozoa or slime molds). Together with Rhizaria, they are molecularly the least well-characterized major group of eukaryotes, including many species of uncertain affinity. Amoebozoa have lobose or tubelike pseudopods, often a single nucleus, and tubular branched mitochondrial cristae (Adl et al. 2005). They range in size from a few microns to several millimeters, and many smaller forms probably remain to be discovered. Cyst formation to survive desiccation or to invade hosts is common. Most taxa are free-living and cosmopolitan in distribution, and they are important as major bacterial predators. In addition, the group includes some animal commensals and opportunistic pathogens.

1. Lobose Amoebae

The lobose amoebae are probably paraphyletic and include the most basal branches of Amoebozoa. They can be roughly divided into two classes: the Tubulinea and Flabellinea (Smirnov et al. 2005). There are also many species of uncertain affinity, which branch at the base of the amoebozoan tree. The Tubulinea share the ability to adopt a monopodial locomotive form with monoaxial cytoplasmic flow. The taxon also includes the arcellinids, the only lobose amoebae to form tests, which they construct from organic material. The Flabellinea are flattened naked amoebae having polyaxial cytoplasmic flow (Smirnov et al. 2005). Among the other lobose amoebae the most distinctive group are Acanthopodida, which differ from other amoebae in having cytoplasmic microtubule-organizing centers (MTOC) and a specific pseudopodial network. The genus Acanthamoeba is very important medically as causative agent of amoebic keratitis, a chronic eye infection (Marciano-Cabral et al. 2000).

2. Archamoebae

The Archamoebae are a tentative grouping of pelobionts and entamoebas (Bapteste et al. 2002; Cavalier-Smith et al. 2004). Most live in low-oxygen environments, and all lack mitochondria (Bapteste et al. 2002; Cavalier-Smith et al. 2004; Smirnov et al. 2005). Instead of mitochondria they have mitosomes, which are small mitochondrial-derived organelles of unknown function (Tovar et al. 1999). Pelobionts are amoeboflagellates, that is, they can assume amoeboid or flagellate forms, the latter with one or many flagella. They vary widely in size; Pelomyxa can be up to 3 mm long with several nonmotile flagella in its tail region. Entamoebae are small nonflagellates and mostly commensals or parasites of animals. Several species live in the mouth and intestinal tract of humans. Entamoeba histolytica causes amoebic dysentery, sometimes invading the liver with more severe consequences. Recent completion of the E. histolytica genome sequence shows that substantial amounts of its novel metabolic repertoire were acquired from bacteria by lateral gene transfer (LGT) (Loftus et al. 2005).

3. Mycetozoa

These are also called Eumycetozoa or social amoebas. They include the plasmodial (myxogastrid), cellular (dictyostelid), and protostelid slime molds, which have strikingly different trophic stage morphologies (Olive and Stoianovitch 1975). However, all possess an amoeboid stage with pointed, smoothly moving pseudopods, and they form similar fruiting bodies consisting of a cellulosic stalk supporting spore-bearing sori. Dictyostelids and myxogastrids have been variously classified as plants, animals, or fungi in the ~150 years since they were first described, and there has been a long running debate as to whether they are even related to each other. However, molecular data confirm that they form a group (Baldauf and Doolittle 1997), and recent evidence suggests that their closest relatives may be the Archamoebae (Nikolaev et al. 2006).

The myxogastrids are also referred to as plasmodial, true, or acellular slime molds, or simply "giant amoebas." The best known is the model organism *Physarum polycephalum*. Myxogastrids are amoeboflagellates, switching between amoeboid and flagellate forms early in their life cycle before maturing into large plasmodia. The latter can grow to a meter or more in diameter and have tens of thousands of synchronously dividing nuclei but no internal cell walls. Plasmodia are mobile and can move substantial distances at rates of ~1 cm per hour, propelled by means of cytoplasmic pulses. Eventually, the plasmodium breaks down to form numerous, often colorful and ornate fruiting bodies (Olive and Stoianovitch 1975).

In contrast, the dictyostelids spend most of their life cycle as single-celled amoebae, foraging for bacteria among detritus or in the soil. Under appropriate conditions, tens of thousands of these amoebae will come together to form a "slug." This is a mobile, macroscopic (2–5 mm) organism surrounded by an outer sheath and with defined head and body regions. Cell fate is determined in the slug such that when it metamorphoses into a fruiting body, the cells in the head of the slug will die during formation of the cellulosic stalk (Strmecki et al. 2005). Much less is known about the almost exclusively microscopic protostelids, which can be either amoeboflagellate or strictly amoeboid and form simple fruiting bodies (Olive and Stoianovitch 1975). They include the sister group to the dictyostelids and myxogastrids (Baldauf and Doolittle 1997) and may be paraphyletic (Spiegel et al. 1995).

C. Rhizaria (Formerly Cercozoa)

The Rhizaria consist largely of testate (shell-forming) amoeboid protists with typically thin finely pointed (filose) pseudopodia. However, the group is morphologically and ecologically very diverse and includes also a number of flagellate or biflagellate species, some naked amoebae, and plasmodial parasites. Rhizarian tests are built from a variety of materials, and the amoebae reside within them through most of their life cycle. The group includes the foraminiferans, the radiolarians, some heliozoans, the euglyphid amoebas, the plasmodiophorids, the chlorarachniophytes, and more. Rhizaria has been defined based exclusively on molecular characters, including phylogenetic analysis of actin, ribosomal RNA, and RNA polymerase II gene sequences (Keeling 2001; Berney and Pawlowski 2003; Longet et al. 2003, Nikolaev et al. 2004) and the presence of a specific insertion in the polyubiquitin gene (Archibald et al. 2003; Bass et al. 2005).

1. Group 1 (Radiolaria)

This group includes two classes of radiolarians, Acantharea and Polycystinea, and one former heliozoan group, Taxopodida. Radiolarians are identified by the combined presence of an internal mineralized "skeleton" and axopodia, which are long, radiating, unbranched processes stiffened by microtubular arrays. All are marine and pelagic, and they can be solitary or colonial. Radiolarians traditionally consist of three divisions—Acantharea, Polycystinea, and Phaeodarea, but Phaeodarea appear to be more closely related to Cercozoa (Polet *et al.* 2004), a different section of Rhizaria (see Section IV.C.). Acantharea have delicate skeletons with radial spicules. These tests are composed of strontium sulfate, joined at the center of the cell and emerging from the cell surface in a regular pattern. Polycystinea have silicious skeletons varying from simple spicules to complex helmet-shaped structures. Taxopodida consist of a single genus, *Sticholonche,* which are large heart-shaped protists with rows of oarlike pseudopodia that it uses for locomotion. The body is covered with silicious spicules and spines, giving the overall appearance of a star (Chachon and Chachon 1977).

2. Group 2 (Foraminifera)

This group includes the foraminiferans and their relatives, the haplosporidians and gromiids. Foraminiferans are a well-studied group with 940 modern genera and a rich fossil record (Lee *et al.* 2000). They are widely distributed in all types of marine environments, and some also occur in freshwater and terrestrial habitats (Pawlowski *et al.* 1999; Meisterfeld *et al.* 2001). They have finely granular reticulated pseudopods with bidirectional cytoplasmic flow. Most foraminiferans possess tests, which may be organic, agglutinated or calcareous, and composed of single or multiple chambers. Many also have complex life cycles consisting of alternating sexual and asexual generations (Lee *et al.* 2000). Haplosporidians are endoparasites that form large multinucleate plasmodia in freshwater and marine invertebrates. The gromiids are widespread marine protists with filose pseudopodia, a large (up to 5 mm) spherical to ovoid organic test with a characteristic layer of honeycomb membranes and a complex life cycle. They appear to be the

sister group to foraminiferans or foraminiferans + haplosporidians (Longet *et al.* 2003).

3. Group 3 (Cercozoa)

This large heterogenous assemblage includes cercomonads, chlorarachniophytes, filose testate amoebae, some former heliozoans (desmothoracids) and radiolarians (Phaeodaria), and plasmodial parasites (Nikolaev *et al.* 2004). Chlorarachniophytes are photosynthetic marine protists with anastomosing, network-like (reticulate) pseudopods and a uniflagellate dispersal stage. They acquired photosynthesis by capturing a green alga, and they retain both the plastid of the green alga and a remnant of its nucleus (nucleomorph), essentially a cell within a cell (McFadden 2001; Hackett *et al.*, Chapter 7, this volume). Chlorarachniophytes are closely related to cercomonads, which are common heterotrophic amoeboflagellates (Keeling 2001). Euglyphid testate amoebae, which build their tests from silica scales, are common in freshwater and in mosses. Plasmodiophorids (Phytomyxea) are endoparasites of plants or heterokont algae. They form multinucleated plasmodia, have a distinctive cruciform nuclear division, and have bi- or tetra-flagellate zoospores (Adl *et al.* 2005). Phaeodaria have siliceous skeletons, usually made of hollow radial spines, and a characteristic thick capsular membrane. Desmothoracids (Nucleohelea) are amoeboflagellates. The trophic amoeboid stage tends to be stalked with a perforated lorica through which multiple long thin pseudopodia protrude, giving it a starlike appearance.

4. Fossils

Both foraminiferan and radiolarian skeletons contribute substantially to the microfossil record in marine sediments extending back to the Cambrian. Their fossilized tests are used in micropaleontology as biostratigraphic markers and as paleoceanographic indicators to determine ancient water temperature, ocean depths, circulation patterns, and the age of water masses.

D. Archaeplastida

Archaeplastida (formerly Plantae) consist of the Rhodophyta (red algae), Glaucophyta, and Chloroplastida (green algae + land plants) (O'Kelly, Chapter 13, this volume). This is almost certainly the group in which eukaryotic plastids arose, by direct acquisition of a cyanobacterium (primary endosymbiosis). This cyanobacterium was converted into a plastid by massive restructuring of its genome, with roughly 95% of its genes being either lost or transferred to the host nucleus (Martin *et al.* 1998). As a result, ~90% of chloroplast proteins are encoded in the nucleus, synthesized in the cytosol, and posttranslationally imported into the plastid (Steiner *et al.* 2005). Thus, one of the primary events in plastid endosymbiosis would have been the evolution of a protein-import machinery (McFadden and van Dooren 2004).

The debate on whether there were one or more origins of eukaryotic photosynthesis is a long running one. Much of this hinges on evidence for the monophyly of (1) Archaeplastida and (2) their plastids. Strong support for the monophyly of these plastids includes the fact that their genomes have a similar derived gene order and composition (Douglas 1998). In addition, all the plastids in this group have only two membranes, whereas all other eukaryote plastids have three or more (Archibald and Keeling 2002; Hackett *et al.,* Chapter 7, this volume). Evidence for the monophyly of archaeplastid nuclear genomes has been slower in coming. However, there is now strong evidence for this from comparative genomics of red and green plants (McFadden and van Dooren 2004) and from multigene datasets, which also place glaucophytes as the first major branch in the group (Rodrigues-Ezpeleta *et al.* 2005).

1. Glaucophyta

The morphology of glaucophytes varies from biflagellates to coccoid nonflagellates to palmelloid forms (nonmotile cells in a mucilaginous matrix). All are small unicells whose light-harvesting complexes share similarities with red algae in that they use only chlorophyll (chl) *a,* which is attached to phycobiliproteins, and their thylakoids are unstacked. One of the most remarkable features of these taxa is their plastids, or "cyanelles." These have a bacterial-like peptidoglycan cell wall in between the inner and outer plastid membranes (Steiner *et al.* 2005). Genera include *Cyanophora, Glaucocystis,* and *Gloeochaete.*

2. Rhodophyta

Rhodophytes vary from large seaweeds to crustose mats that look more like rocks than living plants. Their plastids have two membranes and unstacked thylakoids. Light is harvested primarily with chl *a* and phycoerythrins conjugated to phycobiliproteins. Two major subgroups are recognized, Bangiophyceae and Florideophyceae; the former appears to be older and may have given rise to the latter. However, the taxonomy of the group requires major revision, and many, if not all, of the traditional major divisions may be invalid (Adl *et al.* 2005).

3. Chloroplastida

Chloroplastida include the green algae and the land plants, which were derived from the charaphyte branch of green algae. Thus, members of the Chlorophyta vary from single-celled flagellates to large marine filaments to redwoods. Their plastids have two membranes and stacked thylakoids, and they harvest light with chls *a* and *b* attached to chl*a/b*-binding proteins. For a detailed discussion of Chloroplastida and the origin of land plants, see O'Kelly (Chapter 13, this volume; also Bhattacharya and Medlin 1998; Karol *et al.* 2001).

E. Chromalveolates

Chromalveolates are a large group and morphologically and ecologically extremely diverse. They consist of the alveolates (ciliates, dinoflagellates, and apicomplexans), the stramenopiles (heterokonts), and possibly also the cryptophytes and haptophytes (Cavalier-Smith 2000, but see Hackett *et al.*, Chapter 7, this volume). Secondary endosymbiosis is widespread in the group, and the source in all cases appears to be red algae (Yoon *et al.* 2002b; Hackett *et al.*, Chapter 7, this volume). Dinoflagellates in particular have a tendency to acquire extra plastids on a regular basis (Yoon *et al.* 2002a; Delwiche, Chapter 10, this volume). In cryptophytes, a remnant of the red alga nucleus persists in the form of a nucleomorph, similar to the situation in chlorarachniophytes (see Section II.C.).

1. Stramenopiles

The stramenopiles are morphologically and ecologically one of the most diverse groups of eukaryotes. Nonetheless, they have strong molecular phylogenetic support and phenotypic justification. All major divisions of the group include organisms with a "tinsillated" flagellum, and most have a second, shorter, smooth flagellum (hence the commonly used alternate name "heterokont"). The shorter flagellum is posteriorly directed and often associated with an eyespot. The tinsillated flagellum is anteriorly directed and bears two rows of stiff, tripartite hairs (stramenopiles) along its length. These hairs reverse the flow around the flagellum so that the cell is dragged forward, rather than pushed along. Environmental sampling shows that there may be additional major divisions of the group, so far known only as uncultured ultra-small species (Moreira and Lopéz-Garcia 2002; Massana *et al.* 2006). In fact, there appears to be a huge diversity of very small free-swimming phototrophic, mixotrophic, and heterotrophic stramenopiles in most planktonic systems (Moreira and Lopéz-Garcia 2002).

a. Nonphotosynthetic Stramenopiles

Stramenopiles include a wide diversity of nonphotosynthetic lineages. They are not a monophyletic group and are only presented together here for convenience. Oomycetes are the water molds and downy mildews; they include important plant parasites such as *Phytophthora infestans,* the cause of potato blight. Once classed as fungi, they are now clearly assigned to stramenopiles (Gajadhar *et al.* 1991). The bicosoecids are small heterotrophic biflagellates, such as *Cafeteria,* possibly the world's most abundant predator (Moreira and Lopéz-Garcia 2002). Labyrinthulids or slime nets form filamentous networks and were once thought to be close relatives of slime molds (see Section III.C.; Olive and Stoianovitch 1975). Opalinids, which have cell bodies covered with stramenopile bearing flagella, and the taxonomically enigmatic *Blastocystis* are both commensals in the guts of cold-blooded animals. Although lacking a plastid, recent sequence data suggest that at least some of these taxa may have once been photosynthetic because they retain genes of apparent cyanobacterial origin in their nuclear genomes (Andersson and Roger 2002).

b. Photosynthetic Stramenopiles

Eleven major divisions of photosynthetic stramenopiles are recognized (Adl *et al.* 2005). These are Bacillariophyceae (diatoms), Chrysophyceae, Dictyochophyceae, Eustigmatales, Pelagophyceae, Phaeothamniophyceae (brown algae), Pinguiochrysidales, Raphidophyceae, Schizocladia, Synurales, and Xanthophyceae. Many are purely unicells, but there are also massive multicellular forms, such as the giant kelps. Various extracellular structures are found, such as scales or loricas.

Diatoms are ubiquitous and often the dominant marine photoautotroph. They reside in lidded boxes made of silica (see chapters by Kooistra *et al.*, Chapter 11, and by Hamm and Smetacek, Chapter 14, this volume). There are ~11,000 recognized species, and millions of undescribed ones by some estimates (Norton *et al.* 1996). Phaeophytes are particularly widespread in temperate intertidal and subtidal zones. Some of them have true parenchyma and build "forests" in nearshore environments that support complex ecosystems including fish and marine mammals. The xanthophytes, or yellow-green algae, include primarily freshwater algae. Most of them are unicellular, but a substantial number are colonial and live as naked cells in a gelatinous envelope. The Eustigmatales represent small unicellular coccoid algae, some of which have a very short flagellum and a large orange eyespot that is located outside the plastid. Raphidophytes are relatively large flagellates with the typical stramenopile flagella. They are naked but can bear trichocysts and usually have a variety of brown or green plastids.

2. Haptophytes

Haptophytes get their name from the presence of a unique anterior appendage, the haptonema, used for adhesion and capturing prey. The group includes the coccolithophorids, which build external coverings of calcium carbonate scales (coccoliths) and tend to dominate open oceanic waters worldwide. *Emiliania huxleyi,* in particular, has received considerable attention (see de Vargas *et al.*, Chapter 12, this volume). This is because of its role as a major carbon sink and because its massive blooms affect the temperature and optical qualities of ocean waters and play an important role in cloud production through dimethyl sulfoxide release (Buitenhuis *et al.* 1996). Coccoliths from dead cells accumulate as limestone deposits on the ocean floor, contributing to the largest inorganic reservoirs of carbon on Earth. Noncalcerous haptophyte genera, *Chrysochromulina* and *Prymnesium,* are an important source of toxic blooms.

3. Cryptophytes

The cryptophytes are, perhaps, the least well known of the chromists, being rela-

tively small (mostly 2–10 μm diameter) unicells and particularly good competitors in low light conditions in a wide variety of habitats (Gillot 1989). The group has received considerable attention because of its importance in unraveling the process of secondary endosymbiosis by retaining an intermediate stage in the process. Similar to chlorarachniophytes (see Section II.C.), cryptophyte plastids are accompanied by a remnant of their primary host nucleus (nucleomorph) that still encodes some of the proteins required for plastid function. Analysis of the nucleomorph genome (e.g., Douglas *et al.* 1999, 2001) provided the first phylogenetic evidence for the chimeric nature of algal cells by confirming the red algal origin of the cryptophyte plastid (see Hackett *et al.*, Chapter 7, this volume).

4. Alveolates

The alveolates are another large assemblage of protists with strong molecular and ultrastructural justification. The group includes the dinoflagellates (see Delwiche, Chapter 10, this volume), many of which are algae, the parasitic apicomplexans, and the ciliates (Adl *et al.* 2005). Cortical alveoli, which are saclike structures that lie immediately beneath the plasma membrane, are widely present, although sometimes lost. The alveoli form the pellicle in ciliates, surround the peripheral armor plates in dinoflagellates, and form the pellicular membrane in apicomplexans.

a. Ciliates

Ciliates appear to be the sister group to dinoflagellates + Apicomplexa. They are mostly free-living aquatic unicells characterized by an abundance of flagella (cilia) on their body surface (Hausmann and Hülsmann 1996). Ciliates are also noted for their nuclear dualism, where all cells have one or more of two very different types of nuclei. The smaller micronucleus contains the diploid germ nucleus, and the second much larger macronucleus contains thousands of copies of only the physiologically active genes. Ciliate nuclear genome organization can be truly remarkable; genes may not only be fragmented by introns and often numerous short intervening sequences, but the order of the gene fragments themselves may be scrambled. Therefore, extensive editing is required during generation of the macronucleus in order to produce the active working copy of the gene, and the mechanism by which this occurs is still unknown (Prescott 2000; Dalby and Prescott 2004).

b. *Dinoflagellates*

Dinoflagellates are a diverse, predominantly unicellular group, characterized by having one transverse and one longitudinal flagellum, resulting in a unique rotatory swimming motion. Many are also covered by often elaborate plates or armor. Although the group was probably primitively photosynthetic, only about half of the extant species still are, and many of these are mixotrophs (Stoecker 1999). The latter ingest bacteria and other eukaryotes, sometimes retaining algae or their plastids for varying lengths of time (Jakobsen *et al.* 2000; Tamura *et al.* 2005). The group includes the only known example of tertiary endosymbiosis involving the secondary endosymbiosis of a secondary endosymbiont (Yoon *et al.* 2002a; Delwiche, Chapter 10, this volume). *Symbiodinium* species are endosymbionts of corals and other invertebrates and occasionally other protists. Some dinoflagellates are a common source of phosphorescence in marine waters. Dinoflagellates are an important component of marine ecosystems as primary producers as well as parasites, symbionts, and micrograzers. They also produce some of the most potent toxins known and are the main source of toxic red tides and other forms of fish and shellfish poisoning.

c. *Apicomplexa*

Apicomplexa are the sister group to the dinoflagellates and include some of the most important protozoan disease agents of both invertebrates and vertebrates. All but the co/podellids are obligate and mostly intracellular parasites, and they include the causative agents of malaria and toxoplasmosis. They are characterized by the presence of an intricate apical complex, a system of organelles and microtubules situated at the posterior of the cell that functions in the attachment and initial penetration of the host. Parasitic apicomplexa have complex life cycles that are completed entirely within the host, and they exist outside it only as spores or oocysts. The group appears to have been derived from photosynthetic ancestors. It retains a vestigial plastid (apicoplast), most likely of red algal origin (Fast *et al.* 2001) that may be required for heme, lipid, and/or isoprenoid biosynthesis (Waller and McFadden 2005).

F. Excavates

This is easily the most enigmatic of the major groups of eukaryotes, and, in fact, molecular phylogenetic support for the group as a whole is weak at best. All are single-celled organisms, and many possess some type of conspicuous "excavated" ventral feeding groove (Simpson and Patterson 1999; Cavalier-Smith 2002). There are two recognized subdivisions within the "group." The discicristates include two well-defined subgroups, the Euglenozoa and the Heterolobosea, and possibly also some "jakobid" flagellates (Simpson *et al.* 2006). The latter lack the disc-shaped mitochondrial cristae (O'Kelly 1993) that were originally identified as a unifying character for the group along with strong molecular phylogenetic evidence (Baldauf *et al.* 2000). The second division of excavates, which, for lack of a better name will be referred to here as amitochondriate excavates, is a grab bag of taxa, many of which are largely known as obligate symbionts or parasites. Thus, they tend to have structurally simple cells and fast-evolving gene sequences and appear as the earliest branches in molecular trees

(Sogin *et al.* 1989; Cavalier-Smith and Chao 1996). Although many now consider the latter an artifact (Philippe and Germot, 2000), it seems to be quite a consistent one (Bapteste *et al.* 2002; Hedges *et al.* 2004; Ciccarelli *et al.* 2006).

1. Discicristates

The discicristates include the Euglenozoa and the Heterolobosea. They may also include some of the species known as jakobids, which have bacterial-like mitochondrial genomes (Lang *et al.* 1997). Euglenozoa include kinetoplastids and euglenids. Kinetoplastids are small uniflagellated or biflagellated cells, including the causative agents of sleeping sickness, Chagas' disease, and leishmaniasis. They are also famous for their bizarre mitochondrial genomes, where the genes are essentially encoded in a highly abbreviated shorthand. As a result, oligonucleotide fragments must be posttranscriptionally inserted into the initial messenger RNAs so that they can be properly decoded into protein (Sollner-Webb 1996).

a. Euglenozoa

Euglenids are usually free-living uni- or biflagellate cells enclosed by a thickened pellicle made of proteinaceous strips. Most euglenids are free-living osmotrophs, or phagotrophs, some of which are capable of ingesting whole eukaryotic cells. This is probably how photosynthetic forms, such as *Euglena*, acquired their chloroplasts, through secondary endosymbiosis of a green alga. *Euglena* has an unusual chloroplast genome, particularly the presence of twintrons. These are self-splicing introns within introns, where the inner intron must be spliced out before the outer intron can assume the correct structure for its own splicing (Hallick *et al.* 1993).

b. Heterolobosea

The Heterolobosea are mostly amoebae, although many have flagellate phases in their life cycles (Patterson *et al.* 2000a).

These naked amoebae differ from lobosean amoebae in that their pseudopods develop and move in a sporadic, "eruptive" manner. Most are soil or freshwater bacterivores, although one, *Naegleria fowleri,* is a rare but often fatal facultative human pathogen. The acrasid "slime molds" have been reassigned to this group based on molecular trees (Roger *et al.* 1996; Baldauf *et al.* 2000), which fits with the morphology of their heterolobosean-like pseudopodia (Olive and Stoianovitch 1975).

c. Jakobids

Discicristates probably also include some of the unicells referred to as jakobids (Simpson *et al.* 2006). These are small free-living bacterivores, noted for their bacteria-like mitochondrial genomes. Although most eukaryotes have fewer than 20 genes remaining in their mitochondrial genomes, jakobids retain more than 100. Also, unlike other eukaryotes, these genes are arranged in bacteria-like operons (Lang *et al.* 1997), consistent with the alpha-proteobacterial ancestry of mitochondria (Andersson *et al.* 2003).

2. Amitochondriate Excavates

The true or amitochondriate excavates are a grab bag of unicells, including some real morphological oddities. Many are obligate parasites, and all lack mitochondria. This led to the suggestion that they represented very early branches in the eukaryote tree, preceding the origin of mitochondria (Cavalier-Smith and Chao 1996). However, genes of mitochondrial origin have now been found in excavate nuclear genomes (Roger 1999; Tachezy *et al.* 2001), and some, perhaps even all, have what appear to be mitochondrially derived organelles (Dyall and Johnson 2000; Tovar *et al.* 2003).

The diplomonads typically exhibit a "doubled" morphology, with duplicate nuclei, sets of flagella, and cytoskeletons arranged back-to-back in each cell. The intestinal parasite *Giardia intestinalis* is a major human diarrheal agent, and *Spironucleus* includes some serious fish parasites. Most diplomonads are parasites, and the few free-living species are found only in low-oxygen habitats (Bernard *et al.* 2000). Retortamonads are broadly similar to diplomonads but have a single nucleus, flagellar cluster, and feeding groove per cell (Silberman *et al.* 2002). Most are intestinal commensals. Oxymonads are flagellated symbionts from the intestinal tracts of animals, mostly termites.

Parabasalids are mostly parasites and symbionts that are defined by the presence of a parabasalar apparatus, which is a complex of Golgi stacks and striated cytoskeletal elements. They include hypermastigids and trichomonads. Hypermastigids are huge multiflagellated cells, hundreds of micrometers long and covered in ectosymbiotic bacteria. They are found only in the hindgut of termites, itself a complex ecosystem, and are essential for the breakdown of cellulose. Trichomonads are small teardrop-shaped cells with four to six flagella that cause trichomoniasis, the most common human sexually transmitted protozoan infection (Embley and Hirt 1998).

G. Incertae Sedis

In 1999, Patterson identified 230 protists of uncertain affinity (Patterson 1999). In 2005, this number had only decreased to 204 (Adl *et al.* 2005), so much remains to be done. Most of these are small free-living heterotrophic flagellates or amoebae or are parasites of various kinds. Many, if not most, will undoubtedly turn out to fall within one or more of the groups described previously. Environmental surveys further suggest the existence of major undiscovered eukaryotic lineages (Amaral Zettler *et al.* 2002; Dawson and Pace 2002; Moreira and Lopéz-Garcia 2002). These "nanoeukaryotes," cells less than 2–3μm in diameter, have previously escaped detection because they are all but indistinguishable from bacteria under the light microscope.

Some of the new taxa appear to represent major new subdivisions of established groups (e.g., alveolates and stramenopiles, Section II.E.), including major components of marine ecosystems (Massana *et al.* 2006). Others, however, appear to represent known lineages for which DNA sequences were not previously available (Berney *et al.* 2004; Guillou *et al.* 2004).

III. THE EUKARYOTE ROOT

Probably the single most outstanding question in eukaryote evolution is the location of the root of the tree. For a long time, the predominant theory was that Archezoa (now mostly classified as amitochondriate excavates) arose near the root of eukaryotes, as these tend to form the deepest branches in molecular trees, including those based on large multigene datasets (Philippe and Adoutte 1998; Baldauf *et al.* 2000; Bapteste *et al.* 2002; Ciccarelli *et al.* 2006). However, there has been growing distrust in the ability of molecular phylogeny to resolve the deepest branches in the tree of life, largely due to the problem of LBA (Embley and Hirt 1998; Philippe and Ardoutte 1998).

A radically different placement of the eukaryote root is suggested by the fusion of the genes for dihydrofolate reductase and thymidylate synthase. These genes are adjacent and cotranscribed in bacteria, separate in opisthokonts, and fused in representatives of all other major eukaryote groups except amitochondriate excavates and amoebozoa, which mostly lack the genes entirely (Stechmann and Cavalier-Smith 2002, 2003). Because gene fusions are rare, and gene fissions presumably rarer, this suggests that Archaeplastida, chromalveolates, Rhizaria, and possibly also Excavates (Stechmann and Cavalier-Smith 2003) share a unique common ancestor excluding opisthokonts and possibly also Amoebozoa (Simpson and Roger 2002). Thus, this root potentially divides all eukaryotes into two supergroups—unikonts (opisthokonts and Amoebozoa) and bikonts (everything else).

IV. OXYGENIC PHOTOSYNTHESIS ACROSS THE EUKARYOTE TREE OF LIFE

Photosynthesis is widespread among eukaryotes (see Figure 1) as a result of multiple events and a variety of different mechanisms. Originally, eukaryotes (specifically the Archaeplastida) acquired photosynthesis from cyanobacteria (primary endosymbiosis; see Section II.D.). However, this primary endosymbiont has subsequently been passed around among eukaryotes multiple times, probably ever since the first plastid was established and continuing to the present. This lateral acquisition of eukaryotic photosynthesis can take the form of (1) temporary sequestration of foreign plastids derived from ingested prey (kleptoplastidy); (2) whole algal cells residing within a new host cell in an obligate or, more often, nonobligate manner (algal endosymbiosis); or (3) the plastid being stripped from these endosymbiotic algal cells in stages so that, ultimately, it becomes an obligate organelle of the new host (secondary or tertiary endosymbiotic organelles; see Hackett *et al.*, Chapter 7, this volume). Temporary sequestration of plastids is common in many heterotrophic protists and some invertebrates. This is in many cases "obligate" for the host, but an evolutionary "dead-end" for the plastid because the latter does not divide and hence cannot co-evolve with the host. In contrast, algal endosymbioses are probably related to acquisition of secondary or tertiary endosymbiotic organelles, representing the start and end points, respectively, of a continuum.

Although much attention has been paid to secondary organelle endosymbioses, less is known about transient and hereditary algal endosymbioses. In transient endosymbiosis, the endosymbiont must usually be taken up by the host in each generation. In most cases, these symbionts remain intact inside the host and are capable of existing independently outside of it. Where it has been studied, it appears that the host and symbiont may exchange small molecules, but there is no known gene transfer between them. Uptake of the symbiont by the host takes place in host specific time intervals, and host and sometimes symbiont appear to benefit from the association. The benefits to the host are generally obvious, as in most cases it has been shown that they derive photosynthate from the alga. The benefits to the symbiont, however, may be more complex, ranging from nutrition to protection from predation to improved dispersal and more. In some cases of transient symbiosis, the host digests excess algal symbionts. In contrast, hereditary algal endosymbioses are often obligate and can involve tight synchronization of host and endosymbiont division. In some cases, the endosymbiont is highly modified and probably cannot survive outside the host. In the case of both transient and hereditary endosymbioses (Table 1), the host cell is packed with symbiont cells, and it is heavily dependent on the symbiont such that photosynthesis is its main, and often sole, source of fixed carbon.

Transient endosymbiosis is found in amoebozoans and opisthokonts and is particularly widespread in Rhizaria and alveolates (see Table 1, Figure 1). In all cases where it has been studied, the symbiont contributes to the host's nutrition. Although the hosts are spread across almost the entire breadth of the eukaryote tree, the algal symbionts are, for the most part, quite taxonomically restricted (see Table 1). In most cases they belong to one of two algal groups: (1) dinoflagellates, most commonly of the genus *Symbiodinium* and (2) green algae, mainly from the genus *Chlorella*, or, in the case of lichens, the *Trebouxia* (John *et al.* 2002). Minority algal groups that enter into transient endosymbioses with otherwise nonphotosynthetic organisms include prasinophyte green algae, single-celled red algae, cryptophytes, and diatoms. In addition, there are some, although perhaps surprisingly few, transient endosymbioses

TABLE 1. Documented examples of algal endosymbiosis and temporary plastid sequestration across eukaryotes, including the formal ("Group") and informal ("Organism type") classification of the host, example host species, identity of the endosymbiont, and references

Group	Organism type	Example species	Endosymbiont	References
Unikonts Opisthokonts Fungi: Various fungi	Lichens Other fungi	*Bryoria* spp., *Buellia* spp. *Geosiphon pyriforme*	Fungi plus green algae (e.g., *Trebouxia*) or cyanobacteria green algae	Friedl and Bhattacharya 2001; John et al. 2002, 1996; Schüßler and Wolf 2005; Selosse and Le Tacon 1998;
Animals: Porifera	Fresh water and marine sponges	*Spongilla*	Cyanobacteria, green algae, and *Symbiodinium* spp.	Hill 1996; Hinde et al. 1994; Osinga et al. 2001; Wilkinson 1983, 1987;
Cnidaria	Corals	*Monastrea annularis*	*Symbiodinium* spp.	Savage et al. 2002; Toller et al. 2001; Visram and Douglas 2006;
	Hydrozoa	*Chlorohydra viridissima*	*Chlorella* sp. (green alga)	Bossert and Dunn 1986; Douglas and Smith 1984;
	Various sea anemones	*Anthopleura elegantissima, A. xanthogrammica*	*Symbiodinium*, green algae	Lewis and Muller-Parker 2004; Savage et al. 2002;
Chordata	Various ascidians	*Lissoclinum punctatum*	*Prochloron*	Griffiths and Thinh 1987; Hirose et al. 1998; Newcomb and Pugh 1975; Smith 1935;
Mollusca	Bivalves	*Tridacnidae, Anodonta cygnea, Corculum cardissa*	*Symbiodinium, Chlorella*	Farmer et al. 2001;
	Nudibranches	*Elysia chlorotica, Aeolidia papilosa*	*Symbiodinium*, plastids of various green algae, and others	Rumpho et al. 2000;
Platyhelminthes	Turbellaria	*Amphiscolops langerhansi, Convoluta roscoffensis, C. convoluta, Dalyellia viridis, Phaenocora typhlops, Typhloplana viridata*	Prasinophyte algae, and *Chlorella*, diatoms, *Amphidinium*	Douglas 1985; Douglas 1987; Taylor 1984;
Amoebozoa incertae sedis	lobose amoeba	*Mayorella viridis*	Green algae	Cann 1981;
Rhizaria	Polycystine radiolaria	*Thalassicolla* spp., *Collozoum* spp., *Physematium* spp, *Collosphaera* spp., *Spongodrymus* spp.	Dinoflagellates, prasinophytes	Anderson et al. 1983; Caron and Swanberg 1990; Gast and Caron 1996; Gast et al. 2000, 2003;

Acantharea	Diploconus fusces, Lithoptera mulleri, Amphilonche elongata, Acanthometra pellucida	Dinoflagellates, haptophytes	Febvre and Febvre-Chevalier 1979; Kimor et al. 1992; Taylor 1990;
Euglyphid amoeba	Paulinella chromatophora	Plastids of cyanobacteria	Kepner 1905; Marin et al. 2005; Melkonian and Mollenhauer 2005;
Testate amoeba	Amphitrema flavum, Hyalosphenia papilio	Green algae	Röettger 1995;
Foraminifera	Globigerina bulloides, Globigerinoides spp. Globigerinella siphonifera, Orbulina spp., Peneroplis spp., Sorites spp., Archaias spp., Borelis spp., Calcarina spp.	Dinoflagellates, haptophytes chlorophytes, rhodophytes, diatoms, chrysophytes	Chai and Lee 2000; Faber et al. 1988; Gast and Caron 1996, 2001; Gast et al. 2000; Lee 1998; Norris 1996; Pawlowski et al. 2001a, b; Pochon et al. 2001; Spero and Angel 1991;
	Elphidium spp., Nonion germanicum, Nonionella stella, Bulimina elegantissima	Kleptoplastids from diatoms	Bernard and Bowser 1999; Grzymski et al. 2002; Lopez 1979;
Chromalveolates			
Alveolates			
Ciliates	Climacostomum virens, Euplotes sp., Frontonia spp., Ophrydium versatile, Paramecium bursaria, Prorodon viridis, Stentor sp., Vorticella sp., Maristentor dinoferus	Green algae (Chlorella), dinoflagellates, purple nonsulfur bacteria	Fenchel and Bernard 1993; Esteban and Finlay 1996; Finlay et al. 1987; Gramham and Graham 1980; Hood 1927; Hoshina et al. 2004; Lobban et al. 2002, 2005; Sand-Jensen et al. 1997; Siegel 1960;
	Myrionecta rubra (=Mesodinium rubrum)	Cryptophyte or "incomplete" cryptophyte?	Gustafson et al. 2000; Hansen and Fenchel 2006; Johnson and Stoecker 2005; Johnson et al. 2007, Taylor et al. 1971;
	Perispira ovum	Kleptoplastids from Euglena? Incomplete symbiont?	Johnson et al. 1995;
	Strombidium capitatum, S. viride, S. conicum, S. acutum, S. chlorophilum, Laboea strobila, Tontonia spp.	Kleptoplastids from cryptophytes, chlorophytes, prasinophytes, haptophytes	Blackbourn et al. 1973; Laval-Peuto and Febvre 1986; McManus et al. 2004; Rogerson et al. 1989; Stoecker et al. 1987, 1988/1989, 1988;
Dinoflagellates	Noctiluca scintillans	Prasinophyte	Hansen et al. 2004;

(Continued)

TABLE 1. Documented examples of algal endosymbiosis and temporary plastid sequestration across eukaryotes, including the formal ("Group") and informal ("Organism type") classification of the host, example host species, identity of the endosymbiont, and references—Cont'd

Group	Organism type	Example species	Endosymbiont	References
		Kryptoperidinium foliaceaum (=*Peridinium foliaceum*), *Gymnodinium quadrilobatum*, *Peridinium quinquecorne*, *Dinothrix paradoxa*, *Galeidinium rugatum*, *Durinskia baltica* (=*Peridinium balticum*)	Diatoms	Chesnick *et al.* 1996, 1997; Tamura *et al.* 2005;
		Podolampus bipes, *P. reticulata*	Dictyochophyte	Schweiker and Elbrächter 2004
		Amphidinium acidotum (=*Gymnodinium acidotum*)	"Incomplete" cryptophyte	Fields and Rhodes 1991; Wilcox and Wedmayer 1984;
		Lepidodinium viride	"Vestigial" prasinophyte endosymbiont?	Watanabe *et al.* 1990;
		Amphidinium latum, A. poecilochroum, Cryptoperidiniopsis sp., *Gymnodinium gracilentum, Pfiesteria piscicida*	Kleptoplastids from cryptophytes	Eriksen *et al.* 2002; Feinstein *et al.* 2002; Horiguchi and Pienaar 1992; Jakobsen *et al.* 2000; Larsen 1988; Lewitus *et al.* 1999; Skovgaard 1998;
		Dinophysis acuminata, acuta, norvegica, mitra	Kleptoplastids from cryptophytes and haptophytes or ciliates? Or permanent plastids?	Hackett *et al.* 2003, 2004; Janson 2004; Koike *et al.* 2005; Minnhagen and Janson 2006; Park *et al.* 2006, Takashita *et al.* 2002;
Incertae sedis	Centroheliozoa	*Acanthocystis* sp., *Raphidiophrys* sp.	*Chlorella*	Chroome 1986; Kessler and Huss 1990;
		Clamydaster fimbriatus, Raphidocytis tubifera, Acanthocystis serrata	Kleptoplastids from green and chromophyte algae	Patterson and Durrschmidt 1987;
	Ebrids	*Hermesinum* sp.	Cyanobacteria	Hargraves 2002;
	Katablepharids	*Hatena arenicola*	Prasinophyte	Okamoto and Inouye 2005, 2006.

involving cyanobacteria, particularly pro-chlorophytes. However, in many cases of transient endosymbiosis the descriptions are incomplete and the precise identity of the symbiont is unknown (see Table 1).

Permanent or semipermanent endosymbioses in which the endosymbiont is passed from generation to generation appear to be rarer than transient endosymbioses, but this may be a biased view because the former may be more difficult to investigate experimentally. These often involve highly modified algal endosymbionts that have lost their surface features including flagella and cell wall. In one case the endosymbiont is only passed to one daughter cell (Okamoto and Inouye 2005). Highly evolved symbioses in which the algal endosymbiont persists in culture and is distributed to both daughter cells occur in the dinoflagellates and some ciliates (Reisser 1992; Schweiker and Elbrächter 2004; Tamura *et al.* 2005).

Perhaps most interesting are associations that are difficult to classify. Some appear to be highly successful "incomplete" endosymbioses in which a variety of endosymbiont organelles are maintained, and in some cases replicated, by the host. However, these still require repeated uptake of symbionts for sustained maintenance of the association and proliferation of the host. The blue-green dino-flagellate *Amphidinium acidotum* and red-tide ciliate *Myrionecta rubra* (*Mesodinium rubrum*) are examples of incomplete endosymbiosis (Wilcox and Wedemayer 1984; Johnson and Stoecker 2005). Other types of "incomplete" endosymbionts may represent intermediate steps between algal endosymbiosis and organelle endosymbiosis, for example the vestigial endosymbiont in the dinoflagellate *Lepidodinium viride* (Wantanabe *et al.* 1990).

A. Opisthokonts

Within the opisthokonts, symbioses are found in both fungi and animals. Within fungi, the glomalean *Geosiphon pyriforme* forms bladders containing endosymbiotic *Nostoc* (cyanobacteria). This association could be similar to primitive lichens (Selosse and Le Tacon 1998), which are symbioses between various ascomycete or basidiomycete fungi and green algae (*Chlorella* or *Trebouxia*), or sometimes cyanobacteria. Within animals, symbioses are found in Porifera, Cnidaria, Chordata, Mollusca, and Platyhelminthes. The main symbionts are dinoflagellates (*Symbiodinium* and *Amphidinium*), which are found in corals, various sea anemones, bivalves, nudibranchs, and turbellarians (see Table 1). Colonies of the coral *Montastrea annularis* species complex may contain several different groups of *Symbiodinium* (Toller *et al.* 2001). Green algal endosymbionts are common in Hydrozoa, sea anemones, bivalves, nudibranchs, and turbellarians (see Table 1). Some organisms, such as the nudibranch *Aeolidia papilosa* and the sea anemones *Anthopleura elegantissima* and *A. xanthogrammica*, have either a dinoflagellate (*Symbiodinium*) or a green alga of uncertain taxonomic position (Saunders and Muller-Parker 1997).

In some cases the host only retains the chloroplasts of the symbiont. For example the green sea slug, *Elysia chlorotica* (nudi-branch), acquires chloroplasts by feeding on the xanthophyte *Vaucheria litorea*. Although the bulk of the green alga is digested, its plastids are somehow retained in the slug's digestive tract where they can persist for up to 9–10 months (Rumpho *et al.* 2000). Some turbellarians host diatoms or prasinophytes. The endosymbionts found in ascidians are prochlorophytes, which are cyanobacteria with both chl *a* and *b* (Lewin 2002).

B. Amoebozoa

Transient endosymbiosis is rare in Amoebozoa, probably because they are rarely aquatic. Nonetheless, some lobose amoebae are known to carry endosymbiotic algae. For example, cells of *Mayorella viridis* (synonym *Amoeba viridis*) are packed with green algal unicells of unknown taxonomy (see Table 1). The testate amoebozoan *Hyalosphenia papilio*, which lives in semiaquatic environments, also harbors a green algal symbiont (Röttger 1995).

C. Rhizaria

The Rhizaria include one example of secondary endosymbiosis (chlorarachniophytes, see Section II.C.) and many examples of transient symbioses and plastid retention. Transient endosymbioses are found in various Radiolaria, in euglyphid amoebae and in Foraminifera. Within Radiolaria, symbionts are, so far, only found in the Polycystinae and Acantharea, where they are widespread. These symbionts are mainly dinoflagellates (Spero and Angel 1991; Gast and Caron 1996, 2001; Takahashi *et al.* 2003). However, for each host species, the symbionts are highly specific and must be re-established every generation (Gast and Caron 1996).

About 25% of the 40–50 known species of planktonic Foraminifera harbor internal algal symbionts. These endosymbionts are mostly dinoflagellates, diatoms, or green algae (Gast and Caron 1996; Pawlowski *et al.* 2001a, b). In some cases the foraminifer digests the endosymbiont after a couple of months. However, in most cases the host retains the endosymbionts for much longer, although they are probably lysed before the host reproduces (Norris 1996). Their importance for maintaining the host in low nutrient environments by providing an extra energy source is suggested by the fact that endosymbiont-bearing foraminifers have a wider cosmopolitan distribution than asymbiotic species (Norris 1996).

Algal endosymbionts and plastid retention are very common among benthic Foraminifera. All the large benthic Foraminifera important in carbonate deposition, reef building, and formation of calcerous sands in tropical and semitropical seas have algal endosymbionts (Lee 1992). These symbionts contribute to both carbon acquisition and host calcification. Different families of benthic Foraminifera tend to have different algal types, with one family hosting unicellular rhodophytes, one chlorophytes, one dinoflagellates, and a number of families specializing in diatoms (Lee 1992). Although the diatom-containing

Foraminifera appear to be highly specialized morphologically for harboring endosymbionts, the hosts often harbor more than one species of diatom. In most cases the endosymbionts can survive and grow if they are removed from their host.

Many benthic Foraminifera retain plastids and have morphological features that appear designed to remove and retain the plastids from ingested diatoms (Bernhard and Bowser 1999; Lee 1992). In the dark, plastids can be retained for weeks to months, but in the light the half-life of plastids is much shorter. Thus, foraminiferan hosts need to continually ingest algae to keep a stable plastid population. Plastid retention is particularly common among taxa that inhabit dysoxic environments, and it has been hypothesized that oxygen production during photosynthesis may be important to the host (Bernhard and Browser 1999). Surprisingly, sequestration of plastids is also observed in foraminifera-inhabiting sediments well below the euphotic zone (Bernhard and Bowser 1999). It has been hypothesized that in very low light and aphotic environments the chloroplasts are utilized in assimilation of inorganic nitrogen by the host because nitrate reductase is located in plastids and lacking in most heterotrophic protists (Grzymski *et al.* 2002).

Green algal symbionts are found in a number of sphagnum moss-inhabiting testate amoebae, including the Rhizarian *Amphitrema flavuum* (Mitchell and Gilbert 2004). The euglyphid amoeba *Paulinella chromatophora* is a special case of possible primary endosymbiosis, where evidence now suggests that the symbiont may be a bona fide organelle (Marin *et al.* 2005). Each amoeba contains two permanent kidney-shaped "cyanelles" that appear to be of direct cyanobacterial origin unrelated to modern chloroplasts. The amoeba is dependent on this symbiont, and there is no evidence that *P. chromatophora* feeds, despite long-term maintenance in culture. Repeated attempts to cultivate the endosymbiont outside the host have failed, and the symbiont and host divide synchronously (Marin *et al.* 2005). Whatever its origin, the euglyphid cyanelle appears to be a relatively recent innovation, as the closely related species *P. ovalis* lacks the endosymbiont but preferentially feeds on the cyanobacterium *Synechococcus*. As yet, there is no evidence of any gene transfer from the symbiont to the host nucleus.

D. Archaeplastida

The Archaeplastida is the group in which the plastids that are now so widespread among eukaryotes first arose from a cyanobacterium. All members of the group are photosynthetic, except for a few isolated instances where photosynthesis has been secondarily lost. All chloroplasts in this group are surrounded by two membranes, apparently derived from the membranes of the ancestral cyanobacterium (Archibald and Keeling 2002). All Archaeplastida use chl *a* as a primary photosynthetic pigment of the light-harvesting complex. A variety of other accessory pigments are found, but only the Chloroplastida (green algae and land plants) also have chl *b*.

E. Chromalveolates

1. *Endosymbiotic Organelles*

Members of the chromalveolates, both alveolates and "chromists," possess plastids that they acquired by secondary endosymbiosis, in all cases apparently from a red alga, although possibly not in the same endosymbiotic event (see Hackett *et al.*, Chapter 7, this volume). In addition to the universal chl *a*, their plastids also contain chl *c*, the origin of which is not well understood. Most of these plastids are surrounded by four membranes. Of these, the inner two are derived from the ancestral cyanobacterium, the third from the plasma membrane of the primary host, and the fourth from the phagosomal membrane of the secondary host (Archibald and Keeling 2002). The origin of the plastid in "chromists" is now fairly well understood (Yoon *et al.* 2002b). The situation in

alveolates is slightly less clear, partly because plastid origin in dinoflagellates is complex and variable and because the "plastids" of Apicomplexa are highly reduced and no longer photosynthetic.

Nearly all dinoflagellate plastids are of red algal origin (Hackett *et al.*, Chapter 7, this volume) and mainly contain chl *a* and *c*. Most of them also have three membranes, possibly due to the loss of the outer (phagosomal) one. However, some dinoflagellates have plastids with two to five membranes, and these plastids were acquired through secondary or tertiary endosymbiosis. In the Apicomplexa a remnant plastid with four membranes is found, but it has lost its photosynthetic genes and pigments. It is not clear if ciliates once had plastids and then lost them, which would have to be the case if there was a single origin for alveolate plastids.

2. *Plastid Retention*

Temporary retention of plastids derived from algal prey (kleptoplastidy) occurs in both dinoflagellates and ciliates. Among the dinoflagellates, plastid retention usually, but not always, involves cryptophyte plastids (see Table 1). In dinoflagellate hosts that are not amenable to culture, it can be difficult to distinguish retained plastids from endosymbiotic plastids. For example, the origin of "cryptophyte-like" plastids in photosynthetic *Dinophysis* species is debated (Takashita *et al.* 2002; Hackett *et al.* 2003; Janson 2004). The presence of a haptophyte-like plastid in *Dinophysis mitra* and the heterogeneity in plastid content of many *Dinophysis* species suggests that they are capable of obtaining plastids from prey (Koike *et al.* 2005). In fact, Park *et al.* (2006) reported that *Dinophysis* can be cultured if fed *Myrionecta rubra*—so it probably gets its plastids from a ciliate that gets them from a cryptophyte (Section IV. E. 4).

Most plastid stealing dinoflagellates (except *Dinophysis)* only retain plastids from cryptophytes (see Table 1). Plastid retention is facultative in these species and the host can also grow heterotrophically. In most cases the retained plastids have a short half-life (usually less than a day) (Jakobsen *et al.* 2000; Eriksen *et al.* 2002; Feinstein *et al.* 2002).

Plastid retention is very common among planktonic ciliates in the family Strombidiidae (Stoecker *et al.* 1987, 1998; Rogerson *et al.* 1989; Dolan and Perez 2000). The physiology of plastid retention is different from that in dinoflagellates. Ciliates retain plastids from a variety of algal taxa; including prasinophytes, haptophytes, cryptophytes, and chlorophytes. One host cell can contain several types of plastids (Laval-Peuto and Febvre 1986; Stoecker and Silver 1990). However, some tide pool ciliates specialize in retaining plastids and eyespots from swarmers of green macroalgae (McManus *et al.* 2004). As in the dinoflagellates, the half-life of these kleptoplastids is usually short (less than a day) (Stoecker and Silver 1990). However, unlike dinoflagellates, plastid retention is usually obligate in ciliates; plastid-retaining ciliates usually require both light and a suitable algal source of plastids for growth (Stoecker 1998; Stoecker and Silver 1990).

3. *Algal Endosymbioses*

Examples of endosymbioses are common in some alveolate taxa and range from transient, nonobligatory associations to hereditary endosymbioses in which the host depends on its algal endosymbionts for survival and growth. Some dinoflagellate hosts have a loose relationship with the endosymbiont. For example, some *Noctiluca scintillans* have free-swimming prasinophyte endosymbionts in their central vacuole, but if the endosymbionts are lost, the host can grow as a strict heterotroph (Hansen *et al.* 2004). Other dinoflagellate species always have a highly modified, hereditary diatom endosymbiont (see Table 1) with the single endosymbiont and host cell dividing synchronously. It is thought that this diatom endosymbiont functionally replaced the host plastid, which became the eyespot in these dinoflagellates (Tamura *et al.* 2005). Phylogenetic evidence supports a single

endosymbiotic event as the "ancestor" for several of these endosymbioses (Tamura *et al.* 2005). One hereditary endosymbiosis (*Podolampus* species) involves dictyochophycean endosymbionts (see Table 1) (Schweiker and Elbrächter 2004).

These examples of inheritable endosymbioses in dinoflagellates suggest that an endosymbiotic unit can evolve and radiate into new host-endosymbiont species. An interesting case of what may be a highly evolved endosymbiosis is the "green" dinoflagellate *Lepidodinium viride*. The "vestigial endosymbiont" in this host lacks a nucleus and mitochondria but has cytoplasm, ribosomes, multiple plastids, and vesicles, all surrounded by a double membrane (Watanabe *et al.* 1990). Because *Lepidodinium* can grow without prey, this may be a case of an endosymbiotic alga on its way to becoming an endosymbiotic plastid.

Among marine ciliates, algal endosymbionts are fairly rare, but they are common in freshwater ciliates, particularly in micro-oxic layers (reviewed in Dolan 1992; Stoecker 1998) Most freshwater ciliates have green algal endosymbionts (zoochlorellae). For example, in *Ophrydium versatile,* which forms gelatinous colonies, each individual is packed with endosymbiotic green algae. These algae dominate the ciliate's oxygen metabolism and supply it with photosynthate (Sand-Jensen *et al.* 1997). The free-living ciliate *Paramecium bursaria* hosts a green alga, although the identity of this symbiont seems to vary with the location where the host is found (Hoshina *et al.* 2005). The few cases of "algal" endosymbiosis described in marine ciliates involve either purple nonsulfur bacteria (Fenchel and Bernard 1993) or dinoflagellates (Lobban *et al.* 2002, 2005).

In many of the algal endosymbioses in ciliates, the host picks up the algal endosymbionts from the environment, and the association is nonobligatory for the ciliate. However, the association is specific in that only a few algal strains can form a relationship with a particular host (Reisser 1992). The host may gain carbon by digesting some of the endosymbionts periodically or by utilizing excreted photosynthate. In the case of "heritable" endosymbioses (e.g., *P. bursaria, Climacostomum viridis*), the endosymbionts excrete most of their fixed carbon, grow slowly, and are protected from digestion in perialgal vacuoles (Reisser 1992).

4. Incomplete Endosymbioses

In both dinoflagellates and ciliates, there are some rare relationships that involve "incomplete" algal endosymbionts—usually plastid–mitochondrial complexes embedded in symbiont cytoplasm, separated from host cytoplasm by a membrane. In some cases, symbiont nuclei and other organelles are at least transiently sequestered. The endosymbionts appear to slowly degrade with the plastids usually lasting the longest. Thus, the relationship can resemble algal endosymbiosis if it is observed soon after the ingestion event and can resemble plastid retention in its later stages. These incomplete endosymbioses appear to be distinct, highly specialized, and at least in some cases, highly successful.

An example among the dinoflagellates is the freshwater blue-green species *Amphidinium acidotum* (*Gymnodinium acidotum*), which preys on cryptophytes. The cryptophyte endosymbionts are often incomplete, lacking a nucleus (Wilcox and Wedmayer 1984; Fields and Rhoades 1991). These plastids are retained by the host for about 10 days, which is much longer than "simple" kleptoplastidy in dinoflagellates. Although the endosymbiosis is nonheritable, it is obligate for the host, which can only sustain growth when it is cultured with the cryptophyte (Fields and Rhoades 1991).

Among ciliates, *Perispira ovum*, a specific predator on *Euglena*, may be one example of intermediate endosymbiosis (Johnson *et al*. 1995). *Euglena* chloroplasts, mitochondria, and paramylon are observed in *P. ovum*. Retention of nuclei has not been reported, but that might be because nuclei are more rapidly degraded or lost. More information

is available about another ciliate, *Myrionecta rubra* (*Mesodinium rubrum*). This "pink" ciliate is known for its cosmopolitan distribution in marine waters, high photosynthetic rates, and ability to form red tides. *M. rubra* has variously been described as possessing a permanent cryptophyte endosymbiont, an incomplete endosymbiont, or individual cryptophyte plastids (Taylor *et al*. 1971; Gustafson *et al*. 2000; Hansen and Fenchel 2006). These differences may be due to differences among *M. rubra* populations or to the timing of observations relative to ingestion of cryptophytes. Experimental studies with cultures have shown that chlorophyll synthesis and plastid replication occurs in the host, but that periodic ingestion of cryptophytes is necessary for sustained photosynthesis and population growth (Johnson and Stoecker 2005; Hansen and Fenchel 2006). Temporary sequestration of prey nuclei and utilization of cryptophyte nuclear gene products seem to be necessary for maintenance of plastid function and replication in the host (Johnson *et al*. 2007).

F. Excavates and Incertae Sedis

The only excavate group known to be capable of photosynthesis are the euglenids (see discicristates, Figure 1). Their plastids are of green algal origin, contain chl *a* and *b* and are surrounded by three membranes. Photosynthesis is clearly a derived feature within euglenids as most major divisions of the group are not photosynthetic and many routinely ingest algal prey.

Currently unclassified eukaryotes known to have photosynthetic algal endosymbionts include some centrohelid "Heliozoa." Members of the genera *Acanthocystis* and *Raphidiophrys* have both been reported to harbor green algal endosymbionts (Croome 1986; Kessler and Huss 1990). *Chlamydaster fimbriatus, Raphidocytis tubifera,* and *Acanthocystis serrata* all retain plastids from green and chromophyte algae (Patterson and Dürrschmidt 1987). Another unclassified eukaryote with endosymbionts is the ebridian flagellate

Hermesinum, which harbors endosymbiotic cyanobacteria (Hargraves 2002).

An interesting case of algal endosymbioses occurs in the newly described katablepharid flagellate *Hatena arenicola,* which can harbor a prasinophyte endosymbiont (Okamoto and Inouye 2006). *Hatena* alternates between a colorless feeding stage and an autotrophic stage with a degenerate feeding apparatus and a highly modified, single enlarged prasinophyte endosymbiont. When an autotrophic *Hatena* cell divides, the endosymbiont is passed to one daughter cell; the other, colorless daughter cell reverts to heterotrophy (Okamoto and Inouye 2005, 2006).

V. CONCLUSIONS

The eukaryotes are an ancient and diverse group of organisms. It is perhaps surprising that they have evolved chloroplasts only once, because this original plastid has spread so widely ever since. This suggests that there is either a bottleneck in the process or perhaps some change in modern eukaryotes that makes this more difficult than earlier in their evolution. The latter seems unlikely, because plastids originated after the origin of Archaeplastida. Thus, the protohost cell would already have been a fully developed eukaryote with all major features that define modern eukaryotes.

Nonetheless, the selective advantages of acquiring photosynthesis appear to be strong, as it is widespread and has occurred many times independently (see Table 1, Figure 1). It is interesting to note that, although a wide variety of eukaryotes can enter into transient or heritable symbioses as the host, the taxonomic distribution of endosymbionts is much more restricted (see Table 1). This suggests that there may be some unique characteristic of the symbiont that facilitates these interactions. Clues to this may lie in symbiont membranes, as the ability to exchange materials with the host in a selective and directionally biased manner

may be critical. This might also explain why cyanobacterial–eukaryote symbioses are rare, as such membrane transport systems are lacking in bacteria.

Although hosts for algal endosymbionts are widespread among eukaryotes, some groups are particularly prone to these associations, particularly Rhizaria and alveolates. It has been argued that dinoflagellates have more of their plastid genes transferred to the nucleus (Hackett *et al.* 2004), which may make dinoflagellates good "hosts" for endosymbionts and plastids. Because most heterotrophic dinoflagellates are derived from photosynthetic lineages (Saldarriaga *et al.* 2001), they may have retained some algal nuclear genes or relict plastids and thus also be good hosts. In the case of ciliates it is also possible that some lineages still have plastid-derived genes or relict plastids (Archibald and Keeling 2002).

It is interesting that dinoflagellates are often endosymbiont hosts, but they are rarely endosymbionts themselves. Although "red" plastids are thought to be more "portable" than "green" plastids (Grzebyk *et al.* 2003), this probably does not apply to dinoflagellate plastids because of their reduced plastid genome (Hackett *et al.* 2004). In contrast, cryptophyte plastids are retained by many dinoflagellates and ciliates (see Table 1) and usually persist longer than other plastid types. This may be related to the fact that these plastids retain a reduced nuclear genome (Douglas *et al.* 2001). This may aid plastid survival in the host due to residual enzyme activity in the plastid's periplast, providing photosynthate to the host and extending the survival of the plastid. Genes for about 30 chloroplast proteins are found in the cryptophyte nucleomorph (Douglas *et al.* 2001). It is possible that "runoff" protein synthesis or even limited transcription may continue in the periplast after endosymbiosis, possibly with the aid of host proteins.

Plastids also have other metabolic functions besides photosynthesis. Photosynthesis, by providing oxygen, may allow hosts to survive and grow in suboxic environments in which competition and predation pressure are reduced. Plastids may also serve as food storage organelles. It has been suggested that plastids can play a role in assimilating inorganic nitrogen (Grzymski *et al.* 2002), which could be important in habitats where organic resources have a high C:N ratio. In algae and plants, key precursors in biosynthetic pathways are made in the plastid, for example d-amino-levulinic acid (d-ALA), which is a precursor for the haem component of mitochondrial cytochromes and other oxidative enzymes (Barbrook *et al.* 2006). Essential fatty acids, sterols, and amino acids are also made in the plastid. Some phagotrophic protists may have lost their biosynthetic pathways for these compounds and thus be dependent on algal endosymbionts or retained plastids to supply them. This dependency may explain why some protists that live in the dark retain plastids, why plastid retention is obligatory in some phagotrophs, and why remnant plastids persist in many heterotrophic lineages.

If plastids are so useful, why don't endosymbionts always evolve into them? There are many barriers to the transition from endosymbiont to organelle (Cavalier-Smith 2003) including the requirement for nuclear genes, which encode most of the chloroplast's proteins. These genes would need to be transferred to the new host nucleus and acquire appropriate targeting sequences for transport through chloroplast membranes. For retained plastids, evolution into endosymbiotic organelles faces an additional barrier, the plastids cannot divide in the host (except in the case of incomplete endosymbioses), and therefore cannot evolve.

Algal endosymbiosis, incomplete endosymbioses, and plastid retention may be evolutionarily stable states in that they may have advantages over acquisition and maintenance of permanent photosynthetic machinery by the host. The additional cost of phototrophy in a phagotrophic cell (a mixotrophic protist) has been estimated to be up to 50% of the energy, carbon, nitro-gen, phosphorus, and iron budget of the cell (Raven 1997). The maximum growth rates of heterotrophs are usually higher than of strict autotrophs of similar size. Thus, it may be more efficient for some species to obtain their photosynthetic machinery through ingestion rather than to synthesize and maintain it. When life-cycle events or environmental conditions render algal endosymbionts or plastids a cost rather than a benefit, most hosts can probably rapidly reduce their commitment to autotrophy by digesting or egesting their photosynthetic machinery. This nutritional flexibility may be an important adaptation to food variability while maintaining the ability to grow rapidly.

References

Adl, S.M., Simpson, A.G., Farmer, M.A., Ansersen, R.A., Anderson, O., Barta, J.R., Bowser, S.S., Brugerolle, G. Fensome, R.A., Frederiq, S., James, T.Y., Karpov, S., Kugrens, P., Krug, J., Lane, C.E., Lewis, L.A., Lodge, J., Lynn, D.H., Mann, D.G., McCourt, R.M., Mendoza, L., Moestrup, Ø., Mozley-Standridge, S.E., Nerad, T.A., Shearer, C.A., Smirnov, A.V., Spiegel, F.W., and Taylor, M. (2005). The new higher level classification of eukaryotes with emphasis on the taxonomy of protists. *J. Eukaryot. Microbiol.* **52:** 399–451.

Alberts, B., Johnson, A., Lewis, J., Raff, M., Roberts, K., and Walter, P. (2002). *Molecular Biology of the Cell,* 4th ed. New York, Taylor and Francis.

Amaral Zettler, L.A., Gomez, F., Zettler, E., Keenan, B.G., Amils, R., and Sogin, M.L. (2002). Microbiology: eukaryotic diversity in Spain's River of Fire. *Nature* **417:** 137.

Anderson, O.R., Swanberg, N.R., and Bennett, P. (1983). Assimilation of symbiont-derived photosynthates in some solitary and colonial radiolaria. *Mar. Biol.* **77:** 265–269.

Andersson, J.O., and Roger, A.J. (2002). A cyanobacterial gene in nonphotosynthetic protists—an early chloroplast acquisition in eukaryotes? *Curr. Biol.* **12:** 115–119.

Andersson, S.G., Karlberg, O., Canback, B., and Kurland, C.G. (2003). On the origin of mitochondria: a genomics perspective. *Philos. Trans. R. Soc. Lond. B. Biol. Sci.* **358(1429):** 165–177.

Archibald, J.M., and Keeling, J.P. (2002). Recycled plastids: a "green movement" in eukaryotic evolution. *Trends Genet.* **18:** 577–584.

Archibald, J.M., Longet, D., Pawlowski, J., and Keeling, P.J. (2003). A novel polyubiquitin structure in Cercozoa and Foraminifera: evidence for a new eukaryotic supergroup. *Mol. Biol. Evol.* **20(1):** 62–66.

Baldauf, S.L. (2003). The deep roots of eukaryotes. *Science* **300:** 703–706.

Baldauf, S.L., and Doolittle, W.F. (1997). Origin and evolution of the slime molds (Mycetozoa). *Proc. Natl. Acad. Sci. U S A* **94:** 12007–12012.

Baldauf, S.L., and Palmer, J.D. (1993). Animals and fungi are each other's closest relatives: congruent evidence from multiple proteins. *Proc. Natl. Acad. Sci. U S A* **90:** 11558–11562.

Baldauf, S.L., Roger, A.J., Wenk-Siefert, I., and Doolittle, W.F. (2000). A kingdom-level phylogeny of eukaryotes based on combined protein data. *Science* **290:** 972–977.

Bapteste, E., Brinkmann, H., Lee, J.A., Moore, D.V., Sensen, C.W., Gordon, P., Daruflé, L., Gaasterland, T., Lopez, P., Müller, M., and Philippe, H. (2002). The analysis of 100 genes supports the grouping of three highly divergent amoebae: *Dictyostelium*, *Entamoeba*, and *Mastigamoeba*. *Proc. Natl. Acad. Sci. U S A* **99:** 1414–1419.

Barbrook, A.C., Howe, C.J., and Purton, S. (2006). Why are plastid genomes retained in non-photosynthetic organisms? *Trends Plant Sci.* **11:** 101–108.

Bass, D., Moreira, D., Lopez-Garcia, P., Polet, S., Chao, E.E., von der Heyden, S., Pawlowski, J., and Cavalier-Smith, T. (2005). Polyubiquitin insertions and the phylogeny of Cercozoa and Rhizaria. *Protist* **156(2):** 149–161.

Bernard, C., Simpson, A.G.B., and Patterson, D.J. (2000). Some free-living flagellates (Protista) from anoxic habitats. *Ophelia* **52:** 113–142.

Berney, C., Fahrni, J., and Pawlowski, J. (2004). How many novel eukaryotic "kingdoms"? Pitfalls and limitations of environmental DNA surveys. *BMC Biol.* **4:** 2–13.

Berney, C., and Pawlowski, J. (2003). Revised small subunit rRNA analysis provides further evidence that Foraminifera are related to Cercozoa. *J. Mol. Evol.* **57(1):** 120–127.

Bernhard, J.M., and Bowser, S.S. (1999). Benthic foraminifera of dysoxic sediments: chloroplast sequestration and functional morphology. *Earth Sci. Rev.* **46:** 149–165.

Bhattacharya, D., and Medlin, L. (1998). Algal phylogeny and the origin of land plants. *Plant Physiol.* **116:** 9–15.

Blackbourn, D.J., Taylor, F., and Blackborn, J. (1973). Foreign organelle retention by ciliates. *J. Protozool.* **20:** 286–288.

Borchiellini, C., Manuel, M., Alivon, E., Boury-Esnault, N., Vacelet, J., and Le Parco, Y. (2001). Sponge paraphyly and the origin of Metazoa. *J. Evol. Biol.* **14:** 171–179.

Bossert, P., and Dunn, K.W. (1986). Regulation of intracellular algae by various strains of the symbiotic *Hydra viridissima*. *J. Cell. Sci.* **85:** 187–195.

Buitenhuis, E., Bleijswijk, J., van Bakker, D., and Veldhuis, M. (1996). Trends in inorganic and organic carbon in a bloom of *Emiliania huxleyi* in the North. Sea. *Mar. Ecol. Prog. Ser.* **143:** 271–282.

Cann, J.P. (1981). An ultrastructural study of *Mayorella viridis* (Leidy) (Amoebida, Pramoebidae), a rhizopod containing zoochlorellae. *Arch. Protistenkunde* **124:** 353–360.

Caron, D.A., and Swanberg, N.R. (1990). The ecology of planktonic Sarcodines. *Rev. Aquatic Sci.* **3:** 147–180.

Cavalier-Smith, T. (1998). A revised six-kingdom system of life. *Biol. Rev.* **73:** 203–266.

Cavalier-Smith, T. (2000). Membrane heredity and early chloroplast evolution. *Trends Plant Sci.* **5:** 174–182.

Cavalier-Smith, T. (2002). The phagotrophic origin of eukaryotes and phylogenetic classification of Protozoa. *Int. J. Syst. Evol. Microbiol.* **52:** 297–354.

Cavalier-Smith, T. (2003). Genomic reduction and evolution of novel genetic membranes and protein-targeting machinery in eukaryote-eukaryote chimaeras (meta-algae). *Philos. Trans. R. Soc. Lond. B.* **358:** 109–133.

Cavalier-Smith, T. (2004). Only six kingdoms of life. *Proc. Biol. Sci.* **271:** 1251–1262.

Cavalier-Smith, T., and Chao, E.E. (1996). Molecular phylogeny of the free-living archezoan *Trepomonas agilis* and the nature of the first eukaryote. *J. Mol. Evol.* **43:** 551–562.

Cavalier-Smith, T., and Chao E.E. (2003). Phylogeny of Choanozoa, Apusozoa and other protozoa and early eukaryote megaevolution. *J. Mol. Evol.* **56:** 540–563.

Cavalier-Smith, T., Chao, E.-Y., and Oates, B. (2004). Molecular phylogeny of Amoebozoa and the evolutionary significance of the unikont Phalansterium. *Euk. J. Prot.* **40:** 21–48.

Chachon, J., and Cachon, M. (1977). *Sticholonche zanclea* Hertwig: a reinterpretation of its phylogenetic position based upon new observations on its ultrastructure. *Arch. Protistenkd.* **120:** 148–168.

Chai, J.Y., and Lee, J.J. (2000). Recognition, establishment and maintenance of diatom endosymbiosis in foraminifera. *Micropaleontology* **46:** 182–195.

Chesnick, J.M., Kooistra, W.H.C.F., Wellbrock, U., and Medlin, L.K. (1997). Ribosomal RNA analysis indicates a benthic pennate diatom ancestry for the endosymbionts of the dinoflagellates *Peridinium foliaceum* and *Peridinium balticum* (Pyrrhophyta). *J. Eukaryot. Microbiol.* **44:** 314–320.

Ciccarelli, F.D., Doerks, T., von Mering, C., Creevey, C.J., Snel, B., and Bork, P. (2006). Toward automatic reconstruction of a highly resolved tree of life. *Science* **311:** 1283–1287.

Croome, R. (1986). Observations of the heliozoan genera *Acanthocystis* and *Raphidocystis* from Australia. *Arch. Protistenkd.* **131(3–4):** 189–199.

Dalby, A.B., and Prescott, D.M. (2004). The scrambled actin I gene in *Uroleptus pisces*. *Chromosoma* **112:** 247–254.

Dawson, S.C., and Pace, N.R. (2002). Novel kingdom-level eukaryotic diversity in anoxic environments. *Proc. Natl. Acad. Sci. U S A* **99**: 8324–8329.

Delwiche, C. (2007). The origin and evolution of dino-flagellates. *Evolution of Primary Producers in the Sea.* P.G. Falkowski and A.H. Knoll, eds. Boston, Elsevier, pp. 191–205.

de Vargas, C., Aubry, M.-P., Probert, I., and Young, J. (2007). Origin and evolution of coccolithophores: From coastal hunters to oceanic farmers. *Evolution of Primary Producers in the Sea.* P.G. Falkowski and A.H. Knoll, eds. Boston, Elsevier, pp. 251–285.

Dolan, J. (1992). Mixotrophy in ciliates: a review of *Chlorella* symbiosis and chloroplast retention. *Mar. Microbial Food Webs* **6**: 115–132.

Dolan, J.R., and Perez, M.T. (2000). Costs, benefits and characteristics of mixotrophy in marine oligotrichs. *Freshwater Biol.* **45**: 227–238.

Douglas, A.E. (1985). Growth and reproduction of *Convoluta roscoffensis* containing different naturally occurring algal symbionts. *J. Mar. Biol. Ass. UK* **65**: 871–879.

Douglas, A.E. (1987). Experimental studies on symbiotic *Chlorella* in the neorhabdocoel turbellaria *Dalyellia viridis* and *Typhloplana viridata. Br. Phycol. J.* **22**: 157–161.

Douglas, S.E. (1998). Plastid evolution: origins, diversity, trends. *Curr. Opin. Genet. Dev.* **8**: 655–661.

Douglas, S.E., Murphy, C.A., Spencer, D.F., and Gray, M.W. (1999). Cryptomonad algae are evolutionary chimaeras of two phylogenetically distinct unicellular eukaryotes. *Nature* **350**: 148–151.

Douglas, A.E., and Smith, D.C. (1984). The green hydra symbiosis. VIII Mechanisms in symbiont regulation. *Proc. R. Soc. Lond. B* **221**: 291–319.

Douglas, S., Zauner, S., Fraunholz, M., Beaton, M., Penny, S., Deng, L.T., Wu, X., Reith, M., Cavalier-Smith, T., and Maier, U.G. (2001). The highly reduced genome of an enslaved algal nucleus. *Nature* **410**: 1091–1096.

Dyall, S.D., and Johnson, P.J. (2000). Origins of hydrogenosomes and mitochondria: evolution and organelle biogenesis. *Curr. Opin. Microbiol.* **3**: 404–411.

Embley, T.M., and Hirt, R.P. (1998). Early branching eukaryotes? *Curr. Opin. Genet. Dev.* **8**: 624–629.

Eriksen, N.T., Hayes, K.C., Lewitus, A.J. (2002). Growth responses of the mixotrophic dinoflagellates, *Cryptoperidiniopsis* sp. and *Pfiesteria piscicida*, to light under prey-saturated conditions. *Harmful Algae* **1**: 191–203.

Esteban, G.F., and Finlay, B.J. (1996). Morphology and ecology of the cosmopolitan ciliate *Prorodon viridis. Arch. Protistenkd.* **147**: 181–188.

Farmer, M.A., Fitt, W.K., and Trench, R.K. (2001). Morphology of the symbiosis between *Corculum cardissa* (Mollusca: Bivalvia) and *Symbiodinium corculorum* (Dinophyceae). *Biol. Bull.* **200**: 336–343.

Fast, N.M., Kissinger, J.C., Roos, D.S., and Keeling, P.J. (2001). Nuclear-encoded, plastid-targeted genes suggest a single common origin for apicomplexan and dinoflagellate plastids. *Mol. Biol. Evol.* **18**: 418–426.

Febvre, J., and Febvrechevalier, C. (1979). Ultrastructural study of zooxanthellae of 3 species of Acantharia (Protozoa Actinopoda), with details of their taxonomic position in the Prymnesiales (Prymnesiophyceae, Hibberd, 1976). *J. Mar. Biol. Assoc. UK* **59**: 215–226.

Feinstein, T.N., Traslavina, R., Sun, M.Y., and Lin S.J. (2002). Effects of light on photosynthesis, grazing, and population dynamics of the heterotrophic dinoflagellate *Pfiesteria piscicida* (Dinophyceae). *J. Phycol.* **38**: 659–669.

Felsenstein, J. (1978). Cases in which parsimony or compatibility methods will be positively misleading. *Syst. Zool.* **27**: 401–410.

Fenchel, T., and Bernard, C. (1993). Endosymbiotic purple nonsulfur bacteria in an anaerobic ciliated protozoan. *Ferns Microbiology Lett.* **110(1)**: 21–25.

Fields, S.D., and Rhodes, R.G. (1991). Ingestion and retention of *Chroomonas* spp. (Cryptophyceae) by *Gymnodinium acidotum* (Dinophyceae). *J. Phycol.* **27**: 525–529.

Finlay, B.J., Berninger, U.G., Stewart, L.J., Hindle, R.M., and Davison, W. (1987). Some factors controlling the distribution of 2 pond-dwelling ciliates with algal symbionts (*Frontonia vernalis* and *Euplotes daidaleos*). *J. Protozool.* **34**: 349–356.

Friedl, T., and Bhattacharya, D. (2001). Origin and evolution of eukaryotic lichen algae. *Origin, Evolution and Versatility of Microorganisms.* J. Seckbach, ed. Dordrecht, Kluwer.

Gajadhar, A.A., Marquardt, W.C., Hall, R., Gunderson, J., Ariztia-Carmona, E.V., and Sogin, M.L. (1991). Ribosomal RNA sequences of *Sarcocystis muris, Theileria annulata* and *Crypthecodinium cohnii* reveal evolutionary relationships among apicomplexans, dinoflagellates, and ciliates. *Mol. Biochem. Parasitol.* **45**: 147–154.

Gast, R.J., Beaudoin, D.J., and Caron, D.A. (2003). Isolation of symbiotically expressed genes from the dinoflagellate symbiont of the solitary radiolarian *Thalassicolla nucleata. Biol. Bull.* **204**: 210–214.

Gast, R.J., and Caron, D.A. (1996). Molecular phylogeny of symbiotic dinoflagellates from planktonic foraminifera and radiolaria. *Mol. Biol. Evol.* **13**: 1192–1197.

Gast, R.J., and Caron, D.A. (2001). Photosymbiotic associations in planktonic foraminifera and radiolaria. *Hydrobiologia* **461**: 1–7.

Gast, R.J., McDonnell, T.A., and Caron, D.A. (2000). srDNA-based taxonomic affinities of algal symbionts from a planktonic foraminifer and a solitary radiolarian. *J. Phycol.* **36**: 172–177.

Gillot, M. (1989). Phylum cryptophyte (cryptomonads). *Handbook of Protoctista.* L. Margulis, J.O. Corliss, M. Melkonian, and D.J. Chapman, eds. Boston, Bartlett Publishers, pp. 139–151.

Graham, L.E., and Graham, J.M. (1980). Endosymbiotic *Chlorella* (Chlorophyta) in a species of *Vorticella* (Ciliophora). *Trans. Am. Microsc. Soc.* **99**: 160–166.

Griffiths, D.J., and Thinh, L.V. (1987). Photosynthesis by *in situ* and isolated *Prochloron* (Prochlorophyta) associated with didemnid ascidians. *Symbiosis* **3(2):** 109–122.

Grzebyk, D., Schofield, O., Vetriani, C., and Falkowski, P.G. (2003). The mesozoic radiation of eukaryotic algae: The portable plastid hypothesis. *J. Phycol.* **39:** 259–267.

Grzymski, J., Schofield, O.M., Falkowski, P.G., and Bernhard, J.M. (2002). The function of plastids in the deep-sea benthic foraminifer, *Nonionella stella. Limnol. Oceanogr.* **47(6):** 1569–1580.

Guillou, L., Eikremb, W., Chrétiennot-Dinetc, M.-J., Le Gall, F., Massana, R., Romari, K., Pedros-Alio, C., and Vaulot, D. (2004). Diversity of picoplanktonic prasinophytes assessed by direct nuclear SSU rDNA sequencing of environmental samples and novel isolates retrieved from oceanic and coastal marine ecosystems. *Protist* **155:** 193–214.

Gustafson, D.E., Stoecker, D.K., Johnson, M.D., Van Heukelem, W.F., and Sneider, K. (2000). Cryptophyte algae are robbed of their organelles by the marine ciliate *Mesodinium rubrum. Nature* **405:** 1049–1052.

Hackett, J.D., Maranda, L., Yoon, H.S., and Bhattacharya, D. (2003). Phylogenetic evidence for the cryptophyte origin of the plastid of *Dinophysis* (Dinophysiales, Dinophyceae). *J. Phycol.* **39:** 440–448.

Hackett, J.D., Yoon, H.S., Soares, M.B., Bonaldo, M.F., Casavant, T.L., Scheetz, T.E., Nosenko, T., and Bhattacharya, D. (2004). Migration of the plastid genome to the nucleus in a peridinin dinoflagellate. *Curr. Biol.* **14:** 213–218.

Hackett, J.D., Yoon, H.S., Butterfield, N.J., Sanderson, M.J., and Bhattacharya, D. (2007). Plastid endosymbiosis: Sources and timing of the major events. *Evolution of Primary Producers in the Sea.* P.G. Falkowski and A.H. Knoll, eds. Boston, Elsevier, pp. 109–131.

Halanych, K.M. (2004). The new view of animal phylogeny. *Ann. Rev. Ecol. Evol. Syst.* **35:** 229–256.

Hallick, R.B., Hong, L., Drager, R.G., Favreau, M.R., Monfort, A., Orsat, B., Spielmann, A., and Stutz, E. (1993). Complete sequence of *Euglena gracilis* chloroplast DNA. *Nucl. Acids Res.* **21:** 3537–3544.

Hamm, C., and Smetacek, V. (2007). Armor: Why, when, and how. *Evolution of Primary Producers in the Sea.* P.G. Falkowski and A.H. Knoll, eds. Boston, Elsevier, pp. 311–332.

Hansen, P.J., and Fenchel, T. (2006). The bloom-forming ciliate *Mesodinium rubrum* harbours a single permanent endosymbiont. *Mar. Biol. Res.* **2:** 169–177.

Hansen, P.J., Miranda, L., and Azanza, R. (2004). Green *Noctiluca scintillans*: a dinoflagellate with its own greenhouse. *Mar. Ecol. Prog. Ser.* **275:** 79–87.

Hargraves, P.E. (2002). The ebridian flagellates *Ebria* and *Hermesinum. Plankton Biol. Ecol.* **49:** 9–16.

Harper, J.T., Waanders, E., and Keeling, P.J. (2005). On the monophyly of chromalveolates using a six-protein phylogeny of eukaryotes. *Int. J. Syst. Evol. Microbiol.* **55:** 487–496.

Hausmann, K., and Hülsmann, N. (1996). *Protozoology.* New York, Thieme Medical Publishers.

Hedges, S.B., Blair, J.E., Venturi, M.L., and Shoe, J.L. (2004). A molecular timescale of eukaryote evolution and the rise of complex multicellular life. *BMC Evol. Biol.* **4:** 2–10.

Hill, M.S. (1996). Symbiotic zooxanthellae enhance boring and growth rates of the tropical sponge *Anthosigmella varians* forma *varians. Mar. Biol.* **125:** 649–654.

Hillis, D.M., Huelsenbeck, J.P., and Cunningham, C.W. (1994). Application and accuracy of molecular phylogenies. *Science* **264:** 671–677.

Hinde, R., Pironet, F., and Borowitzka, M.A. (1994). Isolation of *Oscillatoria spongeliae*, the filamentous cyanobacterial symbiont of the marine sponge *Dysidea herbacea. Mar. Biol.* **119:** 99–104.

Hirose, E., Maruyama, T., Cheng, L., and Lewin, R.A. (1998). Intra- and extra-cellular distribution of photosynthetic prokaryotes, *Prochloron* sp., in a colonial ascidian, ultrastructural and quantitative studies. *Symbiosis* **25(1–3):** 301–310.

Hood, C.L. (1927). The zoochlorellae of *Frontonia leucas. Biol. Bull.* **52:** 79–88.

Horiguchi, T., and Pienaar, R.N. (1991). Ultrastructure of a marine dinoflagellate, *Peridinium quinquecorne* Abé (Peridiniales) from South Africa with particular reference to its chrysophyte endosymbiont. *Botanica Marina* **34:** 123–131.

Hoshina, R., Kamako, S.I., and Imamura, N. (2004). Phylogenetic position of endosymbiotic green algae in *Paramecium bursaria* Ehrenberg from Japan. *Plant Biol.* **6:** 447–453.

Hoshina, R., Kato, Y., Kamako, S., and Imamura, N. (2005). Genetic evidence of "American" and "European" type symbiotic algae of *Paramecium bursaria* Ehrenberg. *Plant Biol.* **7:** 526–532.

Huelsenbeck, J.P., Bull, J.J., and Cunningham, C.W. (1996). Combining data in phylogenetic analysis. *Trends Ecol. Evol.* **11:** 152–158.

Jakobsen, H.H., Hansen, P.J., and Larsen, J. (2000). Growth and grazing responses of two chloroplast-retaining dinoflagellates: effect of irradiance and prey species. *Mar. Ecol. Prog. Ser.* **201:** 121–128.

James, T.Y., Lechter, P.M., Longcone, J.E., Mozley-Standridge, S.E., Porter, D., Powell, M.J., Griffith, G.W., and Vilgalys, R. (2006). A molecular phylogeny of the flagellated fungi (Chytridiomycota) and description of a new phylum (Blastocladiomycota). *Mycologia* **98(6):** 860–871.

Janson, S. (2004). Molecular evidence that plastids in the toxin-producing dinoflagellate genus Dinophysis originate from the free-living cryptophyte *Teleaulax amphioxeia. Environ. Microbiol.* **6:** 1102–1106.

John, D.M., Whitton, B.A., and Brook, A.J. (2002). *The Freshwater Algal Flora of the British Isles.* Cambridge, Cambridge University Press.

Johnson, M.D., Oldach, D., Delwiche, C.F., and Stoecker, D.K. (2007). Retention of transcriptionally active cryptophyte nuclei by the ciliate *Myrionceta rubra. Nature* **445:** 426–428.

Johnson, M.D., and Stoecker, D.K. (2005). Role of feeding in growth and photophysiology of *Myrionecta rubra*. *Aquat. Microb. Ecol.* **39**: 303–312.

Johnson, P.W., Donaghay, P.L., Small, E.B., and Sieburth, J.M. (1995). Ultrastructure and ecology of *Perispira ovum* (Ciliophora, Litostomatea)—an aerobic, planktonic ciliate that sequesters the chloroplasts, mitochondria, and paramylon of *Euglena proxima* in a micro-oxic habitat. *J. Eukaryot. Microbiol.* **42**: 323–335.

Karol, K.G., McCourt, R.M., Cimino, M.T., and Delwiche, C.F. (2001). The closest living relatives of land plants. *Science* **294**: 2351–2353.

Keeling, P.J. (2001). Foraminifera and Cercozoa are related in actin phylogeny: two orphans find a home? *Mol. Biol. Evol.* **18**: 1551–1557.

Keeling, P.J., Burger, G., Durnford, D.G., Lang, B.F., Lee, R.W., Pearlman, R.E., Roger, A.J., and Gray, M.W. (2005). The tree of eukaryotes. *Trends Ecol. Evol.* **20**: 670–676.

Keeling, P.J., and Doolittle, W.F. (1996). Alpha-tubulin from early-diverging eukaryotic lineages and the evolution of the tubulin family. *Mol. Biol. Evol.* **13**: 1297–1305.

Keeling, P.J., and Inagaki, Y. (2005). A class of eukaryotic GTPase with a punctate distribution suggesting multiple functional replacements of translation elongation factor 1alpha. *Proc. Natl. Acad. Sci. U S A* **101**: 15380–15385.

Keeling, P.J., Luker, M.A., and Palmer, J.D. (2000). Evidence from beta-tubulin phylogeny that microsporidia evolved from within the fungi. *Mol. Biol. Evol.* **17**: 23–31.

Kepner, W.A. (1905). *Paulinella chromatophora. Biol. Bull.* **9**: 128–129.

Kessler, E., and Huss, V.A.R. (1990). Biochemical taxonomy of symbiotic *Chlorella* strains from *Paramecium* and *Acanthocystis*. *Bot. Acta* **103(2)**: 140–142.

Kimor, B., Gordon, N., and Neori, A. (1992). Symbiotic associations among the microplankton in oligotrophic marine environments, with special reference to the Gulf of Aqaba, Red Sea. *J. Plankton Res.* **14**: 1217–1231.

Koike, K., Sekiguchi, H., Kobiyama, A., Takishita, K., Kawachi, M., Koike, K., and Ogata, T. (2005). A novel type of kleptoplastidy in *Dinophysis* (Dinophyceae): Presence of haptophyte-type plastid in *Dinophysis mitra*. *Protist* **156**: 225–237.

Kooistra, W.H.C.F., Gersonde, R., Medlin, L.K., and Mann, D.G. (2007). The origin and Evolution of the diatoms: Their adaptation. *Evolution of Primary Producers in the Sea*. P.G. Falkowski and A.H. Knoll, eds. Boston, Elsevier, pp. 207–249.

Lang, B.F., Burger, G., O'Kelly, C.J., Cedergren, R., Golding, G.B., Lemieux, C., Sankoff, D., Turmel, M., and Gray, M.W. (1997). An ancestral mitochondrial DNA resembling a eubacterial genome in miniature. *Nature* **387**: 493–497.

Lang, B.F., O'Kelly, C., Nerad, T., Gray, M.W., and Burger, G. (2002). The closest unicellular relatives of animals. *Curr. Biol.* **12**: 1773–1778.

Larsen, J. (1988). An ultrastructural study of *Amphidinium poecilochroum* (Dinophyceae), a phagotrophic dinoflagellate feeding on small species of cryptophytes. *Phycologia* **27**: 366–377.

Laval-Peuto, M., and Febvre, M. (1986). On plastid symbiosis in *Tontonia appendiculariformis* (Ciliophora, Oligotrichina). *Biosystems* **19**: 137–158.

Lee, J.J. (1998). "Living sands"—Larger foraminifera and their endosymbiotic algae. *Symbiosis* **25**: 71–100.

Lee, J.J. (1992). Symbiosis in Foraminifera. *Algae and Symbioses: Plants, Animals, Fungi, Viruses, Interactions Explored*. W. Resisser, ed. Bristol, Biopress Limited, pp. 63–78.

Lee, J.J., Faber, Jr., W.W., Anderson, R.O., and Pawlowski, J. (1991). Life cycles of Foraminifera. *Biology of Foraminifera*. J.J. Lee and O.R. Anderson, eds. London, Academic Press, pp. 285–334.

Lee, J.J., Pawlowski, J., Debenay, J.-P., Whittaker, J., Banner, F., Gooday, A.J., Tendal, O., Haynes, J., and Faber, W.W. (2000). Phylum granuloreticulosa. *The Illustrated Guide to the Protozoa*. J.J. Lee, G.F. Leedale, and P. Bradbury, eds., 2nd ed. Lawrence, KS, Society of Protozoologists, pp. 872–951.

Lewin, R.A. (2002). Prochlorophyta—a matter of class distinctions. *Photosyn. Res.* **73**: 59–61.

Lewis, L.A., and Muller-Parker, G. (2004). Phylogenetic placement of "zoochlorellae" (Chlorophyta), algal symbiont of the temperate sea anemone *Anthopleura elegantissima*. *Biol. Bull.* **2007**: 87–92.

Li, S., Nosenko, T., Hackett, J.D., and Bhattacharya, D. (2006). Phylogenomic analysis identifies red algal genes of endosymbiotic origin in the Chromalveolates. *Mol. Biol. Evol.* **23**: 663–674.

Lobban, C.S., Modeo, L., Verni, F., and Rosati, G. (2005). *Euplotes uncinatus* (Ciliophora, Hypotrichia), a new species with zooxanthellae. *Mar. Biol.* **147**: 1055–1061.

Lobban, C.S., Schefter, M., Simpson, A.G.B., Pochon, X., Pawlowski, J., and Foissner, W. (2002). *Maristentor dinoferus* n. gen., n. sp., a giant heterotrich ciliate (Spirotrichea: Heterotrichida) with zooxanthellae, from coral reefs on Guam, Mariana Islands. *Mar. Biol.* **141**: 207–208.

Loftus, B., Anderson, I., Davies, R., and 51 others (2005). The genome of the protist parasite *Entamoeba histolytica*. *Nature* **433**: 865–868.

Longet, D., Archibald, J.M, Keeling, P.J., and Pawlowski, J. (2003). Foraminifera and Cercozoa share a common origin according to RNA polymerase II phylogenies. *Int. J. Syst. Evol. Microbiol.* **53(Pt 6)**: 1735–1739.

Lopez, E. (1979). Algal chloroplasts in the protoplasm of 3 species of benthic foraminifera—taxonomic affinity, viability and persistence. *Mar. Biol.* **53**: 201–211.

Loytynoja, A., and Milinkovitch, M.C. (2001). Molecular phylogenetic analyses of the mitochondrial ADP-ATP carriers: the Plantae/Fungi/Metazoa trichotomy revisited. *Proc. Natl. Acad. Sci. U S A* **98**: 10202–10207.

Marciano-Cabral, F., Puffenbarger, R., and Cabral, G. A. (2000). The increasing importance of Acanthamoeba infections. *J. Eukaryot. Microbiol.* **47(1):** 29–36.

Marin, B., Nowack, E.C.M., and Melkonian, M. (2005). A plastid in the making: evidence for a second primary endosymbiosis. *Protist* **156:** 425–432.

Martin, W. (2007). Eukaryote and mitochondrial origins: Two sides of the same coin and too much ado about oxygen. *Evolution of Primary Producers in the Sea.* P.G. Falkowski and A.H. Knoll, eds. Boston, Elsevier, pp. 55–73.

Martin, W., Stoebe, B., Goremykin, V., Hansmann, S., Hasegawa, S., and Kowallik, K.V. (1998). Gene transfer to the nucleus and the evolution of chloroplasts. *Nature* **393:** 162–165.

Massana, R., Terrado, R., Forn, I., Lovejoy, C., and Pedros-Alio, C. (2006). Distribution and abundance of uncultured heterotrophic flagellates in the world oceans. *Environ. Microbiol.* **8:** 1515-1522.

McFadden, G.I. (2001). Primary and secondary endosymbiosis and the origin of plastids. *J. Phycol.* **37:** 951–959.

McFadden, G.I., and van Dooren, G.G. (2004). Evolution: red algal genome affirms a common origin of all plastids. *Curr. Biol.* **14:** R514–R516.

McManus, G.B., Zhang, H., and Lin, S. J. (2004). Marine planktonic ciliates that prey on macroalgae and enslave their chloroplasts. *Limnol. Oceanography* **49:** 308–313.

Medina, M., Collins, A.G., Taylor, J.W., Valentine, J.W., Lipps, J.H., Amaral-Zettler, L., and Sogin, M.L. (2005). Phylogeny of Opisthokonta and the evolution of multicellularity and complexity in Fungi and Metazoa. *Int. J. Astrobiol.* **2:** 203–211.

Meisterfeld, R., Holzmann, M., and Pawlowski, J. (2001). Morphological and molecular characterization of a new terrestrial allogromiid species: *Edaphoallogromia australica* gen. et sp. nov. (Foraminifera) from Northern Queensland (Australia). *Protist* **152:** 185–192.

Melkonian, M., and Mollenhauer, D. (2005). Robert Lauterborn (1869–1952) and his *Paulinella chromatophora*. *Protist* **156:** 253–262.

Mendoza, L., Taylor, J.W., and Ajello, L. (2002). The class Mesomycetozoea: a heterogeneous group of microorganisms at the animal-fungal boundary. *Ann. Rev. Microbiol.* **56,** 315–344.

Mitchell, A.D., and Gilbert, D. (2004). Vertical microdistribution and response to nitrogen deposition of testate amoebae in *Sphagnum. J. Euk. Microbiol.* **51:** 480–490.

Moreira, D., Le Guyader, H., and Philippe, H. (2000). The origin of red algae and the evolution of chloroplasts. *Nature* **405:** 69–72.

Moreira, D., and Lopéz-Garcia, P. (2002). The molecular ecology of microbial eukaryotes unveils a hidden diversity. *Trends Microbiol.* **10:** 31–39.

Nash, T.H., ed. (1996). *Lichen Biology.* Cambridge, Cambridge University Press.

Newcomb, E.H., and Pugh, T.D. (1975). Blue-green algae associated with ascidians of the Great Barrier Reef. *Nature* **253:** 533–534.

Nikolaev, S.I., Berney, C., Fahrni, J.F., Bolivar, I., Polet, S., Mylnikov, A.P., Aleshin, V.V., Petrov, N.B., and Pawlowski, J. (2004). The twilight of Heliozoa and rise of Rhizaria, an emerging supergroup of amoeboid eukaryotes. *Proc. Nat. Acad. Sci. U S A* **101:** 8066–8071.

Nikolaev, S.I., Berney, C., Petrov, N.B., Mylnikov, A.P., Fahrni, J.F., and Pawlowski, J. (2006). Phylogenetic position of *Multicilia marina* and the evolution of Amoebozoa. *Int. J. Syst. Evol. Microbiol.* **56:** 1449–1458.

Norris, R.D. (1996). Symbiosis as an evolutionary innovation in the radiation of Paleocene planktic foraminifera. *Paleobiology* **22:** 461–480.

Norton, T.A., Melkonian, M., and Andersen, R.A. (1996). Algal biodiversity. *Phycologia* **35:** 308–326.

Okamoto, N., and Inouye, I. (2005). A secondary symbiosis in progress? *Science* **310:** 287–287.

Okamoto, N., and Inouye, I. (2006). *Hatena arenicola* gen. et sp. nov., a katablepharid undergoing probable plastid acquistion. *Protist* **157(4):** 401–419.

O'Kelly, C.J. (1993). The jakobid flagellates: Structural features of *Jakoba, Reclinomonas* and *Histiona* and implications for the early diversification of eukaryotes. *J. Eukaryot. Microbiol.* **40:** 627–636.

O'Kelly, C.J. (2007). The origin and early evolution of green plants. *Evolution of Primary Producers in the Sea.* P.G. Falkowski and A.H. Knoll, eds. Boston, Elsevier, pp. 287–309.

Olive, L., and Stoianovitch, D. (1975). *The Mycetozoans.* New York, Academic Press.

Osinga, R., Armstrong, E., Burgess, J. G., Hoffmann, F., Reitner, J., and Schumann-Kindel, G. (2001). Sponge-microbe associations and their importance for sponge bioprocess engineering. *Hydrobiologia* **461:** 55–62.

Pace, N.R., Olsen, G.J., and Woese, C.R. (1986). Ribosomal RNA phylogeny and the primary lines of evolutionary descent. *Cell* **45:** 325–326.

Park, M.G., Kim, S., Kim, H.S., Myung, G., Kang, Y.G., and Yih, W. (2006). First successful culture of the marine dinoflagellate *Dinophysis acuminata. Aquat. Microb. Ecol.* **45:** 101–106.

Patterson, D.J. (1999). The diversity of eukaryotes. *Am. Nat.* **154:** S96–S124.

Patterson, D.J., and Dürrschmidt, M. (1987). Selective retention of chloroplasts by algivorous helizoa: Fortuitous chloroplast symbiosis? *Eur. J. Protistol.* **23:** 51–55.

Patterson, D.J., Rogerson, A., and aVørs, N. (2000a). Class heterolobosea. *The Illustrated Guide to the Protozoa.* J.J. Lee, G.F. Leedale, and P. Bradbury, eds., 2nd ed. Lawrence, KS, Society of Protozoologists, pp. 1104–1110.

Patterson, D.J., Rogerson, A., and Simpson, A.G. (2000b). Amoebae of uncertain affinity. *The Illustrated Guide to the Protozoa.* J.J. Lee, G.F. Leedale, and P. Bradbury, eds., 2nd ed. Lawrence, KS, Society of Protozoologists, pp. 804–827.

Pawlowski, J., Bolivar, I., Fahrni, J., de Vargas, C., and Bowser, S.S. (1999). Molecular evidence that *Reticulomyxa filosa* is a freshwater naked foraminifer. *J. Eukaryot. Microbiol.* **46:** 612–617.

Pawlowski, J., Holzmann, M., Fahrni, J.F., and Hallock, P. (2001a). Molecular identification of algal endosymbionts in large miliolid foraminifera: 1. Chlorophytes. *J. Eukaryot. Microbiol.* **48:** 362–367.

Pawlowski, J., Holzmann, M., Fahrni, J.F., Pochon, X., and Lee, J. J. (2001b). Molecular identification of algal endosymbionts in large miliolid foraminifera: 2. Dinoflagellates. *J. Eukaryot. Microbiol.* **48:** 368–373.

Philip, G.K., Creevey, C.J., and McInerney, J.O. (2005). The Opisthokonta and the Ecdysozoa may not be clades: stronger support for the grouping of plant and animal than for animal and fungi and stronger support for the Coelomata than Ecdysozoa. *Mol. Biol. Evol.* **22:** 1175–1184.

Philippe, H., and Adoutte, A. (1998). The molecular phylogeny of protozoa: solid facts and uncertainties. *Evolutionary Relationships Among Protozoa.* G.H. Coombs, K. Vickerman, M. A. Sleigh, and A. Warren, eds. Dordrecht, The Netherlands, Kluwer Academic Publishers, pp. 25–36.

Philippe, H., and Germot, A. (2000). Phylogeny of eukaryotes based on ribosomal RNA: long-branch attraction and models of sequence evolution. *Mol. Biol. Evol.* **17:** 830–834.

Philippe, H., Lartillot, N., and Brinkmann, H. (2005). Multigene analyses of bilaterian animals corroborate the monophyly of Ecdysozoa, Lophotrochozoa, and Protostomia. *Mol. Biol. Evol.* **22:** 1246–1253.

Philippe, H., and Telford, J. (2006). Large-scale sequencing and the new animal phylogeny. *Trends Ecol. Evol.* **21(11):** 614–620.

Pochon, X., Pawlowski, J., Zaninetti, L., and Rowan, R. (2001). High genetic diversity and relative specificity among *Symbiodinium*-like endosymbiotic dinoflagellates in soritid foraminiferans. *Mar. Biol.* **139:** 1069–1078.

Polet, S., Berney, C., Fahrni, J., and Pawlowski, J. (2004). Small subunit ribosomal RNA sequences of Phaeodarea challenge the monophyly of Haeckel's Radiolaria. *Protist* **155:** 53–63.

Prescott, D.M. (2000). Genome gymnastics: unique modes of DNA evolution and processing in ciliates. *Nat. Rev. Genet.* **1:** 191–198.

Ragan, R. A., Goggin, C. L., Cawthorn, R. J., Cerenius, L., Jamieson, A.V., Plourde, S.M., Rand, T.G., Soderhall, K., and Gutell, R.R. (1996). A novel clade of protistan parasites near the animal-fungal divergence. *Proc. Nat. Acad. Sci. U S A* **93:** 11907–11912.

Raven, J.A. (1997). Phagotrophy in phototrophs. *Limnol. Oceanogr.* **42:** 198–205.

Reisser, W. (1992). *Endosymbiotic associations of algae with freshwater protozoa and invertebrates. Algae and Symbioses: Plants, Animals, Fungi, Viruses, Interactions Explored.* W. Reisser, ed. Bristol, Biopress Limited, pp. 1–20.

Richards, T.A., and Cavalier-Smith, T. (2005). Myosin domain evolution and the primary divergence of eukaryotes. *Nature* **436:** 1113–1118.

Rodriguez-Ezpeleta, N., Brinkmann, H., Burey, S.C., Roure, B., Burger, G., Loffelhardt, W., Bohnert, H.J., Philippe, H., and Lang, B.F. (2005). Monophyly of primary photosynthetic eukaryotes: green plants, red algae, and glaucophytes. *Curr. Biol.* **15:** 1325–1330.

Roger, A.J. (1999). Reconstructing early events in eukaryotic evolution. *Am. Nat.* **154:** 146–163.

Roger, A.J., Smith, M.W., Doolittle, R.F., and Doolittle, W.F. (1996). Evidence for the Heterolobosea from phylogenetic analysis of genes encoding glyceraldehyde-3-phosphate dehydrogenase. *J. Eukaryot. Microbiol.* **43:** 475–485.

Rogerson, A., Finlay, B.J., and Berninger, U.G. (1989). Sequestered chloroplasts in the fresh-water ciliate *Strombidium viride* (Ciliophora, Oligotrichida). *Trans. Am. Microscop. Soc.* **108:** 117–126.

Rokas, A., Kruger, D., and Carroll, S.B. (2006). Animal evolution and the molecular signature of radiations compressed in time. *Science* **310:** 1933–1938.

Röttger, R. (1995). *Praktikum der Protozoologie.* Stuttgart, Spektrum Akademischer Verlag.

Ruiz-Trillo, I., Inagaki, Y., Davis, L.A., Sigmund Sperstad, S., Landfald, B., and Roger, A.J. (2004). *Capsaspora owczarzaki* is an independent opisthokont lineage. *Curr. Biol.* **14:** R946–R947.

Rumpho, M.E., Summer, E.J., and Manhart, J.R. (2000). Solar-powered sea slugs. Mollusc/algal chloroplast symbiosis. *Plant Physiol.* **123:** 29–38.

Saldarriaga, J.F., Taylor, F.J., Keeling, P.J., and Cavalier-Smith, T. (2001). Dinoflagellate nuclear SSU rRNA phylogeny suggests multiple plastid losses and replacements. *J. Mol. Evol.* **53:** 204–213.

Sand-Jensen, K., Pedersen, O., and Geertz-Hansen, O. (1997). Regulation and role of photosynthesis in the colonial symbiotic ciliate *Ophrydium versatile*. *Limnol. Oceanogr.* **42:** 866–873.

Saunders, B.K., and Muller-Parker, G. (1997). The effects of temperature and light on two algal populations in the temperate sea anemone *Anthopleura elegantissima* (Brandt, 1835). *J. Exp. Mar. Biol. Ecol.* **211:** 213–224.

Savage, A.M., Goodson, M.S., Visram, S., Trapido-Rosenthal, H., Wiedenmann, J., and Douglas, A.E. (2002). Molecular diversity of symbiotic algae at the latitudinal margins of their distribution: dinoflagellates of the genus *Symbiodinium* in corals and sea anemones. *Mar. Ecol. Prog. Ser.* **244:** 17–26.

Schüßler, A., Schwarzott, D., and Walker, C. (2001). A new fungal phylum, the *Glomeromycota*: phylogeny and evolution. *Mycol. Res.* **105:** 1413–1421.

Schüßler, A., and Wolf, E. (2005). *Geosiphon pyriformis*—a glomeromycotan soil fungus forming endosymbiosis with cyanobacteria. In *In Vitro Culture of Mycorrhizas, Soil Biology* Vol. 4. S. Declerck, D.-G. Strullu, J.A. Fortin, eds. Heidelberg, Springer-Verlag, pp. 271–289.

Schwartz, T.U. (2005). Modularity within the architecture of the nuclear pore complex. *Curr. Opin. Struct. Biol.* **15:** 221–226.

Schweiker, M., and Elbrächter, M. (2004). First ultrastructural investigations of the consortium between a phototrophic eukaryotic endocytobiont and *Podolampas bipes* (Dinophyceae). *Phycologia* **43:** 614–623.

Selosse, M.-A., and Le Tacon, F. (1998). The land flora: a phototroph-fungus partnership? *TREE* **13(1):** 15–20.

Siegel, R.W. (1960). Hereditary endosymbiosis in *Paramecium bursaria*. *Exp. Cell Res.* **19**: 239–252.

Silberman, J.D., Simpson, A.G., Kulda, J., Cepicka, I., Hampl, V., Johnson, P.J., and Roger, A.J. (2002). Retortamonad flagellates are closely related to diplomonads implications for the history of mitochondrial function in eukaryote evolution. *Mol. Biol. Evol.* **19**: 777–786.

Simpson, A.G., Inagaki, Y., and Roger, A.J. (2006). Comprehensive multigene phylogenies of excavate protists reveal the evolutionary positions of "primitive" eukaryotes. *Mol. Biol. Evol.* **23**: 615–625.

Simpson, A.G.B., and Patterson, D.J. (1999). The ultrastructure of *Carpediemonas membranifera* (Eukaryota) with reference to the excavate hypothesis. *Eur. J. Protistol.* **35**: 353–370.

Simpson, A.G., and Roger, A.J. (2002). Eukaryotic evolution: getting to the root of the problem. *Curr. Biol.* **12**: 691–693.

Simpson, A.G., and Roger, A.J. (2004). The real 'kingdoms' of eukaryotes. *Curr. Biol.* **14**: 693–696.

Skovgaard, A. (1998). Role of chloroplast retention in a marine dinoflagellate. *Aquat. Microb. Ecol.* **15**: 293–301.

Smirnov, A., Nassonova, E., Berney, C., Fahrni, J., Bolivar, I., and Pawlowski, J. (2005). Molecular phylogeny and classification of the lobose amoebae. *Protist* **156**: 129–142.

Smith, H.G. (1935). On the presence of algae in certain Ascidiaceae. *Ann. Mag. Nat. Hist.* **15**: 615–626.

Sogin, M.L., Gunderson, J.H., Elwood, H.J., Alonso, R.A., and Peattie, D.A. (1989). Phylogenetic meaning of the kingdom concept: an unusual ribosomal RNA from *Giardia lamblia*. *Science* **243**: 75–77.

Sollner-Webb, B. (1996). Trypanosome RNA editing: resolved. *Science* **273**: 1182–1183.

Spero, H.J., and Angel, D.L. (1991). Planktonic sarcodines: microhabitat for oceanic dinoflagellates. *J. Phycol.* **27**: 187–195.

Spiegel, F.W., Lee, S.B., and Rusk, S.A. (1995). Eumycetozoans and molecular systematics. *Can. J. Bot.* **73**: 738–746.

Stechmann, A., and Cavalier-Smith, T. (2002). Rooting the eukaryote tree by using a derived gene fusion. *Science* **297**: 89–91.

Stechmann, A., and Cavalier-Smith, T. (2003). The root of the eukaryote tree pinpointed. *Curr. Biol.* **13**: R665–666.

Steenkamp, E.T., and Baldauf, S.L. (2004). Origin and evolution of animals, fungi and their unicellular allies (Opisthokonta). *Organelles, Genomes, and Eukaryote Phylogeny*. R.P. Hirt and D.S. Horner, eds. London, Taylor and Francis, pp. 109–129.

Steenkamp, E.T., Wright, J., and Baldauf, S.L. (2006). The protistan origins of animals and fungi. *Mol. Biol. Evol.* **23**: 93–106.

Steiner, J.M., Yusa, F., Pompe, J.A., and Löffelhardt, W. (2005). Homologous protein important machineries in chloroplasts and cyanelles. *Plant J.* **44**: 646–652.

Stoecker, D.K. (1998). Conceptual models of mixotrophy in planktonic protists and some ecological and evolutionary implications. *Eur. J. Protistol.* **34**: 281–290.

Stoecker, D.K. (1999). Mixotrophy among dinoflagellates. *J. Euk. Microb.* **46(4)**: 397–401.

Stoecker, D.K., Michaels, A.E., and Davis, L.H. (1987). Large proportion of marine planktonic ciliates found to contain functional chloroplasts. *Nature* **326**: 790–792.

Stoecker, D.K., and Silver, M.W. (1990). Replacement and aging of chloroplasts in *Strombidium capitatum* (Ciliophora, Oligotrichida). *Mar. Biol.* **107**: 491–502.

Stoecker, D.K., Silver, M.W., Michaels, A.E., and Davis, L.H. (1988). Obligate mixotrophy in *Laboea strobila*, a ciliate which retains chloroplasts. *Mar. Biol.* **99**: 415–423.

Stoecker, D.K., Sliver, M.W., Michaels, A.E., and Davis, L.H. (1988–1989). Enslavement of algal chloroplasts by four *Strombidium* spp. (Ciliophora, Oligotrichida). *Mar. Microb. Food Webs* **3(3)**: 79–100.

Strmecki, L., Greene, D.M., and Pear, C.J. (2005). Developmental decisions in *Dictyostelium discoideum*. *Dev. Biol.* **284**: 25–36.

Tachezy, J., Sanchez, L.B., and Muller, M. (2001). Mitochondrial type iron-sulfur cluster assembly in the amitochondriate eukaryotes *Trichomonas vaginalis* and *Giardia intestinalis*, as indicated by the phylogeny of IscS. *Mol. Biol. Evol.* **18**: 1919–1928.

Takahashi, O., Mayama, S., and Matsuoka, A. (2003). Host-symbiont associations of polycystine Radiolaria: epifluorescence microscopic observation of living Radiolaria. *Mar. Micropaleontol.* **49**: 187–194.

Takishita, K., Koike, K., Maruyama, T., and Ogata, T. (2002). Molecular evidence for plastid robbery (Kleptoplastidy) in *Dinophysis*, a dinoflagellate causing diarrhetic shellfish poisoning. *Protist* **153**: 293–302.

Tamura, M., Shimada, S., and Horiguchi, T. (2005). *Galeidiniium rugatum* gen. et sp. nov. (Dinophyceae), a new coccoid dinoflagellate with a diatom endosymbiont. *J. Phycol.* **41**: 658–671.

Taylor, D.L. (1984). Translocation of carbon and nitrogen from the turbellarian host *Amphiscolops langerhansi* (Turbellaria: Acoela) to its algal endosymbiont *Amphidinium klebsii* (Dinophyceae). *Zool. J. Linnean. Soc.* **80**: 337–344.

Taylor, F.J.R. (1990). Symbiosis in marine protozoa. *Ecology of Marine Protozoa*. G.M. Capriulo, ed. New York, Oxford University Press.

Taylor, F.J.R., Blackbourn, D.J., and Blackbourn, J. (1971). The red-water ciliate *Mesodinium rubrum* and its "incomplete symbionts": a review including new ultrastructural observations. *J. Fish. Res. Bd. Can.* **28**: 391–407.

Toller, W.W., Rowan, R., and Knowltoni, N. (2001). Zooxanthellae of the *Montastraea annularis* species complex: patterns of distribution of four taxa of

Symbiodinium on different reefs and across depths. *Biol. Bull.* **201:** 348–359.

Tovar, J., Fischer, A., and Clark, C.G. (1999). The mitosome, a novel organelle related to mitochondria in the amitochondrial parasite *Entamoeba histolytica*. *Mol. Microbiol.* **32:** 1013–1021.

Tovar, J., Leon-Avila, G., Sanchez, L.B., Sutak, R., Tachezy, J., van der Giezen, M., Hernandez, M., Muller, M., and Lucocq, J.M. (2003). Mitochondrial remnant organelles of *Giardia* function in iron-sulphur protein maturation. *Nature* **426:** 172–176.

Vandenkoornhuyse, P., Baldauf, S.L., Leyval, C., Straczek, J., and Young, J.P. (2002). Extensive fungal diversity in plant roots. *Science* **295:** 2051.

Visram, S., and Douglas, A.E. (2006). Molecular diversity of symbiotic algae (zooxanthellae) in scleractinian corals of Kenia. *Coral Reefs* **25(1):** 172–176.

Waller, R.F., and McFadden, G.I. (2005). The apicoplast: a review of the derived plastid of apicomplexan parasites. *Curr. Issues Mol. Biol.* **7:** 57–79.

Watanabe, M.M., Suda, S., Inouye, I., Sawaguchi, T., and Chihara, M. (1990). *Lepidodiniumviride* gen. et sp. nov. (Gymnodiniales, Dinophyta), a green dinoflagellate with a chlorophyll *a*-containing and *b*-containing endosymbiont. *J. Phycol.* **26:** 741–751.

Whittaker, R.H., and Margulis, L. (1978). Protist classification and the kingdoms of organisms. *Biosystems* **10:** 3–18.

Wilcox, L.W., and Wedemayer, G.J. (1984). *Gymnodinium acidotum* Nygaard (Pyrrophyta), a dinoflagellate with an endosymbiotic cryptomonad. *J. Phycol.* **20:** 236–242.

Wilkinson, C.R. (1983). Net primary productivity in coral reef sponges. *Science* **219:** 410–412.

Wilkinson, C.R. (1987). Significance of microbial symbionts in sponge evolution and ecology. *Symbiosis* **4:** 135–146.

Wolf, Y.I., Rogozin, I.B., and Koonin, E.V. (2004). Coelomata and not Ecdysozoa: evidence from genome-wide phylogenetic analysis. *Genome Res.* **14:** 29–36.

Wolfe, K.H. (2001). Yesterday's polyploids and the mystery of diploidization. *Nat. Rev. Genet.* **2:** 333–341.

Yoon, H.S., Hackett, J., and Bhattacharya, D. (2002a). A single origin of the peridinin-, and fucoxanthin-containing plastids in dinoflagellates through tertiary endosymbiosis. *Proc. Natl. Acad. Sci. U S A* **99:** 11724–11729.

Yoon, H.S., Hackett, J., Pinto, G., and Bhattacharya, D. (2002b). A single, ancient origin of chromist plastids. *Proc. Natl. Acad. Sci. U S A* **99:** 15507–15512.

7

Plastid Endosymbiosis: Sources and Timing of the Major Events

JEREMIAH D. HACKETT, HWAN SU YOON, NICHOLAS J. BUTTERFIELD,
MICHAEL J. SANDERSON, AND DEBASHISH BHATTACHARYA

I. GENERAL INTRODUCTION TO PLASTID ENDOSYMBIOSIS

The acquisition of photosynthesis was a pivotal event in eukaryotic evolution. The ability to capture the sun's energy, once exclusively in the bacterial domain, laid the foundation for the diversification of algae and plants. In this chapter, we review current ideas regarding the origin of plastids (the photosynthetic organelle) in eukaryotes and the timing of these events, with particular emphasis on the initial source of eukaryotic photosynthesis.

There is now a consensus among researchers that photosynthesis was acquired by eukaryotes from bacteria through a process called endosymbiosis: the capture of one organism by another that resulted in the origin of both plastids and mitochondria (see Martin, Chapter 5, this volume). Early in the evolution of eukaryotes, a heterotrophic eukaryote engulfed a cyanobacterium and retained it in its cytoplasm as a permanent acquisition, converting it into the plastid. This event is referred to as "primary" endosymbiosis and gave rise to three extant lineages containing a primary plastid (the Plantae [Moreira *et al.* 2000; Stibitz *et al.* 2000; Matsuzaki *et al.* 2004; Sanchez Puerta *et al.* 2004] or the Archaeplastida [Adl *et al.* 2005]): the Glaucophyta (glaucophyte algae), the Rhodophyta (red algae), and the Viridiplantae (green algae and land plants, Figure 1). The number of independent primary endosymbioses has

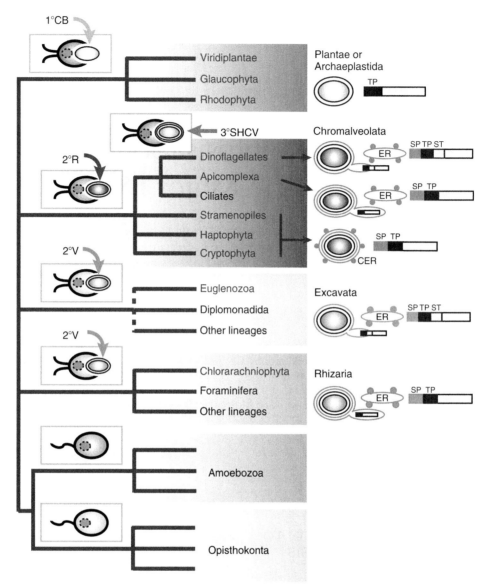

FIGURE 1. Schematic tree of the six eukaryotic supergroups (see Adl *et al.* 2005) showing the different plastid endosymbioses and the origins of the plastid protein import system in photosynthetic eukaryotes. The Viridiplantae, Glaucophyta, and Rhodophyta (Plantae or Archaeplastida) share a plastid that was derived from a putative single primary endosymbiosis (1°) between a nonphotosynthetic protist and a cyanobacterium (CB). Many endosymbiont genes were transferred to the nucleus following the plastid capture. Transit peptides (TP) that were added to the N-terminus of nuclear-encoded proteins allowed Plantae plastid targeting via the TOC/TIC translocon. The single red algal secondary endosymbiosis (2°) that putatively gave rise to the chromalveolate plastid was ancestrally shared by Chromista (Stramenopiles, haptophytes, and cryptophytes) and Alveolata (dinoflagellates, apicomplexans, and ciliates). The chromists contain a four-membrane plastid with a plastid endoplasmic reticulum (CER) and a modified bipartite leader sequence, which targets plastid proteins to the CER via a signal peptide (SP) in addition to the typical transit peptide. Apicomplexan apicoplasts, which have four plastid membranes but lack CER, carry out protein import via the secretory pathway (ER and Golgi apparatus); a system that is also present in chlorarachniophytes that contain a green-algal–derived plastid. Of the three-membrane plastids, the dinoflagellates, which generally contain a chromalveolate plastid of red algal origin, but also have undergone tertiary plastid replacements involving Stramenopile, haptophyte (i.e., 3° SHCV), cryptophyte, or green algae, use a tripartite leader sequence that includes a stop transfer (ST) signal to ensure that plastid proteins in Golgi-derived vesicles are integral membrane proteins that maintain a predominant cytoplasmic component (Nassoury *et al.* 2003). This type of tripartite leader sequence is also found in the Euglenozoa that contain a green-algal–derived secondary plastid. (See color plate.)

profound implications for our understanding of eukaryotic evolution. If it happened as many as three times in different lineages, that would suggest that establishing the combination of a eukaryotic and a prokaryotic cell is relatively easy, in evolutionary terms, the resulting chimera converging on a similar morphology and physiology for each separate occurrence. If it occurred only once, then all extant autotrophic eukaryotes trace their ancestry to this remarkable accident of evolution. Although yet to be definitively proven, the majority of evidence suggests that this primary endosymbiosis occurred a single time in the common ancestor of the Plantae (Gray 1992; Bhattacharya and Medlin 1995; Delwiche 1999; Moreira *et al.,* 2000; McFadden 2001; Palmer 2003; Bhattacharya *et al.* 2004; Rodriguez-Ezpeleta *et al.* 2005). Phylogenetic evidence in support of this idea is presented later in this chapter. The one exception is the filose amoeba *Paulinella chromatophora,* which harbors a plastid from an independent primary endosymbiosis (Bhattacharya *et al.* 1995; Marin *et al.* 2005; Yoon *et al.* 2006). A recent comparative genome analysis of the *Paulinella* endosymbiont demonstrates that it has traits that are typical of cyanobacterial, not plastid genomes, supporting a recent plastid capture (Yoon *et al.* 2006).

Molecular clock analyses using multigene datasets and approaches (e.g., penalized likelihood, Bayesian methods [see later]) that relax the assumption of strict chronometric behavior of genes under study suggest that the primary endosymbiosis is an ancient event in eukaryotic evolution. Although not yet settled (see Douzery et al. 2004), recent analyses suggest that the primary plastid was established ca. 1.5 billion years ago (Ga) in the Mesoproterozoic (e.g., Martin et al. 2003; Hedges et al. 2004; Yoon et al. 2004; Blair et al. 2005). It should be noted that there are many extant examples of intracellular symbioses such as in ciliates; however, these lack several crucial characteristics that distinguish endosymbiosis from intracellular associations between different organisms. These are evolutionary innovations that cement the permanent association between the endosymbiont and the host, making each unable to survive without the other, and allow transport of proteins and metabolites in and out of the organelle. Of these steps, intracellular gene transfer from the captured cell to the host nucleus may be considered a fundamental requirement for endosymbiosis (see Martin, Chapter 5, this volume).

A necessary step in the primary endosymbiosis was the establishment of a reliable connection between the host cell and the ancestral plastid to allow the controlled exchange of metabolic intermediates between the symbiotic partners. Metabolite antiporters are ideally suited for the controlled exchange of solutes between cellular compartments because the antiport function is dependent on the presence of a suitable counter-exchange substrate on the *trans*-site of the membrane. Controlled exchange is critical in this respect because the unregulated flux of metabolites between the host and plastid would have had detrimental effects on the metabolism of both partners and thereby lowered the evolutionary fitness of the symbiosis. As an example, all phosphate translocators catalyze a strict antiport transport of substrates because the export of, for example, triose phosphates (TP transporter, TPT) from plastids does not only remove carbon from the plastid but also phosphate. Phosphate homeostasis of the plastid stroma is extremely important because depletion of the plastidic phosphate pool results in substrate limitation of photophosphorylation (light-driven ATP biosynthesis at the thylakoid membrane from ADP and Pi; Flügge 1998). The TPT imports one Pi into the plastid for each TP exported, thereby maintaining the Pi homeostasis of the plastid stroma. It was recently shown that plastid translocators were established in the common ancestor of the red and green algae (and likely all Plantae/Archaeplastida, supporting their monophyly), allowing this first alga to profit from cyanobacterial carbon fixation (Weber *et al.* 2006). This evolutionary step may have rendered irreversible the association between the plastid and the host cell. The ancestral plastid antiporter evolved from an existing metabolite translocator in the host cell that had evolved due to the pre-existence of mitochondria and an endomembrane system (Figure 2).

Other important steps in plastid establishment included gene transfer from the endosymbiont to the host (e.g., Allen *et al.* 2005). If two organisms live in close association, but each contains its own complete complement of genes, these cells can potentially live apart if necessary. When a gene required for the viability of the endosymbiont is transferred to the nucleus of the host, the endosymbiont may lose the ability to live independently. If these transferred genes code for proteins that function inside the endosymbiont, then a system for importing proteins synthesized in the cytoplasm must also be developed. Both of these have occurred in the lineages that resulted from the primary endosymbiosis thereby enslaving the plastid endosymbiont (sensu Douglas *et al.* 2001). The vacuolar membrane that is presumed to have existed has been lost, resulting in a two-membrane plastid. The majority of the genes that existed on the cyanobacterial nucleoid have been lost or transferred to the nucleus. Most plastids contain a single, circular chromosome of about 200 kilobases (Kb) and encode around 100–120 genes. A free-living cyanobacterium typically has a genome of about 2500 Kb. The genes that remain in the plastid are primarily involved in photosynthesis, transcription and translation of plastid genes, and ATP synthesis via the F-ATPase complex (Nelson and Ben-Shem 2004). However, most genes needed to maintain the plastid are encoded in the nucleus. Plastid-encoded genes are presumably under selection for transfer to the nucleus to avoid the accumulation of deleterious mutations that occur on nonrecom-

bining genomes (Muller's ratchet; Muller 1932). In light of Muller's ratchet and the higher mutation rate caused by reactive oxygen species produced in organelles (Allen and Raven 1996), it is interesting that so many genes remain encoded in the plastid and mitochondria. Several theories have been developed to explain why some genes have not or cannot be transferred to the nucleus (discussed in Martin and Herrmann 1998). Some dinoflagellate algae have the most reduced plastid genomes among photosynthetic eukaryotes and may shed light on this aspect of genome evolution (discussed later). Because most plastid genes are encoded in the nucleus, these organisms must have an efficient system to target and import the products of these genes back into the plastid. All primary plastids are surrounded by two membranes that must be crossed by these proteins (see Figure 1). Plastid proteins that are synthesized in the cytoplasm are first targeted to the plastid with an N-terminal–targeting peptide (McFadden 2001). The targeting peptides show little sequence identity; they are instead rich in basic amino acids. This peptide is recognized by proteins in the outer membrane of the plastid and the protein is pulled in through the two plastid membranes through channels formed by the translocon of the plastid outer membrane (TOC) and translocon of the plastid inner membrane (TIC) protein complexes. These are multiprotein complexes that mediate the translocation of proteins through the plastid membranes (Reumann *et al.* 2005). Once inside, the transit peptide is cleaved, leaving the mature protein. The conservation of many aspects of the plastid-import machinery in Plantae also supports a single plastid origin in these lineages (Matsuzaki *et al.* 2004; McFadden and van Dooren 2004).

The completion of several cyanobacterial and eukaryotic genomes has allowed us to understand more completely the impact of this endosymbiosis on the genome of eukaryotes. An analysis of the *Arabidopsis*

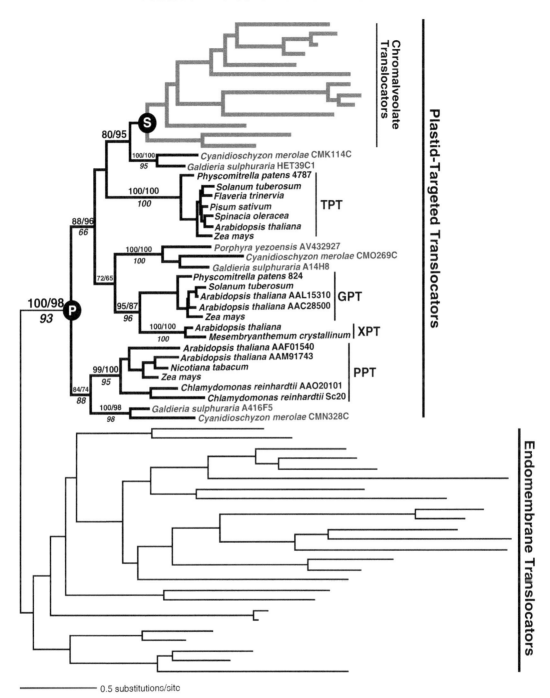

FIGURE 2. Maximum likelihood phylogeny of endomembrane and plastid translocators. The PhyML bootstrap values (200 replications) are shown above the branches on the left of the slash mark, whereas the values to the right are from a neighbor-joining analysis (100 replications). The bootstrap values shown below the branches in italics are from an unweighted maximum parsimony (MP) analysis (2000 replications). Only bootstrap values of interest and >60% are shown. The red algae are shown in grey text. The different plastid translocators are glucose 6-phosphate translocator (GPT), phosphoenolpyruvate translocator (PPT), triose phosphate translocator (TPT), and xylulose 5-phosphate translocator (XPT). The numbers in the filled circles indicate translocators to the right that resulted from primary (P) and secondary (S) endosymbiosis (for details, see Weber *et al.* 2006).

genome has revealed that as much as 18% of the nuclear genome of this land plant may be of cyanobacterial origin (Martin *et al.* 2002). The majority of these transferred genes are not involved in plastid functions. The genomes of two red algae, *Cyanidioschyzon merole* (Matsuzaki *et al.* 2004) and *Galdieria sulfuraria* (Barbier *et al.* 2005), have been completed. These data, together with the wealth of genome information from green algae and land plants (embryophytes), will soon give us a more complete picture of how the primary endosymbiosis impacted the nuclear genome of the host. However, it is clear from present studies that the primary endosymbiosis not only gave eukaryotes the ability to do photosynthesis but also supplied genetic raw material for the evolution of other traits.

Some time after the split between red and green algae, members of these lineages were themselves engulfed by nonphotosynthetic protists in separate endosymbioses and reduced to plastids. Data generated in our lab (Yoon *et al.* 2002), and elsewhere (Fast *et al.* 2001; Harper and Keeling 2003), indicate that one of these so-called secondary endosymbioses putatively gave rise to the photosynthetic ancestor of a protist super-assemblage, the chromalveolates (Cavalier-Smith 1999), comprising the Chromista (cryptophytes, haptophytes, and Stramenopiles; Cavalier-Smith 1986) and the Alveolata (dinoflagellates, ciliates, and apicomplexans; Van de Peer and De Wachter 1997). Again, gene transfer was necessary, but this time from the nucleus of the secondary endosymbiont (red alga) to that of the host. In all groups except the cryptophytes, which maintain a remnant of the red algal nucleus (the nucleomorph; Greenwood 1974), the endosymbiont nucleus and mitochondrion have been eliminated, indicating that all genes necessary to control the plastid must have been transferred to the host nucleus (Douglas *et al.* 2001). In two other separate secondary endosymbioses a plastid of green algal origin was introduced into the euglenophytes and the chlorarachniophyte amoebae (McFadden 2001). The most extreme case in plastid origin is tertiary endosymbiosis (see Figure 1), in which an alga containing a plastid of secondary origin was engulfed and reduced to the photosynthetic organelle. Only dinoflagellates have undergone tertiary endosymbiosis. The chapter now looks in detail at the evidence regarding the source and timing of the plastids that have resulted from primary, secondary, and tertiary endosymbiosis. The greatest attention is paid to the primary endosymbiosis that gave rise to the first eukaryotic plastid from which all (except in *Paulinella chromatophora*) plastids are derived.

II. PRIMARY PLASTID ORIGIN AND PLANTAE MONOPHYLY

A. Generating the Eukaryotic Phylogeny

To place a date on the origin of the primary plastid in the common ancestor of the Plantae, we first generated a multigene tree of eukaryotes. To generate the tree, we chose several "housekeeping" genes that are distributed in a wide diversity of eukaryotes, including our three lineages of interest (the Glaucophyta, Rhodophyta, and Viridiplantae). Our final choice of proteins did not reflect an intentional bias toward those that support Plantae monophyly but rather reflect their availability in databases and ease of unambiguous alignment from a diverse set of target taxa. The 17 proteins used in this study are members of the conserved eukaryotic translation apparatus (elongation factor 1α, elongation factor 2, eukaryotic initiation factor 5A, and the ribosomal proteins rpl13A, rpl9, rpl17, rpl18, and rpl27A), the cytoskeleton (actin, α- and β-tubulin), the α and β subunits of vacuolar ATPase, heat shock proteins 70 and 90, and subunit 8 of the 26S proteasome. Many of these proteins have proven to be reliable in other phylogenetic studies (Baldauf *et al.* 2000; Moreira *et al.* 2000; Arisue *et al.* 2002) and analyses of the individual protein datasets did not provide any clear evidence for lateral gene transfer among the studied taxa (data not shown). Given these observations, we assembled the 17 nuclear-encoded proteins from 40 taxa.

The phylogenetic analysis of the aligned 6458 amino acids was implemented under maximum likelihood (ML) using PhyML (Guindon and Gascuel 2003). The tree of highest likelihood was obtained using the WAG + Γ model of protein sequence evolution (Whelan and Goldman 2001). ML bootstrap analyses (100 replicates) were also done using PhyML. Maximum parsimony (MP) bootstrap analyses (2000 replicates) were done in PAUP* (Swofford 2003) using 10 random additions and TBR branch swapping. Bayesian inference (MrBayes V3.0b4; Huelsenbeck and Ronquist 2001) was used to generate Bayesian posterior probabilities (BPP) for nodes. The MCMC chains (3 cold, 1 heated) were run with the WAG + Γ model for 500,000 generations and sampled every 1000 generations. The first 100 trees were discarded as burn-in. Consistency among the results produced by these different 3 approaches (i.e., ML, MP, BPP) in molecular phylogenetics is generally regarded as indicating a well-supported tree. For bootstrap and Bayesian analyses, robust support for phylogenetic groupings is postulated when values exceed 75–80% and BPP >0.95, respectively. These analyses (Figure 3) show bootstrap and Bayesian support for virtually all of the clades, including the expected monophyly of the fungi and animals (ML = 92%, MP = 100%, BPP = 1.0) and the alveolates and Stramenopiles (ML = 99%, M P= 8 1%, BPP = 1.0). Our analysis also provides moderate support for Plantae monophyly (ML = 76%, MP = 66%, BPP = 1.0) and is, therefore, consistent with a single origin of the primary plastid in eukaryotes (see also Rodriguez-Ezpeleta et al. 2005). Addition of the highly divergent sequences from the amitochondriate taxa (Giardia lamblia, Trichomonas vaginalis) did not disturb Plantae monophyly (or any other topological element in the tree) and these protists were positioned as close relatives of the discicristates (ML = 78%, MP = 100%, BPP = 1.0) as also reported in Arisue et al. (2005). The only node that was not supported in the ML tree was the sister relationship of the Stramenopiles + alveolates and the discicristates + diplomonads. Interestingly, a recent analysis using 30,113 amino acids (aa), positions that included the discicristates and amitochondriate taxa in our tree provided strong ML support (97%) for Plantae monophyly and for the sister group relationship of glaucophytes and Viridiplantae (94%). Our more limited dataset is a subset of that used by Rodriguez-Ezpeleta et al. (2005), except that we included two proteins not in their alignment (heat shock protein 90 and 60S ribosomal protein 9).

Within the Plantae, there is strong support (ML = 93%, MP = 76%, BPP = 1.0) for the sister group relationship of Cyanophora paradoxa (glaucophyte) and the Viridiplantae (see also Rodriguez-Ezpeleta et al. 2005). This result is surprising because only glaucophyte plastids retain the ancestral peptidoglycan wall of the cyanobacterium (Pfanzagl et al. 1996). This would imply that peptidoglycan was lost from plastids on two occasions (i.e., at the base of the red algae and of the Viridiplantae). To investigate this result, we generated alternate topologies and assessed their likelihoods using the "one-sided" Kishino-Hasegawa test (Kishino and Hasegawa 1989; Goldman et al. 2000). This ML test significantly rejected the placement of Cyanophora on the branch uniting the Plantae (P = 0.009), the red algae (P = 0.009), or the embryophytes (P < 0.000) or as sister to Chlamydomonas reinhardtii (P < 0.000). Other positions in the tree were also significantly rejected by the test (e.g., on the branch uniting the excavates + chromalveolates, P = 0.002). Phylogenetic analyses of plastid genes or genomes often suggest an early divergence of glaucophytes among algal plastids (e.g., Martin et al. 1998; Ohta et al. 2003; Yoon et al. 2004). However, analyses of protein sequences

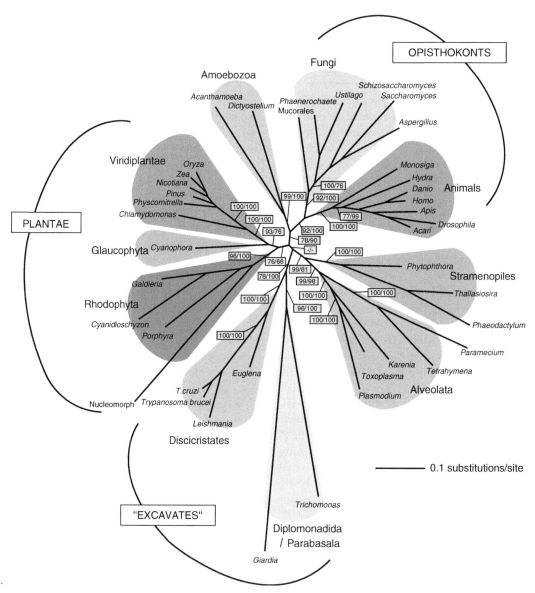

FIGURE 3. Unrooted maximum likelihood tree of eukaryotes inferred from 17 proteins. The numbers indicate the results of PhyML (above) and maximum parsimony (below) bootstrap analyses. Nearly all branches had a probability of 1.0 from Bayesian analysis. Only bootstrap values for the deeper nodes are shown for clarity.

using a broader taxon sampling also support a glaucophyte-green algal relationship (Hagopian *et al.* 2004; Yoon *et al.* 2005) or leave in question the position of this taxon in the plastid tree (Bachvaroff *et al.* 2005). Clearly, more data from the species-poor glaucophytes (ca. 12 species, Kies and Kremer 1990) are needed to assess this important issue (Reyes-Prieto and Bhattacharya, work in progress). In addition, these results suggest to us that larger data sets with more extensive taxon sampling will be critical to conclusively solving the tree of eukaryotes (i.e., including many lineages missing from our trees such as rhizarians, haptophytes, and cryptophytes). This entails genome-wide phylogenetic datasets that incorporate thousands of amino acid positions and include a more complete sampling of lineages.

B. Molecular Clock Analyses

Given our phylogenetic hypothesis shown in Figure 2, we used two different molecular clock methods that relax the requirement for a strict molecular clock (Bayesian inference [Thorne *et al.* 1998] and penalized likelihood [Sanderson 2003]) to estimate the date of primary plastid origin. In the first approach, Bayesian methods that incorporate knowledge of the prior distribution of parameters are used to infer divergence dates under a probabilistic model for protein sequence rate variation (see Kishino *et al.* 2001). In the second approach the ML method is used, in which the molecular rate of divergence is allowed to vary throughout the tree, but closely related species share similar rates (Sanderson 2003). Using the Bayesian approach, the ML tree topology was first used to calculate the branch lengths with the program estbranches (using the JTT protein evolution model; Jones *et al.* 1992) prior to Bayesian estimation of divergence times using the program multidivtime (Kishino *et al.* 2001). The ML tree was rooted on either the branch leading to the animals + fungi or the diplomonads + parabasalids (see discussion later). Because we did not have a true outgroup taxon on these branches, we added a hypothetical outgroup on the appropriate branch, which was coded as missing data in the alignment. This outgroup was removed from the tree when the branch lengths are calculated using estbranches. We placed 8 time constraints on this analysis based on the fossil record: 0.144–0.206 Ga for the monocot–dicot split (Sanderson and Doyle 2001), 0.290–0.320 for the angiosperm–gymnosperm split (Bowe *et al.* 2000), 0.433–0.490 for the origin of embryophytes (land plants; Kenrick and Crane 1997), 0.490–0.543 for the Chelicerata–Insecta split (Benton 1993), 0.290–0.417 for the Diptera–Hymenoptera

split (Benton 1993), 0.354–0.417 for the Actinopterygii–Mammalia split (Benton 1993), and minimum dates of 0.417 for the origin of ascomycetes (Taylor *et al.* 1999). The prior time (for the Bayesian analysis) from the tips to the root was set at 1.8 Ga with a standard deviation of 0.5 Ga to correspond with the oldest known mega-eukaryotic fossils ("*Grypania,*" Han and Runnegar 1992). The mean of the prior distribution of the rate at the root node was calculated according to the multidivtime manual by dividing the median distance from the tips to the root by the prior time from tips to the root. The standard deviation of the prior rate distribution was set equal to the mean. The MCMC chain was run for 1 million generations, sampling every 100 generations, after a burn-in of 100,000 generations. Divergence time estimations were done twice to ensure convergence of the MCMC chain.

We also used the molecular clock software program r8s to estimate divergence times using the penalized likelihood approach. This program relies upon user-supplied branch lengths. Because of this, we placed a hypothetical outgroup (branch length = 0) that bisected the appropriate branch, placing half of the branch length on either side of the branch point. With the new version of the r8s program, a new logarithmic penalty function that penalizes local fractional changes in rate was examined, as well as the older additive penalty used in previous versions of r8s. The logarithmic penalty is expected to perform slightly more effectively in contexts where extrapolation from recent calibrations to deep nodes is required. Optimal smoothing values for both penalty functions corresponded to a constant rate model, which is expected either if rates are constant or if they vary significantly but irregularly across the tree. A clock could be statistically rejected for the dataset, but the modest (although significant) heterogeneity in rate across the tree is apparently distributed in such a way that models that postulate smooth rate variation across the tree do not improve inferences significantly. We were able to calculate confidence intervals using r8s by calculating the branch lengths of the tree shown in Figure 3 from 100 bootstrap datasets with the WAG + Γ model of sequence evolution using tree-puzzle and the puzzleboot script (Holder and Roger 1999), modified to save the trees rather than the distance matrices. The divergence times for each of these trees were calculated using r8s, and the 95% confidence interval of the time distribution for each node was calculated using JMP 4.0 statistical analysis software.

The time of primary plastid origin was assumed to have occurred sometime after the divergence of the Plantae and before the red–green algal split. An important consideration in these analyses was the location of the root of the tree. The placement of the root of the eukaryotic tree has been controversial. Based on the phylogenetic distribution of several gene fusions, it has been proposed that the root lies either on the branch leading to the Opisthokonts (animals + fungi) or the branch between the Opisthokonts + Amoebozoa and other eukaryotes (Stechmann and Cavalier-Smith 2002). However, a recent study examined 22 proteins from eukaryotes in addition to archaebacterial and eubacterial outgroups and determined the most likely position of the root was either between the Opisthokonts and Amoebozoa + others or between the Diplomonads + Parabasalids and other eukaryotes (Arisue et al. 2005). Because it is unclear where the root lies, and because its position could substantially affect our results, we did the molecular clock analyses using the latter two rooting schemes.

Under the conditions used here, Bayesian analysis of the Opisthokont rooted tree showed the origin of the Plantae to be around 1.37 (1.15–1.66) Ga. The split of the red algae from the Viridiplantae + glaucophytes was predicted to have occurred around 1.25 Ga (1.04–1.52; Figure 4A). Using the penalized likelihood approach, the origin of the Plantae was calculated to be around 1.53 (1.17–

1.77) Ga and the split of the Rhodophyta was 1.41 Ga (1.04–1.52). These results are similar to those obtained from the Diplomonad + Parabasalid (DP) rooted tree (Figure 4B) and suggest that the primary endosymbiosis occurred around 1.25 - 1.60 Ga. These dates are consistent with a previous analysis of plastid genes that suggested a minimal date for the primary endosymbiosis of ca. 1.47 Ga (Yoon et al. 2004).

The deeper dates from the PL analysis of the opisthokont-rooted tree and the analyses of the DP-rooted tree are supported by Proterozoic fossil record (not used as calibrations in our analyses). For example, presence of the sexual red alga *Bangiomorpha* in the ca. 1200-million-year-old Hunting Formation, arctic Canada (Butterfield 2001) is consistent with our molecular clock estimate for the divergence of the sexual red alga *Porphyra* after the split of red algae around 1.40 Ga. The >1.0 Ga putative Stramenopile *Palaeovaucheria* (Hermann 1990; Butterfield 2004) also appears after our calculated divergence of Stramenopiles and alveolates between 1.35 and 1.52 Ga, although we consider this finding less robust due to the lack of uniformity in our Stramenopile and alveolate analyses. We note, however, that the molecular clock estimates obtained by Douzery et al. (2004) are less consistent with the fossil data, because they propose a 928 million years ago (Ma) divergence for the red algae and 872 Ma divergence for Stramenopiles. Douzery et al. (2004) also placed the origin of Plantae between 0.892 and 1.162 Ga, some 500 million years later than our point estimate (Douzery et al. 2004). The reasons for these discrepancies are not entirely clear but are likely to include our considerably broader sampling of nonopisthokont (e.g., Plantae) taxa in contrast to the larger number of amino acid positions in their analysis. Both of these factors play important roles in tree topology and branch length estimates, although taxon sampling may be the more significant issue given the breadth of eukaryotic diversity that is being addressed in these studies.

Our dates are, however, consistent with the analyses of Hedges *et al.* (2004) and Blair *et al.* (2005), who used large genome-wide datasets and several relaxed clock methods of analysis. These authors reported dates of 1.414 + 0.121 Ga (+ one standard deviation) for the split of red and green algae and 1.671 + 0.145 Ga for the split of animals/fungi and plants (Blair *et al.* 2005).

The significant bootstrap and Bayesian support for virtually all nodes in our tree suggests that our analyses have sufficient power to address with confidence the interrelationships of the studied taxa. In particular, we take confidence from the strong agreement between the divergence dates using the Bayesian approach and the results from penalized likelihood estimations. We also note that molecular clock estimates for the origin of the Plantae are beginning to converge: within the 95% confidence interval, there is virtually no difference between our latest (Opisthokont root: ~1.16 Ga) and Douzery *et al.'s* earliest (1.162 Ga) dates for Plantae origin.

The evolution of photosynthetic eukaryotes is likely to have had a significant impact on the early biosphere and, as such, should be detectable in the Proterozoic record (Peterson and Butterfield 2005). Interestingly, our 1.38–1.54 Ga molecular clock estimate for the origin of Plantae coincides closely with the first appearance of morphologically differentiated microfossils in ca. 1.43 Ga strata of Roper Group, northern Australia (Javaux *et al.* 2001). Although

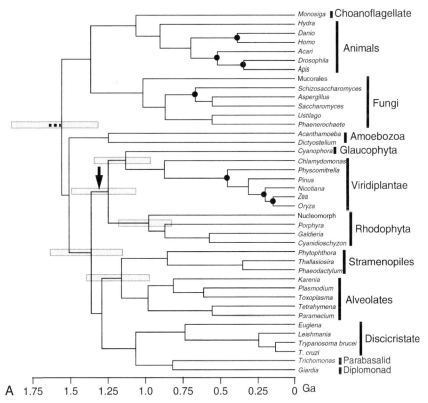

FIGURE 4. Chronograms of the maximum likelihood (ML) tree of eukaryotes from Bayesian molecular clock analysis. The tree is rooted on the branch leading to the **(A)** Opisthokonts (animals + fungi) or

(Continued)

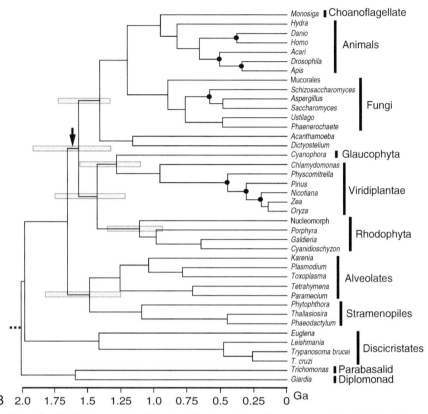

FIGURE 4. (Cont'd)   (B) Diplomonads + Parabasalids. Branch lengths are proportional to time (see scale bar at the bottom). Grey boxes at the nodes indicate the 95% confidence intervals for those nodes. The arrow indicates the branch where the putative primary endosymbiosis occurred.

these are of unknown taxonomic (and metabolic) affinity and offer only a minimum age constraint on first appearance, there are broad indications of a substantial increase in eukaryotic diversity at some point in the early to middle Mesoproterozoic (Knoll 1994; Huntley *et al.* 2006). We interpret this increase to the direct and indirect consequences of eukaryotes gaining the ability to photosynthesize.

C. Conclusions of Plantae Phylogenetic and Molecular Clock Analyses

The unique and relatively late appearance of (extant) photosynthetic eukaryotes has important implications for understanding the early biosphere and its fossil record. Fossil biomarker molecules extracted from

late Archean (2.7 Ga) rocks point to the antiquity of eukaryotes (Brocks *et al.* 1999), but at least one-half of the domain's history appears to have taken place without the advantage of photoautotrophy. As such, the habits and affiliations of putative Palaeoproterozoic eukaryotes need to be reassessed, not least the 1.87 Ga (Schneider *et al.* 2002) Negaunee "Grypania," which have been interpreted as benthic seaweeds (Han and Runnegar 1992), and early sphaeromorphic acritarchs, which are generally assumed to represent the cysts of net phytoplankton (Knoll 1994). If these are indeed eukaryotic (see Cavalier-Smith 2002; Samuelsson and Butterfield 2001), they are better interpreted as the remains of heterotrophic organisms, either protozoans sensu lato or fungi sensu lato (see Butterfield 2005). More importantly, marine biogeochemistry in the absence of plant protists cannot be modeled on simple uniformitarian principles. In the modern oceans, the majority of exported carbon is supplied by relatively large eukaryotic net phytoplankton, whereas most cyanobacterial production is rapidly recycled via a microbial loop (Azam *et al.* 1983). Carbon was certainly exported prior to the Mesoproterozoic but would have been sourced almost entirely by photosynthetic bacteria following distinctive nonuniformitarian transport pathways and imparting distinctive nonuniformitarian biogeochemical signatures. Interestingly, our estimate for the origin of eukaryotic photosynthesis occurs within a billion-year interval of conspicuous biogeochemical stability (Buick *et al.* 1995; Brasier and Lindsay 1998), which, along with fossil biomarker evidence (Summons *et al.* 1999; Brocks *et al.* 2005; Knoll *et al.*, Chapter 8, this volume), points to a continued cyanobacterial hegemony of marine primary productivity. In other words, the evolutionary appearance of photosynthetic eukaryotes substantially preceded their ecological predominance. In summary, our nuclear phylogeny supports the monophyly of photosynthetic eukaryotes containing a primary plastid, and our molecular clock estimates provide a timeline for reconstructing the early evolutionary history of the Plantae. Together with the fossil and geochemical records, these data provide an increasingly resolved view of early eukaryotic photosynthesis and, by extension, important features in evolutionary and Earth history.

Although we do not know why the Plantae primary endosymbiosis ultimately succeeded in a setting where many other cyanobacterial captures must have occurred, an interesting parallel is found when comparing this clade to the only other known case of primary endosymbiosis in *Paulinella chromatophora*. Both of these taxa trace their origin to ancestors that could clearly conduct phagotrophy yet both now rely primarily on photoautotrophy. Phagotrophy was apparently lost in the Plantae ancestor

with only a single prasinophyte green alga known to feed by phagocytosis (Cavalier-Smith 1993). In the case of *Paulinella,* there exists a closely related heterotrophic sister species (*P. ovalis*) that actively feeds on cyanobacteria that are found in vacuoles in the cell (Johnson *et al.* 1988). *P. chromatophora,* however, has lost phagotrophic ability and has become an obligate photoautotroph. This suggests that the gain of photoautotrophy via primary endosymbiosis offers significant selective advantages to the host cell that likely led to its rapid fixation within the Plantae ancestor and more recently, in *P. chromatophora.*

III. SECONDARY PLASTID ENDOSYMBIOSIS

Following Plantae origin, several other groups of eukaryotes acquired plastids, not from cyanobacteria but by capturing one of these primary plastid-containing algae. This "secondary" endosymbiosis has occurred at least three times (Bhattacharya *et al.* 2004). Two groups have acquired the plastids independently from green algae. The Euglenozoa (such as *Euglena gracilis*) are freshwater algae related to the parasitic trypanosomes and contain a three-membrane plastid of green algal origin (Gibbs 1978). The chlorarachniophyte *Bigelowiella natans* is a marine amoeba that contains a four-membrane green algal plastid and retains a remnant of the green algal nucleus between the second and third membranes (starting from the inside) called the nucleomorph (McFadden *et al.* 1994; Gilson and McFadden 1996; Van de Peer *et al.* 1996).

The most abundant groups with secondary plastids acquired them from the red algae. Five algal lineages have plastids of red algal origin. These include the cryptophytes, which are covered in cellulose scales; the haptophytes, which form huge blooms in the open ocean and produce calcium carbonate scales that act as a global carbon sink; and the Stramenopiles, which are a diverse and highly successful group that includes parasitic oomycetes (water molds such as *Phytophthora*), kelps, and the silica-shell–forming diatoms. Together, these groups are called the Chromista, and recent phylogenetic evidence suggests that their plastids are clearly of red algal origin (Figure 5) and closely related, which is consistent with (but does not prove) Chromist monophyly (Yoon *et al.* 2002, 2004). Molecular clock analyses of the plastid gene data suggest that the putatively single chromist red algal secondary endosymbiosis occurred ca. 1.3 Ga (Yoon *et al.* 2004). A less parsimonious alternative to this scenario is that multiple independent, nearly simultaneous secondary endosymbioses involving closely related red algae occurred with different host cells. This would incorrectly suggest chromist monophyly. Recent data that demonstrate a close relationship between nuclear-encoded plastid-targeted proteins of red algal origin in chromists (and in all chromalveolates, see later), however, are consistent with the plastid trees and support the monophyly of these taxa (Li *et al.* 2005; Nosenko *et al.* in press). All chromists contain a plastid with four membranes and the photopigment chlorophyll *c*. The inner two membranes are homologous to the membranes of primary plastids, the third membrane derives from the plasma membrane of the red alga, whereas the outer membrane comes from the phagosomal membrane of the host. As in the chloroarachniophytes, the cryptophytes maintain a remnant of the algal (in this case a red algae) nucleus between the second and third membranes (starting from the inside) of the plastid. The nucleomorph has been lost in the haptophytes and Stramenopiles. These organisms have evolved a modified plastid-targeting signal that includes a signal peptide at the N-terminus, followed by a typical plastid transit peptide. Plastid proteins are first targeted to the endoplasmic reticulum (ER), which is continuous with the outer membrane of the plastid. It is unknown how proteins cross the third membrane. The inner two membranes are spanned using the same TOC/TIC proteins as in primary plastids (see Figure 1 for details).

The remaining two lineages with red algal plastids are related lineages within the alveolates. The Alveolata include the aplastidial ciliates, the parasitic apicomplexans (such as *Plasmodium* and *Toxoplasma*), and the dinoflagellates. Phylogenetic evidence suggests that the common ancestor of the alveolates contained a plastid (Baldauf *et al.* 2000); however, the early branching ciliates are believed to have lost their plastid and no traces of this organelle have yet been found. The parasitic apicomplexans have lost the ability to do photosynthesis, most likely because of their intracellular lifestyle. However, they maintain a vestigial organelle derived from a plastid called the apicoplast. This organelle is surrounded by four membranes, has a small genome, and is retained most likely for fatty acid biosynthesis. There has been controversy over whether the apicoplast is of green or red algal origin (Kohler *et al.* 1997; McFadden *et al.* 1997; Funes *et al.* 2002). The strongest evidence comes from the gene order on the apicoplast genome that most resembles that of red algae (McFadden *et al.* 1997).

All photosynthetic members of the chromists and alveolates contain chlorophyll c_2, a pigment unique to these lineages. For this reason it has been proposed that these organisms share a common ancestor, and they have been termed the chromalveolates. The "chromalveolate hypothesis" posits that there was a secondary endosymbiosis in the common ancestor of the chromists and alveolates and that these two lineages diverged as sister groups. Currently, there is some support for this hypothesis, but the data are incomplete. Phylogenetic analysis of both plastid and nuclear proteins supports the monophyly of the Stramenopiles (chromists) and the alveolates (Van de Peer and De Wachter 1997; Baldauf *et al.* 2000; Yoon *et al.* 2004). In the few nuclear protein analyses that have included them, the position of the cryptophytes and

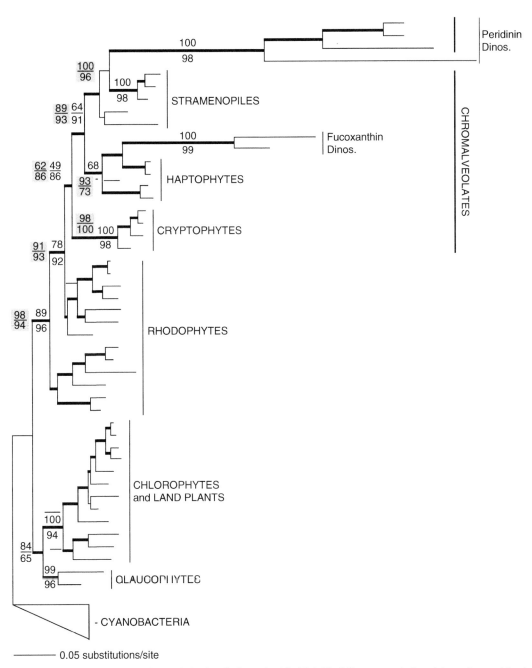

FIGURE 5. Phylogenetic analysis of algal and plant plastids. This PhyML tree was inferred from the combined plastid protein sequences of PsaA, PsaB, PsbA, PsbC, and PsbD. The results of a PhyML bootstrap analysis are shown above the branches, whereas the values below the branches result from a minimum evolution bootstrap analysis. The thick branches represent >95% Bayesian posterior probability. The bootstrap values shown in the grey boxes were calculated after the removal of the highly divergent minicircle sequences in the peridinin dinoflagellates. Only bootstrap values of interest and >60% are shown. The branch lengths are proportional to the number of substitutions per site (see scale in figure). Full details of the phylogenetic analysis are found in Yoon *et al.* (2005).

haptophytes has been poorly resolved (Durnford *et al.* 1999; Tengs *et al.* 2000; Hackett *et al.* 2004b). Interestingly, in these analyses, the Stramenopile–alveolate relationship is still supported, perhaps indicating that the alveolates have a specific relationship to the Stramenopiles, rather than a sister group relationship with a monophyletic Chromista as previously suggested (Cavalier-Smith 2000).

A recent phylogenomic analysis of the chromalveolates provides evidence that members of this lineage share a suite of nuclear genes of red algal origin that encode plastid targeted proteins. Under the chromalveolate hypothesis, it is predicted that, like the plastid genes of red algal origin (Figure 6), genes of plastid function that have been relocated to the nucleus should also have a red algal origin and be monophyletic in phylogenies. This prediction has been met for a number of chromalveolate (except cryptophyte) plastid-targeted nuclear genes (e.g., Miyagishima *et al.* 2004; Li *et al.* 2005). Li *et al.* (2005) used the phylogenomics app-roach with a unigene set of 5081 expressed sequence tags (ESTs) from the haptophyte alga *Emiliania huxleyi* and found that 19 genes encoding plastid-targeted proteins could be unambiguously positioned in phylogenies. Seventeen of these trees supported a red algal origin of chromalveolate plastid-targeted proteins (e.g., see Figure 6), whereas two proteins were of green algal origin (Li *et al.* 2005). The most important evidence for, or against, chromalvoelate monophyly will come from ongoing work in several groups using nuclear genes to position these protists in the eukaryotic tree of life.

IV. TERTIARY PLASTID ENDOSYMBIOSIS

Tertiary endosymbiosis occurs when an alga containing a plastid of secondary endosymbiotic origin (e.g., chromists) is engulfed and reduced to the photosynthetic organelle. Only dinoflagellates are known until now to have tertiary plastids. Plastid-containing dinoflagellates make up approximately one-half of the known taxa and are among the most environmentally and economically important of these protists. The majority of plastid-containing dinoflagellates contain the photopigment peridinin, with the remainder having originated relatively recently through tertiary endosymbioses (Schnepf and Elbrachter 1999). Tertiary plastids in dinoflagellates have been acquired from haptophyte (e.g., *Karenia, Karlodinium*, and *Takayama*) and prasinophyte (*Lepidodinium viride*) algae and from diatoms (*Peridinium foliaceum*) (Watanabe *et al.* 1990, 1991). Currently, there are five plastids known in this group, each with its own evolutionary history, making dinoflagellates the champions of plastid endosymbiosis among eukaryotes (see Hackett *et al.* 2004a). Most photosynthetic species contain chlorophylls *a* and c_2, the carotenoid beta-carotene, and a group of xanthophylls that appears to be unique to dinoflagellates, typically peridinin, dinoxanthin, and diadinoxanthin. These pigments give many dinoflagellates their typical golden-brown color. The peridinin-containing dinoflagellates have evolved a tripartite N-terminal extension containing two hydrophobic domains for targeting nuclear-coded plastid proteins to the organelle (Nassoury *et al.* 2003). The first hydrophobic domain acts like a typical transit peptide, targeting the preprotein to the ER. The second hydrophobic domain acts as a stop-transfer sequence, causing the plastid proteins to be cotranslationally inserted into the ER. Between these two hydrophobic domains is a region rich in serine and threonine that acts as a plastid signal peptide. Once the plastid proteins are inserted into the ER, they are processed through the Golgi apparatus and transported to the outer plastid membrane in Golgi-derived vesicles. These vesicles presumably fuse with the outer membrane of the plastid and the proteins are then pulled through the inner two membranes by a homologous TIC/TOC system.

The plastid genome in peridinin plastids is also remarkably different from that of other photosynthetic eukaryotes (see Delwiche, Chapter 10, this volume). Normally, plastids contain a circular genome that, although varying in complexity and genetic content, is about 150 kilobases (kb) in size and encodes approximately 100 genes. Even the plastid genomes of nonphotosynthetic

eukaryotes (e.g., *Plasmodium falciparum, Epifagus virginiana, Euglena longa*) are single circular molecules with reduced gene content, that is, lacking the genes involved in photosynthesis. In contrast, the plastid genome of peridinin-containing dinoflagellates is reduced and broken up into minicircles. Currently, only 14 proteins encoded on these minicircles have been found, in addition to the large subunit (LSU) and a putative small subunit (SSU) of the plastid ribosomal RNA, "empty"

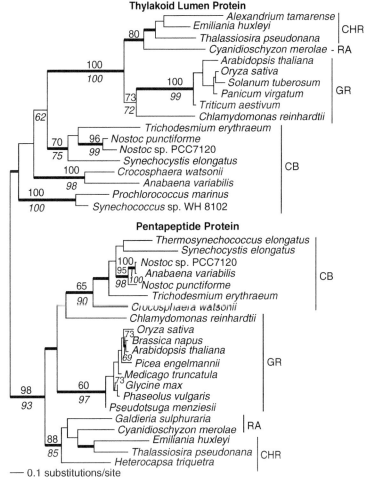

FIGURE 6. Examples of red algal endosymbiotic gene transfer in the chromalveolates (for details, see Li *et al.* 2005). **(A)** PhyML tree of the thylakoid lumen and the closely related pentapeptide proteins. These protein trees have been reciprocally rooted.

(Continued)

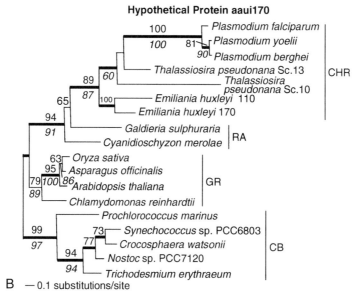

FIGURE 6. (Cont'd) **(B)** PhyML tree of the hypothetical protein "aaui170" from *E. huxleyi* and cyanobacterial and algal/plant homologs rooted with the cyanobacteria. The numbers above and below the branches are the results of PhyML and neighbor joining bootstrap analyses, respectively. The thick branches indicate >0.95 posterior probability from Bayesian inference. Only bootstrap values >60% are shown. Branch lengths are proportional to the number of substitutions per site (see scale bars). CHR, chromalveolates; GR, green algae/plants; RA, red algae; CB, cyanobacteria.

minicircles, and those encoding pseudogenes (Zhang *et al.* 1999, 2001, 2002; Barbrook and Howe 2000; Hiller 2001; Laatsch *et al.* 2004; Nisbet *et al.* 2004). These sequences code for the core subunits of the photosystem, cytochrome b_6f, ATP synthase complex (*atp*A, *atp*B, *pet*B, *pet*D, *psa*A, *psa*B, *psb*A-E, *psb*I) and four other proteins (*ycf*16, *ycf*24, *rpl*28, and *rpl*23). The remaining genes required for photosynthesis have been lost from the plastid and presumably moved to the nucleus. Remarkably, a recent article from Laatsch *et al.* (2004) provided evidence (based on partial sequences) that the minicircles in the peridinin dinoflagellate *Ceratium horridum* are present in the nucleus rather than in the plastid of this species. This raises the possibility that minicircle genes in different dinoflagellates may be found in either, or potentially both, plastids and nuclei. If true, this could reconcile the fact

that no components of the plastid transcription or translation apparatus have been found. However, no plastid targeting sequences are present at the N-terminus of these proteins, raising the question of how these proteins are localized to the plastid if their genes are encoded in the nucleus. Clearly, the extent and type of plastid gene transfer in different dinoflagellates needs to be carefully examined to understand fully plastid evolution in this lineage.

Because of the presence of chlorophyll *c*, it has been proposed that the peridinin plastid of dinoflagellates and the plastids of the chromists (cryptophytes, haptophytes, and Stramenopiles) share a common ancestor through secondary endosymbiosis of a red alga (and the subsequent evolution of chlorophyll *c*; Cavalier-Smith 1999 [see previous]). According to the chromalveolate hypothesis, the ancestral plastid of red algal origin was maintained in the chromist lineage and went through significant changes in the alveolates. The plastid was lost in the ciliates and reduced to the nonphotosynthetic apicoplast in the apicomplexans. The dinoflagellates evolved a tripartite targeting signal to shuttle proteins to the plastid, which was no longer inside the ER. The minicircle genes may have provided the best answer to the question of the origin of the peridinin plastid. Zhang *et al.* (1999) did phylogenetic analyses using a concatenated set of seven minicircle genes and found that the peridinin plastid was sister to the chromists and red algae. A red algal origin remains the most parsimonious solution to the provenance of minicircle genes, a hypothesis that is supported by the recent analysis of a multigene dataset of minicircle and chromist plastid proteins (Yoon *et al.* 2005).

Analyses of nuclear plastid-targeted genes have supported a specific relationship between chromist and alveolate plastid genes (Fast *et al.* 2001; Harper and Keeling 2003). However, the internal relationships among the chromists and dinoflagellates are poorly resolved and do not clearly show chromists and alveolates as sisters. In contrast, analyses of light-harvesting proteins, plastid SSU rDNA, and plastid *atp*I showed a specific relationship between the Stramenopiles and peridinin dinoflagellates to the exclusion of the cryptophytes and/or haptophytes (Durnford *et al.* 1999; Tengs *et al.* 2000; Hackett *et al.* 2004b). This may potentially indicate that the plastids of the alveolates are more closely related to Stramenopiles. It is still unclear, however, if these results stem from phylogenetic artifacts, lateral transfers of Stramenopile genes, or a tertiary endosymbiosis of a Stramenopile that gave rise to the peridinin plastid.

V. SUMMARY

Plastid endosymbiosis has spread photosynthesis to many branches of the eukaryotic tree of life. The timing, frequency, and ecological drivers of these landmark events remain poorly constrained. The presently available evidence suggests that a single primary endosymbiosis gave rise to the Plantae, comprising the glaucophytes, red algae, and Viridiplantae. The singular origin of plastids in eukaryotes appears to have occurred around 1.5 Ga; however, photosynthetic eukaryotes appear to have not become significant primary producers for many millions of years later. The descendants of this first association between a cyanobacterium and a eukaryote have evolved to play a major role in aquatic and terrestrial ecosystems and have further driven the evolution of eukaryotes by contributing their plastids through secondary endosymbiosis. Again, the timing and frequency of these events are unclear; however, it appears the green algae have supplied plastids independently to two lineages (chloroarachniophytes and euglenids). Secondary plastids of red algal origin are found in five lineages (cryptophytes, haptophytes, Stramenopiles, apicomplexans, and dinoflagellates) and may have arisen through a single secondary endosymbiosis. Whereas some of these lineages are relatively rare in nature (chloroarachniophytes, cryptophytes), others have become dominant members of marine ecosystems (haptophytes, stramenopiles, dinoflagellates) or significant pathogens (apicomplexans). Plastid endosymbiosis has clearly been a driving force in eukaryotic evolution and instrumental to the success of many eukaryotic groups and has influenced the evolution of other organisms that utilize these organisms for food or habitat. Whereas the primary endosymbiosis appears to have been a single event, subsequent endosymbioses involving eukaryotes (secondary and tertiary) appear to have been more frequent and ongoing. Members of the dinoflagellates have more recently acquired as many as four plastids from other eukaryotes (Hackett *et al.* 2004a), and often maintain temporary plastids from prey or kleptoplasts (e.g., Koike *et al.* 2005). These tertiary plastids show varying morphology and may represent inter-mediate stages in the reduction of an endosymbiont to an organelle. A recent report describes the flagellate "Hatena," which may be in the process of acquir-ing a secondary plastid from a green alga (Okamoto and Inouye 2005). The filose amoeba *Paulinella chromatophora* may be the only other example of primary endosym-biosis other than the ancient one that gave rise to the Plantae. This amoeba contains a photosynthetic organelle derived from a cyanobacterium that resembles the cyanelle in glaucophytes (Kies 1974; Bhattacharya *et al*. 1995; Marin *et al*. 2005; Yoon *et al*. 2006). Ongoing studies of *Paulinella* may shed light on the early stages of primary endosymbio-sis, a process that has profoundly impacted evolution of life on Earth.

References

Adl, S.M., Simpson, A.G., Farmer, M.A., and 25 others. (2005). The new higher level classification of eukary-otes with emphasis on the taxonomy of protists. *J. Eukaryot. Microbiol*. **52**: 399–451.

Allen, J.F., Puthiyaveetil, S., Strom, J., and Allen, C.A. (2005). Energy transduction anchors genes in organelles. *Bioessays* **27**: 426–435.

Allen, J.F., and Raven, J.A. (1996). Free-radical-induced mutation vs redox regulation: costs and benefits of genes in organelles. *J. Mol. Evol*. **42**: 482–492.

Arisue, N., Hasegawa, M., and Hashimoto, T. (2005). Root of the Eukaryota tree as inferred from combined maximum likelihood analyses of multiple molecular sequence data. *Mol. Biol. Evol*. **22**: 409–420.

Arisue, N., Hashimoto, T., Yoshikawa, H., Nakamura, G., Nakamura, F., Yano, T.A., and Hasegawa, M. (2002). Phylogenetic position of *Blastocystis hominis* and of stramenopiles inferred from multiple molecular sequence data. *J. Eukaryot. Microbiol*. **49**, 42–53.

Azam, F., Fenchel, T., Field, J.G., Gray, J.S., Meyer-Reil, L.A., and Thingstad, F. (1983). The ecological role of water column microbes in the sea. *Mar. Ecol. Prog. Ser*. **10**: 257–264.

Bachvaroff, T.R., Sanchez Puerta, M.V., and Delwiche, C.F. (2005). Chlorophyll *c* containing plastid rela-tionships based on analyses of a multi-gene dataset with all four chromalveolate lineages. *Mol. Biol. Evol*. **22**: 1772–1782.

Baldauf, S.L., Roger, A.J., Wenk-Siefert, I., and Doolittle, W.F. (2000). A kingdom-level phylogeny of eukaryotes based on combined protein data. *Science* **290**: 972–977.

Barbier, G., Oesterhelt, C., Larson, M.D., Halgren, R.G., Wilkerson, C., Garavito, R.M., Benning, C., and Weber, A.P. (2005). Comparative genomics of two closely related unicellular thermo-acidophilic red algae, *Galdieria sulphuraria* and *Cyanidioschyzon merolae*, reveals the molecular basis of the metabolic flexibility of *Galdieria sulphuraria* and significant dif-ferences in carbohydrate metabolism of both algae. *Plant Physiol*. **137**: 460–474.

Barbrook, A.C., and Howe, C.J. (2000). Minicircular plastid DNA in the dinoflagellate *Amphidinium oper-culatum*. *Mol. Gen. Genet*. **263**: 152–158.

Benton, M.J. (1993). *Fossil Record 2*. London, Chapman and Hall.

Bhattacharya, D., Helmchen, T., and Melkonian, M. (1995). Molecular evolutionary analyses of nuclear-encoded small subunit ribosomal RNA identify an independent rhizopod lineage containing the Eug-lyphina and the Chlorarachniophyta. *J. Eukaryot. Microbiol*. **42**: 65–69.

Bhattacharya, D., and Medlin, L.K. (1995). The phyl-ogeny of plastids: a review based on comparisons of small-subunit ribosomal RNA coding regions. *J. Phycol*. **31**: 489–498.

Bhattacharya, D., Yoon, H.S., and Hackett, J D. (2004). Photosynthetic eukaryotes unite: endosymbiosis connects the dots. *Bioessays* **26**: 50–60.

Blair, J.E., Shah, P., and Hedges, S.B. (2005). Evolu-tionary sequence analysis of complete eukaryote genomes. *BMC Bioinformatics* **6**: 53.

Bowe, L.M., Coat, G., and dePamphilis, C.W. (2000). Phylogeny of seed plants based on all three genomic compartments: extant gymno-sperms are monophyletic and Gnetales' closest relatives are conifers. *Proc. Natl. Acad. Sci. U S A* **97**: 4092–4097.

Brasier, M.D., and Lindsay, J.F. (1998). A billion years of environmental stability and the emergence of eukaryotes: new data from northern Australia. *Geol-ogy* **26**: 555–558.

Brocks, J.J., Logan, G.A., Buick, R., and Summons, R.E. (1999). Archean molecular fossils and the early rise of eukaryotes. *Science* **285**: 1033–1036.

Brocks, J.J., Love, G.D., Summons, R.E., Knoll, A.H., Logan, G.A., and Bowden, S.A. (2005). Biomarker evidence for green and purple sulphur bacteria in a stratified Palaeoproterozoic sea. *Nature* **437**: 866–870.

Buick, R., Thornett, J.R., McNaughton, N.J., Smith, J.B., Barley, M.E., and Savage, M. (1995). Record of emergent continental crust similar to 3.5 billion years ago in the Pilbara craton in Australia. *Nature* **375**: 574–577.

Butterfield, N.J. (2001). Paleobiology of the late Meso-proterozoic (ca. 1200 Ma) Hunting Formation, Som-erset Island, arctic Canada. *Precambrian Res*. **111**: 235–256.

Butterfield, N.J. (2004). A vaucheriacean alga from the middle Neoproterozoic of Spitsbergen: implications for the evolution of Proterozoic eukaryotes and the Cambrian explosion. *Paleobiology* **30:** 231–252.

Butterfield, N.J. (2005). Probable Proterozoic fungi. *Paleobiology* **31:** 165–182.

Cavalier-Smith, T. (1986). The kingdom Chromista: origin and systematics. *Progress in Phycological Research.* F. E. Round, D. J. Chapman, eds. Bristol, Biopress, pp. 309–347.

Cavalier-Smith, T. (1993). Kingdom Protozoa and its 18 phyla. *Microbiol. Rev.* **57:** 953–994.

Cavalier-Smith, T. (1999). Principles of protein and lipid targeting in secondary symbiogenesis: euglenoid, dinoflagellate, and sporozoan plastid origins and the eukaryote family tree. *J. Euk. Microbiol.* **46:** 347–366.

Cavalier-Smith, T. (2000). Membrane heredity and early chloroplast evolution. *Trends Plant Sci.* **5:** 174–182.

Cavalier-Smith, T. (2002). The neomuran origin of archaebacteria, the negibacterial root of the universal tree and bacterial megaclassification. *Int. J. Syst. Evol. Microbiol.* **52:** 7–76.

Delwiche, C.F. (1999). Tracing the thread of plastid diversity through the tapestry of life. *Am. Nat.* **154:** S164–S177.

Delwiche, C.F. (2007). The origin and evolution of dinoflagellates. *Evolution of Primary Producers in the Sea.* P.G. Falkowski and A.H. Knoll, eds. Boston, Elsevier, pp. 191–205.

Douglas, S., Zauner, S., Fraunholz, M., Beaton, M., Penny, S., Deng, L.T., Wu, X., Reith, M., Cavalier-Smith, T., and Maier, U.G. (2001). The highly reduced genome of an enslaved algal nucleus. *Nature* **410:** 1091–1096.

Douzery, E.J., Snell, E.A., Bapteste, E., Delsuc, F., and Philippe, H. (2004). The timing of eukaryotic evolution: does a relaxed molecular clock reconcile proteins and fossils? *Proc. Natl. Acad. Sci. U S A* **101:** 15386–15391.

Durnford, D.G., Deane, J.A., Tan, S., McFadden, G.I., Gantt, E., and Green, B.R. (1999). A phylogenetic assessment of the eukaryotic light-harvesting antenna proteins, with implications for plastid evolution. *J. Mol. Evol.* **48:** 59–68.

Fast, N.M., Kissinger, J.C., Roos, D.S., and Keeling, P.J. (2001). Nuclear-encoded, plastid-targeted genes suggest a single common origin for apicomplexan and dinoflagellate plastids. *Mol. Biol. Evol.* **18:** 418–426.

Flügge, U.I. (1998). Metabolite transporters in plastids. *Curr. Opin. Plant Biol.* **1:** 201–206.

Funes, S., Davidson, E., Reyes-Prieto, A., Magallon, S., Herion, P., King, M.P., and Gonzalez-Halphen, D. (2002). A green algal apicoplast ancestor. *Science* **298:** 2155.

Gibbs, S.P. (1978). The chloroplast of Euglena may have evolved from symbiotic green algae. *Can. J. Bot.* **56:** 2883–2889.

Gilson, P.R., and McFadden, G.I. (1996). The miniaturized nuclear genome of eukaryotic endosymbiont contains genes that overlap, genes that are cotranscribed, and the smallest known spliceosomal introns. *Proc. Natl. Acad. Sci. U S A* **93:** 7737–7742.

Goldman, N., Anderson, J.P., and Rodrigo, A.G. (2000). Likelihood-based tests of topologies in phylogenetics. *Syst. Biol.* **49:** 652–670.

Gray, M.W. (1992). The endosymbiont hypothesis revisited. *Int. Rev. Cyt.* **141:** 233–357.

Greenwood, A.D. (1974). The Cryptophyta in relation to phylogeny and photosynthesis. *Electron Microscopy 1974.* J.V. Sanders and D.J. Goodchild, eds. Canberra, Australian Academy of Sciences, pp. 566–567.

Guindon, S., and Gascuel, O. (2003). PHYML—A simple, fast, and accurate algorithm to estimate large phylogenies by maximum likelihood. *Syst. Biol.* **52:** 696–704.

Hackett, J.D., Anderson, D.M., Erdner, D.L., and Bhattacharya, D. (2004a). Dinoflagellates: a remarkable evolutionary experiment. *Am. J. Bot.* **91:** 1523–1534.

Hackett, J.D., Yoon, H.S., Soares, M.B., Bonaldo, M.F., Casavant, T.L., Scheetz, T.E., Nosenko, T., and Bhattacharya, D. (2004b). Migration of the plastid genome to the nucleus in a peridinin dinoflagellate. *Curr. Biol.* **14:** 213–218.

Hagopian, J.C., Reis, M., Kitajima, J.P., Bhattacharya, D., and de Oliveira, M.C. (2004). Comparative analysis of the complete plastid genome sequence of the red alga *Gracilaria tenuistipitata* var. liui provides insights into the evolution of rhodoplasts and their relationship to other plastids. *J. Mol. Evol.* **59:** 464–477.

Han, T.M., and Runnegar, B. (1992). Megascopic eukaryotic algae from the 2.1-billion-year-old negaunee iron-formation, Michigan. *Science* **257:** 232–235.

Harper, J.T., and Keeling, P.J. (2003). Nucleus-encoded, plastid-targeted glyceraldehyde-3-phosphate dehydrogenase (GAPDH) indicates a single origin for chromalveolate plastids. *Mol. Biol. Evol.* **20:** 1730–1735.

Hedges, S.B., Blair, J.E., Venturi, M.L., Shoe, J.L. (2004). A molecular timescale of eukaryotic evolution and the rise of complex multicellular life. *BMC Evol. Biol.* **4:** 2.

Hermann, T.N. (1990). *Organic World Billion Year Ago.* Leningrad, Nauka, pp. 1–49.

Hiller, R.G. (2001). "Empty" minicircles and petB/atpA and psbD/psbE (cytb559 alpha) genes in tandem in Amphidinium carterae plastid DNA. *FEBS Lett.* **505:** 449–452.

Holder, M.E., and Roger, A.J. (1999). PUZZLEBOOT. Available online at http://hades.biochem.dal.ca/Rogerlab/Software/software.html# puzzleboot

Huelsenbeck, J.P., and Ronquist, F. (2001). MrBayes: Bayesian inference of phylogenetic trees. *Bioinformatics* **17:** 754–755.

Huntley, J.W., Xiao, S., and Kowalewski, M. (2006). 1.3 Billion years of acritarch history: An empirical morphospace approach. *Precambrian Res.* **144:** 52–68.

Javaux, E.J., Knoll, A.H., and Walter, M.R. (2001). Morphological and ecological complexity in early eukaryotic ecosystems. *Nature* **412**: 66–69.

Johnson, P.W., Hargraves, P.E., and Sieburth, J.M. (1988). Ultrastructure and ecology of *Calycomonas ovalis* Wulff, 1919, (Chrysophyceae) and its redescription as a testate rhizopod, *Paulinella ovalis* n. comb. (Filosea: Euglyphina). *J. Protozool.* **35**: 618–626.

Jones, D.T., Taylor, W.R., and Thornton, J.M. (1992). The rapid generation of mutation data matrices from protein sequences. *Comput. Appl. Biosci.* **8**: 275–282.

Kenrick, P., and Crane, P.R. (1997). The origin and early evolution of plants on land. *Nature* **389**: 33–39.

Kies, L. (1974). Elektronenmikroskopische Untersuchungen an *Paulinella chromatophora* Lauterborn, einer Thekamöbe mit blaugrünen Endosymbionten (Cyanellen). *Protoplasma* **80**: 69–89.

Kies, L., and Kremer, B.P. (1990). Phylum Glaucocsytophyta. *Handbook of Protoctista*. L. Margulis, J.O. Corliss, M. Melkonian, and D.J. Chapman, eds. Boston, Jones and Bartlett, pp. 152–166.

Kishino, H., and Hasegawa, M. (1989). Evaluation of the maximum likelihood estimate of the evolutionary tree topologies from DNA sequence data, and the branching order in hominoidea. *J. Mol. Evol.* **29**: 170–179.

Kishino, H., Thorne, J.L., and Bruno, W.J. (2001). Performance of a divergence time estimation method under a probabilistic model of rate evolution. *Mol. Biol. Evol.* **18**: 352–361.

Knoll, A.H. (1994). Proterozoic and early Cambrian protists: evidence for accelerating evolutionary tempo. *Proc. Natl. Acad. Sci. U S A* **91**: 6743–50.

Knoll, A.H., Summon, F.E., Waldbauer, J.R., and Zumberge, J.E. (2007). The geological succession of primary producers in the sea. *Evolution of Primary Producers in the Sea*. P.G. Falkowski and A.H. Knoll, eds. Boston, Elsevier, pp. 133–163.

Kohler, S., Delwiche, C.F., Denny, P.W., Tilney, L.G., Webster, P., Wilson, R.J., Palmer, J.D., and Roos, D.S. (1997). A plastid of probable green algal origin in Apicomplexan parasites. *Science* **275**: 1485–1489.

Koike, K., Sekiguchi, H., Kobiyama, A., Takishita, K., Kawachi, M., Koike, K., and Ogata, T. (2005). A novel type of kleptoplastidy in *Dinophysis* (Dinophyceae): Presence of haptophyte-type plastid in *Dinophysis mitra*. *Protist* **156**: 225–237.

Laatsch, T., Zauner, S., Stoebe-Maier, B., Kowallik, K.V., and Maier, U.G. (2004). Plastid-derived single gene minicircles of the dinoflagellate *Ceratium horridum* are localized in the nucleus. *Mol. Biol. Evol.* **21**: 1318–1322.

Li, S., Nosenko, T., Hackett, J.D., and Bhattacharya, D. (2005). Phylogenomic analysis identifies red algal genes of enosymbiotic origin in the chromalveolates. *Mol. Biol. Evol.* **23**: 663–674.

Marin, B., Nowack, E.C.M., and Melkonian, M. (2005). A plastid in the making: evidence for a second primary endosymbiosis. *Protist* **156**: 425–432.

Martin, W. (2007). Eukaryote and mitochondrial origins: Two sides of the same coin and too much ado about oxygen. *Evolution of Primary Producers in the Sea*. P.G. Falkowski and A.H. Knoll, eds. Boston, Elsevier, pp. 55–73.

Martin, W., and Herrmann, R.G. (1998). Gene transfer from organelles to the nucleus: how much, what happens, and why? *Plant Physiol.* **118**: 9–17.

Martin, W., Rotte, C., Hoffmeister, M., Thiessen, U., Gelius-Dietrich, G., Ahr, S., and Henze, K. (2003). Early cell evolution, eukaryotes, anoxia, sulfide, oxygen, fungi first (?), and a tree of genomes revisited. *IUBMB Life* **55**: 193–204.

Martin, W., Rujan, T., Richly, E., Hansen, A., Cornelsen, S., Lins, T., Lesister, D., Stoebe, B., Hasegawa, M., and Penny, D. (2002). Evolutionary analysis of *Arabidopsis*, cyanobacterial, and chloroplast genomes reveals plastid phylogeny and thousands of cyanobacterial genes in the nucleus. *Proc. Natl. Acad. Sci. U S A* **99**: 12246–12251.

Martin, W., Stoebe, B., Goremykin, V., Hansmann, S., Hasegawa, M., and Kowallik, K. (1998). Gene transfer to the nucleus and the evolution of chloroplasts. *Nature* **393**: 162–165.

Matsuzaki, M., Misumi, O., Shin, I.T., and 39 others. (2004). Genome sequence of the ultrasmall unicellular red alga *Cyanidioschyzon merolae* 10D. *Nature* **428**: 653–657.

McFadden, G.I. (2001). Primary and secondary endosymbiosis and the origin of plastids. *J. Phycol.* **37**: 951–959.

McFadden, G.I., Gilson, P.R., Hofmann, C.J.B., Adcock, G.J., and Maier, U.G. (1994). Evidence that an amoeba acquired a chloroplast by retaining part of an engulfed eukaryotic alga. *Proc. Natl. Acad. Sci. U S A* **91**: 3690–3694.

McFadden, G.I., and van Dooren, G.G. (2004). Evolution: Red algal genome affirms a common origin of all plastids. *Curr. Biol.* **14**: R514–516.

McFadden, G.I., Waller, R.F., Reith, M.E., and Lang-Unnasch, N. (1997). Plastids in apicomplexan parasites. *Origin of Algae and Their Plastids*. D. Bhattacharya, ed. New York, Springer-Wein, pp. 261–287.

Miyagishima, S.Y., Nozaki, H., Nishida, K., Matsuzaki, M., and Kuroiwa, T. (2004). Two types of FtsZ proteins in mitochondria and red-lineage chloroplasts: the duplication of FtsZ is implicated in endosymbiosis. *J. Mol. Evol.* **58**: 291–303.

Moreira, D., Le Guyader, H., and Phillippe, H. (2000). The origin of red algae and the evolution of chloroplasts. *Nature* **405**: 69–72.

Muller, H.J. (1932). Some genetic aspects of sex. *Am. Nat.* **66**: 118–138.

Nassoury, N., Cappadocia, M., and Morse, D. (2003). Plastid ultrastructure defines the protein import pathway in dinoflagellates. *J. Cell Sci.* **116**: 2867–2874.

Nelson, N., and Ben-Shem, A. (2004). The complex architecture of oxygenic photosynthesis. *Nat. Rev. Mol. Cell Biol.* **5**: 971–982.

Nisbet, R.E., Koumandou, L.V., Barbrook, A.C., and Howe, C.J. (2004). Novel plastid gene minicircles in the dinoflagellate *Amphidinium operculatum*. *Gene* **331**: 141–147.

Nosenko, T., Lidie, K.L., Van Dolah, F.M., Lindquist, E., Cheng, J.-F., U.S. Department of Energy-Joint Genome Institute, and Bhattacharya, D. (In press). Chimeric plastid proteome in the Florida "red tide" dinoflagellate *Karenia brevis*. *Mol. Biol. Evol.* **23**: 2026–2038.

Ohta, N., Matsuzaki, M., Misumi, O., Miyagishima, S.Y., Nozaki, H., Tanaka, K., Shin, I.T., Kohara, Y., and Kuroiwa, T. (2003). Complete sequence and analysis of the plastid genome of the unicellular red alga *Cyanidioschyzon merolae*. *DNA Res.* **10**: 67–77.

Okamoto, N., and Inouye, I. (2005). A secondary symbiosis in progress? *Science* **310**: 287.

Palmer, J.D. (2003). The symbiotic birth of plastids: How many times and whodunit? *J. Phycol.* **39**: 4–11.

Peterson, K.J., and Butterfield, N.J. (2005). Origin of the Eumetazoa: testing ecological predictions of molecular clocks against the Proterozoic fossil record. *Proc. Natl. Acad. Sci. U S A* **102**: 9547–9552.

Pfanzagl, B., Zenker, A., Pittenauer, E., Allmaier, G., Martinez-Torrecuadrada, J., Schmidt, E.R., Depedro, J.A., and Löffelhardt, W. (1996). Primary structure of cyanelle peptidoglycan of *Cyanophora paradoxa*: a prokaryotic cell wall as part of an organelle envelope. *J. Bacteriol.* **178**: 332–339.

Reumann, S., Inoue, K., and Keegstra, K. (2005). Evolution of the general protein import pathway of plastids. *Mol. Membr. Biol.* **22**: 73–86.

Reyes-Prieto, A., and Bhattacharya, D. Phylogeny of nuclear encoded plastid targeted proteins demonstrates an early divergence of glaucophytes within plantae. In preparation.

Rodriguez-Ezpeleta, N., Brinkmann, H., Burey, S.C., Roure, B., Burger, G., Löffelhardt, W., Bohnert, H.J., Philippe, H., and Lang, B.F. (2005). Monophyly of primary photosynthetic eukaryotes: green plants, red algae, and glaucophytes. *Curr. Biol.* **15**: 1325–1330.

Samuelsson, J., and Butterfield, N.J. (2001). Neoproterozoic fossils from the Franklin Mountains, northwestern Canada: stratigraphic and palaeobiological implications. *Precambrian Res.* **107**: 235.

Sanchez Puerta, M.V., Bachvaroff, T.R., and Delwiche, C.F. (2004). The complete mitochondrial genome sequence of the haptophyte *Emiliania huxleyi* and its relation to heterokonts. *DNA Res.* **11**: 1–10.

Sanderson, M.J. (2003). r8s: inferring absolute rates of molecular evolution and divergence times in the absence of a molecular clock. *Bioinformatics* **19**: 301–302.

Sanderson, M.J., and Doyle, J.A. (2001). Sources of error and confidence intervals in estimating the age of angiosperms from rbcL and 18S rDNA data. *Am. J. Bot.* **88**: 1499–1516.

Schneider, D.A., Bickford, M.E., Cannon, W.F., Schulz, K.J., and Hamilton, M.A. (2002). Age of volcanic rocks and syndepositional iron formations, Marquette Range Supergroup: implications for the tectonic setting of Paleoproterozoic iron formations of the Lake Superior region. *Can. J. Earth Sci.* **39**: 999–1012.

Schnepf, E., and Elbrachter, M. (1999). Dinophyte chloroplasts and phylogeny—a review. *Grana* **38**: 81–97.

Stechmann, A., and Cavalier-Smith, T. (2002). Rooting the eukaryote tree by using a derived gene fusion. *Science* **297**: 89–91.

Stibitz, T.B., Keeling, P.J., and Bhattacharya, D. (2000). Symbiotic origin of a novel actin gene in the cryptophyte *Pyrenomonas helgolandii*. *Mol. Biol. Evol.* **17**: 1731–1738.

Summons, R.E., Jahnke, L.L., Hope, J.M., and Logan, G.A. (1999). 2-Methylhopanoids as biomarkers for cyanobacterial oxygenic photosynthesis. *Nature* **400**: 554–557.

Swofford, D.L. (2003). *PAUP*: Phylogenetic Analysis Using Parsimony (* and Other Methods)*. Sunderland, MA, Sinauer.

Taylor, T.N., Hass, H., and Kerp, H. (1999). The oldest fossil ascomycetes. *Nature* **399**: 648.

Tengs, T., Dahlberg, O.J., Shalchian-Tabrizi, K., Klaveness, D., Rudi, K., Delwiche, C.F., and Jakobsen, K.S. (2000). Phylogenetic analyses indicate that the 19'Hexanoyloxy-fucoxanthin- containing dinoflagellates have tertiary plastids of haptophyte origin. *Mol. Biol. Evol.* **17**: 718–729.

Thorne, J.L., Kishino, H., and Painter, I.S. (1998). Estimating the rate of evolution of the rate of molecular evolution. *Mol. Biol. Evol.* **15**: 1647–1657.

Van de Peer, Y., and De Wachter, R. (1997). Evolutionary relationships among the eukaryotic crown taxa taking into account site-to-site rate variation in 18S rRNA. *J. Mol. Evol.* **45**: 619–630.

Van de Peer, Y., Rensing, S.A., Maier, U.G., and De Wachter, R. (1996). Substitution rate calibration of small subunit ribosomal RNA identifies chlorarachniophyte endosymbionts as remnants of green algae. *Proc. Natl. Acad. Sci. U S A* **93**: 7732–7736.

Watanabe, M.M., Sasa, T., Suda, S., Inouye, I., and Takichi, S. (1991). Major carotenoid composition of an endosymbiont is a green dinoflagellate, *Lepidodinium viride*. *J. Phycol.* **27**: 75.

Watanabe, M.M., Suda, S., Inouye, I., Sawaguchi, T., and Chihara, M. (1990). *Lepidodinium viride* new genus new species Gymnodiniales Dinophyta a green dinoflagellate with a chlorophyll a-containing and chlorophyll b-containing endosymbiont. *J. Phycol.* **26**: 741–751.

Weber, A.P.M., Linka, M., and Bhattacharya, D. (2006). Single, ancient origin of a plastid metabolite translocator family in Plantae from an endomembrane-derived ancestor. *Eukaryot. Cell* **5**: 609–612.

Whelan, S., and Goldman, N. (2001). A general empirical model of protein evolution derived from multiple protein families using a maximum-likelihood approach. *Mol. Biol. Evol.* **18:** 691–699.

Yoon, H.S., Hackett, J.D., Ciniglia, C., Pinto, G., and Bhattacharya, D. (2004). A molecular timeline for the origin of photosynthetic eukaryotes. *Mol. Biol. Evol.* **21:** 809–818.

Yoon, H.S., Hackett, J.D., Pinto, G., and Bhattacharya, D. (2002). The single, ancient origin of chromist plastids. *Proc. Natl. Acad. Sci. U S A* **99:** 15507–15512.

Yoon, H.S., Hackett, J.D., Van Dolah, F.M., Nosenko, T., Lidie, K.L., and Bhattacharya, D. (2005). Tertiary endosymbiosis driven genome evolution in dinoflagellate algae. *Mol. Biol. Evol.* **22:** 1299–1308.

Yoon, H.S., Reyes-Prieto, A., Melkonian, M., and Bhattacharya, D. (2006). Minimal plastid genome evolution in the *Paulinella* endosymbiont. *Curr. Biol.* **16:** R670–R672.

Zhang, Z., Cavalier-Smith, T., and Green, B.R. (2001). A family of selfish minicircular chromosomes with jumbled chloroplast gene fragments from a dinoflagellate. *Mol. Biol. Evol.* **18:** 1558–1565.

Zhang, Z., Cavalier-Smith, T., and Green, B.R. (2002). Evolution of dinoflagellate unigenic minicircles and the partially concerted divergence of their putative replicon origins. *Mol. Biol. Evol.* **19:** 489–500.

Zhang, Z., Green, B.R., and Cavalier-Smith, T. (1999). Single gene circles in dinoflagellate chloroplast genomes. *Nature* **400:** 155–159.

CHAPTER

8

The Geological Succession of Primary Producers in the Oceans

ANDREW H. KNOLL, ROGER E. SUMMONS,
JACOB R. WALDBAUER, AND JOHN E. ZUMBERGE

In the modern oceans, diatoms, dinoflagellates, and coccolithophorids play prominent roles in primary production (Falkowski *et al.* 2004). The biological observation that these groups acquired photosynthesis via endosymbiosis requires that they were preceded in time by other photoautotrophs. The geological observation that the three groups rose to

geobiological prominence only in the Mesozoic Era also requires that other primary producers fueled marine ecosystems for most of Earth history. The question, then, is: what did primary production in the oceans look like before the rise of modern phytoplankton groups?

In this chapter, we explore two records of past primary producers: morphological fossils and molecular biomarkers. Because these two windows on ancient biology are framed by such different patterns of preservational bias and diagenetic selectivity, they are likely to present a common picture of stratigraphic variation only if that view reflects evolutionary history.

I. RECORDS OF PRIMARY PRODUCERS IN ANCIENT OCEANS

A. Microfossils

Microfossils, preserved as organic cell walls or mineralized tests and scales, record the morphologies and (viewed via transmission electron microscopy) ultrastructures of ancient microorganisms. Such fossils can provide unambiguous records of phytoplankton in past oceans—diatom frustules, for example—and they commonly occur in large population sizes, with numerous occurrences that permit fine stratigraphic resolution and wide geographic coverage.

Set against this is a number of factors that limit interpretation. Not all photoautotrophs produce preservable cell walls or scales, and, of those that do, not all generate fossils that are taxonomically diagnostic. Thus, although many modern diatoms precipitate robust frustules of silica likely to enter the geologic record, others secrete weakly mineralized shells with a correspondingly lower probability of preservation. Similarly, whereas dinoflagellates as a group have left a clear record of dinocysts, many extant species do not produce preservable cysts and others form cysts that would not be recognized unambiguously as dinoflagellate in fossil assemblages. (The phylogenetic affinities of fossil dinocysts are established by the presence of an archeopyle, a distinctive excystment mechanism peculiar to but not universally found within dinoflagellates.) Especially in the early history of a group, character combinations that readily distinguish younger members may not be in place. Thus, stem group diatoms

without well-developed frustules might well leave no morphologic record at all in sediments.

By virtue of their decay-resistant extracellular sheaths and envelopes, many cyanobacteria have a relatively high probability of entering the fossil record, and some benthic lineages are both readily preservable and morphologically distinctive (Knoll and Golubic 1992). On the other hand, important picoplankton such as *Prochlorococcus* are unlikely to leave recognizable body (or, as it turns out, molecular) fossils. A number of algal clades include good candidates for fossilization, especially groups with distinctive resting stages (phycomate prasinophytes, dinoflagellates) or mineralized skeletons (diatoms, coccolithophorids, coralline reds, caulerpalean and dasyclad greens). Other primary producers fossilize occasionally but only under unusual depositional or diagenetic circumstances (e.g., Butterfield 2000; Xiao *et al.* 2002, 2004; Foster and Afonin 2006), and still others rarely if ever produce morphologically interpretable fossils.

Diagenesis can obliterate fossils as well as preserve them: organic walls are subject to postdepositional oxidation and mineralized skeletons may dissolve in undersaturated pore waters. The result is that presence and absence cannot be weighted equally in micropaleontology. The presence of a fossil unambiguously shows that the cell from which it derived lived at a certain time in a particular place, but absence may reflect true absence, low probability of fossilization, or obfuscating depositional or diagenetic conditions. For older time intervals, tectonic destruction of the sedimentary record imposes an additional challenge; in particular, subduction inexorably destroys oceanic crust and the sediments that mantle it, so that deep-sea sediments are common only in Jurassic and younger ocean basins.

B. Molecular Biomarkers

The chemical constituents of biomass produced by living organisms can be incorporated into sediments and ultimately into sedimentary rocks that can survive for billions of years. Where these compounds are preserved in recognizable forms, they represent another opportunity for organisms to leave a trace of themselves in the fossil record. Organic biomarkers are the diagenetically altered remains of the products of cellular biosynthesis and may be aptly termed molecular fossils. Most biomarkers are derived from lipids and are potentially stable over billion-year time scales under ideal conditions (Brocks and Summons 2004).

Given the variety of organic compounds produced by cells and the vast quantities of sedimentary organic matter in a rock record that stretches back billions of years, biomarkers are a potentially rich source of information concerning the diversity and ecology of ancient communities. However, the process of organic matter incorporation into rocks and its transformation during deep burial imposes some strong constraints on the kinds of information that can be recovered millions of years after the fact. The classes of molecules that contain molecular sequence information, nucleic acids, and most proteins, do not survive long in the geologic environment. DNA can survive for at least a few hundred thousand years, especially in reducing environments such as euxinic sediments (e.g., Coolen *et al.* 2004) where heterotrophy is curtailed by a lack of electron acceptors, but it is not an option where the aim is to look at changes on million-year or longer time scales. Other kinds of biomolecules, however, prove remarkably resilient in the rock record.

Any molecule with a hydrocarbon skeleton has the potential to be preserved over long periods. For the most part, this refers to the hydrocarbon portions of membrane lipids, which are the major constituent of extractable organic matter (bitumen) in sedimentary rocks. Diagenesis quickly strips these compounds of their reactive polar functionalities, and over longer periods causes stereochemical and structural rearrangements, but hydrocarbon skeletons

can remain recognizable as the products of particular biosynthetic pathways on time scales that approach the age of the Earth (e.g., Brocks and Summons 2004; Peters *et al.* 2005).

The character of the information contained in molecular fossils is variable. Some are markers for the presence and, to the extent they can be quantified relative to other inputs, abundance of particular organisms. The taxonomic specificity of such biomarkers ranges from species to domain level. Others are markers for the operation of a particular physiology or biosynthetic pathway that may have a broad and/or patchy taxonomic distribution. Still other kinds of biomarkers are most strongly associated with specific depositional settings, making their presence more indicative of paleoenvironmental conditions than of any particular biology. Interpretation of the molecular fossil record depends on our ability to recognize biomarker compounds, link them to biosynthetic precursors, and then make inferences about what the presence of those molecules in the rock record tells us about contemporary biology and geochemistry.

Turning to biomarkers that might establish a molecular fossil record of primary production in marine settings, several classes of compounds are promising for their combination of biochemical and/or taxonomic specificity. Pigments are natural candidates, representing markers of the photosynthetic machinery itself. Input of chlorophyll to sediments can result in several kinds of molecular fossils, including porphyrins and the pristane and phytane skeletons of the chlorophyll side chain (Figure 1). It was the recognition of vanadyl porphyrin as the molecular fossil of chlorophyll that led Alfred Treibs (1936) to make the first compelling chemical argument for the biogenic origin of petroleum. Other pigments, such as carotenoids, are subject to very selective preservation, generally requiring the presence of reduced sulfur species; the functional groups that confer many of their biophysical properties and taxonomic

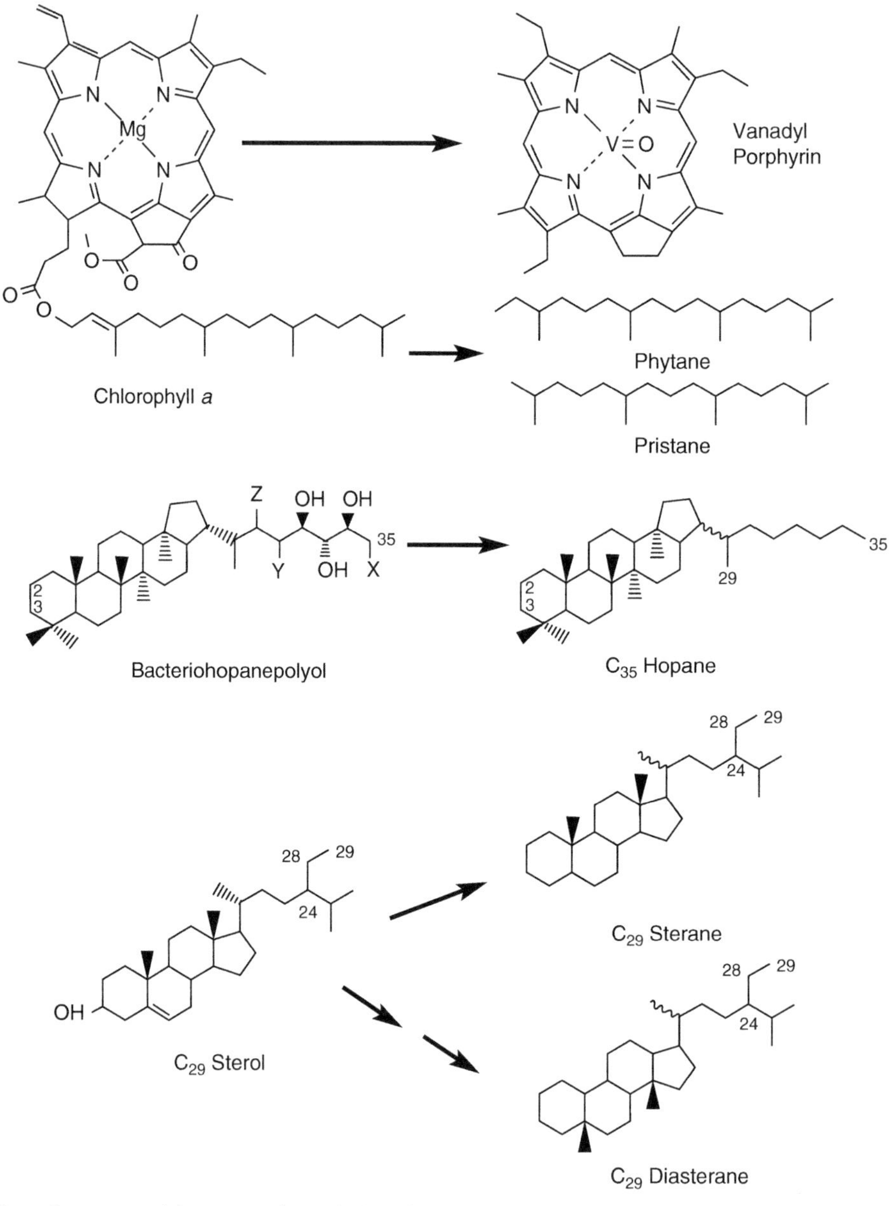

FIGURE 1.   Structures of diagnostic phytoplankton lipids (left) and their fossil counterparts (right).

specificity are lost through chemical reduction processes early in diagenesis (e.g., Kohnen *et al*. 1991, 1993; Hebting *et al*. 2006).

Although the preservation of pigment-derived biomarkers is spotty and information is steadily lost over time as diagenesis proceeds, another class of extraordinarily durable molecules provides us with much

of the molecular fossil record of primary producers, particularly in Paleozoic and older rocks. These are the polycyclic triterpenoids produced by the cyclization of the isoprenoid squalene and found in the membranes of both eukaryotes and bacteria. The main types are the steroids, which are ubiquitous among the Eucarya but known from only a very few bacteria, and the hopanoids, including the bacteriohopanepolyols (BHPs), produced by a wide variety of autotrophic and heterotrophic bacteria. These molecules have the great advantages of a durable polycyclic skeleton that is clearly a biological product and a well-characterized diagenetic fate involving a number of rearrangements that provide information about the postburial history of the organic matter. The structures of some commonly used biomarker lipids and their fossil counterparts are shown in Figure 1. Table 1 summarizes current knowledge of the biological affinities of hydrocarbons commonly found in marine sediment samples, excluding biomarkers derived from terrestrial organisms.

Molecular fossils suffer from some of the same limitations as body fossils. Not all ecologically and biogeochemically important groups leave distinctive molecular fingerprints, making them difficult to follow in time. Moreover, the structures of lipids are not nearly as diverse as body fossils; different, often distantly related or physiologically disparate organisms can produce similar patterns of lipids. Generally, biomarkers will reflect an average of inputs to sediments, which can be influenced by factors including bottom water oxygenation, sediment mineralogy, and grain surface area available for sorptive protection (Hedges and Keil 1995). These inputs are attenuated by remineralization of organic matter as it sinks: this reworking is >95% complete by 3000 m depth (Martin et al. 1987). The high degree of water-column degradation of organic matter in deep basins means that, even where deep-water sediments survive subduction, they commonly contain little organic matter; hence, the biomarker record of open-ocean primary production is poor. The problem of interpreting absence can be acute, because in biomarker analysis, absence can only be defined in terms of detection limits and, hence, is conditionally dependent upon the analytical technology available.

Much of what is known about the diagenesis and preservation of biomarkers derives from studies of the origin and composition of petroleum (e.g., Peters et al. 2005). Petroleum geologists were initially interested in identifying the source rocks from which hydrocarbon accumulations originated. Information on the thermal histories of source rocks is also key for modeling hydrocarbon generation. Biomarkers provide a way to determine both parameters through complementary analyses of sedimentary bitumen in source rocks and their derived oil accumulations. Companies serving the petroleum exploration industry have developed and maintained databases of bitumen and oil composition that can be used to compare oils to bitumen in their source horizons and model the thermal histories of petroleum deposits. These databases can be employed as a predictive tool when exploring in frontier regions. An example is the commercial "Oils" database generated by GeoMark Research, which records geochemical analyses of more than 10,000 crude oil samples from every known petroliferous basin on the globe (www.geomarkresearch.com). The "Oils" data comprise the contents of S, Ni, and V; the carbon isotopic compositions of bulk saturated and aromatic hydrocarbons; and quantitative analysis of approximately 100 individual hydrocarbons, including n-alkanes, acyclic isoprenoids, steroids, and triterpenoids. Abundances of the latter biomarkers, which have been determined using a rigorously reproducible analytical protocol, allow calculation of 23 diagnostic molecular ratios that can be used to predict paleoenvironmental features of an oil's source rock without direct knowledge of the rock itself (Zumberge 1987) or to evaluate hydrocarbon charge histories from field to basin scales (e.g., Zumberge et al. 2005). Averaging of data from numerous oil samples within a well, field, or an entire basin helps to overcome anomalies in individual hydrocarbon samples that reflect differences in maturity and losses from evaporation, water

TABLE 1.   Hydrocarbon biomarkers prevalent in marine sediments and petroleum derived from marine sediments and their known source organisms

Fossil hydrocarbon[a]	Functionalized precursors	Established sources[b,c]	References[d]
C_{27}–C_{35} bacteriohopanes	C_{35} bacteriohopanepolyols (BHPs)	Bacteria although nonspecific	Rohmer et al. 1984 1992
2-Methylhopanes	2-Methyl-BHP	Cyanobacteria although also in methanotrophs and other bacteria	Zundel and Rohmer 1985b, 1985c; Bisseret et al. 1985; Summons et al. 1999
3-Methylhopanes	3-Methyl-BHP	Methanotrophs, other proteobacteria	Zundel and Rohmer 19 85a
Aryl isoprenoids, isorenieratane	Aromatic carotenoids, e.g., isorenieratene, okenone	Green and purple sulfur bacteria	Summons and Powell 1987; Brocks et al. 2005
Gammacerane	Tetrahymanol	Purple nonsulfur bacteria, some protists	Ten Haven et al. 1989; Kleemann et al. 1990
Tricyclic terpanes, cheilanthanes	Unknown	Unknown, probably bacteria	Moldowan et al. 1983
<C_{20} acyclic isoprenoids	Bacterial and algal chlorophylls archaeol	Photosynthetic bacteria and protists Archaea although nonspecific	Peters et al. 2005
>C_{20} acyclic isoprenoids	Glycerol ether lipids; also found as free hydrocarbons	Archaea although nonspecific	Peters et al. 2005
Cholestane	Cholesterol and related C_{27} sterols	Photosynthetic protists, metazoa	Volkman 2003
Ergostane; 24-methylcholestane	Erogosterol and related C_{28} sterols	Photosynthetic protists, prevalent in diatoms	Volkman 2003
Stigmastane; 24-ethylcholestane	Sitosterol, stigmasterol, and related C_{29} sterols	Photosynthetic protists, prevalent in chlorophytes	Volkman 2003
24-n-Propylcholestane	24-n-Propylcholesterol	Marine chrysophytes	Moldowan 1984
Dinosterane, triaromatic dinosteroids	Dinosterol, dinostanol	Dinoflagellates	Summons et al. 1987, 1992; Moldowan and Talyzina 1988
24-Norcholestane	24-Norcholesterol and related 24-nor sterols	Diatoms	Rampen et al. 2005
$C_{20,}$ $C_{25,}$ and C_{30} highly branched isoprenoids (HBIs)	Mono- or polyunsaturated HBIs	Diatoms	Volkman et al. 1994; Belt et al. 2000; Sinninghe Damsté et al. 2004
n-C_{37}–C_{39} alkenones or alkanes	n-C_{37}–C_{39} alkenones	Haptophytes	Volkman et al. 1980; Marlowe et al. 1984

[a]In most cases, the connection between the fossil hydrocarbon and organismic source is supported by detection of diagenetic intermediates in sediments.

[b]Extensive, systematic studies of lipids from microbial cultures are rare. Many lipid–organism relationships remain unknown.

[c]Genomic sequencing will also help identify the metabolic potential of source organisms to produce preservable compounds; such identification depends on (currently incomplete) knowledge of the biosynthetic pathways of biomarker molecules.

[d]Indicative but incomplete list. Reviews that provide extensive citation lists include Brocks and Summons 2004 and Peters et al. 2005.

washing, or biodegradation. Although it is not widely appreciated, the global ubiquity of sedimentary bitumen and oil accumulations of all sizes means that hydrocarbons can also serve as a source of information on trends in the global carbon cycle (e.g., Andrusevich *et al.* 1998, 2000) and patterns in the evolution and environmental distributions of organisms, as discussed further later. The data illustrated in Figures 2, 3, 4, 5, and 6 come from the "Oils" database and represent the averages of numerous samples from a global selection of prominent Cenozoic to Proterozoic petroleum systems, both marine and lacustrine.

Inspection of trends in the "Oils" database suggests that some aspects of the composi-

tion of sedimentary hydrocarbons appear to be relatively invariant with age, changing instead with source rock lithology and sedimentary environment. These features are mostly reflected in the abundance patterns of bacterial and archaeal biomarkers. An example is depicted in Figure 2, which plots the relative abundance of a diagnostic bacteriohopane hydrocarbon (30-norhopane) as a function of paleolatitude. 30-Norhopane can arguably originate in several ways, but one particularly prolific source would be those BHPs with a hydroxyl substituent at position "Z" (see Figure 1); this makes them prone to oxidative cleavage, leading to a C_{29} hydrocarbon (Rohmer *et al.* 1992). Such precursor BHPs occur commonly in proteobacteria,

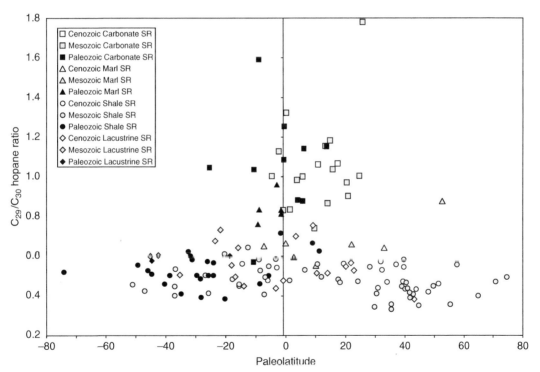

FIGURE 2. This figure depicts a diagnostic biomarker ratio derived from the averaged analyses of numerous oil samples representing important commercial petroleum accumulations plotted versus their paleolatitude. Samples are grouped according to geological era and classified according to the lithology and environment of the source rock, namely marine distal shales, marine marls, marine carbonates, or lacustrine sediments. The ratio of hopanes with 29 carbons to those with 30 carbons tends to be highest in carbonates and marls and in samples from low paleolatitude. Oils sourced from marine distal shales invariably show a C_{29}/C_{30} hopane ratio <0.7, whereas marine carbonates tend to have values of 0.8 or more. This pattern holds irrespective of age over the duration of the Phanerozoic.

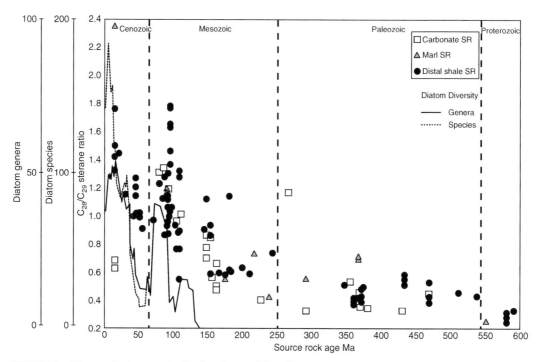

FIGURE 3. The secular increase in the abundance of C_{28} relative to C_{29} steranes over the Phanerozoic Eon, particularly during the last 250 million years, in petroleum systems from the GeoMark Oils database. Overlain are the fossil diatom species and genera diversity curves from Katz *et al.* (2004).

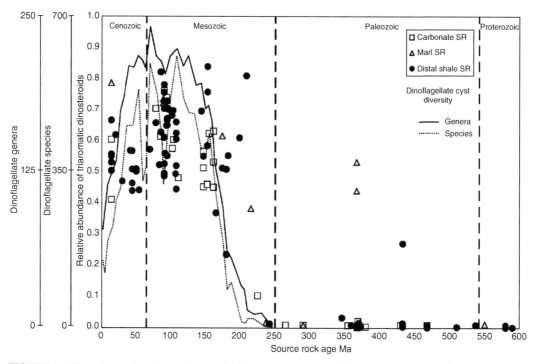

FIGURE 4. The relative abundance of aromatic dinosteranes over the Phanerozoic Eon from the GeoMark Oils database, showing marked increase during the early Mesozoic. Note concordance with genera- and species-level dinoflagellate fossil cyst diversity curves of Katz *et al.* (2004). Paleozoic occurrences of dinosteranes in petroleum sources are infrequent but merit further attention.

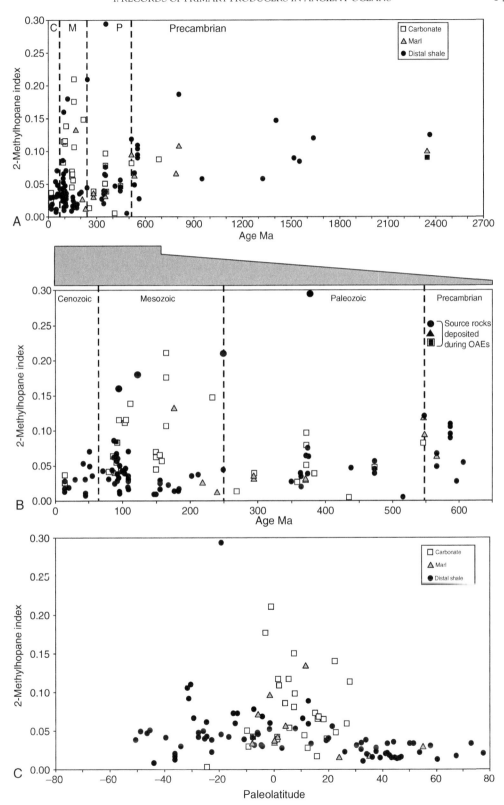

FIGURE 5. For legend see next page.

(Continued)

FIGURE 5. (Cont'd)   (A) C_{30} 2-methylhopane (2-MeH) index ($=C_{30}$ 2-MeH/(C_{30} 2-MeH + C_{30} desmethyl hopane) through geologic time. Values are generally elevated throughout the Precambrian, with distal shales reaching below 0.05 only during the Phanerozoic. Compilation includes data from the GeoMark Oils database and Kuypers et al. (2004) (B) Expansion of the time scale of (A) to focus on the Phanerozoic. Source rocks deposited during oceanic anoxic events (OAEs) are highlighted and often show elevated 2-MeH indices. (C) 2-MeH indices of Phanerozoic petroleum systems by source rock lithology and paleolatitude. The highest 2-MeH values are generally found at low latitudes, often in carbonate depositional environments.

including methylotrophs. They appear to be especially common in carbonate-precipitating sedimentary environments; the ratio of C_{29}/C_{30} hopane tends to be highest in oils from carbonates, intermediate in marls, and lowest in distal shales (Subroto et al. 1991). This relationship holds independent of the geological age of the source rocks. Carbonates accumulate predominantly in low latitudes, explaining the paleogeographic correspondence shown in Figure 2.

In contrast to the trends shown by biomarkers from prokaryotes, acyclic and cyclized terpenoids and steroids (see Figure 1) derived from planktonic algae and vascular plants show strong age-related trends (e.g., Summons and Walter 1990; Brocks and Summons 2004). For example, the oleanoid triterpenoids such as β-amyrin are important in the predation-defense mechanisms of flowering plants. Oleanane, a hydrocarbon derived from these triterpenoids,

is sometimes abundant in oils from rocks deposited on continental margins, showing a marked increase in oils formed from Cenozoic rocks that reflects the Cretaceous radiation of the angiosperms (Moldowan et al. 1994). Low levels of oleanane in rocks as old as Jurassic age, however, suggests either that early flowering plants were sporadic inhabitants of Mesozoic environments or that oleanoid triterpenoid synthesis originated in their phylogenetic precursors (e.g., Peters et al. 2005). Observations concerning algal biomarkers are discussed later.

II. THE RISE OF MODERN PHYTOPLANKTON

A. Fossils and Phylogeny

In modern oceans, three algal groups dominate primary production on continental shelves: the diatoms, dinoflagellates, and

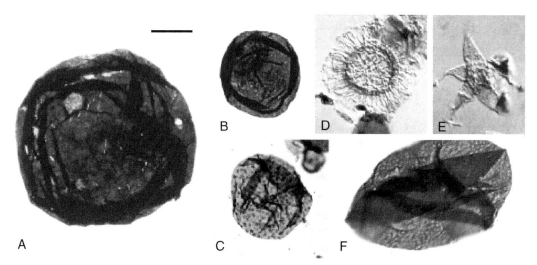

FIGURE 6.   Proterozoic and Early Cambrian protistan microfossils. (A) and (B) are Neoproterozoic leiospherid acritarchs; (C–F) are Early Cambrian acritarchs (C–E) and a prasinophyte phycoma (F; *Tasmanites*). Scale bar = 40 microns.

coccolithophorids. As detailed elsewhere in this volume, fossils clearly suggest that these groups all rose to taxonomic and ecological prominence only during the Mesozoic Era (Delwiche, Chapter 10; de Vargas *et al.*, Chapter 12; Kooistra *et al.*, Chapter 11, this volume). Could there have been, however, an earlier "cryptic" evolutionary history for these groups? For example, might nonmineralizing stem group diatoms or haptophytes have been ecologically important but paleontologically uninterpretable in Paleozoic oceans? Might the significance of Paleozoic dinoflagellates be obscured by fossils that are abundant and diverse but lack archeopyles?

Several reports claim microfossil evidence for Paleozoic diatoms, dinoflagellates, and haptophytes, but such fossils are rare and subject to alternative interpretation as contaminants (the mineralized skeletons) or different taxa (organic fossils). Stratigraphic research indicates that the Paleozoic silica cycle differed substantially from that of the Cretaceous and Tertiary periods, with sponges and radiolarians dominating biological removal of silica from the oceans (Maliva *et al.* 1989). Similarly, sediments preserved in obducted slices of Paleozoic seafloor at best contain only limited evidence for pelagic carbonate deposition. Such observations cannot eliminate the possibility that rare diatoms, coccolithophorids, or calcareous dinoflagellates lived in Paleozoic oceans, but they clearly indicate that these groups did not perform the biogeochemical roles they have played since the Mesozoic Era.

Complementing this, molecular clock estimates for diatom and coccolithophorid diversification, calibrated by well-preserved fossils, suggest that these groups have no long Paleozoic "prehistory" (de Vargas *et al.*, Chapter 12; Kooistra *et al.*, Chapter 11, this volume). On the other hand, some molecular clock analyses suggest divergence times for the plastids in photosynthetic chromalveolates well back into the Proterozoic (Douzery *et al.* 2004; Yoon *et al.* 2004). If these estimates are even broadly cor-

rect, they must be accommodated in one of two ways. Either the molecular clocks date divergence within a closely related group of unicellular red algae that subsequently and individually were incorporated as plastids in chromalveolate algae, or photosynthetic chromalveolates emerged from a single Proterozoic endosymbiosis (Cavalier-Smith 1999) but remained ecologically unimportant or paleontologically unrecognizable until much later.

B. Biomarkers and the Rise of Modern Phytoplankton

The rise to ecological prominence of the three chlorophyll *c*-containing eukaryotic plankton lineages left several imprints in the molecular fossil record, especially in the distributions of steranes with different side-chain alkylation patterns. A secular increase in the ratio of C_{28} to C_{29} steranes (24-methylcholestanes versus 24-ethylcholestanes), first noted by Grantham and Wakefield (1988), has been attributed to increasing production by chlorophyll *c* algae, which dominate C_{28} sterane input, relative to green algae, which synthesize primarily C_{29} steroids (Volkman 2003). An updated plot of the C_{28}/C_{29} sterane ratio versus geological age, based on averages from 123 petroleum systems worldwide, is shown in Figure 3 along with data for diatom diversity. The C_{28}/C_{29} sterane ratio remains below 0.4 in the Neoproterozoic and, with one exception, below 0.7 through the Paleozoic. Corresponding to the diversification of diatoms in the later half of the Mesozoic, there is a rise in the C_{28}/C_{29} sterane ratio to values as high as 1.8, followed by an apparent drop in the Paleocene and Eocene. Values climb again in the Miocene, accompanying a second rise in the numbers of diatom genera and species.

Other biomarkers show marked increases in abundance in the Cretaceous that also likely reflect diatom radiation. These include 24-norcholestanes (Holba *et al.* 1998a, b), the so-called highly branched isoprenoids (HBIs; Sinninghe Damsté *et al.* 1999a, b; Belt *et al.* 2000; Allard *et al.* 2001), long-chain diols, and mid-chain hydroxyl methyl-alkanoates (Sinninghe Damsté *et al.* 2003). The secular increase in 24-norcholestane abundance (Moldowan *et al.* 1991; Holba *et al.* 1998a, b) was observed and linked to the diatom radiation well before a precursor sterol was recognized in a culture of the centric diatom *Thalassiosira aff. antarctica* (Rampen *et al.* 2005). In fact, in the study of Rampen *et al.* (2005) only one, that is *T. aff. antarctica,* of 100 different diatom taxa was found to produce 24-norcholesta-5, 22-dien-3β-ol.

The detail of novel sterol production by diatoms provides an interesting window into the connections among biomarkers, taxonomy, and the physiological roles of lipids. In a recent study of diatom sterols, Suzuki *et al.* (2005) reported that environmental samples of diatoms collected from the North Pacific Ocean and the Bering Sea contained 24-norsterols. Further, samples of the diatom *Coscinodiscus marginatus,* initially devoid of 24-norcholesterol, contained significant amounts of this and the related steroid 27-nor-24-methylcholesta-5,22-dien-3β-ol after storage at 3°C for 30 days. These authors attributed the latter change to bacterial biodegradation. Due to the ubiquity of 24-norsteranes in Mesozoic sediments (Holba *et al.* 1998a, b), and no other evidence for selective side-chain biodegradation of sterols (biodegrading bacteria are unlikely to select for removal of C_{26} and C_{27} of the apparent precursor 24-methylcholesta-5,22-dien-3β-ol and leave other sterols untouched), it seems far more likely that there is a direct, as opposed to diagenetic, source for the 24-norsteroids in sediments. The rarity of 24-norsterols in cultured diatoms (Rampen *et al.* 2005) more likely reflects the fact that sterol biosynthesis responds to physiological conditions and that laboratory culture conditions have as yet not mimicked the natural conditions, such as low temperature and, perhaps, low light under which 24-norsterols are produced by some diatoms. The enigma surrounding the origins of

24-norsteroids, and the ultimate detection of 24-norcholesta-5,22-dien-3β-ol by Rampen *et al.* (2005) in a cold water diatom species, provide a timely reminder that biomarkers not only reflect the presence of particular algal taxa but also reflect the environmental conditions under which they thrive. A corollary to this is that cultured organisms might not always produce the same assemblage of lipids as their counterparts growing under natural conditions.

The other major class of diatom-specific biomarkers is the HBI. So far as is currently known, the occurrence of HBI is confined to four genera, namely *Navicula, Haslea,* and *Pleurosigma* within the pennates, and *Rhizosolenia* among the centrics (Volkman *et al.* 1994; Belt *et al.* 2000; Sinninghe Damsté *et al.* 2004). Both molecular phylogeny and fossils indicate that centric diatoms predate pennates (Kooistra *et al.* 2006). Therefore, the genus *Rhizosolenia* is considered the likely source of the first recorded fossil HBI at about 91 million years ago (Ma) —which predates recorded fossil tests of Rhizosolenid diatoms by about 20 million years (Sinninghe Damsté *et al.* 2004). This discrepancy in timing could reflect incomplete paleontological sampling, which systematically underestimates first appearances, HBI synthesis by a morphologically distinct stem group relative of the rhizosolenids, or both. Biosynthetic pathways are such a fundamental characteristic of organisms that they might be detectable through chemical fossils before the first classical fossils of a clade are ever recognizable.

Dinosteroid biomarkers, derived from the 4-methylsterols of dinoflagellates (Robinson *et al.* 1984), show an analogous pattern of secular increase in the Mesozoic, in accord with microfossil evidence for later Triassic dinoflagellate radiation. As in the case of diatom HBI, however, several reported occurrences of dinosteranes predate fossil cysts, in some cases by hundreds of millions of years (Summons *et al.* 1992; Moldowan and Talyzina 1998; Talyzina *et al.* 2000). These deserve close attention as they may establish a genuine pre-Mesozoic history of dinoflagellates, as predicted by molecular phylogenies and clocks. Some aspects of this putative history are considered further later. In Figure 4, data derived from the "Oils" database show the pattern of secular variation in triaromatic dinosteroid abundances, along with a recent compilation of dinoflagellate cyst diversity (Katz *et al.* 2004).

The third group of modern plankton, the haptophytes, produces distinctive lipids in the form of long-chain (n-C_{37} to C_{39}), unsaturated ketones known as alkenones (Volkman *et al.* 1980; Marlowe *et al.* 1984). These can be abundant and easily recognized in Recent and Cenozoic sediments that have not experienced extensive diagenesis, and they form the basis of a widely used paleotemperature proxy (Brassell *et al.* 1986). Their distinctiveness, however, lies in the carbonyl functionality and one to four unsaturations, which are easily reduced and inherently unstable over geological time scales (Prahl *et al.* 1989). Thus, the oldest reported detection is in Cretaceous sediments (Farrimond *et al.* 1986), and we would not expect to be able to recognize them in Paleozoic or older rocks.

The molecular and morphological records of eukaryotic predominance in shelf primary production are mirrored by indications of relatively low cyanobacterial contributions. Some cyanobacteria are known to biosynthesize BHPs and analogues with an extra methyl group attached to the 2 position of the A ring (2-MeBHP); the hydrocarbon cores of these molecules provide a potentially useful tracer for cyanobacterial input to sedimentary organic matter (Summons *et al.* 1999). Apart from a few notable exceptions, values of the 2-methylhopane (2-MeHI) index fraction of hopanoids (methylated at the 2 position relative to their desmethyl counterparts) tend higher in Proterozoic samples than they are in Paleozoic and, especially, Jurassic and younger oils (Figure 5A). This is especially true of samples from shales. (Examination of the Phanerozoic record [Figure 5B] shows that higher

2-MeHI values are found in oils from carbonate lithologies, formed predominantly at low paleolatitudes [Figure 5C].) The exceptions are associated with widespread anoxia in the oceans (Figure 5B). Kuypers *et al.* (2004a, b) and Dumitrescu and Brassell (2005) studied biomarkers associated with Cretaceous oceanic anoxic events (OAEs) and found that the relative abundances of 2-methylhopanoids, as measured by the 2-MeHI, were distinctively enhanced, along with nitrogen (N), isotopic evidence for cyanobacterial primary productivity (Kuypers *et al.* 2004b). Mass extinction at the Permian–Triassic boundary is also associated with widespread anoxia in shallow oceans (Wignall and Twitchett 2002). In Figure 5, the two data points for the Permian Triassic transition represent unpublished data from the Perth Basin, Australia, and the boundary stratotype in Meishan, China; in these sections, high 2-MeHI correlate with independent molecular, iron speciation, and sulfur isotopic evidence for intense euxinia (Grice *et al.* 2005). The highest 2-MeHI value (0.29) recorded in the GeoMark Oils database comes from the Larapintine Petroleum System, Australia, which includes oils from the Late Devonian reef complex of the Canning Basin (Edwards *et al.* 1997) sourced from black shales deposited near to Frasnian–Famennian boundary, another event characterized by geological and geochemical evidence for pervasive euxinia (e.g., Bond *et al.* 2004). The samples from the Cretaceous OAEs, Permian–Triassic boundary, and Frasnian–Famennian shale all contain abundant isorenieratane and aryl isoprenoids derived from the brown pigmented strains of the green sulfur bacteria (Chlorobiaceae), considered diagnostic for photic zone euxinia (e.g., Summons and Powell 1987; Koopmans *et al.* 1996; Kuypers *et al.* 2004a; Grice *et al.* 2005; van Breugel *et al.* 2005). Kuypers *et al.* (2004b) hypothesize that the N cycle was compromised while euxinic conditions prevailed during the Cretaceous OAEs, creating an unusual opportunity for the proliferation of N-fixing cyanobacteria. The disparate occurrences described previously suggest a more general correlation between high 2-MeHI and photic zone euxinia, a topic we return to in our discussion of Proterozoic primary production.

Although cyanobacteria appear to have been minor contributors to primary production on most Mesozoic and Cenozoic continental shelves, they remain the dominant phytoplankton in open-ocean, oligotrophic environments today. Whether this is a recent or long-standing situation is difficult to discern given the paucity of the deep-sea sedimentary records and the absence of 2Me-BHP in cyanobacterial picoplankton (Summons, unpublished data).

C. Summary of the Rise of Modern Phytoplankton

Fossils, molecular biomarkers, molecular clocks for individual clades, and the sedimentary silica record all tell a consistent story: the modern phytoplankton has Mesozoic roots. How we interpret this transformation depends in no small part on what we think came before.

III. PALEOZOIC PRIMARY PRODUCTION

A. Microfossils

Microfossils of presumptive eukaryotic phytoplankton are both abundant and diverse in Paleozoic marine rocks (Figure 6C–F). A number of forms, including *Tasmanites, Pterospermella,* and *Cymatiosphaera,* have morphologies and ultrastructures that relate them to prasinophyte phycomata (Tappan 1980). Indeed, in well-studied microfossil assemblages from Lower Cambrian shales, at least 20% of described morphotaxa and more than half of all individual fossils are likely prasinophytes (e.g., Volkova *et al.* 1983; Knoll and Swett 1987; Moczydlowska 1991). Others, with regularly distributed processes, tantalizingly

resemble dinocysts, but lack archeopyles. Still other acritarchs (the group name given to closed, organic-walled microfossils of uncertain systematic relationships) (Evitt 1963) do not closely resemble known cysts of modern phytoplankton. Collectively, these microfossils show evidence of marked Cambrian and Ordovician radiations that parallel the two-stage diversification of marine animals (Knoll 1989). For reasons that remain obscure, acritarch diversity drops near the end of the Devonian and remains low for the remainder of the Paleozoic Era (Molyneux *et al.* 1996).

Moldowan and Talyzina (1998) innovatively attempted to break the phylogenetic impasse regarding Cambrian microfossils. Extracts from fossiliferous clays of the Lower Cambrian Lükati Formation, Estonia, contain low abundances of dinosterane and 4a-methyl-24-ethylcholestane, both known to originate from the sterols of dinoflagellates (Robinson *et al.* 1984; Summons *et al.* 1987). Moldowan and Talyzina (1998) divided the microfossil populations in a Lükati sample into three groups—tasmanitids (see Figure 6F; interpreted as prasinophyte phycomata), a low fluorescence group dominated by leiosphaerid acritarchs (also possible phycomata of prasinophytes like the extant *Halosphaera*), and a high fluorescence fraction containing abundant process-bearing acritarchs (e.g., see Figure 6D)—and analyzed the sterane content of these subassemblages. The tasmanitid and low fluorescence fractions contained relatively abundant C_{29} steranes but little or no dinoflagellate lipid. In contrast, the high fluorescence fraction contained relatively high abundances of dinosterane, suggesting that the dominant, process-bearing acritarchs are dinocysts sans archeopyles. It is not clear that sterols play a structural role in cyst walls, making selective adsorption a real possibility. Moreover, whereas dinosterane and 4a-methyl-24-ethylcholestane abundances in these samples are relatively high, their concentrations are absolutely low. Thus, the specific attribution of acritarch taxa to the dinoflagellates remains speculative. Nonetheless, these analyses do clearly suggest that dinoflagellates were present in coastal Cambrian oceans and may have left a morphological as well as biogeochemical record. Given the low abundances of dinoflagellate biomarkers, it is possible that the constituent dinoflagellates were largely heterotrophs, not primary producers.

B. Paleozoic Molecular Biomarkers

The molecular fossil record prior to the rise of the chlorophyll c lineages broadly corroborates the microfossil evidence for the occurrence and potential ecological importance of other eukaryotic phytoplankton in the Paleozoic. In particular, the high abundance of C_{29} steranes relative to C_{27} and

C_{28} homologues suggests a greater role for green algae in marine primary production at this time. This signal is observed in globally distributed rocks and petroleum systems from the latest Neoproterozoic into the Paleozoic and wanes in the later Paleozoic, although the depositional bias for much of this time is toward low paleolatitudes (Figure 7). In a study of tasmanite oil shales from different locations in Tasmania, Revill *et al.* (1994) found that C_{27} and C_{29} steranes were present in roughly equal abundance and dominated over C_{28}. All the samples were shales with a high total organic carbon content, with the visible organic matter primarily comprising *Tasmanites punctatus* microfossils. These early Permian deposits, which were geographically localized, contained abundant dropstones and evidence

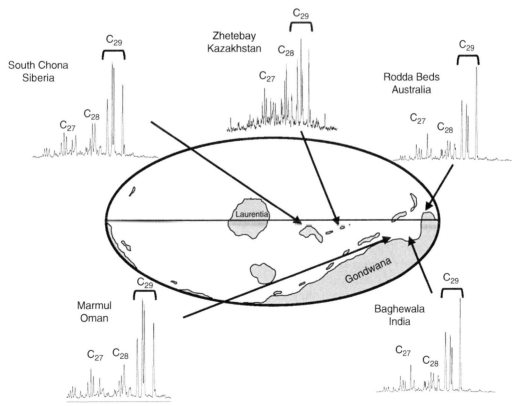

FIGURE 7. Dominance of C_{29} steranes in late Neoproterozoic/Early Cambrian petroleum systems. Paleogeography shows concentration of depositional systems to low paleolatitudes during this time. Continental reconstruction redrawn after C. Scotese, available at www.scotese.com.

of low-temperature minerals, which were clearly glacial in origin. Thus, the *Tasmanites punctatus* may have occupied an ecological niche similar to that occupied by modern sea–ice diatom communities (Revill *et al.* 1994).

Dinosteranes are generally below detection in the Paleozoic marine sediments and oils that have been examined and reported to date. In contrast, triaromatic dinosteroids have been found in significant abundance in several lower Paleozoic sedimentary rocks and petroleum samples. This speaks to the occurrence of either stem or crown group dinoflagellates in Paleozoic oceans. An additional factor in observed dinosterane abundances may be preservational bias. Saturated dinosteroids, dinosteranes, may only be preserved under strongly reducing conditions. A wider search for triaromatic dinosteroids, and authentication of the Paleozoic petroleum data through reanalysis and careful checking of pedigrees, may expose a more extensive pattern of occurrence, and hence, a richer early history for these plankton than is evident from the distribution of fossilized cysts.

C. Paleozoic Summary

Biomarker data for oils and some sediments suggest that dinoflagellates existed in Paleozoic oceans, but with few exceptions, lipids thought to be sourced by dinoflagellates occur in low abundances, raising the question whether Paleozoic dinoflagellates functioned to any great extent as primary producers. The same is true of possible stem-group heterokonts. Thus, although Chl a+c phytoplankton may well have existed in Paleozoic oceans, they do not appear to have played anything like the ecological role they have assumed since the Mesozoic Era began. In contrast, microfossil and biomarker molecules both suggest that green algae played a greater role in marine primary production than they have in the past 100 million years, and biomarkers also suggest a significant role for cyanobacterial production on continental shelves. Macro-

fossils, predominantly of calcareous skeletons, further indicate that red and green algae were ecologically important in the shallow shelf benthos (Wray 1977).

IV. PROTEROZOIC PRIMARY PRODUCTION

Fossils (whether morphological or molecular) are less abundant in Proterozoic rocks than they are in Phanerozoic samples, and Proterozoic sedimentary rocks, themselves, are less abundantly preserved than their younger counterparts. Nonetheless, fossils have been reported from hundreds of Proterozoic localities (Mendelson and Schopf 1992), allowing us to recognize at least broad patterns of stratigraphic and paleoenvironmental distribution. Indeed, Proterozoic micropaleontology has developed to the point where it has become predictive, in the sense that knowledge of age and environmental setting permits reasonable prediction about the fossil content of a given rock sample (e.g., Knoll *et al.* 2006).

A. Prokaryotic Fossils

By the earliest Proterozoic Eon, cyanobacteria must have been important contributors to primary production—there is no other plausible source for the O_2 that began to accumulate in the atmosphere and surface oceans 2.45–2.32 billion years ago (Ga). Consistent with this observation, it has been appreciated since the early days of Precambrian paleontology that cyanobacteria-like microfossils are abundant and widespread constituents of Proterozoic fossil assemblages (Figure 8; Schopf 1968). Not all cyanobacteria have diagnostic morphologies, but some do and others are likely candidates for attribution given knowledge of taphonomy (processes of preservation) and depositional environments represented in the record. By mid-Proterozoic times, if not earlier, all major clades of cyanobacteria existed in marine and near-shore terrestrial

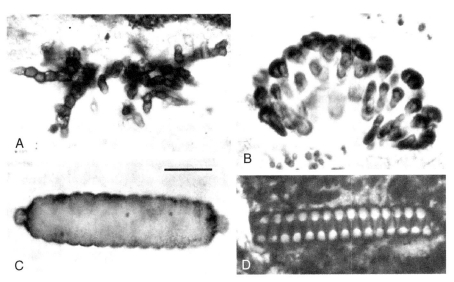

FIGURE 8. Cyanobacteria in Proterozoic sedimentary rocks. **(A)** 700–800-million-year-old endolithic pleur-capsalean fossil. **(B)** 1500-million-year-old mat building cyanobacterium closely related to modern *Entophysalis*. **(C)** 1500-million-year-old short trichome. **(D)** *Spirulina*-like fossil in latest Proterozoic (<600 Ma) phosphorite. Scale bar = 15 microns in **(A–C)**, and = 25 microns in **(D)**.

environments, including those that differentiate akinetes and heterocysts (Tomitani *et al.* 2006). The best-characterized Proterozoic cyan-obacteria come from early diagenetic chert nodules in carbonate successions (e.g., Schopf 1968; Zhang 1981; Knoll *et al.* 1991; Sergeev *et al.* 1995, 2002; Golubic and Seong-Joo 1999). These fossils are largely benthic and largely coastal marine. Stromatolites, however, indicate a much wider distribution of benthic cyanobacteria in the photic zone. (A role for cyanobacteria or of organisms in general is difficult to establish in the precipitated stromatolites found in Earth's oldest well-preserved sedimentary successions; however, the likelihood that cyanobacteria were major architects of Proterozoic stromatolites that accreted primarily by trapping and binding is high) (Grotzinger and Knoll 1999). Microfossils are less useful for evaluating the contributions of cyanobacteria to the phytoplankton of Proterozoic oceans because many were small, nondescript, and likely to settle on the seafloor in places where interpretable preservation was improbable. Given the distribution of planktonic clades

on a phylogenetic tree calibrated by well-documented fossils, however, it is likely that cyanobacteria were important constituents of the phytoplankton in Proterozoic oceans (Sanchez-Baracaldo *et al.* 2005; Tomitani *et al.* 2006; see later).

B. Eukaryotic Fossils

Two problems shadow attempts to understand the Proterozoic history of photosynthetic eukaryotes. Given the polyphyletic evolution of at least simple unicellular and multicellular characters, convergence complicates interpretation of many Proterozoic protistan fossils. In addition, given the oft-observed reality that stem group organisms display only a subset of the characters that collectively identify crown group members of clades, early fossils may challenge finer scale systematic attribution, even though they may be unambiguously eukaryotic.

Despite these problems, a small number of fossil populations provide calibration points for eukaryotic phylogenies. *Bangiomorpha pubescens* (Butterfield 2000) is a large population of multicellular microfossils found in tidal flat deposits of the ca. 1200 Ma Hunting Formation, Arctic Canada. These erect filaments, preserved via rapid burial by carbonate mud and subsequent silicification, display patterns of thallus organization, cell division, and cell differentiation that ally them to the bangiophyte red algae. Complementing this, a moderate diversity of cellularly preserved florideophyte red algal thalli occurs in <600 Ma phosphorites of the Ediacaran Doushantuo Formation, China (Xiao *et al.* 2004). Shifting to another branch of the eukaryotic tree, several taxa of vase shaped microfossils in the ca. 750 Kwagunt Formation, Grand Canyon, Arizona, can be related to lobose testate amoebae, placing a minimum constraint on the timing of amoebozoan divergence (Porter *et al.* 2003).

Accepting the presence of red algae by 1200 Ma, one might expect to observe green algal fossils in younger Proterozoic rocks. Several candidate taxa have been described, of which *Proterocladus,* a branching coenocytic thallus organized much like living *Cladophora,* is most compelling (Butterfield *et al.* 1994). *Palaeovaucheria clavata,* described from >1005 ±4 Ma shales in Siberia (Herman 1990), as well as ca. 750–800 Ma shales from Spitsbergen (Butterfield 2004), has a branching filamentous morphology and pattern of reproductive cell differentiation very similar to that of the extant xanthophyte alga *Vaucheria.* Kooistra *et al.* (Chapter 11, this volume) speculate that this similarity arose via convergence in a green algal clade; either interpretation places the origin of green algae earlier than 1000 Ma.

Fossils show that eukaryotic photoautotrophs were present in the benthos no later than the Mesoproterozoic Era (1600–1000 Ma), but what about the phytoplankton? Unicellular taxa occur in all three divisions of the Plantae, making it likely that such cells existed by the time that *Bangiomorpha* evolved. Of these, however, only the phycomate prasinophytes are likely to have left a tractable fossil record in marine

sedimentary rocks. As noted previously, the three major types of ornamented phycoma known from living prasinophytes have fossil records that extend backward to the Early Cambrian, but there is little evidence of earlier origin. In contrast, extant *Halosphaera* develop smoothly spheroidal phycomata that could easily be represented among leiosphaerid acritarchs in Proterozoic rocks (see Figure 6A and B; Tappan 1980). Ultrastructural and microchemical studies (e.g., Javaux *et al.* 2004; Marshall *et al.* 2005) provide our best opportunity to test this hypothesis.

Chromalveolates may be recorded in a very different way. In 1986, Allison and Hilgert reported small (7–40 μm in maximum dimension), apparently siliceous ovoid scales in cherts of the Tindir Group, northwestern Canada, now judged to be >635 and <710 Ma (Kaufman *et al.* 1992). The scales resemble those formed by living prymnesiophytes and (at a smaller size range) chrysophytes, likely documenting early diversification somewhere within the chromalveolate branch of the eukaryotic tree.

Fossils of any kind are rare in rocks older than about 2000 million years, but unambiguous fossils of eukaryotes occur in shales as old as 1650–1850 Ma (Knoll *et al.* 2006); little is known about their systematic relationships or physiology. Compilations of total diversity (e.g., Knoll 1994; Vidal and Moczydlowska 1997), assemblage diversity (Knoll *et al.* 2006), and morphospace occupation (Huntley *et al.* 2006) through time agree that a moderate diversity of eukaryotic organisms existed in Mesoproterozoic oceans. By ca. 1200 Ma if not earlier, this diversity included photosynthetic eukaryotes. Diversity appears to have increased modestly in the Neoproterozoic, but the major radiations within preservable seaweed and phytoplankton groups took place only at the end of the Proterozoic Era and during the ensuing Cambrian and Ordovician Periods (Knoll *et al.* 2006).

C. Proterozoic Molecular Biomarkers

Rocks containing organic matter amenable to biomarker analysis grow increasingly rare as we sample more deeply into the Proterozoic, and many of those available have undergone extensive heating such that only the most recalcitrant molecules remain. Nevertheless, a molecular fossil record of primary production is emerging for this long interval of Earth history. A major feature of the record is the high relative abundance of 2-MeHI in organic-rich distal shales throughout the Proterozoic (see Figure 5), which, in conjunction with the microfossil record and geochemical evidence for oxic surface waters in oceans, provides strong evidence for the importance of cyanobacterial production. There has been such limited sampling of Paleo- and Mesoproterozoic sediments that it has not been possible to examine these for correlations with lithology, as has been accomplished for the Phanerozoic (see Figure 5C).

Steranes have been reported from a number of Proterozoic successions (e.g., Summons and Walter 1990; Hayes *et al.* 1992; Dutkiewicz *et al.* 2003) and are generally found in low abundance, reflecting at least in part the thermal maturity of their host rocks. These molecular fossils establish the presence of eukaryotes in Proterozoic oceans, but the scarcity of detailed records limits the inferences that can be drawn concerning ecological role or taxonomic affinities (because group-distinctive markers are generally below detection limits). The geologic record of steroid biosynthesis extends into the Late Archean, several hundred million years before the first recognized protistan fossils. However, there continue to be doubts about the syngeneity of these steroids because of the advanced maturity of all the sections studied so far and because of the potential for the bitumens found there to have migrated from younger sequences or to be contaminants from drill and handling (e.g., Brocks *et al.* 2003a, b). In contrast, the Roper and McArthur Basins of northern Australia contain rocks of low-to-moderate thermal maturity, more consistent with the probability of finding genuinely syngenetic biomarkers. Given that studies of the Roper and McArthur Basin sediments and oils consistently show the presence of steroids (Summons *et al.* 1988a, b; Dutkiewicz *et al.* 2003) along with other evidence for the *in situ* (Summons *et al.* 1994) character of the bitumens, there seems little doubt that steroid biosynthesis operated as long ago as 1640 Ma. Preservation is a major limitation for both body and molecular fossil records at this point. Nonetheless, sterane abundances in rocks of this age appear to be low, independent of maturity level, and do not approach Phanerozoic abundances until the Neoproterozoic Era. Based on a few exceptionally well-preserved deposits of organic material in Mesoproterozoic shales, there appear to have been times and places where producer communities were very different from those that characterize later periods. Brocks *et al.* (2005) have reported biomarkers of anaerobic, sulfide-utilizing phototrophs in the carbonate facies of the Barney Creek Formation, Australia, suggesting that euxinic waters extended well into the photic zone. Molecular markers of eukaryotes and cyanobacteria in those portions of the Barney Creek Formation are exceptionally scarce, raising the possibility that, in at least some environments during the Proterozoic, production by anoxygenic photoautotrophs may have been quantitatively important. In fact, the scarcity of steroids and 2-methylhopanoids in samples with most abundant biomarkers for phototrophic sulfur bacteria is also consistent with the highly euxinic conditions they require. The extent to which this scenario reflects global versus local conditions awaits further elucidation, but it is consistent with geochemical proxies for oceanic redox conditions, observed globally (e.g., Logan *et al.* 1995; Arnold *et al.* 2003; Shen *et al.* 2003; Gellatly and Lyons 2005).

In contrast to the scarcity of suitable organic-rich rocks in the Paleoproterozoic and Mesoproterozoic successions, the Neo-proterozoic is replete with well-character-ized organic matter in low maturity sections from Australia, North America, Oman (e.g., Grantham *et al.* 1987) and eastern Sibe-ria (Summons and Powell 1992; see Sum-mons and Walter 1990; Hayes *et al.* 1992, for reviews). Of particular note are the oldest commercial petroleum accumulations in Siberia and Oman. These late Neoprotero-zoic oils display striking biomarker patterns characterized by particularly abundant steroidal hydrocarbons. Predominance of C_{29} steranes over other homologues is a fea-ture of oils from the South Oman Salt Basin that has received much attention since it was first reported by Grantham (1986). Examination of Neoproterozoic petroleum samples worldwide suggests that this is a globally significant feature (see Figure 7) that records the rise of green algae to eco-logical prominence. Further, samples that show the strong predominance of C_{29} ster-anes are also generally characterized by anomalously light carbon isotopic composi-tions, in the range of -33 to $-37°/_{oo}$ PDB. It is likely no coincidence that the oldest com-mercial petroleum deposits bear the promi-nent signature of a green algal contribution to petroleum-prone organic matter and that some green algae are known for their capac-ity to biosynthesize decay-resistant aliphatic biopolymers in their cell wall (algaenans; Derenne *et al.* 1991, 1992; Gelin *et al.* 1996, 1997, 1999), a likely source of acyclic hydro-carbons in these oils (e.g., Höld *et al.* 1999).

D. Summary of the Proterozoic Record

Microfossil and biomarker records are consistent in showing that cyanobacte-ria and eukaryotic microorganisms were both present in Proterozoic oceans. Fossils indicate that the primary endosymbiotic event establishing the photosynthetic Plan-tae took place no later than ca. 1200 Ma, in broad agreement with molecular clock estimates appropriately ornamented by error estimates (Hackett *et al.*, Chapter 7,

this volume). Thus, eukaryotic algae con-tributed to primary production during at least the last 600 million years of the Proterozoic Era. Yet, preserved biomark-ers are dominated by cyanobacteria and other photosynthetic bacteria, suggesting that eukaryotes played a limited quantita-tive role in primary production. Increas-ing amounts of C_{29} steranes appear in later Neoproterozoic samples, typified by the high sterane-to-hopane ratios and strong C_{29} sterane predominances in oils from the South Oman Salt Basin and eastern Siberia (e.g., Grantham 1986; Summons and Powell 1992); this suggests that green algae began to play an increasing role in primary production by 600–700 Ma. The timing of this transition is not well constrained but, in Oman, it begins prior to the Marinoan glaciation and extends to the Neoproterozoic–Cambrian boundary (Grosjean *et al.* 2005, and unpublished data), falsifying the hypothesis that the green algal proliferation was a response to the Acra-man impact event in Australia (McKirdy *et al.* 2006). In short, algae may have emerged as major contributors to global primary pro-duction only during the late Neoproterozoic to Early Paleozoic interval distinguished by marked increases in fossil diversity.

V. ARCHEAN OCEANS

We evaluate the Archean geobiologi-cal record cautiously, as available data are sparse. Sedimentary rocks are limited in volume, especially for the early Archean, and most surviving strata have been altered by at least moderate metamorphism. Thus, any interpretation must be provisional.

The expectation from both phylogeny and the Proterozoic biogeochemical record is that prokaryotic primary producers are likely to have governed early marine ecosystems. Cyanobacteria have the ecological advantage of obtaining electrons from ubiquitous water molecules, but there is no reason to believe that cyanobacteria were the pri-mordial photoautotrophs (see Blankenship *et al.*, Chapter 3, this volume). Indeed, the question of when cyanobacteria, with their coupled photosystems, evolved remains contentious. In an early ocean dominated by anoxygenic photobacteria, the availability of electron donors (Fe^{++}, H$_2$, H$_2$S) would have limited primary production (Kharecha *et al.* 2005).

The four principal lines of evidence used to contract an evolutionary history of Proterozoic oceans apply equally to the Archean record: microfossils, biomarker molecules, sedimentary textures that record microbe/sediment interactions on the ancient seafloor (e.g., stromatolites), and stable isotopic signatures (Knoll 2003b). Few microfossils have been reported from Archean cherts and shales. Somewhat poorly preserved fossils occur in latest Archean cherts from South Africa (Lanier 1986; Klein *et al.* 1987; Altermann and Schopf 1995); these could include cyanobacteria, but other alternatives cannot be rejected. More controversial are the nearly 3500 Ma carbonaceous microstructures interpreted as bacterial, and possibly cyanobacterial, trichomes by Schopf (1993). Recently, not only their systematic interpretation but their fundamental interpretation as biogenic has been called into question (Brasier *et al.* 2002, 2005, 2006). Debate about these structures continues (e.g., Schopf *et al.* 2002a, b, in response to Brasier *et al.* 2002), but few believe that these structures, whatever their origin, provide phylogenetic or physiological insights into early life.

The stromatolite record is similar. At least 40 occurrences of stromatolites have been reported from Archean rocks (Schopf 2006)—not a lot given that the record is 1 billion years long. Those younger than about 3000 Ma include structures that accreted by the trapping and binding of fine particles; such textures are more or less uniformly associated with microbial activity. Bedding surfaces on siliciclastic rocks of comparable age similarly include textures attributable to microbial mat communities (Noffke *et al.* 2006). Older stromatolites are largely precipitated structures whose bio-

genicity is harder to establish. Conoidal forms in ca. 3450 Ma rocks from Western Australia (Hofmann *et al.* 1999; Allwood *et al.* 2006) and "roll-up structures" (sediment sheets that were ripped up and rolled into a cylinder by currents, suggesting microbially mediated cohesion of poorly lithified laminae) in comparably old rocks from South Africa (Tice and Lowe 2004) may well require biological participation to form, but the taxonomic and physiological nature of the participants remains uncertain (see Tice and Lowe 2006, for an argument that anoxygenic photobacteria fueled Early Archean mat ecosystems).

No biogeochemically informative organic molecules are known from Early Archean rocks. Late Archean biomarkers have been reported; controversy surrounding their identification and interpretation has two distinct aspects. The first relates to their provenance and whether or not all the organic matter present in ancient sediments is coeval, as recognized by Brocks *et al.* (2003a, b). This question can best be addressed through studies of cores recently drilled and curated under controlled conditions. For example, the Agouron–Griqualand Paleoproterozoic Drilling Project (AGPDP) and the NASA Astrobiology Institute Drilling Project (ABDP) have recovered fresh cores from South Africa and the Pilbara Craton of Western Australia, respectively, which are being studied for a range of paleobiologic proxies, including analyses of preserved organic matter. One aim of this research is to control or eliminate contamination by hydrocarbons from younger sediments; a second aim is to test for relationships between extractable hydrocarbons and rock properties that could not exist in the case of contamination.

The second aspect of the controversy revolves around the degree to which biosynthetic pathways may have evolved over long time scales. It is fair to state that there must have been evolution in the structure and function of lipids over geological time. However, key enzymes in the biosynthetic pathways leading to sterols (Summons *et al.* 2006) and other triterpenoids in extant organisms are highly conserved. The known geological record of molecular fossils, especially steranes and triterpanes, is notable for the limited number of structural motifs that are recorded. With a few exceptions, the carbon skeletons are the same as those found in the lipids of extant organisms, and no demonstrably extinct structures have been reported. Furthermore, the patterns of occurrence of sterane and triterpane isomers are rigid over billion-year time scales and correlate strongly with environments of deposition, suggesting that diagenetic pathways connecting functional lipids to their fossil biomarker counterparts are also conserved. We also have evidence, through the occurrence of rearranged steranes (diasteranes) and unconventional steroids such as the 2-alkyl and 3-alkyl steranes (Summons and Capon 1988, 1991; van Kaam-Peters *et al.* 1998) and their aromatic counterparts (Dahl *et al.* 1995), that fossil steranes originated from precursors that carried a 3-hydroxyl group and unsaturation in the tetracyclic ring system, as extant sterols do. Thus, there is no evidence for major changes in the known record of chemical fossils that could be attributed to the inception, evolution, or alternative lipid biosynthetic pathways to the 24-alkylated steroids or hopanoids (Kopp *et al.* 2005). Accordingly, if biomarkers that have been identified are confirmed to be indigenous to late Archean rocks, this will constitute robust evidence for the presence of algae and bacteria early in Earth history. The fact that molecular oxygen is an absolute requirement for the biosynthesis of algal sterols also implies that oxygenic photosynthesis must have been present at the time (Summons *et al.* 2006).

The carbon isotopic abundances of Early Archean carbonates and organic matter are comparable to those of younger rocks, indicating fractionation like that imparted by Rubisco-based autotrophy. Indeed, C-isotopic signatures that are consistent with carbon fixation by Rubisco extend backward to nearly 3800 Ma metamorphosed sediments from southwestern Greenland (Rosing and Frei 2004). The question is whether these signatures *require* such an interpretation. Other biochemical pathways for carbon fixation exist and at least some of them impart isotopic signatures that are equally consistent with Archean data (e.g., Knoll and Canfield 1998). It has been known for 2 decades that abiotic syntheses of organic matter, like that demonstrated by Miller (1953), fractionate C-isotopes (Chang *et al.* 1982); the degree of fractionation appears to vary widely as a function of initial conditions. More recently, McCollom and Seewald (2006) have shown that Fischer-Tropsch–type (FTT) synthesis can produce organic compounds depleted in ^{13}C relative to their carbon source to a degree similar to that associated with biological carbon fixation. In these experiments, formic acid was reacted with native iron at 250°C and 325 bar, and a series of *n*-alkanes were produced that were depleted by ~36°/₀₀ relative to the reactant carbon. This isotopic discrimination is in the range observed for the difference in $\delta^{13}C$ values of coexisting carbonate minerals and organic matter in some Archean deposits. This finding emphasizes the importance of understanding the depositional context (e.g., sedimentary versus hydrothermal) of this very ancient carbonaceous matter when assessing its biogenicity.

We conclude that the origin of life predates the known record of preserved sedimentary rocks, but the nature of that life—and, in particular, the nature of primary producers in the oceans—remains uncertain. All known geobiological records from Archean rocks are consistent with an early evolution of cyanobacteria, but few if any require such an interpretation (Knoll 2003a). Indeed, Kopp *et al.* (2005) have hypothesized that cyanobacteria originated only in association with the initial accumulation of free oxygen in the atmosphere, 2320–2450 Ma (Holland 2006). Careful geobiological analyses of well-preserved Archean rocks remain a priority for continuing research.

VI. CONCLUSIONS

In combination, paleontological and organic geochemical data suggest that the second half of Earth history can be divided into three major eras, with respect to marine photosynthesis. Limited data from Paleoproterozoic and Mesoproterozoic rocks suggest that cyanobacteria and other photosynthetic bacteria dominated primary production at that time, with anoxygenic photosynthetic bacteria playing an important role, at least locally, in water masses subtended by a euxinic oxygen-minimum zone. Indeed, available data suggest that cyanobacteria continued as principal photoautotrophs well into the Phanerozoic Eon and long after photosynthesis originated in eukaryotic cells. C_{29} sterane abundances indicate that green algae joined, but did not entirely displace, cyanobacteria as major primary producers during the latest Proterozoic and Cambrian; the second phase of primary production history thus initiated persisted until the Mesozoic radiation of modern phytoplankton dominants. Later Triassic oceans may have been the first in which cyanobacteria played a relatively minor role in continental shelf production. (Of course, they remain important today in the open gyre systems little recorded by pre-Jurassic sedimentary rocks.) The degree to which Chl a+c algae participated in Neoproterozoic and Paleozoic marine ecosystems remains unresolved, but if present their role must have been much smaller than it has been during the past 200 million years.

The observation that the oceans have experienced two major shifts over the past billion years in the composition of primary producers, and the corollary that at least some clades emerged as ecologically dominant primary producers long after their evolutionary origin, invites discussion of possible drivers. The importance of cyanobacteria in Proterozoic primary production can be attributed to at least two circumstances, their early diversification and environmental circumstances in Proterozoic oceans. Prior to the proliferation of eukaryotic algae, cyanobacteria would, of course, have had an open playing field, flourishing in oxygenated surface waters from coastline to mid-ocean gyres, although ceding deeper, at least intermittently euxinic parts of the photic zone to green and purple photosynthetic bacteria. Why, however, does it appear that cyanobacteria continued as dominant features of the photosynthetic biota on continental shelves long after red and green algae entered the oceans? At least in part, the answer may have to do with the nutrient structure of oceans in which, beneath an oxygenated surface layer, the oxygen minimum zone (Brocks *et al.* 2005), if not the entire deep ocean (Canfield 1998), had a high propensity for developing euxinia. Under these conditions, one would expect little fixed N to resurface during upwelling (Anbar and Knoll 2002; Fennel *et al.* 2005), providing strong selective advantage for cyanobacteria able to fix N and scavenge low concentrations of fixed N effectively from seawater.

Increasing oxygenation of the oceans during the Neoproterozoic Era (Canfield *et al.* 2006; Fike *et al.* 2006) would have begun to alleviate the N budget, as the mid-level waters that source upwelling would have been increasingly likely to remain oxic, limiting denitrification and anammox reactions that strip fixed N from ascending anoxic water masses. More ammonium would have been returned to the surface, and nitrate would have begun to accumulate for the first time. In consequence, eukaryotes would have spread more completely across benthic environments and into the phytoplankton, as recorded in the geological record (Knoll *et al.* 2006).

Dinoflagellates, diatoms, and coccolithophorids exhibit many features that collectively account for their ecological success in modern oceans (Delwiche, Chapter 10; de Vargas *et al.*, Chapter 12; Kooistra *et al.*, Chapter 11, this volume). Why, then, do we not see evidence for similar success in Paleozoic seas?

One possibility is that the secondary endosymbioses that led to these groups took place only at the beginning of the Mesozoic Era or shortly earlier. Such a scenario is consistent with clade-specific molecular clocks for diatoms and coccolithophorids but is inconsistent with the hypothesis that secondary endosymbiosis involving red algal photosymbionts occurred only once, in the early history of the chromalveolates (Hackett *et al.*, Chapter 7, this volume). Regardless of the timing of clade origination, however, we need to consider environmental factors, for the simple reason that it is hard to conceive of biological barriers would have prevented secondary endosymbiosis long before the Mesozoic began.

Black shale distributions may provide perspective on this issue. Multiregional to globally widespread black shales are essentially absent from Cenozoic successions but occur at about seven discrete stratigraphic horizons in the Mesozoic record (Jones and Jenkyns 2001). In contrast, there are at least seven black shale horizons in the Devonian record alone and many more in other parts of the Paleozoic, especially the Cambrian and Ordovician (Berry and Wilde 1978). Prior to the dawn of the Cambrian, *most* shales were carbonaceous (e.g., Knoll and Swett 1990; Abbott and Sweet 2000). If the redox structure of the oceans influenced the selective environment of green versus Chl a+c phytoplankton, then it may be that only in Mesozoic oceans did environmental conditions routinely favor the latter. As noted previously, fossils and biomarkers indicate that greens and cyanobacteria transiently re-established themselves as principal primary producers during the Mesozoic OAEs; green sulfur bacteria also proliferated during episodes of photic zone euxinia. Moreover, unlike chromalveolate photoautotrophs, both green algae and cyanobacteria show a pronounced preference for ammonium over nitrate in metabolism (Litchman, Chapter 16, this volume). Thus, the long-term redox evolution of the oceans may govern the composition of marine primary producers through time.

Whatever their drivers, the two observed transitions in the marine photosynthetic biota provide an important framework and stimulus for continuing paleobiological investigations of animal evolution. Latest Proterozoic and Cambrian phytoplankton radiation may not simply be a response to animal evolution (e.g., Peterson and Butterfield 2005) but also a driver. Well-documented (Bambach 1993) increases in body size among Mesozoic (versus Paleozoic) marine invertebrates may reflect the Mesozoic radiation of larger net plankton, while the so-called Mesozoic Marine Revolution among mostly Cretaceous and Cenozoic marine animals may specifically reflect the rise to ecological prominence of diatoms (see Finkel, Chapter 15, this volume). Vermeij (1977) first documented the major evolutionary changes in skeletonized marine fauna during this interval and ascribed it to a late Mesozoic radiation of predators able to penetrate shells. Bambach (1993), however, argued that the required radiation of top predators could only occur as a consequence of increased primary production and, hence, increase nutrient status in the oceans. Bambach (1993) suggested that evolving angiosperms increased nutrient fluxes to the oceans, and although this likely did occur (see Knoll 2003c), the evolution of a high-quality food source and efficient nutrient transporter in the form of diatoms likely played at least an equal role.

A. Directions for Continuing Research

Over the past decade, both paleontologists and organic geochemists have made inroads into problems of photosynthetic history. Nonetheless, there continue to be more questions than answers. Future research will require more and independent studies of fossils and hydrocarbon distribution on Archean and Proterozoic rocks. However, it will also require phylogenetic, biosynthetic, and functional studies of sterols and BHPs (especially 2-Me-BHP) in living organisms that will increase our ability to interpret ancient records. In comparable fashion, continuing research on Proterozoic and Paleozoic microfossils will need to stress wall ultrastructure (Arouri *et al.* 1999, 2000; Talzyina 2000; Javaux *et al.* 2004) and microchemical analysis (e.g., FTIR, hydropyrolysis, x-ray and, perhaps, Raman spectroscopy; Love *et al.* 1995; Schopf *et al.* 2002a, b; Boyce *et al.* 2003; Marshall *et al.*, 2005) interpreted in light of corresponding analyses of living cells and younger, taxonomically unambiguous fossils.

Finally, as noted previously, the evolution of photosynthetic organisms did not take place in a passive or unchanging ocean nor did it occur in an ecological vacuum. Improved understanding of Earth's redox history and the evolutionary record of animals (and land plants) (Falkowski *et al.* 2004) will provide the framework needed to interpret the evolutionary history of marine photoautotrophs as it continues to emerge.

Acknowledgments

AHK, JW, and RES acknowledge research support from NSF grant 0420592. AHK also acknowledges NSF grant OCE-0084032 (Project EREUPT, P. Falkowski, PI). RES is further supported by NASA Exobiology Program (Grant No. NNG05GN62G), NSF (EAR0418619), and PRF grant (41553-AC2) from the American Chemical Society. JRW receives support from an NDSEG Graduate Fellowship from the Office of Naval Research.

References

Abbott, S.T., and Sweet, I.P. (2000). Tectonic control on third-order sequences in a siliciclastic ramp-style basin: an example from the Roper Superbasin (Mesoproterozoic), northern Australia. *Austral. J. Earth Sci.* **47:** 637–657.

Allard, W.G., Belt, S.T., Massé, G., Naumann, R., Robert, J.-M., and Rowland, S. J. (2001). Tetraunsaturated sesterterpenoids (Haslenes) from Haslea ostrearia and related species. *Phytochemistry* **56:** 795–800.

Allison, C.W., and Hilgert, J.W. (1986). Scale microfossils from the Early Cambrian of northwest Canada. *J. Paleontol.* **60:** 973–1015.

Allwood, A.C., Walter, M.R., Kamber, B.S., Marshall, C.P., and Burch, I.W. (2006). Stromatolite reef from the Early Archaean era of Australia. *Nature* **441:** 714–718.

Altermann, W., and Schopf, J.W. (1995). Microfossils from the Neoarchean Campbell Group, Griqualand West sequence of the Transvaal Supergroup, and their paleoenvironmental and evolutionary implications. *Precambrian Res.* **75:** 65–90.

Anbar, A.D., and Knoll, A.H. (2002). Proterozoic ocean chemistry and evolution: a bioorganic bridge? *Science* **297:** 1137–1142.

Andrusevich, V.E., Engel, M.H., Zumberge, J.E., and Brothers, L.A. (1998). Secular, episodic changes in stable isotopic composition of crude oils. *Chem. Geol.* **152:** 59–72.

Andrusevich, V.E., Engel, M.H., and Zumberge, J.E. (2000). Effects of paleolatitude on the stable carbon isotope composition of crude oils. *Geology* **28:** 847–850.

Arnold, G.L., Anbar, A.D., Barling, J., and Lyons, T.W. (2003). Molybdenum isotope evidence for widespread anoxia in mid-Proterozoic oceans. *Science* **304:** 87–90.

Arouri, K., Greenwood P.F., and Walter, M.R. (1999). A possible chlorophycean affinity of some Neoproterozoic acritarchs. *Org. Geochem.* **30:** 1323–1337.

Arouri, K., Greenwood, P.F., and Walter, M.R. (2000). Biological affinities of Neoproterozoic acritarchs from Australia: microscopic and chemical characterisation. *Org. Geochem.* **31:** 75–89.

Bambach, R.K. (1993). Seafood through time—changes in biomass, energetics, and productivity in the marine ecosystem. *Paleobiology* **19:** 372–397.

Belt, S.T., Allard, W.G., Massé, G., Robert, J.-M., and Rowland S. J. (2000). Highly branched isoprenoids (HBIs): identification of the most common and abundant sedimentary isomers. *Geochim. Cosmochim. Acta* **64:** 3839–3851.

Berry, W.B.N., and Wilde, P. (1978). Progressive ventilation of the oceans—an explanation for the distribution of the Lower Paleozoic Black Shales. *Am. J. Sci.* **278:** 257–275.

Bisseret, P., Zundel, M., and Rohmer, M. (1985). Prokaryotic triterpenoids. 2. 2B-Methylhopanoids from *Methylobacterium organophilum* and *Nostoc muscorum*, a new series of prokaryotic triterpenoids. *Eur. J. Biochem.* **150:** 29–34.

Blankenship, R., Sadekar, S., and Raymond, J. (2007). The evolutionary transition from anoxygenic to oxygenic photosynthesis. *Evolution of Primary Producers in the sea.* P.G. Falkowski and A.H. Knoll, eds. Boston, Elsevier, pp. 21–35.

Bond, D., Wignall, P.B., and Racki, G. (2004). Extent and duration of marine anoxia during the Frasnian-Famennian (Late Devonian) mass extinction in Poland, Germany, Austria and France. *Geol. Mag.* **141:** 173–193.

Boyce, C.K., Cody, G.D., Fogel, M.L., Hazen, R.M., and Knoll, A.H. (2003). Chemical evidence for cell wall lignification and the evolution of tracheids in Early Devonian plants. *Int. J. Plant Sci.* **164:** 691–702.

Brasier, M., McLoughlin, N., Green, O., and Wacey, D. (2006). A fresh look at the fossil evidence for early Archaean cellular life. *Phil. Trans. R. Soc. B* **361**: 887–902.

Brasier, M.D., Green, O.R., Jephcoat, A.P., Kleppe, A.K., van Kranendonk, M.J., Lindsay, J.F., Steele, A., and Grassineau, N.V. (2002). Questioning the evidence for Earth's oldest fossils. *Nature* **416**: 76–81.

Brasier, M.D., Green, O.R., Lindsay, J.F., McLoughlin, N., Steele, A., and Stoakes, C. (2005). Critical testing of Earth's oldest putative fossil assemblage from the similar to 3.5 Ga Apex Chert, Chinaman Creek, western Australia. *Precambrian Res.* **140**: 55–102.

Brassell, S.C., Eglinton, G., Marlowe, I.T., Pflaumann, U., and Sarnthein, M. (1986). Molecular stratigraphy: a new tool for climatic assessment. *Nature* **320**: 129–133.

Brocks, J.J., Love, G.D., Snape, C.E., Logan, G.A., Summons, R.E., and Buick, R. (2003a) Release of bound aromatic hydrocarbons from late Archean and Mesoproterozoic kerogens via hydropyrolysis. *Geochim. Cosmochim. Acta* **67**: 1521–1530.

Brocks, J.J., Love, G.D., Summons, R.E., Knoll, A.H., Logan, G.A., and Bowden, S. (2005). Biomarker evidence for green and purple sulfur bacteria in an intensely stratified Paleoproterozoic ocean. *Nature* **437**: 866–870.

Brocks, J.J., and Summons, R.E. (2004). Sedimentary hydrocarbons, biomarkers for early life. *Treatise in Geochemistry*. Vol. 8. H.D. Holland and K. Turekian, eds. Amsterdam, Elsevier, pp. 65–115.

Brocks, J.J., Summons, R.E., Logan, G.A., and Buick, R. (2003b). Molecular fossils in Archean rocks II: composition, thermal maturity and integrity of hydrocarbons extracted from sedimentary rocks of the 2.78 to 2.45 billion years old Mount Bruce Supergroup, Pilbara Craton, Western Australia. *Geochim. Cosmochim. Acta* **67**: 4289–4319.

Butterfield, N.J. (2000). *Bangiomorpha pubescens* n. gen., n. sp.: implications for the evolution of sex, multicellularity and the Mesoproterozoic/Neoproterozoic radiation of eukaryotes. *Paleobiology* **26**: 386–404.

Butterfield, N.J. (2004). A vaucheriacean alga from the middle Neoproterozoic of Spitsbergen: implications for the evolution of Proterozoic eukaryotes and the Cambrian explosion. *Paleobiology* **30**: 231–252.

Butterfield, N.J., Knoll, A.H., and Swett, N. (1994). Paleobiology of the Neoproterozoic Svanbergfjellet Formation, Spitsbergen. *Fossils Strata* **34**: 1–84.

Canfield, D.E. (1998). A new model for Proterozoic ocean chemistry. *Nature* **396**: 450–453.

Canfield, D.E., Poulton, S.W., and Narbonne, G.M. (2007) Late Neoproterozoic deep-ocean oxygenation and the rise of animal life. *Science* **315**: 92–95.

Cavalier-Smith, T. (1999). Principles of protein and lipid targeting in secondary symbiogenesis: euglenoid, dinoflagellate, and sporozoan plastid origins and the eukaryote family tree. *J. Euk. Microbiol.* **46**: 347–366.

Chang, S., Des Marais, D., Mack, R., Miller, S.L., and Strathearn, G.E. (1982). Prebiotic organic syntheses and the origin of life. *Earth's Earliest Biosphere: Its Origin and Evolution*. J.W. Schopf, ed. Princeton, NJ, Princeton University Press, p. 53–92.

Coolen, M.J.L., Muyzer, G., Rijpstra, W.I.C., Schouten, S., Volkman, J.K., and Sinninghe Damsté, J.S. (2004). Combined DNA and lipid analyses of sediments reveal changes in Holocene haptophyte and diatom populations in an Antarctic lake. *Earth Planet. Sci. Lett.* **223**: 225–239.

Dahl J., Moldowan J.M., Summons R.E., McCaffrey M.A., Lipton P.A., Watt D.S., and Hope J.M. (1995). Extended 3β-alkyl steranes and 3-alkyl triaromatic steroids in oils and rock extracts. *Geochim. Cosmochim. Acta* **59**: 3717–3729.

Delwiche, C.F. (2007) The origin and evolution of dinoflagellates. *Evolution of Primary Producers in the Sea*. P.G. Falkowski and A.H. Knoll, eds. Boston, Elsevier, pp. 195–205.

Derenne, S., Largeau, C., Casadevall, E., Berkaloff, C., and Rousseau, B. (1991).Chemical evidence of kerogen formation in source rocks and oil shales via selective preservation of thin resistant outer walls of microalgae: origin of ultralaminae. *Geochim. Cosmochim. Acta* **55**: 1041–1050.

Derenne, S., Le Berre, F., Largeau, C., Hatcher, P. G., Connan, J., and Raynaud, J.-F. (1992). Formation of ultralaminae in marine kerogens via selective preservation of thin resistant outer walls of microalgae. *Org.Geochem.* **19**: 345–350.

de Vargas, C., Aubrey, M.-P., Probert, I., and Young, J. (2006). Origin and evolution of coccolithophores: From coastal hunters to oceanic farmers. *Evolution of Primary Producers in the Sea*. P.G. Falkowski and A.H. Knoll, eds. Boston, Elsevier, pp. 251–285.

Douzery, E.J.P., Snell, E.A., Bapteste, E., Delsuc, F., and Philippe, H. (2004). The timing of eukaryotic evolution: does a relaxed molecular clock reconcile proteins and fossils? *Proc. Nat. Acad. Sci. U S A* **101**: 15386–15391.

Dumitrescu, M., and Brassell, S.C. (2005). Biogeochemical assessment of sources of organic matter and paleoproductivity during the early Aptian oceanic anoxic event at Shatsky Rise, ODP Leg 198. *Org. Geochem.* **36**: 1002–1022.

Dutkiewicz, A., Volk, H., Ridley, J., and George, S. (2003). Biomarkers, brines, and oil in the Mesoproterozoic, Roper Superbasin, Australia. *Geology* **31**: 981–984.

Edwards, D.S., Summons, R.E., Kennard, J.M., Nicoll, R.S., Bradshaw, J., Bradshaw, M., Foster C.B., O'Brien, G.W., and Zumberge, J.E. (1997). Geochemical characterisation of Palaeozoic petroleum systems in northwestern Australia. *Austral. Petrol. Production Explorationists Assoc. J.* **37**: 351–379.

Evitt, W.R. (1963). A discussion and proposals concerning fossil dinoflagellates, hystrichospheres, and acritarchs: 2. *Proc. Nat. Acad. Sci., U S A* **49**: 298–302.

Farrimond, P., Eglinton, G., and Brassell, S.C. (1986). Alkenones in Cretaceous black shales, Blake-Bahama Basin, western North Atlantic. *Org. Geochem.* **10**: 897–903.

Falkowski, P.G., Katz, M.E., Knoll, A.H., Quigg, A., Raven, J.A., Schofield, O., and Taylor, F.J.R. (2004). The evolution of modern eukaryotic phytoplankton. *Science* **305**: 354–360.

Fennel, K., Follows, M., and Falkowski, P.G. (2005). The co-evolution of the nitrogen, carbon and oxygen cycles in the Proterozoic ocean. *Am. J. Sci.* **305**: 526–545.

Fike, D.A., Grotzinger, J.P., Pratt, L.M., and Summons, R.E. (2006). Oxidation of the Ediacaran Ocean. *Nature* **444**: 744–747.

Finkel, Z.V. (2007). Does phytoplankton cell size matter? The evolution of modern marine food webs. *Evolution of Primary Producers in the Sea.* P.G. Falkowski and A.H. Knoll, eds. Boston, Elsevier, pp. 333–350.

Foster, C.B., and Afonin, S.A. (2006). *Syndesmorion* gen. nov. A coenobial alga of Chlorococcalean affinity from the continental Permian-Triassic deposits of Dalongkou section, Xinjiang Province, China. *Rev. Palaeobot. Palynol.* **138**: 1–8.

Gelin, F., Boogers, I., Noordeloos, A.A.M., Sinninghe Damsté, J.S., Hatcher, P.G., and de Leeuw, J.W. (1996). Novel, resistant microalgal polyethers: An important sink of organic carbon in the marine environment? *Geochim. Cosmochim. Acta* **60**: 1275–1280.

Gelin, F., Boogers, I., Noordeloos, A.A.M., Sinninghe Damsté, J.S., Riegman, R., and de Leeuw, J.W. (1997). Resistant biomacromolecules in marine microalgae of the classes Eustigmatophyceae and Chlorophyceae: geochemical implications. *Org. Geochem.* **26**: 659–675.

Gelin, F., Volkman, J.K., Largeau, C., Derenne, S., Sinninghe Damsté, J.S., and de Leeuw, J. W. (1999). Distribution of aliphatic, nonhydrolyzable biopolymers in marine microalgae. *Org. Geochem.* **30**: 147–159.

Gellatly, A.M., and Lyons, T.M. (2005). Trace sulfate in mid-Proterozoic carbonates and the sulfur isotope record of biospheric evolution. *Geochim. Cosmochim. Acta* **69**: 3813–3829.

Golubic, S., and Seong-Joo, L. (1999). Early cyanobacterial fossil record: preservation, palaeoenvironments and identification. *Eur. J. Phycol.* **34**: 339–348.

Grantham, P.J. (1986). The occurrence of unusual C_{27} and C_{29} sterane predominances in two types of Oman crude oil. *Org. Geochem.* **9**: 1–10.

Grantham, P.J., Lijmbach, G.W.M., Posthuma, J., Hughes Clarke, M.W., and Willink, R.J. (1987). Origin of crude oils in Oman. *J. Petroleum Geol.* **11**: 61–80.

Grantham, P.J., and Wakefield, L.L. (1988). Variations in the sterane carbon number distributions of marine source rock derived crude oils through geological time. *Org. Geochem.* **12**: 61–73.

Grice, K., Cao, C., Love, G.D., Bottcher, M.E., Twitchett, R.J., Grosjean, E., Summons, R.E.,

Turgeon, S.C., Dunning, W., and Jin, Y. (2005). Photic zone euxinia during the Permian–Triassic superanoxic event. *Science* **307**: 706–709.

Grosjean, E., Stalvies, C., Love, G.D., Lewis, A.N., Hebting, Y., Fike, D.A., Taylor, P.N., Newall, M.J., Grotzinger, J.P., Farrimond, P., and Summons, R.E. (2005). New oil-source correlations in the South Oman Salt Basin. *Organic Geochemistry: Challenges for the 21st Century.* 22 IMOG Seville, Spain, Abstract OOS-1.

Grotzinger, J.P., and Knoll, A.H. (1999). Proterozoic stromatolites: evolutionary mileposts or environmental dipsticks? *Ann. Rev. Earth Planet. Sci.* **27**: 313–358.

Hackett, J.D., Yoon, H.S., Butterfield, N.J., Sanderson, M.J., and Bhattacharya, D. (2007). Plastid endosymbiosis: Sources and timing of the major events. *Evolution of the Primary Producers in the Sea.* P.G. Falkowski and A.H. Knoll, eds. Boston, Elsevier, pp. 109–131.

Hayes, J.M., Summons, R.E., Strauss, H., Des Marais, D.J., and Lambert, I.B. (1992). Proterozoic biogeochemistry. *The Proterozoic Biosphere: A Multidisciplinary Study.* J.W. Schopf and C. Klein, eds. Cambridge, UK, Cambridge University Press, pp. 81–133.

Hebting, Y., Schaeffer, P., Behrens, A., Adam, P., Schmitt, G., Schneckenburger, P., Bernasconi, S.M., and Albrecht, P. (2006). Biomarker evidence for a major preservation pathway of sedimentary organic carbon. *Science* **312**: 1627–1631.

Hedges, J.I., and Keil, R.G. (1995). Sedimentary organic matter preservation: an assessment and speculative synthesis. *Mar. Chem.* **49**: 81–115.

Herman, N. (1990). *Organic World One Billion Years Ago.* Nauka, Leningrad.

Hofmann, H.J., Grey, K., Hickman, A.H., and Thorpe, R.I. (1999). Origin of 3.45 Ga coniform stromatolites in Warrawoona Group, Western Australia. *Geol. Soc. Am. Bull.* **111**: 1256–1262.

Holba, A.G., Dzou, L.I.P., Masterson, W.D., Hughes, W.B., Huizinga, B.J., Singletary, M.S., Moldowan, J.M., Mello, M.R., and Tegelaar, E. (1998a). Application of 24-norcholestanes for constraining source age of petroleum. *Org. Geochem.* **29**: 1269–1283.

Holba, A.G., Tegelaar, E.W., Huizinga, B.J., Moldowan, J.M., Singletary, M.S., McCaffrey, M.A., and Dzou, L.I.P. (1998b). 24-norcholestanes as age-sensitive molecular fossils. *Geology* **26**: 783–786.

Höld, I.M., Schouten, S., Jellema, J., and Sinninghe Damsté, J.S. (1999). Origin of free and bound mid-chain methyl alkanes in oils, bitumens and kerogens of the marine, Infracambrian Huqf Formation (Oman). *Org. Geochem.* **30**: 1411–1428.

Holland, H.D. (2006). The oxygenation of the atmosphere and oceans. *Phil. Trans. R. Soc. B.* **361**: 903–915.

Huntley, J.W., Xiao, S.H., and Kowalewski, M. (2006). 1.3 billion years of acritarch history: an empirical morphospace approach. *Precambrian Res.* **144**: 52–68.

Javaux, E.J., Knoll, A.H., and Walter, M.R. (2004). TEM evidence for eukaryotic diversity in mid-Proterozoic oceans. *Geobiology* **2:** 121–132.

Jones, C.E., and Jenkyns, H.C. (2001). Seawater strontium isotopes, oceanic anoxic events, and seafloor hydrothermal activity in the Jurassic and Cretaceous. *Am. J. Sci.* **301:** 112–149.

Katz, M.E., Finkel, Z.V., Grzebyk, D., Falkowski, P.G., and Knoll, A.H. (2004). Evolutionary trajectories and biogeochemical impacts of marine eukaryotic phytoplankton. *Ann. Rev. Ecol. Syst.* **35:** 523–556.

Kaufman, A.J., Knoll, A.H., and Awramik, S.M. (1992). Bio- and chemostratigraphic correlation of Neoproterozoic sedimentary successions: the upper Tindir Group, northwestern Canada as a test case. *Geology* **20:** 181–185.

Kharecha, P., Kasting, J., and Siefert, J. (2005). A coupled atmosphere-ecosystem model of the early Archean Earth. *Geobiology* **3:** 53–76.

Kleemann, G., Poralla, K., Englert, G., Kjøsen, H., Liaaen-Jensen, S., Neunlist, S., and Rohmer, M. (1990). Tetrahymanol from the phototrophic bacterium *Rhodopseudomonas palustris*: first report of a gammacerane triterpene from a prokaryote. *J. Gen. Microbiol.* **136:** 2551–2553.

Klein, C., Beukes, N.J., and Schopf, J.W. (1987). Filamentous microfossils in the early Proterozoic Transvaal Supergroup—their morphology, significance, and paleoenvironmental setting. *Precambrian Res.* **36:** 81–94.

Knoll, A.H. (1989). Evolution and extinction in marine realm: some constraints imposed by microplankton. *Phil. Trans. R. Soc. B* **325:** 279–290.

Knoll, A.H. (1994). Proterozoic and Early Cambrian protists: evidence for accelerating evolutionary tempo. *Proc. Nat. Acad. Sci. U S A* **91:** 6743–6750.

Knoll, A.H. (2003a). The geological consequences of evolution. *Geobiology* **1:** 3–14.

Knoll, A.H. (2003b). *Life on a Young Planet*. Princeton, NJ, Princeton University Press.

Knoll, A.H. (2003c). Biomineralization and evolutionary history. *Rev. Mineral. Geochem.* **54:** 329–356.

Knoll, A.H., and Canfield, D.E. (1998). Isotopic inferences on early ecosystems. *Paleontol. Soc. Pap.* **4:** 212–243.

Knoll, A.H., and Golubic, S. (1992). Living and Proterozoic cyanobacteria. *Early Organic Evolution: Implications for Mineral Energy Resources.* M. Schidlowski, S. Golubic, M.M. Kimberley, D.M. McKirdy, and P.A. Trudinger, eds. Berlin, Springer-Verlag, pp. 450–462.

Knoll, A.H., Javaux, E.J., Hewitt, D., and Cohen, P. (2006). Eukaryotic organisms in Proterozoic oceans. *Phil. Trans. R. Soc. B* **361:** 1023–1038.

Knoll, A.H., and Swett, K. (1987). Micropaleontology across the Precambrian–Cambrian boundary in Spitsbergen. *J. Paleontol.* **61:** 898–926.

Knoll, A.H., and Swett, K. (1990). Carbonate deposition during the late Proterozoic era: an example from Spitsbergen. *Am. J. Sci.* **290-A:** 104–132.

Knoll, A.H., Swett, K., and Mark, J. (1991). The Draken Conglomerate Formation: paleobiology of a Proterozoic tidal flat complex. *J. Paleontol.* **65:** 531–569.

Kohnen, M.E.L., Sinninghe Damsté, J.S., Baas, M., Kock-van Dalen, A.C., and de Leeuw, J.W. (1993). Sulphur-bound steroid and phytane carbon skeletons in geomacromolecules: implications for the mechanism of incorporation of sulphur into organic matter. *Geochim. Cosmochim. Acta* **57:** 2515–2528.

Kohnen, M.E.L., Sinninghe Damsté, J.S., and de Leeuw, J.W. (1991). Biases from natural sulphurization in palaeoenvironmental reconstruction based on hydrocarbon biomarker distributions. *Nature* **349:** 775–778.

Kooistra, W.H.C.F., Gersonde, R., Medlin, L.K., and Mann, D.G. (2007). The origin and evolution of the diatoms: Their adaptation to a planktonic existence. *Evolution of Primary Producers in the Sea.* P. G. Falkowski and A.H. Knoll, eds. Boston, Elsevier, pp. 207–249.

Koopmans, M.P., Van Kaam-Peters, H.M.E., Schouten, S., de Leeuw, J.W., Sinninghe Damsté, J.S., Koster, J., Kenig, F., and Hartgers, W.A. (1996). Diagenetic and catagenetic products of isorenieratene: molecular indicators for photic zone anoxia: *Geochim. Cosmochim. Acta* **60:** 4467–4496.

Kopp, R.E., Kirschvink, J.L., Hilburn, I.A., and Nash, C.Z. (2005). The Paleoproterozoic snowball Earth: a climate disaster triggered by the evolution of oxygenic photosynthesis. *Proc. Nat. Acad. Sci. U S A* **102:** 11131–11136.

Kuypers, M.M.M., Lourens, L.J., Rijpstra, W.I.C., Pancost, R.D., Nijenhuis, I.A., and Sinninghe Damsté, J.S. (2004a). Orbital forcing of organic carbon burial in the proto-North Atlantic during oceanic anoxic event 2. *Earth Planet. Sci. Lett.* **228:** 465–482.

Kuypers, M.M.M., van Breugel, Y., Schouten, S., Erba, E., Sinninghe Damsté, J.S. (2004b). N_2-fixing cyanobacteria supplied nutrient N for Cretaceous oceanic anoxic events. *Geology* **32:** 853–856.

Lanier, W.P. (1986). Approximate growth rates of Early Proterozoic microstromatolites as deduced by biomass productivity. *Palaios* **1:** 525–542.

Litchman, E. (2007). Resource competition and the ecological success of phytoplankton. *Evolution of Primary Producers in the Sea.* P.G. Falkowski and A.H. Knoll, eds. Boston, Elsevier, pp. 351–375.

Logan, G.A., Hayes, J.M., Hieshima, G.B., and Summons, R.E. (1995). Terminal Proterozoic reorganisation of biogeochemical cycles. *Nature* **376:** 53–56.

Love, G.D., Snape, C.E., Carr, A.D., and Houghton, R.C. (1995). Release of covalently-bound alkane biomarkers in high yields from kerogen via catalytic hydropyrolysis. *Org. Geochem.* **23:** 981–986.

Maliva, R., Knoll, A.H., and Siever, R. (1989). Secular change in chert distribution: a reflection of evolving biological participation in the silica cycle. *Palaios* **4:** 519–532.

Marlowe, I.T., Green, J.C., Neal, A.C., Brassell, S.C., Eglinton, G., and Course, P.A. (1984). Long chain (*n*-C37-C39) alkenones in the Prymnesiophyceae. Distribution of alkenones and other lipids and their taxonomic significance. *Br. Phycol. J.* **19:** 203–216.

Marshall, C.P., Javaux, E.J., Knoll, A.H., and Walter, M.R. (2005). Combined micro-Fourier transform infrared (FTIR) spectroscopy and Micro-Raman spectroscopy of Proterozoic acritarchs: a new approach to palaeobiology. *Precambrian Res.* **138:** 208–224.

Martin, J.H., Knauer, G.A., Karl, D.M., and Broenkow, W.W. (1987). VERTEX: carbon cycling in the northeast Pacific. *Deep-Sea Res.* **34:** 267–285.

McCollom, T.M., and Seewald, J.S. (2006). Carbon isotope composition of organic compounds produced by abiotic synthesis under hydrothermal conditions. *Earth Planet. Sci. Lett.* **243:** 74–84.

McKirdy, D.M., Webster, L.J., Arouri, K.R., Grey, K., and Gostin, V.A. (2006). Contrasting sterane signatures in Neoproterozoic marine rocks of Australia before and after the Acraman asteroid impact. *Org. Geochem.* **37:** 189–207.

Mendelson, C.V., and Schopf, J.W. (1992). Proterozoic and selected Early Cambrian microfossils and microfossil-like objects. *The Proterozoic Biosphere: A Multidisciplinary Study*. J.W. Schopf and C. Klein, eds. Cambridge, UK, Cambridge University Press, pp. 867–951.

Miller, S.L. (1953). A production of amino acids under possible primitive Earth conditions. *Science* **117:** 527–528.

Moczydlowska, M. (1991). Acritarch biostratigraphy of the Lower Cambrian and the Precambrian–Cambrian boundary in southeast Poland. *Fossils Strata* **29:** 1–127.

Moldowan, J.M. (1984). C30-steranes, novel markers for marine petroleums and sedimentary rocks. *Geochim. Cosmochim. Acta* **48:** 2767–2768.

Moldowan, J.M., Dahl, J.E.P., Huizinga, B.J., Fago, F.J., Hickey, L.J., Peakman, T.M., and Taylor, D.W. (1994). The molecular fossil record of oleanane and its relation to angiosperms. *Science* **265:** 768–771.

Moldowan, J.M., Lee, C.Y., Watt, D.S., Jeganathan, A., Slougui, N.E., and Gallegos, E.J. (1991). Analysis and occurrence of C$_{26}$-steranes in petroleum and source rocks. *Geochim. Cosmochim. Acta* **55:** 1065–1081.

Moldowan, J.M., Seifert, W.K., and Gallegos, E.J. (1983). Identification of an extended series of tricyclic terpanes in petroleum. *Geochimicha et Cosmochimica Acta* **47:** 1531–1534.

Moldowan, J.M., and Talyzina, N.M. (1998). Biogeochemical evidence for dinoflagellate ancestors in the early Cambrian. *Science* **281:** 1168–1170.

Molyneux, S.G., Le Hérissé, A., and Wicander, R. (1996). Paleozoic phytoplankton. *Palynology: Principles and Applications*. J. Jansonius and D.C. McGregor, eds.

Tulsa, OK, American Association of Stratigraphic Palynologists Press, pp. 493–529.

Noffke, N., Eriksson, K.A., Hazen, R.M., and Simpson, E.L. (2006). A new window into Early Archean life: microbial mats in Earth's oldest siliciclastic tidal deposits (3.2 Ga Moodies Group, South Africa). *Geology* **34:** 253–256.

Peters, K.E., Walters, C.C., and Moldowan, J.M. (2005). *The Biomarker Guide,* 2nd ed. Cambridge, UK, Cambridge University Press.

Peterson, K.J., and Butterfield, N.J. (2005). Origin of the Eumetazoa: testing ecological predictions of molecular clocks against the Proterozoic fossil record. *Proc. Nat. Acad. Sci. U S A* **102:** 9547–9552.

Porter, S.M, Meisterfeld, R., and Knoll, A.H. (2003). Vase-shaped microfossils from the Neoproterozoic Chuar Group, Grand Canyon: a classification guided by modern testate amoebae. *J. Paleontol.* **77:** 409–429.

Prahl, F.G., de Lange, G.J., Lyle, M., and Sparrow, M.A. (1989). Postdepositional stability of long-chain alkenones under contrasting redox conditions. *Nature* **341:** 434–437.

Rampen, S.W., Schouten, S., Abbas, B., Panoto, F.E., Muyzer, G., Campbell, C.N., Fehling, J., and Sinninghe Damsté, J.S. (2005). The origin of 24-norcholestanes and their use as age-diagnostic biomarkers. *Organic Geochemistry: Challenges for the 21st Century.* 22 IMOG Seville, Spain, Abstract PB1–10.

Revill, A.T., Volkman, J.K., O'Leary, T., Summons, R.E., Boreham, C.J., Banks, M.R., and Denwer, K. (1994). Depositional setting, hydrocarbon biomarkers and thermal maturity of tasmanite oil shales from Tasmania, Australia. *Geochim. Cosmochim. Acta* **58:** 3803–3822.

Robinson, N., Eglinton, G., and Brassell, S.C. (1984). Dinoflagellate origin for sedimentary 4α-methylsteroids and 5α(H)-stanols. *Nature* **308:** 439–442.

Rohmer, M., Bisseret, P., and Neunlist, S. (1992). The hopanoids: prokaryotic triterpenoids and precursors of ubiquitous molecular fossils. *Biological Markers in Sediments and Petroleum.* J.M. Moldowan, P. Albrecht, and R.P. Philp, eds. New York, Prentice Hall, pp. 1–17.

Rohmer, M., Bouvier-Nave, P., and Ourisson, G. (1984). Distribution of hopanoid triterpenes in prokaryotes. *J. Gen. Microbiol.* **130:** 1137–1150.

Rosing, M.T., and Frei, R. (2004). U-rich Archaean seafloor sediments from Greenland—indications of >3700 Ma oxygenic photosynthesis. *Earth Planet. Sci. Lett.* **217:** 237–244.

Sanchez-Baracaldo, P., Hayes, P.K., and Blank, C.E. (2005). Morphological and habitat evolution in the Cyanobacteria using a compartmentalization approach. *Geobiology* **3:** 145–165.

Schopf, J.W. (1968). Microflora of the Bitter Springs Formation, Late Precambrian, central Australia. *J. Paleontol.* **42:** 651–688.

Schopf, J.W. (1993). Microfossils of the early Archean Apex chert—new evidence of the antiquity of life. *Science* **260:** 640–646.

Schopf, J.W. (2006). Fossil evidence of Archaean life. *Phil. Trans. R. Soc. B* **361:** 869–885.

Schopf, J.W., Kudryavtsev, A.B., Agresti, D.G., Wdowiak, T.J., and Czaja, A.D. (2002a). Laser-Raman imagery of Earth's earliest fossils. *Nature* **416:** 73–76.

Schopf, J.W., Kudryavtsev, A.B., Agresti, D.G., Wdowiak, T.J., and Czaja, A.D. (2002b). Laser-Raman spectroscopy: images of the Earth's earliest fossils? Reply. *Nature* **420:** 477.

Sergeev, V.N., Gerasimenko, L.M., and Zavarzin, G.A. (2002). The Proterozoic history and present state of cyanobacteria. *Microbiology* **71:** 623–637.

Sergeev, V.N., Knoll, A.H., and Grotzinger, J.P. (1995). Paleobiology of the Mesoproterozoic Billyakh Group, northern Siberia. *Paleontol. Soc. Mem.* **39:** 1–37.

Shen, Y., Knoll, A.H., and Walter, M.R. (2003). Evidence for low sulphate and deep water anoxia in a mid-Proterozoic marine basin. *Nature* **423:** 632–635.

Sinninghe Damsté, J.S., Muyzer, G., Abbas, B., Rampen, S.W., Masse, G., Allard, W.G., Belt, S.T., Robert, J.M., Rowland, S.J., Moldowan, J.M., Barbanti, S.M., Fago, F.J., Denisevich, P., Dahl, J., Trindade, L.A.F., and Schouten, S. (2004). The rise of the rhizosolenid diatoms. *Science* **304:** 584–587.

Sinninghe Damsté, J.S., Rampen, S., Rijpstra, W.I.C., Abbas, B., Muyzer, G., and Schouten, S. (2003). A diatomaceous origin for long-chain diols and mid-chain hydroxy methyl alkanoates widely occurring in Quaternary marine sediments: Indicators for high-nutrient conditions. *Geochim. Cosmochim. Acta* **67:** 1339–1348.

Sinninghe Damsté, J.S., Rijpstra, W.I.C., Schouten, S., Peletier, H., van der Maarel, M.J., and Gieskes, W.W.C. (1999a). A C_{25} highly branched isoprenoid alkene and C_{25} and C_{27} n-polyenes in the marine diatom *Rhizosolenia setigera*. *Org. Geochem.* **30:** 95–100.

Sinninghe Damsté, J.S., Schouten, S., Rijpstra, W.I.C., Hopmans, E.C., Peletier, H., Gieskes, W.W.C., and Geenevasen, J.A.J. (1999b). Structural identification of the C25 highly branched isoprenoid pentaene in the marine diatom *Rhizosolenia setigera*. *Org. Geochem.* **30:** 1581–1583.

Subroto, E.A., Alexander, R., and Kagi, R.I. (1991). 30-Norhopanes: their occurrence in sediments and crude oils. *Chem. Geol.* **93:** 179–192.

Summons, R.E., Bradley, A.S., Jahnke, L.L., and Waldbauer, J.R. (2006). Steroids, triterpenoids and molecular oxygen. *Phil. Trans. R. Soc. B* **361:** 951–968

Summons, R.E., Brassell, S.C., Eglinton, G., Evans, E., Horodyski, R.J., Robinson, N., and Ward, D.M. (1988b). Distinctive hydrocarbon biomarkers from fossiliferous sediments of the Late Proterozoic Walcott Member, Chuar Group, Grand Canyon, Arizona. *Geochim. Cosmochim. Acta* **52:** 2625–2637.

Summons, R.E., and Capon, R.J. (1988). Fossil steranes with unprecedented methylation in ring-A. *Geochimicha et Cosmochimica Acta* **52:** 2733–2736.

Summons, R.E., and Capon, R.J. (1991). Identification and significance of 3-ethyl steranes in sediments and petroleum. *Geochim. Cosmochim. Acta* **55:** 2391–2395.

Summons, R.E., Jahnke, L.L., Hope, J.M., and Logan, G.A. (1999). 2-Methylhopanoids as biomarkers for cyanobacterial oxygenic photosynthesis. *Nature* **400:** 554–557.

Summons, R.E., and Powell, T.G. (1987). Identification of aryl isoprenoids in source rocks and crude oils: biological markers for the green sulphur bacteria. *Geochim. Cosmochim. Acta* **51:** 557–566.

Summons, R.E., and Powell, T.G. (1992). Hydrocarbon composition of the Late Proterozoic oils of the Siberian Platform: implications for the depositional environment of the source rocks. *Early Organic Evolution: Implications for Mineral and Energy Resources.* M. Schidlowski, S. Golubic, M.M. Kimberley, D.M. McKirdy, and P.A. Trudinger, eds. Berlin, Springer-Verlag, pp. 296–507.

Summons, R.E., Powell, T.G., and Boreham, C.J. (1988a). Petroleum geology and geochemistry of the Middle Proterozoic McArthur Basin, Northern Australia: III. Composition of extractable hydrocarbons. *Geochim. Cosmochim. Acta* **52:** 1747–1763.

Summons, R.E., Taylor D., and Boreham, C.J. (1994). Geochemical tools for evaluating petroleum generation in Middle Proterozoic sediments of the McArthur Basin, Northern Territory, Australia. *Austral. Petroleum Explor. Assoc. J.* **34:** 692–706.

Summons, R.E., Thomas, J., Maxwell, J.R., and Boreham, C.J. (1992). Secular and environmental constraints on the occurrence of dinosterane in sediments. *Geochim. Cosmochim. Acta* **56:** 2437–2444.

Summons, R.E., Volkman, J.K., and Boreham, C.J. (1987). Dinosterane and other steroidal hydrocarbons of dinoflagellate origin in sediments and petroleum. *Geochim. Cosmochim. Acta* **51:** 3075–3082.

Summons, R.E., and Walter M.R. (1990). Molecular fossils and microfossils of prokaryotes and protists from Proterozoic sediments. *Am. J. Sci.* **290-A:** 212–244.

Suzuki, N., Yasuo, N., Nakajo, T., and Shine, H. (2005). Possible origin of 24-norcholesterol in marine environment. *Organic Geochemistry: Challenges for the 21st Century.* 22 IMOG Seville, Spain, Abstract OB2-2.

Talyzina, N.M. (2000). Ultrastructure and morphology of *Chuaria circularis* (Walcott, 1899) Vidal and Ford (1985) from the Neoproterozoic Visingso Group, Sweden. *Precambrian Res.* **102:** 123–134.

Talyzina, N.M., Moldowan, J.M., Johannisson, A., and Fago, F.J. (2000). Affinities of Early Cambrian acritarchs studied by using microscopy, fluorescence flow cytometry and biomarkers. *Rev. Palaeobot. Palynol.* **108:** 37–53.

Tappan, H. (1980). *The Paleobiology of Plant Protists.* San Francisco, Freeman.

Ten Haven, H.L., Rohmer, M., Rullkotter, J., and Bisseret, P. (1989). Tetrahymanol, the most likely precursor of gammacerane, occurs ubiquitously in marine sediments. *Geochim. Cosmochim. Acta* **53**: 3073–3079.

Tice, M.M., and Lowe, D.R. (2004). Photosynthetic microbial mats in the 3,416-Myr-old ocean. *Nature* **431**: 549–552.

Tice, M.M., and Lowe, D.R. (2006). Hydrogen-based carbon fixation in the earliest known photosynthetic organisms. *Geology* **34**: 37–40.

Tomitani, A., Knoll, A.H., Cavanaugh, C.M., and Ohno, T. (2006). The evolutionary diversification of cyanobacteria: molecular phylogenetic and paleontological perspectives. *Proc. Nat. Acad. Sci. U S A* **103**: 5442–5447.

Treibs, A. (1936). Chlorophyll and hemin derivatives in organic mineral substances. *Angew. Chem.* **49**: 682–686.

van Breugel, Y., Schouten, S., Paetzel, M., Ossebaar, J., Sinninghe Damsté, J.S. (2005). Reconstruction of $\delta^{13}C$ of chemocline CO2 (aq) in past oceans and lakes using the $\delta^{13}C$ of fossil isorenieratene. *Earth Planet. Sci. Lett.* **235**: 421–434.

van Kaam-Peters, H.M.E., Köster, J., van der Gaast, S.J., Dekker, M., de Leeuw, J.W., and Sinninghe Damsté, J.S. (1998). The effect of clay minerals on diasterane/sterane ratios. *Geochim. Cosmochim. Acta* **62**: 2923–2929.

Vermeij, G.J. (1977). The Mesozoic marine revolution; evidence from snails, predators and grazers. *Paleobiology* **3**: 245–258.

Vidal, G., and Moczydlowska, M. (1997). Biodiversity, speciation, and extinction trends of Proterozoic and Cambrian phytoplankton. *Paleobiology* **23**: 230–246.

Volkman, J.K. (2003). Sterols in microorganisms. *Appl. Microbiol. Biotechnol.* **60**: 496–506.

Volkman, J.K., Barrett, S.M., and Dunstan, G.A. (1994). C_{25} and C_{30} highly branched isoprenoid alkanes in laboratory cultures of two marine diatoms. *Org. Geochem.* **21**: 407–413.

Volkman, J.K., Eglinton, G., Corner, E.D.S., and Sargent, J.R. (1980). Novel unsaturated straight-chain C_{37}-C_{39} methyl and ethyl ketones in marine sediments and a coccolithophore *Emiliania huxleyi*. *Advances in Organic Geochemistry 1979*. A.G. Douglas and J.R. Maxwell, eds. London, Pergamon, pp. 219–227.

Volkova, N.G., Kirjanov, V.V., Piscun, L.V., Paškeviciene, L.T., and Jankauskas, T.V. (1983). Plant microfossils. *Upper Precambrian and Cambrian Palaeontology of the East European Platform*. A. Urbanek and A.Y. Rozanov, eds. Warsaw, Wydawnictwa Geologiczne. pp. 7–46.

Wignall, P.B., and Twitchett, R.J. (2002). Extent, duration and nature of the Permian-Triassic superanoxic event. *Geol. Soc. Am. Spec. Paper* **356**: 395–413.

Wray, J.L. (1977). *Calcareous Algae*. Amsterdam, Elsevier, Amsterdam.

Xiao, S., Knoll, A.H., Yuan, X., and Pueschel, C.M. (2004). Phosphatized multicellular algae in the Neoproterozoic Doushantuo Formation, China, and the early evolution of florideophyte red algae. *Am. J. Bot.* **91**: 214–227.

Xiao, S., Yuan, X., Steiner, M., and Knoll, A.H. (2002). Macroscopic carbonaceous compressions in a terminal Proterozoic shale: a systematic reassessment of the Miaohe biota, South China. *J. Paleontol.* **76**: 345–374.

Yoon, H.S., Hackett, J.D., Ciniglia, C., Pinto, G., and Bhattacharya, D. (2004). A molecular timeline for the origin of photosynthetic eukaryotes. *Mol. Biol. Evol.* **21**: 809–818.

Zhang, Y. (1981). Proterozoic stromatolite microfloras of the Gaoyuzhuang Formation (Early Sinian, Riphean), Hebei, China. *J. Paleontol.* **55**: 485–506.

Zumberge, J.E. (1987). Prediction of source rock characteristics based on terpane biomarkers in crude oils: a multivariate statistical approach. *Geochim. Cosmochim. Acta* **51**: 1625–1637.

Zumberge, J.E., Russell, J.A., and Reid, S.A. (2005). Charging of Elk Hills reservoirs as determined by oil geochemistry. *AAPG Bull.* **89**: 1347–1371.

Zundel, M., and Rohmer, M. (1985a). Hopanoids of the methylotrophic bacteria Methylococcus capsulatus and Methylomonas sp. as possible precursors of C_{29} and C_{30} hopanoid chemical fossils. *FEMS Microbiol. Lett.* **28**: 61–64.

Zundel, M., and Rohmer, M. (1985b). Prokaryotic triterpenoids 3. The biosynthesis of 2ß-methylhopanoids and 3ß-methylhopanoids of *Methylobacterium organophilum* and *Acetobacter pasteurianus* ssp. *pasteurianus*. *Eur. J. Biochem.* **150**: 35–39.

Zundel, M., and Rohmer, M. (1985c). Prokaryotic triterpenoids. 1. 3β-Methylhopanoids from Acetobacter sp. and Methylococcus capsulatus. *Eur. J. Biochem.* **150**: 23–27.

9

Life in Triassic Oceans: Links Between Planktonic and Benthic Recovery and Radiation

JONATHAN L. PAYNE AND BAS VAN DE SCHOOTBRUGGE

The Triassic Period, which lasted from 252 to 200 million years ago (Ma) (Gradstein *et al.* 2005), was named by Friedrich von Alberti in recognition of the three distinct units of sedimentary rock that characterize the period in Germany (Carle 1982). The Triassic was an important transitional interval in the history of marine phytoplankton. From the Mesoproterozoic through the Permian (1500–252 Ma), algae with "green" plastids (i.e., plastids with chlorophyll *b* as an accessory pigment) appear to have been the most abundant and diverse eukaryotic phytoplankton (Falkowski *et al.* 2004). The oldest diagnostic body fossils of dinoflagellates and coccolithophorids date from the Late Triassic. Along with diatoms, these phytoplankton lineages with "red" plastids (i.e., plastids with chlorophyll *c* as an accessory pigment) are the dominant eukaryotic phytoplankton in modern oceans (Falkowski *et al.* 2004). Differences in physiology and nutrient requirements between "green" and "red" lineages (Quigg *et al.* 2003) suggest that environmental and ecological changes in Triassic oceans are related to the beginning of the rise of "red" lineages to ecological prominence at this time.

Geographically, the Triassic world was maximally different from the present. The assembly of the supercontinent Pangaea had reached its zenith, leaving more than half of the Earth's surface covered by the Panthalassa Ocean, which spread from pole to pole and spanned nearly three quarters of the globe at the equator (Figure 1). Biologically, the Triassic was a time of transition. Extinction at the end of the preceding Permian Period occurred in less, and perhaps much less, than 500 thousand years ago (Bowring *et al.* 1998; Rampino and Adler 1998; Rampino *et al.* 2000; Twitchett *et al.* 2001) and eliminated approximately half of all marine animal families (Raup 1979) and close to 90% of all marine species (Raup 1979; Stanley and Yang 1994). Calcareous benthic algae were also severely affected by extinction (Flügel 1985, 2002). Direct effects on the diversity and abundance of eukaryotic phytoplankton are poorly known, although cyanobacteria and green sulfur bacteria appear to have increased in abundance at the time of extinction (Grice *et al.* 2005; Xie *et al.* 2005). No consensus has been reached regarding geological triggers and proximal environmental and ecological effects of end-Permian extinction. Primary candidates for a geological trigger are volatile release associated with eruption of the Siberian Traps (Renne and Basu 1991; Wignall 2001; Kamo *et al.* 2003; Grard *et al.* 2005) and bolide impact (Becker *et al.* 2001, 2004; Basu *et al.* 2003). (The Siberian Traps are the world's largest flood basalt province, with an estimated volume of more than 2.5 million km^3 erupted in less than 1 million years spanning the Permian–Triassic boundary [Reichow *et al.* 2002]. Eruption of the Siberian Traps through a thick sequence of coal and carbonate rocks may have caused the eruptions to carry an unusually large volatile load [Svensen *et al.* 2004; Erwin 2006].) Proposed proximal kill mechanisms include

environmental stresses such as anoxia (Wignall and Hallam 1992; Wignall and Twitchett 1996), hypercapnia (CO_2 poisoning) (Knoll *et al.* 1996), hydrogen sulfide poisoning (Kump *et al.* 2003, 2005), and acid rain associated with halogen poisoning (Visscher *et al.* 2004). The selectivity of extinction against animals with heavily calcified skeletons and limited ability to regulate gas exchange (Knoll *et al.* 1996) is unique among Phanerozoic extinctions, suggesting that conditions responsible for this extinction may have been generated only once since the Cambrian radiation of animals more than 500 Ma.

The reorganization of marine ecosystems during the Triassic Period (Figure 2) occurred at the grandest scale since the initial diversification of animals during the Cambrian and Ordovician periods (543–443 Ma). The proportion of marine animal diversity consisting of predators and motile organisms increased substantially (Bambach 2002; Bambach *et al.* 2002). Scleractinian corals started their initial radiation during the Middle Triassic, leading to their dominant role in the formation of coral reefs—now the world's most diverse marine ecosystems. Dinoflagellates and coccolithophorids, two of the most important groups of phytoplankton from the Mesozoic through the Recent, first appeared in the Triassic (Bralower *et al.* 1991; Fensome *et al.* 1996) and the diatoms may have acquired their algal symbiont at this time (Medlin *et al.* 1997). More broadly, the Triassic was a time of transition from a Late Paleozoic diversity plateau to the roughly exponential diversification in the marine realm that has continued from the Triassic up to the present day (Bambach 1999). Through changes in primary productivity and the composition of marine benthic communities, Triassic evolution laid the foundation for the Mesozoic Marine Revolution, an escalation of predatory and defensive interactions first recognized by increased defensive ornamentation in gastropods (Vermeij 1977) that may reflect a long-term increase in food energy supplied to the benthic fauna (Bambach 1993, 1999).

Early-Middle Triassic

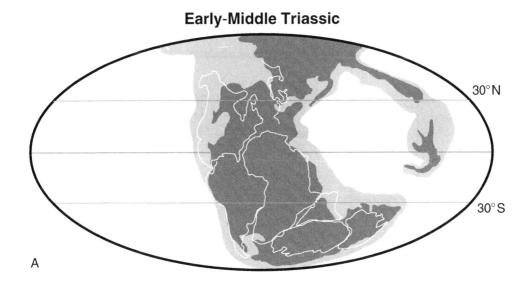

A

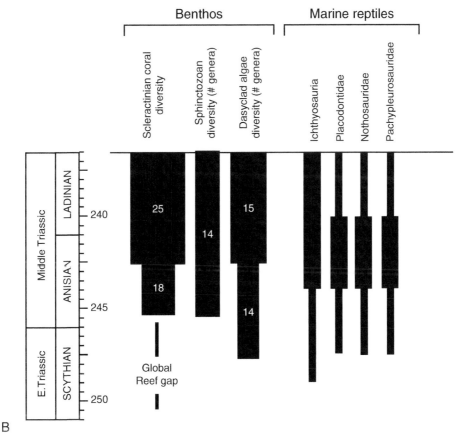

B

FIGURE 1. **(A)** Palaeogeographic reconstruction for the Early-Middle Triassic (modified after Scotese Paleomap Project 2000). **(B)** Diversity trends in Early to Middle Triassic benthic and plankton communities. Diversity of Triassic marine reptiles after Schoch and Wild (1999a, b). Diversity of Scleractinian corals and sphinctozoan sponges from Flügel (2002). Diversity of dasyclad algae from Bucur (1999).

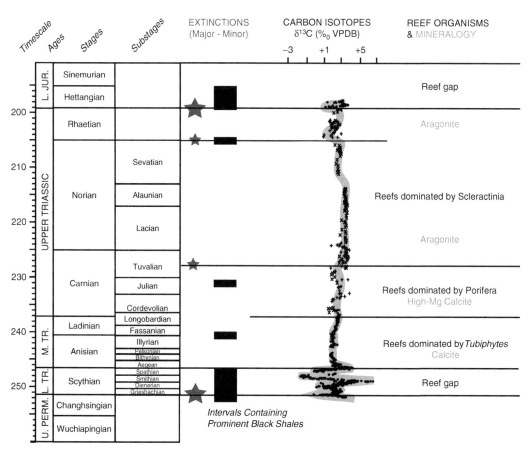

FIGURE 2. Triassic timescale, inorganic carbon isotope record, global diversity, and significant evolutionary events. Timescale is based upon the 2004 IUGS stratigraphic chart and radiometric constraints from (Mundil *et al.* 1996, 2001, 2004; Bowring *et al.* 1998; Muttoni *et al.* 2004a, b). Carbon isotope data for the Permian through Middle Triassic represented by filled circles are whole rock values from Payne *et al.* (2004). Data represented by + and x for the Late Triassic are from Korte *et al.* (2005). + represents data from brachiopod calcite and x represents whole rock values. Data across the Triassic–Jurassic boundary represented by open circles are whole rock values from Galli *et al.* (2005). (See color plate.)

In this chapter, we review the Triassic fossil record of marine animals, algae, and microbes. Specific aspects of this story have been addressed in considerable detail by other authors (Stanley 1988, 2003; Fensome *et al.* 1996; Flügel 2002), and we do not attempt to replicate those efforts. Rather, we first present a summary of Triassic planktic and benthic evolutionary patterns and then follow the lead of Signor and Vermeij (1994) in asking, specifically, how they may have been linked. In this chapter, we focus on marine Triassic life; we also consider conditions on the continents, however, because biolimiting nutrients, such as phosphorus and iron, are primarily delivered to the oceans through weathering and runoff. Because Triassic restructuring of marine ecosystems was so extensive, the period provides a nearly unparalleled view of the environmental and biological interactions that shape marine ecosystems.

I. BENTHOS

A. Benthic Wastelands of the Early Triassic

Whereas lively debate continues regarding the cause(s) of the end-Permian extinction, the consequences are increasingly well known, particularly with respect to benthic macroinvertebrates. The Early Triassic lasted approximately 5 Ma (Bowring et al. 1998; Martin et al. 2001; Mundil et al. 2001, 2004), only 10% of Triassic time (see Figures 1 and 2). Despite its short duration, the interval is biologically distinct from the Middle and Late Triassic. For example, local communities contained extremely low diversity (Schubert and Bottjer 1995; Rodland and Bottjer 2001; Twitchett et al. 2004; Fraiser and Bottjer 2005). Ecological tiering above and below the sediment surface was limited to a few tens of centimeters or less (Twitchett et al. 2004), and maximum size in gastropods, crinoids, and echinoids was greatly reduced (Price-Lloyd and Twitchett 2002; Fraiser and Bottjer 2004; Payne 2005; Twitchett 2005). Entire community types were absent, such as reefs constructed by skeletal animals and algae (Flügel 1994).

Taxonomic selectivity of end-Permian extinction against poorly buffered, heavily calcified suspension feeders (Knoll et al. 1996) is also reflected in the decreased abundance of these organisms in Early Triassic benthic communities. Not only did mollusks weather the mass extinction with comparatively high survival rates, they also became the most significant source of skeletal grains in nearly all Lower Triassic deposits, particularly prior to the Spathian (Figure 3A, B, and D) (Schubert and Bottjer 1995; Fraiser and Bottjer 2004; Twitchett et al. 2004), after playing a minor role in Upper Permian paleocommunities (Payne et al. 2006). Diversity and abundance patterns need not be coupled at this timescale and have been observed to decouple in the aftermath of catastrophic extinction (e.g., McKinney et al. 1998). In the case of the end-Permian event,

however, evidence suggests effects on diversity and abundance were coupled, at least at the highest taxonomic levels. Brachiopods and crinoids, both taxonomically hard-hit by extinction, remained at extremely low relative abundances through most of the Early Triassic (Schubert and Bottjer 1995; Fraiser and Bottjer 2004; Payne et al. 2006). The absence of many genera of gastropods from the Lower Triassic rock record, despite occurrences before and after ("Lazarus" genera) (Batten 1973; Erwin 1996), and the generally smaller sizes of fossils in Lower Triassic assemblages have been interpreted as evidence for reduced animal abundance (or biomass) in Early Triassic oceans (Wignall and Benton 1999, 2000; Twitchett et al. 2000; Twitchett 2001). Compositional analysis of a Late Permian to Late Triassic carbonate platform in southern China provides quantitative support for this interpretation (Payne et al. 2006). Twitchett (2001) suggested low productivity could explain reduced abundance and size of marine animals. Payne et al. (2006; see also Payne and Finnegan 2006) argue that differences in growth efficiency of the dominant organisms in Late Permian and Early Triassic communities (brachiopods, crinoids, and sponges versus mollusks) and changes in food web structure are likely to have been equally important. However, further quantification of changes in skeletal abundance from the Late Permian through the Middle Triassic from other locations is required to confirm the generality of changes in the abundance of skeletal animals through this interval, and the underlying mechanisms remain controversial.

Calcareous benthic algae are nearly absent from Lower Triassic rocks (Flügel 1985, 2002), the exceptions being the phylloid red alga Archaeolithophyllum reported from Lower Triassic (Griesbachian) strata in East Greenland (Wignall and Twitchett 2002b) and a dasyclad green alga from a latest Early Triassic (Upper Spathian) carbonate ramp in Iran (Baud et al. 1991). Calcareous green algae are otherwise unobserved from

Lower Triassic platform carbonates despite the persistence of several dasyclad genera from the Late Permian to the Middle Triassic (Flügel 1985; Bucur 1999).

Dominance of Early Triassic carbonate deposition by precipitated microbialites, ooids, and carbonate mud, a pattern more typical of Precambrian (Knoll and Swett 1990) than post-Ordovician strata, is increasingly recognized as a reflection of biological changes associated with extinction and recovery (Schubert and Bottjer 1992; Baud *et al.* 1997; Wignall and Twitchett 1999; Groves and Calner 2004; Pruss *et al.* 2005). In particular, a microbialite unit, often containing precipitated carbonate fabrics, overlies the extinction horizon in shallow marine carbonate facies in Panthalassa (Sano and Nakashima 1997) and across much of Tethys (Baud *et al.* 1997, 2005; Kershaw *et al.* 1999, 2002; Lehrmann 1999; Lehrmann *et al.* 2003) (see Figure 1C) and may reflect specific conditions associated with extinction.

Persistent low diversity, both locally and globally, appears in the eyes of many researchers to reflect slower than expected recovery, suggesting environmental inhibition of recovery prior to the Middle Triassic (e.g., Hallam 1991; Rodland and Bottjer 2001; Fraiser *et al.* 2005). Persistent carbon cycle instability during the Early Triassic (Payne *et al.* 2004) (see Figure 2) is also suggestive of episodic environmental disturbance through the Early Triassic. Identification of the source of continuing environmental disruption, whether from continued episodes of Siberian Traps volcanism, feedbacks between biological processes and environmental conditions, or other as yet unidentified processes, is required to reach an adequate understanding of Early Triassic evolutionary patterns.

Environmental conditions generally, and water-column chemistry specifically, are widely thought to have governed the rate and biogeographic structure of recovery from end-Permian extinction (Hallam 1991; Wignall and Newton 2003; e.g., Fraiser and Bottjer 2004; Twitchett *et al.* 2004). The occurrence of black shales in deep water sections in Japan (Isozaki 1997) and dark, unbioturbated to weakly bioturbated mudstones in shallow shelf sections from the Italian Dolomites and the western United States (Wignall and Hallam 1992, 1993, 1996; Wignall and Twitchett 1996, 2002b; Wignall *et al.* 1998) suggests persistent low oxygen conditions in deep water and at least intermittent low oxygen conditions in shelf environments (Wignall and Twitchett 2002a). Anoxia appears to have reached its greatest extent during the Griesbachian (Wignall and Twitchett 2002a), after which it was limited to deeper water environments (Isozaki 1997; Woods *et al.* 1999) during the second half of the Early Triassic (Wignall and Twitchett 2002a). Additional evidence for anoxia at the time of end-Permian extinction comes from the occurrence of isorenieratane, a biomolecule synthesized in the photosynthetic apparatus of the sulfide-oxidizing Chlorobiaceae, in sediments straddling the P/Tr boundary in Australia and southern China (Grice *et al.* 2005). Chlorobiaceae require anoxic, H_2S-laden water within the photic zone to perform photosynthesis. The deleterious effects of H_2S have also been suggested as a direct cause of extinction (Kump *et al.* 2005). Observations from the rock record have not been easy to replicate in numerical modeling studies, leading to disagreement regarding potential mechanisms underlying widespread marine anoxia (Hotinski *et al.* 2001; Zhang *et al.* 2001; Winguth and Maier-Reimer 2005).

B. Middle Triassic Recovery of Benthic Ecosystems

Accelerated biotic recovery within marine benthic communities began early in the Middle Triassic. Global diversity increased, as did ecological tiering (Ausich and Bottjer 1982; Twitchett *et al.* 2004). Maximum body size in gastropods increased substantially, both within local communities (Fraiser and Bottjer 2004) and globally (Payne 2005). Calcareous sponges returned to the rock record in the middle of the Anisian (Pelsonian) after their Early Triassic absence

(Figure 3I) (Riedel and Senowbari-Daryan 1991; Flügel 2002). The oldest known scleractinian corals are of approximately the same age (Figure 3F) (Flügel 2002; Stanley 2003). The reappearance of these taxa can be accounted for in several ways: calcifying lineages may have persisted at extremely low abundance through the Early Triassic, calcifying lineages may have survived in as yet unsampled refugia, calcifying lineages may have survived but failed to calcify due to environmental inhibition, or previously noncalcifying lineages that survived the end-Permian extinctions may have acquired the ability to calcify. Other calcifying groups, such as crinoids, recovered from the Spathian to Ladinian and became locally dominant clasts in shell beds (Ramovš 1996; Twitchett and Oji 2005). Rapid diversification of the widely distributed halobiid bivalves (*Halobia, Daonella, Aparimella,* and *Monotis*) (Figure 3J) during the Middle to Late Triassic has been related to the dispersal of halobiid larvae in the plankton (McRoberts 2000), providing one potential connection between changing conditions in the upper water-column and benthic evolution.

The record of bioturbation in Middle Triassic sediments also indicates accelerated recovery of benthic organisms. Although there is some ichnofossil evidence for gradual increase in size and behavioral diversity during the Early Triassic (Twitchett 1999; Pruss and Bottjer 2004), Griesbachian assemblages from middle to high latitudes (Zonneveld *et al.* 2002a) are nearly as diverse as Spathian assemblages from lower paleolatitudes (Pruss and Bottjer 2004). Bioturbation levels as a whole appear to have been depressed throughout this interval, implying a long-term reduction in sediment mixing rates (Twitchett 1999; Pruss and Bottjer 2004). Middle Triassic trace fossil assemblages, on the other hand, exhibit higher diversity and significantly larger maximum burrow diameter than any Early Triassic assemblage (Zonneveld *et al.* 2001; Zonneveld *et al.* 2002b; Pruss and Bottjer 2004).

Despite a host of evolutionary novelties, Middle Triassic reef ecosystems bore stronger taxonomic resemblance to their Permian counterparts than to later Mesozoic reefs (Stanley 1988). Through the Ladinian, scleractinian corals remained less important as reef-builders than holdover Permian-type reef organisms, such as *Tubiphytes* (Figure 3G, H), calcisponges (Figure 3I), bryozoans, and calcareous algae (Figure 3E) (Stanley 1988; Flügel 2002). Volumetrically significant early marine cements in Middle Triassic reefs (Flügel 2002) suggest continued high alkalinity in seawater, although platform geometry and depositional rate probably also played important roles governing reef accumulation and cementation (Seeling *et al.* 2005). The significant lag between the return of reefs on platform margins (early Anisian) and the return of reefs constructed by skeletal metazoans and algae (Ladinian and Carnian) indicates some decoupling of reef formation from the evolution of algae and metazoans. The earliest reefs may have formed in response to changes in some combination of seawater chemistry, carbonate platform geometry, and processes of sedimentation and sediment stabilization on platform margins rather than in response to animal and algal evolution. Once formed, it is possible that reefs served to accelerate diversification by providing a greater range of benthic microenvironments. Such a scenario is supported by the observation that modern reefs are not merely repositories of diversity but also sites of diversification (Briggs 2004). Mapping of environmental distributions of taxa through the Triassic can help to determine whether genera preferentially originated in reef environments or, rather, originated in other environments and then migrated to reefs.

Geochemical proxy records of Middle Triassic oceans are extremely sparse. Pronounced Early Triassic carbon cycle instability yields to persistent Middle Triassic stability early in the Anisian (see Figure 2) (Payne *et al.* 2004). Evidence for deep-water anoxia ends at approximately the same time

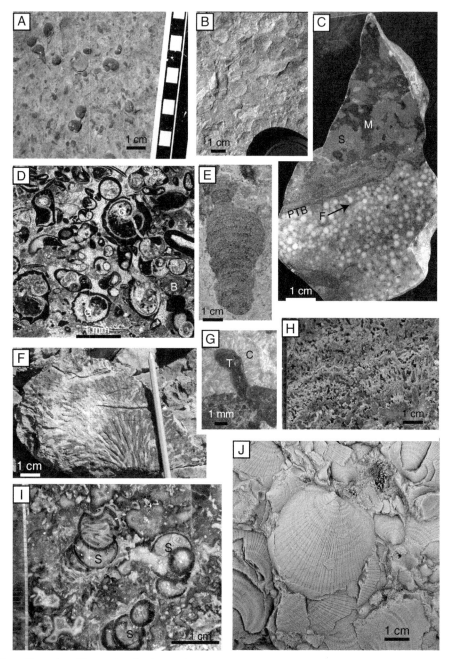

FIGURE 3. Examples of Triassic benthic fauna. (A) Assemblage of Lower Triassic gastropods from the Taskent section, Turkey. (B) Bedding plane assemblage of the pectinid bivalve *Claraia* from the Griesbachian of Guizhou Province, China. (C) Polished slab containing the contact between Upper Permian, fusulinid-bearing packstone and postextinction calcimicrobial boundstone. Sample is from the Longbai section on the Great Bank of Guizhou, China. Fusulinids (F) appear as white grains. The extinction horizon is marked PTB, sediment in the microbialite is marked S and the microbialite framework is marked M. (Courtesy of Daniel Lehrmann.) (D) Photomicrograph of Griesbachian packstone from the Great Bank of Guizhou containing common ornamented gastropods (G) and uncommon articulate brachiopods (B). (E) Ladinian calcareous alga from the Great Bank of Guizhou. (F) Norian colonial scleractinian coral from the Serro Monos Formation of Sonora, Mexico. (G) Photomicrograph of the reef-framework problematicum *Tubiphytes* in a sample from the Great Bank of Guizhou. (H) Polished slab of Anisian reef boundstone from the Great Bank of Guizhou. *Tubiphytes* framework appears white in the polished slab. Sediment and void-filling cements appear dark. (I) Polished slab of Anisian or Ladinian reef limestone from the Yangtze Platform of Guizhou Province, China containing several chambered, calcareous sponges (S). (J) *Halobia stryiaca,* from the Early Norian Hallstatt Limestone, Feuerkogel, Austria (courtesy Christopher McRoberts).

172

(Isozaki 1997; Sugitani and Mimura 1998). It remains unclear whether there is a causal link between carbon cycle stabilization and biotic recovery or simply a correlation, possibly reflecting a common forcing mechanism. Improved temporal resolution of combined paleontological and geochemical proxy records should shed light on whether carbon cycle stabilization was associated with the cause of accelerated recovery or was instead a consequence.

C. Late Triassic Benthic Boom: *Supersize Me*

During the Late Triassic, marine benthic communities developed ecological and taxonomic characteristics that would persist throughout the Mesozoic. Scleractinian corals thrived in deeper, more protected shelf settings during the Ladinian and Early Carnian and were excluded from shallow waters, where other organisms such as dasyclad algae (a commonly preserved group of benthic green algae including numerous calcifying lineages) enjoyed optimal light conditions (Car *et al.* 1981; Stanley 1988; Stanley and Cairns 1988). Following a poorly understood extinction event across the Carnian–Norian boundary (see Figure 2; Benton 1986), scleractinians became the dominant framework-builders of Late Triassic reef ecosystems and remained so for much of the Mesozoic and Cenozoic. At the Carnian–Norian boundary, approximately 90% of all coral species were rapidly replaced by new species distinctive of the so-called Norian–Rhaetian fauna (Stanley 1988; Stanley and Beauvais 1994). In contrast to their Middle Triassic predecessors, Norian and Rhaetian scleractinian corals were capable of building colonies several meters in diameter (Zankl 1969; Bernecker *et al.* 1999), resulting in spectacular accumulations of reef limestones more than 1200 m thick in the Dachstein Mountains of the Northern Calcareous Alps in Austria (Zankl 1977). Reefs blossomed from Europe

to the Indonesian archipelago (Martini *et al.* 2000) and along the southern Tethyan margin from Libya to Australia. Late Triassic reefs largely lacked the widespread, volumetrically significant early marine cementation that provided important structural reinforcement to many Middle Triassic reefs (Flügel 2002).

The Late Triassic rise of reef-building scleractinian corals was likely aided by the first acquisition of symbiotic dinoflagellates that provided corals with energy and enhanced their ability to calcify rapidly (Stanley 1988). Other factors such as oligotrophic surface water conditions and changes in ocean carbonate chemistry also may have played an important role. Riedel (1991) proposed that the symbiotic relationship evolved during the Norian in response to changing trophic conditions.

Megalodontid bivalves (superfamily Megalodontoidea), thick-shelled Paleozoic to early Mesozoic bivalves that likely harbored algal or microbial symbionts (Yancey and Stanley 1999), were ubiquitous inhabitants of Norian back reef areas. They experienced a dramatic increase in maximum body size from the Ladinian to Norian (Végh-Neubrandt 1982) and a concomitant increase in minimum size (Végh-Neubrandt 1982), suggesting a driving mechanism behind the trend (sensu McShea 1994). By Late Norian times, ornamented, high-spired gastropods typical of the Jurassic and Cretaceous dominated some communities, suggesting that Vermeij's (1977) Mesozoic Marine Revolution had its roots late in the Triassic (Nützel and Erwin 2004).

Extinction at the end of the Triassic was less severe than at the end of the Permian but it dramatically impacted marine benthic communities. Reef-building ceased for nearly 6 million years until the Lower Jurassic Pliensbachian and extinction was further associated with platform drowning along the northern and southern margins of the Tethys (Stanley 1988; Hallam and Wignall 1997; Galli *et al.* 2005). Extinction among bivalves was not size-selective, but epifaunal taxa had higher survivorship rates than infaunal taxa (McRoberts and Newton 1995). The causes of the end-Triassic mass extinction remain poorly understood. Terrestrial events, perhaps associated with eruption of the Central Atlantic Magmatic Province (a flood basalt province erupted during early rifting of the central Atlantic Ocean basin that is preserved in eastern North America, northern South America, and western Africa), are currently seen as the likely trigger of extinction (Ward *et al.* 2001, 2004; Hautmann 2004; Galli *et al.* 2005), although bolide impact is also a possibility (Olsen *et al.* 2002).

II. PLANKTON

A. Early Triassic Disaster Species

Body fossil evidence indicates that planktonic primary production during much of the Palaeozoic was dominated by green algae belonging to the Prasinophyceae (Chlorophyta) and Acritarcha—a phylogenetically meaningless group used to lump organic remains of uncertain affinity. Acritarchs are polyphyletic but most likely include members of green algal clades such as the Chlorophyceae (Arouri *et al.* 1999; Marshall *et al.* 2005). Acritarch diversity declined sharply during the Late Devonian to Carboniferous for reasons that are poorly understood, and diversity remained low during the Permian and Triassic. Hence, impoverished assemblages of Early-Middle Triassic phytoplankton were primarily the result of ecosystem changes prior to Permian extinctions. During the Triassic, the most common acritarch genera were *Micrhystridium, Baltisphaeridium*, and *Veryhachium*. Although these acritarch genera share a superficial resemblance with present-day dinoflagellates, they lack morphological characters distinctive of modern dinoflagellate cysts, such as excystment openings, paratabulation patterns on the cyst wall, and indications of a parasulcus and paragirdle. Furthermore, co-occurrences of dinoflagellates and acritarchs in the Early

Jurassic demonstrate substantial differences in environmental preferences (e.g., temperature, salinity, nutrients) (van de Schootbrugge *et al.* 2005).

Conditions in Early Triassic oceans appear to have been favorable for the development of widespread blooms of acritarchs and prasinophytes (Sarjeant 1973; Eshet *et al.* 1995; Twitchett *et al.* 2001). The proliferation of prasinophytes in association with widespread oceanic anoxia, particularly in the aftermath of mass-extinction events, has earned these primitive green algal flagellates the appellation "disaster species" (Tappan 1980). Substantial concentrations of prasinophyte phycomata (i.e., vegetative cysts) are associated with anoxic events during the Jurassic (e.g., Toarcian Posidonia shales) but have become less common since. Early Triassic green algae may have profited from strongly stratified surface ocean waters and elevated pCO_2 possibly generated during eruption of the Siberian Traps. These phytoplankton taxa may also have responded to extreme oligotrophy in the mixed layer due to generally sluggish oceanic circulation (Martin 1996).

Cyanobacteria, photosynthetic prokaryotes that dominate oligotrophic gyres of the present-day oceans, appear to have played a dominant role in primary production in the immediate aftermath of the end-Permian extinction, based on evidence from the molecular fossil record (Grice *et al.* 2005; Xie *et al.* 2005). Isorenieratane, a biomarker for the photoautotrophic, anoxygenic green sulfur bacteria, occurs in abundance near the end-Permian extinction horizon at the GSSP section at Meishan, China, and in the Perth Basin of western Australia (Grice *et al.* 2005), indicating potentially widespread areas of anoxic, sulfide-rich waters in the photic zone.

The lipid biomarker record for the remainder of the Early Triassic is less well known. Isorenieratane and related molecules occur in deep-water samples from southern China of Griebachian through Smithian age (R. Summons, unpublished data). Steranes preserved in Lower Trias-sic strata indicate significant changes in the eukaryotic fraction of planktonic commu-nities as well. In particular, dinosterane and related molecules increased in abundance in Lower Triassic strata after being rare in Late Paleozoic deposits (Summons *et al.* 1992; Moldowan and Jacobson 2000), suggest-ing dinoflagellates or their ancestors were present in moderate abundance at this time. The occurrence of molecular fossils of dino-flagellates predates the first appearance of unequivocal dinoflagellate cysts in the Late Triassic (Carnian) (see Figure 3 and Figure 5).

Medlin *et al.* (1997) hypothesized that the ability to form a resting stage, and the acquisition of a photosymbiont at this time may have allowed diatoms to survive and diversify during environmental disturbance associated with end-Permian extinction. Alternatively, Lee and Kugrens (2000) sug-gested that low pCO_2 during Permo-Car-boniferous glaciations may have favored the survival of lineages with secondary photo-symbionts (e.g., diatoms, dinoflagellates, and coccolithophorids) and the associated ability to concentrate CO_2. Whether or not these hypotheses are correct, the molecular and palynological records suggest that dia-toms, dinoflagellates, and coccolithophorids were not dominant members of the Early or Middle Triassic phytoplankton community. Their rise to ecological dominance appears to have been driven by later events.

B. Middle Triassic Oxygen and Evolution

The taxonomic composition of Middle Triassic phytoplankton is extremely poorly constrained. In the Muschelkalk Basin of Germany, phytoplankton floras occur as low-diversity assemblages that include the prasinophytes *Dictyotidium, Tasmanites,* and *Cymatiosphaera* and the ubiquitous acritarchs *Micrhystridium* and *Veryhachium*. Blooms of prasinophytes in the Muschelkalk presum-ably reflect pulses of fresh water runoff from large rivers draining the Fennoscandian

hinterland (Brocke and Riegel 1996), even though direct evidence for widespread delta deposits in NW Europe during this time is scarce. Middle Triassic phytoplankton was thus composed of taxa with strong Paleo-zoic affinities, a pattern parallel to the one observed in the evolution of benthic commu-nities. Molecular fossil data indicate increas-ing abundance of dinoflagellate biomarkers at this time (Summons *et al.* 1992; Moldowan and Jacobson, 2000); however, dinoflagel-late cysts are unknown from Middle Triassic strata (Fensome *et al.*, 1996).

Radiolarians formed the most important group of Triassic pelagic zooplankton. Their proliferation in Middle to Late Triassic oceans is evident from the widespread deposition of radiolarites and cherty carbonates (e.g., Torlesse Group, New Zealand, Aita and Spörli 1994; Austrian Alps, Rüffer 1999; Japan, Hori *et al.* 2003; Turkey, Tekin and Yurtsever 2003). Although radiolarians are generally inter-preted as indicators of eutrophic conditions, it is of interest to note that some pelagic colo-nial radiolarians in the present-day ocean harbor dinoflagellate symbionts of the spe-cies complex that also inhabits scleractinian corals (Gast and Caron 1996, 2001).

Predatory marine animals, particularly marine reptiles belonging to the Ichthyo-sauria, diversified rapidly during the Middle Triassic, in conjunction with closely related groups such as long-necked predators belong-ing to the Protorosauria (Li *et al.* 2004). The occurrence of derived ichthyosaurs in the Early Triassic, however, suggests that radiation was already under way by this time (Maisch and Matzke 2003). Other diversifying marine rep-tiles included the shell-crushing Placodontia and predatory Nothosauria (Schoch and Wild 1999a). The preservation of ichthyosaurs in black shale facies, such as the Serpiano Oil Shales of the Swiss–Italian border region, epitomize contrasting biological evidence for substantial atmospheric oxygen levels with models and sedimentary evidence indicating low atmospheric and water-column oxygen concentrations. Ichthyosaurs were faster and more continuous swimmers (Masare 1988)

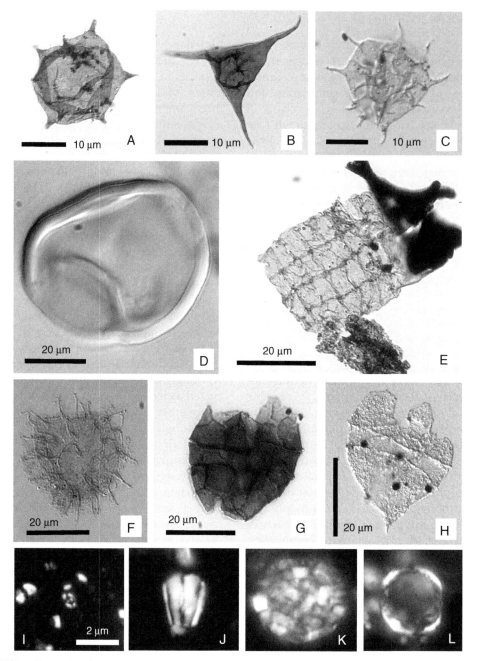

FIGURE 4. Examples of Late Triassic phytoplankton taxa. **(A)** Unidentified acritarch; Late Triassic, uppermost Rhaetian, Mingolsheim, Germany (van de Schootbrugge unpublished data). **(B)** *Veryhachium trispinosum*; Late Triassic, uppermost Rhaetian, Mingolsheim, Germany (van de Schootbrugge unpublished data). **(C)** *Micrhystridium* spp., Late Triassic, Middle Rhaetian, Benken, Switzerland (van de Schootbrugge unpublished data). **(D)** *Tasmanites tardus*; Tr–J boundary, St Audrie's Bay, United Kingdom (van de Schootbrugge *et al.* 2007). **(E)** *Plaesiodictyon mosellanum*; Late Triassic, uppermost Rhaetian, Mingolsheim, Germany (van de Schootbrugge unpublished data). **(F)** *Beaumontella langii*; Tr–J boundary, St Audrie's Bay, United Kingdom (van de Schootbrugge *et al.* 2007). **(G)** *Dapcodinium priscum*; Late Triassic, uppermost Rhaetian, Mingolsheim, Germany (van de Schootbrugge unpublished data). **(H)** *Rhaetogonyaulax rhaetica*; Late Triassic, Middle Rhaetian, Benken, Switzerland (van de Schootbrugge unpublished data). **(I)** *Crucirhabdus minutus*; Late Triassic, Rhaetian, Fischerwiese, Austria (courtesy Paul Bown). **(J)** *Eoconusphaera zlambachensis*; Late Triassic, Rhaetian, Fischerwiese, Austria (courtesy Paul Bown). **(K)** *Prinsiosphaera triassica*; Late Triassic, Rhaetian, Fischerwiese, Austria (courtesy Paul Bown). **(L)** *Orthopithonella geometrica*; Late Triassic, Rhaetian, Fischerwiese, Austria (courtesy Paul Bown).

than other major groups of Mesozoic marine reptiles (mosasaurs, plesiosaurs, and marine crocodiles). Such a lifestyle presumably demanded relatively high blood oxygen levels. Also, numerous finds of female ichthyosaurs and other Triassic marine reptiles with preserved young in the body cavity indicate these reptiles were viviparous (Cheng *et al.* 2004). Evolution of viviparity likewise appears to require high ambient concentrations of atmospheric oxygen (Andrews 2002). These observations contrast with model calculations of extremely low Middle Triassic pO_2 (Berner 2005), although the precise limiting oxygen level for these extinct marine reptiles is not known. More importantly, the temporal mismatch between sedimentary evidence for maximum anoxia during the Early Triassic (Wignall and Twitchett 2002a) and model reconstructions of minimum atmospheric pO_2 during the Middle Triassic (Berner 2005) suggest either poor model resolution and/or decoupling between atmospheric pO_2 and marine anoxia. Oceanographic factors affecting seawater oxygen content such as changing degrees of stratification and consequent bottom-water anoxia can explain much of the discrepancy. Suggestions that low atmospheric oxygen levels played a role in terrestrial extinction at the end of the Permian (Huey and Ward 2005) are likewise complicated by the occurrence of minimum pO_2 during Middle Triassic rediversification and Berner's (2005) calculation that pO_2 did not increase substantially until the Cretaceous, more than 100 million years later. Geochemical constraints from molybdenum, iron, and multiple sulfur isotopes can provide tests of the degree of anoxia in Triassic seawater (Anbar 2004; Johnston *et al.* 2005; Rouxel *et al.* 2005). Empirical constraints on atmospheric oxygen levels through the Triassic are greatly needed but may be more difficult to achieve.

C. Late Triassic Rise of Modern Phytoplankton

The Carnian–Norian boundary was a time of important innovation for calcifying plankton. Calcareous nannoplankton (Haptophyta) started their ascent to become the principal Mesozoic oceanic exporters of carbonate (see Figure 5) (Bown *et al.* 2004). Some of the oldest genera—*Prinsiosphaera* and *Crucirhabdus*—are common in the Austrian Alps (Jafar 1983) and in Australia (ODP leg 122) (Bralower *et al.* 1991). Both of these occurrences are closely related to reef environments, highlighting the importance of shallow shelf environments in early phytoplankton evolution (Katz *et al.* 2004). Although haptophytes do not appear to have played a globally significant role in carbonate export to the open ocean until the Middle Jurassic (Martin 1996), Bellanca *et al.* (1995) have argued that already during the Carnian open ocean micritic limestones in Italy (Sicani Mountains, Sicily) resulted from calcareous nannofossil accumulations. Quantitative data on nannofossil abundance are generally absent for the Late Triassic and much of the evidence is, at present, anecdotal (e.g., Di Nocera and Scandone 1977).

Global diversification of cyst-producing dinoflagellates occurred across the Carnian–Norian boundary. Thecate dinoflagellates are organic-walled phytoplankton, many of which form resting cysts consisting of extremely resistant, sporopollenin-like biopolymers. Although the first unequivocal dinoflagellate cysts occur in the Late Carnian of Arctic Canada (Bujak and Fisher 1976), rapid diversification across the Carnian–Norian boundary is apparent in widely distributed, diverse assemblages from Austria (Morbey 1975, 1978) to Australia (Brenner 1992; Backhouse *et al.* 2002). The Suessiaceae, one of the oldest and probably also most basal lineages of cyst-producing dinoflagellates (LaJeunesse, personal communication 2005), display a characteristic paratabulation pattern consisting of 7 to 10 latitudinal series (Fensome *et al.* 1996; Bucefalo Palliani and Riding 2003). The suessioid tabulation type is only known from the extant Symbiodiniaceae, the most important symbionts of corals. This morphological link between Triassic Suessiaceae and extant *Symbiodinium*

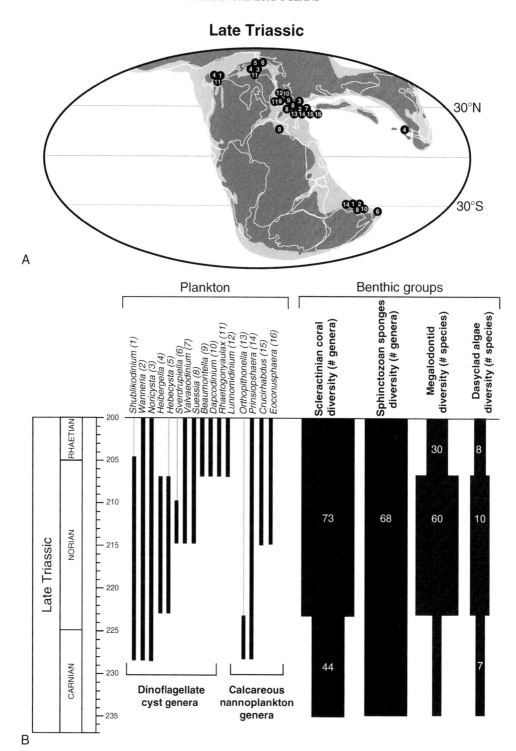

FIGURE 5.   For legend see opposite page.

(Loeblich and Sherley 1979) further supports the hypothesis that the Late Triassic diversification of corals and their success in reef environments were related to the acquisition of photosymbionts (Stanley 1988, 2003).

The biogeographic distribution of suessiacean dinoflagellates matches that of Late Triassic reefs. Norian–Rhaetian Suessiaceae cysts are only known from paleolatitudes between 40°N and 40°S, a range that bracketed reef development during the Late Triassic (Kiessling 2001). The oldest suessiacean genus, *Noricysta,* appears to have originated in low latitudes where it first occurs in sediments intercalated with reef carbonates (Hochuli and Frank 2000). The Norian–Rhaetian Suessiaceae dinoflagellate cyst genera *Suessia, Wanneria,* and *Beaumontella* often occur in sediments that are either intercalated with reef deposits or closely associated with them (Brenner 1992). In Austria, *Beaumontella* has even been found in coral boundstone (Holstein 2004). The close association between suessiacean dinoflagellates and reefs suggests that physical proximity played an important role in the evolution of scleractinian–dinoflagellate symbiosis.

The Late Triassic was also an eventful time for many motile planktonic animals. Aulacoceratids, predecessors of the Belemnitida, a group of squidlike predators that diversified rapidly during the Jurassic and Early Cretaceous, first appeared during the Ladinian–Carnian (Gustomesov 1978).

Despite their initial Middle Triassic radiation, only a few ichthyosaur species are known from the Late Triassic (Motani 2005). Large marine reptiles radiated again during the Early Jurassic and then declined during the Cretaceous, possibly due to competition with sharks (Masare 1988). Conodont animals reached the end of their long evolutionary history, gradually disappearing during the Triassic to become extinct at the end of the period (Hallam and Wignall 1997).

The radiation of coccolithophorids and dinoflagellates was severely disrupted by extinction at the end of the Triassic Period, an event that eliminated as much as 80% of all marine species (Newell 1963). *Crucirhabdus* was the sole coccolithophorid genus and *Beaumontella* and *Dapcodinium* were the only dinoflagellate genera to cross into the Jurassic, indicating extinction substantially affected both groups. The early Jurassic is characterized by the proliferation of green algae, including many prasinophyte taxa (van de Schootbrugge *et al.* 2005, 2007). Van de Schootbrugge *et al.* (2005, 2007) speculate that the predominance of green algal phytoplankton and the general absence of dinoflagellates and coccolithophorids from the Hettangian through Lower Pliensbachian were related to elevated pCO_2 and renewed intensification of oceanic anoxia as a result of stratification. If this is the case, similarities between the aftermaths of end-Permian and end-Triassic extinctions may reflect similar causes, for example environmental changes associated with massive volcanism.

FIGURE 5. **(A)** Palaeogeographic reconstruction for the Late Triassic (modified after Scotese Paleomap Project 2000). **(B)** Red phytoplankton diversity and range chart with Late Triassic dinoflagellate and coccolithophorid genera; (1) *Shublikodinium,* Alaska (Wiggins 1973), Australia (Helby 1987); (2) *Wanneria,* Australia (Helby *et al.* 1987; Brenner 1992), Austria (Holstein 2004); (3) *Noricysta,* Arctic Canada (Bujak and Fisher 1976), Switzerland (Hochuli and Frank 2000); (4) *Heibergella,* Alaska (Wiggins 1973), Arctic Canada (Bujak and Fisher 1976), Timor (Martini *et al.* 2000); (5) *Hebecysta,* Arctic Canada (Bujak and Fisher 1976); (6) *Sverdrupiella,* Arctic Canada (Bujak and Fisher 1976), New Zealand (Helby and Wilson 1988); (7) *Valvaeodinium,* Austria (Morbey 1975); (8) *Suessia,* Austria (Morbey 1975, 1978; Holstein 2004); (9) *Beaumontella,* England (Wall 1965; van de Schootbrugge *et al.* 2007); (10) *Dapcodinium,* Denmark (Evitt 1961), Australia (Backhouse *et al.* 2002); (11) *Rhaetogonyaulax,* England (Sarjeant 1973; Harland *et al.* 1975), Germany (van de Schootbrugge unpublished data); (12) *Lunnomidinium,* Sweden (Lindström 2002); (13) *Tetralithus;* (14) *Prinsiosphaera* (Jafar 1983; Bralower *et al.* 1991); (15) *Crucirhabdus,* Austria (Jafar 1983); (16) *Eoconusphaera,* Austria (Jafar 1983). Diversity of scleractinian corals and sphinctozoan sponges from Flügel (2002). Diversity of dasyclad algae from Bucur (1999). Diversity of megalodontid bivalves from Végh-Neubrandt (1982).

III. BENTHIC–PLANKTONIC COUPLING IN TRIASSIC OCEANS

Both planktonic and benthic ecosystems were severely disrupted by end-Permian events. A few abundant taxa dominated benthic and planktic communities during the Early Triassic. The record of plankton, although limited, suggests low diversity blooms of prasinophytes and acritarchs and possibly an important role for cyanobacteria (Xie *et al.* 2005). Benthic assemblages similarly reflect poorly integrated communities dominated by a few taxa with patchy distributions (e.g., Schubert and Bottjer 1995; Rodland and Bottjer 2001). Persistent carbon cycle instability hints at continued environmental disturbances, although it could alternatively reflect a loss of biological dampening of underlying carbon cycle instability (Payne *et al.* 2004). Accelerated Middle Triassic recovery is most apparent in the benthos. It is indicated by increases in the maximum size and diversity of benthic organisms, the return of calcareous algae and sponges, and the first appearance of scleractinian corals. However, Permian taxonomic affinities characterize the taxa that dominated the initial phase of recovery (Stanley 1988; Erwin 1990). Increased abundance of dinoflagellate biomarkers in the molecular fossil record suggests parallel recovery in the plankton, but the complete absence of preserved dinoflagellate cysts indicates possible decoupling of the abundance of preserved dinoflagellate cysts from the abundance of dinoflagellates in the phytoplankton. At present, it is unclear whether Middle Triassic phytoplankton were dominated by Palaeozoic holdovers or by groups that radiated during the Late Triassic. Regardless, it was not until the Late Triassic that ecosystems took on a distinctly Mesozoic structure. Parallels between benthic and planktonic evolutionary trajectories are clear, but how were the evolution of benthic and planktonic communities connected? Three possibilities exist: (1) the patterns are entirely unrelated—each is driven by independent mechanisms; (2) the patterns are controlled by the same driver(s) but without any direct connections between benthic and planktic evolution; and (3) interactions between planktic and benthic evolution helped to shape Triassic recovery and radiation. Because phytoplankton supply much of the food to benthic ecosystems, we view the first possibility as untenable and address the latter two.

A. Common Driver

One possible environmental control on Triassic evolution in the benthos and plankton is nutrient limitation. For example, expanded anoxia in Early Triassic shallow and deep waters (Wignall and Twitchett 2002a) could have affected the availability of fixed nitrogen and trace metal nutrients (Katz *et al.* 2004; Fennel *et al.* 2005). The occurrence of biomarkers for green sulfur bacteria in Lower Triassic sediments (Grice *et al.* 2005; Summons and Love, unpublished data) and size distributions of pyrite framboids (Wignall *et al.* 2005) are suggestive of photic zone euxinia (anoxia with free hydrogen sulfide) in the Early Triassic water column, at least on a local scale. In a euxinic water column, trace metal nutrients, such as Fe and Mo, required for enzymes involved in key functions such as nitrogen fixation are scarce in the water column, potentially leading to severe nitrogen limitation (Anbar and Knoll 2002). Resulting nitrogen stress would help account for the elevated proportional contribution of cyanobacteria to organic matter in Lower Triassic sediments because cyanobacteria, unlike eukaryotic algae, are able to fix nitrogen and thus compete well under conditions of nitrogen limitation. Regional nitrogen limitation could have reduced levels of primary production and the food supply to benthic ecosystems, thereby limiting biotic recovery. It is unlikely that pervasive euxinia, such as has been proposed for the Proterozoic Ocean (Canfield 1998; Anbar and Knoll 2002), could have existed in Early Triassic oceans given the much higher atmospheric pO_2 during the Triassic. It is more likely that Triassic anoxia and/or euxinia was local or regional, although geological and geochemical evidence is widespread. Waning anoxia in the Middle Triassic, under this scenario, could account for resulting radiations in the planktonic and benthic realms. The possibility of such a scenario can be further constrained through continued investigation of proxies for local and global anoxia through the molecular fossil and geochemical records, as well as modeling of Triassic ocean dynamics.

B. Plankton Control

In addition to food quantity, food quality may also have changed through the Triassic. The molecular fossil and micropaleontological record suggests that cyanobacteria and a low-diversity assemblage of green algae played important roles in primary production, particularly early in the Early Triassic (Grice *et al.* 2005; Xie *et al.* 2005). Some bloom-forming cyanobacteria, such as *Nodularia,* are toxic and have direct deleterious effects on zooplankton and macroinvertebrate communities in marine and freshwater settings (Sellner 1997; White *et al.* 2005). Transfer of carbon from cyanobacteria to invertebrate consumers is generally inefficient (De Bernardi and Guiussani 1990; von Elert *et al.* 2003; Martin-Creuzburg *et al.* 2005), in part because cyanobacteria lack the sterols used by many marine invertebrates as high–molecular-weight precursors in the synthesis of cholesterol (Kanazawa 2001). Consequently, animals such as bivalves and crustaceans fed on cyanobacteria grow more slowly and produce less egg mass than individuals grown on a diet containing eukaryotic algae (von Elert *et al.* 2003; Wacker and von Elert 2003). Thus, increased cyanobacterial contribution to primary production during the Early Triassic (Xie *et al.* 2005) could help to explain low animal abundance (Wignall and Benton 1999; Payne *et al.* 2006) and reduced maximum size (Twitchett 2001; Fraiser and Bottjer 2004; Payne 2005) at this time.

C. Feedback from the Benthos

Burrowing by macrobenthic invertebrates increases the supply of oxygen and other oxidants to sediments, thereby increasing the efficiency of organic remineralization as well as the return of buried nutrients to the water column (Aller 1982; Thayer 1983). Reduced bioturbation during the Early Triassic (Twitchett 1999; Pruss and Bottjer 2004), therefore, could have decreased the nutrient load in the water column by limiting the exposure of organic matter to oxygen and other oxidants, thereby allowing more efficient burial of organic matter and limiting rates of nutrient recycling. Increased depth and intensity of bioturbation during the Middle Triassic would be expected to decrease the burial efficiency of organic matter and thereby increase the quantity of essential nutrients available to phytoplankton. The magnitude of this effect is poorly documented. Based on modeling and observations, however, Thayer (1983) suggested that rates of bioturbation may have increased by as much as an order of magnitude through the Phanerozoic and thereby increased nutrient availability by a comparable amount. Observations of modern sediments suggest the first 100 to 1000 years of oxygen exposure strongly reduce burial efficiency (Hartnett *et al.* 1998), causing burial efficiency to vary by as much as a factor of five. On the other hand, anoxia is generally argued to enhance the efficiency of phosphorus recycling from organic matter (e.g., Van Cappellen and Ingall 1994). Increased consumption of organic carbon by epifaunal or pelagic organisms within the water column as the food web increased in efficiency through the Triassic would also be expected to increase the recycling of nutrients prior to burial. Environmental stabilization during the Middle Triassic may have presented more favorable conditions to both the benthos and plankton, allowing this positive feedback to operate. Further quantification is required in several areas, including the relationship between nutrient availability and export production, the competing effects of bioturbation and oxygen exposure on organic carbon burial efficiency versus the recycling of phosphorus under anoxic conditions and the effect of food-web structure on the fraction of primary production that reaches the sediment surface. The ichnofossil record, itself, also remains poorly quantified in terms of the size and abundance of burrows through the Triassic and the diversity of behaviors represented.

D. Assistance from the Plankton

One of the most direct couplings of plankton and benthic ecosystems during the Triassic was through symbiotic interactions. Based on carbon isotope records from exceptionally preserved Late Triassic corals from Turkey, Stanley and Swart (1995) speculated that the coral–dinoflagellate symbiosis gave these symbiont-bearing corals an edge over other Late Triassic reef-building organisms through faster growth rates. Hence, the evolution of dinoflagellates during the Triassic was of central importance for the evolution of Late Triassic coral reef ecosystems. Reefs are net sources for diversity in the modern ocean (Briggs 2004). Given the close relationship between reef development through the Triassic and benthic diversification, a detailed understanding of the biological and physical aspects of reefs that facilitate diversification could shed light on the role of reefs in the evolution of Triassic benthic communities. If the physical structure and creation of microenvironments by framework organisms is of prime importance, then the acquisition of photosymbionts in scleractinian corals may have played a central role in Triassic diversification. Middle Triassic reefs are characterized by low-growing organisms such as *Tubiphytes* that likely generated little relief on the seafloor. Late Triassic coral reefs probably provided much more physical relief on the shallow seafloor and may have allowed organisms to diversify in a greater variety of microenvironments.

IV. CONCLUSIONS

The Triassic Period was an interval of transition for benthic and planktonic marine ecosystems in terms of taxonomic composition, ecological structure, nutrient requirements, and biogeochemical cycles. The relative importance of common environmental driving forces versus direct or indirect interactions between benthic and planktonic organisms in Triassic evolution remains to be determined. Along with continuing environmental disturbances, changes in the abundance, nutritional value, and availability of food for benthic communities coming from primary producers appears to have played a significant role during the protracted recovery from the end-Permian mass extinction and subsequent Triassic evolution. Nutrient availability, the taxonomic and biochemical composition of the most abundant primary producers, nutrient recycling via bioturbation, and coupling of animal and algal evolution through symbiosis are all potentially significant factors coupling planktonic and benthic evolution through the Triassic. Quantitative assessment of which factors were more or less significant is beyond the resolution of existing data. However, improved resolution of the molecular fossil record, ichnofossil record, proxies for seawater chemistry, and the fossil record of macroinvertebrates all have the potential to refine interpretations of Triassic evolution. Despite the fact that data are not sufficient to provide definitive answers at this time, we see many avenues open for continued exploration, particularly with a multidisciplinary, multiproxy approach.

Events during the Triassic set the stage for the later expansion of coccolithophorids and dinoflagellates and lit the fuse for the benthic escalation of the Mesozoic Marine

Revolution. Improved understanding of life in Triassic oceans is likely to illuminate the general principles underlying much of subsequent Mesozoic and Cenozoic evolution in the marine realm.

Acknowledgments

We thank P. Falkowski and A. Knoll for the invitation to write this essay and for their careful reviews of the manuscript. We also thank A. Bush, D. Erwin, S. Finnegan, W. Oschmann, S. Pruss, and F. Tremolada for constructive comments on the manuscript. We also received generous assistance from several colleagues: C. Korte provided carbon isotope data, R. Summons and G. Love allowed access to unpublished data, and P. Bown, D. Lehrmann, C. McRoberts, and F. Tremolada provided photographs used in the figures.

References

Aita, Y., and Spörli, K.B. (1994). Late Triassic radiolaria from the Torlesse Terrane, Rimutaka Range, North Island, New Zealand. *N Z J. Geol. Geophys.* **37**: 155–162.

Aller, R.C. (1982). The effects of macrobenthos on chemical properties of marine sediment and overlying water. *Animal-Sediment Relations.* P.L. Mccall and M.J.S. Tevesz, ed. New York, Plenum Press, pp. 53–102.

Anbar, A.D. (2004). Molybdenum stable isotopes: observations, interpretations and directions. *Geochemistry of Non-Traditional Stable Isotopes.* C. Johnson, ed. Washington, DC, Mineralogical Society of America, pp. 429–454.

Anbar, A.D., and Knoll, A.H. (2002). Proterozoic ocean chemistry and evolution: a bioinorganic bridge? *Science* **297**: 1137–1142.

Andrews, R.M. (2002). Low oxygen: a constraint on the evolution of viviparity in reptiles. *Physiol. Biochem. Zool.* **75**: 145–154.

Arouri, K., Greenwood, P.F., and Walter, M.R. (1999). A possible chlorophycean affinity of some Neoproterozoic *acritarchs. Org. Geochem.* **30**: 1323–1337.

Ausich, W.I., and Bottjer, D.J. (1982). Tiering in suspension-feeding communities on soft substrata throughout the Phanerozoic. *Science* **216**: 173–174.

Backhouse, J., Balme, B.E., Helby, R., Marshall, N.G., and Morgan, R. (2002). Palynological zonation and correlation of the latest Triassic, Northern Carnarvon Basin. *The Sedimentary Basins of Western Australia.* M. Keep and S.J. Moss, eds. Perth, Petroleum Exploration Society of Australia, pp. 179–201.

Bambach, R.K. (1993). Seafood through time—changes in biomass, energetics, and productivity in the marine ecosystem. *Paleobiology* **19**: 372–397.

Bambach, R.K. (1999). Energetics in the global marine fauna: a connection between terrestrial diversification and change in the marine biosphere. *Geobios* **32**: 131–144.

Bambach, R.K. (2002). Supporting predators: changes in the global ecosystem inferred from changes in predator diversity. *The Fossil Record of Predation.* M. Kowalewski and P.H. Kelley, eds. New Haven, The Paleontological Society, pp. 319–352.

Bambach, R.K., Knoll, A.H., and Sepkoski, J.J. (2002). Anatomical and ecological constraints on Phanerozoic animal diversity in the marine realm. *Proc. Natl. Acad. Sci. U S A* **99**: 6854–6859.

Basu, A.R., Petaev, M.I., Poreda, R.J., Jacobsen, S.B., and Becker, L. (2003). Chondritic meteorite fragments associated with the Permian-Triassic boundary in Antarctica. *Science* **302**: 1388–1392.

Batten, R.L. (1973). The vicissitudes of the Gastropoda during the interval of Guadalupian-Ladinian time. *The Permian and Triassic Systems and Their Mutual Boundary.* A. Logan and L.V. Hills, eds. Calgary, Canadian Society of Petroleum Geologists, pp. 596–607.

Baud, A., Brandner, R., and Donofrio, D.A. (1991). The Sefid Kuh limestone—a late Lower Triassic carbonate ramp (Aghdarband, NE-Iran). *Abhandlungen der Geologisches Bundesanstalt* **38**: 111–123.

Baud, A., Cirilli, S., and Marcoux, J. (1997). Biotic response to mass extinction: the lowermost Triassic microbialites. *Facies* **36**: 238–242.

Baud, A., Richoz, S., and Marcoux, J. (2005). Calcimicrobial cap rocks from the basal Triassic units: western Taurus occurrences (SW Turkey). *Comptes Rendus Palevol* **4**: 501–514.

Becker, L., Poreda, R.J., Basu, A.R., Pope, K.O., Harrison, T.M., Nicholson, C., and Iasky, W. (2004). Bedout: a possible end-Permian impact crater offshore of Northwestern Australia. *Science* **304**: 1469–1476.

Becker, L., Poreda, R.J., Hunt, A.G., Bunch, T.E., and Rampino, M. (2001). Impact event at the Permian Triassic boundary: evidence from extraterrestrial noble gases in fullerenes. *Science* **291**: 1530–1533.

Bellanca, A., DiStefano, P., and Neri, R. (1995). Sedimentology and isotope geochemistry of Carnian deep-water marl/limestone deposits from the Sicani Mountains, Sicily: environmental implications and evidence for a planktonic source of lime mud. *Palaeogeogr. Palaeoclimatol. Palaeoecol.* **114**: 111–129.

Benton, M.J. (1986). More than one event in the Late Triassic mass-extinction. *Nature* **321**: 857–861.

Bernecker, M., Weidlich, O., and Flugel, E. (1999). Response of Triassic reef coral communities to sea-level fluctuations, storms and sedimentation: evidence from a spectacular outcrop (Adnet, Austria). *Facies* **40**: 29–280.

Berner, R.A. (2005). The carbon and sulfur cycles and atmospheric oxygen from middle Permian to middle Triassic. *Geochim. Cosmochim. Acta* **69**: 3211–3217.

Bown, P., Lees, J.A., and Young, J.R. (2004). Calcareous nannoplankton evolution and diversity through time. *Coccolithophores: From Molecular Processes to Global Impact*. H.R. Thierstein and J.R. Young, eds. Berlin, Springer-Verlag, pp. 481–508.

Bowring, S.A., Erwin, D.H., Jin, Y.G., Martin, M.W., Davidek, K., and Wang, W. (1998). U/Pb zircon geochronology and tempo of the end-Permian mass extinction. *Science* **280**: 1039–1045.

Bralower, T.J., Bown, P.R., and Siesser, W.G. (1991). Significance of Upper Triassic Nannofossils from the Southern-Hemisphere (Odp Leg 122, Wombat Plateau, NW Australia). *Mar. Micropaleontol.* **17**: 119–154.

Brenner, W. (1992). First results of Late Triassic palynology of the Wombat Plateau, northwestern Australia. *Proc. Ocean Drilling Progr. Sci. Results* **122**: 413–426.

Briggs, J.C. (2004). Older species: a rejuvenation on coral reefs? *J. Biogeogr.* **31**: 525–530.

Brocke, R., and Riegel, W. (1996). Phytoplankton response to shoreline fluctuations in the Upper Muschelkalk (Middle Triassic) of Lower Saxony (Germany). *Neues Jahrbuch für Geologie und Palaeontologie Abhandlungen* **200**: 53–73.

Bucefalo Palliani, R., and Riding, J.B. (2003). Umbriadinium and Polarella: an example of selectivity in the dinoflagellate fossil record. *Grana* **42**: 108–111.

Bucur, I. (1999). Stratigraphic significance of some skeletal algae (Dasycladales, Caulerpales) of the Phanerozoic. *Depositional Episodes and Bioevents*. A. Farinacci and A.R. Lord, eds. Rome, Università La Sapienza, pp. 53–104.

Bujak, J.P., and Fisher, M.J. (1976). Dinoflagellate cysts from the Upper Triassic of Arctic Canada. *Micropaleontology* **22**: 44–70.

Canfield, D.E. (1998). A new model for Proterozoic ocean chemistry. *Nature* **396**: 450–453.

Car, J., Skaberne, D., Ogorelec, B., Turnsek, D., and Placer, L. (1981). Sedimentological characteristics of Upper Triassic (Cordevolian) circular quiet water coral bioherms in western Slovenia, northwestern Yugoslavia. *European Fossil Reef Models*. D.F. Toomey, ed. Tulsa, Society of Economic Paleontologists and Mineralogists, pp. 233–240.

Carle, W. (1982). Friedrich August von Alberti; Schoepfer des Formationsnamens Trias. *Geologische Rundschau* **71**: 705–710.

Cheng, Y., Wu, X., and Ji, Q. (2004). Triassic marine reptiles gave birth to live young. *Nature* **432**: 383–386.

De Bernardi, R., and Guiussani, G. (1990). Are bluegreen algae a suitable food for zooplankton? An overview. *Hydrobiologia* **200/201**: 29–41.

Di Nocera, S., and Scandone, P. (1977). Triassic nannoplankton limestones of deep basin origin in the Central Mediterranean region. *Palaeogeogr. Palaeoclimatol. Palaeoecol.* **21**: 101–111.

Erwin, D.H. (1990). Carboniferous-Triassic gastropod diversity patterns and the Permo-Triassic mass extinction. *Paleobiology* **16**: 187–203.

Erwin, D.H. (1996). Understanding biotic recoveries: extinction, survival, and preservation during end-Permian mass extinction. *Evolutionary Paleobiology*. D.H. Erwin, D. Jablonski, and J.H. Lipps, eds. Chicago, University of Chicago Press, pp. 398–418.

Erwin, D.H. (2006). *Extinction: How Life on Earth Nearly Ended 250 Million Years Ago*. Princeton, Princeton University Press.

Eshet, Y., Rampino, M.R., and Visscher, H. (1995). Fungal event and palynological record of ecological crisis and recovery across the Permian-Triassic boundary. *Geology* **23**: 967–970.

Evitt, W.R. (1961). Dapcodinium priscum n. gen. n. sp., a dinoflagellate from the Lower Lias of Denmark. *J. Paleontol.* **35**: 996–1002.

Falkowski, P.G., Katz, M.E., Knoll, A.H., Quigg, A., Raven, J.A., Schofield, O., and Taylor, F.J. (2004). The evolution of modern eukaryotic phytoplankton. *Science* **305**: 354–360.

Fennel, K., Follows, M., and Falkowski, P.G. (2005). The co-evolution of the nitrogen, carbon and oxygen cycles in the Proterozoic ocean. *Am. J. Sci.* **305**: 526–545.

Fensome, R.A., MacRae, R.A., Moldowan, J.M., Taylor, F.J.R., and Williams, G.L. (1996). The early Mesozoic radiation of dinoflagellates. *Paleobiology* **22**: 329–338.

Flügel, E. (1985). Diversity and environments of Permian and Triassic Dasycladacean algae. *Paleoalgology: Contemporary Research and Applications*. D.F. Toomey and M.H. Nitecki, eds. Berlin, Springer, pp. 344–351.

Flügel, E. (1994). Pangean shelf carbonates: controls and paleoclimatic significance of Permian and Triassic reefs. *Pangea: Paleoclimate, Tectonics, and Sedimentation During Accretion, Zenith, and Breakup of a Supercontinent*. G.D. Klein, ed. Boulder, CO, Geological Society of America, pp. 247–266.

Flügel, E. (2002). Triassic reef patterns. *Phanerozoic Reef Patterns*. W. Kiessling, E. Flügel, J. Golonka, eds. Tulsa, OK, Society for Sedimentary Geology, pp. 391–464.

Fraiser, M., and Bottjer, D.J. (2005). Restructuring in benthic level-bottom shallow marine communities due to prolonged environmental stress following the end-Permian mass extinction. *Comptes Rendus Palevol* **4**: 515–523.

Fraiser, M.L., and Bottjer, D.J. (2004). The non-actualistic Early Triassic gastropod fauna: a case study of the Lower Triassic Sinbad Limestone member. *Palaios* **19**: 259–275.

Fraiser, M.L., Twitchett, R.J., and Bottjer, D.J. (2005). Unique microgastropod biofacies in the Early Triassic: indicator of long-term biotic stress and the pattern of biotic recovery after the end-Permian mass extinction. *Comptes Rendus Palevol* **4**: 475–484.

Galli, M.T., Jadoul, F., Bernasconi, S.M., and Weissert H. (2005). Anomalies in global carbon cycling and extinction at the Triassic/Jurassic boundary: evidence from a marine C-isotope record. *Palaeogeogr. Palaeoclimatol. Palaeoecol.* **216**: 203–214.

Gast, R.J., and Caron, D.A. (1996). Molecular phylogeny of symbiotic dinoflagellates from planktonic foraminifera and radiolaria. *Mol. Biol. Evol.* **13**: 1192–1197.

Gast, R.J., and Caron, D.A. (2001). Photosynthetic associations in planktonic foraminifera and radiolaria. *Hydrobiologia* **461**: 1–7.

Gradstein, F.M., Ogg, J.G., Smith, A.G., and 37 others. (2005). *A Geologic Time Scale*. Cambridge, Cambridge University Press.

Grard, A., François, L.M., Dessert, C., Dupré, B., and Goddéris, Y. (2005). Basaltic volcanism and mass extinction at the Permo-Triassic boundary: environmental impact and modeling of the global carbon cycle. *Earth Planetary Sci. Lett.* **234**: 207–221.

Grice, K., Cao, C.Q., Love, G.D. et al. (2005). Photic zone euxinia during the Permian-Triassic superanoxic event. *Science* **307**: 706–709.

Groves, J.R., and Calner, M. (2004). Lower Triassic oolites in Tethys: a sedimentologic response to the end-Permian mass extinction. *Geol. Soc. Am. Annu. Meeting Abstr. Progr.* **36**: 336.

Gustomesov, V.A. (1978). The pre-Jurassic ancestry of the Belemnitida and the evolutionary changes in the Belemnoidea at the boundary between the Triassic and the Jurassic. *Paleontol. J.* **3**: 283–292.

Hallam, A. (1991). Why was there a delayed radiation after the end-Paleozoic extinctions? *Hist. Biol.* **5**: 257–262.

Hallam, A., and Wignall, P.B. (1997). *Mass Extinctions and Their Aftermaths*. New York, Oxford University Press.

Harland, R., Morbey, S.J., and Sarjeant, W.A.S. (1975). A revision of the Triassic to lowest Jurassic dinoflagellate Rhaetogonyaulax. *Palaeontology* **18**: 847–864.

Hartnett, H.E., Keil, R.G., Hedges, J.I., and Devol, A.H. (1998). Influence of oxygen exposure time on organic carbon preservation in continental margin sediments. *Nature* **391**: 572–574.

Hautmann, M. (2004). Effect of end-Triassic CO2 maximum on carbonate sedimentation and marine mass extinction. *Facies* **50**: 257–261.

Helby, R., Morgan, R., and Partridge, A.D. (1987). A palynological zonation of the Australian Mesozoic. *Mem. Ass. Australas. Palaeontols.* **4**: 1–94.

Hochuli, P.A., and Frank, S.M. (2000). Palynology (dinoflagellate cysts, spore-pollen) and stratigraphy of the Lower Carnian Raibl Group in the Eastern Swiss Alps. *Eclogae geologicae Helvetiae* **93**: 429–443.

Holstein, B. (2004) Palynologische Untersuchungen der Kossener Schichten (Rhat, Alpine Obertrias). [Unpublished PhD thesis]: Frankfurt, Johann Wolfgang Goethe Universitat Frankfurt.

Hori, R.S., Campbell, J.D., and Grant-Mackie, J.A. (2003). Triassic radiolaria from Kaka Point Structural Belt, Otago, New Zealand. *J. R. Soc. N Z* **33**: 39–55.

Hotinski, R.M., Bice, K.L., Kump, L.R., Najjar, R.G., and Arthur, M.A. (2001). Ocean stagnation and end-Permian anoxia. *Geology* **29**: 7–10.

Huey, R.B., and Ward, P.D. (2005). Hypoxia, global warming, and terrestrial Late Permian extinctions. *Science* **308**: 398–401.

Isozaki, Y. (1997). Timing of Permian-Triassic anoxia. *Science* **277**: 1748–1749.

Jafar, S.A. (1983). Significance of Late Triassic calcareous nannoplankton from Austria and Southern Germany. *Neues Jahrbuch für Geologie und Palaeontologie Abhandlungen* **166**: 218–259.

Johnston, D.T., Wing, B.A., Farquhar, J., Kaufman, A.J., Strauss, H., Lyons, T.W., Kan, L.C., and Canfield, D.E. (2005). Multiple sulfur isotope fractionations in biological systems: a case study with sulfate reducers and sulfur disproportionations. *Am. J. Sci.* **305**: 645–660.

Kamo, S.L., Czamanske, G.K., Amelin, Y., Fedorenko, V.A., Davis, D.W., and Trofimov, V.R. (2003). Rapid eruption of Siberian flood-volcanic rocks and evidence for coincidence with the Permian-Triassic boundary and mass extinction at 251 Ma. *Earth Planetary Sci. Lett.* **214**: 75–91.

Kanazawa, A. (2001). Sterols in marine invertebrates. *Fisheries Sci.* **67**: 997–1007.

Katz, M.E., Finkel, Z.V., Grzebyk, D., Knoll, A.H., and Falkowski, P.G. (2004). Evolutionary trajectories and biogeochemical impacts of marine eukaryotic phytoplankton. *Annu. Rev. Ecol. Evol. System.* **35**: 523–556.

Kershaw, S., Guo, L., Swift, A., and Fan, J.S. (2002). Microbialites in the Permian-Triassic boundary interval in Central China: structure, age and distribution. *Facies* **47**: 83–89.

Kershaw, S., Zhang, T.S., and Lan, G.Z. (1999). A microbialite carbonate crust at the Permian-Triassic boundary in South China, and its palaeoenvironmental significance. *Palaeogeogr. Palaeoclimatol. Palaeoecol.* **146**: 1–18.

Kiessling, W. (2001). Paleoclimatic significance of Phanerozoic reefs. *Geology* **29**: 751–754.

Knoll, A.H., Bambach, R.K., Canfield, D.E., and Grotzinger, J.P. (1996). Comparative Earth history and Late Permian mass extinction. *Science* **273**: 452–457.

Knoll, A.H., and Swett, K. (1990). Carbonate deposition during the Late Proterozoic Era—an example from Spitsbergen. *Am. J. Sci.* **290A**: 104–132.

Korte, C., Kozur, H.W., and Veizer, J. (2005). d13C and d18O values of Triassic brachiopods and carbonate rocks as proxies for coeval seawater and palaeotemperature. *Palaeogeogr. Palaeoclimatol. Palaeoecol.* **226**: 287–306.

Kump, L.R., Pavlov, A., Arthur, M., Kato, Y., and Riccardi, A. (2003). Death by hydrogen sulfide: a kill mechanism for the end-Permian mass extinction.

Seattle, Geological Society of America, Abstract #70–6.

Kump, L.R., Pavlov, A., and Arthur, M.A. (2005). Massive release of hydrogen sulfide to the surface ocean and atmosphere during intervals of oceanic anoxia. *Geology* **33**: 397–400.

Lee, R.E., and Kugrens, P. (2000). Ancient atmospheric CO2 and the timing of evolution of secondary endosymbioses. *Phycologia* **39**: 167–172.

Lehrmann, D.J. (1999). Early Triassic calcimicrobial mounds and biostromes of the Nanpanjiang basin, south China. *Geology* **27**: 359–362.

Lehrmann, D.J., Payne, J.L., Felix, S.V., Dillett, P.M., Wang, H., Yu, Y.Y., and Wei, J.Y. (2003). Permian-Triassic boundary sections from shallow-marine carbonate platforms of the Nanpanjiang Basin, south China: implications for oceanic conditions associated with the end-Permian extinction and its aftermath. *Palaios* **18**: 138–152.

Li, C., Rieppel, O., and LaBarbera, M.C. (2004). A Triassic aquatic protorosaur with an extremely long neck. *Science* **305**: 1931.

Lindström, S. (2002). Lunnomidinium scaniense Lindstrom, gen. et sp. nov., a new suessiacean dinoflagellate cyst from the Rhaetian of Scania, southern Sweden. *Rev. Palaeobot. Palynol.* **120**: 247–261.

Loeblich III, A.R., and Sherley, J.L. (1979). Observations on the theca of the motile phase of free-living and symbiotic isolates of zooxanthella microadriatica (Freudenthal) comb. nov. *J. Mar. Biol. Assoc.* **59**: 195–205.

Maisch, M.W., and Matzke, A.T. (2003). Observations on Triassic ichthyosaurs. Part X: the Lower Triassic Merriamosaurus from Spitzbergen—additional data on its anatomy and phylogenetic position. *Neues Jahrbuch Fur Geologie Und Palaontologie-Abhandlungen* **227**: 93–137.

Marshall, C.P., Javaux, E.J., Knoll, A.H., and Walter, M.R. (2005). Combined micro-Fourier transform infrared (FTIR) spectroscopy and micro-Raman spectroscopy of Proterozoic acritarchs: a new approach to Palaeobiology. *Precambrian Res.* **138**: 208–224.

Martin, M.W., Lehrmann, D.J., Bowring, S.A., Enos, P., Ramezani, J., Wei, J.Y., and Zhang, J. (2001). Timing of Lower Triassic carbonate bank buildup and biotic recovery following the end-Permian extinction across the Nanpanjiang Basin, south China, Geological Society of America Annual Meeting, Volume 33, p. 201.

Martin, R.E. (1996). Secular increase in nutrient levels through the Phanerozoic: implications for productivity, biomass, and diversity of the marine biosphere. *Palaios* **11**: 209–219.

Martin-Creuzburg, D., Wacker, A., and von Elert, E. (2005). Life history consequences of sterol availability in the aquatic keystone species Daphnia. *Oecologia* **144**: 362–372.

Martini, R., Zaninetti, L., Villeneuve, M., Cornee, J.-J., Krystyn, L., Cirilli, S., De Wever, P., Dumitrica, P., and Harsolumakso, A. (2000). Triassic pelagic deposits of

Timor: palaeogeographic and sea-level implications. *Palaeogeogr. Palaeoclimatol. Palaeoecol* **160**: 123–151.

Masare, J.A. (1988). Swimming capabilities of Mesozoic marine reptiles: implications for method of predation. *Paleobiology* **14**: 187–205.

McKinney, F.K., Lidgard, S., Sepkoski, J.J., and Taylor, P.D. (1998). Decoupled temporal patterns of evolution and ecology in two post-Paleozoic clades. *Science* **281**: 807–809.

McRoberts, C.A. (2000). A primitive Halobia (Bivalvia: Halobioidea) from the Triassic of northeast British Columbia. *J. Paleontol.* **74**: 599–603.

McRoberts, C.A., and Newton, C.R. (1995). Selective extinction among End-Triassic European bivalves. *Geology* **23**: 102–104.

McShea, D.W. (1994). Mechanisms of large-scale evolutionary trends. *Evolution* **48**: 1747–1763.

Medlin, L.K., Kooistra, W., Gersonde, R., Sims, P.A., and Wellbrock, U. (1997). Is the origin of the diatoms related to the end-Permian mass extinction? *Nova Hedwigia* **65**: 1–11.

Moldowan, J.M., and Jacobson, S.R. (2000). Chemical signals for early evolution of major taxa: biosignatures and taxon-specific biomarkers. *Int. Geol. Rev.* **42**: 805–812.

Morbey, S. (1975). The palynostratigraphy of the Rhaetian Stage, Upper Triassic in the Kendelbachgraben, Austria. *Palaeontographica B* **152**: 1–75.

Morbey, S.J. (1978). Late Triassic and Early Jurassic subsurface palynostratigraphy in northwestern Europe. *Palinologia* **1**: 355–365.

Motani, R. (2005). Evolution of fish-shaped reptiles (Reptilia: Ichthyopterygia) in their physical environments and constraints. *Annu. Rev. Earth Planetary Sci.* **33**: 395–420.

Mundil, R., Brack, P., Meier, M., Rieber, H., and Oberli, F. (1996). High resolution U-Pb dating of Middle Triassic volcaniclastics: Time-scale calibration and verification of tuning parameters for carbonate sedimentation. *Earth Planetary Sci. Lett.* **141**: 137–151.

Mundil, R., Ludwig, K.R., Metcalfe, I., and Renne, P.R. (2004). Age and timing of the Permian mass extinctions: U/Pb dating of closed-system zircons. *Science* **305**: 1760–1763.

Mundil, R., Metcalfe, I., Ludwig, K.R., Renne, P.R., Oberli, F., and Nicoll, R.S. (2001). Timing of the Permian-Triassic biotic crisis: implications from new zircon U/Pb age data (and their limitations). *Earth Planetary Sci. Lett.* **187**: 131–145.

Muttoni, G., Kent, D.V., Olsen, P.E., Di Stefano, P., Lowrie, W., Bernasconi, S.M., and Hernandez, F.M. (2004a). Tethyan magnetostratigraphy from Pizzo Mondello (Sicily) and correlation to the Late Triassic Newark astrochronological polarity time scale. *Geol. Soc. Am. Bull.* **116**: 1043–1058.

Muttoni, G., Nicora, A., Brack, P., and Kent, D.V. (2004b). Integrated Anisian-Ladinian boundary chronology. *Palaeogeogr. Palaeoclimatol. Palaeoecol.* **208**: 85–102.

Newell, N.D. (1963). Crises in the history of life. *Sci. Am.* **208:** 76–92.

Nützel, A., and Erwin, D.H. (2004). Late Triassic (Late Norian) gastropods from the Wallowa Terrane (Idaho, USA). *Paläontologische Zeitschrift* **78:** 361–416.

Olsen, P.E., Kent, D.V., Sues, H.D., Koeberl, C., Huber, H., Montanari, A., Rainforth, E.C., Fowell, S.J., Szajna, M.J., and Hartline, B.W. (2002). Ascent of dinosaurs linked to an iridium anomaly at the Triassic-Jurassic boundary. *Science* **296:** 1305–1307.

Payne, J.L. (2005). Evolutionary dynamics of gastropod size across the end-Permian extinction and through the Triassic recovery interval. *Paleobiology* **31:** 269–290.

Payne, J.L., and Finnegan, S. (2006). Controls on marine animal biomass through geological time. *Geobiology* **4:** 1–10.

Payne, J.L., Lehrmann, D.J., Wei, J., and Knoll, A.H. (2006). The pattern and timing of biotic recovery from the end-Permian extinction on the Great Bank of Guizhou, Guizhou Province, China. *Palaios* **21:** 63–85.

Payne, J.L., Lehrmann, D.J., Wei, J.Y., Orchard, M.J., Schrag, D.P., and Knoll, A.H. (2004). Large perturbations of the carbon cycle during recovery from the end-Permian extinction. *Science* **305:** 506–509.

Price-Lloyd, N., and Twitchett, R.J. (2002). The Lilliput effect in the aftermath of the end-Permian mass extinction event, Geological Society of America Annual Meeting, Volume 34. Denver, CO, Geological Society of America, p. 355.

Pruss, S., and Bottjer, D.J. (2004). Early Triassic trace fossils of the western United States and their implications for prolonged environmental stress from the end-Permian mass extinction. *Palaios* **19:** 551–564.

Pruss, S.B., Corsetti, F.A., and Bottjer, D.J. (2005). The unusual sedimentary rock record of the Early Triassic: a case study from the southwestern United States. *Palaeogeogr. Palaeoclimatol. Palaeoecol* **222:** 33–52.

Quigg, A., Finkel, Z.V., Irwin, A.J., Rosenthal, Y., Ho, T.Y., and Reinfelder, J.R. (2003). The evolutionary inheritance of elemental stoichiometry in marine phytoplankton. *Nature* **425:** 291–294.

Ramovš, A. (1996). Crinoids in Lower Triassic in Slovenia. *Albertiana* **17:** 22–24.

Rampino, M.R., and Adler, A.C. (1998). Evidence for abrupt latest Permian mass extinction of foraminifera: results of tests for the Signor-Lipps effect. *Geology* **26:** 415–418.

Rampino, M.R., Prokoph, A., and Adler, A. (2000). Tempo of the end-Permian event: high-resolution cyclostratigraphy at the Permian-Triassic boundary. *Geology* **28:** 643–646.

Raup, D.M. (1979). Size of the Permo-Triassic bottleneck and its evolutionary implications. *Science* **206:** 217–218.

Reichow, M.K., Saunders, A.D., White, R.V., Pringle, M.S., Al'Mukhamedov, A.I., Medvedev, A.I., and Kirda, N.P. (2002). Ar-40/Ar-39 dates from the West Siberian Basin: Siberian flood basalt province doubled. *Science* **296:** 1846–1849.

Renne, P.R., and Basu, A.R. (1991). Rapid eruption of the Siberian Traps flood basalts at the Permo-Triassic boundary. *Science* **253:** 176–179.

Riedel, P. (1991). Korallen in der Trias der Tethys: Stratigraphische reichweiten, Diversitätsmuster, Entwicklungstrends und Bedeutung als Rifforganismens. *Mitteilungen der Gesellschaft der Geologie- und Bergbaustudenten in Österreich* **37:** 97–118.

Riedel, P., and Senowbari-Daryan, B. (1991). Pharetronids in Triassic reefs. *Fossil and Recent Sponges.* J. Reitner and H. Keupp, eds. Berlin, Springer, pp. 465–476.

Rodland, D.L., and Bottjer, D.J. (2001). Biotic recovery from the end-permian mass extinction: behavior of the inarticulate brachiopod Lingula as a disaster taxon. *Palaios* **16:** 95–101.

Rouxel, O.J., Bekker, A., and Edward, K.J. (2005). Iron isotope constraints on the Archean and Paleoproterozoic ocean redox state. *Science* **307:** 1088–1091.

Rüffer, T. (1999). Exkurs: Sedimentation und Faziesräume in der nordalpinen Trias. *Trias: Eine ganz andere Welt.* R. Hauschke and V. Wilde, eds. Munich, Dr. Friedrich Pfeil, pp. 175–204.

Sano, H., and Nakashima, K. (1997). Lowermost Triassic (Griesbachian) microbial bindstone-cementstone facies, southwest Japan. *Facies* **36:** 1–24.

Sarjeant, W.A.S. (1973). Acritarchs and Tasmanitids from the Mianwali and Tredian Formations (Triassic) of the Salt and Surghar Ranges, West Pakistan. *The Permian and Triassic Systems and Their Mutual Boundary.* Calgary, Canadian Society of Petroleum Geologists, pp. 35–73.

Schoch, R., and Wild, R. (1999a). Die Wirbeltier-Fauna im Keuper von Süddeutschland. *Trias: Eine ganz andere Welt.* R. Hauschke and V. Wilde, eds. Munich, Dr. Friedrich Pfeil, pp. 395–418.

Schoch, R., and Wild, R. (1999b). Die Wirbeltiere des Muschelkalks unter besonderer berücksichtigung Süddeutschlands. *Trias: Eine ganz andere Welt.* R. Hauschke and V. Wilde, eds. Munich, Dr. Friedrich Pfeil, pp. 331–342.

Schubert, J.K., and Bottjer, D.J. (1992). Early Triassic stromatolites as post mass extinction disaster forms. *Geology* **20:** 883–886.

Schubert, J.K., and Bottjer, D.J. (1995). Aftermath of the Permian-Triassic mass extinction event—paleoecology of Lower Triassic carbonates in the western USA. *Palaeogeogr. Palaeoclimatol. Palaeoecol.* **116:** 1–39.

Seeling, M., Emmerich, A., Bechstadt, T., and Zuhlke, R. (2005). Accommodation/sedimentation development and massive early marine cementation: Latemar vs. Concarena (Middle/Upper Triassic, Southern Alps). *Sediment. Geol.* **175:** 439–457.

Sellner, K.G. (1997). Physiology, ecology, and toxic properties of marine cyanobacteria blooms. *Limnol. Oceanogr.* **42:** 1089–1104.

Signor, P.W., and Vermeij, G.J. (1994). The plankton and the benthos—origins and early history of an evolving relationship. *Paleobiology* 20: 297–319.

Stanley, G.D. (1988). The history of early Mesozoic reef communities: a three-step process. *Palaios* 3: 170–183.

Stanley, G.D. (2003). The evolution of modern corals and their early history. *Earth Sci. Rev.* 60: 195–225.

Stanley, G.D., and Beauvais, L. (1994). Corals from an Early Jurassic coral-reef in British-Columbia—refuge on an oceanic island reef. *Lethaia* 27: 35–47.

Stanley, G.D., and Cairns, S.D. (1988). Constructional azooxanthellate coral communities: an overview with implications for the fossil record. *Palaios* 3: 233–242.

Stanley, G.D., and Swart, P.K. (1995). Evolution of the coral Zooxanthellae symbiosis during the Triassic—a geochemical approach. *Paleobiology* 21: 179–199.

Stanley, S.M., and Yang X. (1994). A double mass extinction at the end of the Paleozoic Era. *Science* 266: 1340–1344.

Sugitani, K., and Mimura, K. (1998). Redox change in sedimentary environments of Triassic bedded cherts, central Japan: possible reflection of sea-level change. *Geol. Mag.* 135: 735–753.

Summons, R.E., Thomas, J., Maxwell, J.R., and Boreham, C.J. (1992). Secular and environmental constraints on the occurrence of dinosterane in sediments. *Geochim. Cosmochim. Acta* 56: 2437–2444.

Svensen, H., Planke, S., Malthe-Sorenssen, A., Jamtveit, B., Myklebust, R., Eidem, T.R., and Rey, S.S. (2004). Release of methane from a volcanic basin as a mechanism for initial Eocene global warming. *Nature* 429: 542–545.

Tappan, H. (1980). *The Paleobiology of Plant Protists*. San Francisco, W.H. Freeman and Company.

Tekin, U.K., and Yurtsever, T.S. (2003). Late Triassic (Early to Middle Norian) radiolarians from the Antalya Nappes, Antalya, SW Turkey. *J. Micropalaeontol.* 22: 147–162.

Thayer, C.W. (1983). Sediment-mediated biological disturbance and the evolution of marine benthos. *Biotic Interactions in Recent and Fossil Benthic Communities*. M.J.S. Tevesz and P.L. McCall, eds. New York, Plenum Press, pp. 479–625.

Twitchett, R.J. (1999). Palaeoenvironments and faunal recovery after the end-Permian mass extinction. *Palaeogeogr. Palaeoclimatol. Palaeoecol.* 154: 27–37.

Twitchett, R.J. (2001). Incompleteness of the Permian-Triassic fossil record: a consequence of productivity decline? *Geol. J.* 36: 341–353.

Twitchett, R.J. (2005). The Lilliput effect in the aftermath of the end-Permian extinction event. *Albertiana* 33: 79–81.

Twitchett, R.J., Krystyn, L., Baud, A., Wheeley, J.R., and Richoz, S. (2004). Rapid marine recovery after the end-Permian mass-extinction event in the absence of marine anoxia. *Geology* 32: 805–808.

Twitchett, R.J., Looy, C.V., Morante, R., Visscher, H., and Wignall, P.B. (2001). Rapid and synchronous collapse of marine and terrestrial ecosystems during the end-Permian biotic crisis. *Geology* 29: 351–354.

Twitchett, R.J., and Oji, T. (2005). Early Triassic recovery of echinoderms. *Comptes Rendus Palevol* 4: 463–474.

Twitchett, R.J., Wignall, P.B., and Benton, M.J. (2000). Discussion on Lazarus taxa and fossil abundance at times of biotic crisis. *J. Geol. Soc.* 157: 511–512.

Van Cappellen, P., and Ingall, E.D. (1994). Benthic phosphorus regeneration, net primary production, and ocean anoxia—a model of the coupled marine biogeochemical cycles of carbon and phosphorus. *Paleoceanography* 9: 677–692.

van de Schootbrugge, B., Bailey, T., Rosenthal, Y., Katz, M., Wright, J.D., Feist-Burkhardt, S., Miller, K.G., and Falkowski, P.G. (2005). Early Jurassic climate change and the radiation of organic-walled phytoplankton in the Tethys Ocean. *Paleobiology* 31: 73–97.

van de Schootbrugge, B., Tremolada, F., Bailey, T.R., Rosenthal, Y., Feist-Burkhardt, S., Brinkhuis, H., Pross, J., Kent, D.V., and Falkowski, P.G. (2007). End-Triassic calcification crisis and blooms of organic-walled disaster species. *Palaeogeogr. Palaeoclimatol. Palaeoecol.* 244: 126–141.

Végh-Neubrandt, E. (1982). *Triassische Megalodontaceae*. Budapest, Akadémiai Kiadó.

Vermeij, G.J. (1977). The Mesozoic marine revolution: evidence from snails, predators and grazers. *Paleobiology* 3: 245–258.

Visscher, H., Looy, C.V., Collinson, M.E., Brinkhuis, H., Cittert, J., Kurschner, W.M., and Sephton, M.A. (2004). Environmental mutagenesis during the end-Permian ecological crisis. *Proc. Natl. Acad. Sci. U S A* 101: 12952–12956.

von Elert, E., Martin-Creuzburg, D., and Le Coz, J.R. (2003). Absence of sterols constrains carbon transfer between cyanobacteria and a freshwater herbivore (Daphnia galeata). *Proc. R. Soc. Lond. Ser. B. Biol. Sci.* 270: 1209–1214.

Wacker, A., and von Elert, E. (2003). Food quality controls reproduction of the zebra mussel (Dreissena polymorpha). *Oecologia* 135: 332–338.

Wall, D. (1965). Microplankton, pollen and spores from the Lower Jurassic in Britain. *Micropaleontology* 11: 151–190.

Ward, P.D., Garrison, G.H., Haggart, J.W., Kring, D.A., and Beattie, M.J. (2004). Isotopic evidence bearing on Late Triassic extinction events, Queen Charlotte Islands, British Columbia, and implications for the duration and cause of the Triassic/Jurassic mass extinction. *Earth Planetary Sci. Lett.* 224: 589–600.

Ward, P.D., Haggart, J.W., Carter, E.S., Wilbur, D., Tipper, H.W., and Evans, T. (2001). Sudden productivity collapse associated with the Triassic-Jurassic boundary mass extinction. *Science* 292: 1148–1151.

White, S.H., Duivenvoorden, L.J., and Fabbro, L.D. (2005). Impacts of a toxic Microcystis bloom on the macroinvertebrate fauna of Lake Elphinstone, Central Queensland, Australia. *Hydrobiologia* **548:** 117–126.

Wiggins, V.D. (1973). Upper Triassic dinoflagellates from arctic Alaska. *Micropaleontology* **19:** 1–16.

Wignall, P.B. (2001). Large igneous provinces and mass extinctions. *Earth Sci. Rev.* **53:** 1–33.

Wignall, P.B., and Benton, M.J. (1999). Lazarus taxa and fossil abundance at times of biotic crisis. *J. Geol. Soc.* **156:** 453–456.

Wignall, P.B., and Benton, M.J. (2000). Discussion on Lazarus taxa and fossil abundance at times of biotic crisis—reply. *J. Geol. Soc.* **157:** 512.

Wignall, P.B., and Hallam, A. (1992). Anoxia as a cause of the Permian Triassic mass extinction—facies evidence from Northern Italy and the Western United States. *Palaeogeogr. Palaeoclimatol. Palaeoecol.* **93:** 21–46.

Wignall, P.B., and Hallam, A. (1993). Griesbachian (Earliest Triassic) Paleoenvironmental changes in the Salt Range, Pakistan and Southeast China and their bearing on the Permo-Triassic mass extinction. *Palaeogeogr. Palaeoclimatol. Palaeoecol.* **102:** 215–237.

Wignall, P.B., and Hallam, A. (1996). Facies change and the end-Permian mass extinction in SE Sichuan, China. *Palaios* **11:** 587–596.

Wignall, P.B., Morante, R., and Newton, R. (1998). The Permo-Triassic transition in Spitsbergen: delta C-13(org) chemostratigraphy, Fe and S geochemistry, facies, fauna and trace fossils. *Geol. Mag.* **135:** 47–62.

Wignall, P.B., and Newton, R. (2003). Contrasting deep-water records from the Upper Permian and Lower Triassic of South Tibet and British Columbia: evidence for a diachronous mass extinction. *Palaios* **18:** 153–167.

Wignall, P.B., Newton, R., and Brookfield, M.E. (2005). Pyrite framboid evidence for oxygen-poor deposition during the Permian-Triassic crisis in Kashmir. *Palaeogeogr. Palaeoclimatol. Palaeoecol.* **216:** 183–188.

Wignall, P.B., and Twitchett, R.J. (1996). Oceanic anoxia and the end-Permian mass extinction. *Science* **272:** 1155–1158.

Wignall, P.B., and Twitchett, R.J. (1999). Unusual intraclastic limestones in Lower Triassic carbonates and their bearing on the aftermath of the end-Permian mass extinction. *Sedimentology* **46:** 303–316.

Wignall, P.B., and Twitchett, R.J. (2002a). Extent, duration, and nature of the Permian-Triassic superanoxic event. *Catastrophic Events and Mass Extinctions; Impacts and Beyond.* C. Koeberl and K.G. MacLeod, eds. Boulder, CO, Geological Society of America, pp. 395–413.

Wignall, P.B., and Twitchett R.J. (2002b). Permian-Triassic sedimentology of Jameson Land, East Greenland: incised submarine channels in an anoxic basin. *J. Geol. Soc.* **159:** 691–703.

Winguth, A.M.E., and Maier-Reimer E. (2005). Causes of the marine productivity and oxygen changes associated with the Permian-Triassic boundary: a reevaluation with ocean general circulation models. *Mar. Geol.* **217:** 283–304.

Woods, A.D., Bottjer, D.J., Mutti, M., and Morrison, J. (1999). Lower Triassic large sea-floor carbonate cements: their origin and a mechanism for the prolonged biotic recovery from the end-Permian mass extinction. *Geology* **27:** 645–648.

Xie, S.C., Pancost, R.D., Yin, H.F., Wang, H.M., and Evershed, R.P. (2005). Two episodes of microbial change coupled with Permo/Triassic faunal mass extinction. *Nature* **434:** 494–497.

Yancey, T.E., and Stanley, G.D. (1999). Giant alatoform bivalves in the Upper Triassic of western North America. *Palaeontology* **42:** 1–23.

Zankl, H. (1969). Der Hohe Goll. Aufbau und Lebensbild eines Dachsteinkalk-Riffes in der Obertrias der nordlichen Kalkalpen. *Abhandlungen der Senkenbergishchen Naturforschenden Gesellschaft* **519:** 1–123.

Zankl, H. (1977). Quantitative aspects of carbonate production in a Triassic reef complex. *Third International Coral Reef Symposium*, Volume 2. R.N. Ginnsburg and D.L. Taylor, eds. Miami, Rosenstiel School of Marine and Atmospheric Sciences, pp. 375–378.

Zhang, R., Follows, M.J., Grotzinger, J.P., and Marshall, J. (2001). Could the Late Permian deep ocean have been anoxic? *Paleoceanography* **16:** 317–329.

Zonneveld, J.-P., Gingras, M.K., and Pemberton, S.G. (2002a). Ichnology and sedimentology of the Lower Montney Formation (Lower Triassic), Kahntah River and Ring Border fields, Alberta and British Columbia. *Canadian Society of Petroleum Geologists Abstracts with Programs*, 355.

Zonneveld, J.P., Gingras, M.K., and Pemberton, S.G. (2001). Trace fossil assemblages in a Middle Triassic mixed siliciclastic-carbonate marginal marine depositional system, British Columbia. *Palaeogeogr Palaeoclimatol. Palaeoecol.* **166:** 249–276.

Zonneveld, J.P., Pemberton, S.G., Saunders, T.D.A., and Pickerill, R.K. (2002b). Large, robust Cruziana from the Middle Triassic of northeastern British Columbia: ethologic, biostratigraphic, and paleobiologic significance. *Palaios* **17:** 435–448.

The Origin and Evolution of Dinoflagellates

CHARLES F. DELWICHE

Dinoflagellates are typically unicellular, but sometimes filamentous or coenocytic, flagellates found in both marine and freshwater environments worldwide. Photosynthetic species are responsible for enormous primary productivity, but many species, whether photosynthetic or not, can also be important predators (e.g., Yang *et al.* 2004). Some species produce potent toxins that can be a cause of morbidity and mortality from direct exposure or indirectly as a result of bioaccumulation in top predators (Fensome *et al.* 1993; Tindall and Morton 1988). Consequently, dinoflagellates constitute a vital link in aquatic ecosystems and, depending upon the context, can be strongly beneficial or detrimental to human uses of these environments.

The typical dinoflagellate cell displays two flagella, one of which wraps the cell in an equatorial groove called the "cingulum,"

with the other projecting toward the posterior of the cell and lying in, but often extending from, a groove termed the "sulcus" (Figure 1). The name "dinoflagellate" refers to the distinctive whirling motion of the swimming cells, which is a function of this flagellar arrangement. Not all dinoflagellates display this cell morphology but most show it in at least one stage of the life history, and many are of this form during vegetative growth. The cells can, however, be spectacularly modified, and in several highly derived species the identity of the organism as a dinoflagellate was recognized only after observation of a "dinospore" life stage (Fensome *et al.* 1993) or on the basis of other properties.

Also highly characteristic of dinoflagellates is a nucleus with chromosomes that remain condensed throughout the life history (although this is not in and of itself

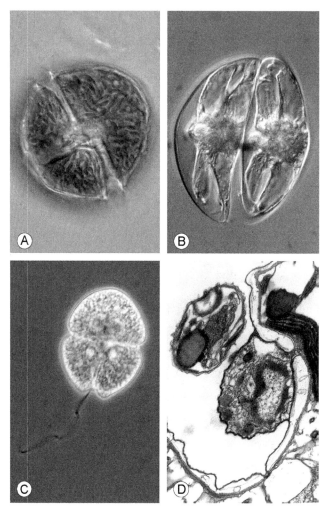

FIGURE 1.   Representative dinoflagellates. **(A)** *Peridinium* sp., with the cingulum (extending from 11 o'clock to 5 o'clock), sulcus (8 o'clock), and peduncle visible near the junction of cingulum and sulcus. **(B)** *Pyrocystis lunula,* which is one of several bioluminescent dinoflagellates; it is not flagellate in its vegetative state but releases "dinokont" zoospores under some conditions. **(C)** *Akashiwo sanguinea,* one of many unarmored species once lumped as "*Gymnodinium,*" with the trailing flagellum visible extending from the sulcus. **(D)** *Amoebophrya* sp., in the early stages of infecting *Akashiwo sanguinea*; also visible are the cortical alveolae of the host, and a portion of the host chloroplast. (Parts **[A]**, **[B]**, and **[C]** from Charles Delwiche; **[D]** from John Miller.) (See color plate.)

unique to dinoflagellates), unusual histone-like proteins and lack of typical eukaryotic nucleosomes (Chudnovsky *et al.* 2002), and nuclear division with a spindle apparatus that consists of several massive bundles of microtubules that penetrate the intact nuclear envelope (mitosis is "closed," i.e., the nuclear envelope remains intact throughout mitosis). The DNA in many cells is thought to be modified with 5' hydroxymethyluracil (Rae and Steele 1978), and standard molecular biological techniques are often difficult to apply in the group. There is a series of vesicles ("cortical alveolae") that lie beneath the cell surface. In many species, these are occupied by cellulosic plates that make the cell rigid and confer a distinctive appearance reminiscent of medieval armor (see Hamm and Smetacek, Chapter 14, this volume). This armor can be quite spectacular in appearance, and morphological classification relied heavily on it, particularly in the fossil record (Fritsch 1965; Fensome *et al.* 1993). Thus, classical taxonomy divided species into "armored" and "naked" clades and easily identified a number of natural groups with distinctive plate structure. An elaborate system of notation has been developed to describe the arrangement of the armor plates or "thecae." Naked species often proved much more difficult to place. Although classification based on plate tabulation has a number of drawbacks, not the least of them being the difficulty of placing species with thin or nonexistent thecae, this system remains the most reliable and effective means for morphological identification of field material.

About half of all dinoflagellates are photosynthetic, the majority of these relying on a chloroplast that is surrounded by three membranes and pigmented with a highly modified carotenoid, peridinin (Hofmann *et al.* 1996; Figure 2). This "peridinin-type" plastid is not the only type of plastid found in dinoflagellates (see later), but it is the most widespread and is widely thought to be the ancestral plastid type. Many dinoflagellates—whether photosynthetic or not—are also capable of heterotrophic feeding, often making use of a tubelike peduncle (or "tentacle") that extends from near the junction between the cingulum and sulcus to take in food via phagocytosis or myzocytosis. As a consequence, dinoflagellates are often mixotrophic and can rapidly shift from an environmental role as a primary producer to that of a predator (Bockstahler and Coats 1993; Li *et al.* 1996). This mixotrophy also seems to have made the association of the plastid with the host cell more facultative than in most other algal groups, and several dinoflagellates seem to have lost their peridinin-type plastid, only to independently acquire a different type of plastid. A major outstanding question in the evolution of dino-

flagellates is how the history of plastid acquisition, retention, and loss correlates with the phylogeny of the host cells.

I. PALEONTOLOGICAL DATA

The earliest undisputed, structural fossils of dinoflagellates are cysts dating from the Triassic (e.g., *Suessia swabiana*, ~200 million years ago [Ma]), with a few likely Permian records (Fensome *et al.* 1996). Some Silurian specimens (~410 Ma) have been attributed to the group, but the affinity is uncertain (Tappan 1980). By the lower Jurassic organisms with nearly modern morphologies were abundant, and there was a great diversification of dinoflagellates during this period, with diversity reaching a maximum (substantially exceeding modern diversity) in the Cretaceous. Although dinoflagellates are widespread and environmentally important in the modern environment, they appear to have neither the diversity nor the absolute abundance achieved in the Cretaceous. Thus, although structural specimens resembling modern dinoflagellates may extend as far back as the Silurian, the primary diversification of the group occurred in the Jurassic and Cretaceous (Tappan 1980; Falkowski *et al.* 2004). It is worth noting that there are significant preservational biases in almost any type of fossil, and many well-documented modern dinoflagellates would not be expected to preserve well, particularly given that only around 13% of modern dinoflagellates are known to form cysts (Fensome *et al.* 1996). It is possible that the apparent Cretaceous diversity reflects an overabundance of easily fossilized forms but more likely that it represents only a fraction of the true Cretaceous diversity.

Sedimentary biomarkers imply that the dinoflagellates may be much older than their structural fossils suggest (Moldowan and Talyzina 1998; Knoll *et al.*, Chapter 8, this volume). Triaromatic dinosteroids, which are thought to be characteristic of dinoflagellates, have been shown to be locally abundant in Ordovician sediments (over 450 Ma) and are detectable well into the Proterozoic (Brocks and Summons 2003).

One factor complicating the study of fossil dinoflagellates is the relatively poor ability of the vegetative cell wall to survive fossilization. Some, but not all, species form resistant cysts that are impregnated with dinosporin, a sporopollenin-like, highly durable compound. These can lie dormant in the sediment for many years (Anderson *et al.* 2002; McGillicuddy *et al.* 2003) and are thought to account for the ability of toxic-bloom–forming dinoflagellates to appear in vast numbers almost overnight when appropriate conditions prevail. Such cysts are formed internally in the dinoflagellate cell, and although they often bear a variety of processes reminiscent of wings and arms, the relationship between these structures and that of the vegetative cell is often difficult to discern. Linking the morphology of a fossil cyst to that of a living dinoflagellate requires detailed knowledge of the structure and life history of the organism.

Acritarchs are, by definition, microfossils with no known affinity (Evitt 1963). There is a considerable diversity of acritarchs in Ordovician, Silurian, and Devonian sediments, and several authors have attempted to link these to dinoflagellates. Some later (Jurassic and Cretaceous) acritarch-like cysts (i.e., certain hystrichospheres) have been clearly shown to be dinoflagellate cysts and consequently are no longer treated as acritarchs (Colbath and Grenfell 1995). A correlation has been noted between the presence of triaromatic dinosteroids and acritarch abundance (Moldowan *et al.* 1996; Meng *et al.* 2005), implying that these acritarchs may be the cysts of ancestral dinoflagellates. However, a number of important questions remain unanswered, including whether the acritarchs themselves are the source of the dinosteroids, the chemical basis for this association (are triaromatic dinosteroids incorporated into dinosporin?), and the details of the metabolic pathway responsible for the production of dinosteroids.

Accurate interpretation of biomarkers depends upon the taxonomic specificity of the biomarker, its geological stability, and low mobility or containment within specific strata (Knoll *et al.*, Chapter 8, this volume). These considerations are added to the broader concerns of stratigraphy, including chronological constraints, correlation with other strata, and modification subsequent to deposition (Moldowan *et al.* 1996; Brocks *et al.* 1999; Brocks and Summons 2003). Concurrent with addressing these issues will be the long, hard work of making correlations between cyst morphology and the structure of the vegetative cell. This is difficult enough with extant dinoflagellates but is challenging indeed with Ordovician acritarchs that likely have no extant equivalents. However, if these early acritarchs are indeed affiliated with dinoflagellates, the prospects are good that subsequent work will clarify the nature of the relationship.

It is clear, in any case, that the Jurassic radiation reflects a genuine biological process and very likely constitutes the diversification of modern dinoflagellates. Cysts and thecal plates are primarily formed by photosynthetic dinoflagellates, and these species are also much more likely to occur in large populations. There is, therefore, a strong bias in favor of photosynthetic forms within the fossil record. One plausible hypothesis is that the Jurassic radiation reflects the acquisition of the peridinin-type plastid by dinoflagellates (although this would have to be reconciled with the chromalveolate hypothesis; see later).

II. PHYLOGENY OF DINOFLAGELLATES

A. Sources of Information

Despite the efforts of several groups of very talented scientists, there is no wholly credible consensus phylogeny of dinoflagellates. Molecular phylogenetic analyses of most photosynthetic organisms have relied heavily on chloroplast data, particularly plastid genome maps, as well as the analysis of gene sequences from the plastid genes *rbcL* and *atpB* (which encode the large subunit of form I ribulose-1,5-bisphosphate carboxylase/oxygenas, or "rubisco," and the beta subunit of ATPase, respectively). None of these is appropriate for phylogenetic analysis of dinoflagellates. First, because many dinoflagellates are not photosynthetic, and several of the photosynthetic species rely on plastids acquired independently, plastid phylogeny in dinoflagellates does not track that of the host cells (Falkowski and Raven 2007). Second, even among peridinin-pigmented dinoflagellates, the plastid genome is highly modified, with an extremely high rate of sequence evolution (Zhang *et al.* 2000). Furthermore, in peridinin-pigmented dinoflagellates, the rubisco is a nuclear-encoded form II rubisco, unique among oxygenic phototrophs (Morse *et al.* 1995; Rowan *et al.* 1996; Delwiche 1999). Consequently, several robust tools for molecular phylogenetics have been unavailable for use in the study of dinoflagellates.

The most widely used loci for phylogenetic purposes are various portions of the ribosomal (rDNA) operon, particularly the regions encoding the small (SSU) and large (LSU) subunit of the ribosome, and to a lesser extent the internal transcribed spacer regions (ITS). The SSU rDNA is the dominant locus for "universal" phylogenies and has been sequenced from a very wide variety of organisms, including many dinoflagellates (Pace 1997). Phylogenies based on these molecules often have very short internal branches, with a diversity of short and long branches within the dinoflagellates (Lenaers *et al.* 1991; Daugbjerg *et al.* 2000; Saldarriaga *et al.* 2001), and are sensitive to the analytical method (Saldarriaga *et al.* 2001).

If molecular evolution were truly clocklike, then one would expect all taxa to have a similar distance to the root of the tree. It is clear that real sequence evolution is rarely clocklike, but the dinoflagellate rDNA data are particularly variable (see Figure 2). Dinoflagellate rDNA phylogenies typically show clusters of long-branch taxa and a poorly resolved "backbone" of short-branch taxa. Among these short branches there is little phylogenetic information, and the GPP complex of Saunders (1997) is a well-recognized difficult phylogenetic problem. However, a more fundamental question is whether it is possible to correctly place long-branch taxa in such a tree. The long-branch dinoflagellates often cluster together to at least some degree, but it is difficult to be confident whether this reflects phylogeny or "long-branch attraction," and dinoflagellate molecular phylogenies are in some ways difficult to reconcile with traditional phylogeny (Fensome *et al.* 1999; Taylor 2004). The situation is further complicated by the possibility of more complex variation in the mode of sequence evolution such as those noted for dinoflagellate plastid genes (Shalchian-Tabrizi *et al.* 2006a).

Only a few studies have examined nuclear loci other than the rDNA operon (Leander and Keeling 2004; Shalchian-Tabrizi *et al.* 2006a), but these suggest that nuclear-protein–coding genes may avoid some of the difficulties of

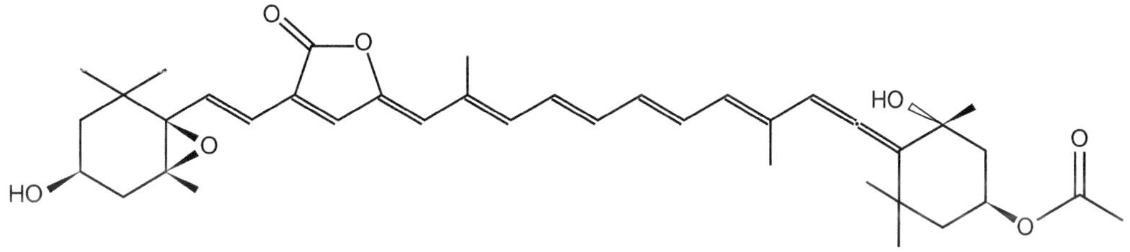

FIGURE 2. Structure of peridinin, the characteristic pigment of the most common type of plastid in dinoflagellates. (Strain *et al.* 1971).

the ribosomal operon, albeit with the added complications of large gene families embedded in potentially enormous genomes. The mitochondrial genome is apparently very small, but cytochrome b has been explored for phylogenetic purposes (Zhang *et al.* 2005).

Significant studies have emphasized specific lineages within the dinoflagellates, such as the strains with symbiotic associations with animals such as coral and clams, as well as with other protists (Sadler *et al.* 1992; McNally *et al.* 1994; Gast and Caron 1996; Takishita *et al.* 2003; Santos 2004; Takabayashi *et al.* 2004; LaJeunesse *et al.* 2005). Others have used phylogenetic methods to examine specific evolutionary phenomena, notably the diversity of dinoflagellate chloroplasts and their phylogenetic origins (Chesnick *et al.* 1997; Tengs *et al.* 2000; Yoon *et al.* 2002; López-García *et al.* 2001).

B. The Phylogeny

Despite the relative difficulty of dinoflagellate molecular phylogenetics, some fairly consistent features are apparent (Figure 3). In most cases there is a clade of photosynthetic dinoflagellates that consists primarily of peridinin-pigmented species (the Dinophyceae *sensu stricto*), but that also includes some (presumably secondary) heterotrophs and species with secondary and tertiary plastids with other pigmentation types (Saunders *et al.* 1997; Shalchian-Tabrizi *et al.* 2006a). In a few cases the nonperidinin species within the Dinophyceae *s.s.* retain an eyespot that has been interpreted as a remnant peridinin-type plastid. Branching below the Dinophyceae *s.s.* are several lineages of heterotrophs, typically including the predatory *Noctiluca; Amoebophrya,* which is a parasite of other dinoflagellates; the arthropod parasites *Syndinium* and *Hematodinium;* and the predatory *Oxyrrhis* (López-García *et al.* 2001; Moon-van der Staay *et al.* 2001; Skovgaard *et al.* 2005).

A striking discrepancy between the two trees shown in Figure 3 is the placement of *Noctiluca*, a large, predatory dinoflagellate. *Noctiluca* is distinctly different from Dinophyceae *sensu stricto* and has often been placed as an outgroup to the rest of the dinoflagellates. Most molecular analyses place it among the Syndinea, as shown in Figure 1A, but in Figure 1B it is deeply embedded within the Dinophyceae *sensu stricto*. A similarly dramatic rearrangement was observed for the putative outgroup species *Oxyrris marina* by Saldarriaga *et al.* (2003) as a function of analytical method. Many phenomena could be responsible for such highly discordant phylogenies, ranging from alignment errors to phylogenetic artifacts and chimeric genes. This variability may largely be explained by the relatively small amount of phylogenetic information carried by rDNA in dinoflagellates; Shalchian-Tabrizi *et al.* (2006a) compared phylogenies of dinoflagellates based on the protein-coding gene *hsp90* to those based on SSU rDNA and found that in combined analyses the hsp90 topology dominated. Unfortunately, at present few protein-coding gene sequences are available from a sufficient diversity of dinoflagellates to permit detailed phylogenetic analyses.

Also branching as an outgroup to the phototrophs are some as-yet unidentified environmental sequences (López-García *et al.* 2001; Moon-van der Staay *et al.* 2000) that very likely come from species that have been described morphologically but for which no molecular data have yet been determined (see Figure 3B). Molecular research has proceeded more quickly on photosynthetic dinoflagellates than on heterotrophic species, presumably because of the relative difficulty of culturing heterotrophs, and it is entirely likely that many of these mysterious sequences will eventually be found to derive from described organisms. Such environmental molecular studies illustrate the great diversity of dinoflagellates, help reveal diversity that is not easily studied with morphological methods,

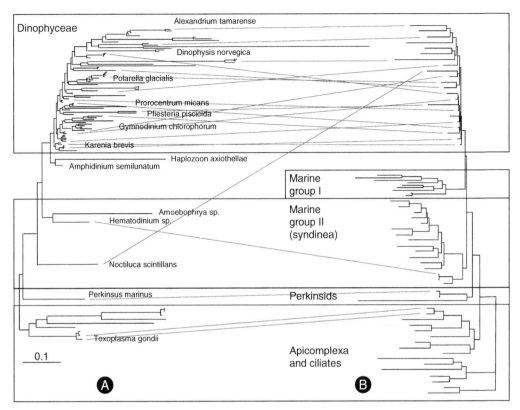

FIGURE 3.   Comparison of two recent SSU rDNA phylogenetic trees of dinoflagellates. The tree on the left comes from a study of diversity of dinoflagellate plastid types (Shalchian-Tabrizi *et al.* 2006a), whereas the tree on the right is from a study of the arthropod parasite *Syndinium* (Skovgaard *et al.* 2005). The trees have been placed on a common scale (scale bar = 0.1 substitutions/site) and have been arranged to facilitate comparison of taxon sampling. Those taxa that are present in both analyses are linked by grey lines. Both trees were found using likelihood methods (Bayesian and maximum likelihood, respectively), albeit with different models of sequence evolution.

and suggest that there are many species of dinoflagellate yet to be described.

C. Reconciling Molecular and Morphological Phylogenies

Traditional classifications emphasized the gross morphology of the cell. They made a strong distinction between the "desmokont" taxa, which have thecal plates divided into two distinct halves, and "dinokont" taxa, which, when armored, have numerous thecal plates arranged in a complex pattern (Fritsch 1965). An excellent overview of dinoflagellate morphology was provided by Taylor (1989), who provided a conceptual diagram (not

a formal phylogenetic tree) suggesting relationships among the known forms. This treatment followed earlier authors and placed near the base of the tree structurally simple desmokont taxa such as *Prorocentrum*. Nonphotosynthetic predatory and parasitic species such as *Noctiluca* and *Amoebophrya* were treated as derived, but Taylor has explicitly noted that this was intended essentially to be an unrooted tree and could be interpreted equally well inverted, with the superficially simple *Prorocentrum* seen as a derived form.

Classification of dinoflagellates has been carried out largely independently by scientists working with fossil data and those working with living organisms. A major effort to integrate these two data types was made by Fensome *et al.* (1993), and this remains the most thorough recent review of dinoflagellate diversity and classification, although molecular phylogenetic data are leading to reinterpretation and modification of that classification. Another form of systematic bias comes from the fact that non-photosynthetic species typically do not have armor and thus can be difficult to identify. As a consequence, photosynthetic and non-photosynthetic dinoflagellates have often been studied almost independently, and taxonomy of dinoflagellates can be handled under either the botanical or zoological code (at the discretion of the author).

Where molecular phylogenies have proven to be extremely helpful is in recognizing close relationships among morphologically divergent taxa. This has led to the ongoing reclassification of taxa previously placed in the form-genus *Gymnodinium* and similar reclassification of other naked species. Analysis of rDNA data has also allowed the placement of highly modified species such as parasites (Skovgaard *et al.* 2005), has largely resolved the debate about the ancestral and derived characteristics for dinoflagellates (placing *Noctiluca* and *Oxyrris* as ancestral forms and prorocentroids as derived), and has excluded the hypothesis that the nuclear organization of dinoflagellates is the ancestral condition for eukaryotes. However, despite this progress, molecular phylogenies for dinoflagellates remain preliminary, and morphological analysis remains a vital tool for dinoflagellate phylogeny.

III. THE PLASTIDS OF DINOFLAGELLATES

The vast majority of photosynthetic dinoflagellates incorporate a "peridinin-type" plastid that is pigmented by an unusual, modified carotenoid, peridinin (see Figure 2), along with chlorophylls *a* and *c*. The pigments are bound to a protein and together form a characteristic water-soluble complex. The peridinin-type plastid is surrounded by three membranes, all of which lie relatively closely around the plastid, do not have attached ribosomes, and do not form an obvious chloroplast endoplasmic reticulum (ER) like that of heterokonts, haptophytes, and cryptomonads (see Kooistra *et al.*, Chapter 11, this volume). The plastids of dinoflagellates (like those of all algae other than reds, greens, and glaucocystophytes) are thought to be of secondary origin, that is, acquired from another eukaryote (Delwiche *et al.* 2004; Gibbs 1981). The source is presumed to have been a red alga, although the timing of this event and the relationship of this plastid to those of other algae with secondary plastids derived from red algae remains an area of active study.

The similarity in pigmentation between dinoflagellates, heterokonts, haptophytes, and cryptomonads (all of which contain chlorophyll *c*) has long been noted, and it has long been suspected that these taxa should be classified together as "chromophytes" (Christensen 1989). However, the similarity among these organisms is primarily a function of their plastid properties. Consequently, the chromophyte concept fell into disfavor when the endosymbiotic nature of plastids made clear the possibility that these properties reflected endosymbiont, rather than host, phylogeny. It was also recognized that there were many nonphotosynthetic close relatives of both dinoflagellates and heterokonts. Cavalier-Smith (1999) postulated that the evolution of the polypeptide-import apparatus found in chloroplasts and mitochondria (and consequently the endosymbiotic origin of organelles) must be an extremely rare event and on that basis proposed the existence of a eukaryotic supergroup, the "chromalveolates" (see Fehling *et al.*, Chapter 6, this volume). If the chromalveolate hypothesis (Cavalier-Smith 1999) is correct, then dinoflagellates, along with their relatives the Apicomplexa and ciliates, share a common ancestry with the other chlorophyll *c* containing algae (heterokonts, haptophytes, and cryptomonads). The implications of this hypothesis are far-reaching, and the data bearing on it are contradictory and confusing (for excellent reviews, see Bodyl 2005; Keeling 2004).

A critical element of the chromalveolate hypothesis is that not only are the host-cell lineages presumed to be monophyletic but also their common ancestor is presumed to have had a plastid (Cavalier-Smith 1999; Grzebyk *et al.* 2003). This is striking, because although it minimizes the number of secondary endosymbiotic events, it implies a relatively large number of secondary losses of photosynthesis. Several significant chromalveolate groups are nonphotosynthetic and were not suspected of having previously been photosynthetic. Among these are nonphotosynthetic heterokonts, including oomycetes, labyrinthulomycetes, bicosoecids, and opalinids, and alveolates (the last of which includes ciliates and nonphotosynthetic dinoflagellates). All of these would be predicted to have lost photosynthesis. One obvious corollary of this hypothesis is that there should be substantial numbers of endosymbiont genes remaining in the nuclear genomes of these organisms. This is predicted because tremendous numbers of endosymbiont genes are found in the nuclear genome of phototrophs; these were apparently transferred early in the evolution of plastids (Palmer and Delwiche 1998), and many of these play roles in cellular function that are not directly related to the plastid itself. Thus, even when photosynthesis is lost, one would not predict loss of all endosymbiont-derived genes. Study of the nuclear genomes of nonphotosynthetic chromalveolates should help evaluate this prediction.

The peridinin-type plastid does resemble those of heterokonts, haptophytes, and cryptomonads in several ways. Aside from the presence of chlorophyll *c*, which is unknown among red algae, these plastids have thylakoids in groups of three, and the storage polysaccharide (starch in dinoflagellates) is deposited outside of the plastid. However,

unlike those groups, the plastid lumen is separated from the cytosol by only three membranes, and there is no obvious chloroplast ER (a membrane system thought to correspond to the food vacuole of the host cell). Thus, a relationship among these plastids seems plausible but is by no means certain.

The chromalveolate hypothesis may, however, be incompatible with the hypothesis that the Jurassic diversification of dinoflagellate cysts stemmed from the acquisition of plastids by the group. Yoon *et al.* (2004) attempted to establish a timeline for major events in the evolution of photosynthetic eukaryotes using molecular data. Their analyses did not include dinoflagellates but placed the divergence among other members of the chromalveolata in the Proterozoic, well over 540 Ma and perhaps as much as 1200 Ma, long before the Jurassic diversification. Such estimates should be viewed with caution, given the highly stochastic nature of molecular sequence evolution and the small number of calibration points currently available. If Ordovician acritarchs are, indeed, attributable to ancestral photosynthetic dinoflagellates, then a Proterozoic origin for the group is reasonable, but the Jurassic diversification would then remain. As an alternative hypothesis, Katz *et al.* (2004) suggest that it may be attributable to the opening of shallow, continental seas during the breakup of Gondwana that favor cyst-producing dinoflagellates. Jurassic fossils attributable to the Suessiaceae, Gonyaulacales, and Peridiniales span much of the diversity of the single photosynthetic clade as it is presently understood. If the peridinin-type plastid is homologous to that of other chromalveolates, and the analyses of Yoon *et al.* (2004) are even close to being correct, then the photosynthetic dinoflagellate clade must have persisted for hundreds of millions of years prior to its Jurassic diversification, without leaving any other photosynthetic descendants during that time.

That would be a strange possibility but not necessarily more so than other features of the peridinin-type plastid. Unlike all other photosynthetic eukaryotes, carbon fixation is carried out by a form II rubisco encoded in the nuclear genome (Morse *et al.* 1995; Whitney *et al.* 1995; Rowan *et al.* 1996). Rubisco of this type is unique to peridinin-containing dinoflagellates among eukaryotes and is elsewhere associated with anoxygenic photosynthesis (Morse *et al.* 1995). In those organisms for which the enzyme kinetics have been characterized, form II rubisco can fix carbon very rapidly but is highly sensitive to the presence of oxygen. Dinoflagellates are the only known organisms that use this enzyme in oxygenic photosynthesis, almost certainly with the assistance of a carbon-concentrating mechanism (Whitney and Andrews 1998; Kaplan and Reinhold 1999). This form II rubisco is apparently a spectacular example of the rampant horizontal gene transfer that has influenced rubisco evolution (Delwiche and Palmer 1996; Paoli *et al.* 1998).

Another peculiarity of the peridinin-type plastid is its apparent lack of a coherent, canonical plastid genome. Many plastid genes seem to be encoded on single-gene minicircles (Zhang *et al.* 1999; Barbrook *et al.* 2001), and no typical plastid genome has ever been detected. Although messenger RNA has been localized to the plastid, and DNA has been shown to be present, there is also evidence that the minicircles are encoded in the nuclear genome of at least some species (Laatsch *et al.* 2004). It may be that the plastid is only weakly dependent upon the plastid genome; many typically plastid-encoded genes have been detected as apparent nuclear transcripts, suggesting that the plastid genome has nearly been lost (Bachvaroff *et al.* 2004; Hackett *et al.* 2004). However, relatively few species have been examined, and none of these has been studied in the detail expected in molecular biological studies of plants and animals. It is important to bear in mind that considerable work is needed to determine whether these features are variable within the group.

It is perhaps not surprising that several dinoflagellates are pigmented with plastids other than the peridinin-type plastid, given that many, if not most, dinoflagellates are also capable of phagotrophy (Bhattacharya *et al.* 2004; Delwiche *et al.* 2004; Keeling 2004). Some species (e.g., *Pyrocystis lunula*) have vegetative stages with thick cell walls that presumably preclude phagocytosis, but even most photosynthetic species can also function as predators. Non–peridinin-type plastids were evidently acquired via secondary (or tertiary) endosymbiosis and include plastids derived from green algae, haptophytes, and cryptomonads, as well as a nearly intact heterokont, a diatom (Chesnick *et al.* 1997; Jenks and Gibbs 2000). It is unclear why dinoflagellates, alone among mixotrophic algae, have acquired several different types of plastids. One hypothesis is that the highly reduced nature of the plastid genome combined with the unusually large number of nuclear-encoded, plastid-expressed genes has made the plastid genome nearly unnecessary. If this is the case, and the dinoflagellate nuclear genome has a highly flexible ability to support endosymbiotic plastids of any origin, then one might predict that non–peridinin-type plastids in dinoflagellates would import proteins that in the lineage from which they are derived are encoded within the plastid genome. One obvious flaw with this hypothesis is that these "minor" plastid types in dinoflagellates retain distinctive characters of the lineage from which they are derived, suggesting that key genes determining of biochemistry, pigmentation, and ultrastructure have traveled with the plastid. Consequently the high diversity of plastid types in dinoflagellates compared with other taxa remains unexplained.

IV. DINOFLAGELLATES IN THE PLANKTON

Dinoflagellates occupy a variety of environmental niches, but the majority of the photosynthetic species are plank-tonic. They occur in both freshwater and marine environments but cannot tolerate highly turbulent waters and consequently are most common in bays and estuaries, on the open ocean where there is relatively lit-tle upwelling, and in lakes and ponds. They benefit from a diurnal cycle in which they migrate from the euphotic zone during the day to deeper, more nutrient-rich waters below the pycnocline at night (Harding and Coats 1988). As a result they are relatively minor constituents of the phytoplankton in areas with extensive upwelling, and under conditions with substantial wind-driven mixing. Two fascinating and important aspects of dinoflagellates in nature would require more space than is available here but have recently received excellent reviews to which I refer the reader: their potential role in harmful algal blooms (Anderson *et al.* 2002; Luckas *et al.* 2005; Edwards *et al.* 2006), and their role as the symbionts in associa-tion with corals, clams, and other animals (Stanley 2003; Van Oppen *et al.* 2005; Van Oppen and Gates 2006; Rowan 1998; Souter and Linden 2000).

The cyst-forming dinoflagellates are of particular importance because of their ability to appear in the sudden, high-density blooms known as "red tides" (or "mahogany" or "brown" tides) and the role of cysts in the fossil record. The cysts can lie dormant in onshore sediments for years until conditions favor germina-tion and suddenly a huge population can appear, literally overnight. The population dynamics of such blooms are fascinating and complex (Bockstahler and Coats 1993; Park *et al.* 2004). For example, in the Chesa-peake Bay, blooms of *Akashiwo sanguinea* (see Figure 1C) can develop quite quickly. Over time, as the bloom develops, competi-tion from cryptomonads, which have a rela-tively high growth rate, can lead to shading of the *A. sanguinea*, which can then shift to a predatory mode and ingest the competing cryptomonads. The situation is further com-plicated by *Amoebophrya* spp (see Figure 1D),

which are parasitic dinoflagellates that attack *A. sanguinea*. As the population den-sity of *A. sanguinea* increases, so does its risk of attack from *Amoebophrya*, which can lead to a sudden decline in the *A. sanguinea* pop-ulation. There are further complications, with potential interactions by viruses, fungi, and parasites of *Amoebophrya*. Understand-ing such population dynamics is an active area of research.

Cyst formation is not, however, a suc-cessful strategy for dinoflagellates in the open ocean. When the seafloor is too dis-tant from the surface, it is neither practi-cal for a cyst to receive germination signals from the surface nor feasible for a motile cell to migrate to the surface. Consequently cysts are characteristic of the coastal envi-ronment (although by no means do all coasta dinoflagellates make cysts), and a different life strategy is necessary for the open ocean.

One of the most dramatic and famous features of dinoflagellates is the biolumines-cence that can illuminate the wakes of ships and footprints on the beach at night (see, e.g., Abrahams and Townsend 1993). Although dinoflagellate bioluminescence can produce dramatic displays in onshore environments such as Bioluminescent Bay in Puerto Rico, bioluminescence appears to be fundamen-tally an adaptation to life in the open ocean. In dinoflagellates, bioluminescence typi-cally occurs in specialized organelles called scintillons and involves a regulated, three-component process (Hastings 1959; Li *et al.* 1997; Okamoto *et al.* 2001; Okamoto and Hastings 2003). During the daytime, when they are photosynthetic and lie close to the surface, bioluminescent dinoflagellates do not emit light. At night, as they migrate below the pycnocline, they begin to emit flashes of light. The genetic regulation of biolumines-cence has been an area of active study and has provided important insights into circa-dian rhythms and the genetic mechanisms that control them (Hastings and Johnson 2003; Okamoto and Hastings 2003; Rossini *et al.* 2003).

The light is quite bright, and under fairly common conditions can be the major source of light in the environment. Exactly why dinoflagellates emit light remains hotly debated, but one rather surprising possibility is that the light emitted by dinoflagellates may help provide illumination for the predators that consume the herbivores that might otherwise eat the dinoflagellates. As far-fetched as this hypothesis may sound, it has received some empirical support. For example, predatory cephalopods have been shown to feed more effectively (Fleisher and Case 1995), and herbivorous copepod populations have been shown to decline when dinoflagellate bioluminescence is present. However, much about bioluminescence (Esaisas and Curl 1972) remains poorly understood, not least of which is why a process that requires substantial energetic input confers a selective advantage.

Dinoflagellates are remarkable and unfamiliar organisms in almost every way. This is in part because they are distantly related to the more familiar animals, fungi, and plants and in part because they inhabit environments that are often remote and inhospitable to human occupation. That unfamiliarity would, by itself, be an important reason to study these organisms (by studying them it is possible to gain a more generalized understanding of how life is structured), but they are also profoundly important in the environment in a diversity of roles, from primary producer to predator and parasite.

References

Abrahams, M.V., and Townsend, L.D. (1993). Bioluminescence in dinoflagellates: A test of the burglar alarm hypothesis. *Ecology* **74**: 258–260.

Anderson, D.M., Glibert, P.M., and Burkholder, J.M. (2002). Harmful algal blooms and eutrophication: nutrient sources, composition, and consequences. *Estuaries* **25**: 704–726.

Bachvaroff, T.R., Concepcion, G.T., Rogers, C.R., Herman, E.M., and Delwiche, C.F. (2004). Dinoflagellate expressed indicate massive transfer to the nuclear genome sequence tag data of chloroplast genes. *Protist* **155**: 65–78.

Barbrook, A.C., Symington, H., Nisbet, R.E.R., Larkum, A., and Howe, C.J. (2001). Organisation and expression of the plastid genome of the dinoflagellate *Amphidinium operculatum*. *Mol. Genet. Genomics* **266**: 632–638.

Bhattacharya, D., Yoon, H.S., and Hackett, J.D. (2004). Photosynthetic eukaryotes unite: endosymbiosis connects the dots. *Bioessays* **26**: 50–60.

Bockstahler, K.R., and Coats, D.W. (1993). Grazing of the mixotrophic dinoflagellate *Gymnodinium-Sanguineum* on ciliate populations of Chesapeake Bay. *Mar. Biol.* **116**: 477–487.

Bodyl, A. (2005). Do plastid-related characters support the chromalveolate hypothesis? *J. Phycol.* **41**: 712–719.

Brocks, J.J., Logan, G.A., Buick, R., and Summons, R.E. (1999). Archean molecular fossils and the early rise of eukaryotes. *Science* **285**: 1033–1036.

Brocks, J.J., and Summons, R.E. (2003). Sedimentary hydrocarbons, biomarkers for early life. *Treatise in Geology*. H.D. Holland, and K. Turekian, eds. Oxford, Elsevier, pp. 63–115.

Cavalier-Smith, T. (1999). Principles of protein and lipid targeting in secondary symbiogenesis: euglenoid, dinoflagellate, and sporozoan plastid origins and the eukaryote family tree. *J. Eukaryot. Microbiol.* **46**: 347–366.

Chesnick, J.M., Kooistra, W., Wellbrock, U., and Medlin, L.K. (1997). Ribosomal RNA analysis indicates a benthic pennate diatom ancestry for the endosymbionts of the dinoflagellates *Peridinium foliaceum* and *Peridinium balticum* (Pyrrhophyta). *J. Eukaryot. Microbiol.* **44**: 314–320.

Christensen, T. (1989). The chromophyta, past and present. *The Chromophyte Algae: Problems and Perspectives*. J.C. Green, B.S.C. Leadbeater, and W.L. Diver, eds. Oxford, Clarendon Press, pp. 1–12.

Chudnovsky, Y., Li, J.F., Rizzo, P.J., Hastings, J.W., and Fagan, T.F. (2002). Cloning, expression, and characterization of a histone-like protein from the marine dinoflagellate *Lingulodinium polyedrum* (Dinophyceae). *J. Phycol.* **38**: 543–550.

Colbath, G.K., and Grenfell, H.R. (1995). Review of biological affinities of Paleozoic acid-resistant, organic-walled eukaryotic algal microfossils (including Acritarchs). *Rev. Palaeobot. Palynol.* **86**: 287–314.

Daugbjerg, N., Hansen, G., Larsen, J., and Moestrup, O. (2000). Phylogeny of some of the major genera of dinoflagellates based on ultrastructure and partial LSU rDNA sequence data, including the erection of three new genera of unarmoured dinoflagellates. *Phycologia* **39**: 302–317.

Delwiche, C.F. (1999). Tracing the thread of plastid diversity through the tapestry of life. *Am. Nat.* **154**: S164–S177.

Delwiche, C.F., Andersen, R.A., Bhattacharya, D., Mishler, B., and McCourt Richard, M. (2004). Algal evolution and the early radiation of green plants. *Assembling the Tree of Life*. J. Cracraft, and M.J. Donoghue, eds. London, Oxford University Press, pp. 121–137.

Delwiche, C.F., and Palmer, J.D. (1996). Rampant horizontal transfer and duplication of rubisco genes in eubacteria and plastids. *Mol. Biol. Evol.* **13**: 873–882.

Edwards, M., Johns, D.G., Leterme, S.C., Svendsen, E., and Richardson, A.J. (2006). Regional climate change and harmful algal blooms in the northeast Atlantic. *Limnol. Oceanogr.* **51**: 820–829.

Esaisas, W., and Curl, Jr., H.C. (1972). Effect of dinoflagellate bioluminescence on copepod ingestion rates. *Limnol. Oceanogr.* **16**: 901–906.

Evitt, W.R. (1963). A discussion and proposals concerning fossil dinoflagellates, hystrichospheres, and acritarchs. 1. *Proc. Natl. Acad. Sci. U S A* **49**: 158–164.

Falkowski, P.G., Katz, M.E., Knoll, A.H., Quigg, A., Raven, J.A., Schofield, O., and Taylor, F.J.R. (2004). The evolution of modern eukaryotic phytoplankton. *Science* **305**: 354–360.

Falkowski, P.G., and Raven, J.A. (2007). *Aquatic Photosynthesis*, 2nd ed. Princeton, Princeton University Press, pp. 484.

Fehling, J., Stoecker, D., and Baldauf, S. (2007). Photosynthesis and the eukaryote tree of life. *Evolution of Primary Producers in the Sea*. P.G. Falkowski and A.H. Knoll, eds. Boston, Elsevier, pp. 75–107.

Fensome, R.A., MacRae, R.A., Moldowan, J.M., Taylor, F.J.R., and Williams, G.L. (1996). The early Mesozoic radiation of dinoflagellates. *Paleobiology* **22**: 329–338.

Fensome, R.A., Saldarriaga, J.F., and Taylor, F. (1999). Dinoflagellate phylogeny revisited: reconciling morphological and molecular based phylogenies. *Grana* **38**: 66–80.

Fensome, R.A., Taylor, F.J.R., Norris, G., Sarjeant, W.A.S., Wharton, D.I., and Williams, G.L. (1993). A classification of fossil and living dinoflagellates. *Micropaleontology* Special Paper #7: 1–351.

Fleisher, K.J., and Case, J.F. (1995). Cephalopod predation facilitated by dinoflagellate luminescence. *Biol. Bull.* **189**: 263–271.

Fritsch, F.E. (1965). *The Structure and Reproduction of Algae*, Vol 1. Cambridge, UK, Cambridge University Press.

Gast, R.J., and Caron, D.A. (1996). Molecular phylogeny of symbiotic dinoflagellates from plankton foraminifera and radiolaria. *Mol. Biol. Evol.* **13**: 1192–1197.

Gibbs, S.P. (1981). The chloroplasts of some algal groups may have evolved from endosymbiotic eukaryotic algae. *Ann. N Y Acad. Sci.* **361**: 193–208.

Grzebyk, D., Schofield, O., Vetriani, C., and Falkowski, P.G. (2003). The Mesozoic radiation of eukaryotic algae: The portable plastid hypothesis. *J. Phycol.* **39**: 259–267.

Hackett, J.D., Yoon, H.S., Soares, M.B., Bonaldo, M.F., Casavant, T.L., Scheetz, T.E., Nosenko, T., and Bhattacharya, D. (2004). Migration of the plastid genome to the nucleus in a peridinin dinoflagellate. *Curr. Biol.* **14**: 213–218.

Hamm, C., and Smetacek, V. (2007). Armor: why, when, and how. *Evolution of Primary Producers in the Sea*. P.G. Falkowski and A.H. Knoll, eds. Boston, Elsevier, pp. 311–332.

Harding, L.W., and Coats, D.W. (1988). Photosynthetic physiology of Prorocentrum-Mariae-Lebouriae (Dinophyceae) during its subpycnocline transport in Chesapeake Bay. *J. Phycol.* **24**: 77–89.

Hastings, J.W. (1959). Bioluminescence in marine dinoflagellates. Paper presented at First National Biophysics Conference, Yale University, Yale University Press.

Hastings, J.W., and Johnson, C.H. (2003). Bioluminescence and chemiluminescence. *Biophotonics*, Part A. Boston, Elsevier, pp. 75–104.

Hofmann, E., Wrench, P.M., Sharples, F.P., Hiller, R.G., Welte, W., and Diederichs, K. (1996). Structural basis of light harvesting by carotenoids: peridinin-chlorophyll-protein from *Amphidinium carterae*. *Science* **272**: 1788–1792.

Jenks, A., and Gibbs, S.P. (2000). Immunolocalization and distribution of Form II Rubisco in the pyrenoid and chloroplast stroma of *Amphidinium carterae* and Form I Rubisco in the symbiont-derived plastids of *Peridinium foliaceum* (Dinophyceae). *J. Phycol.* **36**: 127–138.

Kaplan, A., and Reinhold, L. (1999). CO2 concentrating mechanisms in photosynthetic microorganisms. *Annu. Rev. Plant Physiol. Plant Mol. Biol.* **50**: 539–570.

Katz, M.E., Finkel, Z.V., Grzebyk, D., Knoll, A.H., and Falkowski, P.G. (2004). Evolutionary trajectories and biogeochemical impacts of marine eukaryotic phytoplankton. *Annu. Rev. Ecol. Evol. System.* **35**: 523–556.

Keeling, P.J. (2004). Diversity and evolutionary history of plastids and their hosts. *Am. J. Bot.* **91**: 1481–1493.

Knoll, A.H., Summons, R.E., Waldbauer, J.R., and Zumberge, J.E. (2007). The geological succession of primary producers in the oceans. *Evolution of Primary Producers in the Sea*. P.G. Falkowski and A.H. Knoll, eds. Boston, Elsevier, pp. 133–163.

Kooistra, W.C.H.F., Gersonde, R., Medlin, L.K., and Mann, D.G. (2007). The Origin and evolution of the diatoms: Their adaptation to a planktonic existence. *Evolution of Primary Producers in the Sea*. P.G. Falkowski and A.H. Knoll, eds. Boston, Elsevier, pp. 207–249.

Laatsch, T., Zauner, S., Stoebe-Maier, B., Kowallik, K.V., and Maier, U.-G. (2004). Plastid-derived single gene minicircles of the dinoflagellate *Ceratium horridum* are localized in the nucleus. *Mol. Biol. Evol.* **21**: 1318–1322.

LaJeunesse, T.C., Lambert, G., Andersen, R.A., Coffroth, M.A., and Galbraith, D.W. (2005). *Symbiodinium* (Pyrrhophyta) genome sizes (DNA content) are smallest among dinoflagellates. *J. Phycol.* **41**: 880–886.

Leander, B.S., and Keeling, P.J. (2004). Early evolutionary history of dinoflagellates and apicomplexans (Alveolata) as inferred from hsp90 and actin phylogenies. *J. Phycol.* **40**: 341–350.

Lenaers, G., Scholin, C., Bhaud, Y., Sainthilaire, D., and Herzog, M. (1991). A molecular phylogeny of dinoflagellate protists (Pyrrhophyta) inferred from the sequence of 24s ribosomal-RNA divergent domain-D1 and domain-D8. *J. Mol. Evol.* **32:** 53–63.

Li, L., Hong, R., and Hastings, J.W. (1997). Three functional luciferase domains in a single polypeptide chain. *Proc. Natl. Acad. Sci. U S A* **94:** 8954–8958.

Li, A.S., Stoecker, D.K., Coats, D.W., and Adam, E.J. (1996). Ingestion of fluorescently labeled and phycoerythrin-containing prey by mixotrophic dinoflagellates. *Aquat. Microb. Ecol.* **10:** 139–147.

López-García, P., Rodríguez-Valera, F., Pedrós-Alió, C., and Moreira, D. (2001). Unexpected diversity of small eukaryotes in deep-sea Antarctic plankton. *Nature* **409:** 603–607.

Luckas, B., Dahlmann, J., Erler, K., Gerdts, G., Wasmund, N., Hummert, C., and Hansen, P.D. (2005). Overview of key phytoplankton toxins and their recent occurrence in the North and Baltic seas. *Environ. Toxicol.* **20:** 1–17.

McGillicuddy, D.J., Signell, R.P., Stock, C.A., Keafer, B.A., Keller, M.D., Hetland, R.D., and Anderson, D.M. (2003). A mechanism for offshore initiation of harmful algal blooms in the coastal Gulf of Maine. *J. Plankton Res.* **25:** 1131–1138.

McNally, K.L., Govind, N.S., Thome, P.E., and Trench, R.K. (1994). Small-subunit ribosomal DNA-sequence analyses and a reconstruction of the inferred phylogeny among symbiotic dinoflagellates (Pyrrhophyta). *J. Phycol.* **30:** 316–329.

Meng, F.W., Zhou, C.M., Yin, L.M., Chen, Z.L., and Yuan, X.L. (2005). The oldest known dinoflagellates: Morphological and molecular evidence from Mesoproterozoic rocks at Yongji, Shanxi Province. *Chinese Sci. Bull.* **50:** 1230–1234.

Moldowan, J.M., Dahl, J., Jacobson, S.R., Huizinga, B.J., Fago, F.J., Shetty, R., Watt, D.S., and Peters, K.E. (1996). Chemostratigraphic reconstruction of biofacies: molecular evidence linking cyst-forming dinoflagellates with pre-Triassic ancestors. *Geology* **24:** 159–162.

Moldowan, J.M., and Talyzina, N.M. (1998). Biogeochemical evidence for dinoflagellate ancestors in the early Cambrian. *Science* **281:** 1168–1170.

Moon-van der Staay, S.Y., De Wachter, R., and Vaulot, D. (2001). Oceanic 18S rDNA sequences from picoplankton reveal unsuspected eukaryotic diversity. *Nature* **409:** 607–610.

Morse, D. (1995). A nuclear-encoded Form-II Rubisco in dinoflagellates (Vol. 268, Pg. 1622, 1995). *Science* **269:** 17–17.

Morse, D., Salois, P., Markovic, P., and Hastings, J.W. (1995). A nuclear-encoded form II rubisco in dinoflagellates *Science* **268:** 1622–1624.

Okamoto, O.K., and Hastings, J.W. (2003). Novel dinoflagellate clock-related genes identified through microarray analysis. *J. Phycol.* **39:** 519–526.

Okamoto, O.K., Liu, L.Y., Robertson, D.L., and Hastings, J.W. (2001). Members of a dinoflagellate luciferase gene family differ in synonymous substitution rates. *Biochemistry* **40:** 15862–15868.

Pace, N. (1997). A molecular view of microbial diversity and the biosphere. *Science* **276:** 734–740.

Palmer, J.D., and Delwiche, C.F. (1998). The origin and evolution of plastids and their genomes. *Molecular Systematics of Plants II.* D.E. Soltis, P.S. Soltis, and J.J. Doyle, eds. Boston, Kluwer Academic Publishers, pp. 375–409.

Paoli, G.C., Soyer, F., Shively, J., and Tabita, F.R. (1998). Rhodobacter capsulatus genes encoding form I ribulose-1,5-bisphosphate carboxylase/oxygenase (cbbLS) and neighbouring genes were acquired by a horizontal gene transfer. *Microbiology* **144:** 219–227.

Park, M.G., Yih, W., and Coates, D.W. (2004). Parasites and phytoplankton, with special emphasis on dinoflagellate infections. *J. Eukaryot. Microbiol.* **51,** 145–155.

Rae, P.M., and Steele, R.E. (1978). Modified bases in the DNAs of unicellular eukaryotes: an examination of distributions and possible roles, with emphasis on hydroxymethyluracil in dinoflagellates. *Biosystems* **10:** 37–53.

Rossini, C., Taylor, W., Fagan, T., and Hastings, J.W. (2003). Lifetimes of mRNAs for clock-regulated proteins in a dinoflagellate. *Chronobiol. Int.* **20:** 963–976.

Rowan, R. (1998). Diversity and ecology of zooxanthellae on coral reefs. *J. Phycol.* **34:** 407–417.

Rowan, R., Whitney, S.M., Fowler, A., and Yellowlees, D. (1996). Rubisco in marine symbiotic dinoflagellates: form II enzymes in eukaryotic oxygenic phototrophs encoded by a nuclear multigene family. *Plant Cell* **8:** 539–553.

Sadler, L.A., McNally, K.L., Govind, N.S., Brunk, C.F., and Trench, R K. (1992). The nucleotide-sequence of the small subunit ribosomal-RNA gene from *Symbiodinium-Pilosum,* a symbiotic dinoflagellate. *Curr. Genet.* **21:** 409–416.

Saldarriaga, J.F., McEwan, M.L., Fast, N.M., Taylor, F.J.R., and Keeling, P.J. (2003). Multiple protein phylogenies show that *Oxyrrhis marina* and *Perkinsus marinus* are early branches of the dinoflagellate lineage. *Int. J. Syst. Evol. Microbiol.* **53:** 355–365.

Saldarriaga, J.F., Taylor, F.J.R., Keeling, P.J., and Cavalier-Smith, T. (2001). Dinoflagellate nuclear SSU rRNA phylogeny suggests multiple plastid losses and replacements. *J. Mol. Evol.* **53:** 204–213.

Santos, S.R. (2004). Phylogenetic analysis of a free-living strain of *Symbiodinium* isolated from Jiaozhou Bay, PR China. *J. Phycol.* **40:** 395–397.

Saunders, G.W., Hill, D.R.A., Sexton, J.P., and Anderson, R.A. (1997). Small-subunit ribosomal RNA sequences from selected dinoflagellates: testing classical evolutionary hypotheses with molecular systematic methods. *Pl. Syst. Evol. Suppl.* **11:** 237–260.

Shalchian-Tabrizi, K., Minge, M.A., Cavalier-Smith, T., Nedreklepp, J.M., Klaveness, D., and Jakobsen, K.S. (2006a). Combined heat shock protein 90 and ribosomal RNA sequence phylogeny supports multiple replacements of dinoflagellate plastids. *J. Eukaryot. Microbiol.* **53:** 217–224.

Shalchian-Tabrizi, K., Skanseng, M., Ronquist, F., Klaveness, D., Bachvaroff, T.R., Delwiche, C.F., Botnen, A., Tengs, T., and Jakobsen, K.S. (2006b). Heterotachy processes in rhodophyte-derived secondhand plastid genes: implications for addressing the origin and evolution of dinoflagellate plastids. *Mol. Biol. Evol.* **23:** 1504–1515.

Skovgaard, A., Massana, R., Balagué, V., and Saiz, E. (2005). Phylogenetic position of the copepod-infesting parasite *Syndinium turbo* (Dinoflagellata, Syndinea). *Protist* **156:** 413–423.

Souter, D.W., and Linden, O. (2000). The health and future of coral reef systems. *Ocean Coastal Manage.* **43:** 657–688.

Stanley, G.D. (2003). The evolution of modern corals and their early history. *Earth Sci. Rev.* **60:** 195–225.

Strain, H.H., Svec, W.A., Aitzetmu, K., Grandolf, M.C., Katz, J.J., Kjosen, H., Norgard, S., Liaaenje, S., Haxo, F.T., Wegfahrt, P., and Rapoport, H. (1971). Structure of peridinin, characteristic dinoflagellate carotenoid. *J. Am. Chem. Soc.* **93:** 1823–1825.

Takabayashi, M., Santos, S.R., and Cook, C.B. (2004). Mitochondrial DNA phylogeny of the symbiotic dinoflagellates (*Symbiodinium*, Dinophyta). *J. Phycol.* **40:** 160–164.

Takishita, K., Ishikura, M., Koike, K., and Maruyama, T. (2003). Comparison of phylogenies based on nuclear-encoded SSU rDNA and plastid-encoded psbA in the symbiotic dinoflagellate genus *Symbiodinium*. *Phycologia* **42:** 285–291.

Tappan, H. (1980). *The Paleobiology of Plant Protists*. San Francisco, W.H. Freeman and Co.

Taylor, F.J.R. (1989). Dinoflagellate morphology. *The Biology of Dinoflagellates*. F.J.R. Taylor, ed. Oxford, Blackwell Scientific Publications, pp. 24–91.

Taylor, F.J.R. (2004). Illumination or confusion? Dinoflagellate molecular phylogenetic data viewed from a primarily morphological standpoint. *Phycol. Res.* **52:** 308–324.

Tengs, T., Dahlberg, O.J., Shalchian-Tabrizi, K., Klaveness, D., Rudi, K., Delwiche, C.F., and Jakobsen, K.S. (2000). Phylogenetic analyses indicate that the 19′ hexanoyloxy-fucoxanthin-containing dinoflagellates have tertiary plastids of haptophyte origin. *Mol. Biol. Evol.* **17:** 718–729.

Tindall, D.R., and Morton, S.L. (1998). Community dynamics and physiology of epiphytic/benthic dinoflagellates associated with ciguatera. *Physiological Ecology of Harmful Algal Blooms*. D.M. Anderson, A.D. Cembella, and G.M. Hallegraeff, eds. Berlin, Springer-Verlag, pp. 292–312.

Van Oppen, M.J.H., and Gates, R.D. (2006). Conservation genetics and the resilience of reef-building corals. *Mol. Ecol.* **15:** 3863–3883.

Van Oppen, M.J.H., Mieog, J.C., Sanchez, C.A., and Fabricius, K.E. (2005). Diversity of algal endosymbionts (zooxanthellae) in octocorals: the roles of geography and host relationships. *Mol. Ecol.* **14:** 2403–2417.

Whitney, S.M., and Andrews, T.J. (1998). The CO2/O-2 specificity of single-subunit ribulose-bisphosphate carboxylase from the dinoflagellate, *Amphidinium carterae*. *Aust. J. Plant Physiol.* **25:** 131–138.

Whitney, S.M., Shaw, D.C., and Yellowlees, D. (1995). Evidence that some dinoflagellates contain a ribulose-1,5-bisphosphate carboxylase oxygenase related to that of the alpha-proteobacteria. *Proc. R. Soc. Lond. Ser. B Biol. Sci.* **259:** 271–275.

Yang, E.J., Choi, J.K., and Hyun, J.H. (2004). Distribution and structure of heterotrophic protist communities in the northeast equatorial Pacific Ocean. *Mar. Biol.* **146:** 1–15.

Yoon, H.S., Hackett, J.D., and Bhattacharya, D. (2002). A single origin of the peridinin- and fucoxanthin-containing plastids in dinoflagellates through tertiary endosymbiosis. *Proc. Natl. Acad. Sci. U S A* **99:** 11724–11729.

Yoon, H.S., Hackett, Y.D., Ciniglia, C., Pinto, G., and Bhattacharya, D. (2004). A molecular timeline for the origin of photosynthetic eukaryotes. *Mol. Biol. Evol.* **21:** 809–818.

Zhang, H., Bhattacharya, D., and Lin, S. (2005). Phylogeny of dinoflagellates based on mitochondrial cytochrome b and nuclear small subunit rDNA sequence comparisons. *J. Phycol.* **41:** 411–420.

Zhang, Z., Green, B.R., and Cavalier-Smith, T. (1999). Single gene circles in dinoflagellate chloroplast genomes. *Nature* **400:** 155–159.

Zhang, Z.D., Green, B.R., and Cavalier-Smith, T. (2000). Phylogeny of ultra-rapidly evolving dinoflagellate chloroplast genes: a possible common origin for sporozoan and dinoflagellate plastids. *J. Mol. Evol.* **51:** 26–40.

11

The Origin and Evolution of the Diatoms: Their Adaptation to a Planktonic Existence

WIEBE H.C.F. KOOISTRA, RAINER GERSONDE, LINDA K. MEDLIN,
AND DAVID G. MANN

Diatoms are unicellular eukaryotes with cell sizes that usually range between 10 and 200 μm. Their hallmark is the ornamented compound silica cell wall called a frustule (Round *et al*. 1990; also see Hamm and Smetacek, Chapter 14, this volume). Frustule architecture is extremely diverse (e.g., Round *et al*. 1990). Yet, variation is also apparent in other features, such as organelle arrangement (Schmid 1988), the shape, number, and ultrastructure of the chloroplasts (Mann 1996), and the type of sexual reproduction (Kaczmarska *et al.* 2001; Chepurnov *et al.* 2004).

Diatoms constitute one of the major lineages of photosynthetic eukaryotes on Earth and there are probably well over ca. 10^5 species (Mann and Droop 1996), the vast majority of which are not uniquely identified as such. They are ubiquitous in the plankton and benthos of oceanic, coastal, and freshwater habitats, and they abound even in temporarily wet terrestrial environments. Not surprisingly, their contribution to global productivity and nutrient cycling is significant (Rau *et al.* 1997; Falkowski *et al.* 1998; Mann 1999; Smetacek 1999; Boyd *et al.* 2000). Planktonic diatoms are apparently extremely well adapted to growth in deeply mixed turbulent water, where cells are only intermittently exposed to high light levels (Mitchell and Holm-Hansen 1991; Falkowski and Raven 2007). Their central vacuole can store nutrients when in plentiful supply for later use, their light-harvesting system is able to protect itself rapidly against high light intensity, and their CO_2 uptake mechanisms are highly efficient. In addition, diatoms can form resting stages to overcome periods adverse to growth.

Results from ultrastructural, biochemical, molecular phylogenetic, and comparative genomic approaches have revealed that diatoms belong to the heterokontophytes, a group of algae with golden-brown to brown chloroplasts originating from a red algal endosymbiont. The heterokontophytes form a clade within heterotrophic heterokonts; the latter are paraphyletic. The heterokonts share heterodynamic flagella (van den Hoek *et al.* 1995; Graham and Wilcox 2000). Within the diatoms themselves, the radial centrics (shaped like Petri dishes) appear to constitute the most ancient group, multipolar centrics (shaped like gaudy perfume bottles) are slightly younger, and pennates (shaped like boat hulls) form a relatively young group (see early morphological phylogeny in Simonsen 1979). Within the latter, the raphid pennates, those with a slit called a raphe, represent

the most recent diversification and are the most species-rich group of extant diatoms (e.g., Kooistra *et al.* 2003a, 2006; Medlin and Kaczmarska 2004; see also Figures 1 and 2). Marine planktonic diatoms are found, however, predominantly in the radial and multipolar centrics, though there are also a few important pennate representatives.

Insights into the evolution of the diatoms can also be gleaned from their extensive fossil record. Frustule elements are preserved in various concentrations and in various states, and often in all their resplendence, over millions of years (e.g., Gersonde and Harwood 1990), and organic molecular fossils specific for particular diatom lineages, such as *Rhizosolenia,* are detectable in ocean sediments (Rospondek *et al.* 2000; Sinninghe Damsté *et al.* 2004). Radial centrics first appear in the Jurassic, multipolar centrics are present in the Early Cretaceous, and pennates appear in the Late Cretaceous. Diatoms come relatively unscathed through the Cretaceous–Tertiary (K/T) boundary and show from then onward an ever-increasing diversity (e.g., Schrader and Fenner 1976; Fenner 1977, 1985; Gombos and Ciesielski 1983; Baldauf 1984; Harwood and Maruyama 1992; Gladenkov and Barron 1995; Stoermer and Smol 1999; Harwood *et al.* 2004; Sims *et al.* 2006).

Here we review the evolution of the diatoms and of various groups within the diatoms and evaluate the significance of particular shared derived traits. Questions discussed include the following: Why do the diatoms dominate the modern marine phytoplankton? Why did they as members of the chromist lineages (here including dinoflagellates, heterokontophytes, and haptophytes) win over red and green unicellular phytoplankton? Why did heterokontophytes win over dinoflagellates and haptophytes? Why did diatoms win over other heterokontophytes? Such insights would be valuable to climatologists trying to discover if there are any historical analogues for the current Earth system, with atmospheric CO_2 now at levels unknown since the mid-Tertiary

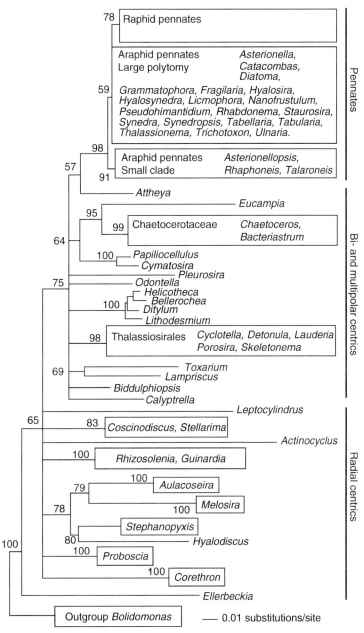

FIGURE 1. Neighbor joining (NJ) phylogeny inferred from maximum likelihood pair-wise distances among nuclear SSU rDNA sequences of various diatom species. Sequences of *Bolidomonas* spp. were included as outgroup. Maximum likelihood calculations were constrained with substitution rate parameters: $A\Delta C = 0.9$, $A\Delta G = 2.4$, $A\Delta T = 1.2$, $C\Delta G = 1.1$, $C\Delta T = 3.8$ versus $G\Delta T$ set to 1.0. Assumed proportion of invariable sites along the alignment $= 0.4$ and a gamma distribution of rates at variable sites with shape parameter (alpha) $= 0.5$. Assumed nucleotide frequencies: $A = 0.25$, $C = 0.18$, $G = 0.25$, $T = 0.32$. Bootstrap values (1000 replicates) have been generated using the same settings; values associated to minor clades to the right have been omitted. The tree is an abridged version of Figure 36A, B, and C in Kooistra *et al.* (2006).

(see, e.g., Falkowski *et al.* 2004a, b) and an ocean pH that within a few centuries may be lower than at any time since the end of the Carboniferous (Doney 2006).

I. THE HALLMARK OF THE DIATOMS: THE SILICA FRUSTULE

The compound silica cell wall of the diatoms is composed of two valves (see Figure 2), each accompanied by a series of girdle bands (also called cingular bands, see Figure 2B and F). The valves are generally the more complex structures. A typical valve looks like a Petri dish with a flat area called the valve face and a rim called the mantle. The girdle bands underlap the mantle. One valve and its girdle bands are, together, called the epitheca, and they overlap a similar but slightly narrower set of frustular elements called the hypotheca. Thus, the frustule completely encloses the protoplasm while permitting growth by expansion.

A. Frustule Shape and Ornamentation and Their Bearings on Diatom Taxonomy

The valve face generally shows a central ring (annulus), central area, or a thickened midrib (sternum, see Figure 2H) from which ribs (interstriae, see Figure 2H) extend. Rows of pores are present between the interstriae and each pore is itself partly occluded by fine silica structures in the form of strips, flaps, elaborate meshes, or sievelike plates. In many species, various silica elements for further structural reinforcement are laid down inside the valves (see, e.g., Figure 2C and I). Many species have clusters of differently shaped, tightly packed slits or pores, called pore fields, located at the valve apices or poles (see Figure 2H). These structures exude polysaccharides to form mucilage pads for attachment to sister cells or other substrata. Most diatoms possess one or two types of tubes (processes) that run through their valve face. Labiate processes (rimoportulae,

see Figure 2D) are the most widespread process in the diatoms and are probably involved in mucilage excretion and nutrient uptake (Medlin *et al.* 1986; Pickett-Heaps *et al.* 1986; Kühn and Brownlee 2005). Their name originates from the liplike compressed shape of the end of the tube on the inside of the valve. Strutted processes (fultoportulae, see Figure 2F) are encountered exclusively in Thalassiosirales (see Round *et al.* 1990) and are involved in the excretion of β-chitin fibrils (see Figure 2E). These fibrils extend into the environment or connect adjacent sister cells into a chain (Walsby and Xypolyta 1977). A third kind of process, the raphe (see Figure 2I), consists of two v-shaped slits located on one side of the midrib that allow the cells to slide actively over the substratum by means of an ingenious and cost-effective system of traction (Edgar and Pickett-Heaps 1984). The raphe creates a structural weakness in the valve. To overcome this problem, many raphid pennate species possess silica struts, called fibulae (see Figure 2I), bridging the raphe canal. Other tubular processes exist but are relatively rare; for example, a simple large pore next to the annulus in *Leptocylindrus* or a series of minute ones on the valve mantle of *Ellerbeckia*.

Taxonomists have divided the diatoms into several groups of convenience based on combinations of valve structures and processes (Round *et al.* 1990; Medlin and Kaczmarska 2004). Radial centrics possess radially organized valves with an annulus (ring), ribs radiating from the center, and labiate processes, if present, usually located in a ring in the valve mantle or secondarily repositioned near the center (note small labiate processes in valve mantle in Figure 2A). Multipolar centrics also show a radial pore organization, but their cell outline is usually elongate (see Figure 2G), triangular (see Figure 2C), or starlike. Their valve poles are adorned with pore fields, and their labiate or tubular processes, if present, are situated within the annulus or slightly off-center, although there are exceptions. *Biddulphiopsis* possesses large labiate processes in the valve mantle (see Figure 2C). Seemingly radial centric Thalassiosirales (see e.g., Figure 2E and F) also belong in the bipolar centrics (Medlin *et al.* 1993) based on molecular phylogenetic data (see Figure 1) and possibly also on shared internal features, such as the arrangement of their Golgi bodies (Medlin and Kaczmarska 2004). Pennate diatoms are elongated as well, with a midrib, from which ribs extend out more or less perpendicularly (see Figure 2H and I). They are subdivided into raphid pennates, which have raphe slits (see Figure 2I) and araphid pennates, which lack such slits but instead have apical pore fields and apical labiate processes (see Figure 2H). The most recent taxonomic revision of the diatoms (Medlin and Kaczmarska 2004) gives class status to the radial centrics, the bipolar centrics (including the Thalassiosirales), and the pennates.

B. Frustule Construction

Construction of frustule elements takes place in silica deposition vesicles (SDVs) within the cell under mildly acidic conditions (Vrieling *et al.* 1999) and lasts only a few hours at most (see Hazelaar *et al.* 2005). Ortho-silicic acid $Si(OH)_4$, the most abundant form of soluble silica in the sea, is taken up through the plasmalemma and concentrated in the SDV by means of silica transporter proteins. Their source, a family of silicon transporter genes (SITs), has been identified in *Cylindrotheca fusiformis* and other diatoms (Hildebrand *et al.* 1998; Martin-Jezequel *et al.* 2000). Silica is not hoarded in the cell; instead, SIT gene expression is induced at the beginning of cell division. The mechanism of intracellular silicic acid accumulation to supersaturated levels and its transport to the SDV is currently unknown. Within the SDV, the silicic acid encounters an organic matrix of silaffins and long-chain polyamines, which have been implicated in accelerating silicic acid polycondensation (i.e., silica formation) and controlling the silica nanostructure (Kröger and Sumper 1998; Kröger *et al.* 1999, 2000, 2001, 2002; Poulsen *et al.* 2003; Sumper and Kröger 2004).

Construction of valves follows immediately upon mitotic nuclear division, and girdle bands are formed either during brief periods in the interphase (e.g., *Navicula pelliculosa*) or more or less continuously throughout interphase (e.g., *Cylindrotheca fusiformis*). In valve-SDVs, silica deposition generally commences along the annulus or midrib and then proceeds along the ribs to the edges (Pickett-Heaps *et al.* 1990; Round *et al.* 1990). Often, a second layer of silica is deposited underneath or on top of the primary one, commonly in a hexagonal pattern, thus giving rise to reinforced chambered valves (Hamm and Smetacek, Chapter 14, this volume).

Calcium-binding glycoproteins, called frustulins, coat the frustule elements (Kröger *et al.* 1996, 1997; van de Poll *et al.* 1999). When completed, the elements are exocytosed via fusion of the SDV-membrane with the plasmalemma. A third category of proteins, called pleuralins, is located in the zone where the ultimate girdle band of the epitheca overlaps the hypotheca (Kröger *et al.* 1997). Pleuralins seem to guide the next plane of mitotic cell division. Both frustulins and silaffins are synthesized as precursor forms, which contain aminoterminal sequences; the most terminal ones resemble signal sequences required for cotranslational import of proteins in the endoplasmic reticulum, whereas the subsequent ones may function to transport the proteins from the Golgi apparatus to the SDV, possibly in vesicles, to keep the enzymes separate from dissolved silica in the cytoplasm.

II. DIATOM PHYLOGENY

The nuclear small subunit (SSU) ribosomal DNA (18S rDNA) gene region is used universally for phylogeny reconstruction, and its sequence is now available for several thousands of planktonic organisms, including diatoms and other heterokontophytes. The use of other gene regions has been explored, for example, the nuclear-encoded 28S (LSU) rRNA gene (Sörhannus *et al.* 1995),

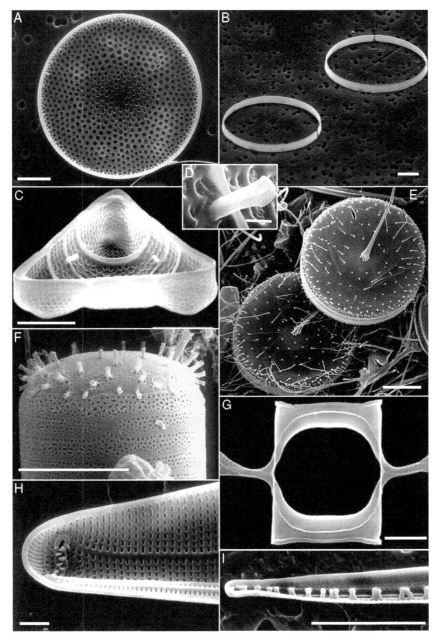

FIGURE 2. Scanning electron micrographs of extant diatom frustule elements. **(A)** Valve interior view of radial centric *Coscinodiscus*. **(B)** girdle band of *Coscinodiscus* in tilted view. **(C)** Valve interior view (tilted) of multipolar centric diatom *Biddulphiopsis alternans*. **(D)** Detail of C showing labiate process. **(E)** Two adjacent cells in a chain of *Thalassiosira rotula*. Chitin fibrils exuded through strutted processes extend into the exterior or connect sister cells. **(F)** *Lauderia annulata* showing strutted processes in valve and girdle bands. **(G)** Exterior side view of two valves belonging to adjacent sister cells of multipolar centric *Chaetoceros decipiens*. The two valves are connected by means of their setae (see Section V.D.4). **(H)** Valve interior view of araphid pennate *Pseudohimantidium atlanticum*. Note the midrib, the striae, and interstriae perpendicular upon the midrib, as well as the composite labiate process and the apical pore field near the valve apex. **(I)** Valve interior view of raphid pennate *Bacillaria paxilifer*. Small struts (fibulae) bridge the raphe canal. Scale bars 10 micrometers; 2D, 1 micrometer.

the plastid-encoded markers *rbc*L (Mann *et al.* unpublished results), 16S (Medlin and Kaczmarska 2004) and tufA (Medlin *et al.* 1997), and the mitochondrial markers COX 1 (Ehara *et al.* 2000; Evans *et al.* in press) and rpoA (Fox and Sörhannus 2004), and has revealed relationships in accordance with those of SSU rDNA phylogenies. Yet, datasets for these markers lack the taxonomic coverage of SSU data. However, the sheer number of SSU rDNA sequences is a mixed blessing. This marker contains only ca. 1800 base pairs, setting a maximum on the number of variable positions. With thousands of distinct sequences included, there may not be enough sequence positions to resolve all relationships. In a phylogeny inferred from a restricted number of sequences, any internode may obtain proper support, but with many new sequences added, internodes will break up into many badly supported ones (e.g., Goertzen and Theriot 2003).

A. The Heterokont Ancestry of the Diatoms

Diatoms belong to the heterokontophytes (also called ochrophytes or stramenopile algae, see Cavalier-Smith and Chao 1996), traditionally referred to as the golden-brown microalgae and the brown macroalgae (see history of names in Medlin *et al.* 1997). Their principal character is the heterodynamic flagellar apparatus, with a forward swimming long flagellum adorned with two rows of tripartite glycoproteinaceous hairs (mastigonemes), and a backwards swimming, shorter, naked flagellum. This flagellar system defines the heterokont lineage, which also includes heterotrophic groups.

Within the autotrophic heterokontophytes, there seems to be a basal divergence between a clade with diatoms and bolidomonads (Guillou *et al.* 1999) and a clade with the remaining golden-brown algae (see Cavalier-Smith and Chao 1996). The latter clade includes the eustigmatophyceans, chrysophyceans, and raphidophyceans, with both freshwater and coastal marine

taxa, the freshwater synurophyceans, and the marine pelagomonads, silicoflagellates, rhizochromulinids, pedinellids, tribophyceans, and phaeophytes.

The ancestry of the heterokontophytes can be examined from two perspectives. From the perspective of the host cell, the earliest divergences in the lineage are heterotrophic heterokonts (e.g., bicoecids, thraustochytrids, labyrinthulids, and oomycetes). From the perspective of the plastids, relatives must be sought among the plastids of other chromist lineages (i.e., cryptophytes, haptophytes, and dinoflagellates), all of which gained their plastids from red algae and the red algal plastids themselves. Endosymbiotic events provide the explanation for this interwoven phylogenetic pattern (see Graham and Wilcox 2000; Grzebyk *et al.* 2003; Stiller *et al.* 2003).

What is not resolved yet is whether the chromist lineages acquired their endosymbionts independently (Bhattacharya and Medlin 1995; Medlin *et al.* 1997; Yoon *et al.* 2002a) or only once (Fast *et al.* 2001; Yoon *et al.* 2002b; Cavalier-Smith 2003; Harper and Keeling 2003). If the heterokontophytes acquired their endosymbiont independently from the others, then endosymbiosis probably took place in the Mesozoic and not before the late Permian (Kooistra and Medlin 1996; Medlin *et al.* 1997). If, however, a single endosymbiosis event occurred in the common ancestry of all of these lineages, then the event must have happened in the Precambrian (Yoon *et al.* 2004; Bhattacharya and Medlin 2005). Independent acquisitions may appear more parsimonious because no secondary losses of photosynthetic capacity are needed in the heterotrophic heterokont clades. Yet, secondary loss is easy, whereas independent acquisition of endosymbionts needs subsequent independent losses of similar sets of plastid genes, transfer of similar sets of phylogenetically related genes from the endosymbiont nucleus and plastids to the host nucleus, and an independent retention of a similar set of phylogenetically related genes in the plastid (Grzebyk *et al.* 2003; see also review by Palmer 2003).

B. Diatom Phylogenies

Recent published phylogenies (Sörhannus *et al.* 1995; Medlin *et al.* 1996a, b, 2000, 2004; Ehara *et al.* 2000; Kooistra *et al.* 2003a, 2006; Medlin and Kaczmarska 2004; Sinninghe Damsté *et al.* 2004; Fox and Sörhannus 2004; Sörhannus 2004) reveal the following patterns. The radial centric and the multipolar centrics form clades (Figure 1 in Medlin and Kaczmarska 2004; Sinninghe Damsté *et al.* 2004; Sims *et al.* 2006), grades, or they show ill-supported basal ramifications with many well-supported terminal lineages (all other published trees). One particularly diverse clade, sprouting from the sometimes ill-resolved basal clades, contains all of the multipolar centrics and pennates. The multipolar centrics, in turn, group into several well-supported lineages emerging from a clade or grade with ill-resolved basal ramifications. All pennates form a well-supported clade emerging from the bipolar diversity, probably rendering multipolar centrics paraphyletic. The pennates, likewise, include several well-supported lineages that emerge from ill-resolved basal ramifications. The raphid pennates constitute the most diverse clade within the pennate diversity, rendering the araphid pennates paraphyletic. Figure 1 provides an oversight of these tree topologies.

1. Basal Radial Centrics

In several trees, *Ellerbeckia sol* (= *Paralia sol*) is sister to the radial centrics or even sister to all other diatoms (see Figure 1). This diatom and a few relatives possess characters found nowhere else in the extant diatom diversity. Using the similar taxon, *E. arenaria,* as a morphological example, we find robust, massive frustules that possess elaborately chambered valve walls and linking structures between adjacent valve faces that consist of interlocking, radial ridges (Crawford 1988). Valves lack labiate processes but instead have unique tubular processes in the valve mantle that open on the inside at the top of a small dome (Round *et al.* 1990, p. 168). *Ellerbeckia* and morphologically similar *Paralia* include only a few extant species, but many similar forms are known from the fossil record. These diatoms are too heavy for the plankton; they occur in shallow turbulent coastal environments where they drift just above the substratum.

2. Radial Centrics

The badly resolved basal ramifications in the radial centrics probably result from rapid adaptive radiation in marine and freshwater planktonic habitats and epiphytic communities. From these ramifications emerge a few well-supported and relatively species-poor radial centric lineages (see Figure 1). One of them contains the chain-forming *Leptocylindrus*; another includes predominantly solitary planktonic cells with lightly silicified, chambered valves and radially arranged labiate processes (*Coscinodiscus* [see Figure 2A and B] and *Stellarima*); a third has cells with an elongated pervalvar axis, with many girdle bands, its members often form chains (*Rhizosolenia* and *Guinardia*). A fourth clade contains more heavily silicified diatoms, forming chains with interlocking spines (*Aulacoseira, Stephanopyxis*) or with central mucilage pads (*Melosira, Podosira, Hyalodiscus*); the former are planktonic, the latter epiphytic or epilithic. Freshwater forms of these chain-forming diatoms often have their valve faces pressed against one another rendering them unavailable for exchange with the environment.

3. Multipolar Centrics

The highly diverse clade with multipolar centrics and pennates emerges from the basal centric ramification (see Figure 1). Apparently, multipolarity was a success, because multipolar centrics also show a series of well-supported lineages that emerge from a basal sprawl of tightly packed ramifications, suggesting that they, too, went through a period of rapid adaptive radiation. All the deeper lineages in the multipolar centrics (and the lineage with pennates) are epiphytic and epilithic suggesting that this was the ancestral habitat, but several lineages adopted a planktonic lifestyle (e.g., *Eucampia, Chaetoceros,* Lithodesmiales, and Thalassiosirales). The mucilage pads excreted by the apical pore fields permitted these diatoms to retain a connection with their sister cells while also allowing connection with the substratum. An additional advantage of these apically oriented mucilage connections is that they permit exposure of most of the valve face to the environment. The newly acquired properizonial bands (see Section II.C.3) permitted anisometric expansion of the auxospore into multipolar shapes. The presence and positions of labiate processes in, or near, the valve centers of bipolar centrics defines individual clades, but this trait varies wildly all over this group of diatoms. For example, the ancestral multipolar centric *Biddulphiopsis* (see Figure 2C) possesses labiate processes (see Figure 2D) in the valve mantle, although most bipolar centrics possess a single central process or a few pericentral ones. The Thalassiosirales (*Lauderia, Skeletonema, Thalassiosira*; see Figure 2E and F) with their radial valve perimeter also belong to the clade containing the multipolar centrics. Their radial state must be a reversal from multipolarity to radial centricity. Notably, this group does not form a properizonium during auxospore formation (see Section II.C.3). Thalassiosirales is now one of the most successful groups of planktonic diatoms.

4. Pennates

The araphid pennates constitute a morphologically far more homogeneous group than the multipolar centrics. They have an elongate shape, a midrib, striae perpendicular to it, apical labiate processes, and apical pore fields (see Figure 2H). Nonetheless, araphid pennates are paraphyletic (see Figure 1), and none of the traits are uniquely shared among them; midrib and striae orientation are shared with raphid pennates, and apical pore fields are found also in multipolar centrics. Because the focus of this review is on the plankton, we do not discuss this group here in detail. A paper on the evolution of this group is forthcoming (Kooistra *et al.* submitted).

The primary innovation within the araphid pennate group is the midrib (see Figure 2H), which provides mechanical support in extremely elongated diatoms. The advantage of such elongated cells in combination with the already existing apical mucilage pads might be the opportunity to form clusters or tufts, or even complex branched stalks (e.g., in *Licmophora*). Another possible advantage is improved light capturing. Moreover, these tufts or stalks generally stick through the boundary layer, improving nutrient capture but also exposing the cells to drag. A thin, elongate shape and a flexible mucilage pad have a cutting edge in that the cells can fine-tune their orientation instantly, depending on intensity *and* direction of the current (Medlin 1991).

The araphid pennates show a basal dichotomy between a well-supported but small clade, containing the planktonic genus *Asterionellopsis* as well as *Rhaphoneis* and *Talaroneis,* and a diverse clade (see Figure 1). The small araphid clade (in, e.g., Sörhannus 2004) is a strange amalgam of dissimilar taxa exhibiting considerable diversity of colony shape and frustule structure, although they are at least all strictly marine. *Rhaphoneis* and *Talaroneis* are representatives of relatively species-poor lineages in the extant diversity, but the lineages they represent have a rich fossil record with both planktonic and benthic members. *Rhaphoneis* is one of the first extant araphids to appear in the fossil record (Sims *et al.* 2006).

The diverse clade shows a poorly resolved basal sprawl from which emerge several well-supported araphid pennate lineages and a clade with all of the raphid pennates. Apparently, this second group went through a period of rapid diversification. The various araphid lineages can be described by frustule structure and by habitat, but very few of them have unique traits (Kooistra *et al.* in prep). Most araphid lineages abound in benthic epiphytic communities, and that is where the ancestors of rapid diatoms evolved. Yet, a few have developed a planktonic lifestyle (e.g., Thalassionematales).

Raphid pennates are monophyletic and, therefore, the raphe has been acquired only once, probably through elongation of apical labiate processes in an ancestral araphid pennate (Hasle 1974; Mann 1984; Round *et al.* 1990, p. 41). The raphe (see Figure 2I) permitted diatoms to move actively, for example, in search of mating partners (Bates and Davidovich 2002), nutrition, light, or to move away from adversaries or excessive light. It allowed raphid pennates to colonize unstable environments, such as intertidal mud flats where they stabilize the sediments (e.g., Paterson and Hagerthey 2001). Not surprisingly, the acquisition of the raphe enabled a rapid and massive radiation, and raphid pennates now constitute the largest group among the diatoms, in terms of genera and species, even though it is the youngest of the major lineages.

C. The Life Cycle and Its Bearings on Phylogeny

Diatoms are diploid organisms in the vegetative stage. Their life cycle consists of a prolonged period of mitotic divisions followed by a brief period of sexual reproduction, which comprises meiosis followed immediately by cell fusion and formation of a specialized zygote, called the auxospore. Typical life cycles of radial centrics, multipolar centrics, and pennates are illustrated in Figure 3A, B, and C, respectively.

1. The Need for Sex in Diatoms

The rigid frustule imposes two evolutionary constraints on the diatoms. During mitotic division, new valves generally form back-to-back inside the parental cell to produce two more or less equal daughter cells (Günther and Folke 1993). Daughter cells can thus occur solitary or form uniseriate filaments, but true multicellularity (in the sense of cell differentiation) is impossible. In addition, the hypotheca must fit inside the epitheca. So, with each mitotic division, the daughter cell inheriting the parental epitheca forms a new hypotheca of the same size as the original one, but the daughter inheriting the parental hypotheca now uses it as an epitheca and produces a slightly narrower hypotheca inside it. So, with each mitotic division, average cell size usually decreases, and the variance in size increases (Figure 3). The slow decrease in the average size of the frustule in a diatom population is known as the MacDonald–Pfitzer hypothesis (Rao and Desikachary 1970). The most common way to escape from this miniaturization trap is by means of sexual reproduction and auxospore formation, which restores the original cell size. Only a few centric and pennate diatoms can enlarge themselves during their vegetative phase (Von Stosch 1965; Gallagher 1983; Chepurnov *et al.* 2004).

2. Gamete Formation and Conjugation

Sexuality can be induced when vegetative cells reach a particular range of frustule sizes and/or when they are exposed to specific day length–temperature windows, when key nutrients run out, or, simply, when a mate presents itself (Armbrust and Chisholm 1990; Mann 1999; Hiltz *et al.* 2000; Chepurnov *et al.* 2004; Amato *et al.* 2005). Armbrust (1999) identified a gene family specifically expressed during the onset of sexual reproduction in *Thalassiosira weissflogii.* Three of the genes (*Sig1, Sig2,* and *Sig3*) encode proteins containing epidermal growth factor-like domains present in extracellular proteins that promote cell–cell interactions during different developmental stages of animals. The domains are absent in green plants. The SIG-polypeptides may be involved in sperm–egg recognition in animals and diatoms alike (Armbrust 1999; Falciatore and Bowler 2002).

Centrics generate nonflagellated macrogametes ("eggs") and flagellated microgametes ("sperms"), whereas pennates normally

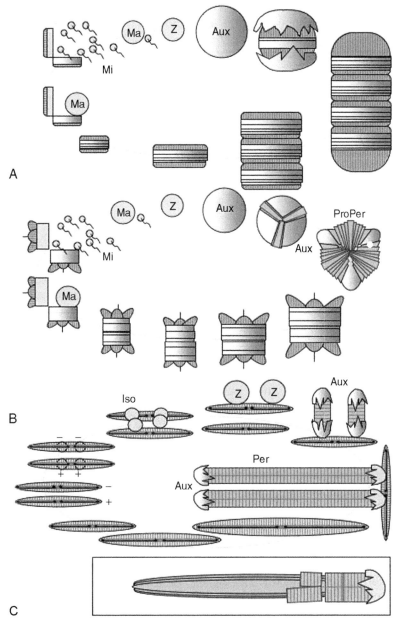

FIGURE 3. Life cycles of **(A)** radial centrics, **(B)** multipolar centrics, and **(C)** pennates. Aux, auxospore; Iso, isogamete; Ma, macrogamete; Mi, microgamete; Per, perizonium composed of perizonial bands; ProPer, properizonium composed of properizonial bands; Z, zygore; +, cell of plus-mating type-strain; –, cell of minus-mating type-strain. Perizonial and properizonial bands are formed inside a primary organic cell wall. Many multipolar centrics form small silica discs in this primary layer. **Inset:** Fully developed pennate auxospore as in **(C)** but with the primary layer of (vertically oriented) transverse perizonial bands partially removed to show secondary layer of (horizontally oriented) longitudinal perizonial bands. The initial cell forms inside the perizonium.

produce nonflagellated amoeboid gametes of equal size, called isogametes (see Figure 3; Round *et al.* 1990; Edlund and Stoermer 1997; Jensen *et al.* 2003; Chepurnov *et al.* 2004; Medlin and Kaczmarska 2004; Kooistra *et al.* 2006). Formation of isogametes represents a major innovation for pennate diatoms, but this mode of reproduction works best when partner cells can remain in close contact with one another during gamete formation, for example, on surfaces. There, potential mates locate one another, signal each other to go through meiosis, and, in the meantime, stay together. Then, the active "male" gamete can move in an amoeboid fashion and locate the nearby passive "female" gamete (or both gametes are active in fully isogamous diatoms), because pheromones move slower in the boundary layer than in open water. The advantage is that the "male" does not need to invest in many small flagellated gametes, most of which will go to waste. The system of partner location works especially well in raphid pennates, which can search actively for a partner, align with it, commence meiosis together, and then let their gametes conjugate. However, this mode of gamete formation may be disadvantageous in the plankton (see Section V.D.4.) and may explain why relatively few pennate lineages have become planktonic.

3. Auxospore Formation

After gamete fusion, the auxospore expands to restore the original cell size (for the probable mechanism, see Section V.D.1). Zygotes first form an outer organic layer to which silica scales can be added. Usually, these scales are flat with an annulus, like simplified centric valves, but in a few cases three-dimensional spiny scales are formed (see Round *et al.* 1990). In all radial centrics studied to date and in Thalassiosirales, which belongs to the multipolar centric clade, the auxospore wall is composed exclusively of the organic wall with its scales, and the initial valves formed in the auxospore are simply domed or hemispherical (see Figure 3A). In auxospores of all other multipolar centrics

studied to date, a second layer is laid down underneath the primary one. This layer, the properizonium, consists of silica bands, which force the auxospore to develop anisometrically to generate the outline of the future multipolar vegetative cell. The silica frustule is then formed inside the properizonium (see Figure 3B).

The apparent absence of this system of hoops in Thalassiosirales is suggestive of secondary loss. The lack of phylogenetic resolution among the major lineages in molecular analyses of multipolar centric diatoms leaves it open as to whether multipolarity was acquired once or several times. However, the fact that members of different multipolar lineages produce properizonia and the complexity of these structures (von Stosch 1982) suggest that multipolarity itself was probably acquired only once and that fossil multipolar diatoms are the ancestors of extant multipolar diatoms.

In pennate auxospores, the initial organic wall often ruptures, and a stack of silica hoops is laid down to force a bipolar expansion, carrying the ruptured initial zygote wall aloft as caps on the ends of the auxospore (see Figure 3C). When elongation is complete, a second series of bands is often added beneath the first layer (see inset, Figure 3C), but these bands are orientated longitudinally. The layer of longitudinal perizonial bands inside the layer of pericentral transverse ones adds to the rigidity of the highly elongated zygote. Together, the first and second series of bands constitute the "perizonium." Then, the initial valves are laid down, and the new cell is complete. In all diatoms, valve formation is intricately coupled with nuclear division. So, in order to form the initial two diatom thecae (two valve plus bands) inside the auxospore, the nucleus must divide mitotically twice without cytokinesis, and the two superfluous nuclei degenerate.

The properizonium and the perizonium are probably different states of the same character; Kaczmarska *et al.* (2000, 2001) review the known distribution of these structures among diatoms. Molecular trees indicate that the properizonium appeared at the origin of the multipolar centrics, whereas the slightly more differentiated perizonium was acquired along the internode leading to the pennates. However, there is still no information on life cycles in the *Asterionellopsis-Asterioplanus-Rhaphoneis-Talaroneis* clade.

Properizonial and perizonial bands are relatively lightly silicified and fragile structures, and chances are low that they can be found in the fossil record. And even if these bands were found in the record, it would be challenging to link them to vegetative frustules.

The auxospore needs to go through two nuclear divisions, without cell division, in order to produce the two initial thecae (e.g., Chepurnov *et al.* 2004). The two superfluous nuclei disappear. These nuclear divisions seem a waste of energy, but there might be an economically sound explanation for it. Dissolved silica cannot be stored far above saturation levels, and once it precipitates, it redissolves only excessively slowly. Thus, the cell must upgrade its intake of silica during mitotic cell division and during auxospore formation. Therefore, regulator genes that initiate the mitotic division might also regulate the silica uptake. This could explain the absolute requirement diatoms have for silica before the nucleus can divide (Darley and Volcani 1969), as well as the need, during auxospore development, to divide the nucleus before new thecae can be formed.

III. THE ORIGIN OF THE FRUSTULE

Many hypotheses have been put forward as to why diatoms have encased themselves in a silica cell wall (Mann and Marchant 1989; Nikolaev and Harwood 2000; Hamm *et al.* 2003; Harwood *et al.* 2004; Raven and Waite 2004, see also Hamm and Smetacek, Chapter 14, this volume). To answer this question one must tease it apart into several related ones: why, how, and at what stage in the evolution did the silica sequestering process originate? Why were silica elements released to the exterior? Why don't they have a simple case consisting of scales like, for example, the calcareous ones in *Emiliania*? How and why did the compound silica cell wall composed of valves and girdle bands evolve? And why did diatoms select silica for wall construction?

A. The Origin of Silica Sequestering and Metabolism

Hydroxylated quartz surfaces are some of the most abundant substrates for selective polymerization and aggregation of biomers (Hench 1989). Such surfaces have been implicated as playing an important role in the origin of life because organic molecules, that is, amino acids, can be selectively absorbed onto their surface for further condensation into proteins. Medlin (2002) reviewed the evidence from mammalian cell lines that silica seems needed in order to prevent cells from aging, to allow normal formation of connective tissues, to reduce the effects of toxic metals, to inhibit fungal attack, and to place cells in a prolonged resting state. So, intracellular silica concentration needs to be fine-tuned according to the needs of the cell, and genes are needed to produce the regulators and enzymes to facilitate uptake and excretion of silica. In case of overabundance, silica can be removed directly through the plasmalemma or via a vesicle. The first silica scales may have developed through silica polymerization in mildly acidic vesicles, and these scales were then discarded through fusion of the vesicle membrane with the plasmalemma.

Mann and Marchant (1989) hypothesized that the first cells using of the discarded silica scales in their cell wall were resting stages. Once the polymerized silica conferred an adaptive advantage to the cell, enzyme systems may have been developed or modified to (a) facilitate silica excretion into specialized silica deposition vesicles,

(b) construct organic compounds that guide construction of the silica scales, and (c) cover these scales to protect them against dissolution (see Section I.B). For the next step, cells would have modified membrane-mediated systems and macromorphogenetic mechanisms involving organelles and the cytoskeleton to mold cell wall features (Kröger and Sumper 1998), thereby creating a myriad array of valve morphologies.

Mayama and Kuriyama (2002) hypothesized that early heterokontophytes inherited or at least improved the ability to sequester silica into cell wall structures from polycystinean hosts by means of horizontal gene transfer. Extant polycystineans produce siliceous shells and usually have endosymbiotic algae, among them are pennate diatoms (Anderson 1992; Polet *et al.* 2004). The diatom endosymbionts do not make a frustule, but they develop one as soon as they are set free. Early heterokontophytes may have lacked cell walls or have been similarly enticed by the host not to make them, thus potentially facilitating horizontal gene transfer. Fossil polycystineans are known from the Late Precambrian onward (Anderson 1983), and so their origin coincides (Yoon *et al.* 2004; Bhattacharya and Medlin 2005) or predates (Kooistra and Medlin 1996; Medlin *et al.* 1997, 2000) the hypothesized emergence of the heterokontophytes. It is improbable that the whole genetic toolbox to produce silica elements was transferred wholesale, but key genes may have been exchanged step by step. The hypothesis is testable by comparing phylogenies of genes involved in silica sequestering with organismal phylogenies among heterokontophytes and other protistan groups.

B. The Evolution of the Frustule in Vegetative Cells

Several heterokontophytes (silicoflagellates, synurophytes, chrysophytes, parmophytes, xanthophytes) produce silica structures, for example, siliceous scales in the cell wall of resting stages or vegetative cells, or silica spines (van den Hoek *et al.* 1995; Graham and Wilcox 2000). Such silica structures are found across the heterokontophytan phylogeny (see Cavalier-Smith and Chao 1996; Guillou *et al.* 1999; Medlin *et al.* 2000; Kawachi *et al.* 2002) but not among the heterotrophic heterokonts in the sister lineages of the heterokontophytan clade. The ability to use silica in cell wall elements is also encountered in other autotrophic eukaryotic lineages but then only in a few species, for example, scales in the haptophyte *Hyalolithus neolepis* (Yoshida *et al.* 2006; see also de Vargas *et al.*, Chapter 12, this volume) and in some dinoflagellates (Harding and Lewis 1994).

Within heterokonts, the ability to make silica elements was probably acquired in the last common ancestor of the extant heterokontophytes. At first sight, the actinophryid heliozoans appear to be an exception because these organisms are heterotrophic heterokonts *and* they are adorned with silica spines. Yet, these heliozoans are derived pedinellids that have lost the ability to photosynthesize secondarily (Mikrjukov and Patterson 2001; Nikolaev *et al.* 2004). *Bolidomonas*, the apparent closest relative of diatoms, lacks any silica structures and has likely lost the ability to make silica elements. In each of the groups that possess silica metabolism, the end product of the silica metabolism is unique for that group.

Mann and Marchant (1989) hypothesized that diatoms adopted a life cycle in which features of the resting stage (no flagella, siliceous cell wall) were retained in the vegetative stage. The key innovation to retain the silica cell wall also during vegetative growth must have been the reshaping of radial silica scales into two valves and two series of encircling girdle elements. Several centric diatoms possess silica scales in their auxospore wall, which show an essentially radial centric architecture with a central annulus and striae radiating from it, reminiscent of the architecture of radial centric valves. The arrangement of silica elements of the diatom wall persists throughout the vegetative cycle, even during cell division, while permitting cell growth though addition of new girdle bands to the hypotheca.

The diplont life cycle of diatoms has also been invoked to support the hypothesis that diatoms developed from a "resting cyst gone vegetative." In most other heterokontophytes, vegetative stages are haploid, and only the resting cysts, which result from the sexual phase, are diploid. There are exceptions, but these are apparently derived (Graham and Wilcox 2000). Diatoms with their silica frustule could then be seen as diploid "resting cells" changed into vegetative ones. Yet, Medlin *et al.* (1993) argue that the diplont life cycle of diatoms might be an ancestral state because two heterotrophic heterokont lineages are diplont as well (Alexopoulos *et al.* 1996).

Parmophytes seem to have developed a similar cell wall organization of interlocking silica elements of different shapes. However, during their cell division, parts of the cell membrane must be exposed, thus exhibiting a weak spot in the defenses. Whether Parmophytes are related to diatoms or not remains to be uncovered.

IV. THE FOSSIL RECORD

Fossils have lost their DNA, rendering it impossible to determine their position in molecular phylogenies. Therefore, inferences of their phylogenetic position depend on alternative sources of information. Generally, two types of fossils are available in heterokontophytes: silica remains and biochemical compounds indicative of the existence of particular organismal groups (Sinninghe Damsté *et al.* 2004; Rampen *et al.* 2007; Knoll *et al.*, Chapter 8, this volume). Most of the heterokontophytan silica fossils consist of diatom frustules, but silicoflagellates have added their basket-like, spinulose exoskeletons (the silica structures are situated outside the plasmalemma).

Fossil frustules can be placed in the phylogeny of the extant diversity based on

several morphological and biochemical character states that each were acquired only once during the evolution of the diatoms. For instance, if a fossil valve shows a midrib, a pennate arrangement of striae, and a raphe, then there is strong evidence that the valve belonged to a raphid pennate diatom. If the fossil valve shows a centric organization of striae, a multipolar outline, and apically oriented pore fields, then we can assume that auxospore formation was guided by properizional bands and that this valve belonged to a multipolar centric diatom.

Unfortunately, even the extant diversity shows many cases of convergence and parallelism (e.g., Fryxell and Medlin 1981). Many ultrastructural traits have been acquired multiple times independently or have gone lost secondarily. Some multipolar centrics, such as Thalassiosirales, have lost their properizional bands and have reversed to a radial centric valve outline. In addition, the needle-like shape of *Rhizosolenia* is also encountered in another but unrelated radial centric *Proboscia* and in the multipolar centric *Neocalyptrella* (Hasle 1975; van de Meene and Pickett-Heaps 2004; Hernandez-Becerril and Maeve-del Castillo 1996). Similar cases exist in araphid and raphid pennates (Kooistra *et al.* 2003a, b, 2006; Medlin and Kaczmarska 2004). Thus, one should be aware of such deceptions when comparing information in the fossil record with that in the extant diversity.

A. The Early Fossil Record of the Heterokontophytes

The biochemical and morphological fossil records hint at the existence of heterokontophytes from the Late Precambrian period onwards. The most ancient morphological fossils are ambiguous. Casts of what has been interpreted as *Ralfsia*-like brown alga in the 570 million years ago (Ma) Doushantuo shales (P.R. China) (Xiao *et al.* 1998) and *Vaucheria*-like fossils in 900–1000 Ma Siberian strata (Knoll 1996; Xiao *et al.* 1998) could be interpreted as red and siphonaceous green algae, respectively. Silicoflagellates have extensive fossil record beginning at 120 Ma (McCartney *et al.* 1990) with peak diversity in the Miocene (Desikachary and Prema 1996). Biochemical fossils (biomarkers) suggest that heterokontophytes were in existence from the Paleozoic times onward. Sarcinochrysidalean algae (related to Pelagomonadales) produce two sterols: 24-n-propylidene-cholesterol and 24-n-propylcholesterol. These sterols are thought to be precursors of C_{30} steranes, which have been found in 500 Ma Ordovician and 360 Ma Devonian oils (Moldowan *et al.* 1990). 24-Nor-diacholestanes are known from the late Proterozoic (Figure 4); these compounds were believed to be diatom specific (Moldowan and Jacobson 2000), but they might have been present in the ancestral lineages as well. Recently, these compounds have

been detected also in dinoflagellates and thus are not diatom specific (Rampen *et al.* 2007). Nonetheless, their increased importance from the Cretaceous, and especially from the Early Tertiary onward, is probably related to the rise of the diatoms.

B. The Fossil Record of the Diatoms

Diatoms possess a splendid fossil record based on fossil silica frustules and on biochemical markers, from which much can be deduced concerning their evolutionary history (Trappan and Loeblich 1973; Strel'nikova 1974, 1975, 1992; Gersonde and Harwood 1990; Moldowan *et al.* 1990; Barron and Mahood 1993; Sinninge Damsté *et al.* 2004; Sims *et al.* 2006). Nikolaev and Harwood (1997, 2000) and Medlin and

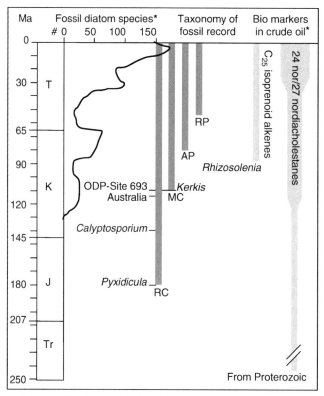

FIGURE 4. Overview of diatom fossil record; # fossil diatom species redrawn from Katz *et al.* (2005). Taxonomy of fossils: RC, radial centrics; MC, multipolar centrics; AP, araphid pennates; RP, raphid pennates. Antarctic ODP-Site 693 with *Kerkis bispinosa* from Gersonde and Harwood (1990). Australia (Nikolaev and Harwood 1997), *Calyptosporium* from Korea (Harwood *et al.* 2004), *Pyxidicula* from Liassic sponge reefs in Germany (Rothpletz 1896). Biomarkers in crude oil: record of C25 heavily branched isoprenoid alkenes (Sinninghe Damsté *et al.* 2004) indicative for origin of rhizosolenoid diatoms; 24 nor/27 nordiacholestanes redrawn from Katz *et al.* (2005).

Kaczmarska (2004) have proposed links between fossil frustules and modern ones based on ultrastructural similarities. In general, the order in which the major groups arose, as inferred from molecular phylogenies, that is, radial centrics first, then also multipolar centrics, then araphid pennates, then raphid pennates, agrees with the stratigraphic evidence of the fossil record (see Figure 4).

1. The Pre-Cretaceous Fossil Record of Diatoms

Diatom fossils of Precambrian to Jurassic age have been reported, but virtually all these claims are plagued with interpretation difficulties (e.g., Kwiecinska 2000; Sieminska 2000). Such ancient fossils often consist of fragments that could belong to other organisms (e.g., Strel'nikova 1974), or they resemble modern diatoms that could have been washed in from younger strata or from extant diversity. An exception from this may be siliceous hardparts reported to occur at high abundance in Jurassic sponge reef deposits (180 Ma) in southern Germany and described as *Pyxidicula bollensis* and *P. liasica* (Rothpletz 1896; see Figure 4). The outline of both taxa closely resembles *Amblypyrgus campanellus*, a diatom described from the late Early Cretaceous (Gersonde and Harwood 1990).

Scarcity of pre-Cretaceous diatom fossils might be because early diatoms were rare or lived in habitats adverse to fossilization or because frustules were lightly silicified. In coastal habitats, diatom frustules have a high probability of being dissolved, pyritized, or crushed and diagenetically transformed from biogenic opal into chert (see von Rad *et al.* 1977) or minerals, such as zeolites. The latter is the main reason why Cretaceous and Paleogene diatom fossils are rare. Diatom frustules preserve much better on the ocean floor because of the slow sedimentation pace. Yet, even the sample reaching the ocean floor is biased toward heavily silicified forms because frustular silica dissolves slowly in seawater undersaturated for silica as soon as bacteria have decomposed the frustule's organic coating (Biddle and Azam 1999). Even the ocean floor is not permanent because seafloor spreading and subduction rework the ocean crust on average every 110 million years (Worsley *et al.* 1986) rendering chances of finding undisturbed pre-Cretaceous deep-water facies low.

Harwood *et al.* (2004) reported diatom fossils (*Calyptosporium*) of Late Jurassic age from Korea (see Figure 4). The frustules seem to have been deposited in terrestrial habitats, leading Harwood (personal communication) to propose that diatoms arose in ephemeral water bodies. Arguments can be made for freshwater or saline pools. "Ur-diatoms," abundant as nonsilicifying unicells in coastal waters or freshwater lakes, may have adapted to life in ephemeral ponds by acquiring features that allowed the cell to go in dormancy in an instant. The key innovations may have been the retention of the silica metabolism to put cells in resting stage in the vegetative cells (Medlin 2002), the retaining of the silica cell wall of the resting stage in the vegetative stage, and modifying its shape to render growth possible while being encased (Mann and Marchant 1989). Valves are rigid, but the sets of girdle bands permit freedom of cell volume adjustment. In ephemerally wet habitats, a heavy silica wall could help prevent desiccation and could put the cells into a resting state at short notice until the next inundation (Medlin 2002). Turgor maintenance may be another reason for the encasement, that is, if diatoms have a freshwater origin. Encasing to counter turgor pressure seems relevant in freshwater and terrestrial diatoms (Ubeleis 1957), but not in marine ones; Kühn and Brownlee (2005) argue that turgor pressure of the protoplasm on the wall of the marine diatom *Coscinodiscus wailesii* is low.

The early diatoms probably recolonized coastal waters, but their massive frustules would have made them too heavy for a planktonic existence. In this respect, *Paralia sol* (transferred to *Ellerbeckia* [Crawford and Sims 2006]) as the first diatom lineage in the radial centric clade in the molecular trees is a reminder of these early heavily silicified diatoms that are benthic or abound tumbling among the detritus over the substratum

(Round *et al.* 1990). Extant members of *Ellerbeckia*, as originally described, mainly occur in freshwater or terrestrial habitats, whereas *Paralia* is a marine genus. So, the first divergence in the radial centrics is a marine representative of what originally may have been a terrestrial genus.

2. The Cretaceous Fossil Record of the Diatoms

The first rich diatom assemblages are of late Early Cretaceous age at 115 Ma (Forti and Schulz 1932; Nikolaev and Harwood 1997) and 110 Ma B.P. (Gersonde and Harwood 1990; Harwood and Gersonde 1990) from Northern Germany, Australia, and the Antarctic margin (see Figure 4). Most productive are deposits recovered at the Antarctic margin. Gersonde and Harwood (1990) and Harwood and Gersonde (1990) described 24 diatom species attributed to 14 genera from very well-preserved assemblages from a ca. 10-m thick interval. Unfortunately, the layer was not probed all the way through. So, it is unclear when diatoms actually first appeared in this paleoenvironment. All fossils in these records are diatoms in that they have valves and girdle bands, and an annulus is visible in several cases (e.g., Gersonde and Harwood 1990, pl. 14, Figure 3), indicating a centric type of organization, but none of these could be assigned to any modern diatom genus or family. All the valves show a very similar ultrastructure of the wall revealing a principal layer with radially organized striae composed of simple pores and, in most specimens, an external secondary, heavily silicified layer with a honeycomb pattern, forming pseudoloculate valves (Figure 5).

Virtually all valves in the 110 Ma record possess a radial valve outline, but a few fossil species, for example, *Bilingua rossii* and *Kerkis bispinosa* (see Figure 5A), have bipolar valves. These species must have produced some kind of properizonial bands (see Figure 3B) during auxosporulation to achieve multi-(bi)-polarity. Thus, the 110-million-year-old layer already contains bipolar forms reminiscent of modern multipolar centrics (Sims *et al.* 2006).

Several fossil frustules, for example, those of *Archepyrgus* (see Figure 5H) and *Amblypyrgus*, show linking structures, poroid organization in the valve, and short cingular bands comparable to those of living genera, *Aulacoseira* or *Stephanopyxis*, respectively. Yet, the fossils are heavily silicified and with massive connections between their valves. Modern *Aulacoseira* and *Stephanopyxis* possess labiate processes, whereas *Archepyrgus* and *Amblypyrgus* do not. Then there is similarly heavy *Rhynchopyxis* (see Figure 5E) with a tubelike structure emanating from the annulus and a marginal fringe. Both structures recall the morphology of modern multipolar centric *Ditylum*, but central tubes are found also in other multipolar centrics (e.g., *Neocalyptrella*). According to Medlin and Kaczmarska (2004) and Kaczmarska *et al.* (2006) the tube of *Rhynchopyxis* may have evolved into a central strutted process and into the central labiate process of the multipolar centrics.

Gladiopsis modica (see Figure 5D; Gersonde and Harwood 1990) also possesses a central process in the valve, but some of the illustrations show scale-shaped girdle bands typical for some of the radial centrics. Sims *et al.* (2006) point out difficulties associated with the interpretation of the 110-million-year-old fossil genera *Gladius* in Gersonde and Harwood (1990, plates 7 and 8) and *Archaegladiopsis* in Nikolaev and Harwood (1997). All modern diatoms divide by constructing a pair of daughter valves inside the parental ones. The two fossil genera show clublike, often extremely elongated, tubelike valves into which daughter valves do not fit. Sims *et al.* (2006) suggest that members of these genera possessed modes of cell division unlike that observed in contemporary diatoms and, thus, could be viewed as a now extinct lineage as sister to all extant diatoms. Alternatively, these organisms could be diatoms with extended girdle areas in which valve formation took place, like in modern *Rhizosolenia*. The presence of such extended girdle areas has been observed in the fossil record (Gersonde and Harwood 1990).

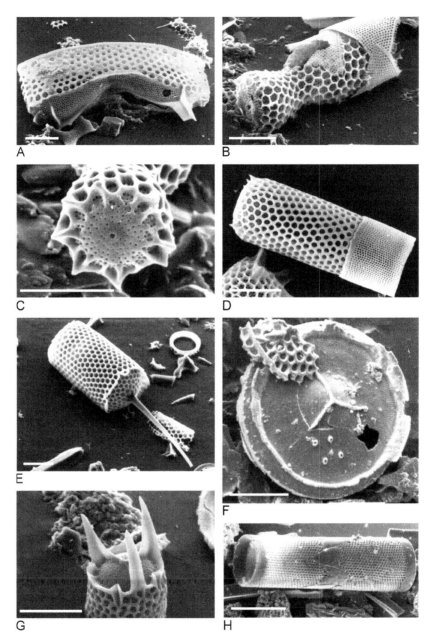

FIGURE 5. Scanning electron micrographs of diatom valves from the 110 Ma B.P. Antarctic diatom fossil record ODP-Leg 693 (Gersonde and Harwood 1990). **(A)** Multi (bi) polar centric diatom *Kerkis bispinosa.* **(B)** *Gladiopsis lagenoides.* Note diamond-shaped girdle bands and heavily silicified, semiloculate, secondary layer. **(C)** Detail of **(B)** showing central pore and primary layer with radial pattern of striae. **(D)** *Gladiopsis modica,* with girdle band. **(E)** *Rhynchopyxis siphonoides* showing central tubelike structure. **(F)** *Trochus elegantulus.* The small pores in the valve face, the collar along the valve perimeter, and the Y-shaped ridges in the valve center resemble similar structures in some extant *Melosira* species (see Round *et al.* 1990, p. 154). **(G)** *Basilicostephanus ornatus.* Spines on valve of adjacent sister cell must have locked in grooves in valve shown. **(H)** *Archepyrgus melosiroides,* valves of adjacent sister cells showing heavy linking structures completely enclosing the valve face. Similar, but less heavy linking structures are encountered in extant *Aulacoseira.* Scale bars 10 micrometers.

Trochus (see Figure 5F), *Microorbis,* and *Prae-thalassiosiropsis* do not have the secondary honeycomb layer (pseudoloculate) and are, therefore, more lightly silicified. The valve structure and the presence of a central perforate process of *Praethalassiosiropsis hasleae* makes this taxon a potential precursor of the Late Cretaceous and Paleocene-lower Eocene *Thalassiosiropsis wittiana* (Hasle and Syvertsen 1985), which resembles species of the extant genus *Thalassiosira* (Round *et al.* 1990 pp. 132, 133). Such a process might have developed into the strutted processes of Thalassiosirales (see Kaczmarska *et al.* 2006).

Unfortunately, there are some large gaps in the diatom fossil record. One particularly frustrating gap occurs just after the occurrence of the Lower Cretaceous flora described previously. Diatoms were present, but they are poorly preserved (Sims *et al.* 2006). Well-preserved fossil diatoms reappear at ca. 90 Ma, in the Late Cretaceous (e.g., Strel'nikova 1974, 1975, 1992; Tapia and Harwood 2002), providing the next clear window into the evolving diatom diversity. It reveals the persistence of species from the 110-Ma fossil record, such as *Basilicostephanus* (see Figure 5G), *Gladius,* and *Gladiopsis* (see Figure 5B–D), but many other fossils show remarkable similarities in shape and valve structure to modern genera (e.g., *Triceratium, Hemiaulus, Actinoptychus, Stephanopyxis, Paralia*), and for that reason they have often been classified as belonging to these modern genera. Most of these diatoms are without doubt radial or multipolar centrics.

Araphid pennates appear only after 90 Ma in the fossil record, and these pennates only show a simple midrib (e.g., Strel'nikova 1974). Diatoms in the small araphid clade (*Rhaphoneis* and relatives) are especially well represented in these early pennate fossil records, but it seems that they have diminished in importance relative to the diatoms in its sister clade with the remainder of the araphid pennates and the raphid pennates. There is a recent report (Chacón-Baca *et al.* 2002) of a freshwater diatom flora, including pennate diatoms, from 70-million-year-old deposits in Mexico. The material (preserved in chert) does not allow confident identification of the diatom lineages represented, but if accurately dated and without contamination, then the report of Chacón-Baca *et al.* (2002) would extend the fossil record of nonmarine diatoms to the latest Cretaceous. Chacón-Baca *et al.* (2002) indicate that some of the pennate diatoms bear a raphe. However, this is not convincingly documented in their figures. Raphid pennates are not known before the Palaeocene (Strel'nikova 1992), but from that time onward they seem to diversify rapidly. This diversification is in accordance with the poorly resolved basal sprawl of clades visible in the raphid pennate diversity.

Biochemical signals for planktonic diatoms are also available in the Mesozoic fossil record. The previously mentioned 24-nordiacholestanes become more prominent from the Early Cretaceous and especially the Early Tertiary onward (Moldowan and Jacobson 2000; Rampen *et al.* 2007), more or less in concordance with the appearance of fossil frustules. Rhizosolenoid diatoms produce C_{25} highly branched isoprenoid alkenes, and these biochemical compounds have been detected in marine sediments of 91.5 million years and younger (see Figure 4), which is well before their appearance in the silica fossil record. These centric diatoms represent a relatively deep lineage in the diatom diversity (Sinninge-Damsté *et al.* 2004).

3. The Tertiary Fossil Record of the Diatoms

The asteroid impact at the K/T boundary coincided with the extinction of some diatom taxa, but it was by far not as dramatic as in many other marine groups. Harwood (1988) estimated that 84% of the species survived the K/T event in a section of the Sobral Formation (North-Seymour Isl., Antarctica), whereas Chambers (1996), using a broader sample of diatom records through this period, estimated a survival of 37%, which is still considerable. Changes in the Tertiary diatom flora, as well as the events that shaped this flora, have been reviewed extensively in Sims *et al.* (2006). Here we focus on changes

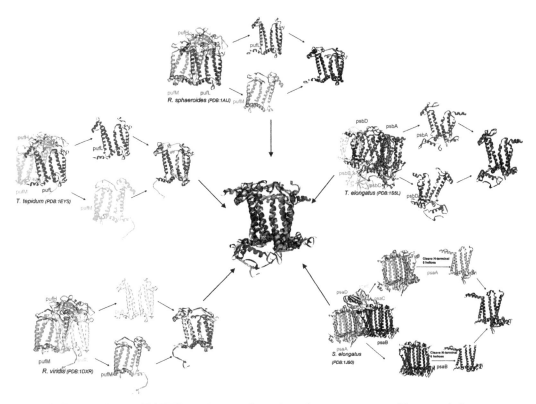

BLANKENSHIP *ET AL.*, FIGURE 2. Structural overlay of reaction centers. The central figure portrays α-proteobacteria: *Rhodobacter sphaeroides* (1AIJ), L, M chains; *Rhodopseudomonas viridis* (1DXR), L, M chains; *Thermochromatium tepidum* (1EYS), L, M chains; Cyanobacteria: *Thermosynechococcus elongatus* (1S5L), D1, D2 chains of photosystem II; *Synechococcus elongatus* (1JB0), A1, A2 chains of photosystem I. The figures along the boundaries exhibit similar structure of the two heterodimers of each reaction center.

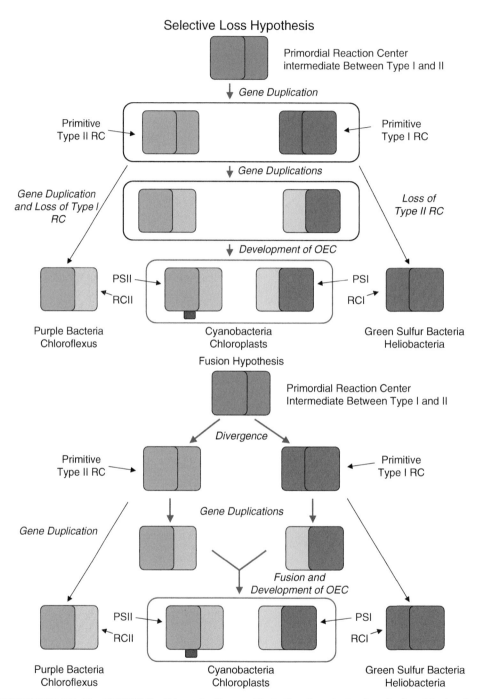

BLANKENSHIP *ET AL.*, FIGURE 5. Schematic diagram illustrating the selective loss (top) and fusion (bottom) hypotheses for the evolutionary development of photosynthetic reaction centers. The core protein subunits of the various types of reaction centers are shown as colored figures. Homodimeric complexes have two identical subunits, whereas heterodimeric complexes have two similar yet distinct subunits, represented by slightly different colors. The gene duplication, divergence, and loss events that led to existing organisms are indicated. Cells containing two types of reaction centers are shown enclosed within an outline. Black outline indicates an anoxygenic organism and red outline indicates an oxygenic organism. The oxygen evolving complex (OEC) is shown as a red figure. Time flows from the top to the bottom of the diagram, with the primordial homodimeric reaction center at the top and the five existing groups of phototrophic prokaryotes as well as the eukaryotic chloroplast at the bottom. PS, photosystem; RC, reaction center.

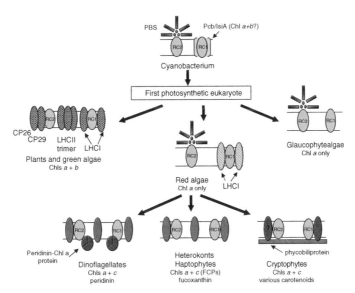

GREEN, FIGURE 1. Light-harvesting antennas of cyanobacteria and chloroplasts. The cyanobacterial ancestor of the chloroplast had phycobilisomes (PBSs) and an antenna of the Pcb/IsiA family that may have been associated with PS I. It may have synthesized chlorophyll (Chl) *b*. Descendants of the first photosynthetic eukaryote diversified into the green line (plants and green algae) with the loss of PBSs, the red line (rhodophyte algae) with the loss of Chl *b*, and the glaucophyte algae. Green line and red line chloroplasts ave antennas of the light-harvesting complex (LHC) superfamily (long green ovals), binding Chl *a+b* (hatched green ovals) and Chl *a* (hatched yellow ovals), respectively. One or more secondary endosymbioses involving a red algal endosymbiont gave rise to the four major algal groups with Chl *a/c* (LHC) antennas (orange, brown, red ovals). The peridinin-Chl *a* protein is found only in dinoflagellates (Macpherson and Hiller 2003).

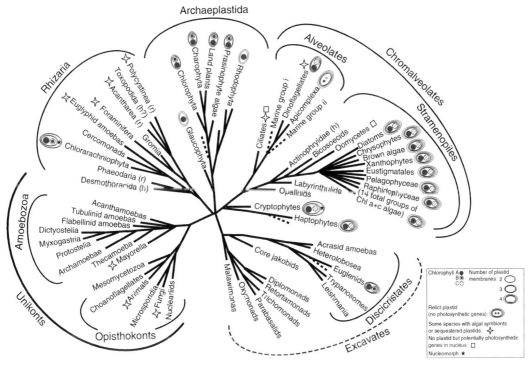

FEHLING *ET AL.*, FIGURE 1. Consensus phylogeny of eukaryotes showing the distribution of photosynthesis among the major groups. The tree shown is a consensus phylogeny of eukaryotes based on a combination of molecular phylogenetic and ultrastructural data. The tree is further annotated to indicate the distribution of photosynthesis across the tree. Groups harboring endosymbiotic organelles are indicated by schematic plastids with variable numbers of membranes and chlorophyll composition, as indicated in the key to the lower right of the figure. Groups with members harboring transient algal endosymbionts are indicated by stars. Taxa classified as Radiolaria are indicated by an "r" following their names, and former "Heliozoa" are indicated by an "h."

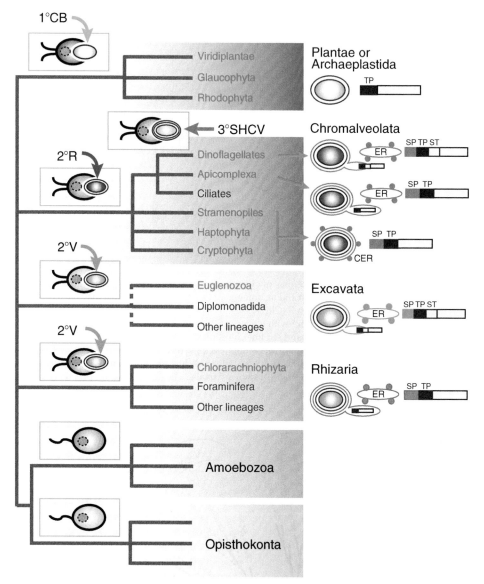

HACKETT ET AL., FIGURE 1. Schematic tree of the six eukaryotic supergroups (see Adl *et al*. 2005) showing the different plastid endosymbioses and the origins of the plastid protein import system in photosynthetic eukaryotes. The Viridiplantae, Glaucophyta, and Rhodophyta (Plantae or Archaeplastida) share a plastid that was derived from a putative single primary endosymbiosis (1°) between a nonphotosynthetic protist and a cyanobacterium (CB). Many endosymbiont genes were transferred to the nucleus following the plastid capture. Transit peptides (TP) that were added to the N-terminus of nuclear-encoded proteins allowed Plantae plastid targeting via the TOC/TIC translocon. The single red algal secondary endosymbiosis (2°) that putatively gave rise to the chromalveolate plastid was ancestrally shared by Chromista (stramenopiles, haptophytes, and cryptophytes) and Alveolata (dinoflagellates, apicomplexans, and ciliates). The chromists contain a four-membrane plastid with a plastid endoplasmic reticulum (CER) and a modified bipartite leader sequence, which targets plastid proteins to the CER via a signal peptide (SP) in addition to the typical transit peptide. Apicomplexan apicoplasts, which have four plastid membranes but lack CER, carry out protein import via the secretory pathway (ER and Golgi apparatus); a system that is also present in chlorarachniophytes that contain a green-algal–derived plastid. Of the three-membrane plastids, the dinoflagellates, which generally contain a chromalveolate plastid of red algal origin, but also have undergone tertiary plastid replacements involving stramenopile, haptophyte (i.e., 3° SHCV), cryptophyte, or green algae, use a tripartite leader sequence that includes a stop transfer (ST) signal to ensure that plastid proteins in Golgi-derived vesicles are integral membrane proteins that maintain a predominant cytoplasmic component (Nassoury *et al*. 2003). This type of tripartite leader sequence is also found in the Euglenozoa that contain a green-algal–derived secondary plastid.

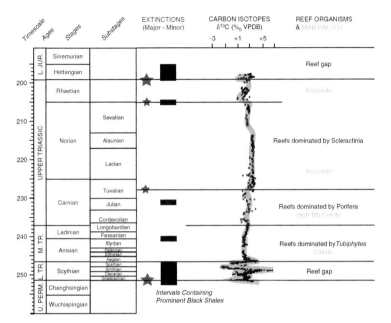

PAYNE AND VAN DE SCHOOTBRUGGE, FIGURE 2.   Triassic timescale, inorganic carbon isotope record, global diversity, and significant evolutionary events. Timescale is based upon the 2004 IUGS stratigraphic chart and radiometric constraints from (Mundil *et al.* 1996, 2001, 2004; Bowring *et al.* 1998; Muttoni *et al.* 2004a, b). Carbon isotope data for the Permian through Middle Triassic represented by filled circles are whole rock values from Payne *et al.* (2004). Data represented by + and x for the Late Triassic are from Korte *et al.* (2005). + represents data from brachiopod calcite and x represents whole rock values. Data across the Triassic–Jurassic boundary represented by open circles are whole rock values from Galli *et al.* (2005).

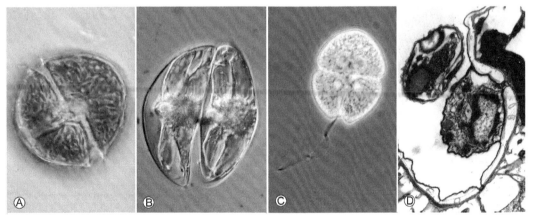

DELWICHE, FIGURE 1.   Representative dinoflagellates. **(A)** *Peridinium* sp., with the cingulum (extending from 11 o'clock to 5 o'clock), sulcus (8 o'clock), and peduncle visible near the junction of cingulum and sulcus. **(B)** *Pyrocystis lunula,* which is one of several bioluminescent dinoflagellates; it is not flagellate in its vegetative state but releases "dinokont" zoospores under some conditions. **(C)** *Akashiwo sanguinea,* one of many unarmored species once lumped as "*Gymnodinium,*" with the trailing flagellum visible extending from the sulcus. **(D)** *Amoebophrya* sp., in the early stages of infecting *Akashiwo sanguinea;* also visible are the cortical alveolae of the host, and a portion of the host chloroplast. (Parts **[A]**, **[B]**, and **[C]** from Charles Delwiche; **[D]** from John Miller.)

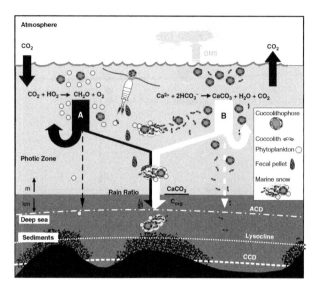

DE VARGAS *ET AL.*, **FIGURE 1.** Role of coccolithophores in biogeochemical cycles. Through the production of their $CaCO_3$ coccoliths, coccolithophores play a key role in the global carbon cycling. Although they thrive in the photic layer of the world ocean, the coccolithophores actively participate in gas exchange (CO_2, DMS) between seawater and the atmosphere and the export of organic matter and carbonate to deep-oceanic layers and deep-sea sediments. They are the main actors of the *carbonate counter-pump* **(B)**, which, through the *calcification* reaction, is a short-term source of atmospheric CO_2. Via the ballasting effect of their coccoliths on marine snow, coccolithophores are also a main driver of the *organic carbon pump* **(A)**, which removes CO_2 from the atmosphere. Thus, organic and carbonate pumps are tightly coupled through coccolithophore biomineralization. Ultimately, certain types of coccoliths particularly resistant to dissolution are deposited at the seafloor, where they have built a remarkable fossil archive for the last 220 million years. The three main carbonate dissolution horizons are depicted: ACD, aragonite compensation depth; Lysocline (complete dissolution of planktic foraminifera); and CCD, calcite compensation depth. See text for more details. (Inspired from Rost and Riebesell 2004.)

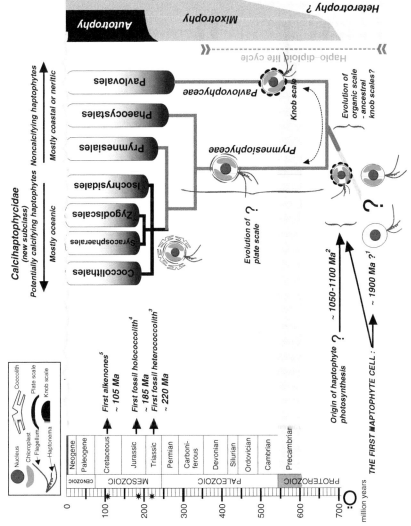

DE VARGAS ET AL., FIGURE 4. Benchmarks in the evolutionary history of the haptophytes and the Calcihaptophycidae. Major innovations are shown along a geological time scale on the left side of the figure, and a synthesis of recent *molecular* phylogenetic data using representative species of the seven extant haptophyte orders (this chapter and Saez *et al.* 2004) is depicted on the right side. Biological, phylogenetic, and paleontological data tend to support a scenario according to which the haptophytes have broadly evolved from coastal or neritic heterotrophs/mixotrophs to oceanic autotrophs since their origination in the Proterozoic. Carbonate biomineralization in the Calcihaptophycidae may have been a key evolutionary step for the stepwise invasion of the oligotrophic pelagic realm, starting ~220Ma. See text for further details. 1, this chapter; Figure 5; 2, Yoon et al. 2004; 3, Bown 1987; 4, Bown 1983; 5, Farrimond *et al.* 1986.

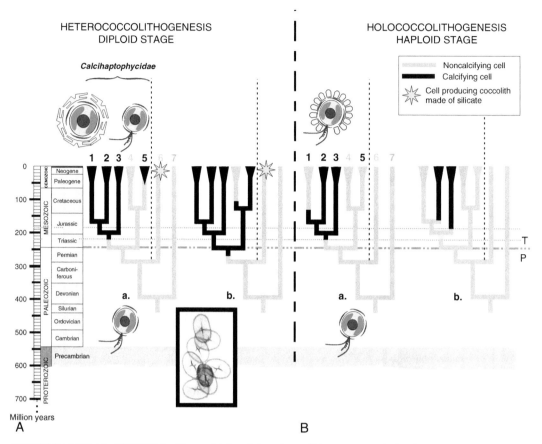

DE VARGAS *ET AL.*, FIGURE 6. The multiple origins of coccolithogenesis in the haploid and diploid stages of calcihaptophytes. These schematic trees (based on SSU rDNA data, see Figure 3) illustrate the evolutionary history of biomineralization (within the Prymnesiophyceae **1**, Pleurochrysidaceae and Hymenomonadaceae; **2**, Coccolithaceae and Calcidiscaceae; **3**, Helico-, Ponto-, Rhabdo-, and Syraco-sphaeraceae; **4**, Isochrysidaceae; **5**, Noelaerhabdaceae; **6**, Prymnesiaceae; **7**, Phaeocystaceae). Carbonate coccolithogenesis first evolved in the diploid phase (A) of primitive Calcihaptophycidae after the Permo–Triassic boundary. It was later reinvented, possibly on multiple occasions, in the haploid phase of certain families **(B)**, and we suggest that even diploid lineages within the family Noelaerhabdaeae may have reinvented calcification in the early Cenozoic **(Aa.)**. Biomineralization based on silicate precipitation independently evolved within the Prymnesiaceae (star symbol). Note that all Prymnesiophyceae, skeletonized or not, secrete and assemble at the cell surface organic plate-scales like the one shown in the black-framed SEM picture in **(A)**. The horizontal dotted lines in the middle of the figure indicate, from bottom to top, the P/Tr boundary, the first fossil heterococcolith, and the first fossil holococcolith.

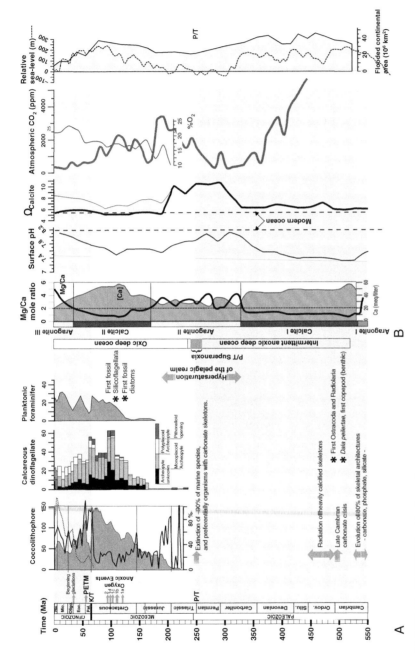

DE VARGAS *ET AL.*, FIGURE 8.   For legend see next page.

FIGURE 8. Abiotic global forcing on the oceanic carbonate system and the evolution of morphological species richness in pelagic microcalcifiers. Concerning the carbonate system, the Phanerozoic can be divided into two periods of equal length: a first during the Paleozoic, when the biotic part of the system was essentially coastal benthic and the abiotic components were relatively unstable; a second in the Meso-Cenozoic, when the new planktonic calcifiers originated, diversified, shifted a significant part of carbonate precipitation into the pelagic realm, and at the same time stabilized the concentration of atmospheric CO_2. On this chart, along a geological time scale highlighting a few major paleoceanographic events, several essential biotic **(A)** and abiotic **(B)** components of the carbonate system are represented with, from left to right: coccolithophore morphospecies diversity and their rate of turnover (= rate of extinction + rate of "speciation") (Bown *et al.* 2004, 2005); morpho-species diversity of calcareous dinoflagellates (Streng *et al.* 2004); morpho-species diversity in planktonic foraminifers (Tappan and Loeblich 1988); secular variations in absolute concentration of Ca^{2+} and Mg:Ca ratio of seawater (Stanley and Hardie 1999); mean surface saturation with respect to calcite according to Ridgwell (2005); atmospheric CO_2 concentration (Royer *et al.* 2004) and atmospheric O_2 concentration (Falkowski *et al.* 2005); and relative sea level (Haq *et al.* 1987) and flooded continental areas (Ronov 1994). In addition, a few major ecological events in Paleozoic carbonate biota as described in Knoll (2003) are indicated in **(A)**. P/Tr, Permian/Triassic; K/T, Cretaceous/Tertiary; PETM, Paleocene-Eocene Thermal Maximum.

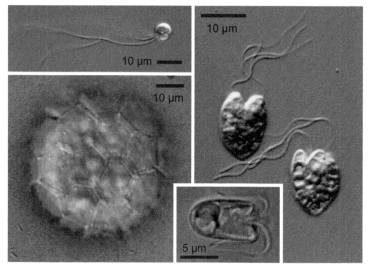

O'KELLY, FIGURE 6. Light micrographs of selected prasinophyte algae. Counterclockwise from top left: flagellate cells of *Pterosperma* sp.; phycoma stage of *Pterosperma* sp. (cf. *P. cristatum*); flagellate cell of *Pyramimonas parkeae*; flagellate cells of *Halosphaera minor*.

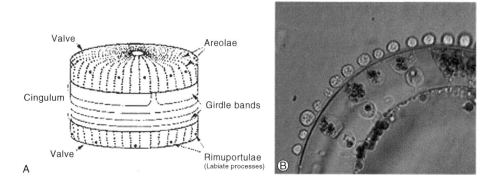

HAMM AND SMETACEK, FIGURE 2. **(A)** Potential sites of infection on the surface of a diatom frustule. **(B)** Numerous cells of the parasitoid *Pirsonia diadema* feeding on the diatom *Coscinodiscus wailesii* (Kühn 1995).

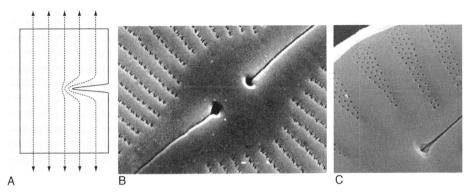

HAMM AND SMETACEK, FIGURE 6.   Sketch displaying stress concentration around a crack (from Gordon 1978) , and the "blunted" ends of the raphe of two pennate diatoms. Scanning electron microscope (SEM) pictures from C. Kages/F. Hinz.

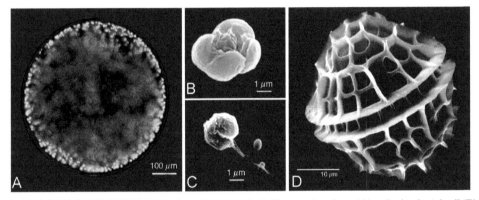

HAMM AND SMETACEK, FIGURE 8.   Organic covers. Left: *Phaeocystis* colony **(A)**, naked colonial cell **(B)**, and free-living flagellate **(C)**. Right: The stiff, lightweight cellulose armor of a dinoflagellate *Protoceratium spinulosum* **(D)**. Part **(D)** from Hallegraeff (1998).

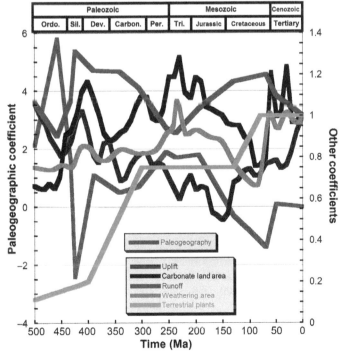

GUIDRY *ET AL.*, FIGURE 2.   Model weathering coefficients.

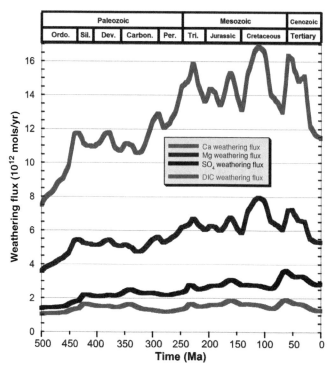

GUIDRY *ET AL.*, FIGURE 3.   Ca, Mg, SO₄, and dissolved inorganic carbon (DIC) weathering fluxes.

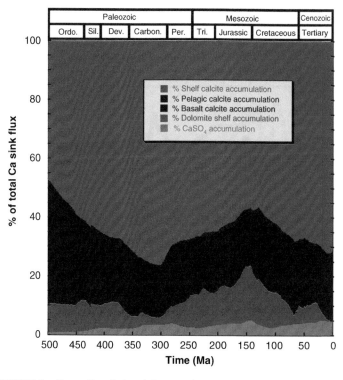

GUIDRY *ET AL.*, FIGURE 5.   Ocean Ca relative sink strength.

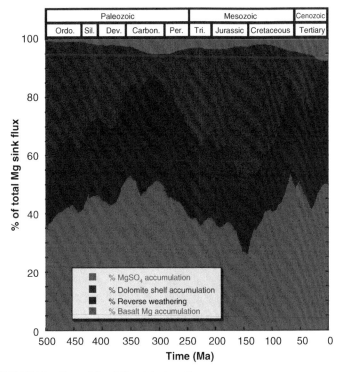

GUIDRY *ET AL.*, FIGURE 6.   Ocean Mg relative sink strength.

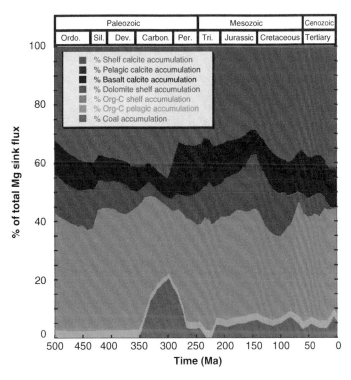

GUIDRY *ET AL.*, FIGURE 7.   Ocean carbon relative sink strength.

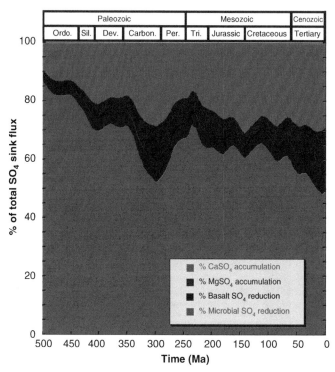

GUIDRY ET AL., FIGURE 8. Relative SO_4 sink flux strengths (percentage of the total SO_4 flux) for the past 500 million years.

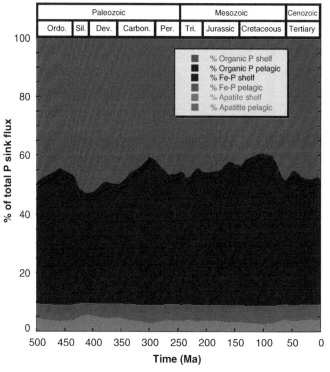

GUIDRY ET AL., FIGURE 9. Ocean P relative sink strength.

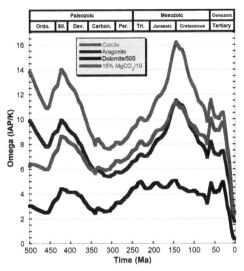

GUIDRY *ET AL.*, FIGURE 10. Ocean saturation state for calcite, aragonite, 15% MgCO₃, and dolomite for the past 500 million years.

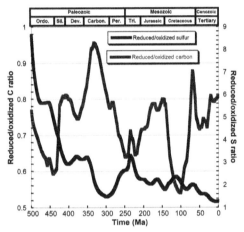

GUIDRY *ET AL.*, FIGURE 11. Reduced/oxidized C and S flux ratio for the past 500 million years.

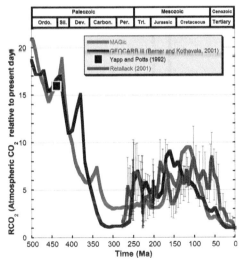

GUIDRY *ET AL.*, FIGURE 13. Atmospheric CO₂ as calculated from the GEOCARB and *MAGic* models and compared with the proxy paleosol data of Yapp and Potts (1992) and the stomatal index date of Retallack (2001).

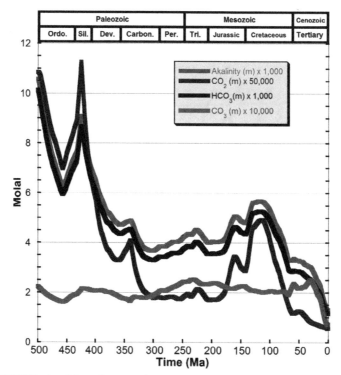

GUIDRY ET AL., FIGURE 14. CO_2–carbonic acid system parameters of seawater.

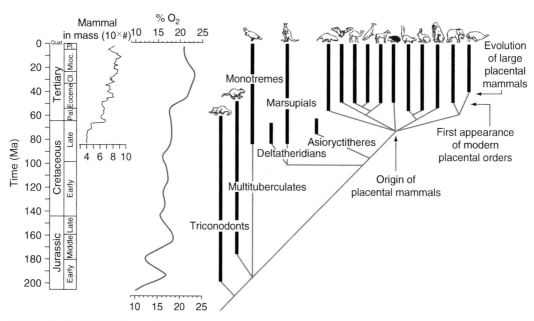

KATZ ET AL., FIGURE 4.   Mammal evolutionary events based on fossil, morphological (Murphy *et al.* 2001), and molecular (Murphy *et al.* 2001; Springer *et al.* 2003) evidence compared with oxygen concentrations in Earth's atmosphere modeled over the past 205 million years using carbon and sulfur isotope datasets; O_2 levels approximately doubled over this time from 10% to 21%, punctuated by rapid increases in the Early Jurassic and from the Eocene to the present. Changes in average mammalian body mass is taken from Alroy (1999). Vertical black bars represent known fossil ranges; dashed lines represent inferred phylogenetic branching. Only some of the ordinal-level placental mammals are shown. Reproduced from Falkowski *et al.* (2005).

in the marine plankton (Trappan and Loeblich 1973; Schrader and Fenner 1976; Gombos and Ciesielski 1983; Fenner 1985; Harwood and Maruyama 1992; Gladenkov and Barron 1995; Stoermer and Smol 1999; Harwood *et al*. 2004).

In the plankton, a major shift in diversity took place in the Oligocene. Many lineages present since the Cretaceous disappeared, or became far less common, whereas new forms emerged, or became more common, such as the radial centrics *Actinocyclus, Coscinodiscus,* and Rhizosoleniaceae, the bipolar centrics Thalassiosirales and Chaetocerotales (based on resting spores), the araphid pennates *Thalassionema, Thalassiothrix, Bogorovia,* and *Synedra*-like diatoms, and raphid pennates belonging to the genera *Rouxia, Fragilariopsis,* and *Nitzschia*. However, several of these diatom lineages did not evolve in the plankton; their ancestors were epiphytic or bottom dwellers. Instead, they adopted a planktonic lifestyle and then diversified in the plankton, but these diversifications are minor in comparison to the radiations of the araphid pennates and the raphid pennates that all happened well before the Oligocene. Thus, one cannot infer radiation events by looking at the fossil record of planktonic diatoms alone.

V. THE SUCCESS OF THE DIATOMS IN THE PLANKTON

Of all the groups of eukaryotic phytoplankton in the contemporary oceans, the diatoms are by far the most successful, not only in numbers of species but also in amount of biomass and primary production. Graham and Wilcox (2000) list estimated numbers of marine species for the major groups of extant phytoplankton organisms. There are about 5880 red algal species, but only a few of those (e.g., *Rhodosorus, Rhodella, Porphyridium*) are planktonic unicells. Of the green lineage, the majority is also benthic, and there are only ca. 100 known species of marine prasinophytes.

Then, there are about 200 cryptophytes (all planktonic; 100 marine and 100 freshwater). In the chromist lineages (here defined as heterokontophytes, haptophytes, and dinoflagellates), there are ca. 1800 to 2000 dinoflagellates (most of these marine), 500 haptophytes (all marine planktonic), and roughly 6000–11,000 heterokontophytes.

Within the heterokontophytes, there are an estimated 800–1000 freshwater chrysophytes (benthic and planktonic), ca. 120 freshwater synurophytes (planktonic), and ca. 15 freshwater eustigmatophytes (planktonic). There are about 600 tribophyceans (xanthophytes), occurring primarily in freshwater and on soil. In the marine realm, there are about 10 raphidophytes (few freshwater), a few pelagomonads, sarcinochrysidaleans (benthic), silicoflagellates, rhizochromulinids, pedinellids, and bolidophytes, and most of these are planktonic. Most of the ca. 1500 brown algae are marine, and all are benthic. All these heterokontophytes taken together are not as diverse as the ca. 10^5 diatoms estimated by Mann and Droop (1996). If the comparison is restricted to marine planktonic species, the difference is even stronger. The ca. 20–30 heterokontophytes other than diatoms are no match for the ca. 5000 (Katz *et al*. 2005) to 10,000 (Sournia, 1995) marine planktonic diatoms. It should be noted, however, that all these estimated numbers are based generally on morphologically delineated species and that they might be on the low side. Results of recent biodiversity studies have uncovered considerable hidden and cryptic diversity in all the phytoplankton groups (e.g., Massana *et al*. 2004; see also Section VII). Proportional diversities will probably remain the same, even if absolute diversities of each group are seriously underestimated.

A. The Paleo-Environmental Settings and the Fates of the Various Phytoplankton Lineages

Paleozoic phytoplankton evolution is difficult to assess because many of the microfossils (e.g., acritarchs, *Tasmanites*) of those times cannot be linked securely to extant lineages (Knoll 1994; see also Knoll *et al.*, Chapter 8, this volume). Biomarkers suggest that ancestors of diatom and/or dinoflagellate ancestors already existed in the plankton throughout the Paleozoic (Moldovan and Jacobson 2000; Rampen 2007). Yet, there are no reliable Paleozoic records of diatom frustules, whereas other silica microfossils, such as those of radiolarians and sponge spicules, have been recorded. So, it appears that diatoms did not exist yet and that their heterokont ancestors, or even unrelated organisms, produced these biomarkers.

The extinction event at the Permo-Triassic (P/Tr) boundary also affected the phytoplankton and with it most other marine life (Raup and Sepkoski 1982), thus permitting open niches for new phytoplankton groups to diversify. Dinoflagellates with fossilizable hard parts appeared, at least easily recognizable as such, in the mid-Triassic (Delwiche, Chapter 10, this volume), and haptophytes first appeared as recognizable coccolith fossils in the Late Triassic (Katz *et al.* 2005; de Vargas *et al.*, Chapter 12, this volume). The heterokontophytes (mainly diatoms and silicoflagellates) were present unambiguously in the Early Cretaceous (Gersonde and Harwood 1990; Harwood and Gersonde 1990; McCartney *et al.* 1990; Desikachary and Prema 1996). The K/T impact event caused a major mass extinction of coccolithophorids, whereas the diatoms (see Figure 4) and dinoflagellates were less affected (Katz *et al.* 2005). All three groups recovered to pre-K/T diversity by the beginning of the Eocene (ca. 55 Ma), but from then on, dinoflagellate and coccolithophore diversity declined steadily (see references in Katz *et al.* 2005; see also Delwiche, Chapter 10, and de Vargas *et al.*, Chapter 12, this volume). The general trend from the Triassic onward is that dinoflagellate and haptophyte biodiversity waxed with eustatic sea level rises correlated with continental rifting, and waned subsequently with sea level drops related to the coming of age of oceanic crusts (e.g., Worsley *et al.* 1986). Silicoflagellates and planktonic diatoms did not follow this trend;

they diversified continuously following the Eocene. Diatoms show two major pulses, one near the end of the Eocene (40–35 Ma) and one in the Middle to Late Miocene (15–8 Ma), and are now by far the most diverse group in the plankton (Trappan and Loeblich 1973). Silicoflagellates are, as the diatoms, known from the Early Cretaceous onward and show their maximum diversity in the Miocene (20–5 Ma) with ca. 100 morphologically defined species. A few widespread species are all that remain (Desikachary and Prema 1996).

The cyclic process of continental rifting and reassembly is known as the Wilson cycle (Wilson 1966; Worsley *et al.* 1986). Continental rifting leads to a rapid increase in the amount of tectonically passive coastline fringing lowlands with wet climates, hence increased runoff, and increased vegetation-mediated weathering in previously arid regions in the continent interior. Moreover, during such a breakup, young, buoyant ocean crust causes sea level to rise, flooding continental shelves and increasing the extent of shallow, epicontinental seas. Increased space, habitat heterogeneity, and nutrient availability are assumed to have resulted in a radiation of the coastal phytoplankton. The Wilson cycle also affects ocean chemistry because during breakup, new mid-oceanic ridges are formed, and seawater percolating through these ridges becomes depleted in Mg and enriched in Ca. Assuming that rates of spreading do not vary through time, higher Mg/Ca ratios in seawater prevail toward supercontinent assembly when total ridge length is shortest, whereas lower Mg/Ca ratios abound following continent breakup. Aragonite and high-Mg calcite oceanic conditions are typical for assembly periods, whereas low-Mg calcite conditions prevail during breakup. Not surprisingly, coccolithophorids, which precipitate calcite, waxed and waned with the unfolding of the present Wilson cycle. Diatoms need calcium, too, but not to the same extent as coccolithophorids and are, therefore, less affected by the Mg/Ca-ratio of seawater (Falkowski *et al.* 2004a, b; Katz *et al.* 2005).

The latest Wilson cycle also affected CO_2 and O_2 content of the ocean and atmosphere. First, with the continental mass in pieces scattered over the world ocean, terrestrial climates were wetter, and more CO_2 was sequestered in vegetation. Moreover, large amounts of the primary production by the newly emerging plankton were buried along the passive continental margins and in epicontinental basins. Both resulted in a decrease in CO_2 and an increase in O_2 levels. Low CO_2 levels and high O_2 levels decrease the efficiency of RuBisCo in fixing CO_2 (Falkowski et al. 2004a, b; Katz et al. 2005). The diatoms and possibly other heterokontophytes may have overcome this problem if they truly have a C-4 mode of carbon fixation (Reinfelder et al. 2000) (see Sections V.C.2 and VII).

An important factor believed to have boosted the diversification of diatoms is the evolution of grasses in the terrestrial ecosystems. Grasses absorb considerable amounts of dissolved silica in the groundwater and sequester it in their tissues in the form of phytoliths, which are typical for grasses. Phytoliths dissolve more easily than the silica in rocks and sediment and can be washed away with the runoff via rivers into the sea. Phytoliths first appear and diversify in the Late Cretaceous (Piperno and Sues 2005), then expand markedly during the Late Eocene/Early Oligocene to become common since the Middle Miocene. So, silica fluxes to the oceans may have increased because of the expansion of grasses. All of these changes correlate with major increases in the abundance and diversity of the marine planktonic diatoms across this period (Retallack 2001; Falkowski et al. 2004a, b).

Finally, during the last 2.5 million years, climate variability ranging at Milankovitch time scales (Milankovitch 1920; Hays et al. 1976) may have affected diversity of coastal and oceanic plankton. The change in the dynamics of climate cycles characterized by more pronounced glacial/interglacial variability is a consequence of the onset of Northern hemisphere glaciation (Lisiecki and Raymo 2007). During a cold period, sea level is low,

and a considerable part of the continental shelves is dry land. Moreover, dryer climates lead to decreased river outflow into coastal seas but increased aeolian dust input in the open ocean. Such a period may depress coastal diatom diversity but probably boosts production and diversity of open ocean diatoms. During warm periods, the system might be the reverse, with flooding of continental shelves, increased nutrient input in coastal regions, and decreased dust input in the open ocean. Hence, coastal diversity rises and open ocean diversity decreases, especially in the southern ocean, which is currently Fe-starved (Boyd et al. 2000). It would be interesting to see if such trends could be tracked in the oceanic fossil record.

B. Why Did Chromists Win Over Prasinophytes or Red Microalgae?

All chromist lineages share a series of characteristics not found in planktonic chlorophytes, prasinophytes, and red algae that may give them a selective edge over these three algal groups.

1. The Photosystem

The chromist lineages absorb wavelengths over a broad light spectrum. Plastids of redalgae possess chlorophyll a and the accessory pigments β-carotene and phycobilins, with which they absorb wavelengths from 480–580 nm; green algal plastids have chlorophyll a and b and a set of carotenoids, with which they catch a more energetic part of the spectrum (400–520 nm). This would render red algae suited for coastal environments, whereas green algae are better at living in the deep blue waters of the open ocean. Peculiarly, prasinophytes, which belong to the green algae, dominate modern coastal picoplankton communities (Medlin et al. 2006). Perhaps they occupy a size niche where diatoms cannot thrive. The chromist chloroplasts lost phycobilins but retained chlorophyll a and β-carotene and acquired various forms of chlorophyll c and fucoxanthin (see Jeffrey and Wright

2005). These pigments permitted light, harvesting outside the absorption spectrum of green and red algae (400–580 nm).

The xanthophylls diadinoxanthin and diatoxanthin protect the photosystem of the chromist lineages from harmful effects of light saturation (Falkowski and Raven 2007). With these compounds, heterokontophytes, haptophytes, and most dinoflagellates can photoacclimate their light capturing system rapidly to the quantity of the incoming light, especially from low light conditions into high light. Diadinoxanthin, the inactive precursor of diatoxanthin, is always in stock and can be transferred into the active compound within a minute when exposed to high light stress. Diatoxanthin dissipates energy by means of nonphotochemical quenching (Lavaud *et al.* 2004). The reverse reaction back to diadinoxanthin is much slower. The content of chlorophyll and accessory light capturing pigments can also be fine-tuned to the amount of incoming light, but these responses are much slower.

2. Uptake and Recycling of Nutrients and Organic Material

Diatoms, together with most other chromist lineages, have a highly efficient uptake and internal recycling system for N, P, and other nutrients, making use of enzymatic pathways inherited from the original heterokont host as well as from the endosymbiont (Takabayashi *et al.* 2005). Although nutrient uptake is energy demanding, and therefore partly light dependent, there are pathways that function in the dark and which might have been retained from the heterokont ancestor. Some diatoms and dinoflagellates sink below the pycnocline to stock nutrients and then rise into the photic zone to photosynthesize, using up the nutrient stock (Villareal *et al.* 1993). This daily migratory behavior is opposite to the migration of their grazers, which move up in the evening and down in the morning (Banse 1964). In addition, those species of multipolar centric *Hemiaulus* and radial centric *Rhizosolenia,* occurring in oligotrophic

(tropical) conditions, commonly maintain N_2 fixing cyanobacterial endosymbionts (Villareal 1990, 1991).

Uptake of organic material is common in chromists, and several species feed on other microorganisms (e.g., Lewin and Hellebust 1975; Nilsson and Sundbäck 1996; Tuchman 1996; see also Van den Hoek *et al.* 1995; Graham and Wilcox 2000). Chromists take up dissolved organic nitrogen (DON; Bronk and Flynn 2005) and dissolved organic carbon (DOC; Lewitus 2005), and they even need essential vitamins (see Lewitus 2005). They can also grow on organic compounds in the dark, if concentrations are high enough. Diatoms have the complete enzyme system needed for a urea cycle (see Section VII). Such a cycle is unknown in green and red algae, implying that the genetic toolbox behind the enzyme system must have been inherited from heterotrophic ancestors.

The presence of the frustule, and the lack of flagella in the vegetative stages, prohibit a diatom from hunting down and ingesting moving prey. Nevertheless, diatoms take up organic material (see Graham and Wilcox 2000). Extensive membrane trafficking has been shown in the diatoms, demonstrating that both endocytosis of particulate organic matter (POC), and exocytosis are common features of the diatoms (Kühn and Brownlee 2005), likely taking place through portules, such as the labiate processes.

3. Resting Stages

Most chromist lineages have some form of resting stage, specialized dormant cells surrounded by a heavy cell wall. The possibility to form resting stages has likely been inherited from the heterotrophic host(s) involved in the secondary endosymbiosis event(s). Resting stages of dinoflagellates (cysts) are an integral part of the sexual cycle, whereas the resting stages of diatoms (spores or resting cells) are not; auxospores of diatoms are generally *not* resting stages. Resting cells are common in diatoms, whereas spores that differ markedly from the morphology of the vegetative cells are only known from a few taxa scattered among radial and multipolar centrics (e.g., *Eucampia* and *Chaetoceros*) (French and Hargraves 1985; McQuoid and Hobson 1996; Montresor and Lewis 2005). Resting stages rain down on the sediment. If they are swirled up into the water column and encounter conditions favorable for germination, then they can seed a new bloom. However, many will probably be eaten or become buried. Resting stages of dinoflagellates and diatoms feature prominently in the fossil record (e.g., Harwood and Gersonde 1990).

Resting stages, such as cysts, are unknown in modern red algae (van den Hoek *et al.* 1995), and might, therefore, explain why the group is so poorly represented in the modern plankton. The situation among the green algae is very different. Resting cysts are encountered in many chlorophytes (see, e.g., Van den Hoek *et al.* 1995; Graham and Wilcox 2000) and prasinophytes (e.g., *Nephroselmis* [Suda *et al.* 1989], *Pterosperma* [Inouye *et al.* 1990], and *Pyramimonas* [Van den Hoff *et al.* 1989; Daugbjerg *et al.* 2000]) and might be more widespread among the latter. Thus, as the chromists, prasinophytes can participate in coastal phytoplankton blooms and spend the rest of the year in dormancy in the sediment. This might partly explain their abundance in *coastal* picoplankton (Not *et al.* 2004; Medlin *et al.* 2006). Such a strategy would not function in mid-ocean because the bottom is simply too far down.

C. Why Did Heterokontophytes Win Over Haptophytes and Dinoflagellates?

Diatoms, and possibly all marine planktonic heterokontophytes, share two characters not found in haptophytes and dinoflagellates. One of these is their low quota for essential trace metals; the other one is their apparently highly efficient CO_2 uptake mechanism.

1. *Quotas for Trace Elements*

Trace elements, such as Fe, Mn, Zn, Cu, Co, Mo, and even Cd, are essential as coenzymes in biochemical reactions in phytoplankton. As an example, the diatom, *Thalassiosira weissflogii*, has a carbonic anhydrase that needs Cd (Lane *et al.* 2005). Yet, the various red and green algal lineages differ markedly in their need for these metals. Falkowski *et al.* (2004a) note that members of the green plastid lineage have relatively high Fe:P, Zn:P, and Cu:P ratios, whereas members of the chromist lineage have higher Ma:P, Cd:P, and Co:P ratios. Ho (2005) shows that prasinophytes, which are dominant in coastal picoplanktonic communities (Not *et al.* 2004; Medlin *et al.* 2006) and which are among the ancestral lineages in the green diversity, need relatively high quotas not only for Fe, Zn, and Cu but also for Cd and Co.

Ho (2005) provides trace metal quotas for four dinoflagellates, two prymnesiophytes (coccolithophorids), and four diatoms. The dinoflagellates show relatively high quotas for the trace metals tested (Fe, Mn, Zn, Cu, Co, and Cd). The coccolithophorids have high quota for Mn, Co, and Cd, whereas they have low quotas for Fe, Zn, and Cu. Strikingly, the diatoms have very low quota for all the six trace metals screened. The values are presented relative to the quotas for P and might, therefore, give a slightly distorted view of the real needs because the green lineages have slightly higher N:P ratios than chromists and red algae (Falkowski *et al.* 2004b; see also www.marinebiology.edu/Phytoplankton/profiles.htm accessed 05/31/06).

Diatoms (and probably all heterokontophytes) possess the Fe-containing cytochrome c_6 in their photosynthetic electron carrier complex instead of the Cu-containing plastocyanin encountered in the photosystems of other algae (Katz *et al.* 2005). Falkowski and Raven (2007) argue that such a replacement would be advantageous in the oxygen-starved oceans of the P/Tr boundary period because in such an environment solubility of Fe would have increased, whereas that of Cu would have decreased. However, such a replacement would have been a disadvantage in the 250 million years since then. Nonetheless, replacement of Cu for Fe makes sense in modern coastal regions. In coastal North Atlantic waters, Fe is more common (21 nM) than Cu (8.5 nM). In contrast, mid-oceanic waters often show very low Fe concentrations, for example, in the mid-North Pacific, Fe is rarer (0.8 nM) than Cu (2.0 nM) and in the Southern Ocean the situation is even worse (Ho 2005).

It is not clear if the low trace metal quotas are a feature of diatoms or of all heterokontophytes. Knowledge of trace metal composition of various non-diatom heterokontophytes is urgently needed.

2. Carbon Fixation

Diatoms are responsible for a large part of the inorganic carbon fixed each year in the ocean. The most common carbohydrate storage product is chrysolaminarian, which is found in storage vacuoles. Lipid droplets are also common.

In the marine environment with a pH of ca. 8.2–8.1, $[CO_2]$ is extremely low, and inorganic carbon is mainly available as HCO_3^-. Moreover, the hydration/dehydration reaction $CO_2 + OH^- \Delta HCO_3^-$ proceeds much slower than the CO_2 uptake in marine algae. Carbon fixation rates can only be maintained if this reaction is catalyzed. A phylogenetically unrelated group of enzymes called carbonic anhydrases perform this function. They are located in the plasmalemma, and most of them need hydroxylated Zn as coenzyme (though see Lane et al. 2005). The enzymes rapidly replace CO_2 used by the Calvin cycle, but they cannot elevate the $[CO_2]$ concentration relative to that of $[HCO_3^-]$ (Falkowski and Raven 2007).

Microalgae fix CO_2 into the C-3 compound phosphoglyceric acid. Ribulose-1, 5-bisphosphate carboxylase (RuBisCo), the enzyme responsible for C-3 carbon fixation, requires CO_2, which is present in seawater at about 10 μM. However, the RuBisCo enzyme of diatoms has a half-saturation constant for CO_2 of 30–60 μM, implying that marine diatoms should be CO_2-limited. Instead, kinetic and growth studies have

shown that photosynthesis in marine diatoms is $[CO_2]$-insensitive (see references in Reinfelder et al. 2000).

It has been hypothesized that the diatoms have a C-4 concentrating mechanism (Reinfelder et al. 2000, 2004). Granum et al. (2005) discuss enzymes and structural components for a C-4-type CO_2-concentrating mechanism in marine diatoms that include the enzymes phosphoenolpyruvate (PEP) carboxylase and PEP carboxykinase. Reinfelder et al. (2004) demonstrated that inhibition of PEP carboxylase resulted in a 10-fold decrease in whole cell photosynthesis in the diatom Thalassiosira weissflogii acclimated to low CO_2 (10 μM), but it had little effect on photosynthesis in a C-3 marine chlorophyte. Elevated CO_2 (150 μM) or low O_2 (80–180 μM) restored photosynthesis, thus linking PEP carboxylase inhibition with CO_2 supply in T. weissflogii. Short-term ^{14}C uptake and CO_2 release experiments using a PEP carboxykinase inhibitor revealed that much of the carbon taken up by diatoms during photosynthesis is stored as organic carbon before being fixed in the Calvin cycle, showing that the C-4 pathway functions as a CO_2 concentrating mechanism. If C-4 photosynthesis is present among marine diatoms, it may account for a significant portion of carbon fixation and export in the ocean and would explain the greater enrichment of ^{13}C in diatoms compared with other classes of phytoplankton (Reinfelder et al. 2000, 2004; Granum et al. 2005). Significantly, unicellular C-4 carbon assimilation may even have predated the appearance of multicellular C-4 plants (Katz et al., Chapter 18, this volume).

In any case, these adaptations may have given diatoms, and possibly other heterokontophytes, a selective advantage in oceanic primary production above their competitors. Geider et al. (1997) demonstrated that maximum growth rates under light saturation in the marine diatoms Phaeodactylum, Skeletonema, and Thalassiosira and the freshwater chrysophyte Olisthodiscus were roughly 4 to 10 times those of five dinoflagellates and two haptophytes tested. Yet, it is still unclear if C-4 metabolism is a characteristic typical of the diatoms, or only for some of them, or for the heterokonto-phytes in general.

The relative inaccessibility of carbon for photosynthesis is a problem in contemporary oceans, but matters are expected to change dramatically over the next centuries. It is likely that the pH of the surface seawater will decrease by almost a full degree over the next millennium as a consequence of ongoing anthropogenic CO_2 emission (see, e.g., Doney 2006). Increased availability of CO_2 may seem good news for phytoplankton, diatoms in particular, but it spells bad news for some marine organisms making use of calcite (coccolithophorids, foraminifera) and aragonite (e.g., corals, calcifying green algae, pteropods) in their cell wall or exoskeleton.

D. Why Did Diatoms Win Over Other Heterokontophytes?

Diatoms possess several characteristics not found in other heterokontophytes. One is the large central vacuole in which nutrients can be stored for later use; another one is the silica frustule. In addition, diatoms deter grazers by means of secondary metabolites, whereas secondary compounds produced by other plankton organisms seem to have less effect on diatoms than on other phytoplankton groups. Last but not least diatoms have diversi fied in many different habitats, and some of the new traits acquired in these various habitats were suited for life in the plankton, as well. Several marine planktonic diatom lineages evolved from benthic ancestry.

1. The Central Vacuole

Ofallheterokontophycean,haptophycean, and dinoflagellate planktonic algae, only the diatoms possess a large central vacuole. This vesicle provides multiple advantages to the diatom. It permits a luxury uptake and stor-

age of nutrients in moments of plenty (Tozzi et al. 2004). With a full store of nutrients, a diatom cell can divide up to four times (generating 16 cells) to bridge periods of nutrient shortage when growth in other phytoplankton organisms is impaired (except when the limiting nutrient is silicon, because large amounts of silicate cannot be stored by diatom cells). This storage capacity renders diatoms particularly well adapted to environments where nutrients are available in short pulses (Raven 1997; Tozzi et al. 2004; Katz et al. 2005). In addition, it has been suggested that diatoms use their vacuole to help regulate their buoyancy. Thus, they sink until they reach the optimum nutrient level and then stop sinking to uptake nutrients (Villareal et al. 1993).

In coastal eutrophic ecosystems, especially larger sized phytoplankton cells sustain higher carbon-specific photosynthesis than smaller cells (Cermeño et al. 2005). A large central vacuole allows a cell to have an increased plasmalemma surface given a volume of protoplasm. Most of the protoplasm is close to the plasmalemma, fostering exchange with the environment. In addition, in centric diatoms with a large central vacuole, the numerous small chloroplasts are distributed along the plasmalemma and do not shade one another. If the same volume of protoplasm were packed in a cell without vacuole, the average distance of the cytoplasm to the plasmalemma would be larger, and the chloroplasts would shade one another.

Being large, with a large central vacuole, also constitutes a strategy to avoid grazing. Smaller cells are heavily grazed by quickly responding microzooplankton, whereas large cells are eaten mainly by mesozooplankton with longer generation times and therefore slower responses to a sudden bonanza of food (Kiorboe 1993). Moreover, a cell largely composed of a vacuole is not the most nutritious choice of food. But nonetheless, diatoms are grazed upon.

The large vacuole also allows cells to control expansion (Mann and Chepurnov

2004). Changes in cell volume, documented in several species, are much more difficult to achieve in nonvacuolate species, where there are fewer opportunities for segregating away metabolites into osmotically inactive forms and then mobilizing them again to drive cell expansion during discrete phases of volume increase, for example, during vegetative enlargement and during auxospore swelling (see Section II.C.3).

2. Advantages of Silica Encasing

One reason for the use of silica for cell wall construction is that it is in ample supply in most of the ocean, and, for a given wall volume, it costs less energy to make a wall of silica than one of carbohydrate (Raven 1983). It has been argued that abundance of silica and the low energy costs of incorporating it into a wall was a reason for the evolution of silica metabolism. Nonetheless, in a seasonal time scale, diatom blooms may deplete available silica such that silica becomes a limiting factor for primary production (Martin-Jezequel et al. 2000).

Diatoms have no flagella in their vegetative phase and, consequently, cannot escape their enemies. Nevertheless, loss of mobility must have been counterbalanced by advantages for a silica cell wall (Smetacek 1999; Hamm and Smetacek, Chapter 14, this volume). Several hypotheses have been put forward. Defense is often invoked because of the compressive strength of silica walls (Hamm et al. 2003) and its possible immunity to enzymatic attack. Very few pathogens have been reported to attack diatoms. Notable exceptions are chitrids and unicellular oomycetes (Sparrow 1960) and cercomonads (Kühn et al. 2004), which find entry into the cell via the processes or at times the areolae. Bacteria are usually only present on senescent cells. There is quite a body of literature on viral attacks on microalgae, but there are only two reports of viruses attacking diatoms (Nagasaki et al. 2004, 2005). In response to impalement with a micropipette, diatoms produce numerous membrane profiles at the wound area to seal it and to attempt to expel the micropipette; this has been suggested as

a mechanism by which diatoms can avoid pathogens (Kühn et al. 2004). Raven and Waite (2004) suggest that the silica wall permits parasitized cells to sink, enhancing survival of the population. One could argue that selection does not work on species but on individuals; however, diatoms divide clonally during their vegetative phase, implying that nearby cells have a high probability of belonging to the same clone. Individual cells or chains sacrifice themselves to increase the survival chances of their clone. In fact, there are records of apparently "altruistic" hypersensitive responses in freshwater diatoms, in which the host cell dies before the parasite can complete its life cycle (Canter and Jaworski 1979). Grazer defense has been invoked, but, nevertheless, there exists a plethora of herbivores that graze on diatoms. Grazing itself is unavoidable because there will always be organisms able to crack the cell wall. Yet, defenses do not need to be perfect, they only have to raise the costs for grazers to get at the food to a point where the grazing effort is too costly (Smetacek 1999).

3. Secondary Compounds and Allelopathy

Diatoms possess an effective biochemical defense system against grazers. When the cell is damaged, C_{16} and C_{20} polyunsaturated fatty acids are liberated from membrane storage sites and enzymatically converted, in a matter of seconds, into bioactive unsaturated aldehydes or related oxylipins (Pohnert 2000). These secondary compounds have the same effect as alkaloids produced by many higher plants in response to cell damage; they have a teratogenic effect, that is, they induce abortion and birth defects in the offspring of predator animals, such as copepods, that are exposed to them during gestation (Miralto et al. 1999; Ianora et al. 2004). Of the 51 marine planktonic diatom species screened for these compounds to date, 20 produce detectable amounts, implying that this antigrazer defense system is ecologically important (Wichard et al. 2005).

These compounds are not produced with the goal of antagonizing other microalgae, because healthy diatoms do not excrete them into the medium. Nonetheless, recent work has shown that the compounds in high doses do have allelopathic effects on *Thalassiosira weissflogii* in that they damage their DNA, inhibit cell growth (Casotti *et al.* 2005), and cause apoptosis.

Whether diatoms produce allelopathic compounds to outcompete other microalgae remains to be seen. The focus of research into allelopathic compounds has been on harmful algal bloom-forming species of cyanobacteria, dinoflagellates, haptophytes, and various flagellated heterokontophytes, such as raphidophytes and pelagophytes (Cembella 2003; Legrand *et al.* 2003; Fistarol 2004; Fistarol *et al.* 2005; Arzul and Gentien 2005). Diatoms have simply not yet been screened thoroughly for allelopathic effects (but see, Arzul and Gentien 2005; Lundholm *et al.* 2005). Yet, allelopathic compounds produced by dinoflagellates seem to have less effect on diatoms than on other algal and protozoan groups (Tillmann *et al.* in press). Why this is so is not yet clear, but, in general, healthy cells that have an internal nutrient store in their vacuole can grow, whereas those under nutrient stress are more susceptible to allelopathic substances (Fistarol 2004). This is where diatoms might have a cutting edge above other phytoplankton.

4. Ecological Diversity

An important reason why diatoms are so successful in the modern plankton is that they have managed to diversify in so many different habitats: in freshwater, coastal habitats, the open ocean, upwelling zones, extremely oligotrophic environments, polar sea-ice communities, the tropics, hypersaline, alkaline, and acidic pools, drifting bottom communities, epiphytic and epizoic communities, and so on. Estimates of the number of diatom species in all these habitats together vary wildly, but one striking fact is that of the more than 10^5 species of diatoms (Mann and Droop 1996), "only" ca. 10,000 (Sournia 1995) inhabit the marine plankton. In other words, the bulk of the diatom biodiversity is not planktonic.

The major innovations in morphology, ecology, and life cycle, that is, the evolution of multipolar centrics, araphid pennates, and raphid pennates, appear to have occurred in benthic communities and on unstable substrata. These innovations could then be tried out also in the plankton. Thus, a very important reason why diatoms are now so diverse in marine planktonic habitats is that their planktonic diversity was enriched, again and again, by new lineages with new designs from benthic habitats (Figure 6).

None of the other algal groups have managed to establish themselves in such a range of niches and therefore lack the backup. Of the ca. 500 haptophytes, almost all are planktonic, but several have benthic stages in their life cycles, and only two are freshwater. Dinoflagellates have fared better; of the ca. 1800–2000 dinoflagellates at least some are benthic, and about 200 live in freshwater habitats (van den Hoek *et al.* 1995). Of the other heterokontophytes, raphidophytes constitute a small group of closely related coastal and freshwater species, and all but one of them are planktonic. Of the brown algae, only the gametes spend time in the plankton. Chrysophytes and synurophytes have diversified into benthic and planktonic niches, but they are all freshwater species. The few extant silicoflagellates, pelagomonads, rhizochromulinids, and pedinellids are all marine planktonic.

Although the major morphological innovations in diatoms happened in benthic communities, it does not mean that nothing happened in the plankton. For instance, the plankton is the place where several new modes of chain formation have been acquired. Chains must provide a net advantage to planktonic diatoms. Only very small and very large planktonic diatoms do not form chains. What exactly the advantages are remains to be debated, but the shape and length of chains seem a compromise between various environmental factors, such as effectiveness of grazer defense, adjustment of buoyancy, optimization

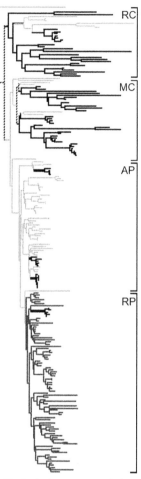

FIGURE 6. Ecological strategies of diatoms: black, typical planktonic; light grey, benthic, attached to substratum, or free-living, tumbling over the bottom, can incidentally be swirled up, but not typically planktonic; dark grey, moving actively over substratum by means of raphe system. State associated to basal ramifications ambiguous (hatchet lines). *Ellerbeckia sol* is not planktonic, but *Bolidomonas*, the nearest sister of the diatoms, is and several members of the heterokontophytan sister lineage possess planktonic stages or are planktonic.

of nutrient uptake through micro shear (Reynolds 1984; Karp-Boss *et al.* 1996), and, in pennate planktonic diatoms, a possibility for entanglement of chains to retain proximity for sexual reproduction. In any case, chain length in *Skeletonema* and *Chaetoceros* is inversely related to turbulence in the surface layer (Schöne 1979).

One type of chain formation is by interlocking silica projections, spines, or tubes from the valves. This type of linkage is found in the radial centrics *Paralia*, *Stephanopyxis*, and *Aulacoseira*, in the multipolar centrics *Detonula* and *Skeletonema*, in the unrelated araphid pennates *Staurosira*, *Fragilariforma*, and *Fragilaria*, and even in raphid pennates such as *Diadesmis*. It must thus have been acquired independently multiple times. It is also the obvious mode of chain formation in most diatoms in the 110-million-year-old fossil record in Gersonde and Harwood (1990), suggesting that this is among the oldest modes of chain formation in the plankton (see, e.g., Figure 5G and H).

Another widespread type of chain formation is by mucilage pads. Pads are encountered across the diatom diversity; some epiphytic radial centrics have them in the valve center (*Arachnoidiscus*), whereas many basal lineages in the multipolar centrics and the majority of the araphid pennates have apical pads. Apical pads often lead to the formation of characteristic zigzag chains or stellate patterns found in many araphid pennates and bipolar centrics.

The multipolar centric radiation probably occurred in the benthos. There, sister cells formed chains by means of apical mucilage pads. Yet, several lineages within this group adopted a planktonic lifestyle, and it is there that they acquired new ways of forming chains. Several of these novelties resulted in modest diversifications within the plankton (e.g., Thalassiosirales, see Kaczmarska *et al.* 2006). Thalassiosirales acquired strutted processes (see Figure 2F) through which chitin fibrils (see Figure 2E) are exuded. The latter maintain connections with sister cells or extend in the surrounding water. The diverse genus *Chaetoceros* developed thin hollow spines, called setae, which interlock with those of sister cells (see Figure 2G; Pickett-Heaps 1998a, b). The setae are extended into the surroundings and cytoplasm, and in some species even chloroplasts can move through these structures. *Eucampia* and *Hemiaulus*, the planktonic sister genera of *Chaetoceros*, connect only through mucilage pads on elevated apical pore fields and are by far not as diverse as *Chaetoceros*. The closely related genera *Bellerochea*, *Helicotheca*, *Ditylum*, and *Lithodesmium* possess ridges along the rim of the valve face that underlap and overlap with those on the sister valve. *Cymatosira* forms ribbons through interlocking spines on the rim of the valve face. These diatoms have little use for mucilage pads and, hence, have drastically reduced apical pore fields or lost them altogether (for electron microscopical [EM] illustrations of genera, see Round *et al.* 1990).

In the small araphid clade, *Asterionellopsis* and *Asterioplanus* are planktonic, forming helical winding colonies. Similar colony shape is encountered in some planktonic cyanobacteria, in rhizosolenoid diatoms, and in many bipolar centrics including *Eucampia*, some species of *Chaetoceros,* and some populations of *Skeletonema*. Winding chain formation can be inferred from diatom species as old as 110 million years, with an elongated and curved pervalval axis (e.g., *Ancylopyrgus* and *Basilicostephanus* [see Figure 5G], Gersonde and Harwood 1990). Helices allow rapid buoyancy-regulated movements; both sinking and floating are facilitated (Reynolds and Walsby 1975; Booker and Walsby 1979).

A few members of the large araphid pennate group have developed a planktonic lifestyle secondarily from epiphytic ancestry. *Thalassionema* and its close relatives (*Trichotoxon, Thalassiothrix, Lioloma*) form zigzag or stellate colonies. These taxa constitute a sister clade of the heavily silicified synedroid epiphytic genera *Tabularia* and *Catacombas*. A few other lineages (*Fragilaria, Diatoma, Asterionella*) are planktonic as well, but these are exclusively in the freshwater. Of all the raphid pennate diversity, only a few lineages developed a truly marine planktonic lifestyle (see Figure 6). Among these are *Fragilariopsis* and its sister, *Pseudo-nitzschia*. The former occurs as ribbons, but the latter forms needle-like chains appearing superficially like the needle-like chains of the radial centric diatom *Rhizosolenia*. They achieve such a shape by attaching to their sister cells at the valve apices. Some

apparently benthic raphid pennates can be found in the plankton as well, but these seem to use the marine snow as a substratum. Others live in the brine channels of seasonal sea ice.

A possible reason why so few pennate diatoms are truly planktonic may have to do with their mode of sexual reproduction (see Figure 3C). Their amoeboid isogametes (see Section II.C.2.) are a liability in the phytoplankton because they need to crawl over a surface to one another and lack the agility of flagellated microgametes of centrics in finding a partner. In order to go through sexual reproduction, planktonic pennate diatoms must increase their encounter rate by sinking to the bottom, being lifted up into the sea ice, or forming a bloom. Not surprisingly, most planktonic pennates are typical for coastal regions in which seasonal blooms can form. Nonetheless, some pennates have adapted to oceanic conditions and are now important species in the southern ocean. The raphid pennates *Pseudo-nitzschia* and *Fragilariopsis* and the araphid pennates *Thalassiothrix and Thalassionema* can form blooms and go through sexual reproduction in the plain open ocean.

VI. CRYPTIC DIVERSITY IN PLANKTONIC DIATOMS AND ITS BEARING ON EVOLUTION

A controversial idea in microbial research is that species are ubiquitous and cosmopolitan and that for these reasons their speciation rates are low (Finlay *et al.* 2002, 2004). That idea is, however, based on morphologically delineated species. A number of molecular genetics studies of microbes have shown that the morphological entities currently described as single taxa are, in fact, assemblages of cryptic species; that is, assemblages of genetically and biologically distinct, albeit often closely related species (e.g., Medlin *et al.* 1991, 2000; Sàez *et al.* 2003; John *et al.* 2003; Montresor *et al.* 2003; Sarno *et al.* 2005, 2007; Armato *et al.* 2007). Cryptic diversity has been discovered throughout the diatom diversity (Lundholm *et al.* 2002a, b, 2003; Behnke *et al.* 2004; Mann *et al.* 2004; Beszteri *et al.* 2005a, b; Sarno *et al.* 2005). Microsatellite data suggest the *Pseudo-nitzschia pungens* appears to be single, cosmopolitan interbreeding species (Evans *et al.* 2005), but most other so-called cosmopolitan species investigated show otherwise (Rynearson and Ambrust 2004). Amato *et al.* (2007) demonstrated that even within the Gulf of Naples, *Pseudo-nitzschia delicatissima* and *P. pseudodelicatissima* each consist of at least three genetically distinct entities, all of them occurring together in a bloom. Crossing experiments revealed that all these entities are reproductively isolated. The existence of all this cryptic diversity suggests that speciation rates are much higher than previously believed.

Nevertheless, each of the cryptic species could in its turn be cosmopolitan and ubiquitous. However, results of a recent global biogeographic survey of coastal diatom genus *Skeletonema* (Kooistra *et al.* submitted) suggest otherwise. The results have shown large, although disjunct, distribution patterns in most species but also revealed peculiar gaps in the patterns: *S. japonicum* was found in all temperate coastal zones except in Europe, and *S. grethae* was encountered in the Northern Caribbean and the U.S. East Coast but nowhere else. Such gaps might result from local extinction and difficulties with reinoculation from elsewhere. Results from counts at phytoplankton time series stations demonstrate that species can, indeed, appear (Nehring 1998) and disappear (Smayda 1998) on a regional scale.

How quickly do phytoplankton species go globally extinct? A glance at the tree shows that the more thoroughly sampled genera (e.g., *Skeletonema, Aulacoseira*) consist of shallow and densely packed clades on top of long internodes. Of course, this does not mean that these genera were monospecific during most of their evolution and then suddenly diversified. They probably diversified throughout their evolutionary history but lost diversity at a more or less equal pace by means of background extinction, weeding away species, and ultimately whole lineages. Such dynamic speciation and extinction may be apparent in the fossil record, but morphometric techniques are needed to track the anagenesis and cladogenesis of the different fossil species.

How can a phytoplankton species go extinct? Competition is unlikely because interactions among more than a few species tend to become chaotic (Hutchinson 1961; Huisman *et al.* 2001), and plankton blooms are usually composed of a plethora of species. We believe that especially for coastal species, to persist somewhere they must participate in blooms to counterbalance relentless losses through burial, sinking, and grazing during the period in which they do not bloom. In case a species misses the opportunity to bloom in a couple of consecutive years, its "seed bank" may be obliterated, and the species is regionally gone. If that happens everywhere, then the species goes extinct. A species can start missing the opportunity for bloom formation at a given locality in case of changing environmental conditions (Hays *et al.* 2005), or the changes may erode its genetic diversity, and therewith its ability to cope with further changes. Anything affecting bloom formation of phytoplankton species could thus obliterate them, and maybe even in a matter of years.

What is the significance of this microevolutionary process for the evolution of the diatoms as a whole? As stated before, raphid pennates are highly diverse, but evolutionary young, whereas radial centrics are far less diverse but show deep, that is, ancient lineages. The differences in species richness between pairs of sister clades are clearly correlated with, and probably result from, whether one of them acquired evolutionary novelties. The obvious reason why raphid pennates are far more speciose than their nearest araphid pennate sister lineages is that the newly acquired raphe allowed the raphid pennates to fill a plethora of niches unavailable to its araphid relatives and, hence, diversify rapidly. The often poorly resolved basal ramifications in the radial centrics, the multipolar centrics, and the araphid pennates probably result from similar brief periods of adaptive radiation. Subsequently, the background extinction pruned back the initial diversity into a series of well-supported lineages on top of the badly resolved "rake of booming diversity." The possible result of such random weeding has been that with time, fewer and fewer lineages survived. The radiation of the radial centrics happened longer ago than the one that sprouted the diversity of the raphid pennates, and hence, the radial centrics are phylogenetically deep but poorly diverse whereas the raphid pennates are, instead, highly diverse but phylogenetically shallow. Moreover, with time, morphological evolution keeps on changing the survivors so that they differentiate more and more from their ancestors and from one another. For example, the pre-Cretaceous centric diatoms are morphologically distinct from modern forms, whereas the early pennates resemble their supposed relatives in the modern pennates to such an extent that taxonomists readily place these fossils in modern genera.

The fact that the raphid pennates are now the most diverse group of diatoms in existence does not mean that the other lineages failed to acquire novelties. Innovative design leading to booming diversity is observed in the centrics as well, but it is by far not at the magnitude of the adaptive radiation of the raphid pennates. The lineage, including *Chaetoceros* with its interconnecting setae (see Figure 2G), is more diverse than its neighbors *Eucampia* and *Hemiaulus,* which lack such setae. The latter two form chains the ancestral way, by means of their mucilage pads. Thalassiosirales, with their strutted processes and chitin mechanisms for chain formation, are another centric success story (see Figure 2E and F). They have diversified profusely and their speies are now dominant constituents of the phytoplankton.

VII. THE DAWNING FUTURE OF DIATOM RESEARCH: GENOMICS

Through the use of comparative genomics we have information on a species' evolutionary history to gain insights into the mechanisms of molecular evolution. The study of molecular evolution among closely related organisms or groups of organisms helps us to provide the foundations for the application of the use of genes and their products for understanding ecosystem resilience, to understand the key mechanisms that have given rise to the similarities and differences within and between species' genomes, to understand the consequences of variation in genome sizes, gene families, and the exploitation of sequence polymorphisms in organisms, to appreciate how combinations of a diversity of kinds of mutations, genetic drift and selection have given rise to today's organisms, to see how the comparative biological approaches within and between species give additional insight into genome structure, evolution and function, and to design new research approaches to address questions in genetics, evolution, and gene function using comparative genomics. A number of comparative genomic studies of algae have been made through EST libraries of gene expression, often under stress conditions. John *et al.* (2005) compared the percentage of identifiable genes from cDNA libraries in the raphid pennate diatom *Fragilariopsis cylindrus* (Mock *et al.* 2006), the toxic dinoflagellate *Alexandrium ostenfeldii,* and the toxic haptophyte *Chrysochromulina polylepis.* The percentage of identifiable genes in the two chromophytes was similar and in the same range of two other chromophytes (*Laminaria digitata* [Phaeophyta] Crépineau *et al.* 2000; *Phaeodactylum tricornutum* [raphid pennate diatom] Scala *et al.* 2002) but lower than those from higher plants (Meinke *et al.* 1998). The percentage of identifiable genes in the dinoflagellate was about 12%, which is significantly lower than in other algae, higher plants, or animals. The most interesting finding from the diatom, *Fragilariopsis*, whose EST library was induced under cold and temperature stress conditions, was that the genes for photosystem II were only induced in this ice alga when light and temperature were reduced, thus making the diatom perfectly adapted for survival in sea ice, its natural habitat (Mock and Valentin 2004). Such studies offer great insight into how organisms are adapted to their natural habitat, and we can expect them to reveal much more about adaptive strategies in the future.

Recently, the first genome sequence of a heterokont organism was completed (Armbrust *et al.* 2004). The diatom was selected because of its small genome size. There were several novel findings in the genome of *Thalassiosira pseudonana* relative to other eukaryotic organisms. It contains more than 11,000 protein-coding genes, of which about half could not be assigned to any known gene. The majority of the genes with a known homology to animal, red, or green nuclear genes were more similar to those from animals than to those from plants (see Figure 3 in Armbrust *et al.*, 2004), and a similar finding has been reported from EST analysis of the raphid pennate *Phaeodactylum tricornutum* (Montsant *et al.* 2005). However, comparisons with animals are not very productive, because these belong to a quite different group of eukaryotes (the opisthokonts: Keeling *et al.* 2005), and comparisons with other organisms are hampered by the fact that few whole genomes are yet available for the "chromalveolate" group of eukaryotes to which the heterokont host cell belongs (Keeling *et al.* 2005). Overall, the nuclear genome of *Thalassiosira* must be a chimera between (1) components, inherited from the heterotrophic ancestor of the host cell, (2) components from the nuclear, plastid, and perhaps mitochondrial genomes of the red algal endosymbiont, and probably also (3) a minority of components acquired through lateral transfers from other eukaryotic and prokaryotic cells (cf. the diverse origins of genes in the chlorarachniophyte *Bigelowiella*, which also represents a secondary endosymbiosis, in this case between an amoeba and a green alga: Archibald *et al.* 2003).

The major categories of transcription factors in *Thalassiosira* were of the heat-shock family, which is rare in other eukaryotes (Armbrust *et al.* 2004). Diatoms also do not have many receptor kinases or leucine-rich repeat containing receptors, so the diatoms likely have a novel class of receptors. The most unusual finding in the *Thalassiosira* genome and in the EST library of *Phaeodactylum* is that of a complete urea cycle. This cycle is totally unknown for any green plant; it is curious as to why the diatoms should have this cycle because they certainly do not secrete urea, although they can use it as a food source. Urea could be used as an osmolyte or a long-term N store, but more likely one of the intermediates in the urea cycle, ornithine, feeds into the polyamine cycle in the diatoms to make the unusual proteins involved in silica precipitation. Arginine, another intermediate in the urea cycle, is likely funneled into a nitric oxide cycle to play a role in pathogen defense. The controversy regarding C-4 or C-3 photosynthesis (Reinfelder *et al.* 2000, 2004; see V.C.2) for diatoms was not resolved, despite the identification of genes coding for all the enzymes in the C-4 pathway. There was no clear evidence for the plastid localization of the decarboxylating enzymes, so further work is needed to clarify carbon fixation pathways in the diatoms. Two pathways for fatty acid synthesis were found in the diatoms. One pathway was localized in the mitochondria, like animals, and the other pathway was in the peroxisomes, like plants. The latter pathway explains how diatoms can withstand long periods of darkness because they can shift stored fatty acids into carbohydrate production, when needed.

Sequencing and annotation of the genomeing of the pennate *Phaeodactylum tricornutum* is now complete. The annotation is backed up with EST libraries of about 90,000 sequences, so annotation of the genome has been easier. Many of the libraries were made under environmentally stressed conditions, which proved very informative for studying adaptive mechanisms at the genetic level (see www.biologie.ens.fr/diatomics/EST/). Such comparative genome projects and expression analyses offer exciting starting points for experimental investigations into the adaptive strategies of the marine phytoplankton, not just the diatoms.

Acknowledgments

We thank Christophe Brunet, Raffaella Casotti, David Harwood, Adrianna Ianora, Marina Montresor, Maurizio Ribera, and Urban Tillman for suggestions and ideas and for checking the text in their areas of expertise. Ugo Sacchi, Isabella Percopo, and Richard Crawford are thanked for help with photos of extant diatoms. The project has been carried out in the frame of the MarBEF Network of Excellence "Marine Biodiversity and Ecosystem Functioning" (contract no. GOCE-CT-2003-505446), which is funded in the EU's Sixth Framework Program.

References

Alexopoulos, C.J., Mims, C.W., and Blackwell, M. (1996). *Introductory Mycology*. New York, Wiley.

Amato, A., Orsini, L., D'Alelio, D., and Montresor, M. (2005). Life cycle, size reduction patterns, and ultrastructure of the pennates planktonic diatom *Pseudo-nitzschia delicatissima* (Bacillariophyceae). *J. Phycol.* **41**: 542–556a.

Amato, A., Kooistra, W.H.C.F., Levialdi Ghiron, J.H., Mann, D.G., Pröschold, T., and Montresor, M. (2007). Reproductive isolation among sympatric cryptic species in marine diatoms. *Protist* **158**: 193–207.

Anderson, O.R. (1983). *Radiolaria*. New York, Springer.

Anderson, O.R. (1992). Radiolarian algal symbioses. *Algae and Symbioses*. W. Reisser, ed. Bristol, UK, Biopress, pp. 92–102.

Archibald, J.M., Rogers, M.B., Toop, M., Ishida, K., and Keeling, P.J. (2003). Lateral gene transfer and the evolution of plastid-targeted proteins in the secondary plastid-containing alga *Bigelowiella natans*. *Proc. Natl. Acad. Sci. U S A* **100**: 7678–7683.

Armbrust, E.V. (1999). Identification of a new gene family expressed during the onset of sexual reproduction in the centric diatom *Thalassiosira weissflogii*. *Appl. Env. Microbiol.* **65**: 3121–3128.

Armbrust, E.V., Berges, J.A., Bowler, C., Green, B.R., and 41 others. (2004). The genome of the diatom *Thalassiosira pseudonana*: ecology, evolution, and metabolism. *Science* **306**: 79–86.

Armbrust, E.V., and Chisholm, S.W. (1990). Role of light and the cell cycle on the induction of spermatogenesis in a centric diatom. *J. Phycol.* **26**: 470–478.

Arzul, G., and Gentien, P. (2005). Allelopathic interactions among marine Microalgae. *Algal Cultures, Analogues of Blooms and Applications*, Vol. 1. D.V. Subba Rao, ed. Enfield, NH, Science Publishers, pp. 131–162.

Baldauf, J.G. (1984). Cenozoic diatom biostratigraphy and paleoceanography of the Rockall Plateau region, North Atlantic, Deep Sea Drilling Project Leg 81. *Init. Rep. Deep Sea Drill. Proj.* **81**: 439–478.

Banse, K. (1964). On the vertical migration of zooplankton in the sea. *Prog. Oceanogr.* **2**: 56–125.

Barron, J., and Mahood, A. (1993). Exceptionally well-preserved early Oligocene diatoms from glacial sediments of Prydz Bay, East Antarctica. *Micropaleontology* **39**: 29–45.

Bates, S.S., and Davidovich, N.A. (2002). Factors affecting the sexual reproduction of diatoms, with emphasis on *Pseudo-nitzschia* spp. *Proceedings of the LIFEHAB Workshop: Life History of Microalgal Species Causing Harmful Algal Blooms*. E. Garces, A. Zingone, B. Dale, M. Montresor, and B. Reguera, eds. Office for official publications of the European Communities, Luxembourg, pp. 27–30.

Behnke, A., Friedl, T., Chepurnov, V.A., and Mann, D.G. (2004). Reproductive compatibility and rDNA sequence analyses in the *Sellaphora pupula* species complex (Bacillariophyta). *J. Phycol.* **40**: 193–208.

Beszteri, B., Acs, E., and Medlin, L.K. (2005a). Conventional and geometric morphometric studies of valve ultrastructural variation in two closely related *Cyclotella* species (Bacillariophyceae). *Eur. J. Phycol.* **40**: 73–88.

Beszteri, B., Acs, E., and Medlin, L.K. (2005b). Ribosomal DNA sequence variation among sympatric strains of the *Cyclotella meneghiniana* complex (Bacillariophyceae) reveals cryptic diversity. *Protist* **156**: 317–333.

Bhattacharya, D., and Medlin, L.K. (1995). The phylogeny of plastids, a review based on comparison of small subunit ribosomal RNA coding regions. *J. Phycol.* **31**: 489–498.

Bhattacharya, D., and Medlin, L.K. (2005). Dating algal origin using molecular clock methods. *Protist* **155**: 9–10.

Biddle, K.D., and Azam, F. (1999). Accelerated dissolution of diatom silica by marine bacterial assemblages. *Nature* **397**: 508–512.

Booker, M.J., and Walsby, A.E. (1979). The relative form resistance of straight and helical blue-green algal filaments. *Br. Phycol. J.* **14:** 141–150.

Boyd, P.W., Watson, A.J., Law, C.S., Abraham, E.R., Trull, T., Murdoch R., Bakker, D.C.E., Bowie, A.R., Buesseler, K.O., Chang, H., Charette, M., Croot, P., and 31 others. (2000). A mesoscale phytoplankton bloom in the polar Southern Ocean stimulated by iron fertilization. *Nature* **407:** 695–702.

Bronk, D.A., and Flynn, K.J. (2005). Algal cultures as a tool to study the cycling of dissolved organic nitrogen. *Algal Cultures, Analogues of Blooms and Applications,* Vol. 1. D.V. Subba Rao, ed. Enfield, NH, Science Publishers, pp. 301–342.

Canter, H.M., and Jaworski, G.H.M. (1979). The occurrence of a hypersensitive reaction in the planktonic diatom *Asterionella formosa* Hassall parasitized by the chytrid *Rhizophydium planktonicum* Canter Emend., in culture. *New Phytol.* **82:** 187–206.

Casotti, R., Mazza, S., Brunet, C., Vantrepotte, V., Ianora, A., and Miralto, A. (2005). Growth inhibition and toxicity of the diatom aldehyde 2-trans, 4-trans-decadienal on *Thalassiosira weissflogii* (Bacillariophyceae). *J. Phycol.* **41:** 7–20.

Cavalier-Smith, T. (2003). Genomic reduction and evolution of novel genetic membranes and protein-targeting machinery in eukaryote-eukaryote chimaeras (meta-algae). *Phil. Trans. R. Soc. London Ser. B* **358,** 109–133.

Cavalier-Smith, T., and Chao, E.E. (1996). Molecular phylogeny of the free-living archeozoan *Trepomonas agilis* and the nature of the first eukaryote. *J. Mol. Evol.* **43:** 551–562.

Cembella, A.D. (2003). Chemical ecology of eukaryotic microalgae in marine ecosystems. *Phycologia* **42:** 420–447.

Cermeño, P., Emilio Marañón, E., Rodríguez, J., and Fernández, E. (2005). Large-sized phytoplankton sustain higher carbon-specific photosynthesis than smaller cells in a coastal eutrophic ecosystem. *Mar. Ecol. Prog. Ser.* **279:** 51–60.

Chacón-Baca, E., Beraldi-Campesi, H., Cevallos-Ferriz, S.R.S., Knoll, A.H., and Golubic, S. (2002). 70 Ma nonmarine diatoms from northern Mexico. *Geology* **30:** 279–281.

Chambers, P.M. (1996). *Late Cretaceous and Palaeocene Marine Diatom Floras.* PhD thesis, University College London.

Chepurnov, V.A., Mann, D.G., Sabbe, K., and Vyverman, W. (2004). Experimental studies on sexual reproduction in diatoms. *Int. Rev. Cytol.* **237:** 91–154.

Crawford, R.M. (1988). A reconsideration of *Melosira arenaria* and *M. teres*: resulting in a proposed new genus *Ellerbeckia. Algae and the Aquatic Environment.* F.E. Round, ed. Bristol, UK, Biopress, Bristol, pp. 413–433.

Crawford, R.M., and Sims, P.A. (2006). The diatoms *Radialiplicata sol* (Ehrenb.) Gleser and *R. clavigera* (Grun.) Gleser and their transfer to *Ellerbeckia*, thus a genus with freshwater and marine representatives. *Nova Hedwigia.* In press.

Crépineau, F., Roscoe, T., Kaas, R., Kloareg, B., and Boyen, C. (2000). Characterization of complementary cDNAs from the expressed sequence tag analysis of life cycle stages of *Laminaria digitata* (Phaeophyceaes). *Plant Mol. Biol.* **43:** 503–513.

Darley, W.M., and Volcani, B.E. (1969). Role of silicon in diatom metabolism: a silicon requirement for deoxyribonucleic acid synthesis in the diatom *Cylindrotheca fusiformis* Reimann and Lewin. *Exp. Cell Res.* **58:** 334–342.

Daugbjerg, N., Marchant, H.J., and Thomson, H.A. (2000). Life history stages of *Pyramimonas tychotreta* (Prasinophyceae, Chlorophyta), a marine flagellate from the Ross Sea, Antarctica. *Phycol. Res.* **48:** 199–209.

Delwiche, C. (2007). The origin and evolution of dinoflagellates. *Evolution of Primary Producers in the Sea.* P.G. Falkowski and A.H. Knoll, eds. Boston, Elsevier, pp. 191–205.

Desikachary, T.V., and Prema, P. (1996). Silicoflagellates (Dictyochophyceae). *Bibliotheca Phycol.* **100:** 1–298.

de Vargas, C., Aubry, M-P., Probert, I., and Young, J. (2007). Origin and evolution of coccolithophores: From coastal hunters to oceanic farmers. *Evolution of Primary Producers in the Sea.* P.G. Falkowski and A.H. Knoll, eds. Boston, Elsevier, pp. 251–285.

Doney, S.C. (2006). The dangers of ocean acidification. *Sci. Am.* **294:** 58–65.

Edgar, L.A., and Pickett-Heaps, J.D. (1984). Diatom locomotion. *Progress in Phycological Research,* Vol. 3. F.E. Round, D.J. Chapman, eds. Bristol, UK, Biopress, pp. 47–88.

Edlund, M.B., and Stoermer, E.F. (1997). Ecological, evolutionary, and systematic significance of diatom life histories. *J. Phycol.* **33:** 897–918.

Ehara, M., Inagaki, Y., Watanabe, K.I., and Ohama, T. (2000). Phylogenetic analysis of diatom *cox*I genes and implications of a fluctuating GC content on mitochondrial genetic code evolution. *Curr. Genet.* **37:** 29–33.

Evans, K.M., Wortley, A.H., and Mann, D.G. (2007). An assessment of potential diatom "barcode" genes (*cox*1, *rbc*L, 18S and ITS rDNA) and their effectiveness in determining relationships in *Sellaphora* (Bacillariophyta). *Protist.* In press.

Evans, K.M., Kühn, S.F., and Hayes, P.K. (2005). High levels of genetic diversity and low levels of genetic differentiation in North Sea *Pseudo-nitzschia pungens* (Bacillariophyceae) populations. *J. Phycol.* **41:** 506–514.

Falciatore, A., and Bowler, C. (2002). Revealing the molecular secrets of marine diatoms. *Ann. Rev. Plant Biol.* **53:** 109–130.

Falkowski, P.G., Barber, R.T., and Smetacek, V. (1998). Biogeochemical controls and feedbacks on ocean primary production. *Science* **281:** 200–206.

Falkowski, P.G., Katz, M.E., Knoll, A.H., Quigg, A., Raven, J.A., Schofield, O., and Taylor, F.J.R. (2004a). The evolution of modern eukaryotic phytoplankton. *Science* **305:** 354–360.

Falkowski, P.G., and Raven, J.A. (2007). *Aquatic Photosynthesis*. Princeton, NJ, Princeton University Press.

Falkowski, P.G., Schofield, O., Katz, M.E., van de Schootbrugge, B., and Knoll, A.H. (2004b). Why is the land green and the ocean red? *Coccolithophores—From Molecular Processes to Global Impact*. H. Thierstein, J.R. Young, eds. Amsterdam, Elsevier, pp. 429–453.

Fast, N.M., Kissinger, J.C., Roos, D.S., and Keeling, P.J. (2001). Nuclear-encoded, plastid-targeted genes suggest a single common origin for apicomplexan and dinoflagellate plastids. *Mol. Biol. Evol.* **18**: 418–426.

Fenner, J. (1977). Cenozoic diatom biostratigraphy of the equatorial and southern Atlantic Ocean. *Init. Rep. Deep Sea Drill. Proj.* **39**: 491–623.

Fenner, J.M. (1985). Late Cretaceous to Oligocene planktic diatoms. Plankton Stratigraphy. H.M. Bolli, J.B. Saunders, and K. Perch-Nielsen, eds. Cambridge, UK, Cambridge University Press, pp. 713–762.

Finlay, B.J., Monaghan E.B., and Maberly, S.C. (2002). Hypothesis: the rate and scale of dispersal of freshwater diatom species is a function of their global abundance. *Protist* **153**: 261–273.

Finlay, B.J., Esteban, G.F., and Fenchel, T. (2004). Protist diversity is different? *Protist* **155**: 15–22.

Fistarol, G.O. (2004). *The Role of Allelopathy in Phytoplankton Ecology*. PhD thesis. Department of Biology and Environmental Science. University of Kalmar, Kalmar, Sweden.

Fistarol, G.O., Legrand, C., and Granéli, E. (2005). Allelopathic effect on a nutrient-limited phytoplankton species. *Aquat. Microb. Ecol.* **41**: 153–161.

Fox, M., and Sörhannus, U. (2004). The usefulness of the Rpo mitochondrial gene in assessing diatom evolution. *J. Euk. Microbiol.* **50**: 471–475.

Forti, A., and Schulz, P. (1932). Erste Mitteilung über Diatomeen aus dem Hannoverschen Gault. *Beih. Bot. Zentralbl.* **50**: 241–246.

French, F.W. III, and Hargraves, P.E. (1985). Spore formation in the life cycles of the diatoms *Chaetoceros diadema* and *Leptocylindrus danicus*. *J. Phycol.* **21**: 477–483.

Fryxell, G.A., and Medlin, L.K. (1981). Chain forming diatoms: evidence of parallel evolution in *Chaetoceros*. *Cryptogamie (Algologique)* **2**: 3–29.

Gallagher, J.C. (1983). Cell enlargement in *Skeletonema costatum* (Bacillariophyceae). *J. Phycol.* **19**: 539–542.

Geider, R.J., MacIntyre, H.L., and Kana, T.M. (1997). Dynamic model of phytoplankton growth and acclimation: responses of the balanced growth rate and the chlorophyll *a*: carbon ratio to light, nutrient limitation and temperature. *Mar. Ecol. Prog. Ser.* **148**: 187–200.

Gersonde, R., and Harwood, D.M. (1990). Lower Cretaceous diatoms from ODP Leg 113 site 693 (Weddell Sea) Part 1: vegetative cells. *Proc ODP Sci. Results* **113**: 365–402.

Gladenkov, A.Y., and Barron, J.A. (1995). Oligocene and early middle Miocene diatom biostratigraphy of Hole 884B. *Proc. ODP Sci. Results* **145**: 21–41.

Goertzen, L.R., and Theriot, E.C. (2003). Effects of outgroup selection, taxonomic sampling, character weighting, and combining data on interpretation of relationships among the heterokont algae. *J. Phycol.* **39**: 423–439.

Gombos, A.M., and Ciesielski, P.F. (1983). Late Eocene to Early Miocene diatoms from the Southwest Atlantic. *Init. Rep. Deep Sea Drill. Proj.* **71**: 583–634.

Graham, L.E., and Wilcox, L.W. (2000). *Algae*. Upper Saddle River, NJ, Prentice Hall.

Granum, E., Raven, J.A., and Leegood, R.C. (2005). How do marine diatoms fix 10 billion tonnes of inorganic carbon per year? *Can. J. Bot.* **83**: 898–908.

Grzebyk, D., Schofield, O., Vetriani, C., and Falkowski, P.G. (2003). The Mesozoic radiation of eukaryotic algae: the portable plastid hypothesis. *J. Phycol.* **39**: 259–267.

Guillou, L., Chrétiennot-Dinet, M.-J., Medlin, L.K., Claustre, H., Loiseaux-de Goër, S., and Vaulot, D. (1999). *Bolidomonas*: a new genus with two species belonging to a new algal class, the Bolidophyceae (Heterokonta). *J. Phycol.* **35**: 368–381.

Günther, F., and Folke, C. (1993). Characteristics of nested living systems. *J. Biol. Syst.* **1**: 257–274.

Hamm, C.E., Merkel, R., Springer, O., Jurkojc, P., Maier, C., Prechtel, K., and Smetacek, V. (2003). Architecture and material properties of diatom shells provide effective mechanical protection. *Nature* **421**: 841–843.

Hamm, C.E., and Smetacek, V. (2007). Armor: why, when, and how. *Evolution of Primary Producers in the Sea*. P.G. Falkowski and A.H. Knoll, eds. Boston, Elsevier, pp. 311–332.

Harding, I.C., and Lewis, J. (1994). Siliceous dinoflagellate thecal fossils from the Eocene of Barbados. *Paleontology* **37**: 825–840.

Harper, J.T., and Keeling, P.J. (2003). Nucleus-encoded, plastid-targeted glyceraldehyde-3-phosphate dehydrogenase (GAPDH) indicates a single origin for chromalveolate plastids. *Mol. Biol. Evol.* **20**: 1730–1735.

Harwood, D.M. (1988). Upper Cretaceous and lower Paleocene diatom and silicoflagellate biostratigraphy of Seymour Island, eastern Antarctic Peninsula. *Geol. Soc. Am. Mem.* **169**: 55–129.

Harwood, D.M., Chang, K.H., and Nikolaev, V.A. (2004). Late Jurassic to earliest Cretaceous diatoms from Jasong Synthem, Southern Korea: evidence for a terrestrial origin. *Abstracts, 18th International Diatom Symposium, Miedzyzdroje, Poland*. A. Witkowski, T. Radziejewska, B. Wawrzyniak-Wydrowska, G. Daniszewska-Kowalczyk, and M. Bak, eds. Szczecin, Poland, Szczecin University Press, p. 81.

Harwood, D.M., and Gersonde, R. (1990). Lower Cretaceous diatoms from ODP Leg 113 Site 693 (Weddell Sea). Part 2: resting spores, chrysophycean cysts, an endoskeletal dinoflagellate, and notes on the origin of diatoms. *Proc. ODP Sci. Results* **113**: 403–425.

Harwood, D.M., and Maruyama, T. (1992). Middle Eocene to Pleistocene diatom biostratigraphy of ODP Leg 120, Kerguelen Plateau. *Proc. ODP Sci. Results* **120**: 683–733.

Hasle, G.R. (1974). The "mucilage pore" of pennate diatoms. *Nova Hedwigia Beih.* **45**: 167–194.

Hasle, G.R. (1975). Some living marine species of the diatom family rhizosoleniaceae. *Nova Hedwigia Beih.* **53**: 99–152.

Hasle, G.R., and Syvertsen, E.E. (1985). *Thalassiosiropsis,* a new diatom genus from the fossil records. *Micropaleontology* **31**: 82–91.

Hays, J.D., Imbrie, J., and Shackleton, N.J. (1976). Variations in the earth's orbit: Pacemaker of the ice ages. *Science* **194**: 1121–1132.

Hays, G.C., Richardson, A.J., and Robinson, C. (2005). Climate change and marine plankton. *Trends Ecol. Evol.* **20**: 337–344.

Hazelaar, S., van der Strate, H.J., Gieskes, W.W.C., and Vrieling, E.G. (2005). Monitoring rapid valve formation in the pennate diatom *Navicula salinarum* (Bacillariophyceae). *J. Phycol.* **41**: 354–358.

Hench, L.L. (1989). Bioceramics and the origin of life. *J. Biomed. Mat. Res.* **23**: 685–703.

Hernández-Becerril, D.U., and Maeve-Del Castillo, M.E. (1996). The marine planktonic diatom *Rhizosolenia robusta* (Bacillariophyta): morphological studies support its transfer to a new genus, *Calyptrella* gen. nov. *Phycologia* **35**: 198–203.

Hildebrand, M., Dahlin, K., and Volcani, B.E. (1998). Characterization of silicon transporter gene family in *Cylindrotheca fusiformis:* sequences, expression analysis and identification of homologs in other diatoms. *Mol. Gen. Genet.* **260**: 480–486.

Hiltz, M., Bates, S.S., and Kaczmarska, I. (2000). Effect of light:dark cycles and cell apical length on the sexual reproduction of the pennate diatom *Pseudonitzschia multiseries* (Bacillariophyceae) in culture. *Phycologia* **39**: 59–66.

Ho., T.-Y. (2005). The trace metal composition of marine microalgae in cultures and natural assemblages. *Algal Cultures, Analogues of Blooms and Applications,* Vol. 1. D.V. Subba Rao, ed. Enfield, NH, Science Publishers, pp. 271–300.

Huisman, J., Johansson, A.M., Folmer, and Weissing, F.J. (2001). Towards a solution of the plankton paradox: the importance of physiology and life history. *Ecol. Lett.* **4**: 408–411.

Hutchinson, G.E. (1961). The paradox of the plankton. *Am. Naturalist* **95**: 137–145.

Ianora, A., Miralto, A., Poulet, S.A., Carotenuto, Y., Buttino, I., Romano, G., Casotti, R., Pohnert, G., Wichard, T., Colucci-D'Amato, L., Terrazzano, G., and Smetacek, V. (2004). Aldehyde suppression of copepod recruitment in blooms of a ubiquitous planktonic diatom. *Nature* **429**: 403–407.

Inouye, I., Hori, T., and Chihara, M. (1990). Absolute configuration analysis of the flagellar apparatus of *Pterosperma* christatum (Prasinophyceae) and consideration of its phylogenetic position. *J. Phycol.* **26**: 329–344.

Jeffrey, S.W., and Wright, S.W. (2005). Photosynthetic pigments in marine macroalgae: insights from cultures and the sea. *Algal Cultures, Analogues of Blooms and Applications,* Vol. 1. D.V. Subba Rao, ed. Enfield, NH, Science Publishers, pp. 33–90.

Jensen, K.G., Moestrup, Ø., and Schmid, A.-M.M. (2003). Ultrastructure of the male gametes of *Coscinodiscus wailesii* and *Chaetoceros laciniosus* (Bacillariophyceae). *Phycologia* **42**: 98–105.

John, U., Fensome, R.A., and Medlin, L.K. (2003). The application of a molecular clock based on molecular sequences and the fossil record to explain biogeographic distributions within the *Alexandrium tamarense* 'species complex' (dinophyceae). *Mol. Biol. Evol.* **20**: 1015–1027.

John, U., Mock, T., Valentin, K., Cembella, A.D., and Medlin, L.K. (2005). Dinoflagellates come from outer space but haptophytes and diatoms do not, Harmful Algae, 2002. *Proceedings of the Xth International Conference on Harmful Algae.* K.A. Steidinger, J.H. Landsberg, C.R. Tomas, and G.A. Vargo, eds. Florida Fish and Wildlife Conservation Commission and Intergovernmental Oceanographic Commission of UNESCO, 2003. St. Petersburg, FL.

Kaczmarska, I., Bates, S.S., Ehrman, J.M., and Leger, C. (2000). Fine structure of the gamete, auxospore and initial cell in the pennate diatom *Pseudo-nitzschia multiseries* (Bacillariophyta). *Nova Hedwigia* **71**: 337–357.

Kaczmarska, I., Beaton, M., Benoit, A.C., and Medlin, L.K. (2006). Evolution of the Thalassirosirales. *J. Phycol.* **42**: 121–138.

Kaczmarska, I., Ehrman, J.M., and Bates, S.S. (2001). A review of auxospore structure, ontogeny and diatom phylogeny. *Proceedings of the 16th International Diatom Symposium.* A. Economou-Amilli, ed. Athens, Greece, University of Athens, pp. 153–168.

Karp-Boss, L., Boss, E., and Jumars, P.A. (1996). Nutrient fluxes to planktonic osmotrophs in the presence of fluid motion. *Oceanogr. Mar. Biol. Ann. Rev.* **34**: 71–107.

Katz, M.E., Fennel, K., and Falkowski, P.G. (2007). Geochemical and biological consequences of phytoplankton. *Evolution of Primary Producers in the Sea.* P.G. Falkowski and A.H. Knoll, eds. Boston, Elsevier, pp. 405–429.

Katz, M.E., Finkel, Z.V., Grzebyk, D., Knoll, A.H., and Falkowski, P.G. (2005). Evolutionary trajectories and biogeochemical impacts of marine eukaryotic phytoplankton. *Ann. Rev. Ecol. Syst.* **35**: 523–556.

Kawachi, M., Inouye, I., Honda, D., O'Kelly, C.J., Bailey, J.C., Bidigare, R.R., and Andersen, R.A. (2002). The Pinguiophyceae classis nova, a new class of chromophyte algae whose members produce large amounts of omega-3 fatty acids. *Phycol. Res.* **50**: 31–47.

Keeling, P.J., Burger, G., Durnford, D.G., Lang, B.F., Lee, R.W., Pearlman, R.E., Roger, A.J., and Gray, M.W. (2005). The tree of eukaryotes. *Trends Ecol. Evol.* **20**: 670–676.

Kiorboe, T. (1993). Turbulence, phytoplankton cell size, and the structure of pelagic food-webs. *Adv. Mar. Biol.* **29**: 1–72.

Knoll, A.H. (1994). Proterozoic and Early Cambrian protists: evidence for accelerating evolutionary tempo. *Proc. Natl. Acad. Sci. U S A* **91**: 6743–6750.

Knoll, A.H. (1996). Archean and proterozoic paleontology, palynology: principles and applications. *Am. Assoc. Stratigr. Palynol. Found*. **1**: 51–80.

Knoll, A.H., Summons, R.E., Waldbauer, J.R., and Zumberge, J.E. (2007). The geological succession of primary producers in the oceans. *Evolution of Primary Producers in the Sea*. P.G. Falkowski and A.H. Knoll, eds. Boston, Elsevier, pp. 133–163.

Kooistra, W.H.C.F., Chepurnov, V., Medlin, L.K., De Stefano, M., Sabbe, K., and Mann, D.G. (2006). The phylogeny of the diatoms. *Plant Genome: Biodiversity and Evolution*. A.K. Sharma and A. Sharma, eds. Enfield, NH, Science Publishers, pp. 117–178.

Kooistra, W.H.C.F., De Stefano, M., Mann, D.G., Salma, N., and Medlin, L.K. (2003b). Phylogenetic position of *Toxarium*, a pennate-like lineage within centric diatoms (Bacillariophyceae). *J. Phycol*. **39**: 185–197.

Kooistra, W.H.C.F., Mann, D.G., and Medlin, L.K. (2003a). The phylogeny of the diatoms: a review. *Silica in Biological Systems*. W.E.G. Müller, ed. London, Elsevier Press, pp. 59–97.

Kooistra, W.H.C.F., and Medlin, L.K. (1996). Evolution of the diatoms (Bacillariophyta). IV. A reconstruction of their age from small subunit rRNA coding regions and fossil record. *Mol. Phylogenet. Evol*. **6**: 391–407.

Kröger, N., Bergsdorf, C., and Sumper, M. (1996). Frustulins: domain conservation in a protein family associated with diatom cell walls. *Eur. J. Biochem*. **239**: 259–264.

Kröger, N., Deutzmann, R., Bergsdorf, C., and Sumper, M. (2000). Species specific polyamines from diatoms control silica morphology. *Proc. Natl. Acad. Sci. U S A* **97**: 14133–14138.

Kröger, N., Deutzmann, R., and Sumper, M. (1999). Polycationic peptides from diatom biosilica that direct silica nanosphere formation. Science **286**: 1129–1132.

Kröger, N., Deutzmann, R., and Sumper, M. (2001). Silica precipitating peptides from diatoms: the chemical structure of silaffin 1A from *Cylindrotheca fusiformis*. *J. Biol. Chem*. **276**: 26066–26070.

Kröger, N., Lehnmann, G., Rachel, R., and Sumper, M. (1997). Characterization of a 200-kDa diatom protein that is specifically associated with a silica-based substructure of the cell wall. *Eur. J. Biochem*. **250**: 99–105.

Kröger, N., Lorenz, S., Brunner, E., and Sumper, M. (2002). Biosilica morphogenesis requires silaffin phosphorylation. *Science* **298**: 584–586.

Kröger, N., and Sumper, M. (1998). Diatom cell wall proteins and the cell biology of silica biomineralisation. *Protist* **149**: 213–219.

Kühn, S.F., and Brownlee, C. (2005). Membrane organisation and dynamics in the marine diatom *Coscinodiscus wailesii* (Bacillariophytceae). *Bot. Mar*. **48**: 297–305.

Kühn, S.F., Medlin, L., and Eller, G. (2004). Phylogenetic position of the parasitoid nanoflagellate *Pirsonia* inferred from nuclear-encoded Small subunit Ribosomal DNA and a description of *Pseudopirsonia* n. gen. and *Pseudopirsonia mucosa* (Drebes) comb. nov. *Protist* **155**: 143–156.

Kwiecinska, B. (2000). How the diatoms were found in the Proterozoic marbles at Przeworno. *The Origin and Early Evolution of Diatoms: Fossil, Molecular and Biogeographical Approaches*. A. Witkowski and J. Sieminska, eds. Cracow, Poland, W. Szafer Institute of Botany, Polish Academy of Sciences, pp. 281–289.

Lane, T.W., Saito, M.A., George, G.N., Pickering, I.J., Prince, R.C., and Morel, F.M. (2005). Biochemistry: a cadmium enzyme from a marine diatom. *Nature* **435**: 42.

Lavaud, J., Rousseau, B., and Etienne, A.L. (2004). General features of photoprotection by energy dissipation in planktonic diatoms (Bacillariophyceae). *J. Phycol*. **40**: 130–137.

Legrand, C., Regefors, K., Fistarol, G.O., and Granéli, E. (2003). Allelopathy in phytoplankton—biochemical, ecological and evolutionary aspects. *Phycologia* **42**: 406–419.

Lewin, J., and Hellebust, J.A. (1975). Heterotrophic nutrition of the marine pennate diatom *Navicula pavillardi* Hustedt. *Can. J. Microbiol*. **21**: 1335–1342.

Lewitus, A.J. (2005). Osmotrophy in marine microalgae. *Algal Cultures, Analogues of Blooms and Applications*, Vol. 1. D.V. Subba Rao, ed. Enfield, NH, Science Publishers, pp. 343–384.

Lisiecki, L.E., and Raymo, M.E. (2007). Plio-Pleistocene climate evolution: trends and transitions in glacial cycle dynamics. *Quatern. Sci. Rev*. **26**: 56–69.

Lundholm, N., Daugbjerg, N., and Moestrup, Ø. (2002b). Phylogeny of the Bacillariaceae with emphasis on the genus *Pseudo-nitzschia* (Bacillariophyceae) based on partial LSU rDNA. *Eur. J. Phycol*. **37**: 115–134.

Lundholm, N., Hansen, P.J., and Kotaki, Y. (2005). Lack of allelochemical effects of the domoic acid-producing marine diatom *Pseudo-nitzschia multiseries*. *Mar. Ecol. Prog. Ser*. **288**: 21–33.

Lundholm, N., Hasle, G.R., Fryxell, G.A., and Hargraves, P.E. (2002a). Morphology, phylogeny and taxonomy of species within the *Pseudo-nitzschia americana* complex (Bacillariophyceae) with descriptions of two new species, *Pseudo-nitzschia brasiliana* and *Pseudo-nitzschia linea*. *Phycologia* **41**: 480–497.

Lundholm, N., Moestrup, Ø., Hasle, G.R., and Hoef-Emden, K. (2003). A study of the *P. pseudodelicatissima/ cuspidata* complex (Bacillariophyceae): what is *P. pseudodelicatissima*? *J. Phycol*. **39**: 797–813.

Mann, D.G. (1984). An ontogenetic approach to diatom systematics. *Proceedings of the 7th International Diatom Symposium*. D.G. Mann, ed. Koenigstein, Germany, O Koeltz, pp. 113–144.

Mann, D.G. (1996). Chloroplast morphology, movements and inheritance in diatoms. *Cytology, Genetics and Molecular Biology of Algae*. B.R. Chaudhary and S.B. Agarwal, eds. Amsterdam, SPB Publishing, pp. 249–274.

Mann, D.G. (1999). The species concept in diatoms. *Phycologia* **38**: 437–495.

Mann, D.G., and Chepurnov, V.A. (2004). What have the Romans ever done for us? The past and future contribution of culture studies to diatom systematics. *Nova Hedwigia* **79**: 237–291.

Mann, D.G., and Droop, S.J.M. (1996). Biodiversity, biogeography and conservation of diatoms. Biogeography of freshwater algae. *Hydrobiologia* **336**: 19–32.

Mann, D.G., and Marchant, H.J. (1989). The origins of the diatom and its life cycle. *The Chromophyte Algae: Problems and Perspectives*. J.C. Green, B.S.C. Leadbeater, and W.L. Diver, eds. Oxford, Clarendon Press, pp. 307–323.

Mann, D.G., McDonald, S.M., Bayer, M.M., Droop, S.J.M., Chepurnov, V.A., Loke, R.E., Ciobanu, A., and du Buf, J.M.H. (2004). Morphometric analysis, ultrastructure and mating data provide evidence for five new species of *Sellaphora* (Bacillariophyceae). *Phycologia* **43**: 459–482.

Martin-Jezequel, V., Hildebrand, M., and Brzezinski, M.A. (2000). Silicon metabolism in diatoms: Implications for growth. *J. Phycol.* **36**: 821–840.

Massana, R., Balagué, V., Guillou, L., and Pedrós-Alió, C. (2004). Picoeukaryotic diversity in an oligotrophic coastal site studied by molecular and culturing approaches. *FEMS Microbiol. Ecol.* **50**: 231–243.

Mayama, S., and Kuriyama, A. (2002). Diversity of mineral cell coverings and their formation process: a review focused on the siliceous coverings. *J. Plant. Res.* **115**: 289–295.

McCartney, K., Wise, E.W., Harwood, D.H., and Gersonde, R. (1990). Enigmatic lower Albian silicoflagellates from ODP Site 693: Progenitors of the Order Silicoflagellata? *Proc. ODP Sci. Results* **113**: 427–442.

McQuoid, M.R., and Hobson, L.A. (1996). Diatom resting stages. *J. Phycol.* **32**: 889–902.

Medlin, L.K. (1991). Evidence for parallel evolution in some pennate diatom genera. *Diat. Res.* **6**: 109–124.

Medlin, L.K. (2002). Why silica or better yet why not silica? Speculations as to why the diatoms utilize silica. *Diat. Res.* **17**: 453–459.

Medlin, L.K., Crawford, R.M., and Andersen, R.A. (1986). Histochemical and ultrastructural evidence of the labiate process in the movement of centric diatoms. *Br. Phycol. J.* **21**: 297–301.

Medlin, L.K., Elwood, H.J., Stickel, S., and Sogin, M.L. (1991). Morphological and genetic variation within the diatom *Skeletonema costatum* (Bacillariophyta)—evidence for a new species, *Skeletonema pseudocostatum*. *J. Phycol.* **27**: 514–524.

Medlin, L.K., Gersonde, R., Kooistra, W.H.C.F., and Wellbrock, U. (1996a). Evolution of the diatoms (Bacillariophyta). II. Nuclear-encoded small-subunit

rRNA sequence comparisons confirm a paraphyletic origin for the centric diatoms. *Mol. Biol. Evol.* **13**: 67–75.

Medlin, L.K., and Kaczmarska, I. (2004). Evolution of the diatoms: V. Morphological and cytological support for the major clades and a taxonomic revision. *Phycologia* **43**: 245–270.

Medlin, L.K., Kooistra, W.H.C.F., Gersonde, R., and Wellbrock, U. (1996b). Evolution of the diatoms (Bacillariophyta). III. Molecular evidence for the origin of the Thalassiosirales. *Nova Hedwigia* **11**: 221–234.

Medlin, L.K., Kooistra, W.H.C.F., Mann, D.G., Muyzer, G., Chepurnov, V., and Sabbe, K. (2004). Evolution of the diatoms: VI. Insight into the pennates. *Abstracts, 18th International Diatom Symposium, Miedzyzdroje, Poland*. A. Witkowski, T. Radziejewska, B. Wawrzyniak-Wydrowska, G. Daniszewska-Kowalczyk, and M. Bak, eds. Szeczecin, Poland, Szczecin University Press, p. 101.

Medlin, L.K., Kooistra, W.H.C.F., Potter, D., Saunders, G.W., and Andersen, R.A. (1997). Phylogenetic relationships of the "golden algae" (hepatophytes, heterokont chrysophytes) and their plastids. *Plant Syst. Evol.* Suppl. **11**: 187–210.

Medlin, L.K., Kooistra, W.H.C.F., and Schmid, A.-M.M. (2000). A review of the evolution of the diatoms - a total approach using molecules, morphology and geology. *The Origin and Early Evolution of the Diatoms: Fossil, Molecular and Biogeographical Approaches*. A. Witkowski, and J. Siemimska, eds. Cracow, Poland, Szafer Institute of Botany, Polish Academy of Science, pp. 13–35.

Medlin, L.K., Metflies, K., Mehl, H., Wiltshire, K., and Valentin, K. (2006). Picoplankton diversity at the Helgoland Time Series site as assessed by three molecular methods. *Microbial Ecol.* **167**: 1432–1451.

Medlin, L.K., Williams, D.M., and Sims, P.A. (1993). The evolution of the diatoms (Bacillariophyta). I. Origin of the group and assessment of the monophyly of its major divisions. *Eur. J. Phycol.* **28**: 261–275.

Meinke, D.W., Cherry, J.M., Dean, C., Rounsley, S.D., and Koornneef, M. (1998). *Arabidopsis thaliana*; a model plant for genome analysis. *Science* **282**: 662–682.

Mikrjukov, K.A., and Patterson, D.J. (2001). Taxonomy and phylogeny of Heliozoa. III. Actinophryids. *Acta Protozool.* **40**: 3–25.

Milankovitch, M. (1920). *Theorie mathematique des phenomenes thermiques produits par la radiation solaire*. Paris, France, Gauthier-Villars.

Miralto, A., Barone, G., Romano, G., Poulet, S.A., Ianora, A., Russo, G.L., Buttino, I., Mazzarella, G., Laabir, M., Cabrini, M., and Giacobbe, M.G. (1999). The insidious effect of diatoms on copepod reproduction. *Nature* **402**: 173–176.

Mitchell, B.G., and Holm-Hansen, O. (1991). Observations and modeling of the Antarctic phytoplankton crop in relation to mixing depth. *Deep-Sea Res.* **38**: 981–1008.

Mock, T., Krell, A., Glöckner, G., Kolukisaoglu, Ü., and Valentin, K. (2006). Analysis of expressed sequence tags (ESTs) from the polar diatom *Fragilariopsis cylindrus*. *J. Phycol.* In press.

Mock, T., and Valentin, K. (2004). Photosynthesis and cold acclimation—molecular evidence from a polar diatom. *J. Phycol.* **40:** 732–741.

Moldowan, J.M., Fago, F.J., Lee, C.Y., Jacobson, S.R., Watt, D.S., Slougui, N.-E., Jeganathan, A., and Young, D.C. (1990). Sedimentary 24-n-propylcholestanes, molecular fossils diagnostic of marine algae. *Science* 247: 309–312.

Moldowan, J.M., and Jacobson, S.R. (2000). Chemical signals for early evolution of major taxa: biosignatures and taxon-specific biomarkers. *Int. Geol. Rev.* 42: 805–812.

Montresor, M., and Lewis, J. (2005). Phases, stages and shifts in the life cycles of marine phytoplankton. *Algal Cultures, Analogues of Blooms and Applications*, Vol. 1. D.V. Subba Rao, ed. Enfield, NH, Science Publishers, pp. 91–130.

Montresor, M., Sgrossi, S., Procaccini, G., and Kooistra, W.H.C.F. (2003). Intraspecific diversity in *Scrippsiella trochoidea* (Dinophyceae): evidence for cryptic species. *Phycologia* 42: 56–70.

Montsant, A., Jabbari, K., Maheswari, U., and Bowler, C. (2005). Comparative genomics of the pennate diatom *Phaeodactylum tricornutum*. *Plant Physiol.* **137:** 500–513.

Nagasaki, K., Tomaru, Y., Katanozaka, N., Shirai, Y., Nishida, K., Itakura, S., and Yamaguchi, M. (2004). Isolation and characterization of a novel single-stranded RNA virus infecting the bloom-forming diatom *Rhizosolenia setigera*. *Appl. Environ. Microbiol.* **70:** 704–711.

Nagasaki, K., Tomaru, Y., Tomaru, Y., Takao, Y., Nishida, K., Shirai, Y., Suzuki, H., and Nagumo, T. (2005). Previously unknown virus infects marine diatom. *Appl. Environ. Microbiol.* **71:** 3528–3535.

Nehring, S. (1998). Non-indigenous phytoplankton species in the North Sea: supposed region of origin and possible transport vector. *Arch. Fish. Mar. Res.* **46:** 181–194.

Nikolaev, S.I., Berney, C., Fahrni, J.F., Bolivar, I., Polet, S., Mylnikov, A.P., Aleshin, V.V., Petrov, N.B., and Pawlowski, J. (2004). The twilight of Heliozoa and rise of Rhizaria, an emerging supergroup of amoeboid eukaryotes. *Proc. Natl. Acad. Sci. U S A* **101:** 8066–8071.

Nikolaev, V.A., and Harwood, D.M. (1997). New process, genus and family of Lower Cretaceous diatoms from Australia. *Diat. Res.* **12:** 43–53.

Nikolaev, V.A., and Harwood, D.M. (2000). Diversity and system of classification in centric diatoms. *The Origin and Early Evolution of the Diatoms: Fossil, Molecular and Biogeographical Approaches.* A. Witkowski and J. Siemimska, eds. Cracow, Poland, W. Szafer Institute of Botany, Polish Academy of Sciences, pp. 37–53.

Nilsson, C., and Sundbäck, K. (1996). Amino acid uptake in natural sediment-living microalgal assemblages studied by microautoradiography. *Hydrobiologia* 332: 119–129.

Not, F., Latasa, M., Marie, D., Cariou, T., Vaulot, D., and Simon, N. (2004). A single species, *Micromonas pusilla* (Prasinophyceae), dominates the Eukaryotic picoplankton in the western English Channel. *Appl. Env. Microbiol.* **70:** 4064–4072.

Palmer, J.D. (2003). The symbiotic birth and spread of plastids: how many times and whodunit? *J. Phycol.* **39:** 4–11.

Paterson, D.M., and Hagerthey, S.E. (2001). Microphytobenthos in contrasting coastal ecosystems: biology and dynamics. *Ecological Comparisons of Sedimentary Shores. Ecological Studies: Analysis and Synthesis, 151.* K. Reise, ed. Berlin, Springer-Verlag, pp. 105–125.

Pickett-Heaps, J.D. (1998a). Cell division and morphogenesis of the centric diatom *Chaetoceros decipiens* (Bacillariophyceae) I. Living cells. *J. Phycol.* **34:** 989–994.

Pickett-Heaps, J.D. (1998b). Cell division and morphogenesis of the centric diatom *Chaetoceros decipiens* (Bacillariophyceae) II. Electron microscopy and a new paradigm for tip growth. *J. Phycol.* **34:** 995–1004.

Pickett-Heaps, J.D., Hill, D.R.A., and Wetherbee, R. (1986). Cellular movement in the centric diatom *Odontella sinensis*. *J. Phycol.* **22:** 334–339.

Pickett-Heaps, J.D., Schmid, A-M.M., and Edgar, L.A. (1990). The cell biology of diatom valve formation. *Progr. Phycol. Res.* **7:** 1–168.

Piperno, D.P., and Sues, H.-D. (2005). Dinosaurs dined on grass. *Science* 310:1126–1128.

Pohnert, G. (2000). Wound-activated chemical defense in unicellular planktonic algae. *Angew. Chem. Int. Ed.* **39:** 4352–4354.

Polet, S., Berney, C., Fahrni, J., and Pawlowski, J. (2004). Small-subunit ribosomal RNA gene sequences of *Phaeodarea* challenge the monophyly of Haeckel's Radiolaria. *Protist* 155: 53–63.

Poulsen, N., Sumper, M., and Kröger, N. (2003). Biosilica formation in diatoms: characterization of native silaffin-2 and its role in silica morphogenesis. *Proc. Natl. Acad. Sci. U S A* **100:** 12075–12080.

Rampen, S.W., Schouten, S., Abbas, B., Elda Pannoto, F., Muyzer, G., Campbell, C.N., Fehling, J., and Sinninghe Damsté, J.S. (2007). On the origin of 24-norcholestanes and their use as diagnostic biomarkers. *Geology* 35: 419–422.

Rao, V.N.R., and Desikachary, T.V. (1970). MacDonald-Pfitzer hypothesis and cell size in diatoms. *Nova Hedwigia Beih.* 31: 485–493.

Rau, G.H., Riebesell, U., and Wolf-Gladrow, D.A. (1997). CO2-dependent photosynthetic 13C fractionation in the ocean: A model versus observations. *Global Biogeochem. Cycles* 11: 267.

Raup, D.M., and Sepkoski, J.J. Jr. (1982). Mass extinctions in the marine fossil record. *Science* **215**: 1501–1503.

Raven, J.A. (1983). The transport and function of silicon in plants. *Biol. Rev.* **58**: 179–207.

Raven, J.A. (1997). The vacuole: a cost-benefit analysis. *Adv. Bot. Res. Adv. Plant Pathol.* **25**: 9–86.

Raven, J.A., and Waite, A.M. (2004). The evolution of silicification in diatoms: inescapable sinking and sinking as escape? *New Phytol.* **162**: 45–61.

Reinfelder, J.R., Kraepiel, A.M.L., and Morel, F.M.M. (2000). Unicellular C-4 photosynthesis in a marine diatom. *Nature* **407**: 996–999.

Reinfelder, J.R., Milligan, A.J., and Morel, F.M.M. (2004). The role of the C4 pathway in carbon accumulation and fixation in a marine diatom. *Plant Physiol.* **135**: 2106–2111.

Reynolds, C.S. (1984). Phytoplankton periodicity—the interactions of form, function and environmental variability. *Freshw. Biol.* **14**: 111–142.

Reynolds, C.S., and Walsby, A.E. (1975). Water-blooms. *Biol. Rev.* **50**: 437–481.

Retallack, G.J. (2001). Neogene expansion of the North American prairie. *Palaios* **12**: 380–390.

Rospondek, M.J., Köster, J., and Sinninghe Damsté, J.S. (2000). Organic molecular fossils of diatoms. *The Origin and Early Evolution of the Diatoms: Fossil, Molecular and Biogeographical Approaches.* A. Witkowski and J. Siemimska, eds. Cracow, Poland, W. Szafer Institute of Botany, Polish Academy of Sciences, pp. 123–135.

Rothpletz, A. (1896). Über die Flysch-Fucoiden und einige andere fossile Algen, sowie über liasische Diatomeen führende Hornschwämme. *Zeitschrift der Deutschen Geologischen Gesellschaft* **48**: 854–915.

Round, F.E., Crawford, R.M., and Mann, D.G. (1990). *The Diatoms. Biology and Morphology of the Genera.* Cambridge, Cambridge University Press.

Rynearson, T.A., and Armbrust, E.V. (2004). Genetic differentiation among populations of the planktonic diatom *Ditylum brightwellii* (Bacillariophyceae). *J. Phycol.* **40**: 34–43.

Sáez, A.G., Probert, I., Geisen, M., Quinn, P., Young, J.R., and Medlin, L.K. (2003). Pseudo-cryptic speciation in coccolithophores, *Proc. Natl. Acad. Sci. U S A* **100**: 7163–7168.

Sarno, D., Kooistra, W.H.C.F., Hargraves, P.E., Zingone, A. (2007). Diversity in the genus *Skeletonema* (Bacillariophyceae). III. Phylogenetic position and morphology of *Skeletonema costatum* and *Skeletonema grevillei*, with the description of *Skeletonema ardens* sp. nov. *J. Phycol.* **43**: 156–170.

Sarno, D., Kooistra, W.H.C.F., Medlin, L.K., Percopo, I., and Zingone, A. (2005). Diversity in the genus *Skeletonema* (Bacillariophyceae): II. An assessment of the taxonomy of *S. costatum*-like species, with the description of four new species. *J. Phycol.* **41**: 151–176.

Scala, S., Carels, N., Falciatore, A., Chiusano, M.L., and Bowler, C. (2002). Genome properties of the diatom *Phaeodactylum tricornutum. Plant Physiol.* **129**: 993–1002.

Schrader, H.-J., and Fenner, J. (1976). Norwegian sea Cenozoic diatom biostratigraphy and taxonomy. Part 1. *Init. Rep. Deep Sea Drill. Proj.* **38**: 921–1099.

Schmid, A.-M.M. (1988). The special Golgi-ER-mitochondrium unit in the diatom genus *Coscinodiscus. Plant Syst. Evol.* **158**: 211–233.

Schöne, H. (1979). Untersuchungen zur ökologischen Bedeutung des Seegangs für das Plankton mit besonderer Berücksichtigung mariner Kieselalgen. *Int. Revue ges. Hydrobiol.* **55**: 595–677.

Sieminska, J. (2000). The discoveries of diatoms older than the Cretaceous. *The Origin and Early Evolution of Diatoms: Fossil, Molecular and Biogeographical Approaches.* A. Witkowski, and J. Sieminska, eds. Cracow, Poland, Szafer Institute of Botany, Polish Academy of Sciences, pp. 55–74.

Simonsen, R. (1979). The diatom system: ideas on phylogeny. *Bacillaria* **2**: 9–71.

Sims, P.A., Mann, D.G., and Medlin, L.K. (2006). Evolution of the diatoms: insights from fossil, biological and molecular data. *Phycologia* **45**. In press.

Sinninghe Damsté, J.S., Muyzer, G., Abbas, B., Rampen, S.W., Masse, G., Allard, W.G., Belt, S.T., Robert, J.-M., Rowland, S.J., Moldowan, J.M., Barbanti, S.M., Fago, F.J., Denisevich, P., Dahl, J., Trindade, L.A.F., and Schouten, S. (2004). The rise of the rhizosolenoid diatoms. *Science* **304**: 584–587.

Smayda, T.J. (1998). Patterns of variability characterizing marine phytoplankton, with examples from Narragansett Bay. *ICES J. Mar. Sci.* **55**: 562–573.

Smetacek, V. (1999). Diatoms and the ocean carbon cycle. *Protist* **150**: 25–32.

Sörhannus, U., Gasse, F., Perasso, R., and Baroin-Tourancheau, A. (1995). A preliminary phylogeny of diatoms based on 28s ribosomal RNA sequence data. *Phycologia* **34**: 65–73.

Sörhannus, U. (2004). Diatom phylogenetics inferred based on direct optimization of nuclear-encoded SSU rRNA sequences. *Cladistics*, **20**: 487–497.

Sournia, A. (1995). Red tide and toxic marine phytoplankton of the world ocean: an inquiry into biodiversity. *Harmful Marine Algal Blooms.* P. Lassus, G. Arzul, E. Erard-Le-Denn, P. Gentien, and C. Marcaillou-Le-Baut, eds. Paris, Technique et Documentation, Lavoisier, pp. 103–112.

Sparrow, F.K. Jr. (1960). *Aquatic Phycomycetes.* Ann Arbor, University of Michigan Press.

Stiller, J.W., Reel, D.C., and Johnson, J.C. (2003). A single origin of plastids revisited: convergent evolution in organellar genome content. *J. Phycol.* **39**: 95–105.

Stoermer, E.F., and Smol, J.P. (1999). *The Diatoms. Applications for the Environmental and Earth Sciences.* Cambridge, UK, Cambridge University Press.

Strel'nikova, N.I. (1974). *Diatomei pozdnego mela (zapadnaya Sibir').* [*Diatoms of the late Cretaceous (western Siberia)*]. Moscow, Izdatel'stvo Nauka.

Strel'nikova, N.I. (1975). Diatoms of the Cretaceous Period. *Nova Hedwigia, Beiheft* **53**: 311–321.

Strel'nikova, N.I. (1992). *Paleogenovye diatomovye vodorosli. [Paleogene diatom algae].* St. Petersburg, St. Petersburg University.

Suda, S., Watanabe, M.M., and Inouye, I. (1989). Evidence for sexual reproduction in the primitive green alga *Nephroselmis olivacea* (Prasinophyceae). *J. Phycol.* **25:** 596–600.

Sumper, M., and Kroger, N. (2004). Silica formation in diatoms: the function of long-chain polyamines and silaffins. *J. Mat. Chem.* **14:** 2059–2065.

Takabayashi, M., Wilkerson, F.P., and Robertson, D. (2005). Response of glutamine synthetase gene transcription and enzyme activity to external nitrogen sources in the diatom *Skeletonema costatum* (Bacillariophyceae). *J. Phycol.* **41:** 84–94.

Tapia, P.M., and Harwood, D.M. (2002). Upper Cretaceous diatom biostratigraphy of the Arctic Archipelago and northern continental margin, Canada. *Micropaleontology* **48:** 303–342.

Tillmann, U., Alpermann, T., John, U., and Cembella, A. (2007). Allelochemical interactions and short-term effects of the dinoflagellate *Alexandrium* on selected photoautotrophic and heterotrophic protists. *Harmful Algae.* In press.

Tozzi, S., Schofield, O., and Falkowski, P. (2004). Historical climate change and ocean turbulence as selective agents for key phytoplankton functional groups. *Mar. Ecol. Prog. Ser.* **274:** 123–132.

Trappan, H., and Loeblich, A.R. (1973). Evolution of the oceanic plankton. *Earth-Sci. Rev.* **9:** 207–240.

Tuchman, N.C. (1996). The role of heterotrophy in algae. *Algal Ecology: Freshwater Benthic Ecosystems.* R.J. Stevenson, M.L. Bothwell, and R.L. Lowe, eds. San Diego, CA, Academic Press, pp. 299–319.

Ubeleis, I. (1957). Osmotischer Wert, Zucker-und-Harnstoffpermeabilität einiger Diatomeen. *Österr. Akad. Wiss. Math.-Naturw. Kl. Sitzber. Abt.* **166:** 395–433.

Van de Meene, A.M.L., and Pickett-Heaps, J (2004). Valve morphogenesis in the centric diatom *Rhizosolenia setigera* (Bacillariophyceae, Centrales) and its taxonomic implications. *Eur. J. Phycol.* **39:** 93–104.

Van de Poll, W.H., Vrieling, E.G., and Gieskes, W.W.C. (1999). Location and expression of frustulins in the pennate diatoms *Cylindrotheca fusiformis, Navicula pelliculosa,* and *Navicula salinarum* (Bacillariophyceae). *J. Phycol.* **35:** 1044–1053.

Van den Hoek, C., Mann, D.G., and Jahns, H.M. (1995). *Algae: An Introduction to Phycology.* Cambridge, UK, Cambridge University Press.

Van den Hoff, J., Buton, H.R., and Vesk, M. (1989). An encysting stage bearing a new scale type, of the Antarctic prasinophyte *Pyramimonas gelidicola* and its paleolimnological and taxonomic significance. *J. Phycol.* **25:** 446–454.

Villareal, T.A. (1990). Laboratory cultivation and preliminary characterization of the *Rhizosolenia* (Bacillariophyceae)-*Richelia* (Cyanophyceae) symbiosis. *Mar. Ecol.* **11:** 117–132.

Villareal, T.A. (1991). Nitrogen-fixation by the cyanobacterial symbiont of the diatom genus *Hemiaulus*. *Mar. Ecol. Prog. Ser.* **76:** 201–204.

Villareal, T.A., Altabet, M.A., and Culver-Rymsza, K. (1993). Nitrogen transport by vertically migrating diatoms mats in the North Pacific Ocean. *Nature* **363:** 709–712.

Von Rad, U., Riech, V., and Rosch, H. (1977). Silica diagenesis in continental marginal sediments off Northwest Africa. *Init. Rep. Deep Sea Drill. Proj.* **41:** 879–905.

Von Stosch, H.A. (1965). Manipulierung der Zellgrösse von Diatomeen im Experiment. *Phycologia* **5:** 21–44.

Von Stosch, H.A. (1982). On auxospore envelopes in diatoms. *Bacillaria* **5:** 127–156.

Vrieling, E.G., Gieskes, W.W.C., and Beelen, T.P.M. (1999). Silicon deposition in diatoms: Control by the pH inside the silica deposition vesicle. *J. Phycol.* **35:** 548–559.

Walsby, A.E., and Xypolyta, A. (1977). The form resistance of chitan fibres attached to the cells of *Thalassiosira fluviatilis* Hustedt. *Br. Phycol. J.* **12:** 215–223.

Wichard, T., Poulet, S., Halsband-Lenk, C., Albaina, A., Harris, R.P., Liu, D., and Pohnert, G. (2005). Survey of the chemical defence potential of diatoms: screening of fifty-one species for a, b, c, d-unsaturated aldehydes. *J. Chem. Ecol.* **31:** 949–958.

Wilson, J.T. (1966). Did the Atlantic close and then re-open? *Nature* **211:** 676–681.

Worsley, T.R., Nance, R.D., and Moody, J.B. (1986). Tectonic cycles and the history of the earth's biogeochemical and paleoceanographic record. *Paleoceanography* **1:** 233–263.

Xiao, S., Knoll, A.H., and Yuan, X. (1998). Morphological reconstruction of *Maiohephyton bifurcatum*. A possible brown alga from the Doushantuo Formation (Neoproterozoic), South China, and its implication for Stramenopile evolution. *J. Paleont.* **72:** 1072–1086.

Yoon, H.S., Hackett, J.D., and Bhattacharya, D. (2002a). A single origin of the peridinin and fucoxanthin-containing plastids in dinoflagellates through tertiary endosymbiosis. *Proc. Natl. Acad. Sci. U S A* **99:** 11724–11729.

Yoon, H.S., Hackett, J., Ciniglia, C., Pinto, G., and Bhattacharya, D. (2004). A molecular timeline for the origin of photosynthetic eukaryotes. *Mol. Biol. Evol.* **21:** 809–818.

Yoon, H.S., Hackett, J.D., Pinto, G., and Bhattacharya, D. (2002b). The single ancient origin of chromid plastids. *Proc. Natl. Acad. Sci. U S A* **99:** 15507–15512.

Yoshida, M., Noël, M.-H., Nakayama, T., Naganuma, T., and Inouye, I. (2006). A haptophyte bearing siliceous scales: ultrastructure and phylogenetic position of *Hyalolithus neolepis* gen. et sp. nov. (Prymnesiophyceae, Haptophyta). *Protist* **157:** 213–234.

12

Origin and Evolution of Coccolithophores: From Coastal Hunters to Oceanic Farmers

COLOMBAN DE VARGAS, MARIE-PIERRE AUBRY,
IAN PROBERT, AND JEREMY YOUNG

I. COCCOLITHOPHORES AND THE BIOSPHERE

The coccolithophores are calcifying protists that have formed a significant part of the oceanic phytoplankton since the Jurassic. Their role in regulating the Earth system is considerable. Through their secretion of a tiny composite exoskeleton (the *coccosphere* made of multiple *coccoliths*), the coccolithophores are estimated to be responsible for about half of *all* modern precipitation of $CaCO_3$ in the oceans (Milliman 1993). Coccolithophores thus play a primary role in the global carbon cycle (Figure 1). The ecological and biogeochemical impacts of their skeletons are multiple and act on a wide range of ecological to geological time scales. On ecological time scales, coccolithophore biomineralization plays a major role in controlling the alkalinity and carbonate chemistry of the photic zone of the world ocean. Counterintuitively, the precipitation of carbonate is a source of CO_2 for the upper ocean and atmosphere (see Figure 1B). On the other hand, the biogenic carbonate produced by coccolithophores constitutes an ideal material for aggregating with the huge reservoir of particulate organic carbon

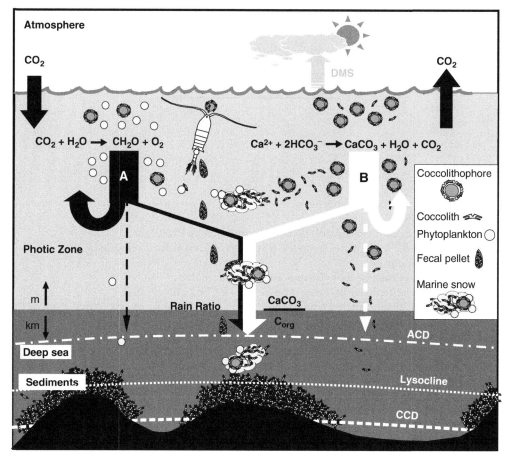

FIGURE 1. Role of coccolithophores in biogeochemical cycles. Through the production of their $CaCO_3$ coccoliths, coccolithophores play a key role in the global carbon cycling. Although they thrive in the photic layer of the world ocean, the coccolithophores actively participate in gas exchange (CO_2, DMS) between seawater and the atmosphere and the export of organic matter and carbonate to deep-oceanic layers and deep-sea sediments. They are the main actors of the *carbonate counter-pump* (**B**), which, through the *calcification* reaction, is a short-term source of atmospheric CO_2. Via the ballasting effect of their coccoliths on marine snow, coccolithophores are also a main driver of the *organic carbon pump* (**A**), which removes CO_2 from the atmosphere. Thus, organic and carbonate pumps are tightly coupled through coccolithophore biomineralization. Ultimately, certain types of coccoliths particularly resistant to dissolution are deposited at the seafloor, where they have built a remarkable fossil archive for the last 220 million years. The three main carbonate dissolution horizons are depicted: ACD, aragonite compensation depth; Lysocline (complete dissolution of planktic foraminifera); and CCD, calcite compensation depth. See text for more details. (Inspired from Rost and Riebesell 2004). (See color plate.)

created by photosynthesis in the upper oceanic layers. The accumulation of coccoliths into marine snow ballasts organic matter that otherwise would not sink to deep-oceanic layers and, potentially, to the deep seafloor. According to Honjo *et al.* (in press), coccoliths are the main driver of the open ocean organic carbon pump (see Figure 1A), which removes CO_2 from the atmosphere. In fact, the effect of coccolith ballasting on atmospheric CO_2 concentration could outweigh CO_2 output from biomineralization. Overall, on geological time scales, certain types of coccoliths particularly resistant to dissolution (~30% of the modern diversity, according to Young *et al.* [2005]) slowly accumulate in deep-oceanic sediments, at rates of less than 10 mm/thousand years (Ka) to more than 100 mm/Ka (Baumann *et al.* 2004). In the Cretaceous, when they started proliferating in the open oceans, the coccolithophores were responsible for switching the major site of global carbonate deposition from shallow seas to the deep ocean for the first time in the history of the Earth (Hay 2004), thus revolutionizing the regulation of ocean carbon chemistry (Ridgwell and Zeebe 2005). Since this time, coccoliths have been the prime contributors to the kilometers-thick accumulation of calcareous ooze covering ~35% of the ocean floor. This carbonate deposit is one of the main stabilizing components of the Earth system via the mechanism of *carbonate compensation* (Broecker and Peng 1987); its fate is eventually to be subducted into the mantle of the Earth, thus depleting carbon from its surface for millions of years.

Overall, the evolutionary and ecological success of coccolithophores for the last 220 million years have literally transformed the fate of inorganic and organic carbon in the Earth system, leading to a global decrease in the saturation state of seawater with respect to carbonate minerals (Ridgwell 2005) and participating in the long-term increase of atmospheric O_2 (Falkowski *et al.* 2005). The biological revolution underlying these long-term biogeochemical changes occurred when certain *haptophyte* protists evolved the ability to genetically control the intracellular nucleation and growth of $CaCO_3$ crystals on pre-existing organic scales, forming tiny, exquisitely sculptured skeletal plates: the *coccoliths*. Since this invention, the coccolithophores have diversified into more than 4000 morphological species, most of which are now extinct. Although apparently complex, coccolithophore biomineralization (or *coccolithogenesis*) appears to be a rather versatile process that can be quantitatively and qualitatively regulated, depending on, for instance, environmental conditions or the stage in the life cycle of the cells. Coccolithogenesis has been modulated multiple times in the course of coccolithophore evolution, either through the continuing innovation of remarkable morphostructures or, in some lineages, even via the complete shut down and possible reinvention of the process itself (see later). Here, we offer an up-to-date summary of coccolithophore evolution, integrating recent stratophenetic, molecular phylogenetic, biogeochemical, and biological data. We discuss the origin and nature of the haptophyte ancestors of coccolithophores, the origin of coccolithophores, and the onset(s) of calcification and illustrate different evolutionary trajectories that succeeding lineages have followed. This evolutionary scheme is then correlated to abiotic and biotic records of historical change in the Earth system, allowing us to evaluate the various extrinsic and possibly intrinsic genomic forces that have driven coccolithophore evolution and the resulting feedbacks of their evolution on the ecosystem. Finally, based on our interpretations of coccolithophore evolutionary history, we envision an uncertain future for this clade in the high CO_2 and high Mg/low Ca world of the Anthropocene.

II. WHAT IS A COCCOLITHOPHORE?

The term *coccolith* was coined by Huxley (1857) for mineral bodies resembling coccoid cells, which he observed in deep-sea sediments with a relatively low magnification microscope. This word, meaning literally "spherical stone," is a rather unsubtle description for remarkably diverse and exquisitely sculptured calcite platelets (Figure 2). Wallich (1861, 1877) first described the association of coccoliths on single spherical structures, which he termed *coccospheres*, and Lohmann (1902) introduced the word *coccolithophore* for the organisms producing these structures, which appeared to be part of a larger entity of eukaryotic life, the haptophyte microalgae (see later). Biological observations later revealed that many species within the coccolithophores do not calcify during part of their life cycle (Billard and Inouye 2004), and certain taxa such as *Isochrysis galbana* or *Dicrateria inornata* have

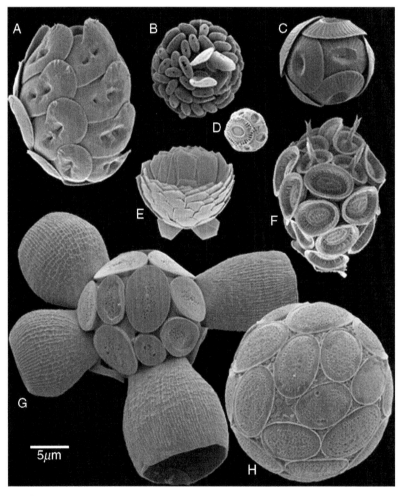

FIGURE 2. Morphostructural diversity in extant coccospheres and their heterococcoliths (diploid stage of the life cycle, see Figure 3). This plate illustrates the astounding calcareous morphostructures observed in modern Calcihaptophycidae (new subclass introduced in this chapter). **(A)** *Helicosphaera carteri*, **(B)** *Algirosphaera robusta*, **(C)** *Coccolithus pelagicus*, **(D)** *Emiliania huxleyi*, **(E)** *Florisphaera profunda*, **(F)** *Syracosphaera pulchra*, **(G)** *Scyphosphaera apsteinii*, and **(H)** *Pontosphaera japonica*. Independent of their size (< 1 to 30 μm) and even though their shape has varied mostly between circular to long elliptical, heterococcoliths exhibit a vast array of morphostructures that have helped reconstructing the phylogenetic history of the coccolithophores (Aubry 1998; Bown *et al.* 2004; Bown 2005).

no known calcifying stages (de Vargas and Probert 2004b). Recently, molecular phylogenetic surveys of environmental and cultured samples have unveiled a growing number of noncalcifying haptophytes, which may partly join the clade comprising more traditional, skeletonized coccolithophores. From geology to biology, and more recently to genetics, the group containing *calcifying haptophytes* is rapidly changing, embracing an increasing diversity of noncalcifying, partly calcifying, or differentially calcifying species. This emerging monophyletic entity of *potentially* calcifying haptophytes has no formal scientific name, and we propose here the erection of a new subclass, the *Calcihaptophycidae*.[1] Further justification of this new taxonomic designation is provided through this chapter.

A. Coccoliths and Coccolithogenesis

Coccolith is a collective term that designates all of the *biomineralized, calcified scales* produced by extant and extinct haptophytes. Although morphologically highly diverse (see Figure 2), the singularity of coccoliths

lies in their size (a few μm), rather homogeneous calcitic composition (although aragonitic coccoliths are known [Manton and Oates 1980; Cros and Fortuno 2002]), optical characteristics, and remarkable symmetry. Haptophyte biomineralization is unusual among eukaryotes as it occurs *intracellularly*. Intracellular *coccolithogenesis* requires the maintenance of sustained net fluxes of Ca^{2+} and inorganic carbon from the external medium to the intracellular Golgi-derived vesicle in which calcification occurs (Brownlee and Taylor 2004). Calcium is the main anion used for signal transduction in eukaryotes; consequently, its intracellular concentration and compartmentalization need to be under *extreme control*. Marine organisms must also regulate calcium concentration to prevent the intracellular precipitation of apatite (Constantz 1986).

The complex crystallographic structure of most coccoliths suggests that coccolithogenesis is a *highly organized* process under significant biological control. It is believed to involve proteinic templates to support the nucleation of $CaCO_3$ crystals (Corstjens *et al.* 1998; Schroeder *et al.* 2005), complex

---

[1]Taxonomic note: within the division Haptophyta and the class Prymnesiophyceae, the new subclass Calcihaptophycidae comprises currently four modern orders (the Isochrysidales, Syracosphaerales, Zygodiscales, and Coccolithales) and 3 extinct orders (Hay 1977). It defines the monophyletic group containing the last, probably noncalcifying common ancestor of all calcifying haptophytes, commonly named "coccolithophores," and all of its descendants including those that do not calcify or calcify in a single stage of their life cycle. Nakayama et al. (2005) discovered a noncalcifying prymnesiophyte, *Chrysoculter rhomboideus*, which looks like *Chrysochromulina* sp. but appears as an ancestral lineage within the Calcihaptophycidae according to genetic analyses of the 18S rDNA and rbcL gene sequences (note, however, that no representatives of the orders Syracosphaerales or Zygodiscales were included in this analysis). A number of ultrastructural characters of the flagellar/haptonematal complex appear to be synapomorphies of the clade including *Chrysoculter* and the coccolithophores: the fibrous root (F1), the electron-dense plate on the Rl microtubular sheet, the R2 of four microtubules with appendages, and the generally low number of microtubules in the emergent part of the haptonema (when present) (Nakayama et al. 2005). Another feature shared by *Chrysoculter* and coccolithophores is the presence of a single transitional plate in the transition region of the flagellum (other prymnesiophytes have two). In coccolithophores, this single transitional plate that may or may not have helical bands (Beech and Wetherbee 1988; Kawachi and Inouye 1994; Sym and Kawachi 2000) has been interpreted as homologous to the proximal transitional plate of other, more primitive prymnesiophytes, based on its position in the flagellum and general morphological features (a perforated septum with an axosome [Nakayama et al. 2005]). Note that these authors interpreted the single transitional plate of *Chrysoculter* as homologous with the distal transitional plate of other prymnesiophytes.

These five ultrastructural features appear to be reliable diagnoses to define the new subclass Calcihaptophycidae Probert et Young, whose formal, Latin description is *Prymnesiophyceae plerumque ferens coccolithos ut minimum ex parte cursus vitae. Species sine crystallis carbonatis calcii descriptae. Apparatus flagellaris plerumque cum radice fibrosa (F1), structura tabulari opaca juxta fasciculum tubularem R1, R2 microtubulis quatuor appendiculatus, et typice microtubulis paucis in parte emergenti haptonemis (si adest). Regio transitiva flagellorum cum tabula transitiva una.*

polysaccharides to control their growth and thus sculpt the coccolith (Marsh 2003), and certainly many other, as yet unknown, proteins, enzymes, and transcription factors needed for the formation, shaping, transport, and cellular addressing of the different vacuoles enclosing the liths and their basic components (Ca^{2+}, CO_3^-). To date, ~100,000 partial messenger RNAs (expressed sequence tags, or ESTs) have been sequenced from a few strains of the coccolithophore species *Emiliania huxleyi* by different research groups (mainly the Joint Genome Institute, California). In total, ~20,000 unique sequences (unigenes) were identified (Betsy Read, personal communication). Quinn *et al.* (2006) used a microarray holding ~2300 unique oligonucleotide sequences to identify a *minimum* number of *46 genes* displaying overexpression associated with coccolithogenesis. These upregulation patterns were confirmed by quantitative polymerase chain reaction (PCR). Even though the function of most of these proteins is currently unknown, these preliminary data support the idea that coccolithogenesis is a rather complex phenomenon, involving multiple structural and regulatory molecules under the control of a significant genetic network.

Coccoliths have long been classified into two broad structural groups, *heterococcoliths* and *holococcoliths* (Braarud *et al.* 1955). Heterococcoliths are complex morphostructures that consist of strongly modified calcite crystals arranged in interlocking cycles. They form within cytoplasmic Golgi-derived vesicles by a process that begins with *nucleation* of a protococcolith ring of simple crystals arranged around the rim of an organic base plate in alternating vertical (V) and radial (R) crystallographic orientations (Young *et al.* 1992). The crystals subsequently grow in various directions to form the final structure (Young *et al.* 2004). In some cases, additional nucleation and growth of calcite crystals occurs in the central area (Young *et al.* 1999, 2004). Holococcoliths, by contrast,

are simpler assemblies of noninterlocking rhombohedral crystallites of uniform size (~0.1 μm across) and are thought to be at least partly formed extracellularly (e.g., Rowson *et al.* 1986; Sym and Kawachi 2000; Young *et al.* 2003). Interestingly, heterococcoliths and holococcoliths are produced by *single* species in different stages of their life cycle (Figure 3). Although likely based on a common genetic background, heterococcolithogenesis and holococcolithogenesis should recruit different cellular pathways for at least part of the calcification process.

Other biogenic calcareous structures of similar size, but lacking the characteristic features of either heterococcoliths or holococcoliths, have long been *incertae sedis* classified into a category *nannoliths* (Haq and Boersma 1978). Examples include the pentagonal plates of *Braarudosphaera*, the horseshoe-shaped *ceratoliths*, and numerous fossil groups of uncertain affinities such as *discoasters* and *sphenoliths*. However, recent observations have systematically shown that nannoliths are indeed secreted by species within the Calcihaptophycidae. For instance, the coccolithophore *Ceratolithus cristatus* has been shown to produce both *ceratoliths* and two types of heterococcoliths (Alcober and Jordan 1997; Sprengel and Young 2000). The aragonitic cup-shaped nannoliths of *Polycrater* are formed by the haploid stage of the genus *Alisphaera* (Cros and Fortuno 2002). Recent molecular data have even shown that *Braarudosphaera* with its unique pentagonal nannoliths is a deep-branching Calcihaptophycidae (Takano *et al.* 2006). Thus, since their origin, the Calcihaptophycidae appear to have mastered widely different modes of calcification, resulting in several distinctive biomineralized skeletal structures that can all conveniently be referred to as coccoliths.

III. THE HAPTOPHYTES

Today, ~280 different, morphologically defined coccosphere types (Young *et al.* 2003) inhabit the photic zone of the global

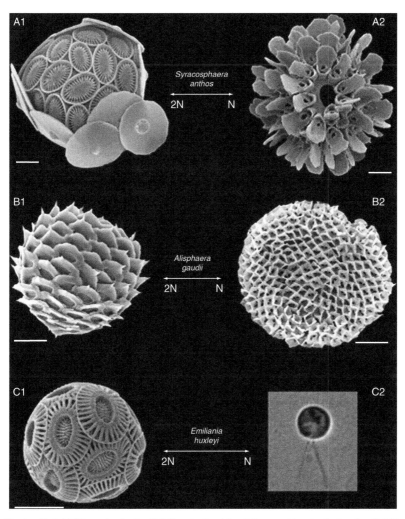

FIGURE 3.   Haplo–diploid life cycles in the Calcihaptophycidae. Like most haptophytes, the Calcihaptophycidae are capable of independent asexual reproduction in both the diploid (2N) and haploid (1N) stages of their life cycle. Diploid and haploid stages of a single species express radically different phenotypes and can be either calcifying or noncalcifying. Scale bars are 2 μm. **(A)** *Syracosphaera anthos;* A1: diploid phase producing heterococcoliths; A2: haploid phase with holococcolith, a form previously described as a discrete species, *Periphyllophora mirabilis.* **(B)** *Alisphaera gaudii.* B1 as in A1; B2: Aragonitic-nannolith bearing phase, inferred to be haploid (ex-genus *Polycrater*). This life cycle is based on observations of combination coccospheres (Cros and Fortuno 2002). **(C)** *Emiliania huxleyi.* C1 as in A1; C2: Noncalcifying haploid phase, which occurs repeatedly in culture of initially diploid calcifying cells (e.g., Houdan *et al.* 2004a).

ocean. Another 6 noncalcifying coastal morphospecies (the family Isochrysidaceae) are included in the Calcihaptophycideae (new subclass, this chapter), which comprises 4 orders and 10 families (Jordan *et al.* 2004). They belong to the class Prymnesiophyceae (6 orders, ~360 species) that, together with the Pavlovophyceae (1 order, 12 species), constitute the division Haptophyta (Jordan *et al.* 2004). Haptophytes are unicellular chlorophyll a + c containing ("red lineage") algae. They occur principally as solitary free-living motile cells that possess two smooth flagella, unequal in length in the Pavlovophyceae and more or less equal in the Prymnesiophyceae. Other forms include colonies of motile cells, and nonmotile cells that may be solitary and form pseudo-filaments or mucus-bound aggregations. Most haptophytes are marine organisms, inhabiting littoral, coastal, and oceanic waters. In the noncoccolithophore haptophytes, however, brackish and freshwater species are common and a single freshwater coccolithophore, *Hymenomonas roseola*, has been documented (Manton and Peterfi 1969). Haptophytes have also been reported in symbiotic relationships with foraminifers and acantharians (Gast *et al.* 2000), and a coccolithophore was found in skin lesions of dogfishes (Leibovitz and Lebouitz 1985). The production of exotoxins and allelopathic activity has been documented or inferred in various taxa from across the phylogeny of the prymnesiophytes, including the noncalcifying genera *Prymnesium*, *Chrysochromulina*, and *Phaeocystis* and certain members of the coccolithophore genus *Pleurochrysis* (Houdan *et al.* 2004b).

The *Haptophyta* are distinguished by the presence of a unique organelle called a *haptonema* (from the Greek *hapsis*—touch), which is superficially similar to a flagellum but differs in the arrangement of its microtubules and in its use for *prey capture* or *attachment*. The haptonema varies in length among haptophytes and in many coccolithophore species is reduced to a vestigial structure. The use of the haptonema to capture prey has been illustrated for some members of the noncoccolithophore genus *Chrysochromulina* (e.g., Kawachi *et al.* 1991), whereas in the related genus *Prymnesium* ingestion is by means of pseudopodal development at the nonflagellar pole with no involvement of the short haptonema (Tillmann 1998). Particle ingestion has been documented in one species of Calcihaptophycidae (Parke and Adams 1960), and Billard and Inouye (2004) noted that coccolithophores with less rigid coccolith coverings or noncalcifying life cycle stages are likely candidates for *phagotrophy*. The uptake and use of dissolved organic carbon has also been demonstrated for several species

of *Chrysochromulina* (Pintner and Pravasoli 1968). *Mixotrophy* thus appears to be widespread in the noncalcifying haptophytes and may also be a significant physiological trait of the coccolithophores. Indeed, although nearly all known haptophytes contain chloroplasts, a number of coccolithophore species found in polar waters (notably from the family Papposphaeraceae) have been reported to be aplastidial heterotrophic organisms (Marchant and Thomsen 1994). Whether these taxa are genuinely heterotrophic or forms that have secondarily lost the photosynthetic apparatus remains to be verified.

Haptophytes typically produce nonmineralized, *organic scales* in Golgi-derived vesicles. These scales are subsequently extruded onto the cell surface and in rare cases onto the haptonema or one of the flagella. In the Pavlovophyceae, these scales are relatively simple *knoblike* structures, whereas in the Prymnesiophyceae more elaborate and usually ornamented *plate* scales, reminiscent of the coccoliths, are produced. The prymnesiophyte ancestor of the coccolithophores evolved the ability to control the intracellular precipitation of calcite onto such organic plate scales and the assembly of the mature carbonate scales at the cell surface (typically exterior to a layer of nonmineralized plate scales).

An increasing number of species from across the phylogeny of the Prymnesiophyceae have been shown to exhibit *haplo–diploid life cycles* (e.g., Vaulot *et al.* 1994; Green *et al.* 1996; Larsen and Edvardsen 1998; Houdan *et al.* 2004a). In a haplo–diploid life cycle both stages, haploid and diploid, are capable of *independent asexual reproduction*. This life cycle strategy, possibly a synapomorphy for all haptophytes, is arguably the most significant biological feature differentiating the haptophytes from the diatoms (diploid life cycle) and dinoflagellates (haploid life cycle). In the prymnesiophytes, the morphology of the cell covering (scales and coccoliths) differs between haploid and diploid stages. In the coccolithophores, in particular, current evidence indicates that diploid stages typically produce heterococcoliths, whereas haploid stages produce holococcoliths or nannoliths or do not calcify at all (see Figure 3). The absence of calcification or the production of nannoliths may also occur in as yet undiscovered diploid stages (de Vargas and Probert 2004b).

IV. TOOLS AND BIASES IN THE RECONSTRUCTION OF COCCOLITHOPHORE EVOLUTION

In the pelagic realm, most of the functional and biological diversity is found in unicellular organisms (prokaryotes, viruses, protists) that are rapidly remineralized in the water column after death. Coccolithophore skeletons thus represent a rare opportunity to reconstruct the tempo and mode of evolution in a group of marine planktonic microbes. Despite its superior quality in terms of completeness and continuity, the fossil record of coccolithophores is difficult to interpret for several preservational and biological reasons recently reviewed in Young *et al.* (2005). First, the assemblages of living coccolithophores dwelling in the photic oceanic layers are significantly altered before they reach the ocean floor. Most, if not all, coccoliths sink to the ocean bottom attached to marine snow or packed into the fecal pellets of copepods (see Figure 1). In addition to the potential damage caused by their passage through copepod mandibles and digestive tracts, the coccoliths may be dissolved by metabolic CO_2 produced by the degradation of organic matter concentrated in both marine snow and feces. The increased acidity of deep-oceanic waters further influences dissolution of coccoliths, which are ultimately subject to diagenesis at the sediment–water interface. A recent detailed study of sinking planktonic assemblages (Andruleit *et al.* 2004) has shown that most of the mor-

phological diversity is entirely dissolved in the upper water column. Among the ~280 types of coccosphere (morphospecies) known from the modern plankton, only 57 are common to rare in Holocene sediments (Young *et al.* 2005). Up to 70% of the diversity is thus erased from the readily accessible recent fossil record.

In modern nannoplankton, the coccoliths lost to dissolution are typically tiny (< 3 μm long) and consist of delicate structures. However, dissolution appears to be more taxon- than size-specific, such that coccoliths are either almost entirely preserved in a few *morphostructural* groups (e.g., the *placoliths*) or largely dissolved or even totally erased in others (e.g., holococcoliths and Syracosphaeraceae, except for a few particularly large species). Overall, censuses of paleodiversity are mostly dependent on the abundance of available strata with ideal preservation conditions, and obviously the time spent by experts on these samples. Another major bias of the sediment record is the relative absence of *coccospheres*. The carbonate ooze consists essentially of detached coccoliths. However, these are merely single building blocks of the coccolithophore skeleton, and their number, type (*polymorphism* or *varimorphism*), and arrangement (imbrication, multiple layers, *dithecatism*, etc.) onto a complete coccosphere is likely ecologically and physiologically more relevant than the morphostructure of the lith itself (Aubry in press-b).

Recent advances in the biological knowledge of the coccolithophores further challenge fossil-based assessments of their paleobiodiversity. The broad phylogenetic distribution of haplo–diploid life cycles in prymnesiophytes (including coccolithophores) suggests that this is the ancestral state for this group. Maintaining the physiological ability to grow vegetatively under both haploid and diploid genomes expressing radically different phenotypes is not frequent among eukaryotes, which have mostly been channeled into either the haploid or the diploid mode of life. This haplo–diploid "double life" should therefore be of primary evolutionary and/or ecological significance. It is likely a strategy to rapidly escape negative selection pressures exerted on one stage, such as grazing, parasite or virus infection, or abrupt environmental changes. However, the factors triggering shifts from diploid to haploid stages in coccolithophores are virtually unknown, as are the ecology and physiology of the *haploid* stages. In fact, it is not an exaggeration to state that research on coccolithophores has almost entirely ignored half of their life cycle up to now. This significant gap is even more pronounced in the fossil record. Although most haploid stages are covered with delicate, tiny holococcoliths with high dissolution potential, others, like the haploid cell of *Emiliania,* are simply naked.

Finally, molecular phylogenetic data have highlighted two additional problems. First, a few studies of bulk ribosomal DNA sequences in natural communities of eukaryotic picoplankton (cells smaller than 3 μm) have revealed a potentially important and ancient diversity of unknown pico-haptophytes (e.g., Moon-Van Der Staay *et al.* 2000; Diez *et al.* 2001). Despite limited sequencing efforts, the new clades are widely divergent and dispersed within the haptophyte phylogeny, which suggests that tiny, noncalcifying haptophytes and, potentially, Calcihaptophycidae form a significant component of the group. Further field investigations using DNA fluorescent probes to identify and *quantify* picoplankton have shown that the pico-haptophytes form a major component of the assemblages of this size fraction in Atlantic waters (up to 35% of the pico-eukaryotes, Not *et al.* 2005) and along a basin-wide Indian Ocean transect from oligotrophic to mesotrophic conditions (F. Not, personal communication). This ghost diversity of naked and tiny haptophytes represents likely a fundamental ecological strategy followed by some lineages, which are clearly inaccessible in the sediment record. Last but not least, molecular data seriously challenge the morphological

species concepts classically used in coccolithophore taxonomy. Using both nuclear and chloroplastic genetic markers, five classical morphological species have been revealed to be, in fact, *monophyletic groups of sibling species,* isolated by several million years of evolution according to molecular clock estimations (Sáez *et al.* 2003). This disconnection between slow, morphological differentiation and more rapid genetic evolution has been revealed in all kinds of skeletonized eukaryotic plankton (e.g., dinoflagellates [John *et al.* 2003], diatoms [Orsini *et al.* 2004; Amato *et al.* 2005], foraminifera [de Vargas *et al.* 1999]) and synthesized into a concept of "planktonic super-species" (de Vargas *et al.* 2004). The genetic data indicate that the morphological criteria currently used to define coccolithophore species are too broad and most, if not all, current morphospecies are clusters of a few sibling species, often, but not always, distinguished by subtle structural characters of the coccolith. Characters such as the size of coccoliths, minor morphological details, and their number and arrangement on the coccosphere are likely to prove critical to distinguish species. This means that assessment of true *species-level diversity* in the fossil record is currently unfeasible. The morphological interpretations of species paleoecology are equally biased, as tiny and discrete morphological differences may reflect isolated biological species adapted to different spatio-temporal ecological niches, as has been demonstrated in several morphospecies of foraminifers (de Vargas *et al.* 2002).

To sum up, the classical, morphological view of coccolithophore biodiversity and evolution is largely oversimplified. Naked or poorly calcified cells, coccospheres, haploid stages, biological species, and characters such as motility and phagotrophy are fundamental information that is difficult—sometimes impossible—to retrieve from the sediments. A common practice in coccolithophore research is to merge the concept of coccoliths with the considerably more complex protists responsible for their production. In our view, coccolithophore studies based on coccoliths primarily reflect the ecology, physiology, or evolution of the function *calcification* in coccolithophores, that is, the genetic network involved in *coccolithogenesis*. Although the fossil record remains the main key to the origin and ancient history of coccolithophore calcification, biological and physiological data are clearly needed to anchor the function *calcification* into other equally fundamental processes of the coccolithophore cells and thus broaden the interpretations based on fossil coccoliths. On the other hand, molecular phylogenetics and comparative genomics give access to a detailed, independent understanding of modern diversity and functions, as well as a coarse evolutionary framework to calibrate and interpret the extinct and fossilized diversities. Unfortunately, both biological and molecular data are still very scarce in coccolithophores and largely focused on the recently evolved and atypical species *Emiliania huxleyi* (see later). However, preliminary exploration of the interfaces among coccolithophore palaeontology, geochemistry, and genomics unveils the forthcoming power of such an approach.

V. THE EVOLUTION OF HAPTOPHYTES UP TO THE INVENTION OF COCCOLITH: FROM COASTAL HUNTERS TO OCEANIC FARMERS?

A. The Origin of the Haptophytes and Their Trophic Status

Although the haptophytes are one of the deepest branching groups in the phylogeny of the eukaryotes (Baldauf 2003), the first reliably identified fossil coccolith appears only ~220 million years ago (Ma) (Bown *et al.* 2004). Given the current absence of any other specific biomarkers for the group (except the alkenones characterizing the Isochrysidales, Figure 4), questions concerning the origins of haptophytes and haptophyte photosynthesis cannot be solved by analysis of the

geological record. Comparison of cytology and biochemical homologies, integrated into molecular phylogenetics and molecular clocks, is the only tool available for reconstructing the early evolution of the group.

The origin of haptophytes is the subject of considerable debate. The chlorophyll-c–containing algae comprise four major lineages (the alveolates, cryptophytes, haptophytes, and heterokonts, the three latter being sometimes called the chromists), which, based on similarities among their plastids, were grouped together into a super-cluster, the chromalveolates (Cavalier-Smith 1999). According to this theory, the chromalveolate clade originated through a *single* secondary symbiotic event when a biciliate anterokont host enslaved a red alga. Cavalier-Smith (2002) speculates that this event occurred after the Varangerian snowball Earth melted, ~580 Ma. The resulting eukaryote chimaera evolved chlorophyll c_2 prior to diverging into the alveolates and the chromists. If this scenario is true, the first haptophyte would have originally diverged as a photosynthetic protist in the latest Neoproterozoic.

Recent molecular evidence from plastid-targeted and plastid-encoding protein structures (Fast *et al.* 2001; Harper and Keeling 2003), from phylogenies based on multiple plastid genes (Yoon *et al.* 2004; Bachvaroff *et al.* 2005), and from comparison of complete plastid genome sequences (Sánchez Puerta *et al.* 2005), are consistent with a single origin of chlorophyll-c–containing plastids from red algae. However, these data, exclusively based on chloroplastic genes, do not preclude the possibility of multiple transfers of related chlorophyll-c–containing plastids among distantly related heterotrophic hosts (Bachvaroff *et al.* 2005). There is growing support for the monophyly of an alveolate/heterokont clade from individual and multigene phylogenies based on *nucleus*-encoded proteins (e.g., Baldauf *et al.* 2000; Ben Ali *et al.* 2001; Stechmann and Cavalier-Smith 2003; Harper *et al.* 2005). In nuclear gene phylogenies, however, the

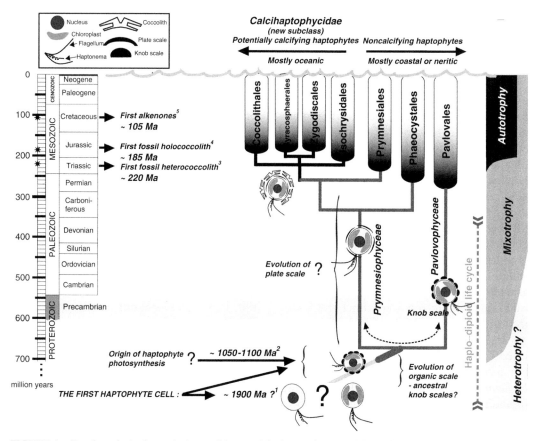

FIGURE 4. Benchmarks in the evolutionary history of the haptophytes and the Calcihaptophycidae. Major innovations are shown along a geological time scale on the left side of the figure, and a synthesis of recent *molecular* phylogenetic data using representative species of the seven extant haptophyte orders (this chapter and Saez *et al.* 2003) is depicted on the right side. Biological, phylogenetic, and paleontological data tend to support a scenario according to which the haptophytes have broadly evolved from coastal or neritic heterotrophs/mixotrophs to oceanic autotrophs since their origination in the Proterozoic. Carbonate biomineralization in the Calcihaptophycidae may have been a key evolutionary step for the stepwise invasion of the oligotrophic pelagic realm, starting ~220 Ma. See text for further details. 1, this chapter, Figure 5; 2, Yoon et al. 2004; 3, Bown 1987; 4, Bown 1983; 5, Farrimond *et al.* 1986. (See color plate.)

haptophytes have an ambiguous, deep and unsolved position, jumping from one sister clade to the other, depending on the analysis. The largest multigene analysis to date places them as the earliest branching group within the Chromalveolata (Hackett *et al.*, Chapter 7, this volume). Nonetheless, even if the chromist and alveolate hosts indeed form a monophyletic group, the hypothesis of multiple secondary endosymbioses via serial transfer is perfectly reasonable, and it is possible that the ancestral haptophyte lineage was originally aplastidial.

In a molecular clock analysis based on multiple chloroplastic genes, Yoon *et al.* (2004) dated the time of divergence between haptophyte and heterokont plastids at ~1050–1100 Ma. Assuming that the plastid phylogeny equals the host phylogeny, the authors proposed this date for the origin of the haptophytes. We have recently sequenced the SSU and LSU rDNA of a wide range of haptophytes, including many species of coccolithophores with an excellent fossil record. The phylogenetic consistency between the two ribosomal datasets, and between the molecular trees and the morphological taxonomy and stratigraphic ranges of the analyzed taxa, allowed us to infer molecular clocks using multiple maxi-minimum time constraints in the Cenozoic (Figure 5). Our maximum likelihood analyses point toward a significantly earlier origin of the haptophytes, in the Early Proterozoic ~1900 Ma. This time maybe artificially pushed backward by the ML algorithm (Peterson and Butterfield 2005), or if quantum evolution accelerated the transformation of haptophyte rDNA at the origin of the group (the stem branch) (Simpson 1944). However, it fits the multigene analyses by Hedges *et al.* (2004) and the first geological record of putative alveolates ~1100 Ma (Summons and Walter 1990). A Paleoproterozoic origin of the haptophytes matches also the basal position of the group in the eukaryotic tree together with multiple Paleoproterozoic records of eukaryote life as a whole (Knoll *et al.*, Chapter 8, this volume). Note that the recent attempt by Berney and Pawlowski (2006) to date the eukaryotic tree using supposedly accurate microfossil records and relaxed molecular clocks was biased by a miscalibration of the most important node of their tree: the maximum time constraints within the haptophytes. The authors, who estimated a founding date for the haptophyte at ~900 Ma, arbitrarily imposed a post Cretaceous–Tertiary (K/T) divergence time between the branches leading to *Calcidiscus* (Calcidiscaceae) and *Pleurochrysis* (Pleurochrysidaceae, a family without a fossil record). According to our data, which include more species and use accurate geological calibrations within entirely fossilized groups, the split between the two families occurred much earlier in the Mesozoic.

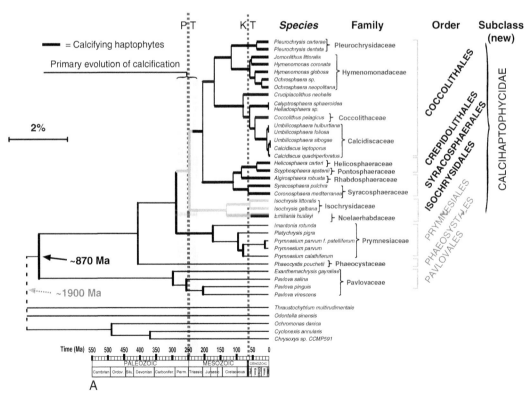

FIGURE 5. The origin and evolution of coccolithophores according to ribosomal DNA. SSU (**A**) and LSU

(Continued)

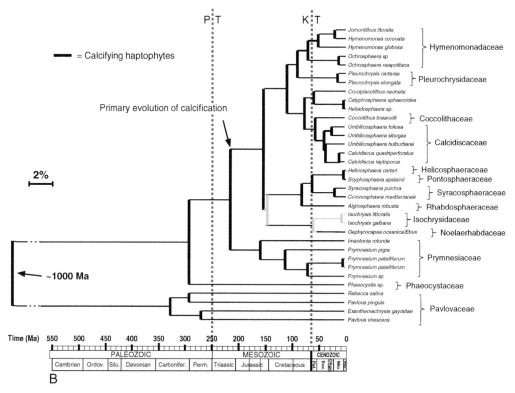

B

FIGURE 5. Cont'd   (B) rDNA trees including 34 haptophyte taxa, representatives of the 7 extant orders within the Haptophyta. The SSU rDNA were fully sequenced (alignment of 1750 sites), whereas a 5' fragment containing the D1 and D2 domains was sequenced for the LSU rDNA (950 sites). Unambiguously aligned sites were used for maximum likelihood analyses enforcing a molecular clock, as implemented in *PAUP* (Swofford 2000). Models of DNA substitution that best fit our datasets were selected through a hierarchical likelihood ratio test (Posada and Crandall 1998), and maximum time constraints were based on strato-phenetic events (details presented elsewhere). Despite small topological differences, both trees are in relative good congruence with both classical, morphological taxonomy and interpretation of the fossil record. The monophyly of the haptophytes with potential for calcification (the Calcihaptophycidae, new subclass) is obvious, and the origin of coccolithogenesis is located somewhere between the molecular origin of the group and the first apparent coccolith in the fossil record (~220 Ma).

If the hypothesis of an Early Proterozoic origin of the haptophytes is true, then haptophytes must have been primarily heterotrophic and acquired a red plastid via secondary endosymbiosis hundreds of million of years after their origin (see Figure 4). The proposed primary function as a *hunting* apparatus of the single most important autapomorphy of the group, the haptonema, fits this heterotrophic scenario, as does the fact that many haptophytes from early diverging, noncalcifying lineages (Prymnesiales, Phaeocystales, see Figure 4) have conserved heterotrophic behavior (mixotrophy). Strikingly, the early-branching lineages in nuclear molecular phylogenies of extant heterokonts, cryptophytes, and alveolates are systematically represented by heterotrophic (aplastidial) taxa. This observation lends further support to a hypothesis of a wide primordial diversity of heterotrophic predators in *all* ancestral lineages of the modern chlorophyll-c–containing algae, which would have acquired photosynthetic abilities only later via independent serial acquisitions of related plastids (for instance from a red algal prey that was highly efficient at establishing polyphyletic endocytosymbioses and transferring genes to its host nucleus). In fact, the oldest *firm* evidence of photosynthesis in haptophytes can only be dated from the divergence time between the Pavlovophyceae and Prymnesiophyceae, whose modern representatives are almost exclusively mixotrophic or autotrophic (see Figure 4). This node was estimated at ~805 Ma using a chloroplastic molecular clock (Yoon *et al.* 2004) and, respectively, at ~780 and 1000 Ma according to our new SSU and LSU rDNA clocks (see Figures 5 and 6). For once, chloroplastic and nuclear data are in good agreement, so that it seems reasonable to assume that photosynthetic haptophytes were present *at least* from the Neoproterozoic Era (1000–542 Ma). These primordial photosynthetic haptophytes were likely phagoautotrophs, a feature apparently shared by most extant noncalcifying prymnesiophytes.

B. Paleozoic Haptophytes and the Ancestors of the Coccolithophores

Again, without fossil record evidence or specific biogeochemical signatures, it is difficult to infer the morphological, physiological, and ecological features of haptophytes during the Paleozoic, prior to the evolution of the coccolithophores. An intuitively appealing hypothesis, which would account for the fact that coccoliths did not evolve for at least the first half of the evolutionary history of the prymnesiophytes, would be that plate scales, that is, the organic matrices that appear necessary for intracellular coccolith-type biomineralization, originated only shortly after the Permo–Triassic (P/Tr) boundary event as the group underwent rapid radiation into environmental niches vacated by the massive marine extinction. In fact, proto-prymnesiophytes must have evolved the organic plate scales at least before the Phaeocystales/Prymnesiales split that, according to our molecular clock analyses (see Figure 5), occurred in the mid-Paleozoic. It is thus likely that relatively derived and diverse prymnesiophytes were

present in the oceans well before the onset of haptophyte calcification. The molecular trees indicate that pavlovophytes were also present in Paleozoic oceans. The deep branches of most analyzed *Pavlova* spp. further suggest that extant lineages survived both the P/Tr and K/T mass extinctions, probably finding refuge in the coastal ecosytems they typically inhabit nowadays.

Fossils in Paleozoic shales indicate that phytoplankton with morphological features similar to members of extant green algal lineages were abundant and diverse in waters overlying contemporary continental shelves, leading Falkowski *et al.* (2004) and others to suggest that green phytoplankton were taxonomically and ecologically dominant throughout this period. The recurring presence of black shales through much of the Paleozoic is suggestive of frequent anoxia in the deep global ocean (Anbar and Knoll 2002). Basin-wide anoxia would correspond to fundamentally different oceanic chemistry, particularly in terms of the bioavailability of various trace elements, metals, and nutrients (Quigg *et al.* 2003). In particular, the increased availability of Fe, P, and ammonium in reduced oceanic conditions would have given strong ecological advantage to the green microalgae (Falkowski *et al.* 2004) and may have restricted the red lineages, including the haptophytes, to the better oxygenated coastal regions where rivers delivered essential metals (Katz *et al.* 2004). Among modern haptophytes, the morphologically relatively simple pavlovophytes are apparently restricted to near-shore, brackish, or freshwater environments often with semibenthic modes of life, and this may mirror the ancestral ecological strategy of Paleozoic haptophytes, including scale-bearing photophagotrophic proto-prymnesiophytes. In coastal environments, the cells may have complemented their nitrogen and trace metal needs via mixotrophy, a way to offset the presumed competitive advantage of green algae. Modern members of the noncalcifying prymnesiophytes (Phaeocystales and Prymnesiales) are found in both

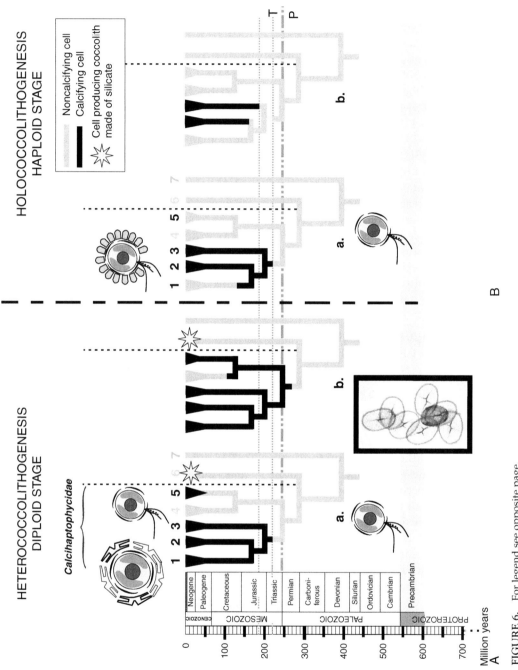

FIGURE 6.   For legend see opposite page.

coastal and oceanic environments. This may be indicative of Cenozoic colonization of oceanic realms from coastal pools of diversity across prymnesiophyte phylogeny. This evolutionary pattern is somehow reminiscent of the foraminifers, which became fully planktonic only ~180 Ma. Some time after the P/Tr catastrophic event, certain prymnesiophytes within the Calcihaptophycidae evolved the ability to biomineralize their organic scales and secrete calcareous coccoliths: they were starting out on the long evolutionary pathway leading to their supremacy in the pelagic realm.

VI. THE ORIGIN OF CALCIFICATION IN HAPTOPHYTES: WHEN, HOW MANY TIMES, AND WHY?

The first *reliable* coccoliths and non-dino-flagellate nannoliths in the fossil record are present from the Late Triassic Norian stage (217–204 Ma) (Bown 1988). The nannofossils present earlier in Carnian sediments (228–217 Ma) are nannoliths of uncertain affinity and calcareous dinoflagellates. Minute (2–6 μm), finely structured, multicrystalline fossils are present in Paleozoic Pennsylvanian and Permian limestones but subject to different interpretations: coccoliths (see review in Tappan 1980) or inorganic calcareous objects or cases of contamination (Bown et al. 2004). Triassic coccoliths

have simple *murolith* morphologies (i.e., with narrow, wall-like rims) of very small size, 2–3 μm (Bown et al. 2004), at the lower limit for preservation in the fossil record, according to Young et al. (2005). Modern murolith-producing coccolithophores in the family Hymenomonadaceae are exclusively found in coastal environments and, as a consequence, have not left a fossil record. Should the Norian coccoliths have evolved from even smaller and/or coastal forms, the origin of coccolithogenesis could significantly predate the first fossil appearance of coccoliths. However, coccolithophores are one of the rare cases in which molecular clocks tend to confirm the first appearance of a group of organisms based on fossil data. SSU and LSU rDNA clocks respectively give ~270–240 and ~200 Ma as the earliest possible dates for the origin of the Calcihaptophycidae (see Figure 5). As discussed later, these dates correspond to the origin of *potentially* calcifying haptophytes and not necessarily to the onset of biomineralization per se. Thus, molecular data indicate that the ancestral lineage of the Calcihaptophycidae originated around the time of the P/Tr boundary and evolved coccolithogenesis very soon after their genetic differentiation, as witnessed by the clear heterococcoliths found in ~220-million-year-old sediments (Bown et al., 2004). As prymnesiophytes with organic plate scales were present in the Paleozoic, why did they start to calcify only between 250 and 220 Ma

FIGURE 6. The multiple origins of coccolithogenesis in the haploid and diploid stages of calcihaptophytes. These schematic trees (based on SSU rDNA data, see Figure 5) illustrate the evolutionary history of biomineralization (within the Prymnesiophyceae **1**, Pleurochrysidaceae and Hymenomonadaceae; **2**, Coccolithaceae and Calcidiscaceae; **3**, Helico-, Ponto-, Rhabdo-, and Syraco-sphaeraceae; **4**, Isochrysidaceae; **5**, Noelaerhabdaceae; **6**, Prymnesiaceae; **7**, Phaeocystaceae). Carbonate coccolithogenesis first evolved in the diploid phase (**A**) of primitive Calcihaptophycidae after the Permo–Triassic boundary. It was later reinvented, possibly on multiple occasions, in the haploid phase of certain families (**B**), and we suggest that even diploid lineages within the family Noelaerhabdaeae may have reinvented calcification in the early Cenozoic (**Aa.**). Biomineralization based on silicate precipitation independently evolved within the Prymnesiaceae (star symbol). Note that all Prymnesiophyceae, skeletonized or not, secrete and assemble at the cell surface organic plate-scales like the one shown in the black-framed SEM picture in (**A**). The horizontal dotted lines in the middle of the figure indicate, from bottom to top, the P/Tr boundary, the first fossil heterococcolith, and the first fossil holococcolith. (See color plate.)

and not before? Was this triggered by the randomness of rare and complex genomic innovation or rather dominantly driven by extrinsic biotic or abiotic selection pressures?

A. Genetic Novelties?

Unfortunately, too little is known about the structure and functions of genomes of calcifying and noncalcifying haptophytes, and the minimum network of genes required to build a coccolith, to assess which genetic change(s) allowed coccolithogenesis. However, the protein(s) and polysaccharides directly recruited for calcite nucleation and growth are likely *few* (although potentially highly variable) and may not be new but rather be derived from ancestral biochemical pathways. Marin *et al.* (1996) proposed that the polysaccharides involved in the control of skeleton growth first evolved as calcification inhibitors in Proterozoic oceans. In the Precambrian, extreme oceanic supersaturation (Ω) resulted in rapid and sometimes massive abiotic precipitation of $CaCO_3$ (the *Strangelove* ocean mode of Zeebe and Westbroek 2003). Primitive eukaryotes would have thus evolved *anticalcifying* molecules to avoid spontaneous carbonate encrustation of cells and tissue surfaces. Therefore, the polysaccharides recruited for coccolithogenesis in the Triassic may have existed for hundreds of millions of years before their use as microskeleton architects. Moreover, one of the most typical features of prymnesiophytes, the plate scales made of cellulosic microfibrils and arranged in specific patterns at the cell surface, represent an ideal *pre-evolved* material for $CaCO_3$ crystal nucleation and growth. In this view, first promoted by Westbroek and Marin (1998), coccoliths, and carbonate skeletons in general, are the "simple" product of *new associations between ancestral biochemical processes* and thus may not be as original as they appear (in other words, they are not dependent on major cellular and biochemical inventions). Supporting this theory, Knoll (2003)

estimated that carbonate skeletons evolved independently *at very least* 28 times within the eukaryotes.

B. Multiple Origins for Coccolithogenesis?

The clustering of all calcifying haptophytes into a monophyletic entity (the Calcihaptophycidae, see Figure 5) and the early origin of *heterococcolithogenesis,* a highly distinctive biomineralization mode, may argue for a single origin of calcification in haptophytes (Young *et al.* 1992). However, genera branching at the base of the Calcihaptophycidae, such as *Chrysoculter* (Nakayama *et al.* 2005) or *Braarudosphaera* (Nakayama *et al.* 2005; Takano *et al.* 2006), are noncalcifying or secreting relatively simple carbonate scales (nannoliths). Thus, the Calcihaptophycidae ancestor may have been noncalcifying, and its diversification may have given rise to various, nonheterococcolith modes of calcification early in the evolution of the group (such as holococcolithogenesis and diverse nannolithogeneses).

We suggest that coccolithophore biomineralization is a game of *bricolage,* that is, that the assembly of a few necessary, but not necessarily new, materials and cellular processes allows the building of a coccolith. In this game, each individual module of the entire process may be recruited on a different mode, and the potential (re)invention or switch-off of a single or a few module(s) may respectively generate or prevent calcification in a particular taxon. The most striking example supporting this model are *holococcoliths* (see Figure 3). Holococcolithogenesis is clearly different from heterococcolith formation in terms of crystal nucleation, growth regulation, and locus of calcification (Young *et al.* 1999; Sym and Kawachi 2000), proving that calcification in haploid stages of the Calcihaptophycidae required some form of *reinvention* of the initial process. Despite the delicate nature and limited preservation potential of holococcoliths in deep-sea sediments, their earliest evidence in the fossil record dates from ~185 Ma. The ~35-Ma delay between the first records of heterococcoliths and holococcoliths in the fossil archive (see Figure 4) does suggest that this reinvention occurred many millions of years after the evolution of heterococcolithogenesis by diploid Calcihaptophycidae. Molecular phylogenies seem to confirm that holococcolithogenesis occurred for the first time after the divergence of the Isochrysidales (whose extant members, such as *Emiliania*, do not calcify in their haploid stage; see Figure 6B) and that the Pleurochrysidaceae and Hymenomonadaceae may have secondarily lost the ability to produce holococcoliths (see Figure 6Ba). Note that it is also possible that holococcolith biomineralization, a rather flexible and simple process compared to heterococcolithogenesis, evolved independently in at least two and possibly more lineages (see Figure 6Bb).

Which of the genetic modules involved in heterococcolith biomineralization were recycled in holococcolithogenesis, or, in other words, which molecules and/or cellular processes were newly recruited in holococcolith formation, is a fundamental question to which there is presently no answer. Similarly, the biomineralization of strikingly different nannoliths (Figure 7) or particular structures of certain heterococcoliths (such as the central process in *Algirosphaera*) likely represent partial reinventions or even *independent origins* of the biomineralization process within the Calcihaptophycidae. Here, the case of by far the most famous species of Calcihaptophycidae, *Emiliania huxleyi*, is worth examining. *Emiliania* is the laboratory rat of coccolithophore research, to such a degree that the concept of coccolithophores itself is regularly (and unfortunately!) confounded with this species. However, *Emiliania* and its sister and ancestral species (the Noelaerhabdaceae) are strongly atypical in many ways and may result from an *independent Cenozoic origin* of calcification (heterococcolith-like). In molecular phylogenies based on both rDNA (see Figure 5) and chloroplastic genes (Fujiwara *et al.* 2001; Sáez *et al.* 2003), the

Noelaerhabdaceae typically branch at the base of the Calcihaptophycidae tree. Moreover, their sister clade, the Isochrysidaceae, contains *noncalcifying* taxa and one species, *Chrysotila lamellosa*, which induces extracellular calcification (see later). Secondary loss of heterococcolith formation in the Isochrysidaceae is a seemingly parsimonious interpretation of this phylogenetic pattern (Figure 6Ab). However, another hypothesis is that the Isochrysidales diverged from the presumably noncalcifying Calcihaptophycidae ancestor prior to the first evolution of heterococcolithogenesis. They remained noncalcifying and possibly coastal (like modern Isochrysidaceae) during the Mesozoic and through the K/T crisis, and independently evolved heterococcolithogenesis ~50 Ma when the first fossil Noelaerhabdaceae appeared in the geological record (see Figure 6Aa). Since then, this new mode of calcification dominated Calcihaptophycidae biomineralization. *Emiliania* is the baby of the group; it diverged only ~250,000 years ago from the gephyrocapsids and has played a prominent role in Quaternary oceans (Thierstein *et al.* 1977).

Several ecological, stratophenetic, physiological, and cellular features of *Emiliania* and its sisters favor the hypothesis of an independent invention of calcification in this group. Unlike most other coccolithophore species, which are K-strategists in oligotrophic waters, most of the Noelaerhabdaceae prefer mesotrophic conditions where they can produce atypical massive blooms. The living representatives of the group exhibit unusual life cycles with a *noncalcifying* haploid phase. Most importantly, they possess several ultrastructural features that distinguish them from other coccolithophores, such as the apparent absence of any vestige of a *haptonema,* very thin and unique flakelike scales underlying the coccoliths, the X-body in the haploid phase, and an atypical *reticular body* (a system of membranes separated from the Golgi-body; see Paasche [2002] and references therein) around the coccolith

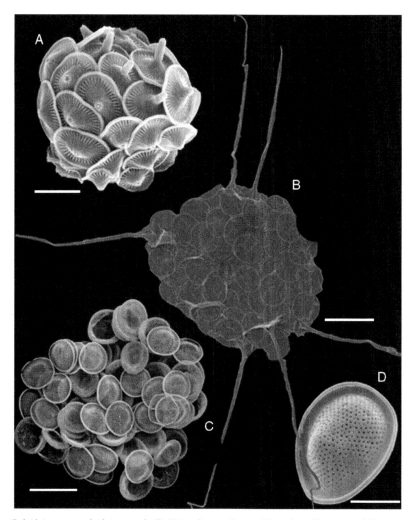

FIGURE 7. Calcifying, noncalcifying, and silicifying haptophytes. This plate illustrates different morphostrategies used by the haptophytes to cover the cell membrane: calcite coccoliths in *Cyrtosphaera lecaliae* **(A)**, cellulosic scales in a noncalcifying *Chrysochromulina* sp. **(B)**, and siliceous scales in *Hyalolithus neolepis* (**[C]**, specimen courtesy of Rick Jordan; **[D]**, a single silicoscale). Scale bars are respectively 2, 3, 10, and 2 μm. These remarkable morphological convergences are likely based on common cellular and molecular processes recruiting orthologous and homologous biochemical processes but also convergent and even novel molecules. As in coccolithogenesis, the silicification of *Hyalolithus* occurs within intracellular vacuoles, which are shaped and dispatched around the cell via cytoskeletal forces.

vesicles. Finally, there is no definite fossil link when they first appear in Early Eocene sediments. The proposed derivation from the extinct Prinsiaceae, based on morphological similarities (Romein 1979; Gallagher 1989; Young *et al.* 1992), has not been proven by stratophenetic studies and is challenged in terms of morphostructural

evolution by Aubry (in press-a). Although some of these exceptional features are not directly linked to calcification and may be due to derived rather than ancestral evolutionary processes, the combination of them supports the idea that coastal, noncalcifying Isochrysidales, survivors of the K/T extinction, have evolved *de novo* the necessary conditions for heterococcolithogenesis. (Note that the embryonic V/R pattern of biomineralization observed in the proto-coccolith ring of *Emiliania* [Young *et al*. 1992] implies that this potential independent origin of heterococcolithogenesis was based on homologous [but not necessarily orthologous] proteinic templates).

In this context, it is worth noting that living members of the noncalcifying Isochrysidales may provide clues as to the first evolutionary steps leading toward intracellular biomineralization in the Calcihaptophycideae and may, in fact, be considered to be in the process of reinventing coccolithogenesis in modern oceans. Although *Chrysotila lamellosa* (a species within the Isochrysidaceae, the sister family of *Emiliania* and the Noelaerhabdaceae) does not produce coccoliths, it does produce small organic plate scales and it is typically found as aggregations of nonmotile cells, which are invested with a thick layer of hyaline mucilage. In old cultures, *calcified* deposits are observed to form in interstices within the mucilage mass, presumably partly driven by modifications of the environment (increased pH) due to photosynthetic activity (Green and Course 1983). The chemical nature of the mucilage produced by *C. lamellosa* is not known, but Green and Course (1983) speculate that it is formed of polysaccharides with sulphated and uronic acid residues and perhaps complexed proteins providing a matrix upon which deposition of calcium salts may occur. If chemical analogies between this system and intracellular coccolith formation are confirmed, it can be postulated that extracellular calcification in benthic prymnesiophytes of the type observed in *C. lamellosa* may be a direct evolutionary precursor to coccolithogenesis, the critical step "simply" being the intracellularization of this process. Were this the case, it can further be speculated that the first coccoliths may not have been as complex in crystallographic and structural terms as hetero- and holococcoliths, perhaps

being relatively unregulated nannolith-like structures.

Last but not least, biomineralization is actually not restricted to the Calcihaptophycideae and has evolved independently within the Prymnesiales on a remarkably similar mode but using silica rather than carbonate precipitation (see Figure 7). Different authors (Pienaar 1980; Green *et al*. 1982) recorded certain species of *Prymnesium* producing layers of scales with electron-dense siliceous material, interpreted as resting cysts. More recently, Yoshida *et al*. (2006) used transmission electron microscopy and molecular phylogenetics to show that *Hyalolithus neolepis,* a haptophyte taxon branching within the Prymnesiales, secretes silica scales by a mode surprisingly similar to the one employed in carbonate coccolithogenesis. The cell controls the intracellular precipitation and growth of silica onto organic templates within vesicles probably derived from the peripheral endoplasmic reticulum, and the silica scales are extruded and arranged into a composite test, strikingly similar to those produced by the organic scales of other Prymnesiales or the coccospheres of the Calcihaptophycidae. A significant part of the genetic network controlling the formation of siliceous scales in haptophytes is likely homologous (and partly orthologous) to the machinery used in haptophyte calcification. This spectacular morphological convergence (see Figure 7) provides strong support for a hypothesis of haptophyte biomineralization as a game of *bricolage* and justifies the introduction of the new subclass Calcihaptophycidae for the monophyletic group of potentially *calcifying* haptophytes.

The question of the ease with which Calcihaptophycidae can switch on and off, adapt and modify, and/or independently re-evolve intracellular calcification is crucial for understanding the past and present and for predicting the future of these organisms. It highlights the need for further genetic and biochemical investigations on a wider range of taxa. If coccolithogenesis is indeed an interplay between pre-evolved modules around a few crucial elements that the Calcihaptophycidae acquired at their origin (most probably specific protein(s) and polysaccharides), then haptophyte calcification may be much more adaptable than previously thought. Coccolithogenesis may have more origins than assumed, and some lineages could stop calcifying and subsequently reactivate the process, even on a different mode, as has been proposed for coral biomineralization (see the "naked coral hypothesis" of Stanley [2005]).

C. Environmental Forcing on the Origin of Haptophyte Calcification

The Late Triassic appears to have been a founding period for the primary diversification of various, unrelated marine calcifiers, including *coccolithophores, calcareous dinoflagellates,* and *scleractinian corals* (Figure 8). It thus seems that a global environmental forcing stimulated or at least opened the possibility for the simultaneous development of meters-wide aragonitic coral colonies on the shore and the earliest calcitic microalgae in the plankton. Which factors may have either hindered such biological innovation earlier in the Paleozoic or specifically triggered them in the Triassic?

Atmospheric CO_2 rise has been invoked as a major threat for modern corals (Leclercq *et al.* 2000) and coccolithophores (Riebesell *et al.* 2000). Because the atmosphere is in equilibrium with the upper ocean, high pCO_2 decreases pH and therefore lowers the carbonate ion concentration of surface waters. This in turns lowers the saturation state (Ω) of the ocean, which can strongly decrease the potential for organisms to secrete calcareous skeletons. However, coccolithophores originated in a high pCO_2 world, where concentrations of atmospheric carbon dioxide appear to have been four to six times higher than today (see Figure 8, and Katz *et al.*, Chapter 18, this volume). Other factors must have buffered the Triassic oceans with high alkalinity to keep Ω hig h enough for aragonite

and calcite deposition. Paleoceanic sea levels and Mg/Ca chemistry have been proposed as driving forces for the colonization of the pelagic realm by calcifying organisms (Ridgwell and Zeebe 2005). The Triassic was characterized by very low sea levels (and therefore reduced area of shallow water carbonate platforms) and seawater with low $Ca^{2}+$ concentration and a high Mg:Ca ratio (Mg inhibits calcite precipitation) (see Figure 8). These conditions of extremely reduced biocalcification on the shelves are thought to have given rise to a highly oversaturated ocean. Weathering of exposed carbonates would have enhanced the effect of decreasing shelf area to further boost the carbonate saturation of the ocean (Walker *et al.* 2002). The very high open ocean Ω inferred for the Late Triassic was certainly a necessary condition for calcite precipitation; however, such conditions (low sea levels, decreased shelf area, high Mg:Ca ratio) seem to have prevailed for around 100 million years prior to the origin of coccolithogenesis (see Figure 8). Other factors may have hindered planktonic biocalcification in the late Paleozoic, and these may be related to the reasons suggested to have led to the supremacy of green algae in Paleozoic oceans, while pushing the red algae to near-shore niches. Deep-water anoxia was particularly pronounced near the end of the Permian and persisted through the Early Triassic (Isozaki 1997). Ocean anoxia alter redox chemistry of the oceans and especially increase the concentration of the cations Fe and Mn, which are known to inhibit the precipitation of $CaCO_3$ (Ridgwell and Zeebe 2005). Thus, we propose that the *relaxation* of anticalcifying Paleozoic conditions in the *Aragonite II* open oceans created an environmental matrix favorable for pelagic calcification.

D. Why Were Coccoliths Invented?

The partial reinvention of calcification in the haploid phase of coccolithophore life cycles and the possible multiple origins of coccolithogenesis indicate that biomineralization has been positively selected at least a few times in the evolutionary history of the haptophytes. In fact, widely different groups of protists, in particular the haptophytes, dinoflagellates, and foraminifers, adopted calcification and started to proliferate in this favorable Late Triassic ocean (see Figure 8). However, the establishment of oceanic conditions permitting calcification was not the evolutionary force that *directly* drove the invention and maintenance of biomineralization. Strong ecological and/or physiological advantage(s) impacting the survival and fitness of the cells must have selected those individuals that started calcifying and at the same time became clearly planktonic (note that the foraminifers were calcifying since the early Cambrian but invaded the planktonic realm only about 180 Ma). It may be that each group of pelagic microcalcifiers followed these evolutionary steps (calcification and migration to the pelagic realm) for different reasons.

Multiple hypotheses to explain the primary function of covering the cell with a carbonate crust have been proposed for each group. In coccolithophores, the debate centers mainly on whether the function of coccoliths is principally *ecological biotic* (protection against grazing and/or viruses, see Hamm and Smetacek, Chapter 14, this volume), *ecological abiotic* (light concentration, protection against ultraviolet [UV], dissipation of light energy under high irradiances, control of sinking rate), or *cellular biochemical* (carbon concentration mechanism for photosynthesis, phosphorus metabolism, maintenance of a balance between high external and low intracellular Ca concentration) (see reviews by Young 1994; Paasche 2002). Despite a plethora of hypotheses, relatively few experiments on living coccolithophores have been conducted to actually test these ideas, and the large majority of experiments have been performed on one of the most atypical calcihaptophytes, *Emiliania huxleyi*. The lack of knowledge of the physiology and ecology of divergent coccolithophore lineages, having sometimes drastically

different ways of building their coccoliths and coccospheres, is arguably the main barrier to interpretation of their evolution. Key selection pressures, such as predation, light intensity, or infection-parasitism, that may act very differently on naked versus calcified cells have not yet been rigorously tested. Even the generally accepted *trash-can* hypothesis, which claims that the transformation of HCO_3^- to CO_2 during calcification (see Figure 1) acts as a carbon concentration mechanism (CCM) providing carbon dioxide to photosynthesis, has been seriously challenged in *E. huxleyi* (Herfort *et al.* 2004; Rost and Riebesell 2004; Scarlett Trimborn, personal communication). The fact that most planktonic and benthic foraminifera are nonphotosynthetic (i.e., have no symbionts) further confirms that calcification and photosynthesis can be fully uncoupled processes. We are left with a simple fact: It is obviously less nutritional and harder for zooplankton to eat armored protists rather than naked cells, and this "grazing" hypothesis has the advantage of applying to all sorts of skeletonized microplankton that radiated in the early Mesozoic (Hamm and Smetacek, Chapter 14, this volume). Note, however, that the kilometer-thick carbonate ooze at the bottom of the ocean is to some extent the product of the export of coccoliths via copepod fecal pellets, which argues in fact for intense grazing!

Whereas most of the hypothetical functions of coccoliths are testable using living species, it is much harder to know which ones prevailed in the Late Triassic and were primarily selected for at the onset of coccolithogenesis. First, all potentially involved selection pressures, chemical or biological, are *inferred* from paleoproxies and have thus a significant degree of inaccuracy. But mostly, the modern coccolithophores are the products of more than 200 million years of evolution, which is equivalent to ~75 billion generations of daily-dividing phytoplankton! Their genomes have had a tremendous amount of time to randomly drift in multiple isolated lineages and evolve under strong

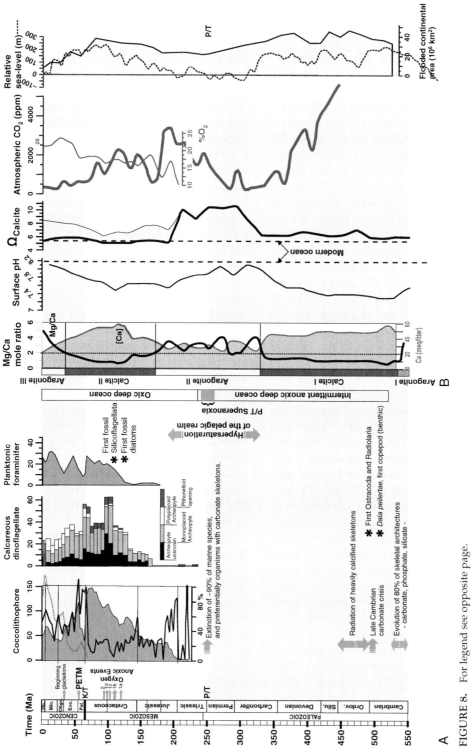

FIGURE 8.   For legend see opposite page.

and changing climatic and biotic pressures. The coccolithophores have survived one major extinction event, underwent four important biological turnovers, and have adapted to different oceans since their origin. The relative importance of the forces that originally selected coccolithogenesis in the founding population of naked haptophytes must have changed over time and among their evolving diversity. In the following sections we explore how the morphological diversity of coccolithophores has changed over time and examine which forces may have shaped the different trajectories and ecological strategies that the main lineages have adopted.

VII. MACROEVOLUTION OVER THE LAST 220 MILLION YEARS

A. Forces Shaping the Evolution of Coccolithophores and Coccolithogenesis

The forces driving the ecological and evolutionary success of coccolithophores are multiple, act on different time scales, and affect various taxonomic ranges, from a single population to the entire group. We can classify them into three categories: *extrinsic biotic, intrinsic biotic, and extrinsic abiotic*. Initially, the species endure constant negative selection pressures due to predation,

parasitism, viral infection, and competition with other photoautotrophs for nutrients. The biotic arms race is particularly acute in the pelagic realm (Hamm and Smetacek, Chapter 14, this volume), where organisms are in perpetual motion and have generally very short generation times. In fact, these extrinsic biotic forces largely control population dynamics on ecological time scales and certainly drive a significant part of genome evolution through natural selection. On longer time scales, new genetic—and in particular *morphogenetic*—inventions can drive novel adaptations, adaptive speciations, and even radiations of a particular taxon into a new ecological niche (de Vargas and Probert 2004). Finally, many abiotic forces, such as the bioavailability of essential elements or the physical stability of ecological niches, may shape the evolution of the group over even larger taxonomic and time scales. These abiotic factors are typically controlled primarily by the evolution of the Earth system (oceanic currents, tectonics, climate changes) and may ultimately control the fate of planktonic life, particularly sensitive to physico-chemical changes of their environment.

In coccolithophores, the main difficulty is to untie the multiple forces shaping the evolution of the group as a whole, including its entire biological and functional complexity, from those acting on the function

FIGURE 8. Abiotic global forcing on the oceanic carbonate system and the evolution of morphological species richness in pelagic microcalcifiers. Concerning the carbonate system, the Phanerozoic can be divided into two periods of equal length: a first during the Paleozoic, when the biotic part of the system was essentially coastal benthic and the abiotic components were relatively unstable; a second in the Meso-Cenozoic, when the new planktonic calcifiers originated, diversified, shifted a significant part of carbonate precipitation into the pelagic realm, and at the same time stabilized the concentration of atmospheric CO_2. On this chart, along a geological time scale highlighting a few major paleoceanographic events, several essential biotic **(A)** and abiotic **(B)** components of the carbonate system are represented with, from left to right: coccolithophore morphospecies diversity and their rate of turnover (= rate of extinction + rate of "speciation") (Bown *et al.* 2004, 2005); morpho-species diversity of calcareous dinoflagellates (Streng *et al.* 2004); morpho-species diversity in planktonic foraminifers (Tappan and Loeblich 1988); secular variations in absolute concentration of Ca^{2+} and Mg:Ca ratio of seawater (Stanley and Hardie 1998); mean surface saturation with respect to calcite according to Ridgwell (2005); atmospheric CO_2 concentration (Royer *et al.* 2004) and atmospheric O_2 concentration (Falkowski *et al.* 2005); and relative sea level (Haq *et al.* 1987) and flooded continental areas (Ronov 1994). In addition, a few major ecological events in Paleozoic carbonate biota as described in Knoll (2003) are indicated in **(A).** P/Tr, Permian–Triassic; K/T, Cretaceous–Tertiary; PETM, Paleocene–Eocene Thermal Maximum. (See color plate.)

calcification. In oceanography and palaeontology, the coccolithophores are often considered as a single *functional group* and their success is scaled on their ability to produce thick, abundant, and diversified calcareous liths. The coccolithophores actually encompass wide morphological, functional, and ecological diversities and have adopted various strategies at different epochs, most probably in reaction to different forcings.

B. Broad Patterns of Morphological Diversity

Bown *et al.* (2004) recently reviewed current understanding of calcareous nannoplankton evolution through their ~220-Ma history, using a synthesis of *morphospecies* diversity data over time and various inferred rates of evolutionary change. They show that rates of speciation, extinction, and turnover were markedly more variable in the Cenozoic than in the Mesozoic and promote the idea that the main force driving increases in coccolithophore diversity (in fact, strictly speaking, coccolith morphological diversity) over geological time is the *long-term stability* of *oligotrophic* to *mesotrophic* water masses (see also Aubry 1992, 1998). The presence of such environments in the greenhouse world during the Mesozoic and Paleogene promoted long-term diversification of the K-strategist phytoplankton, and in particular the coccolithophores. The shift into an icehouse world in the Oligocene, with the establishment of new, cold and vertically mixed water masses at high latitudes, led to long-term diversity decline and higher rates of both speciation and extinction of morphospecies. Furthermore, these latter environmental conditions, where nutrient delivery is pulsed, became largely advantageous for the diatoms, which may have excluded many coccolithophore species through competition (Bown 2005).

As discussed previously, it is currently not possible to define the species-level based solely on coccolith morphology, and so Bown's diversity curve (see Figure 8)

should be interpreted with caution. In fact, its general pattern roughly fits previously published curves of morphological diversity at the genus level, and thus it mostly depicts trends of *morphostructural invention* (and the subsequent evolutionary success of these morphostructures) within the 26 coccolithophore families that have built *dissolution-resistant coccoliths* over the last ~220 Ma. Because most of the diversity within the Calcihaptophycidae (species with null or poor preservation potential, noncalcifying species) has been erased from the sediment record, the "real" coccolithophore biodiversity over geological time may actually be radically different. For example, the calcifying species within the families *Pleurochrysidaceae* and *Hymenomonadaceae* are significant components of the modern coastal oceans, and molecular clocks suggest that they have been present for most of the Mesozoic, surviving the K/T mass extinction (see Figure 5). Despite this long evolutionary history, no traces of their delicate coccoliths are observed in the fossil record. Generally speaking, the coastal or neritic coccolithophores are not the best calcifiers. They often produce small and poorly calcified coccoliths and their haploid stages are frequently noncalcifying (e.g., the *Pleurochrysidaceae* and *Hymenomonadaceae,* see Figure 5). Thus, the Bown curve of fossil coccolith diversity may be largely skewed toward *oceanic heavy calcifiers* and therefore *intrinsically* reflects long-term stability of oligotrophic water masses. Evolution of the coastal biodiversity may even be driven by forces opposite to those acting on oceanic diversity. Should this be the case, the counts of total Calcihaptophycidae diversity over geological time would tend to be flattened.

C. Oligotrophy and Water Chemistry

Whatever the past diversity of poorly calcified or naked coccolithophores was, a few overarching abiotic forcings are likely to have influenced the *intensity of coccolithogenesis,* reflected in the broad diversity pattern of coccoliths resistant to dissolution (Bown's curve in Figure 8). Note that these global forcings likely acted on all pelagic microcalcifiers, independent of their taxonomic or trophic status. Indeed, the morphodiversity curves of *calcareous dinoflagellates* and *planktonic foraminifera* show trends of fluctuation through time very similar to the coccolithophore curve (see Figure 8).

In coccolithophores in general, there does appear to be a broad connection between *biomineralization* and *pelagic oligotrophy*. The first calcified haptophytes appeared in eutrophic shelf and epeiric environments, where they were confined throughout most of the Jurassic. However, their size, diversity, and impact on carbon cycling was relatively limited until they invaded the *oligotrophic open oceans* in the late Jurassic, where they steadily diversified into an astonishing morphological diversity of larger forms up to the end of the Cretaceous, producing massive amounts of deep-sea chalk. Still today, the large majority of coccolithophore biodiversity displaying strong calcification features is found in oligotrophic stratified water masses, while the shelf or coastal species have a general tendency to produce smaller and more fragile coccoliths or sometimes not to calcify at all. Patterns in haptophyte life cycles may provide additional clues. The rDNA trees indicate that haploid and diploid phases have undergone increased differentiation through the evolutionary history of the haptophytes (see Figure 5). Being essentially isomorphic in the ancestral Pavlovales (if haplo-diploidy indeed exists in this lineage), the haploid phases began their morphological (and most probably ecological) differentiation in the Prymnesiales and Phaeocystales, until they reinvented coccolithogenesis under a different mode in the Calcihaptophycidae. However, early branching and predominantly coastal to neritic coccolithophores (Isochrysidaceae, Noelaerhabdaceae, Hymenomonadaceae, Pleurochrysidaceae) are either noncalcifying or do not calcify in their haploid stage (see Figure 6). In contrast, all open ocean cal-cihaptophytes seem to secrete *holococcoliths* or *nannoliths* during this stage (see Figures 5 and 6). It seems, thus, that calcification of both diploid and haploid stages has been a *key adaptation for the successful invasion of the oligotrophic pelagic realm by the calcihaptophytes*, and that the success of coccolithophore *biomineralization* (i.e., high diversity of large and robust coccoliths) is indeed roughly positively correlated to the extent and stability of such environmental conditions (see Figure 8).

Secular changes in seawater chemistry may also broadly constrain coccolithogenesis. In particular, a seminal study by Sandberg (1983) showed that changes in oceanic conditions promoted abiotic precipitation of alternatively calcite and aragonite over periods of hundreds of millions of years (see *Aragonite* and *Calcite* oceans in Figure 8). Stanley and Hardie (1999) built upon these data and proposed that hydrothermal ridge activity related to rates of seafloor spreading controls the relative seawater concentration of Mg^{2+} and Ca^{2+}. At times of low spreading rates, the release of Ca^{2+} to seawater and the capture of Mg^{2+} that occurs during the transformation of rocks at active ridges are significantly reduced, so that the Mg:Ca ratio of open ocean waters increases and favors the precipitation of high-Mg calcite and *aragonite* (Stanley and Hardie 1999). Conversely, the Mg:Ca ratio decreases when seafloor spreading rates are high, which would promote the precipitation of low-Mg calcite, such as that found in coccoliths. The transition between *calcite* and *aragonite* oceans (see Figure 8) appears to have been a strong, overarching evolutionary pressure on organisms building carbonate skeletons (especially those with limited physiological control on biomineralization), as shown by the extinction of calcitic lineages and their replacement by aragonitic taxa in different groups such as corals, sponges, and green and red algae.

The status of the Calcihaptophycidae within this theoretical framework is still ambiguous. Their relatively complex mechanism of *intracellular* crystal nucleation and growth suggests that these protists exert a significant control on biomineralization, which would thus *escape,* at least to some degree, environmental influence. In particular, coccolithophores should have a strong capacity to buffer the fluids from which they precipitate their calcite skeletons. However, the curve of coccolithophore morphospecies diversity (which roughly represents the ecological and evolutionary success of heavy, dissolution-resistant coccoliths) broadly *correlates* (inversely) with the seawater Mg:Ca ratio, with maxima and minima respectively occurring in the late Cretaceous (see Figure 8). The increasingly calcitic oceans in the Cretaceous would have thus selected for the diversification of heavy calcifiers, leading to the formation of massive chalk deposits known from this time. In fact, recent experiments mimicking the ionic composition of Cretaceous seawater seem to confirm that lower Mg:Ca ratio (and high Ca^{2+}) sustains significantly higher rates of calcification and growth in three species within the Calcihaptophycidae (Stanley *et al.* 2005). The authors suggest that increased calcification in low Mg:Ca waters promotes, via the production of CO_2, photosynthesis and thus growth rate. However, several studies have demonstrated that calcification in fact does not promote photosynthesis in *E. huxleyi* (e.g., Herfort *et al.* 2004), and further data using a wider range of taxa are clearly needed. *Emiliania* could, once again, prove to be the exception. It should be noted that, contrary to the generally accepted view that coccoliths are low-Mg calcitic structures, Stanley and collaborators reported that the two coastal species used in their experiments (*Pleurochrysis* and *Ochrosphaera*) secreted liths of high-Mg calcite under modern, high-Mg:Ca conditions. This would mean that at least these coastal taxa, which secrete abundant, tiny and relatively simple coccoliths, do not strictly control cation incorporation during intracellular biomineralization. However, our analyses of calcite from a variety of

coccolithophore species cultured in present day seawater (Probert, Stoll, and Young unpublished results) have to date never revealed a Mg content as high as those reported by Stanley *et al.* (2005).

D. Changes in Morphostructural Strategies

In their ~220-million-year evolutionary history, the Calcihaptophycidae have produced a large array of morphologies and structures. Much of the history of coccolithophore calcification is locked in the *heterococcoliths,* whose *morphostructures* (i.e., distinctive organization of individual crystals into *cycles*) make them both resistant to dissolution and remarkably useful for macroevolutionary studies. In heterococcolith evolution, the temporal distribution of morphostructures exhibits a pattern that correlates with the main geological events punctuating the history of life. Morphostructural losses and innovations occur at chronostratigraphic boundaries associated with mass extinctions (e.g., the K/T boundary, ~65 Ma) and major biological turnovers (e.g., the Paleocene–Eocene [P/E] boundary, ~55 Ma) (Aubry 1998). This correlation suggests that certain types of calcification/crystallographies may no longer be possible or beneficial under the new ecological circumstances created by the event, thereby causing the extinction or rarefaction of the lineages that produced them, unless survival without calcifying has occurred. It must be appreciated that the loss of numerous morphostructures at the K/T boundary (extinction of 62% of the families [Bown *et al.* 2004]) is evolutionarily more significant than the associated 93% reduction in species richness. However, because coccolith species are mere morphotypes, a marked decrease in species richness per family is essentially a measure of the evolutionary success of a particular morphostructure.

Using a morphostructural approach at both the coccolith and coccosphere levels, the macroevolution of coccolithophores can be divided into two modes, separated by the P/E boundary ~55 Ma. The first, the Mesozoic-Paleocene (*MP*) mode, is one of great morphologic diversification. Simply stated, it seems that *MP* coccolithophores expressed their full morphogenetic potential, with little apparent selection pressure on morphostructures. The *MP* mode was abruptly, yet only momentarily, interrupted by the K/T event, which eliminated all taxa that were common through the Late Cretaceous (Perch-Nielsen 1981). Taxa that were uncommon in the Cretaceous, but which can be traced back to an Early Jurassic ancestry, survived the K/T boundary and rapidly evolved into new Cenozoic morphostructures. The rate of Paleocene diversification that produced 11 families (or 60% of the Cenozoic families) in 10 million years compares well with the Early Jurassic radiation of the group, when 56% of the Mesozoic families evolved in 17 million years (Bown *et al.* 2004).

The second mode, Eocene-Oligocene-Neogene (*EON*), is fundamentally different. In the *EON*, coccolith and coccosphere morphologies seem to have been under significantly stronger selection pressure, as independent, polyphyletic lineages have been channeled into similar morphostructural trajectories, or *strategies* (Aubry in press-b). In other words, the paths of morphogenetic diversification became apparently restricted after the P/E event, and this restriction seems to have only grown stronger with time. Different morphological strategies characterize the *EON* mode, such as the increase in coccolith size during the Eocene, but these are still poorly studied. Only the youngest strategy is currently described (Aubry in press-b). It became established ~2.8 Ma around the Middle/Late Pliocene boundary. At this time, most large coccoliths became extinct in various *unrelated* lineages (even within the successful family Noelaerhabdaceae), and a polyphyletic shift occurred from morphostructures with overlapping elements to morphostructures characterized by jointive or even disjunct elements. The extant coccolithophores are part of this new strategy. It must be noted that the change from the *MP* to the *EON* mode and the establishment of the youngest strategy did not occur abruptly and thus cannot be compared to massive extinctions or any other biotic changes related to catastrophic environmental change. It appears rather that global extrinsic abiotic or biotic pressures have been driving the selection of morphostructures over long time periods in coccolithophores.

Finally, clear taxonomic shifts have been occurring over time beneath these global morphogenetic trajectories. The evolutionary and/or ecological success of a particular morpho-taxonomic unit may reflect new genomic inventions providing a higher fitness to the group. As an example, the Coccolithales dominated early Paleogene communities until ~40 Ma, both in terms of diversity and amount of biomineralization. They remained an important component of the coccolithophores through the early Oligocene, but the Isochrysidales became increasingly diverse and abundant after the early Eocene, except at high latitudes where *Coccolithus pelagicus* continued to intermittently dominate (Beaufort and Aubry 1992). This important taxonomic transition may be related, as we have seen previously, to a reinvention of biomineralization in the Isochysidales.

VIII. THE FUTURE OF COCCOLITHOPHORES

Over the last years, several studies have demonstrated the rapid impact of rising anthropogenic CO_2 on the carbonate system in the oceans. The surface ocean is acting as a sink for CO_2 (Sabine *et al.* 2004), where the water pH and concentration of CO_3^{2-} ions are predicted to drop by respectively 0.4 units and 50% by the end of this century. Carbonate is thermodynamically less stable under such conditions. In addition, Orr and collaborators have recently predicted, using a range of models of the ocean-carbon cycle, an imminent and dramatic shoaling of the carbonate compensation depth (see Figure 1) by several hundreds, if not thousands, of meters (Orr *et al.* 2005). Surface waters at high latitudes will become undersaturated with respect to aragonite within decades, and calcite undersaturation will only lag that of aragonite by 50 to 100 years, which may lead to massive extinction of pelagic calcifiers, including the Calcihaptophycidae. In 2000, Riebesell and collaborators tuned the pH (using HCl and NaOH) of mono-specific cultures of *Emiliania* and *Gephyro-capsa* to preindustrial and future high CO_2 world values and reported reduced calcite production and coccolith malformations at increased CO_2 concentrations (Riebesell *et al.* 2000). Furthermore, the seawater Mg: Ca ratio has shown a significant increase since the K/T, with the onset of a new, Neogene, Aragonite ocean (see Figure 8). The modern Mg:Ca ratio of ~5 is higher than ever before in the Phanerozoic and may significantly increase the metabolic cost to the Calcihaptophycidae of producing calcite coccoliths.

Obviously, extant coccolithophores are facing a fast-changing ocean imposing strong pressures on their calcification. However, recent culture and mesocosm studies mimicking predicted pCO_2 conditions (e.g., Delille *et al.* 2005; Riebesell *et al.* 2000) have major limitations. In these studies, the carbonate chemistry was modified *abruptly* in *short-term* experiments involving a *single clone* of a *single species*. This is basically testing the physiological response (or acclimation potential) of an individual to abnormal change, ignoring the ongoing evolutionary adaptation of species and communities. In fact, natural populations of pelagic species are immense, occupying circum-global biogeographic ranges, and thus genetically highly polymorphic (Medlin *et al.* 1996). Pelagic genomes are dividing on daily time scales, thus adapting (i.e., slightly modifying their fitness and ecological range) at exceptionally high pace through the constant and rapid reset of the worldwide population.

The intense genetic turnover characterizing pelagic biodiversity may be a key evolutionary strategy for survival in this unstable and climatically responsive environment, which is obviously difficult to test in laboratory conditions.

To sum up, we have shown in this chapter that there is not a singular coccolithophore (and certainly not *Emiliania*!) but several, widely divergent groups of potentially calcifying haptophytes, the Calcihaptophycidae. Our journey through their fossil record and molecular evolution has shown that their biomineralization was originally selected in a high CO_2, low pH, aragonite ocean (see Figure 8), whose conditions may actually resemble the future Anthropogenic world after 2100. They radiated into an astounding morphological diversity of highly productive species in the Cretaceous Calcite II Ocean, which was relatively acidic under a high CO_2 atmosphere (see Figure 8). And they were bigger than ever, producing thicker and large coccoliths across the Paleocene-Eocene Thermal Maximum (probably the best geologic analogue for future global change), when a massive increase in atmospheric CO_2 over a 10,000-year period caused rapid $CaCO_3$ dissolution at the seafloor and shoaling of the CCD by at least 2 km (Zachos *et al.* 2005). Thus, representing the ultimate haptophyte adaptation to the pelagic realm, the Calcihaptophycidae may in fact be strongly equipped against extinction, capable of multiplying in both haploid and diploid, calcifying or noncalcifying phases of their life cycle and having the potential to reinvent biomineralization at any time from coastal pools of noncalcifying taxa. Future palaeontogenomic approaches (de Vargas and Probert 2004b) will certainly help unveil the biological and functional complexity of the calcareous flowers of the oceans.

Acknowledgments

We express our warm thanks to Miguel Frada, Hui Liu, and Swati Narayan-Yadav for their great energy and help in reconstructing the molecular evolution of the coccolithophores. In addition, many thanks to Paul Falkowski and Andy Knoll for their invitation to participate in this volume; Chantal Billard for her help with the Latin diagnosis of the Calcihaptophycidae; Betsy Read, Fabrice Not, Isao Inouye, and Bjoern Rost for sharing unpublished data, and Ghebail Araia and Daniel Vaulot for their patience. This work was supported by the U.S. NSF grant DEB-0415351 and an ATIP fellowship from the Centre National de la Recherche Scientifique (CdV). It is part of the pluridisciplinary project BOOM (Biodiversity of Open Ocean Microcalcifiers), funded by the Institut Français de la Biodiversité via the Agence National de la Recherche, grant ANR-05-BDIV-004.

References

Alcober, J., and Jordan, R.W. (1997). An interesting association between *Neosphaera coccolithomorpha* and *Ceratolithus cristatus* (Haptophyta). *Eur. J. Phycol.* **32:** 91–93.

Amato, A., Orsini, L., D'alelio, D., and Montresor, M. (2005). Life cycle, size reduction patterns, and ultrastructure of the pennate planktonic diatom *Pseudo-nizschia delicatissima* (Bacillariophyceae). *J. Phycol.* **41:** 542–556.

Anbar, A.D., and Knoll, A.H. (2002). Proterozoic ocean chemistry and evolution: a bioinorganic bridge? *Science* **297:** 1137–1142.

Andruleit, H., Rogalla, U., and Stager, S. (2004). From living communities to fossil assemblages: origin and fate of coccolithophores in the northern Arabian Sea. *Micropaleontology* **50:** 5–21.

Aubry, M.-P. (1992). Late Paleogene calcareous nannoplankton evolution: a tale of climatic deterioration. *The Eocene-Oligocene Climatic and Biotic Changes.* D. Prothero and W. A. Berggren, eds. Princeton, Princeton University Press, pp. 272–309.

Aubry, M.-P. (1998). Early Paleogene calcareous nannoplankton evolution: a tale of climatic amelioration. *Late Paleocene-Early Eocene Climatic and Biotic Events in the Marine and Terrestrial Records.* M.-P. Aubry, S. Lucas and W. A. Berggren, eds. New York, Columbia University Press.

Aubry, M.-P. (In press-a). *Handbook of Cenozoic Calcareous Nannoplankton. Micropress, the Micropaleontology Project.*

Aubry, M.-P. (In press-b). A major Mid-Pliocene calcareous nannoplankton turnover: change in life strategy in the photic zone. *Geol. Soc. Am. Special Paper.*

Bachvaroff, T.R., Sanchez Puerta, M.V., and Delwiche, C.F. (2005). Chlorophyll c-containing plastid relationships based on analyses of a multigene data set with all four chromalveolate lineages. *Mol. Biol. Evol.* **22:** 1772–1782.

Baldauf, S.L. (2003). The deep roots of eukaryotes. *Science* **300:** 1703–1706.

Baldauf, S.L., Roger, A.J., Wenk-Siefert, I., and Doolittle, W.F. (2000). A kingdom-level phylogeny of eukaryotes based on combined protein data. *Science* **290:** 972–977.

Baumann, K.-H., Bockel, B., and Frenz, M. (2004). Coccolith contribution to South Atlantic carbonate sedimentation. *Coccolithophores—From Molecular Processes to Global Impact.* H.R. Thierstein and J.R. Young, eds. New York, Springer, pp. 367–402.

Beech, P.L., and Wetherbee, R. (1988). Observations on the flagellar apparatus and peripheral endoplasmic-reticulum of the coccolithophorid, *Pleurochrysis carterae* (Prymnesiophyceae). *Phycologia* **27:** 142–158.

Ben Ali, A., De Baere, R., Van Der Auwera, G., De Wachter, R., and Van De Peer, Y. (2001). Phylogenetic relationships among algae based on complete large subunit rRNA sequences. *Int. J. Syst. Evol. Microbiol.* **51:** 737–749.

Berney, C., and Pawlowski, J. (2006). A molecular timescale for eukaryote evolution recalibrated with the continuous microfossil record. *Proc. R. Soc. B Biol. Sci.* **273:** 1867–1872.

Billard, C., and Inouye, I. (2004). What's new in coccolithophore biology? *Coccolithophores—From Molecular Processes to Global Impact.* H.R. Thierstein and J.R. Young, eds. Springer Verlag, pp. 1–29.

Bown, P.R. (2005). Calcareous nannoplankton evolution: a tale of two oceans. *Micropaleontology* 299–308.

Bown, P.R., Lees, J.A., and Young, J.R. (2004). Calcareous nannoplankton evolution and diversity through time. *Coccolithophores—From Molecular Processes to Global Impact.* H. R. Thierstein and J. R. Young, eds. Springer Verlag, pp. 481–505.

Braarud, T., Deflandre, G., Halldal, P., and Kamptner, E. (1955). Terminology, nomenclature, and systematics of the Coccolithophoridae. *Micropaleontology* **1:** 157–159.

Broecker, W.S., and Peng, T.-H. (1987). The role of CaCO3 compensation in the glacial to interglacial atmospheric CO2 change. *Global Biogeochemical Cycles* **1:** 15–26.

Brownlee, C., and Taylor, A.R. (2004). Calcification in coccolithophores. *Coccolithophores—From Molecular Processes to Global Impact.* H. R. Thierstein and J. R. Young, eds. Springer, pp. 31–49.

Cavalier-Smith, T. (1999). Principles of protein and lipid targeting in secondary symbiogenesis: euglenoid, dinoflagellate, and sporozoan plastid origins and the eukaryote family tree. *J. Eukary. Microbiol.* **46:** 347–366.

Cavalier-Smith, T. (2002). The phagotrophic origin of eukaryotes and phylogenetic classification of protozoa. *Int. J. Syst. Evol. Microbiol.* **52:** 297–354.

Constantz, B.R. (1986). Coral skeleton construction: a physiochemically dominated process. *Palaios* **1:** 152–157.

Corstjens, P.L.A.M., Van Der Kooij, A., Linschooten, C., Brouwers, G.-J., Westbroek, P., and Jong, E.W.D.V.-D. (1998). GPA, a calcium-binding protein in the coccolithophorid *Emiliania huxleyi* (Prymnesiophyceae). *J. Phycol.* **34:** 622–630.

Cros, L., and Fortuno, J.M. (2002). Atlas of northwestern Mediterranean coccolithophores—preface. *Scientia Marina* **66:** 1–70.

Delille, B., and others (2005). Response of primary production and calcification to changes of pCO(2) during experimental blooms of the coccolithophorid *Emiliania huxleyi*. *Global Biogeochemical Cycles* **19,** GB2023.

de Vargas, C., Bonzon, M., Rees, N., Pawlowski, J., and Zaninetti, L. (2002). A molecular approach to biodiversity and ecology in the planktonic foraminifera *Globigerinella siphonifera* (d'Orbigny). *Mar. Micropaleontol.* **45:** 101–116.

de Vargas, C., and Probert, I. (2004b). New keys to the past: current and future DNA studies in coccolithophores. *Micropaleontology* **50:** 45–54.

de Vargas, C., Sáez, A.G., Medlin, L.K., and Thierstein, H.R. (2004). Super-species in the calcareous plankton. *Coccolithophores—From Molecular Processes to Global Impact*. H.R. Thierstein and J.R. Young, eds. Springer Verlag, pp. 271–298.

Diez, B., Pedros-Alio, C., and Massana, R. (2001). Study of genetic diversity of eukaryotic picoplankton in different oceanic regions by small-subunit rRNA gene cloning and sequencing. *Appl. Environ. Microbiol.* **67:** 2932–2941.

Falkowski, P.G., and others (2005). The rise of oxygen over the past 205 million years and the evolution of large placental mammals. *Science* **309:** 2202–2204.

Falkowski, P.G., Schofield, O., Katz, M.E., Van De Schootenbruggem, B., and Knoll, A.H. (2004). Why is land green and the ocean red? *Coccolithophores: From Molecular Processes to Global Impact*. H.R. Thierstein and J.R. Young, eds. Springer Verlag.

Farrimond, P., Eglinton, G., and Brassell, S.C. (1986). Alkenones in Cretaceous black shales, Blake-Bahama Basin, western North Atlantic. *Org. Geochem.* **10:** 897–903.

Fast, N., Kissinger, J.C., Roos, D.S., and Keeling, P.J. (2001). Nuclear-encoded, plastid-targeted genes suggest a single common origin for apicomplexan and dinoflagellate plastids. *Mol. Biol. Evol.* **18:** 418–426.

Fujiwara, S., Tsuzuki, M., Kawachi, M., Minaka, N., and Inouye, I. (2001). Molecular phylogeny of the haptophyta based on the rbcL gene and sequence variation in the spacer region of the RUBISCO operon. *J. Phycol.* **37:** 121–129.

Gallagher, L.T. (1989). Reticulofenestra: a critical review of taxonomy, structure and evolution. *Nanofossils and Their Applications*. J. A. Crux and S. E. V. Heck, eds. New York, Ellis Harwood, pp. 41–75.

Gast, R.J., Mcdonnell, T.A., and Caron, D.A. (2000). srDNA-based taxonomic affinities of algal symbionts from a planktonic foraminifer and a solitary radiolarian. *J. Phycol.* **36:** 172–177.

Green, J.C., and Course, P.A. (1983). Extracellular calcification in *Chrysotila lamellosa* (Prymnesiophyceae). *Br. Phycol. J.* **18:** 367–382.

Green, J.C., Course, P.A., and Tarranb, G.A. (1996). The life-cycle of *Emiliania huxleyi*: a brief review and a study of relative ploidy levels analysed by flow cytometry. *J. Mar. Syst.* **9:** 33–44.

Green, J.C., Hibberd, D.J., and Pienaar, R.N. (1982). The taxonomy of *Prymnesium* (Prymnesiophyceae) including a description of a new cosmopolitan species, *P. patellifera* sp. nov., and further observations on *P. parvum* N. Carter. *Br. Phycol. J.* **17:** 363–382.

Hackett, J.D., Yoon, H.S., Butterfield, N.J., Sanderson, M.J., and Bhattacharya, D. (2007). Plastid endosymbiosis: sources and timing of the major events. *Evolution of Primary Producers in the Sea*. P.G. Falkowski and A.H. Knoll, eds. Boston, Elsevier, pp. 109–131.

Hamm, C. and Smatacek, V. (2007). Armor: why, when, and how. *Evolution of Primary Producers in the Sea*. P.G. Falkowski and A.H. Knoll, eds. Boston, Elsevier, pp. 311–332.

Haq, B.U., and Boersma, A. (1978). *Introduction to Marine Micropaleontology*. Elsevier.

Haq, B.U., Hardenbol, J., and Vail, P.R. (1987). Chronology of fluctuating sea levels since the Triassic. *Science* **235:** 1156–1167.

Harper, J.T., and Keeling, P.J. (2003). Nucleus-encoded, plastid-targeted glyceraldehyde-3-phosphate dehydrogenase (GAPDH) indicates a single origin for chromalveolate plastids. *Mol. Biol. Evol.* **20:** 1730–1735.

Harper, J.T., Waanders, E., and Keeling, P.J. (2005). On the monophyly of chromalveolates using a six-protein phylogeny of eukaryotes. *Int. J. Syst. Evol. Microbiol.* **55:** 487–496.

Hay, W.W. (1977). Calcerous nannofossils. *Oceanic Micropaleontology*, vol. 2. A.T. Ramsay, ed. New York, Academic Press, pp. 1055–1200.

Hay, W.W. (2004). Carbonate fluxes and calcareous nannoplankton. *Coccolithophores—From Molecular Processes to Global Impact*. H. Thierstein and J. R. Young, eds. New York, Springer, pp. 509–527.

Hedges, S., Blair, J., Venturi, M., and Shoe, J. (2004). A molecular timescale of eukaryote evolution and the rise of complex multicellular life. *BMC Evol. Biol.* **4:** 2.

Herfort, L., Loste, E., Meldrum, F., and Thake, B. (2004). Structural and physiological effects of calcium and magnesium in *Emiliania huxleyi* (Lohmann) Hay and Mohler. *J. Struct. Biol.* **148:** 307–314.

Houdan, A. and others (2004a). Flow cytometric analysis of relative ploidy levels in holococcolithophore-heterococcolithophore (Haptophyta) life cycles. *Syst. Biodiversity* **1:** 14–28.

Houdan, A., Bonnard, A., Fresnel, J., Fouchard, S., Billard, C., and Probert, I. (2004b). Toxicity of coastal coccolithophores (Prymnesiophyceae, Haptophyta). *J. Plankton Res.* **26:** 875–883.

Huxley, T.H. (1857). *Deep Sea Soundings in the North Atlantic Ocean, Between Ireland and New Foundland.* London, Lords Commissioners of the Admiralty, pp. 63–68.

Isozaki, Y. (1997). Permo-Triassic boundary superanoxia and stratified superocean: records from lost deep sea. *Science* **276:** 235–238.

John, U., Fensome, R.A., and Medlin, L. (2003). The application of a molecular clock based on molecular sequences and the fossil record to explain biogeographic distributions within the *Alexandrium tamarense* species complex (Dinophyceae). *Mol. Biol. Evol.* **20:** 1015–1027.

Jordan, R.W., Cros, L., and Young, J.R. (2004). A revised classification scheme for living haptophytes. *Micropaleontology* **50:** 55–79.

Katz, M.E., Finkel, Z.V., Grzebyk, D., Knoll, A.H., and Falkowski, P.G. (2004). Evolutionary trajectories and biogeochemical impacts of marine eukaryotic phytoplankton. *Annu. Rev. Ecol. Evol. Syst.* **35:** 523–556.

Katz, M.E., Fennel, K., and Falkowski, P.G. (2007). Geochemical and biological consequences of phytoplankton evolution. *Evolution of Primary Producers in the Sea.* P.G. Falkowski and A.H. Knoll, eds. Boston, Elsevier, pp. 405–429.

Kawachi, M., and Inouye, I. (1994). Ca2+−mediated induction of the coiling of the haptonema in *Chrysochromulinahirta* (Prymnesiophyta = Haptophyta). *Phycologia* **33:** 53–57.

Kawachi, M., Inouye, I., Maeda, O., and Chihara, M. (1991). The haptonema as a food-capturing device: observations on *Chrysochromulina hirta* (Prymnesiophyceae). *Phycologia* **30:** 563–573.

Knoll, A.H. (2003). Biomineralization and evolutionary history. *Rev. Mineral. Geochem.* **54:** 329–356.

Knoll, A.H., Summons, R.E., Waldbauer, J.R., and Zumberge, J.E. (2007). The geological succession of primary producers in the oceans. *Evolution of Primary Producers in the Sea.* P. G. Falkowski and A.H. Knoll, eds. Boston, Elsevier, pp. 133–163.

Larsen, A., and Edvardsen, B. (1998). Relative ploidy levels in *Prymnesium parvum* and *P. patelliferum* (Haptophyta) analyzed by flow cytometry. *Phycologia* **37:** 412–424.

Leclercq, N., Gattuso, J.P., and Jaubert, J. 2000. CO_2 partial pressure controls the calcification rate of a coral community. *Global Change Biol.* **6:** pp. 329–334.

Leibovitz, L., and Lebouitz, S.S. (1985). A coccolithophorid algal dermatitis of the spiny dogfish, *Squalus acanthias* L. *J. Fish Dis.* **8:** 351–358.

Lohman, H. (1902). Die Coccolithophoridae, eine Monographie der Coccolithen bildenden Flagellaten, zugleich ein Beitrag zur Kenntnis des Mittelmeerauftriebs. *Archiv für Protistenkunde* **1:** 89–165.

Manton, I., and Oates, K. (1980). Polycrater galapagensis gen. et sp. nov., a putative coccolithophorid from the Galapagos Islands with an unusual aragonitic periplast. *Br. Phycol. J.* **15:** 95–103.

Manton, I., and Peterfi, L.S. (1969). Observations on the fine structure of coccoliths, scales and the protoplast of a freshwater coccolithophorid, *Hymenomonas roseola* Stein, with supplementary observations on the protoplast of *Cricosphaera carterae. Proc. R. Soc.* **172:** 1–15.

Marchant, H.J., and Thomsen, H.A. (1994). Haptophytes in polar waters. *The Haptophyte Algae.* J.C. Green and B.S.C. Leadbeater, eds. Oxford, Clarenden Press, pp. 209–228.

Marin, F., Smith, M., Isa, Y., Muyzer, G., and Westbroek, P. (1996). Skeletal matrices, muci, and the origin of invertebrate calcification. *PNAS* **93:** 1554–1559.

Marsh, M.E. (2003). Regulation of CaCO3 formation in coccolithophores. *Comp. Biochem. Physiol. B Biochem. Mol. Biol.* **136:** 743–754.

Medlin, L.K. and others (1996). Genetic characterization of *Emiliania huxleyi* (Haptophyta). *J. Mar. Syst.* **9:** 13–31.

Milliman, J.D. (1993). Production and accumulation of calcium carbonate in the ocean—budget of a nonsteady state. *Global Biogeochemical Cycles* **7:** 927–957.

Moon-Van Der Staay, S.Y., Van Der Staay, G.W.M., Guillou, L., Vaulot, D., Claustre, H., and Medling, L.K. (2000). Abundance and diversity of prymnesiophytes in picoplankton community from the equatorial Pacific Ocean inferred from 18S rDNA sequences. *Limnol. Oceanogr.* **45:** 98–109.

Nakayama, T., Yoshida, M., Noel, M.H., Kawachi, M., and Inouye, I. (2005). Ultrastructure and phylogenetic position of *Chrysoculter rhomboideus* gen. et sp nov (Prymnesiophyceae), a new flagellate haptophyte from Japanese coastal waters. *Phycologia* **44:** 369–383.

Not, F., and others (2005). Late summer community composition and abundance of photosynthetic picoeukaryotes in Norwegian and Barents seas. *Limnol. Oceanogr.* **50:** 1677–1686.

Orr, J.C., Fabry, V.J., Aumont, O., and others (2005). Anthropogenic ocean acidification over the twenty-first century and its impact on calcifying organisms. *Nature* **437:** 681–686.

Orsini, L., Procaccini, G., Sarno, D., and Montresor, M. (2004). Multiple rDNA ITS-types within the diatom *Pseudo-nitzschia delicatissima* (Bacillariophyceae) and their relative abundances across a spring bloom in the Gulf of Naples. *Mar. Ecol. Prog. Ser.* **271:** 87–98.

Paasche, E. (2002). A review of the coccolithophorid *Emiliania huxleyi* (Prymnesiophyceae), with particular reference to growth, coccolith formation, and calcification-photosynthesis interactions. *Phycologia* **40:** 503–529.

Parke, M., and Adams, I. (1960). The motile (Crystallolithus hyalinus Gaarder and Markali) and nonmotile phases in the life history of Coccolithus pelagicus (Wallich) Schiller. *J. Mar. Biol. Assoc. U K* **39:** 263–274.

Perch-Nielsen, K. (1981). Nouvelles observations sur les nannofossiles calcaires à la limite Crétacé/Tertiaire près de El Kef, Tunisie. *Cahiers de Micropaleontologie* **3**: 25–36.

Peterson, K.J., and Butterfield, N.J. (2005). Origin of the Eumetazoa: testing ecological predictions of molecular clocks against the Proterozoic fossil record. *Proc. Natl. Acad. Sci. U S A* **102**: 9547–9552.

Pienaar, R.N. (1980). Observations on the structure and composition of the cyst of *Prymnesium* (Prymnesiophyceae). Annual Conference Proceedings of Electron Microscopy Society of South Africa, pp. 73–74.

Pintner, I.J., and Pravasoli, L. (1968). Heterotrophy in subdued light of three *Chrysochromulina* species., p. 25–31. U.S.-Japan seminar on marine microbiology. *Bull. Miaki Mar. Biol. Inst. Kyoto University*.

Posada, D., and Crandall, K.A. (1998). MODELTEST: testing the model of DNA substitution. *Bioinformatics* **14**: 817–818.

Quigg, A., and 8 others (2003). The evolutionary inheritance of elemental stoichiometry in marine phytoplankton. *Nature* **425**: 291–294.

Quinn, P., Bowers, R.M., Zhang, X., Wahlund, T., Fanelli, M., Olszova, D., and Read, B. (2006). cDNA microarrays as a tool for identification of biomineralization proteins in the coccolithophorid *Emiliania huxleyi* (Haptophyta). *Appl. Environ. Microbiol.* **72**: 5512–5526.

Ridgwell, A. (2005). A mid Mesozoic revolution in the regulation of ocean chemistry. *Mar. Geol.* **217**: 339–357.

Ridgwell, A., and Zeebe, R.E. (2005). The role of the global carbonate cycle in the regulation and evolution of the Earth system. *Earth Planetary Sci. Lett.* **234**: 299–315.

Riebesell, U., Zondervan, I., Rost, B., Tortell, P.D., Zeebe, R.E., and Morel, F.M. (2000). Reduced calcification of marine plankton in response to increased atmospheric CO_2. *Nature* **407**: 364–367.

Romein, A.J.T. (1979). Lineages in Early Paleogene calcareous nannoplankton. *Utrecht Micropaleontol. Bull.* **22**: 231.

Ronov, A.B. (1994). Phanerozoic transgression and regressions on the continents—A quantitative approach based on areas flooded by the sea and areas of marine and continental deposition. *Am. J. Sci.* **294**: 777–801.

Rost, B., and Riebesell, U. (2004). Coccolithophores and the biological pump: responses to environmental changes. *Coccolithophores—From Molecular Processes to Global Impact*. H.R. Thierstein and J.R. Young, eds. Springer Verlag, pp. 99–125.

Rowson, J.D., Leadbeater, B.S.C., and Green, J.C. (1986). Calcium carbonate deposition in the motile (*Crystallolithus*) phase of *Coccolithus pelagicus* (Prymnesiophyceae). *Br. Phycol. J.* **21**: 359–370.

Royer, D.L., Berner, R.A., Montañez, I.P., Tabor, N.J., and Beerling, D.J. (2004). CO_2 as a primary driver of Phanerozoic climate. *GSA Today* **14**: 4–10.

Sabine, C.L., and 14 others (2004). The oceanic sink for anthropogenic CO_2. *Science* **305**: 367–371.

Sáez, A.G., Probert, I., Quinn, P., Young, J.R., Geisen, M., and Medlin. L.K. (2003). Pseudocryptic speciation in coccolithophores. *Proc. Natl. Acad. Sci. U S A* **100**: 7163–7168.

Sánchez Puerta, V., Bachvaroff, T.R., and Delwiche, C.F. (2005). The complete plastid genome sequence of the haptophyte *Emiliania huxleyi*: a comparison to other plastid genomes. *DNA Res.* **12**: 151–156.

Sandberg, P.A. (1983). An oscillating trend in Phanerozoic non-skeletal carbonate mineralogy. *Nature* **305**: 19–22.

Schroeder, D.C., and 8 others (2005). A genetic marker to separate *Emiliania huxleyi* (Prymnesiophyceae) morphotypes. *J. Phycol.* **41**: 874–879.

Sprengel, C., and Young, J.R. (2000). First direct documentation of associations of *Ceratolithus cristatus* ceratoliths, hoop-coccoliths and *Neosphaera coccolithomorpha* planoliths. *Mar. Micropaleontol.* **39**: 39–41.

Stanley, J., and George D. (2003). The evolution of modern corals and their early history. *Earth Sci. Rev.* **60**: 195–225.

Stanley, S.M., and Hardie, L.A. (1999). Hypercalcification: paleontology links plate tectonics and geochemistry to sedimentology. *GSA Today* **9**: 2–7.

Stanley, S.M., Ries, J.B., and Hardie, L.A. (2005). Seawater chemistry, coccolithophore population growth, and the origin of Cretaceous chalk. *Geology* **33**: 593–596.

Stechmann, A., and Cavalier-Smith, T. (2003). Phylogenetic analysis of eukaryotes using heat-shock protein Hsp90. *J. Mol. Evol.* **57**: 408–419.

Streng, M., Hildebrand-Habel, T., and Willems, H. (2004). A proposed classification of archeopyle types in calcareous dinoflagellate cysts. *J. Paleontol.* **78**: 456–483.

Summons, R.E., and Walter, M.R. (1990). Molecular fossils and microfossils of procaryotes and protists from Proterozoic sediments. *Am. J. Sci.* **290**: 212–244.

Swofford, D. (2000). *PAUP*. Phylogenetic Analysis Using Parsimony (* and other methods)*. Version 4.

Sym, S., and Kawachi, M. (2000). Ultrastructure of *Calyptrosphaera radiata*, sp. nov. (Prymnesiophyceae, Haptophyta). *Eur. J. Phycol.* **35**: 283–293.

Takano, Y., Hagino, K., Tanaka, Y., Horiguchi, T., and Okada, H. (2006). Phylogenetic affinities of an enigmatic nannoplankton, *Braarudosphaera bigelowii* based on the SSU rDNA sequences. *Mar. Micropaleontol.* **60**: 145–156.

Tappan, H. (1980). *The Paleobiology of Plant Protists*. San Francisco, Freeman.

Thierstein, H.R., Geitzenauer, K.R., Molfino, B., and Shackleton, N.J. (1977). Global synchroneity of late Quaternary coccolith datum levels Validation by oxygen isotopes. *Geology* **5**: 400–404.

Tillmann, U. (1998). Phagotrophy by a plastidic haptophyte, *Prymnesium patelliferum*. *Aq. Microb. Ecol.* **14**: 155–160.

Vaulot, D., Birrien, J-L., Marie, D., Casotti, R., Veldhuis, M., Kraay, G., and Chretiennot-Dinet, M-J. (1994). Morphology, ploidy, pigment composition, and genome size of cultured strains of *Phaeocystis* (Prymnesiophyceae). *J. Phycol.* **30**: 1022–1035.

Walker, L.J., Wilkinson, B.H., and Ivany, L.C. (2002). Continental drift and Phanerozoic carbonate accumulation in shallow-shelf and deep-marine settings. *J. Geol.* **110:** 75–87.

Wallich, G.C. (1861). Remarks on some novel phases of organic life and on the boring powers of minute annelids at great depths in the sea. *Ann. Magazine Natural History* **3:** 52–58.

Wallich, G.C. (1877). Observations on the coccosphere. *Ann. Magazine Natural History* **4:** 342–350.

Westbroek, P., and Marin. F. (1998). A marriage of bone and nacre. *Nature* **392:** 861–862.

Yoon, H.S., Hackett, J.D., Ciniglia, C., Pinto, G., and Bhattacharya, D. (2004). A molecular timeline for the origin of photosynthetic eukaryotes. *Mol. Biol. Evol.* **21:** 809–818.

Yoshida, M., Noel, M.-H., Nakayama, T., Naganuma, T., and Inouye, I. (2006). A haptophyte bearing siliceous scales: ultrastructure and phylogenetic position of *Hyalolithus neolepis* gen. et sp. nov. (Prymnesiophyceae, Haptophyta). *Protist* **157:** 213–234.

Young, J.R. (1994). Functions of coccoliths. *Coccolithophores*. A. Winter and W.G. Seisser, eds. Cambridge University Press, pp. 63–82.

Young, J.R., Davis, S.A., Bown, P.R., and Mann, S. (1999). Coccolith ultrastructure and biomineralisation. *J. Struct. Biol.* **126:** 195–215.

Young, J.R., Didymus, J.M., Bown, P.R., Prins, B., and Mann, S. (1992). Crystal assembly and phylogenetic evolution in heterococcoliths. *Nature* **356:** 516–518.

Young, J.R., Geisen, M., Cros, L., Kleijne, A., Sprengel, C., and Probert, I. (2003). A guide to extant calcareous nannoplankton taxonomy. *J. Nannoplankton Res.* **28(1):** 1–125.

Young, J.R., Geisen, M., and Probert, I. (2005). Review of selected aspects of coccolithophore biology with implications for paleobiodiversity estimation. *Micropaleontology* **51:** 267–288.

Young, J.R., Henriksen, K., and Probert, I. (2004). Structure and morphogenesis of the coccoliths of the CODENET species. *Coccolithophores—From Molecular Processes to Global Impact*. H. Thierstein and J. R. Young, eds. Springer, 191–216.

Zachos, J.C., and 11 others (2005). Rapid acidification of the ocean during the Paleocene-Eocene thermal maximum. *Science* **308:** 1611–1615.

Zeebe, R.E., and Westbroek, P. (2003). A simple model for the CaCO3 saturation state of the ocean: the "Strangelove," the "Neritan," and the "Cretan" Ocean. *Geochem. Geophys. Geosyst.* **4:** 1104.

13

The Origin and Early Evolution of Green Plants

CHARLES J. O'KELLY

With the possible exception of lions, tigers, and bears, green plants are the most familiar of nonhuman living things. We eat them, write on them, build with them (or make it look like we are), burn them, decorate with them, wrap ourselves in them, breathe their waste products. They are the palm trees waving over South Pacific beaches, the crimson-and-gold maples of New Hampshire in October, the amber waves of grain. To Linnaeus, they were one of the three kingdoms of matter (vegetal, animal, mineral), vital to understanding God's good Earth. And on that good Earth, at least the drier parts, the adjective is not necessary, for practically every plant is green.

In today's salt water, however, it's a different story, as readers of this volume have gathered by now. The oceans grow little wood and not much "grass." Among the seaweeds and the phytoplankton, green is a minority color, struggling with the reds and browns for a place in the sun, and it has been doing so since the Proterozoic (Katz *et al*. 2004). The green plants found that place in the sun, not in the briny deep but emerging from the freshwater vernal pools of the Ordovician—some 1000 million years after the green plants are thought to have first appeared, as one of the three descendant lineages of the first photosynthetic eukaryotes.

Because of the obvious practical importance of green plants to humankind, efforts to understand their origin and evolutionary history are many, including those now underway under the "Deep Green" umbrella (http://ucjeps.berkeley.edu/TreeofLife/related.php). A comprehensive review is well beyond the scope of this contribution. Instead, the text touches briefly on these topics:

1. Green plants defined: how we know a green plant when we see one.
2. Green plant body plans: what we know of the evolutionary history of some of the basic types.
3. The core structure of the green plant phylogenetic tree.
4. Difficulties (unresolved nodes) in the phylogenetic tree, with special reference to those areas that are most relevant to larger issues of phototroph evolution in aquatic (especially marine) environments.
5. A brief note on green plants in today's marine environment.

I. GREEN PLANTS DEFINED

Not all of the photosynthetic eukaryotes that are green in color are "green plants." Euglenoids, chlorarachniophytes, and "green" dinoflagellates are green because they have chloroplasts derived from green plants by secondary or tertiary endosymbiosis (Hackett et al., Chapter 7, this volume). Tribophytes (also called "xanthophytes") and some raphidophytes, two groups of "red" plants, are green because their chloroplasts lack the orange xanthophylls, such as fucoxanthin, that are present in most of their relatives, for instance the diatoms (Andersen 2004). Forms of the name "heterokonts," a name that is often used for the lineage that includes tribophytes, raphidophytes, and diatoms (Andersen 2004; Fehling et al., Chapter 6; Hackett et al., Chapter 7, this volume), were originally used to separate green-colored algae with flagellate cells bearing unequal-length flagella (i.e., the

tribophytes) from *bona fide* green algae with flagellate cells bearing equal-length flagella, long before it was understood that these two groups of "green" plants had entirely different evolutionary histories (e.g., Blackman 1900).

The plants that we *do* refer to as the "green plants" are monophyletic in practically all of the phylogenetic trees that have been generated from DNA data (Lewis and McCourt 2004; Fehling et al., Chapter 6; Hackett et al., Chapter 7, this volume). The name "Viridiplantae" (Cavalier-Smith 1981) is often used for this lineage (as in Falkowski et al. 2004a) instead of "the Plant Kingdom," or "Plantae," which, historically, has either included anything remotely plantlike or has been restricted to the descendants of the first plant with an archegonium (sterile jacket of cells around the female gamete, an egg) and sporopollenin-walled spores (products of meiosis). The former definition includes all the "red" plants, and the latter excludes the green algae. Some recent authors have objected to Viridiplantae and have suggested replacements such as "Chlorobionta" (Lewis and McCourt 2004) or "Chloroplastida" (Adl et al. 2005).

The green plants share several morphological, ultrastructural, and biochemical characters (Lewis and McCourt 2004). Ironically, however, nearly all of these characters are associated with the plastid, the ancestrally cyanobacterial component of the chimera that is the green plant cell. Fortunately, no green plant is without a plastid, although in the few colorless species the plastid may never express chlorophyll (i.e., it never becomes a chloroplast). There are three sets of plastid-associated characters.

1. In the chloroplasts, thylakoids are stacked, but the number of thylakoids in a stack is variable, not fixed. In land plant chloroplasts, the thicker stacks ("grana") are often distinctive, but among green algae the stacking is usually less prominent (Figure 1). Nonetheless, in all other groups, the chloroplasts have no thylakoid stacking

(as in red algae and glaucophytes), have fixed numbers of thylakoids in a stack (e.g., two in cryptomonads and three in the golden algal lineages), or have been "stolen" from green plants.

2. True starch is stored within the plastids (see Figure 1). In other eukaryotic photoautotrophs, the polysaccharide storage product either is not starch or it is stored outside the chloroplast.

3. A unique primary pigment set in the chloroplasts, including chlorophylls a and b, α- and β-carotene, and xanthophylls including the β-carotenoid zeaxanthin and its derivates violaxanthin and neoxanthin, plus the α-carotenoid lutein series (lutein, loroxanthin, prasinoxanthin, siphonaxanthin, siphonein).[1]

The green plant pigment set suggests one possible reason why the land is green and the ocean is red, in addition to those postulated by Falkowski *et al.* (2004a and this entire volume). As is well known, the action spectrum for land plant photosynthesis has prominent peaks in the red and blue parts of the spectrum and a trough in the green. Recent work indicates that the principal xanthophylls in land plant chloroplasts (lutein, neoxanthin, the "xanthophyll cycle" pigments) have light harvesting as well as photoprotective and structural roles in the light-harvesting complexes to which they belong (e.g., Gradinaru *et al.* 2000; Croce *et al.* 2001; Liu *et al.* 2004) and that their roles are, as needed, somewhat interchangeable (e.g., Polle *et al.* 2001). They are therefore contributing to the "classical" action spectrum,

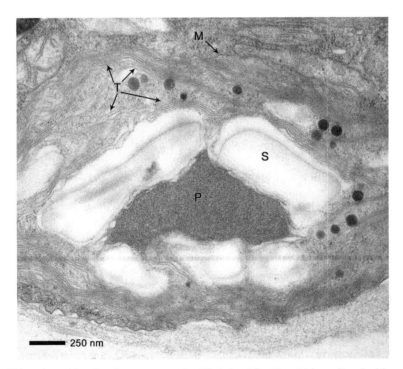

FIGURE 1.   Chloroplast of the ulvophycean green alga *Ulothrix subflaccida,* with bounding double membrane (M), thylakoids (T) with irregular stacking, a crystal of the photosynthetic enzyme ribulose bisphosphate carboxylase/ oxygenase, forming a structure called a *pyrenoid* (P) that is surrounded by starch grains (S).

[1]Many of these carotenoids, especially those of the lutein series, have multiple biologically active derivatives (e.g., Egeland *et al.* 1997; Yoshii *et al.* 2005), so it may be more appropriate to speak of, for instance, not "siphonaxanthin" but "siphonaxanthins."

especially in the blue to blue-green part of the spectrum. The orange xanthophylls (prasinoxanthin, siphonoxanthin, siphonein) of the lutein series also have a light-harvesting function; they extend the photosynthetic action spectrum of algae that possess them into the green range but to a lesser extent than red algal phycobilins or "brown" algal carotenoids such as fucoxanthin (Yokohama 1981; Anderson 1985; Fawley *et al.* 1986).

The orange xanthophylls are absent from land plants and nearly all freshwater green algae, including the "lab rat green flagellate" *Chlamydomonas reinhardtii,* although some are present in the phylogenetically pivotal freshwater flagellate *Mesostigma viride* (Yoshii *et al.* 2003). They are, instead, common among marine planktonic species and in benthic species from subtidal habitats (Yokohama 1981; O'Kelly 1982; Egeland *et al.* 1997; Fawley *et al.* 2000; Yoshii *et al.* 2002, 2004, 2005), where the enhanced green light-harvesting capacity may play a role in helping these algae hold their places. Nevertheless, it is, perhaps, no surprise that the overwhelming majority of green algae have evolved roots, stems, and leaves and crawled out onto the dry land, so that they no longer have to compete with their more photosynthetically efficient "red" counterparts in the sea.

The principal character that is thought to be derived from the eukaryotic ancestor of green plants, is found only in the green plants, and does not vary within the green plants except in fine details (Melkonian 1984) is the "stellate structure" found in the transition zone between flagella and basal bodies, a structure resembling a tube of corrugated cardboard that, under the electron microscope, is a nine-pointed star in cross section and is H-shaped in longitudinal section (Figures 2 and 3).

The terms "basal body" and "centriole" apply to two forms of the same cytoskeletal structure, a cylindrical array of microtubules. The canonical arrangement is nine blades of three microtubules each. There are numerous variations on this theme across the tree of life, although variants are not common among the protists. Basal bodies appear at

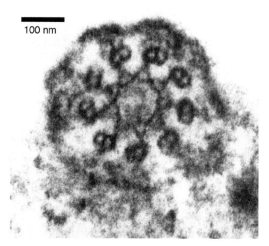

FIGURE 2.   Stellate structure in the transition zone between the basal body and the flagellum in a zoospore of the ulvophycean alga *Bolbocoleon piliferum,* viewed in cross section of both the structure and the containing transition zone.

the base of flagella and are required for the formation, although not the subsequent maintenance, of flagella. Centrioles do not appear at the base of flagella. In many eukaryotes, they appear at the poles of the mitotic and meiotic spindles.

In many green algae, especially more ancestral species, basal bodies or centrioles

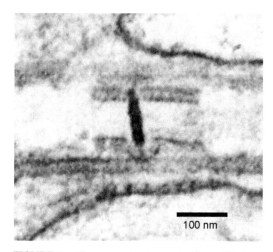

FIGURE 3.   The same structure as in Figure 2 in longitudinal section, showing the typical "H" shape, seen in a zoospore of the ulvophycean alga *Chlorocystis cohnii.*

are always present. In more derived species, centrioles are absent from vegetative cells, forming, apparently *de novo,* only during cell division or flagellate cell (zoospore, gamete) formation. In angiosperms, which make no flagellate sperm, centrioles have never been observed. Obviously, where flagella and basal bodies/centrioles are absent, so are stellate structures. So even this character is not universal for green plants.

The other "host" characters often cited as being diagnostic for green plants have to do with the number, structure, and arrangement of flagella on flagellate cells. Again, these characters do not help those green plants that do not express flagellate cells, and there is more variation within the lineage than is sometimes credited.

In a few green algae, the vegetative cell is flagellate; *Chlamydomonas* is a familiar example. Most of the rest of the green plants have vegetative cells that are nonflagellate, but they may produce flagellate zoospores or gametes.

The zoospore is a flagellate, usually swimming, reproductive cell produced by green plant species that are nonflagellate in the vegetative state. Zoospores differ from gametes because they do not undergo syngamy (gamete fusion). Depending on the life history of the species involved (see below), zoospores may be the products of meiosis and therefore be an essential part of the sexual process, or they may be produced by mitosis, and serve to clone the vegetative phase from which they arose. Most green algae that are nonflagellate in the vegetative state produce zoospores; an exception is the Zygnematales (*Spirogyra* and its relatives) in which neither flagellate zoospores nor gametes are produced. Zoospores are absent from the stoneworts and brittleworts (*Chara, Nitella*) and all of the archegoniate plants.

The basic flagellar number is two—the green plants are part of the "bikont" assemblage of eukaryotes (Fehling *et al.*, Chapter 6, this volume). Uniquely among the photosynthetic bikonts, many green algal flagellate cells express $2n$ flagella, with n = a small integer (five or less, most often two). In cells

with this configuration, the flagella are normally mounted at or near the apical end of the cell. The flagella lack hairs, although they may bear scales, especially on the more ancestral forms, and are the same length (they are "isokont") and have the same motion (they are "isodynamic") (Lewis and McCourt 2004). They drive the cell forward or backward depending on the shape of the flagellar beat (Inouye and Hori 1991; Hoops 1997; Sym *et al.* 2000).

To these general rules, there are several exceptions, especially in flagellar number. Several small flagellates, from a number of different lineages, express only one flagellum (O'Kelly 1992; Sym and Pienaar 1993), and in a few of the larger-celled prasinophytes, uniflagellate cells have been observed (Stewart and Mattox 1978; Daugbjerg *et al.* 2000; Moestrup *et al.* 2003). In some small flagellates with two flagella, the flagella are of unequal length (Sym and Pienaar 1993; Nakayama *et al.* 2000; Zingone *et al.* 2002). A single species of flagellate, *Trichloridella paradoxa,* was reported by Scherffel and Pascher (in Pascher 1927, as *Trichloris paradoxa*) to be triflagellate. This organism has not been seen again. Well documented, however, are the triflagellate zoospores of a marine microfilamentous species, *Acrochaete wittrockii* (Nielsen 1983). Zoospores of some species of Bryopsidaceae (Hoek *et al.* 1995), and the zoospores and gametes of Oedogoniales (Pickett-Heaps 1975), have a ring of flagella around an apical dome (the "stephanokont" condition). Male gametes of horsetails (Renzaglia *et al.* 2006), ferns (Renzaglia *et al.* 2001), and the gymnospermous cycads and *Ginkgo* (Norstog *et al.* 2004), the only flagellate cells in these groups, are likewise multiflagellate, with the flagella arising from a complex blepharoplast that forms a spiral band around at least the apical end of the cell.

II. GREEN PLANT BODY PLANS

Green plants express by far the most versatility in body plans of any lineage of autotrophs. Perhaps this is compensation for the fact that the basic building block, the uninucleate cell, is flat boring by comparison with the elaborate cellular morphologies and extracellular excretions found in other groups of eukaryotic photoautotrophs, particularly the diatoms, dinoflagellates, and haptophytes that have supplanted the green plants in marine plankton (Figure 4).

The unicellular flagellate is considered to be the most ancestral green plant body. In modern representatives, the flagellate cell may have no structural coating (such a cell is often called "naked," although there is usually some sort of glycocalyx on the outer surface of the cell membrane), or bear one or more layers of self-assembling polysaccharide units (Leadbeater and Green 1993; Becker *et al.* 1994; Okuda 2002). These units may be detached from each other (the "scales" of the ultrastructural literature) or fused into a theca (as in *Tetraselmis*) or cell wall (as in *Chlamydomonas*). To date, mineralized scales, such as occur on chrysophyte and haptophyte algae, have not been found on green plants.

From the unicellular flagellate, a bewildering array of body plans has arisen, including colonial flagellates, unicellular and colonial nonflagellate cells (with or without scales, theca, or cell wall; the "coccoidal" morphology), cellular structures forming branched or unbranched filaments, blades, tubes, or solid parenchymatous masses with or without tissue differentiation, or acellular structures forming erect stalks with radially symmetrical branches, or massive thalli composed of a network of siphons (Hoek *et al.* 1995; Graham and Wilcox 2000; Lewis and McCourt 2004). In most of the larger or more elaborate body plans, the cells have chemically complex walls whose principal structural units are interlocking microfibrils. The principal structural polysaccharide in these walls is usually cellulose, but in several groups is, instead, a mannan or xylan (reviewed by Okuda 2002). In some of the larger seaweeds and in the stoneworts, the cell walls may become impregnated with calcium carbonate, but otherwise mineralization of

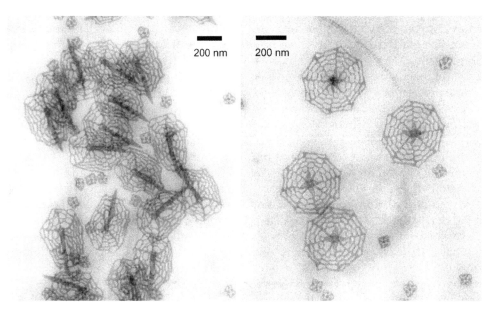

FIGURE 4. Scales of the prasinophyte *Pterosperma* sp., detached from the cells and viewed via transmission electron microscopy. Left, small underlayer and larger shield-shaped scales from the flagella. Right, small underlayer and larger spiderweb scales from the cell body.

walls seldom occurs. Size and complexity of cells and thalli range from the naked aflagellate picoalga *Ostreococcus tauri*, the smallest known autotrophic eukaryote, to fronds of *Ulva* more than a meter in length, or single thalli of *Codium* weighing, fresh, more than a kilogram. Not to mention giant sequoias.

Until the 1970s, algal morphologists tended to assume that the evolution of thallus form was relatively conservative, and they often drew elaborate trees to determine how major lineages typified by one type of thallus could have arisen from lineages typified by another, usually simpler, body type (e.g., Fritsch 1916, 1935, 1945). We now know that most of the simpler body plans have evolved multiple times, sometimes multiple times within a single clade of green algae (Lewis and McCourt 2004), and that simple body plans, including unicellular flagellates and coccoids, may evolve from more complex types (e.g., Friedl and O'Kelly 2002). Commonly, genera considered to be "morphological archetypes," such as *Ulothrix*, "the" exemplar of the unbranched, uniseriate filament morphology, have proven to be polyphyletic—sometimes with the segregate taxa placed in separate green algal classes (Floyd and O'Kelly 1984; O'Kelly and Floyd 1984a, b; Lewis and McCourt 2004).

The polyphyly of thallus forms, especially of the simpler ones, in green algae complicates the interpretation of fossils. This difficulty is made worse by the fact that few green algae are conveniently fossilized and often are so nondescript when fossilized that identities are hard to establish. The relatively few species that are calcified, mostly tropical "seaweeds," do tend to make relatively good fossils, and the orders Dasycladales and Bryopsidales ("Caulerpales" of O'Kelly and Floyd 1984a; Floyd and O'Kelly 1990; Graham and Wilcox 2000) are well represented by these remains (which also include some non-calcified species, e.g., LoDuca *et al.* 2003). A few genera of modern-looking green algae are now being found through the study of borings in Paleozoic and Mesozoic shells and other calcium-carbonate substrata (Glaub and Vogel 2004). Sporopollenin-containing cyst walls are formed in relatively few green algae, but similar forms, such as the leiospherids (oftentimes identifiable by their simplicity relative to other acritarchs), occur through long stretches of the fossil record, from the mid-Proterozoic onward (e.g., Tappan 1980; Javaux *et al.* 2003, 2004; Falkowski *et al.* 2004b). One recent prasinophyte genus (*Pachysphaera*) has been merged with a Paleozoic palynomorph genus (*Tasmanites*) on the basis of the close similarity of cyst wall ornamentation in fossil and living representatives (Guy-Ohlson and Boalch 1992).[2]

A. Green Plant Life Histories

Interpretation of green plant body plans is further complicated by the fact that the body plan may vary in any given species during the course of reproduction, especially sexual reproduction. It is not infrequent to find that two elements within the life history of a single species to have very different body plans and that taxonomists have given those elements separate names.

There are three main sexual life history patterns expressed in green plants. Each type is determined both by the evolutionary history of the species and the environment in which it has evolved.

Zygotic, or haploid-dominant life history has a haploid vegetative plant body, and because it produces gametes (by mitosis), it is called the *gametophyte*. The gametophyte

[2]Boalch and Guy-Ohlson (1992) called the combined genus *Tasmanites* because the name of the fossil genus was published earlier than (and therefore, it had priority over) the name of the extant genus. Subsequently, however, the International Code of Botanical Nomenclature (ICBN) changed the rules, giving names of extant taxa priority over names of fossil taxa regardless of publication date. Thus, the correct genus name for both extant and fossil species is now *Pachysphaera* (ICBN St. Louis, Art. 11.7, Example 28; Greuter *et al.* 2000).

may also reproduce itself, outside of the sexual cycle, by zoospores or other types of propagules, as well as by simple fragmentation. The only diploid cell is the zygote, which typically secretes a thick wall around itself. Typically, on germination, the zygote undergoes meiotic division and releases one to four haploid products, as zoospores or some other propagule type. This life history pattern seems adapted to predictable environmental extremes, such as seasonal freezing or drying of water courses, in which both short-term bursts of growth and long-term resting stages are advantageous. Practically all freshwater green algae have this life history pattern.

The archegoniate plant life history is derived from the zygote-dominant life history by the insertion of mitotic divisions between the formation of the zygote and the onset of meiosis, producing a separate vegetative plant that, because it produces spores by meiosis, is named the *sporophyte*. Initially, the sporophyte is parasitic on the gametophyte, and in the more ancestral plants, it remains so throughout its life span, as is the case in modern mosses, liverworts, and hornworts. In more derived species, the sporophyte becomes independent, and eventually it is the *gametophyte* that becomes the dependent, parasitic generation, as in the ovules of gymnosperms and angiosperms.

Gametic, or diploid-dominant is the familiar life history pattern of multicellular animals. The vegetative plant is diploid, and gametes are produced by meiosis. Specialized cells for asexual reproduction (e.g., zoospores) are seldom seen. The pattern is thought to be adaptive for long-lived species in stable habitats. It is a rare pattern for green plants, found only among species of the order Bryopsidales, such as the reef-building members of the genus *Halimeda*.

Alternation of phases is a life history pattern in which there are separate, free-living vegetative sporophytic and gametophytic plants. Neither phase is ever nutritionally dependent on the other, which distinguishes this pattern from that of the archegoniate plants. The sporophyte and gametophyte may be morphologically identical (*isomorphic*) or dissimilar (*heteromorphic*). These patterns are found only among the ulvophytes and are thought to be adapted to life in less environmentally stable habitats. Predictable disturbances are thought to favor heteromorphic patterns (most ulvophytes in freshwater habitats have this pattern), and *unpredictable* disturbances are thought to favor isomorphic phase alternations.

III. THE CORE STRUCTURE OF THE GREEN PLANT PHYLOGENETIC TREE

Pickett-Heaps and Marchant (1972; see also Pickett-Heaps 1975), primarily on the basis of ultrastructural studies on the patterns of mitosis and cytokinesis in several green algae, developed a two-lineage hypothesis for green plant evolution that was, at the time, a radical departure from previous thought. After three decades of subsequent research on green algal morphology, ultrastructure, and gene/genomic sequences, their proposal has been remarkably well supported (Lewis and McCourt 2004). There are two principal lineages, plus several smaller ones that are collectively termed "prasinophytes" or sometimes (Mattox and Stewart 1984) "micromonadophytes" (Figure 5).

A. The Archegoniate Line

This lineage, which in much modern literature is termed the Streptophyta (Lewis and McCourt 2004), contains all the archegoniate and postarchegoniate plants plus a relatively small number of algal groups.

A number of common names have been applied to the group of plants here called "archegoniate and postarchegoniate": the bryophytes, ferns, gymnosperms, and angiosperms. In addition to "archegoniate," these plants have been dubbed "higher plants," "land plants," and "embryophytes,"

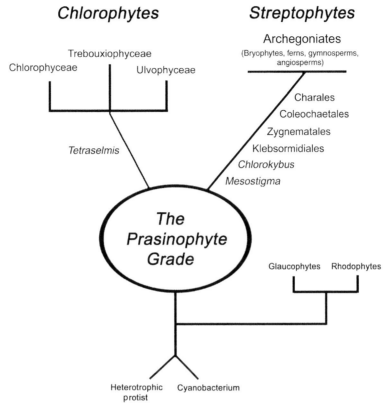

FIGURE 5. Diagrammatic representation of the main lines (Chlorophytes and Streptophytes) of green plant evolution. The "Prasinophyte Grade" incorporates at least six lineages of green algae (see Figure 7) with ancestral morphological characteristics that, on present data, cannot be placed in either of the main lines. Two additional "prasinophyte" lineages, the Chlorodendrales (genus *Tetraselmis*) and *Mesostigma*, have been placed robustly at the bases of, respectively, the Chlorophyte and Streptophyte phylogenetic trees. Each of the six algal taxa within the Streptophytes has been raised to class rank (e.g., Mesostigmatophyceae, Chlorokybophyceae, Klebsormidophyceae, etc.; Lewis and McCourt 2004).

as well as "archegoniates." All of these names trouble some pedant somewhere. The term higher plants implies that these plants are somehow "better" than the rest, and it is now avoided except as the basis for the pun "drier plants," used by phycologists. Land plants overlooks the large number of terrestrial and subaerial green algae, most of which belong to the chlorophyte line (Lewis and McCourt 2004). Embryophyte alludes to the formation of an embryo by the zygote, a diagnostic feature for this lineage of plants (e.g., Kenrick and Crane 1997), but a few chlorophyte algae, *Volvox* in particular, also produce embryos. Archegoniate refers to the *archegonium*, a sterile jacket of cells around the egg. This structure is *the* evolutionary innovation that is unique to the common ancestor of bryophytes and the plants with stems, roots, and leaves. Unfortunately, in angiosperms and gnetophytes, archegonia no longer develop in the female gametophytes, which are tiny in both size and cell number (e.g., Carmichael and Friedman 1996). Nevertheless, archegoniate, especially if it is understood to encompass the "postarchegoniate" gnetophytes and angiosperms, seems to be the best "tag" for the clade that dominates the vegetation of nearly all terrestrial ecosystems.

The plants of the archegoniate line are united in gene-sequence–based phylogenies and by a small number of ultrastructural and biochemical characters (Graham and Wilcox 2000; Lewis and McCourt 2004), including a "halved" (Stewart and Mattox 1978; Melkonian 1984; O'Kelly and Floyd 1984b; O'Kelly 1992) flagellar apparatus that incorporates a multilayered structure (MLS) and an open, persistent mitotic spindle. Although the algal representatives are relatively few, they nevertheless include many of the typical morphologies, including flagellate (*Mesostigma*), coccoidal (Zygnematales, in part), sarcinoid colonial (*Chlorokybus*), uniseriate filamentous (orders Klebsormidiales and, in part, Zygnematales), and branched filamentous with uninucleate (order Coleochaetales) or multinucleate (order Charales) cells. Most of the algal groups, including two with exactly one species each, are currently treated as classes, for example, Mesostigmatophyceae and Chlorokybophyceae (Lewis and McCourt 2004).

Plants belonging to the archegoniate lineage are, obviously, widely distributed in all kinds of habitats throughout the globe, but there is one striking peculiarity: There are almost no modern marine representatives. Of the more than 250,000 species of flowering plants, there may be 50 truly marine ones, almost all of which are mangroves or seagrasses. Perhaps 5 species of stoneworts (Charales) grow in brackish waters (e.g., Blindow 2000). Valentin *et al.* (2005) identified 2 strains of marine picoplankton (out of 63) as "Streptophyta."

For all the algal representatives of this clade, the sexual life history is "zygotic." The life history of the archegoniate plants features a vegetative diploid spore-producing generation (the "sporophyte") that is, at least initially, parasitic on the gametophyte (Graham 1993). The evolutionary origin of this life history was once vigorously debated (e.g., Fritsch 1916, 1945), but it is now firmly established from neontological evidence that it arose from a streptophyte algal ancestor with a zygotic life history, through intercalations of mitotic divisions between the events of syngamy and meiosis (Graham 1993; Graham and Gray 2001; Lewis and McCourt 2004).

Reconciling this conclusion with the fossil record has proved problematic—a problem made worse by the fact that only extant Charales (stoneworts and brittleworts) undergo any significant mineralization (typically, the partial calcification of vegetative cells and the more thorough calcification of oogonia) and thus are the only streptophyte algae to have an interpretable fossil record. The earliest such fossils are from the Silurian, some tens of millions of years later than the Ordovician microfossils interpreted as the first remains of archegoniate plants, and are all from marine sediments. Graham and Gray (2001) argued that these marine remains are unlikely to have originated in freshwater habitats, because they lack the eroded appearance of similar fossils that demonstrably have been transported from freshwaters. Graham and Gray postulated that charophytes and the first archegoniates evolved in the Paleozoic equivalent of vernal pools, periodically dry bodies of water that would have accumulated little sediment, and hence scant opportunity for the preservation of plant remains.

B. The Chlorophyte Line

This lineage, often named the Chlorophyta (Lewis and McCourt 2004), contains the great majority of green algal species and morphological diversity. These plants, again, are united in gene-sequence–based phylogenies and a small number of ultrastructural and biochemical characters, most notably a flagellar apparatus that is nearly radially symmetrical, with usually four microtubular roots subtending two or four basal bodies, and without an MLS (Melkonian 1984; O'Kelly and Floyd 1984b; O'Kelly 1992). Mitotic characters are variable within the lineage, but the open, persistent spindle typical of the streptophytes does not occur (Mattox and Stewart 1984). Several investigators follow Mattox and Stewart (1984) in recognizing three classes within the lineage, Ulvophyceae (before 1982, "Ulvaphyceae"), Trebouxiophyceae (formerly Pleurastrophyceae; Friedl 1995), and Chlorophyceae, defined primarily on the basis of molecular sequence trees, flagellar apparatus absolute configurations, and, in some cases, details of morphology, reproduction, and life history (e.g., Graham and Wilcox 2000; Lewis and McCourt 2004). Other workers, however, recognize as many as seven classes (Hoek *et al.* 1995). Lewis and McCourt (2004) summarize subgroupings in each of the three classes that they recognize.

The Chlorophyceae and Trebouxiophyc-eae are widely represented in freshwaters and on land but are poorly represented in the sea. A few halotolerant species of *Chlamydomonas* exist (e.g., Sasa *et al.* 1992; Kim *et al.* 1994). The best-known species of the trebouxiophycean genera *Prasiola* and *Rosenvingiella* are found in the highest inter-tidal zones of rocky shores, but these coastal forms are marginally marine and are atypi-cal of the group. Most of the species in these genera are from freshwater or terrestrial habitats (Rindi *et al.* 2004). In the picoplank-ton, especially in inshore waters, numerous cells of coccoidal trebouxiophytes related to the genus *Nannochloris* may be found. The taxonomy of this complex is, however, still under investigation (Henley *et al.* 2004).

The Ulvophyceae are relatively poorly represented in *fresh* water. Nearly all species in the orders Bryopsidales (=Caulerpales) and Dasycladales and a majority of the species in the Cladophorales (=Siphonocladales), Ulotrichales, and Ulvales are marine (O'Kelly and Floyd 1984a; Floyd and O'Kelly 1990; Graham and Wilcox 2000; Yoshii *et al.* 2004).

The ulvophycean orders Bryopsidales and Dasycladales also contain the only truly siphonous thalli in the green plants, in fact no species in either order that is *not* of sipho-nous construction has yet been identified. These are also the only orders in the chloro-phyte lineage that contain heavily calcified species. Consequently, most of the body fos-sils that are attributable to green algae have been placed in one or the other of the two, although the three species of *Proterocladus* are commonly assigned to the Cladopho-rales (Butterfield *et al.* 1994). All other prin-cipal morphological types are represented

among the orders of the Chlorophyceae, Trebouxiophyceae, and Ulvophyceae. The ulvophycean order Ulotrichales is excep-tional in that it encompasses every morpho-logical type except the vegetative flagellates and the true siphons (O'Kelly and Floyd 1984a; Floyd and O'Kelly 1990; Friedl and O'Kelly 2002; O'Kelly *et al.* 2004a).

In the Chlorophyceae, all known represent-atives, including the famous *Chlamydomonas*, the sexual life history is zygotic. In the Ulvo-phyceae, however, sexual life history patterns may be zygotic, gametic, or an alternation of diploid and haploid phases. Different sexual life history types define orders, and sometimes families, within the Ulvophyceae (O'Kelly and Floyd 1984a; Floyd and O'Kelly 1990; Graham and Wilcox 2000). Sexual life histories in the Ulotrichales and Dasycla-dales are unusual variants of the zygotic type, with the zygote long-lived, photosyn-thetically active, and undergoing noticeable growth and, in Dasycladales, profound dif-ferentiation prior to the first, meiotic, nuclear division. Among the Trebouxiophyceae, only *Prasiola* and *Rosenvingiella* (Prasiola-les) are known to have sexual life histories, and the pattern is unique and incompletely characterized (Hoek *et al.* 1995).

Major lineages of the Ulvophyceae had differentiated at least by the earliest Paleozoic, and probably much earlier, especially if the tubular *Proterocladus* body fossils from 700–800-million-year-old strata are correctly assigned to the Cladophorales (Butterfield *et al.* 1988, 1994). Calcified Dasycladales were present during the early Cambrian (Berger and Kaever 1992). The trace fossils *Ichnoreticulina elegans*, closely resembling modern *Ostreobium* (Bryopsi-dales; Vogel *et al.* 2000, as *Reticulina elegans*), and *Cavernula pedunculata*, closely resem-bling the sporophyte (elaborated zygote) stage of modern shell-boring Ulotrichales (O'Kelly *et al.* 2004a), have been recorded continuously from the Ordovician to the Recent, whereas *Rhopalia catenata*, equated with modern *Phaeophila dendroides* (Ulva-les; Vogel *et al.* 2000; O'Kelly *et al.* 2004b), has been continuously recorded since the

Triassic (Glaub and Vogel 2004; I. Glaub, personal communication) (Figure 6).

C. The Prasinophytes

The prasinophytes are the green algae that "do not fit." Originally, the taxon "Prasinophyceae" was created to accommodate green phytoflagellates that had chlorophylls a and b and stored starch in the chloroplast but were markedly different from *Chlamydomonas*-like cells in their morphology, especially in having their flagella mounted in an apical pit or groove rather than on a prominence ("papilla") at the anterior end of the cell (Chadefaud 1977). It now encompasses a fairly large number of flagellate and coccoidal species; recent research has uncovered a significant amount of previously unrecorded diversity, mostly among coccoidal, picoplanktonic forms (Fawley *et al.* 2000; Zingone *et al.* 2002; Guillou *et al.* 2004). The new forms include *Ostreococcus tauri*, an abundant and ecologically diverse alga (Rodríguez *et al.* 2005) with both the smallest cell size (Courties *et al.* 1994) and the smallest nuclear genome (Courties *et al.* 1998) of any photosynthetic eukaryote. Members of the group have a wide variety of ultrastructural features (O'Kelly 1992; Sym and Pienaar 1993; Zingone *et al.* 2002; Moestrup *et al.* 2003) and photosynthetic pigment signatures (Egeland *et al.* 1997; Yoshii *et al.* 2002, 2003, 2005; Zingone *et al.* 2002; Latasa *et al.* 2004) and are united only by an absence of similarities to each other and to the streptophyte or chlorophyte clades.

The absence of "positive" characters suggests that "the Prasinophyceae" represents a grade rather than a clade, and DNA data support this suggestion (Lewis and McCourt 2004). Nuclear gene sequences (Bhattacharya

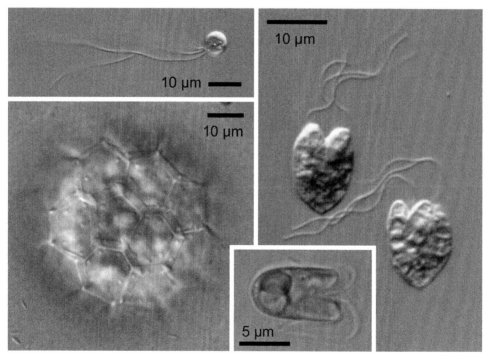

FIGURE 6.   Light micrographs of selected prasinophyte algae. Counterclockwise from top left: flagellate cells of *Pterosperma* sp.; phycoma stage of *Pterosperma* sp. (cf. *P. cristatum*); flagellate cell of *Pyramimonas parkeae*; flagellate cells of *Halosphaera minor*. (See color plate.)

et al. 1998; Marin and Melkonian 1999) and organelle genomes (Lemieux *et al.* 2000; Turmel *et al.* 2002), together with flagellar apparatus ultrastructure (Melkonian 1989), all indicate that the freshwater flagellate *Mesostigma viride* is a basal member of the streptophyte lineage. *Tetraselmis* and *Scherffelia* (family Chlorodendraceae) are placed at or near the base of the chlorophyte lineage on the basis of both mitotic ultrastructure and nuclear gene sequence data (Friedl 1995; Zingone *et al.* 2002; Guillou *et al.* 2004). Pedinomonads (genera *Marsupiomonas*, *Pedinomonas*,[3] and *Resultor*; Moestrup 1991) have similarities in organellar genome content and organization with members of the Chlorophyceae, including *Chlamydomonas* (Turmel *et al.* 1999a). The remaining prasinophytes, with the exception of *Monomastix*, for which ultrastructural and molecular data are not yet sufficient for taxonomic placement (O'Kelly 1992), form a cluster of unresolved lineages at or near the base of the green plant phylogenetic tree; Guillou *et al.* (2004) recognized six such lineages. Organellar genome structure and content in a species of *Nephroselmis,* which belongs to one of these six clades, suggests that they are in fact among the earliest-diverging lineages of green plants, and perhaps among the earliest-diverging lineages of eukaryotes (Gray *et al.* 1998; Turmel *et al.* 1999a, b).

Prasinophytes are almost exclusively planktonic and are predominantly marine. A few species, for instance those adapted to tide pools, may exhibit a rhythmic settling behavior, the nearest approach to a benthic stage (Griffin and Aken 1993). For most species, the only form of reproduction is vegetative cell division. Only one species, *Nephroselmis olivacea,* is known to undergo sexual reproduction, which is of the zygotic type; the zygote forms a cyst (Suda *et al.* 1989, 2004). Cyst formation has been observed in a few other prasinophytes (e.g., Daugbjerg *et al.* 2000; Moro *et al.* 2002), but in all cases, with the possible exception of *Pyramimonas tychotreta* (Daugbjerg *et al.* 2000) and *Cymbomonas tetramitiformis* (Moestrup *et al.* 2003), it appears to be an asexual process.

In most prasinophytes, cysts are nearly the same size as the cells that produce them (e.g., Suda *et al.* 1989, 2004; Daugbjerg *et al.* 2000; Moro *et al.* 2002; Moestrup *et al.* 2003). It appears (data are limited) that they are mostly polysaccharide in nature, lacking sporopollenins or other biodegradation-resistant compounds (Aken and Pienaar 1985; Suda *et al.* 1989). In three extant genera, however, *Halosphaera* (Boalch and Mommaerts 1969), *Pachysphaera* (Parke 1966; Boalch and Parke 1971), and *Pterosperma* (Parke *et al.* 1978), the cyst, which is planktonic, inflates to a considerable size, often approaching 1.0 mm, and consists of multiple wall layers, the outermost of which contains a sporopollenin-like substance. These cysts, or "phycomata," closely resemble the leiospherids and tasmanitids of especially Proterozoic and Paleozoic marine sediments (Boalch and Parke 1971; Tappan 1980; Falkowski *et al.* 2004b; Javaux *et al.* 2003, 2004). Moestrup *et al.* (2003) postulated that the cyst of *Cymbomonas* is structurally a phycoma, although it does not inflate significantly relative to the size of vegetative cells (Figure 7).

These four (counting *Cymbomonas*) "phycomate prasinophyte" genera form a clade, together with the described genus *Pyramimonas*, the species of which form nonphycomate cysts (Aken and Pienaar 1985; Daugbjerg *et al.* 2000; Moro *et al.* 2002), and the as-yet-undescribed genus *Prasinopapilla* (Guillou *et al.* 2004). These genera are unique among the prasinophytes in that they combine, in their flagellar apparatuses, aspects of both the "cruciate" system present

---

[3]Tracking the evolutionary history of *Pedinomonas* is complicated by the fact that some of the early DNA phylogenies (e.g., Kantz *et al.* 1990) were constructed with an 18s rDNA sequence from "*Pedinomonas minutissima,*" which is not a green alga but a chlorarachniophyte, *Bigelowiella natans* (Moestrup and Sengco 2001).

in the chlorophyte line and the asymmetry present in the streptophyte line (O'Kelly 1992; Moestrup *et al.* 2003). Two of these genera, *Halosphaera* and *Pterosperma,* have MLSs as part of their flagellar apparatuses; in one, *Pyramimonas,* MLSs are absent (O'Kelly 1992); for the remaining three genera, data are lacking (Figures 8 and 9).

In all genera for which data are available (features of *Prasinopapilla* are not published), the species contain an elaborate duct/vacuole system extending from one side of the flagellar pit to a large vacuole in the posterior end of the cell, the aperture of which

is associated with a number of striated, and potentially contractile, fibers (O'Kelly 1992; Moestrup *et al.* 2003; see Figures 4 and 5). Circumstantial evidence that this duct/vacuole complex, which appears in no other group of green plants, is a feeding mechanism has been accumulating since the 1960s (O'Kelly 1992; Moestrup *et al.* 2003), although actual phagotrophic mixotrophy is well documented only for the Antarctic species *Pyramimonas gelidicola* (Bell and Laybourn-Parry 2003). Nevertheless, the combination of a long fossil record, a flagellar apparatus configuration from which

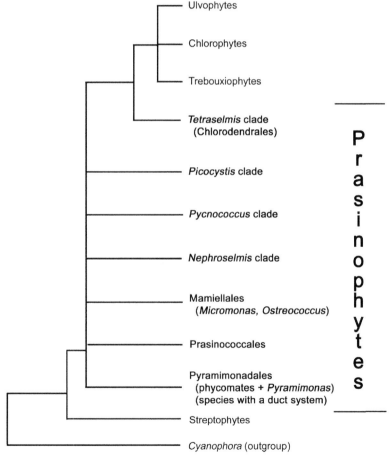

FIGURE 7. Simplified phylogenetic tree based on available 18s rRNA gene sequence data (cf. Guillou *et al.* 2004), showing the clades within the prasinophyte grade. The duct-containing, potentially (in one case actually) mixotrophic prasinophytes (phycoma-producing species plus *Pyramimonas*) form one of the six lineages whose branching order is not yet resolved.

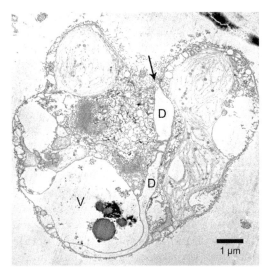

FIGURE 8. Longitudinal section of a flagellate cell of *Halosphaera* aff. *minor*, showing the duct (D) system, the presumed cytostome/cytopharynx complex, extending from the apical pit (arrow) to the posterior vacuole (V).

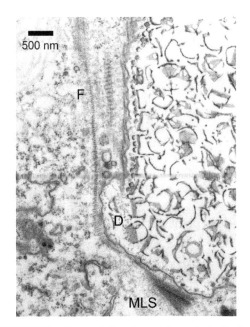

FIGURE 9. Tangential section of the apical pit of a flagellate cell of *Halosphaera* aff. *minor* at the opening of the duct system (D), showing some of the striated fibers (F) extending from the multilayered structure (MLS) to the duct aperture.

all other patterns in the green plants can be plausibly derived by reduction, and the only documented instances of phagotrophic mixotrophy, and structures for that purpose, in the green plants, affords the phycomate prasinophyte clade a special place in our understanding of the early history of velvet pastures and sylvan groves.

IV. DIFFICULTIES IN THE GREEN PLANT PHYLOGENETIC TREE

Although the core structure of the green plant phylogenetic tree is now reasonably well understood, from both morphological and molecular data, there remain several major unresolved nodes. One of these, the "prasinophyte grade," which presumably contains the modern relics of the earliest diversification events in green plants, is discussed earlier in the chapter. Of the many other unresolved nodes, including the key one describing the early diversification of the land flora (Kenrick and Crane 1997; Graham and Gray 2001), two are relevant to questions of origin and early diversification of aquatic chlorobionts: the identity of the lineage ancestral to green plants, and the early diversification of the "seaweed" orders Cladophorales, Dasycladales, and Bryopsidales.

A. The Identity of the Lineage Ancestral to Green Plants

The molecular evidence is overwhelming that the green plant plastid results from a primary endosymbiotic event between a host eukaryotic cell and a cyanobacterium and that the green plant lineage is one of three descendants of that initial endosymbiosis, the other two being the red algae (rhodophytes) and the glaucophytes (Fehling *et al.*, Chapter 6; Hackett *et al.*, Chapter 7, this volume). A prediction of this interpretation of "plant" ancestry is that these three photosynthetic lineages sprung from a phagotrophic ancestor that would have swallowed a cyanobacterium and "chosen" to keep it rather than digest it. This putative ancestor has not been clearly identified in molecular sequence trees, and efforts to infer potential candidates from morphological data are complicated by the apparent rapid and profound diversification of structure in the descendant lineages. For example, red algae lack any trace of flagellar systems or centrioles, and they have highly ordered mitotic spindles (reviewed by Broadwater *et al.* 1992; Sluiman 1993), whereas both glaucophytes and the green plants have flagella, centrioles (reviewed by O'Kelly 1992), and unordered mitotic spindles (Pickett-Heaps 1972, 1975; Sluiman 1993). Cytoskeletal ("flagellar apparatus") configurations in glaucophytes and green algae share some similarities but also differ in many respects (O'Kelly 1992). If these three lineages did indeed share a common ancestor, then there must have been a rapid divergence in both plastid characters (Delwiche 1999) and cytoskeletal organization that soon became fixed in the patterns now common to each lineage.

Also, phagotrophy, common among members of other eukaryotic photoautotroph lineages, such as the dinoflagellates, cercomonads (*Chlorarachnion*), haptophytes, and heterokonts ("stramenopiles"), is known or suspected, among the primary plastid bearers, only for members of the phycomate prasinophyte clade, as detailed in the previous section, and the structures associated with this phagotrophy are not obviously similar to those in any known group of heterotrophs. Phagotrophy is unknown in the red algae and glaucophytes, and in neither group are there amoeboid cells or flagellate cells with structures that might serve a food-capture function (Mignot *et al.* 1969; Broadwater *et al.* 1992; Kugrens *et al.* 1999).

The microtubular root most closely associated with the presumed cytostome in the phycomate prasinophyte clade is the one formed on the right side of the eldest basal body in the flagellar apparatus, is often elaborated into a broad spline, and bears the multilayered structure; it is therefore the homolog of the MLS-bearing root (*roots* in *Mesostigma,* because the flagellar apparatus is partially "doubled") in the flagellate cells of streptophyte plants (O'Kelly 1992). The MLS is positionally homologous and, in many species structurally most similar, to the "C fibre" subtending microtubular root R2 in many groups of "excavates" (Simpson 2003), especially the jakobid and retortamonad flagellates (O'Kelly 1993, 1997; Simpson and Patterson 2001; Simpson 2003). On the basis of mitochondrial genome evidence, the jakobids are the most ancient group of eukaryotes (Gray *et al.* 1998), and the mitochondrial genome of the prasinophyte *Nephroselmis* shares many ancestral features with the jakobids (Gray et al. 1998; Turmel *et al.* 1999a). Conceivably, the jakobids and the green plants share a common ancestor that possessed an MLS-bearing cytoskeleton. However, DNA-sequence-based molecular phylogenies do not show any clear-cut relationships among the excavate protists or between any excavate group and the green plants (Simpson 2003).

Very recently, a group of coccoidal picoplanktonic algae has been identified that is sister to the "primary plastid endosymbiosis" clade in molecular-sequence–based phylogenetic trees (Not *et al.* 2007). As yet, nothing is known of the ultrastructural features of these algae. If future work supports the preliminary assessment of the phylogenetic position of this previously unknown taxon of photosynthetic eukaryotes, the group could provide important new information leading to a better assessment of green plant origins.

B. The Early Diversification of the "Seaweed" Orders

As noted earlier, most recognizable fossil green algal thalli from the latest Proterozoic and the Paleozoic are assigned to one of three extant orders, the Cladophorales, Dasycladales, and Bryopsidales. These orders are usually placed with the

Ulotrichales, Ulvales, and Oltmannsiellopsidales in the class Ulvophyceae (Mattox and Stewart 1984; O'Kelly and Floyd 1984a; Floyd and O'Kelly 1990; Friedl and O'Kelly 2002); together, these orders encompass practically all the extant nonprasinophyte marine algae, including all the macroscopic forms ("seaweeds") except the marginally marine *Prasiola*.

Ironically, these most anciently attested green macrophytes have, what are believed to be, the most derived characters of all chlorophyte algae, in terms of thallus morphology and life history (O'Kelly and Floyd 1984a; Floyd and O'Kelly 1990) as well as in DNA-sequenced–based phylogenetic trees. In these trees, long evolutionary distances separate members of these three orders from their presumed outgroups (Ulotrichales and Ulvales) as well as from other green plants, and their relationships are, consequently, obscured (Zechman *et al.* 1990; Olsen *et al.* 1994; Sherwood *et al.* 2000; Watanabe *et al.* 2001; Famà *et al.* 2002). In the Dasycladales especially, sparse taxon sampling, complicated by the large number of extinctions known to have occurred in this clade, negatively influence phylogenetic reconstructions (Olsen *et al.* 1994; Zechman 2003). Consequently, some workers have elevated all three orders to class status (Hoek *et al.* 1995; Watanabe *et al.* 2001; Leliaert *et al.* 2003).

López-Bautista and Chapman (2003) demonstrated the consequences of inadequate taxon sampling on phylogenetic assessments of the "seaweed" orders when they included representatives in the Trentepohliales in their molecular phylogenies. Trentepohlialean algae, as López-Bautista and Chapman (2003) describe, are so unusual in their habitat, morphology, reproduction, and ultrastructure that their phylogenetic position could not be adequately assessed. Ribosomal RNA gene sequence phylogenies not only placed the Trentepohliales in the Ulvophyceae but indicated that the Trentepohliales is sister to the Cladophorales, Dasycladales, and Bryopsidales. This analysis is further supported (C.J. O'Kelly,

B. Wysor, and W.K. Bellows unpublished results) by inclusion of sequences derived from certain coccoidal green algae that have ultrastructural features of the Ulotrichales but lie outside that order in gene sequence phylogenies, and from *Blastophysa rhizopus*, which has morphological and ultrastructural features of the Cladophorales and, to a lesser extent, the Dasycladales (Chappell *et al.* 1991) but are sister to the coccoidal taxa in gene sequence (cf. Zechman *et al.* 1990). It appears from these results that the relationships of the "seaweed" orders of green algae to other chlorophytes are obscured by marked discontinuities in the rates of morphological *versus* molecular evolution in the earliest-diverging extant species, which up until now have been difficult to find and identify, perhaps because their morphological features are not what investigators had been expecting in taxa sister to the likes of *Acetabularia*.

V. GREEN PLANTS IN THE MODERN MARINE ENVIRONMENT

Ever since the Paleozoic, the land, and the lake have been green, the ocean red. Even the phylogenetic distribution of green plants reflects this dichotomy. Only the prasinophytes and ulvophytes are well represented in salt water. And these groups are poorly represented in freshwater and on land, where the streptophytes, chlorophytes, and trebouxiophytes hold sway.

Pick up a flora of red, brown, and green seaweeds (e.g., Abbott and Hollenberg 1976). The section on green algae is inevitably the smallest. And those green species are usually found in the marginal places, like the upper intertidal regions of rocky shores, or the lowered salinity areas of estuaries, or, in the tropics, the shifting coral sands where, ironically, they compete for space and light with most of the few species of truly marine angiosperms, places where few other seaweeds can find toeholds (Graham and Wilcox 2000). The instances where green algae are numerically dominant, such as on "*Halimeda* reefs" (Wood 1999), are rare; and often when they do occur, we call them "green tides," take the blame for them, and try to get rid of them (e.g., Blomster *et al.* 2002; Nelson *et al.* 2003).

Drag a net through the phytoplankton. Practically everything you get is "red"; a diatom, a haptophyte, a dinoflagellate. Only if you are in exactly the right place at the right time will you find a few green things in the haul, and even then you have to look carefully to be sure that what you have is a green plant and not a green dinoflagellate.

There is, however, one important exception to the "red-dominant" rule of the sea that is emerging, in the domain of the picoplankton. Especially in more eutrophic, nearshore waters, prasinophyte picoplankton species can be abundant and may even form nearly monospecies stands (e.g., Vaquer *et al.* 1996; Díez *et al.* 2001; Not *et al.* 2004; Zhu *et al.* 2005). Population densities vary over time and depth in the water column (Zhu *et al.* 2005; Countway and Caron 2006) and may reach quite high values quickly (and collapse just as quickly; O'Kelly *et al.* 2003). From the dynamics of these algae, the smallest of the small, are emitted, perhaps, the echoes of the tasmanite blooms of the Ordovician.

VI. CONCLUSIONS

The overall shape of the green plant evolutionary tree is now fairly well established. Certain branch points, especially at the base of the tree and at important nodes such as the ones associated with the evolution of the "seaweed" genera and with the early radiations of the archegoniate plants, remain to be resolved. Approaches to resolving these nodes include acquiring a fuller understanding of pivotal taxa within each node, including comprehensive morphological studies and a molecular dataset that includes, at the least, organellar genomes as well as multiple nuclear genes plus a better understanding of the extant diversity at those nodes, which may include cryptic taxa such as the picoeukaryotes that seem to be sister to the primary plastid clade or the coccoidal taxa that seem to reside at the base of the "seaweed" assemblages. As nuclear genome sequencing becomes more prevalent, comparative data from this source will need to be employed as well; for example, it will be possible to determine whether the special features of *Ostreococcus* genomic (Derelle *et al.* 2006) and cellular organization (e.g., Six *et al.* 2005; Farinas *et al.* 2006) are due to its near-ancestral position in the green plant phylogenetic tree or are a consequence of its extreme reduction in size and complexity. It is the least we can do for what is arguably the single lineage of photosynthetic eukaryotes that is the most crucial for life on Earth.

Acknowledgments

Original research reported here, and the writing of this review, were supported by grants from NSF. Grateful thanks to Brent Mishler and the University of California at Berkeley for supporting my sabbatical leave, during which this chapter was written, and to Linda K. Medlin and Ingrid Glaub for sharing unpublished information.

References

Abbott, I.A., and Hollenberg, G.J. (1976). *Marine Algae of California*. Stanford University Press.

Adl, S.M., Simpson, A.G.B., Farmer, M.A., Andersen, R.A., Anderson, O.R., Barta, J.R., Bowser, S.S., Brugerolle, G., Fensome, R.A., Fredericq, S., James, T.Y., Karpov, S., Kugrens, P., Krug, J., Lane, C.E., Lewis, L.A., Lodge, J., Lynn, D.H., Mann, D.G., McCourt, R.M., Mendoz, L., Moestrup, Ø., Mozley-Standridge, S.E., Nerad, T.A., Shearer, C.A., Smirnov, A.V., Spiegel, F.W., and Taylor, M.F.J.R. (2005). The new higher level classification of eukaryotes with emphasis on the taxonomy of protists. *J. Eukar. Microbiol.* **52**: 399–451.

Aken, M.E., and Pienaar, R.N. (1985). Preliminary investigations on the chemical composition of the scale-boundary and cyst wall of *Pyramimonas pseudoparkeae* (Prasinophyceae). *S. Afr. J. Bot.* **51:** 408–416.

Andersen, R.A. (2004). Biology and systematics of heterokont and haptophyte algae. *Am. J. Bot.* **91:** 1508–1522.

Anderson, J.M. (1985). Chlorophyll-protein complexes of a marine green alga, *Codium* species (Siphonales). *Biochim. Biophys. Acta* **806:** 145–153.

Becker, B., Marin, B., and Melkonian, M. (1994). Structure, composition, and biogenesis of prasinophyte cell coverings. *Protoplasma* **181:** 233–244.

Bell, E.M., and Laybourn-Parry, J. (2003). Mixotrophy in the Antarctic phytoflagellate, *Pyramimonas gelidicola* (Chlorophyta: Prasinophyceae). *J. Phycol.* **39:** 644–649.

Berger, S., and Kaever, M.D. (1992). *Dasycladales: An Illustrated Monograph of a Fascinating Algal Order.* Stuttgart, Thieme.

Bhattacharya, D., Weber, K., An, S.S., and Berning-Koch, W. (1998). Actin phylogeny identifies *Mesostigma viride* as a flagellate ancestor of the land plants. *J. Mol. Evol.* **47:** 544–550.

Blackman, F.F. (1900). The primitive algae and the Flagellata. An account of modern work bearing on the evolution of the algae. *Ann. Bot.* **14:** 647–688.

Blindow, I. (2000). Distribution of charophytes along the Swedish coast in relation to salinity and eutrophication. *Int. Rev. Hydrobiol.* **85:** 707–717.

Blomster, J., Bäck, S., Fewer, D.P., Kiirikki, M., Lehvo, A., Maggs, C.A., and Stanhope, M.J. (2002). Novel morphology in *Enteromorpha* (Ulvophyceae) forming green tides. *Am. J. Bot.* **89:** 1856–1763.

Boalch, G.T., and Guy-Ohlson, D. (1992). *Tasmanites,* the correct name for *Pachysphaera* (Prasinophyceae, Pterospermataceae). *Taxon* **41:** 529–531.

Boalch, G.T., and Mommaerts, J.P. (1969). A new punctate species of *Halosphaera. J. Mar. Biol. Assoc. U K* **49:** 129–139.

Boalch, G.T., and Parke, M. (1971). The prasinophycean genera (Chlorophyta) possibly related to fossil genera, in particular the genus *Tasmanites. Proceedings of the Second Planktonic Conference Roma 1970.* A. Farinacci, ed. Rome, Edizioni Tecnoscienza, Rome, pp. 99–105.

Broadwater, S.T., Scott, J.L., and Garbary, D.J. (1992). Cytoskeleton and mitotic spindle in red algae. In *The Cytoskeleton of the Algae.* D. Menzel, ed. Boca Raton, FL, CRC Press, pp. 93–112.

Butterfield, N.J., Knoll, A.H., and Swett, K. (1988). Exceptional preservation of fossils in an Upper Proterozoic shale. *Nature* **334:** 424–427.

Butterfield, N.J., Knoll, A.H., and Swett, K. (1994). Paleobiology of the Neoproterozoic Svanbergfjellet Formation, Spitsbergen. *Fossils Strata* **34:** 1–84.

Carmichael, J.S., and Friedman, W.E. (1996). Double fertilization in *Gnetum gnemon* (Gnetaceae): its bearing on the evolution of sexual reproduction within the Gnetales and the anthophyte clade. *Am. J. Bot.* **83:** 767–780.

Cavalier-Smith, T. (1981). Eukaryote kingdoms: seven or nine? *BioSystems* **14:** 461–481.

Chadefaud, M. (1977). Les Prasinophycées. Remarques historiques, critiques et phylogénétiques. *Bulletin Société Phycologique France* **22:** 1–18.

Chappell, D.F., O'Kelly, C.J., and Floyd, G.L. (1991). Ultrastructure of biflagellate zoospores in the enigmatic marine green alga *Blastophysa rhizopus. J. Phycol.* **27:** 423–428.

Countway, P.D., and Caron, D.A. (2006). Abundance and distribution of *Ostreococcus* sp. in the San Pedro Channel, California, as revealed by quantitative PCR. *Appl. Environ. Microbiol.* **72:** 2496–2506.

Courties, C., Perasso, R., Chrétiennot-Dinet, M.J., Gouy, M., Guillou, L., and Troussellier, M. (1998). Phylogenetic analysis and genome size of *Ostreococcus tauri* (Chlorophyta, Prasinophyceae). *J. Phycol.* **34:** 844–849.

Courties, C., Vaquer, A., Troussellier, M., Lautier, J., Chrétiennot-Dinet, M.J., Neveux, J., Machado, M.C., and Claustre, H. (1994). Smallest eukaryotic organism. *Nature* **370:** 255.

Croce, R., Müller, M.G., Bassi, R., and Holwarth, A.R. (2001). Carotenoid-to-chlorophyll energy transfer in recombination of the major light-harvesting complex (LHCII) of higher plants. I. Femtosecond transient absorbant measurements. *Biophys. J.* **80:** 910–915.

Daugbjerg, N., Marchant, H.J., and Thomsen, H.A. (2000). Life history stages of *Pyramimonas tychotreta* (Prasinophyceae, Chlorophyta), a marine flagellate from the Ross Sea, Antarctica. *Phycol. Res.* **48:** 199–209.

Delwiche, C.E. (1999). Tracing the thread of plastid diversity through the tapestry of life. *Am. Naturalist* **154:** S164-S177.

Derelle, E., Ferraz, C., Rombauts, S., Rouze, P., Worden, A.Z., Robbens, S., Partensky, F., Degroeve, S., Echeynie, S, Cooke, R., Saeys, Y., Wuyts, J., Jabbari, K., Bowler, C., Panaud, O., Piegu, B., Ball, S.G., Ral, J.P., Bouget, F.Y., Piganeau, G., De Baets, B., Picard, A., Delseny, M., Demaille, J., Van de Peer, Y., and Moreau, H. (2006). Genome analysis of the smallest free living eukaryote *Ostreococcus tauri* unveils many unique features. *Proc. Natl. Acad. Sci. U S A* **103:** 11647–11652.

Díez, B., Pedrós-Alió, C., and Massana, R. (2001). Study of genetic diversity of eukaryotic picoplankton in different oceanic regions by small-subunit rRNA gene cloning and sequencing. *Appl. Environ. Microbiol.* **67:** 2932–2941.

Egeland, E.S., Guillard, R.R.L., and Liaaen-Jensen, S. (1997). Additional carotenoid prototype representatives and a general chemosystematic evaluation of carotenoids in Prasinophyceae (Chlorophyta). *Phytochemistry* **44:** 1087–1097.

Falkowski, P.G., Katz, M.E., Knoll, A.H., Quigg, A., Raven, J.A., Schofield, O., and Taylor, F.J.R. (2004b). The evolution of modern eukaryotic phytoplankton. *Science* **305:** 354–360.

Falkowski, P.G., Schofield, O., Katz, M.E., van de Schootbrugge, B., and Knoll, A. (2004a). Why is the land green and the ocean red? *Coccolithophorids.* H. Thierstein and J. Young, eds. Berlin, Springer-Verlag, pp. 429–453.

Famà, P., Wysor, B., Kooistra, W.H.C.F., and Zuccarella, G.C. (2002). Molecular phylogeny of the genus *Caulerpa* (Caulerpales, Chlorophyta) inferred from chloroplast *tufA* gene. *J. Phycol.* **38**: 1040–1050.

Farinas, B., Mary, C., de O Manes, C.-L., Bhaud, Y., Peaucellier, G., and Moreau, H. (2006). Natural synchronization for the study of cell division in the green unicellular alga *Ostreococcus tauri*. *Plant Mol. Biol.* 60: 277–292.

Fawley, M.W., Stewart, K.D., and Mattox, K.R. (1986). The novel light-harvesting pigment-protein complex of *Mantoniella squamata* (Chlorophyta): phylogenetic implications. *J. Mol. Evol.* **23**: 168–176.

Fawley, M.W., Yun, Y., and Qin, M. (2000). Phylogenetic analyses of 18s rDNA sequences reveal a new coccoid lineage of the Prasinophyceae (Chlorophyta). *J. Phycol.* **36**: 387–393.

Fehling, J., Stoecker, D., and Baldauf, S.L. (2007). Photosynthesis and the eukaryote tree of life. *Evolution of Primary Producers in the Sea.* P.G. Falkowski and A.H. Knoll, eds. Boston, Elsevier, pp. 75–107.

Floyd, G.L., and O'Kelly, C.J. (1984). Motile cell fine structure and the circumscription of the Ulotrichales and Ulvales (Ulvophyceae, Chlorophyta). *Am. J. Bot.* **71**: 111–120.

Floyd, G.L., and O'Kelly, C.J. (1990). Phylum Chlorophyta: class Ulvophyceae. *Handbook of Protoctista.* L. Margulis, J.O. Corliss, M. Melkonian and D.J. Chapman, eds. Boston, Jones and Bartlett, pp. 617–635.

Friedl, T. (1995). Inferring taxonomic positions and testing genus level assignments in coccoid green lichen algae: a phylogenetic analysis of 18S ribosomal RNA sequences from *Dictyochloropsis reticulata* and from members of the genus *Myrmecia* (Chlorophyta, Trebouxiophyceae cl. nov.). *J. Phycol.* **31**: 632–639.

Friedl, T., and O'Kelly, C.J. (2002). Phylogenetic relationships of green algae assigned to the genus *Planophila* (Chlorophyta): evidence from 18S rDNA sequence data and ultrastructure. *Eur. J. Phycol.* **37**: 373–384.

Fritsch, F.E. (1916). The algal ancestry of the higher plants. *New Phytol.* **15**: 233–250.

Fritsch, F.E. (1935). *The Structure and Reproduction of the Algae,* vol. 1. Cambridge University Press.

Fritsch, F.E. (1945). Algae and archegoniate plants. *Ann. Bot. Lond. (New System)* **9**: 1–29.

Gray, M.W., Lang, B.F., Cedergren, R., Golding, B.G., Lemieux, C., Sankoff, D., Turmel, M., Brossard, N., Delage, E., Littlejohn, T.G., Plante, I., Rioux, P., Saint-Louis, D., Zhu, Y., and Burger, G. (1998). Genome structure and gene content in protist mitochondrial DNAs. *Nucleic Acids Res.* **26**: 865–878.

Glaub, I., and Vogel, K. (2004). The stratigraphic record of microborings. *Fossils Strata* **51**: 126–135.

Gradinaru, C.C., Stokkum, I.H.W.V., Pascal, A.A., Grondelle, R.V., and Amerongen, H.V. (2000). Identifying the pathways of energy transfer between carotenoids and chlorophylls in LHCII and CP29. A multicolour femtosecond pump-probe study. *J. Phys. Chem. B* **104**: 9330–9342.

Graham, L.E. (1993). *The Origin of Land Plants.* New York, John Wiley.

Graham, L.E., and Gray, J. (2001). The origin, morphology, and ecophysiology of early embryophytes: neontological and paleontological perspectives. *Plants Invade the Land: Evolutionary and Environmental Perspectives,* P.G. Gensel and D. Edwards, eds. New York, Columbia University Press, pp. 140–158.

Graham, L.E., and Wilcox, L.W. (2000). *Algae.* Upper Saddle River, NJ, Prentice Hall.

Griffin, N.J., and Aken, M.E. (1993). Rhythmic settling behavior in *Pyramimonas parkeae* (Prasinophyceae). *J. Phycol.* **29**: 9–15.

Greuter, W., McNeill, J., Barrie, F.R., Burdet, H.-M., Demoulin, V., Filgueiras, T.S., Nicolson, D.H., Silva, P.C., Skog, J.E., Trehane, P., Turland, N.J., and Hawksworth, D.L., eds. (2000). *International Code of Botanical Nomenclature (St. Louis Code).* Regnum Vegetabile 138. Königstein, Koeltz Scientific Books.

Guillou, L., Eikrem, W., Chrétiennot-Dinet, M.J., Le Gall, F., Massana, R., Romari, K., Pedrós-Alió, C., and Vaulot, D. (2004). Diversity of picoplanktonic Prasinophyceae assessed by direct SSU rDNA sequencing of environmental samples and novel isolates retrieved from oceanic and coastal marine ecosystems. *Protist* **155**: 193–214.

Guy-Ohlson, D., and Boalch, G.T. (1992). Comparative morphology of the genus *Tasmanites* (Pterospermatales, Chlorophyta). *Phycologia* **31**: 523–528.

Hackett, J.D., Yoon, H.S., Butterfield, N.J., Sanderson, M.J., and Bhattacharya, D. (2007). Plastid endosymbiosis: sources and timing of the major events. *Evolution of Primary Producers in the Sea.* P.G. Falkowski and A.H. Knoll, eds. Boston, Elsevier, pp. 109–131.

Henley, W.J., Hironaka, J.L., Guillou, L., Buchheim, M.A., Fawley, M.W., and Fawley, K.P. (2004). Phylogenetic analysis of the "*Nannochloris*-like" algae and diagnoses of *Picochlorum oklahomensis* gen. et sp. nov. (Trebouxiophyceae, Chlorophyta). *Phycologia* **43**: 641–652.

Hoops, H.J. (1997). Motility in the colonial and multicellular Volvocales: structure, function and evolution. *Protoplasma* **199**: 99–112.

Inouye, I., and Hori, T. (1991). High speed video analysis of the flagellar beat and swimming patterns of algae: possible evolutionary trends in green algae. *Protoplasma* **164**: 59–64.

Javaux, E.J., Knoll, A.H., and Walter, M.R. (2003). Recognizing and interpreting the fossils of early eukaryotes. *Origins Life Evol. Biosphere* **33**: 75–94.

Javaux, E.J., Knoll, A.H., and Walter, M.R. (2004). TEM evidence for eukaryotic diversity in mid-Proterozoic oceans. *Geobiology* **2:** 121–132.

Kantz, T.S., Theriot, E.C., Zimmer, E.A., and Chapman, R.L. (1990). The Pleurastrophyceae and Micromonadophyceae: a cladistica analysis of nuclear rRNA sequence data. *J. Phycol.* **26:** 711–721.

Katz, M.E., Finkel, Z.V., Grzebyk, D., Knoll, A.H., and Falkowski, P.G. (2004). Evolutionary trajectories and biogeochemical impacts of marine eukaryotic phytoplankton. *Annu. Rev. Ecol. Evol. Syst.* **35:** 523–556.

Kenrick, P., and Crane, P.R. (1997). The origin and early evolution of plants on land. *Nature* **389:** 33–39.

Kim, Y.-S., Oyaizu, H., Matsumoto, M.M., and Nozaki, H. (1994). Chloroplast small-subunit ribosomal RNA gene sequence from *Chlamydomonas parkeae* (Chlorophyta): molecular phylogeny of a green alga with a peculiar pigment composition. *Eur. J. Phycol.* **29:** 213–217.

Kugrens, P., Clay, B.L., Meyer, C.J., and Lee, R.E. (1999). Ultrastructure and description of *Cyanophora biloba*, sp. nov., with additional observations on *C. paradoxa* (Glaucophyta). *J. Phycol.* **35:** 844–854.

Latasa, M., Scharek, R., Le Gall, F., and Guillou, L. (2004). Pigment suites and taxonomic groups in Prasinophyceae. *J. Phycol.* **40:** 1149–1155.

Leadbeater, B.S.C., and Green, J.C. (1993). Cell coverings of microalgae. *Ultrastructure of Microalgae.* T. Berner, ed. Boca Raton, FL, CRC Press, pp. 71–98.

Leliaert, F., Rousseau, F., de Reviers, B., and Coppejans, E. (2003). Phylogeny of the Cladophorophyceae (Chlorophyta) inferred from partial LSA rRNA gene sequences: is the recognition of a separate order Siphonocladales justified? *Eur. J. Phycol.* **38:** 233–246.

Lemieux, C., Otis, C., and Turmel, M. (2000). Ancestral chloroplast genome in *Mesostigma viride* reveals an early branch of green plant evolution. *Nature* **403:** 649–652.

Lewis, L.A., and McCourt, R.M. (2004). Green algae and the origin of land plants. *Am. J. Bot.* **91:** 1535–1556.

Liu, Z., Yan, H., Wang, K., Kuanng, T., Zhang, J., Gui, L., An, X., and Chang, W. (2004). Crystal structure of spinach major light-harvesting complex at 2.7Å resolution. *Nature* **428:** 287–292.

LoDuca, S.T., Kluessendorf, J., and Mikulic, D.G. (2003). A new noncalcified dasycladalean alga from the Silurian of Wisconsin. *J. Paleontol.* **77:** 1152–1158.

López-Bautista, J.M., and Chapman, R.L. (2003). Phylogenetic affinities of the Trentepohliales inferred from small-subunit rDNA. *Int. J. Syst. Evol. Microbiol.* **53:** 2099–2106.

Marin, B., and Melkonian, M. (1999). Mesostigmatophyceae, a new class of streptophyte green algae revealed by SSU rRNA sequence comparisons. *Protist* **150:** 399–417.

Mattox, K.R., and Stewart, K.D. (1984). Classification of the green algae: a concept based on comparative cytology. *The Systematics of the Green Algae.* D.E.G. Irvine and D.M. John, eds. London, Academic Press, pp. 29–72.

Melkonian, M. (1984). Flagellar apparatus ultrastructure in relation to green algal classification. *The Systematics of the Green Algae.* D.E.G. Irvine and D.M. John, eds. London, Academic Press, pp. 73–120.

Melkonian, M. (1989). Flagellar apparatus ultrastructure in *Mesostigma viride* (Prasinophyceae). *Plant Syst. Evol.* **164:** 93–122.

Mignot, J.P., Joyon, L., and Pringsheim, E.G. 1969. Quelques particularités structurales de *Cyanophora paradoxa* Korsch., protozoaire flagellé. *J. Protozool.* **16:** 138–145.

Moestrup, Ø. (1991). Further studies of presumedly primitive green algae, including the description of Pedinophyceae class. nov. and *Resultor* gen. nov. *J. Phycol.* **27:** 119–133.

Moestrup, Ø., Inouye, I., and Hori, T. (2003). Ultrastructural studies on *Cymbomonas tetramitiformis* (Prasinophyceae). I. General structure, scale microstructure, and ontogeny. *Can. J. Bot.* **81:** 657–671.

Moestrup, Ø., and Sengco, M. (2001). Ultrastructural studies on *Bigelowiella natans*, gen. et sp. nov., a chlorarachniophyte flagellate. *J. Phycol.* **37:** 624–646.

Moro, I., La Rocca, N., Dalla Valle, L., Moschin, E., Negrisolo, E., and Andreoli C. (2002). *Pyramimonas australis* sp. nov. (Prasinophyceae, Chlorophyta) from Antarctica: fine structure and molecular phylogeny. *Eur. J. Phycol.* **37:** 103–114.

Nakayama, T., Kawachi, M., and Inouye, I. (2000). Taxonomy and the phylogenetic position of a new prasinophycean alga, *Crustomastix didyma* gen. & sp. nov. (Chlorophyta). *Phycologia* **39:** 337–348.

Nelson, T.A., Lee, D.J., and Smith, B.C. (2003). Are "green tides" harmful algal blooms? Toxic properties of water-soluble extracts from two bloom-forming macroalgae, *Ulva fenestrata* and *Ulvaria obscura*. *J. Phycol.* **39:** 874–879.

Nielsen, R. (1983). Culture studies of *Acrochaete leptochaete* comb. nov. and *A. wittrockii* comb. nov. (Chaetophoraceae, Chlorophyceae). *Nordic J. Bot.* **3:** 689–694.

Norstog, K.J., Gifford, E.M., and Stevenson, D.W. (2004). Comparative development of the spermatozoids of cycads and *Ginkgo biloba*. *Botanical Rev.* **70:** 5–15.

Not, F., Latasa, M., Marie, D., Cariou, T., Vaulot, D., and Simon, N. (2004). A single species, *Micromonas pusilla* (Prasinophyceae), dominates the eukaryotic picoplankton in the western English Channel. *Appl. Environ. Microbiol.* **70:** 4064–4072.

Not, F., Valentin, K., Romari, K., Lovejoy, C., Massana, R., Tobe, K., Vaulot, D., and Medlin, L.K. (2007). Picobiliphytes: a marine picoplanktonic algal group with unknown affinities to other eukaryotes. *Science* **315:** 192–193.

O'Kelly, C.J. (1982). Chloroplast pigments in selected marine Chaetophoraceae and Chaetosiphonaceae (Chlorophyta): the occurrence and significance of siphonaxanthin. *Bot. Mar.* **25:** 133–137.

O'Kelly. C.J. (1992). Flagellar apparatus architecture and the phylogeny of "green" algae: chlorophytes, euglenoids, glaucophytes. *The Cytoskeleton of the Algae.* D. Menzel, ed. Boca Raton, FL, CRC Press, pp. 315–345.

O'Kelly, C.J. (1993). The jakobid flagellates: structural features of *Jakoba*, *Reclinomonas* and *Histiona* and implications for the early diversification of eukaryotes. *J. Eukar. Microbiol.* **40:** 627–636.

O'Kelly, C.J. (1997). Ultrastructural features of *Reclinomonas americana* Flavin & Nerad, 1993 (Histionidae): trophic cells, zoospores, and cysts. *Eur. J. Protistol.* **33:** 337–348.

O'Kelly, C.J., and Floyd, G.L. (1984a). Correlations among patterns of sporangial structure and development, life histories, and ultrastructural features in the Ulvophyceae. *The Systematics of the Green Algae.* D.E.G. Irvine and D.M. John, eds. London, Academic Press, pp. 121–156.

O'Kelly, C.J., and Floyd, G.L. (1984b). Flagellar apparatus absolute orientations and the phylogeny of the green algae. *BioSystems* **16:** 227–251.

O'Kelly, C.J., Sieracki, M.E., Thier, E.C., and Hobson, I.C. (2003). A transient bloom of *Ostreococcus* (Chlorophyta, Prasinophyceae) in West Neck Bay, Long Island, New York. *J. Phycol.* **39:** 850–854.

O'Kelly, C.J., Wysor, B., and Bellows, W.K. (2004a). *Collinsiella* (Ulvophyceae, Chlorophyta) and other ulotrichalean taxa with shell-boring sporophytes form a monophyletic clade. *Phycologia* **43:** 41–49.

O'Kelly, C.J., Wysor, B., and Bellows, W.K. (2004b). Gene sequence diversity and the phylogenetic position of algae assigned to the genera *Phaeophila* and *Ochlochaete* (Ulvophyceae, Chlorophyta). *J. Phycol.* **40:** 789–799.

Okuda, K. (2002). Structure and phylogeny of cell coverings. *J. Plant Res.* **115:** 283–288.

Olsen, J.L., Stam, W.T., Berger, S., and Menzel, D. (1994). 18s rDNA and evolution in the Dasycladales (Chlorophyta): modern living fossils. *J. Phycol.* **30:** 729–744.

Parke, M. (1966). The genus *Pachysphaera. Some Contemporary Studies in Marine Science.* H. Barnes, ed. London, Allen and Unwin, pp. 555–563.

Parke, M., Boalch, G.T., Jowett, R., and Harbour, D.S. (1978). The genus *Pterosperma* (Prasinophyceae): species with a single equatorial ala. *J. Mar. Biol. Assoc. U K* **58:** 239–276.

Pascher, A. (1927). Volvocales = Phytomonadinae. In *Die Süsswasser-Flora Deutschland, Österreichs und der Schweiz.* A. Pascher, ed. Jena, Gustav Fischer, pp. 20–505.

Pickett-Heaps, J.D. (1972). Cell division in *Cyanophora paradoxa. New Phytol.* **71:** 561–567.

Pickett-Heaps, J.D. (1975). *Green Algae: Structure, Reproduction and Evolution in Selected Genera.* Sunderland, MA, Sinauer.

Pickett-Heaps, J.D., and Marchant, H.J. (1972). The phylogeny of the green algae: a new proposal. *Cytobios* **6:** 255–264.

Polle, J.E.W., Niyogi, K.K., and Melis, A. (2001). Absence of lutein, violaxanthin and neoxanthin affects the functional chlorophyll antenna size of photosystem-II but not that of photosystem-I in the green alga *Chlamydomonas reinhardtii. Plant Cell Physiol.* **42:** 482–491.

Renzaglia, K.S., Dengate, S.B., Schmitt, S.J., and Duckett, J.G. (2006). Spermatozoid of *Equisetum arvense* L: a correlated TEM and SEM investigation. *New Phytol.* In press.

Renzaglia, K.S., Johnson, T.H., Gates, H.D., and Whittier, D.P. (2001). Architecture of the sperm cell of *Psilotum. Am. J. Bot.* **88:** 1151–1163.

Rindi, F., McIvor, L., and Guiry, M.D. (2004). The Prasiolales of Atlantic Europe: an assessment based on morphological, molecular, and ecological data, including the characterization of *Rosenvingiella radicans* (Kützing) comb. nov. *J. Phycol.* **40:** 977–997.

Rodríguez, F., Derelle, E., Guillou, L., Le Gall, F., Vaulot, D., and Moreau H. (2005). Ecotype diversity in the marine picoeukaryote *Ostreococcus. Environ. Microbiol.* **7:** 853–859.

Sasa, T., Suda, S., Watanabe, M.M., and Takaichi, S. (1992). A yellow marine *Chlamydomonas*: morphology and pigment composition. *Plant Cell Physiol.* **33:** 527–534.

Sherwood, A.R., Garbary, D.J., and Sheath, R.G. (2000). Assessing the phylogenetic position of the Prasiolales (Chlorophyta) using *rbc*L and 18s rRNA gene sequence data. *Phycologia* **39:** 139–146.

Simpson, A.G.B. (2003). Cytoskeletal organization, phylogenetic affinities and systematics in the contentious taxon Excavata (Eukaryota). *Int. J. Syst. Evol. Microbiol.* **53:** 1759–1777.

Simpson, A.G.B., and Patterson, D.J. (2001). On core jakobids and excavate taxa: the ultrastructure of *Jakoba incarcerata. J. Eukar. Microbiol.* **48:** 480–492.

Six, C., Worden, A.Z., Rodríguez, F., Moreau, H., and Partensky, F. (2005). New insights into the nature and phylogeny of prasinophyte antenna proteins: *Ostreococcus tauri*, a case study. *Mol. Biol. Evol.* **22:** 2217–2230.

Sluiman, H.J. (1993). Nucleus, nuclear division, and cell division. *Ultrastructure of Microalgae.* T. Berner, ed. Boca Raton, FL, CRC Press, pp. 221–267.

Stewart, K.D., and Mattox, K.R. (1978). Structural evolution in the flagellated cells of green algae and land plants. *BioSystems* **10:** 145–152.

Suda, S., Watanabe, M.M., and Inouye, I. (1989). Evidence for sexual reproduction in the primitive green alga *Nephroselmis olivacea* (Prasinophyceae). *J. Phycol.* **25:** 596–600.

Suda, S., Watanabe, M.M., and Inouye, I. (2004). Electron microscopy of sexual reproduction in *Nephroselmis olivacea* (Prasinophyceae, Chlorophyta). *Phycol. Res.* **52:** 273–283.

Sym, S.D., Kawachi, M., and Inouye, I. (2000). Diversity of swimming behavior in *Pyramimonas* (Prasinophyceae). *Phycol. Res.* **48:** 149–154.

Sym, S.D., and Pienaar, R.M. (1993). The class Prasinophyceae. *Prog. Phycol. Res.* **9**: 281–376.

Tappan, H. (1980). *The Paleobiology of Plant Protists*. San Francisco, W. H. Freeman.

Turmel, M., Lemieux, C., Burger, G., Lang, B.F., Otis, C., Plante, I., and Gray, M.W. (1999a). The complete mitochondrial DNA sequences of *Nephroselmis olivacea* and *Pedinomonas minor*: two radically different evolutionary patterns within green algae. *Plant Cell* **11**: 1717–1730.

Turmel, M., Otis, C., and Lemieux, C. (1999b). The complete chloroplast DNA sequence of the green alga *Nephroselmis olivacea*: insights into the architecture of ancestral chloroplast genomes. *Proc. Natl. Acad. Sci. U S A* **96**: 10248–10253.

Turmel, M., Otis, C., and Lemieux, C. (2002). The complete mitochondrial DNA sequence of *Mesostigma viride* identifies this green alga as the earliest green plant divergence and predicts a highly compact mitochondrial genome in the ancestor of all green plants. *Mol. Biol. Evol.* **19**: 24–38.

Valentin, K., Mehl, H., and Medlin, L. (2005). Picoplankton culture assessment using single strand conformation polymorphism and partial 18S sequencing. *J. Plankton Res.* **27**: 1149–1154.

van den Hoek, C., Mann, D.G., and Jahns, H.M. (1995). *Algae: An Introduction to Phycology*. Cambridge, Cambridge University Press.

Vaquer, A., Troussellier, M., Courties, C., and Bibent B. (1996). Standing stock and dynamics of picophytoplankton in the Thau Lagoon (Northwest Mediterranean Coast). *Limnol. Oceanogr.* **41**: 1821–1828.

Vogel, K., Gektidis, M., Golubic, S., Kiene, W.E., and Radtke, G. (2000). Experimental studies on microbial bioerosion at Le Stocking Island, Bahamas and One Tree Island, Great Barrier Reef, Australia: implications for paleoecological reconstructions. *Lethaia* **33**: 190–204.

Watanabe, S., Kuroda, N., and Maiwa, F. (2001). Phylogenetic status of *Helicodictyon planctonicum* and *Desmochloris halophila* gen. et comb. nov. and the definition of the class Ulvophyceae (Chlorophyta). *Phycologia* **40**: 421–434.

Wood, R. (1999). *Reef Evolution*. Oxford University Press.

Yokohama, Y. (1981). Green light-absorbing pigments in marine green algae, their ecological and systematic significance. *Jpn. J. Phycol.* **29**: 209–222 [in Japanese].

Yoshii, Y., Hanyuda, T., Wakana, I., Miyaji, K., Arai, S., Ueda, K., and Inouye, I. (2004). Carotenoid compositions of *Cladophora* balls (*Aegagropila linnaei*) and some members of the Cladophorales (Ulvophyceae, Chlorophyta): their taxonomic and evolutionary implication. *J. Phycol.* **40**: 1170–1177.

Yoshii, Y., Tanaishi, S., Maoka, T., Hanada, S., and Inouye, I. (2002). Characterization of two unique carotenoid fatty acid esters from *Pterosperma cristatum* (Prasinophyceae, Chlorophyta). *J. Phycol.* **38**: 297–303.

Yoshii, Y., Tanaishi, S., Maoka, T., and Inouye, I. (2003). Photosynthetic pigment composition in the primitive green alga *Mesostigma viride* (Prasinophyceae): phylogenetic and evolutionary implications. *J. Phycol.* **38**: 297–303.

Yoshii, Y., Takaichi, S., Maoka, T., Suda, S., Sekiguchi, H., Nakayama, T., and Inouye, I. (2005). Variation of siphonoxanthin series among the genus *Nephroselmis* (Prasinophyceae, Chlorophyta), including a novel primary methoxy carotenoid. *J. Phycol.* **41**: 827–834.

Zechman, F.W. (2003). Phylogeny of the Dasycladales (Chlorophyta, Ulvophyceae) based on analyses of rubisco large subunit (*rbc*L) gene sequences. *J. Phycol.* **39**: 819–827.

Zechman, F.W., Theriot, E.C., Zimmer, E.A., and Chapman, R.L. (1990). Phylogeny of the Ulvophyceae (Chlorophyta): cladistic analysis of nuclear-encoded rRNA sequence data. *J. Phycol.* **26**: 700–710.

Zhu, F., Massana, R., Not, F., Marie, D., and Vaulot, D. (2005). Mapping of picoeukaryotes in marine ecosystems with quantitative PCR of the 18s rRNA gene. *FEMS Microbial Ecol.* **52**: 79–92.

Zingone, A., Borra, M., Brunet, C., Forlani, G., Kooistra, W.C.F., and Procaccini, G. (2002). Phylogenetic position of *Crustomastix stigmatica* sp. nov. and *Dolichomastix tenuilepis* in relation to the Mamiellales (Prasinophyceae, Chlorophyta). *J. Phycol.* **38**: 1024–1039.

14

Armor: Why, When, and How

CHRISTIAN HAMM AND VICTOR SMETACEK

"Der Welten Kleines auch ist wunderbar und groß, und aus dem Kleinen bauen sich die Welten"—Gottlieb Christian Ehrenberg (1795–1876), motto of his Ph.D. thesis and the inscription on his gravestone.

Armor implies physical defense against attack by other organisms and is hence distinct from mechanical structures that confer protection or provide support against nonbiological environmental stressors. A comparison between armor-plated, military and streamlined, civilian vehicles illustrates this difference. Human history teaches us that the arms race is a powerful driving force in the evolution of technology. This principle—the evolving interaction between attack and defense mechanisms and techniques—also applies to Darwinian evolution, except that natural selection takes the place of intelligent design.

The fitness of an individual organism is expressed in the degree of its ability to gather resources and avoid becoming a resource itself—bottom-up and top-down selection, respectively. In both cases, the organism is competing with its neighbors, either for resources with those of the same trophic level or for avoidance of becoming a resource with those sharing the same enemy ranging from specific pathogens and parasites to predators. The co-evolution of specific attack and defense systems is well documented in terrestrial and benthic biota (Ehrlich and Raven 1964; John *et al.* 1992; Rausher 2001).

In contrast, planktologists have traditionally focused on bottom-up competition within the growth environment and neglected organismal properties that promote survival in the mortality environment, defined by Smetacek *et al.* (2004). Chemical defense systems are currently attracting attention (Cembella 2003), but mechanical defenses are still poorly appreciated. Indeed, the relationship between form and function in protistan plankton remains largely mysterious (Sournia 1982).

In this chapter, we provide a brief history of the conceptual framework of plankton evolutionary ecology to explain why the role of defense in the evolution of unicellular plankton has been neglected so far. The aim of this historical overview is also to point out what can be learned from studying the arms race: the other side of the coin. Given the range of attack techniques to which protists are exposed—from viruses to zooplankton—we define armor as all forms of mechanical defense against pathogens, parasites, and ingestors. These in turn will range from slimy or tough cell walls that hamper purchase, bar entry, or withstand puncturing to long spines that deter ingestion. Following a brief comparison with terrestrial systems we expand on the argument, first broached by Smetacek (2001), that evolution of eukaryotes in the plankton is driven by the arms race. To this end, we survey the range of attack systems evolved by pathogens, parasites, and predators in the plankton, align them with their respective defense systems, and speculate on their evolutionary history. In the final section, we examine how the various types of armor (cell walls, scales, frustules, and colony skin) of selected groups provide protection against specific forms of attack.

I. WHY ARMOR

Early life forms assembled organic molecules from the environment to build themselves and fuel their growth. Evolution at this stage will have been driven by resource competition and death caused by resource deprivation: ultimately, the energy required to maintain cellular structures. Dead organisms will have represented a new type of resource whose utilization could be hastened by deployment of exo-enzymes. The arms race began when living organisms were killed by exo-enzymes—the transition from scavenger to predator. The origin of protective layers of slime and, ultimately, cell walls, that is, armor, was an inevitable result. Indeed, the universality of cell walls in unicellular plankton might well be a reason why their role in selection has been taken for granted and hence overlooked.

To understand the role of armor in individual selection, one needs to observe its performance under various forms of attack in the mortality environment. In the case of larger organisms, the relationships between form and function can be assessed by visual observation and tested with straight-forward experiments. In contrast, interactions among unicellular organisms are difficult to observe because microscopes, in contrast to telescopes and binoculars, do not reveal the required breadth and depth of focus. Because nobody has watched protistan interactions in the wild, our assumptions of plankton behavior are based on inference colored by the concepts applied.

A. History of the Concept "Armor" Applied to Plankton

Like any concept transplanted from the realm of human sensory experience to a realm outside of it (in this case the pelagial or pelagic environment inhabited by the plankton), the word "armor" carries with it roots, the subliminal connotations that influence the ramifications of the concept in its new, theoretical environment. We illustrate this effect by contrasting the connotations of the English word armor with its German counterpart "Panzer." Whereas armor is associated with steel, whether the coat of mail worn by knights or the projectile-proof plates of military vehicles and ships still in use today (dictionary definition), Panzer has broader connotations reflected in its wider usage: Panzernashorn: one-horned rhinoceros, Schildkrötenpanzer: the shells of a tortoise, but also Panzerglas: bullet-proof glass and simply Panzer: tank (armored vehicle).

So it is not surprising that the 19th-century German-speaking scientists, who were the first to systematically study protists, freely used the word Panzer to describe protistan cell walls that their English-speaking colleagues substituted with the more ambiguous "shell." In English "armored plankton" does have the ring of scientific hyperbole if not overkill. In the few instances where armor has entered English terminology, for example, dinoflagellates, the connotations are of ornamentation rather than defense. However, the pioneers of protistan ecology clearly had defense on their mind, otherwise they would have chosen other widely used German terms with differing connotations: *Gehäuse* derived from *Haus* (house), which implies protective structure in a broader sense, that is, also against the physical environment (as in snails), or *Außenskelett* (exoskeleton), which implies a shape-giving, supportive function (as in arthropods) but does not exclude defense.

The impression of the early researchers who interpreted apparently tough outer walls as defense is exemplified by Ehrenberg's (1838) description of diatoms as Panzertierchen (armored little animals) implying that their silica shell protected them against a range of potential predators. In his Latin description of the taxon "Bacillaria" he also used "lorica" (cuirass or corselet worn by Roman soldiers), the French description contains the word "carapace."

Defense is defined by attack. However, attack and defense systems were not studied systematically by any of the early microbial naturalists. Methodological constraints were one of the reasons; another, the rise of taxonomy accompanied by its increasing attention to details of the armor as criteria to differentiate species. Not surprisingly, the large armored forms were the first to receive attention by taxonomists and the numbers of species required specialization. Because species have to be defined on the basis of objective criteria, taxonomists reveled in the custom of using dead languages to coin new names and concepts bereft of subjective roots. This sterile jargon is aimed at curbing imagination instead of stimulating it. Ernst Haeckel is an outstanding example of this trend: He described hundreds of Radiolaria species (Haeckel 1862) and was apparently so overwhelmed by the sheer diversity of forms that he was unable to imagine an equivalent range of functions to which these forms might be attributed.

The way out of the dilemma was to declare that there was no function, that the forms were produced by an underlying natural law that automatically generated pattern expressed in the variety of shapes. In a textbook of general zoology, Haeckel (1866) introduced this concept and termed the mysterious driving force the morphogenetic basic law (*Grundgesetz*), which he compared to an organic crystallography. Haeckel was a bold thinker looking for rules to organize the bewildering diversity of organisms being described at the time. He was the champion of Darwinian evolution in Germany and ferociously antireligious. Nevertheless, his concept of an internal programing of organic matter that manifests itself in the shape of whole organisms is ultimately the internalization of intentional design. Natural selection is the quality control of this exuberant, prodigious shape maker. Because random mutation within the genome was not known at the time, Haeckel can be forgiven for focusing on appreciation rather than explanation of planktonic forms. Haeckel's famous and influential coffee table book *Art Forms of Nature* (1904) was strongly influenced by *A Handbook of Ornament* (Meyer 1888), which probably strengthened the perception that biogenic forms are often "artistic" rather than functional (Figure 1).

However, the etymology of "ornamental forms" reflects its deeper roots. Ornament is derived from the Latin ornamentum, meaning "equipment, trappings, embellishment." Ornament's original function was understood to exceed mere decoration and to serve as a way of equipping a person for ceremony or battle. Similarly, the Sanskrit term for ornament, alamkara, encompasses meanings that include invigoration and making one fit.

Haeckel's influence was so pervasive that attributing functions to the diversity of planktonic forms appeared a hopeless task (Smetacek et al. 2002). This was in contrast to invertebrate biologists who took delight in relating form to function in their graphic descriptions of the feeding behavior of the various types of meroplanktonic larvae (Hardy 1956). The last systematic attempt at relating form and function in unicellular plankton is that of Sournia (1982), who gave up in despair and appealed to planktologists to ask their children for clues. But were not these the early innocent observers exemplified by Ehrenberg who looked at the plankton with childlike eyes and whose visionary descriptions (Ehrenberg 1838) were later buried in the plethora of overwhelming detail?

A different note was struck by Hensen (1887), who coined the term plankton and also launched biological oceanography on the agricultural paradigm. He insisted that diatoms were not part of the food chain leading to fish because they consisted of little more than the worthless silica ("wertlose Kieselerde"), that is, they were all shell with little content and not worth eating. This can also be interpreted as a form of mechanical defense in which volume and not necessarily strength of the coating is significant.

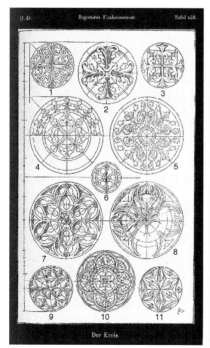

FIGURE 1.   Striking similarity between a plate taken from *A Handbook of Ornament* published by Meyer in 1888 and a plate from the much better known Haeckel's *Art Forms of Nature* (1904). One is left with the impression that planktonic shapes are mere ornament.

His belief was based on general observations of dense diatom blooms subjected to little grazing pressure and his claim of the scarcity of diatoms in the copepod guts he examined.

Such a radical view provoked a number of studies that showed that diatoms were indeed eaten by a broad range of organisms (Smetacek *et al.* 2002). The accent of these studies was on proving that diatoms were not only edible but good food to boot on which a number of organisms thrived. Interestingly, the term grazing, with its connotations of sheep grazing on the lush grass of a meadow, was applied to copepod feeding. So phytoplankton, particularly diatoms, became the pastures of the sea, described with terms like yield and crop, all concepts that thrived in the fertile ground of the agricultural paradigm established by Hensen.

By the middle of the 20th century, the agricultural paradigm with its diatom-copepod-fish food-chain conceptual framework had developed into the cornerstone of biological oceanography (Raymont 1963) till it was side-tracked by discovery of the microbial "loop" (Pomeroy 1974; Azam *et al.* 1983). In the 1970s, ultrastructural studies of the mouthparts of the copepods revealed the presence of numerous chemoreceptors, which allowed a new insight into the potential of selective feeding of zooplankters (Friedmann and Strickler 1975).

During the 1980s, the application of high-speed cinematography revealed that copepod feeding movements were faster than the human eye could follow. The footage demonstrated that copepods actually fed selectively on a broad range of particles and exhibited complex handling techniques depending on the shape and size of the prey (Alcaraz *et al.* 1980; Paffenhöfer and Lewis 1990). The complex feeding behavior of the ubiquitous dinoflagellate genus *Protoperidinium* was another major discovery of the period (Jacobson 1999). These observations of selective grazing showed that there was ample scope for the evolution of defense mechanisms in the plankton but the implications were not immediately realized by mainstream pelagic ecology.

B. Why Should Protists and the Pelagial Be Different?

1. The Terrestrial Analogy

Biological oceanography was launched on a terrestrial paradigm at a time when agricultural thinking with its promise of easy quantification held sway. The aim of agriculture is to maximize production of desired properties (increase yield) and minimize that of undesired ones. In practice, crop defenses, both mechanical and chemical, are bred out by domestication because their function is fulfilled by fences and pesticides. The focus of bio-oceanography has accordingly been on the production and fate of biomass, that is, on the primary metabolites, reflected in the much greater effort spent on studying the growth as compared to the mortality environment (Smetacek *et al.* 2002).

To appreciate the efficiency of any defense, one has to first understand the form of attack against which it evolved. In the macrophyte realm, the fact that a plant, or a part of it, is eaten by some animal or infected by some pathogen or parasite does not mean that this is accomplished easily or that the plant is not effectively defended against a host of other potential enemies. Clearly, mechanical defenses such as the cuticles of leaves protect against desiccation, but they also deter the mycelia of parasitic fungi and the piercing mouth parts of many, but not all, insects and they are poor protection against ungulate grazing. The same can be said of the silica phytoliths of grasses and bamboos that aid in stiffening the plant (skeleton function) but also deter a host of herbivores including many, but not all, ungulates. Clearly, appropriate measurements can be made to discern when the degree of toughness of a protecting agent goes well beyond the requirements of a life-supporting function (preventing evaporation or maintaining optimum rigidity against fluctuating environmental pressure) and becomes part of a life-protecting function, that is, defense. In terms used previously, when does a water tank become a Panzer?

In the pelagic realm, any photo-autotrophic population with a large number of small cells will be more competitive in resource acquisition than a species with equivalent biomass comprising fewer, hence larger, cells. This is a fact dictated by the physicochemical environment via the surface to volume ratio (s/v). So the evolution of eukaryote phytoplankton can only be explained in the light of the mortality environment where survival of both genomes, that of the photoautotrophic endosymbiont as well as its exosymbiont ingestor, results in a new organism with a novel combination of properties and with a smaller s/v ratio. However, growth rates of the autotroph in an endosymbiotic relationship will be lower than that of its free-living cousins not only because uptake of dissolved nutrients will be hampered by the host cell but also because a significant proportion of photosynthetic products will be diverted for its reproduction.

A possible exception would be under nutrient-limiting conditions if the host cell provisions its endosymbionts with nutrients derived from digestion of ingested particles. Such a mixotrophic relationship, although widespread, does not explain the dominance of eukaryote phytoplankton under the nutrient-replete conditions in which blooms develop. The more obvious advantage accruing to the endosymbiont is protection from ingestion by the predatory cousins of the ingesting cell, implying reduction in mortality rate of the endosymbiont (Smetacek 2001). In this connection we consider it more appropriate to refer to the form-giving organism of the endosymbiotic relationship as the exosymbiont rather than the host cell as the latter term has connotations of a parasitic or temporary condition rather than an integrated, mutually beneficial symbiosis.

Another reason why the evolution of defenses in the marine plankton has not received the attention it deserves is based on a widespread belief that an arms race cannot evolve among protists because the target of evolutionary selection is the individual cell. This is a serious misconception: in an asexually reproducing organism, the target of natural selection is the sum total of all the cells comprising the clone (Mayr 2001). In protistan plankton, the individual is a cloud of cells. So there is ample scope for predators to sample their food, learn to select easy-to-handle cells, and avoid others, thus driving evolution of mechanical and chemical defenses in populations of their prey species. So there is no reason why an arms race, equivalent to that in biomes dominated by multicellular plants, should not also be raging in the plankton.

C. Form and Function in Sessile and Drifting Photoautotrophs

The diversity of forms present in protistan plankton has always amazed the eye of the terrestrial beholder, because we are used to plants that come in standard, understandable shapes dictated by the physical environment and constrained by phylogeny. Thus, land plant lineages evolved trees in characteristic shapes but with a common principle: competition for light in the air and water and nutrients in the ground. The gravitational field coupled with wind energy selected the architecture and material properties of terrestrial plants (roots, trunks, and crown, cellulose fortified with lignin, respectively) as an optimal solution. The degree of freedom of shape ranges from that of the unbranched palm tree, the symmetrically branched conifer, to that of the standard branched tree. The phylogeny and function of each of these morphotypes are not only easily understood by us, we also use the tree as an abstract symbol to depict conceptual, organizational frameworks. Armor, in the form of thorns, tough cell walls, or thick bark, is clearly of secondary importance in determining the overall shape. Even German-speaking botanists did not use the term Panzer when describing land plants.

The shapes of marine macrophytes differ, with the exception of sea grass, from those of land plants, but again, the range of shapes is restricted (from ribbons to filaments) and reflects the environmental energy of the habitat: the gradient from exposed rocky shores to secluded coral reefs. Interestingly, vascular plants have not succeeded in recolonizing the sea except for sea grasses, which seem to have colonized a habitat (sandy coasts) not occupied by macroalgae. The latter did not develop a root system to "hold on" to sandy beaches because nutrients are taken up by the photosynthetic tissue directly. This also applies to sea grasses, whose roots primarily function as holdfasts in soft sediment and secondarily for taking up interstitial nutrients. Although macrophytes, including sea grasses, occur along the entire range of nutrient concentrations prevalent in the sea, sophisticated structures for increasing dissolved nutrient uptake do not seem to have evolved, in contrast to the manifold mechanisms such as cilia and pumps developed by zoobenthos to collect suspended particles. Apparently there is no optimal-solution shape dictated by the physical environment for taking up dissolved nutrients or gases at low Reynold's numbers but just a rule: the greater the s/v ratio, the more efficient the uptake of both photons and molecules.

For phytobenthos organisms below the Kolmogorow scale of turbulent diffusion (< ca. 1000 μm), mechanical stress exerted by shear becomes irrelevant, and shape should no longer matter as long as it does not interfere with holding on to the substrate. The various protistan photoautotrophic lineages represented among the benthic microbiota do not appear to have evolved adaptive shapes that clearly differentiate them from phytoplankton. Nevertheless, the preponderance of pennate diatoms in the phytobenthos does suggest that their streamlined

morphology is particularly well suited to life on or within sediment and obviously connected to their motility; so it is surprising that the same shape, including the raphe that enables locomotion on the surface of particles, is also widespread in pelagic pennates. On the other hand, various centric diatom species have also adapted to life on, or in close association with, the benthos as in mud flats and in the surf zone. The other protistan group that thrives on the sediment surface is Foraminifera. The flattened shells of benthic forms are distinct from those of their more spherical pelagic counterparts, but it is of interest to note that only calcifying and agglutinated heterotrophic protists have colonized the benthos. Silicifying heterotrophs such as Radiolaria are absent even in the deep sea where competition with diatoms for silicic acid should not occur. The significance of this absence is worth pondering.

It follows from this brief comparison of sessile and free-living plants that a relationship between form and function aimed at maximizing light and nutrient harvesting is no longer apparent at the protistan-size scale. In the picoplankton-size range occupied by prokaryotes, cell shape is more or less spherical or rod-shaped, the result of optimal packaging rather than a response to the environment. Diversity of shape increases rapidly in the nanoplankton-size range and reaches its maximum in the micro-size range occupied by bizarrely shaped dinoflagellates, radiolarians, and diatoms. Interestingly, in groups with autotrophic and heterotrophic representatives, in particular dinoflagellates and ciliates, the nutritional mode is not reflected in morphotype, suggesting that the latter serves a function other than resource uptake. Thus, exosymbionts have retained their shape even after changing their life style. Survival in the mortality environment, rather than competition in the growth environment, seems to play the crucial role is determining shape of protists in contrast to that of higher organisms.

As pointed out previously, a school of thought, spearheaded initially by Ernst Haeckel and more recently by Stephen Jay Gould, holds that thinking up a function for every type of form—the "adaptionist approach" ("if this organism is the answer, what is the question?")—is a wild goose chase. The supposedly nonfunctional, roughly triangular wall space between adjacent arches of the San Marco cathedral in Venice (spandrels) have been used as an analogy (Gould and Lewontin 1979). Spandrels do have a mechanical function, although it is not optimized for weight reduction. In addition, they were retained for decoration and their shape will have been dictated by the particular style of architecture in vogue rather than mechanical considerations. They are clearly the product of intentional design that becomes intelligible in its historical context. However, in connection with natural structures, Darwin was of a different opinion: "Natural selection is continually trying to economize in every part of the organisation." "Thus, as I believe, natural selection will always succeed in the long run in reducing and saving every part of the organisation, as soon as it is rendered superfluous" (Darwin 1859). The implication is that natural selection is parsimonious and that the evolution of form is ruled by the role of its function in maintaining fitness of the individual and ultimately the species.

D. Attacking Organisms/Attacking Tools

To appreciate the efficiency of any form of defense one has to first understand the form of attack—whether mechanical or chemical—against which it evolved. The size of the attacker relative to the prey organism is also crucial in the unicellular world, so we follow familiar usage and differentiate attack systems in the three size categories: pathogens, parasitoids, and predators. All three categories have in common that they must gain access to the

plasma of their prey, whether mechanically or via enzymes. However, the techniques employed are very different, as will be the defenses against them. In the following, we present a brief description of the mortality environment sensu Smetacek *et al.* (2004) in which phytoplankton have to grow.

1. Pathogens

Although the presence of pelagic pathogens has been known for a long time, interest in their occurrence and function has blossomed only fairly recently (Suttle 2005). However, most of the interest is focused on viruses; reports on pathogenic bacteria are few (Stewart and Brown 1969; Nagai and Imai 1998; Córdova *et al.* 2002). There is, however, no reason why they should not pose an equivalent threat to protists as they do to multicellular organisms.

It is now well established that viruses and heterotrophic nanoflagellates (HNFs) together prevent bacterial populations from reaching concentrations much above 10^6 cells/ml (Pernthaler 2005). This is a reflection of viral infection potential in the pelagial, implying that other organisms, in particular those that routinely attain high densities, such as bloom-forming phytoplankton, are better defended than the bacteria. However, the ability to ward off viral attack appears to vary considerably among phytoplankton lineages, although more dedicated studies (with publication of negative results) are required before firm statements can be made.

Rampant infection of coccolithophorids and decimation of their blooms by viruses have been reported (Suttle 2005). However, large-scale infection of a diatom bloom has yet to be reported and only two viruses have been isolated so far from diatoms. Both are exceptionally small and infectivity was species-specific and in one case clone-specific. The high degree of specificity indicates that diatoms as a group are well defended against viruses. The possible nature of the defense can be deduced from another bloom-forming group—colonial species of *Phaeocystis*. Hamm *et al.* (1999) have shown that colony shape is maintained by an inelastic but plastic skin with a pore diameter of less than 4nm, which is highly permeable to dissolved substances. The cells inside the colony are never observed in contact with the skin, which raises some intriguing questions as to how the skin expands with growth of the colony. But that is another matter; the point here is that the cells cannot be contacted by viruses, which is why viral infection of colonial cells has not been observed, although it is common among solitary flagellates (Jacobsen *et al.* 1996). So only the colonial stage of this genus forms blooms that rival the biomass attained by diatoms.

The fact that both diatoms and *Phaeocystis* colonies can build up blooms in a variety of environments stretching from the coast to the open ocean and from the tropics to the poles suggests that they are not adapted to a specific range of light or nutrient environments. It is probably also not a coincidence that diatoms and *Phaeocystis* colonies reach high biomass levels within sea ice. The skin then takes the shape of the brine channel within which the colony grows. Because the stiff diatom frustules cannot adapt to the spatial constraints within the sea ice, small-celled species predominate. These "sea-ice species" also thrive in the water column following ice melt so it does not appear that they have adapted their physiology to the specific conditions prevalent in either habitat. It is tempting to suggest that immunity against viral attack, hence reduction in mortality rate, is the common property enabling biomass accumulation to bloom proportions in these species and that the diatom frustule serves a similar role in warding off viral infection as does the colony skin.

Viruses can only generate small forces. Smith *et al.* (2001) have shown that DNA confinement can build up an internal force to ~50pN; this force may be available for initiating the ejection of the DNA from the capsid during infection. In addition, the contraction of the tail sheath of bacteriophages generates a force that suffices to puncture bacterial cell walls. However, most viruses need a direct contact to the cell membrane in order to infect a cell. This may be possible if phytoplankton organisms have only organic covers directly adjacent to the protoplast or if biomineralized covers have temporary loopholes, such as shown for coccolithophorids (Bratbak *et al.* 1993).

2. Pelagic Parasitoids

As in the case of pathogens, planktonic parasitoids have been known for a long time but their role in pelagic ecosystems has been underrated so far. The term is widely used in limnology (Sommer 1994) and is derived from terrestrial ecology, where it is applied to the special type of predation where a small organism feeds on a much larger one while growing or multiplying concomitantly. The parasitoid either forcibly enters its prey and eats it from inside or settles on its surface and feeds on the prey plasma through a tube. In either case, force is required to puncture the prey and the size of the parasitoid sets an upper limit to the force it can generate. Parasitoid attack can be warded off by strengthening the cell wall although the degree of toughness required will be far below that required of a defense against larger predators. The fact that naked species such as amoebae are rare to absent in the marine plankton is indicative of the ubiquity of the threat posed by parasitoids. To exert force on the cell surface of a much larger organism, the parasitoid has to first gain purchase on it. As in the case of pathogens, this can be deflected by a layer of mucus adhering to the cell wall. However, ultimately, effective protection will depend on the strength of the armor relative to the force the parasitoid can generate.

Parasitoid–prey relationships have been reported from a number of organism groups including dinoflagellates that feed on nauplii and ciliates that prey on euphausiids.

Interestingly, parasitoid attack of *Phaeocystis* colonies has yet to be observed, indicating that the mechanical properties of the colony skin provide perfect protection. The best known parasitoids from the pelagic realm are nanoflagellates ("zoospores") from several lineages that feed on large diatom cells (Kühn 1995; Tillmann and Reckermann 2002). Most parasitoids feed only on one or a few diatom species, indicating co-evolution of armor and attack techniques based on the mode of entering the frustule (Kühn 1998). This implies that all diatom species not attacked by a given parasitoid species are effectively protected against penetration by it. No cases of puncturing the silica frustule have been reported, suggesting that even the thinnest frustules are resistant to parasitoid attack. Rather, parasitoids "squeeze" into the frustule through specific sites: either between the girdle bands or through pores in the valves (Figure 2). This implies that the surface of the silica frustule is an effective armor against this class of organisms and it is only the chinks in the armor that render them vulnerable.

In contrast to viral attack, there are reports of mass mortality of diatom blooms due to parasitoids that, because they superficially resemble the ubiquitous bacterivorous HNF, are overlooked in the free-living stage and only recorded when feeding on diatom cells. The literature is anecdotal, but where they have been systematically surveyed over the annual cycle, many species are reported to have recurrent seasonal cycles geared to those of their prey (Drebes 1974). The fact that blooms can be decimated by parasitoids indicates that their growth and infection of new cells can potentially keep up with that of the prey. Therefore, because most diatom blooms are not accompanied by mass parasitoid infection, we infer that these organisms are generally kept under control together with the HNF by grazers such as ciliates and zooplankton larvae. Absence of a frustule would shorten handling time (finding the chink and squeezing through it), thereby increasing parasitoid growth rates and their likelihood of overtaking, hence decimating, growing diatom populations. We suggest that the obligate requirement of diatoms for silicon reflects their vulnerability to pathogens and parasitoid attack, which can be met by a minimum frustule thickness.

Large armored dinoflagellates such as *Ceratium* and *Alexandrium* that also form blooms appear to be even less susceptible to parasitoid attack than diatoms, as, to our knowledge, there are no reports of decimation of a dinoflagellate bloom comparable to the reports for diatoms. Parasitoids of these species have been described (Drebes

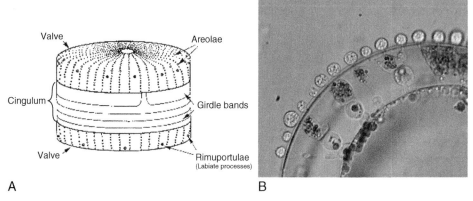

FIGURE 2.   **(A)** Potential sites of infection on the surface of a diatom frustule. **(B)** Numerous cells of the parasitoid *Pirsonia diadema* feeding on the diatom *Coscinodiscus wailesii* (Kühn 1995). (See color plate.)

1979), but the rate of infection must be low, suggesting that the cellulose plates of dinoflagellates and their mode of attachment are superior to those of diatoms. The superiority must come at a price: cellulose armor will be more expensive in energy terms than silica, so diatoms can maintain high growth rates at low light levels.

E. Ingestors or Predators

All the regular herbivores, including protistan ingestors of similar size or larger than their prey, are included here. The largest are centimeter-sized euphausiids and the even larger Peruvian anchoveta, and the smallest are in the same size range as their prey. The difference from the previous categories is that prey shape, and not just its size and surface properties, matters. We argue that the range of shapes in armored phytoplankton larger than 10 μm represent responses to specific types of ingestors. The armor impregnable to small ingestors is mere crispness to the gizzard of a krill. However, the nonselective feeding pressure exerted by large, mobile ingestors is by nature patchy because they feed on swathes of the environment: A dense krill swarm might ingest almost all the potential food in its path but it is only several meters deep and so covers only a portion of the water column. A lot of potential food escapes, so selection exerted by nondiscriminatory filter feeders is weak when compared with the impact of small, more uniformly distributed ingestors. However, not all ingested cells are digested and growth experiments have shown that individual cells of a range of taxa survive euphausiid gut passage (Fowler and Fisher 1983).

Protistan predators share the water column with their phytoplankton prey, actively search for food by swimming with flagellae or cilia, and have growth rates comparable to those of phytoplankton. Their selective pressure will hence be potentially large. Because of their larger size, they can exert more force than the parasitoids and more

elaborate armor is required to deter ingestion. Phagocytic protists have developed a range of techniques to overcome prey defenses (Tillmann 2004). Some species of dinoflagellates have a powerful peduncle with which they pierce their prey and suck in the contents (Jacobson 1999). Species of soft-bodied ciliates appear to be their preferred food, and we know of no reports of large peduncle-feeders preying on diatoms. Apparently, the silica frustules withstand puncturing or crushing by organic-tipped weapons of protists. The skin of *Phaeocystis* colonies appears to be equally effective.

However, the contents of ingested diatoms can be digested without apparent damage to the frustule, as demonstrated by a broad range of diatom-ingesting dinoflagellates. The spines of diatoms deter ingestion but can be overcome in the pallium-feeding mode. The prey item, chains, spines, and all, is covered with a feeding veil (pallium), which is subsequently retracted by the predator leaving behind empty frustules. This spectacular, modified form of ingestion has a major drawback: The deployed pallium will be vulnerable to attack, so prey handling time is crucial. Protuberances of the cell wall will increase the time required to envelope the prey and hence slow growth rates. Although pallium feeders are ubiquitously distributed, they are preferentially grazed by copepods and hence only attain large population sizes sufficient to control blooms under exceptional circumstances.

Not surprisingly, many phagocytic protists (tintinnid ciliates, armored dinoflagellates, Radiolaria, Acantharia, and Foraminifera) also carry lightweight armor based on similar construction principles as in the case of phytoplankton. Some of the protists use phytoplankton shells to construct their own armor (Young and Geisen 2002).

Copepods are the dominant grazers in the ocean in terms of biomass but also impact because they occupy an intermediate position between the large, nonselective, swathe-feeding euphausiids and salps and the selective protists. They tend to feed selectively, and the different species have clear preferences (Koehl 1984). All have powerful mandibles lined with elaborate teeth reinforced with silica with which they crush their prey, including diatom frustules (Figure 3). Clearly these have co-evolved with diatoms (Beklemishev 1954), and a broad range of feeding techniques and abilities are represented in marine copepod assemblages. Thus, feeding experiments with two small, coastal copepod species (*Acartia clausi* and *Temora longicornis*) have shown that, whereas both could successfully tackle chains of the spiny genus *Chaetoceros*, only *Temora* was able to bite out chunks of the equally large frustules of the genus *Coscinodiscus* and suck out the plasma (Jansen 2006). Cells of the armored dinoflagellate *Dinophysis* also appeared to survive copepod gut passage and were the main constituent of the feces of copepods feeding on a natural, summer phytoplankton assemblage (Wexels Riser *et al.* 2003). It should be pointed out that the contents of copepod feces only reflect

undigested food items, which explains why diatom frustules, whether intact or crushed, appear so prominently in them.

The coastal copepods *Calanus helgolandicus* and *Temora longicornis,* when offered a culture of the heavily silicified oceanic diatom species *Fragilariopsis kerguelensis,* fed voraciously but only cracked less than half of the ingested cells (Jansen 2006). The cells within whole frustules in the feces appeared to have survived gut passage as indicated by vital stains. Large copepods such as *Rhincalanus gigas* and *Calanus similimus,* that co-occur with *F. kerguelensis,* manage to crush a much larger percentage of ingested cells; nevertheless, many still appear to survive gut passage (Schultes 2004). The remarkable mechanical strength of the frustules of this diatom has been demonstrated, using micromanipulators and finite element crash models, by Hamm *et al.* (2003). Thus, grazing by the selectively feeding protistan and copepod assemblages will result in distribution of pressure over a wide range of armor types, as reflected in the concomitantly

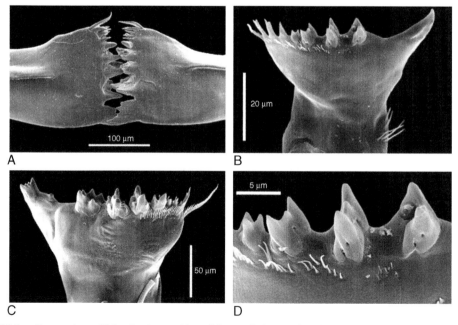

FIGURE 3.   Copepod mandibles: In vivo position of the gnathobases of *Methidia gerlachei* **(A)**, and diverse specific morphologies of *Calanoides acutus,* **(C)**, and *Stephos longipes,* overview and detail (Michels 2003).

occurring diversity in natural phytoplankton assemblages.

A very effective armor-crushing mechanism is undoubtedly the euphausiid gizzard, which is lined with comblike teeth reminiscent of the mandibles of copepods (Figure 4). However, it is not known whether these, too, are reinforced with silica. The crushed contents of euphausiid feces bear witness to the efficiency of this gastric mill, but, nevertheless, intact individuals of various armored species are commonly observed in krill feces. On the other hand, salps have not developed crushing mouth parts or gizzards but nevertheless indiscriminately ingest phytoplankton assemblages. However, their efficiency of digestion has not been systematically demonstrated and live *Synechococcus* cells have been found in their feces (Pfannkuche and Lochte 1993). In the Southern Ocean, the inverse correlation between high concentrations of salps and diatoms has been attributed to avoidance rather than grazing pressure. Smetacek *et al.* (2004) have argued that the long barbed spines and needle-shaped cells of dominant diatoms characteristic of the Antarctic Circumpolar Current are adaptations to deter salp feeding.

It follows that, as can be expected from any arms race, a variety of attack systems will co-evolve with an equal variety of defense systems. Given the heavy grazing pressure prevalent in the pelagial, one can assume an equivalent selection for defenses, whether by deterring ingestion or surviving gut passage. No single armor type can provide universal protection against the armies of pathogens, parasitoids, and ingestors operating in the water column.

II. WHEN

It is highly likely that the origins of armor date back to the early Proterozoic when prokaryotes developed cell walls to protect themselves against chemical attack by other prokaryotes even prior to the evolution of phagocytosis. Indeed, a durable cell outer layer is a prerequisite for the evolution of endosymbiosis, as otherwise the endosymbiont could not survive within the exosymbiont. As mentioned previously, the various instances where new phytoplankton lineages were started when an efficient photosynthesizer sought shelter in, or was taken over by, a well-protected exosymbiont resulted in species radiations, which had profound implications for ocean and hence planetary biogeochemistry (Knoll 2003).

The advent of mineral armor probably extends well back into the Precambrian, but during the Cambrian Explosion the

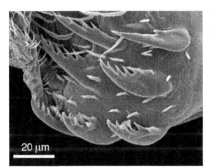
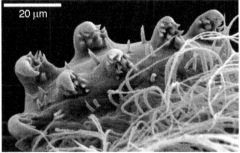

FIGURE 4. View of the interior of the gizzard of *Euphausia superba* showing the comblike internal teeth of the gastric mill adapted to crush diatom frustules. (J. Michels, unpublished).

evolutionary arms race intensified as evidenced by the prominence of mineral armor in the fossil record. As the compressive strength of armor increased, crushing and piercing tools or weapons increased in efficiency concomitantly. However, mineral armor is not necessarily superior to organic armor, as demonstrated by the presence of both calcareous and organic types in dinoflagellates and prymnesiophytes. In fact, calcifying dinoflagellates were prolific in the Cretaceous but are represented by only a few species today that are of minor importance. Dinoflagellates with cellulose armor plating are the dominant types today. Similarly, coccolithophorids are very prominent in chalk deposits from the Cretaceous, whereas the highly effective organic skin of *Phaeocystis* is reported to have evolved comparatively recently (about 30 million years ago [Ma] [Lange *et al.* 2002]).

The case of silica armor is most illuminating in this respect as it demonstrates that it is not just the presence or absence of minerals that is crucial but, in particular, the geometry and architecture in relation to the life cycle. Thus, silicoflagellates with an internal skeleton of silica are reported to have played a more important role in Mesozoic plankton than the diatoms (Parsons *et al.* 1977), although the origin of the latter goes back to the Jurassic. There has been speculation regarding the success of diatoms during the Cenozoic, exemplified by the massive deposits of diatom silica during the Neogene. Whether this is due to a greater supply of silicic acid or to an improvement in the efficiency of the silica frustule is unclear. In any case, Mesozoic diatom frustules look heavy and crudely constructed compared to the elegant lightweight construction of most modern diatoms (Gersonde and Harwood 1990). A similar development from heavy-duty to lightweight calcareous scales from the Mesozoic to Cenozoic can be observed in coccolithophorids. This trend to lightweight construction today is probably due to both sophistication in herbivore mouth parts and a shortage of the building material.

III. HOW

In this section we examine some aspects of physical barriers, that is, "armor" in its broadest sense, in terms of material properties and construction principles. Thus, a coat of mucilage is probably the simplest and perhaps earliest form of armor as it can ward off various attack systems from those of viruses to ingestors and continues to be deployed across all size ranges from bacteria to fish and amphibians. Many phytoplankton species secrete vast amounts of mucus or "exopolymer particles" (Decho 1990) as defense against grazing (Malej and Harris 1993). However, mucus is structurally unspecific ("messy") and may have unfavorable side effects, such as supporting aggregation and sedimentation of algae (Passow 2002). Possibly the major disadvantage of mucus is that it has to be produced continuously as it sloughs off at the outer surface and is hence metabolically more costly than a rigid armor that does not have to be renewed in order to be effective.

Ideally, efficient armor needs to combine optimal stability with a minimum of weight, so it is not surprising that the armor but also endoskeletons of modern plankton exhibit the typical properties of stable lightweight constructions developed by engineers. Because material cross-section and (external) pressure both scale with the square of the length scale, stable lightweight constructions are characterized by typical, well-matching geometries and material properties.

Although the overall geometric properties of phytoplankton are well described, mainly by light and electron microscopy, the material properties of phytoplankton armor are not well known. Several properties are crucial for understanding the use of specific materials as armor. The geometries of stable lightweight constructions may vary, but they all comply with relatively few, fundamental rules. Here we show some basic principles of lightweight engineering that are reflected in the structures of many plankton shells. Gordon (1978) and Mattheck (2004) have provided introductory information on the mechanics of structural engineering.

A. Material

A very clear concept in solid mechanics is that of stiffness of a given material defined by the modulus of elasticity or Young's Modulus E or Y. It describes the deformation (strain e) of a material as a function of a certain stress S. This can be experimentally determined from the slope of a stress–strain curve (Figure 5).

$$Y = \frac{stress}{strain} = \frac{F/A}{\Delta l / l_0} = \frac{Fl}{A\Delta l_0}$$

In brittle materials this function is almost linear (elastic region of Figure 5), but in many materials, such as metals or elastomeres, it contains or is a nonlinear function (plastic region in Figure 5). The higher E becomes, the stiffer is a certain material. Typical values are 2 GPa for polymers, 20 GPa for bone or hardwood, 70 GPa for glass, and 200 GPa for steel.

The ultimate strength of a material is defined by the highest stress a material can resist without breaking (see Figure 5). The values of maximal compressive stress and tensile stress may differ, so that the ultimate tensile strength of a material may have different values from the ultimate compressive strength. Although some metals and ceramics may reach strengths of well over 1000 MPa, high-strength steels range between 400 and 600 MPa. A similar value has been calculated for the silica of diatoms (Hamm *et al.* 2003). In contrast, most polymers have values of less than 100 MPa.

Toughness results from strength and stiffness. It is defined as the amount of energy that a material can absorb before rupturing and can be quantified by calculating the area (i.e., by taking the integral) underneath the stress–strain curve (see Figure 5). The modulus of toughness is measured in units of joules per cubic meter (J/m^3); however, this is not often used to describe material properties.

$$U_e = \int \frac{YA}{l_0} dl = \frac{YA\Delta l^2}{2l_o}$$

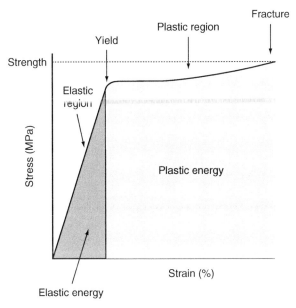

FIGURE 5. Stress–strain diagram for a material subject to a classic tension test.

Typical tough materials include many metals and polymers, because they can dissipate a lot of energy due to a large degree of plastic deformation before failure; brittle materials include glass/opal and ceramics and most inorganic crystalline minerals.

It seems paradoxical, at first sight, that many phytoplankton groups use apparently brittle materials such as silica (i.e., opal) or calcite (coccolithophores) to create armor. On the other hand, biomineralized materials are, in fact, almost always composites (Jackson *et al.* 1990), in which small amounts of organics improve the strength and toughness of the dominating mineral by a large factor (e.g., up to 3000 in the case of nacre compared to aragonite; Okumura and de Gennes 2001). There is evidence that biomineralized phytoplankton shells such as diatoms, in analogy to nacre, are also made of functional composites (Hamm *et al.* 2003).

B. The Geometry

In addition to the material properties, the geometry of a structure defines its use as mechanical protection. Certain geometries (e.g., sharp angles, notches, and crack tips) create stress concentrations (Figure 6). If the stress exceeds the strength of a material, failure occurs. If the crack length cannot be

shortened, it is necessary to reduce stress concentrations by blunting crack tips. This is expressed by the simplified formula:

$$s\left(1+2\sqrt{\frac{L}{r}}\right),$$

where L is the length of the crack, and r the radius of the crack tip. Examples for stress reduction can be seen in many pennate diatoms, which need a slit (a crack) in their shell for locomotion. The tips of these structures are always blunted, bent, and/or reinforced (see Figure 6).

The stiffness contributed by the geometry results from the value of the second moment of area (also called moment of inertia). It measures the efficiency of a shape in respect to its resistance to bending. If a beam or a shell is bent, the external regions of the structure are more strongly deformed than are the inner regions. A certain plane of the structure, which passes through the centroid, is not deformed at all and therefore does not help the structure to resist deformation; it is, therefore, called "neutral axis or fiber." For a light but stiff structure, it is necessary to move the material as far away from the neutral axis as possible, which results in i-beams, honeycomb sandwich constructions, and corrugated materials in the technical world and

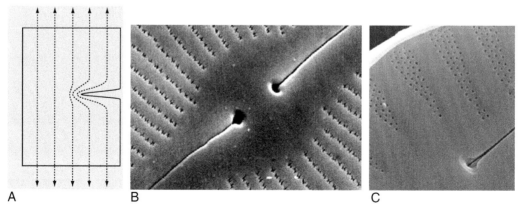

FIGURE 6. Sketch displaying stress concentration around a crack (from Gordon 1978) and the "blunted" ends of the raphe of two pennate diatoms. Scanning electron microscope (SEM) pictures from C. Kages/F. Hinz. (See color plate.)

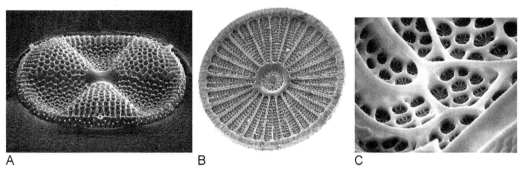

FIGURE 7. **(A, B, C)** Frustules of the marine diatoms *Actinoptychus, Arachnoidiscus,* and *Isthmia* (detail), respectively, displaying different ways of building a stiff lightweight construction by generating high moments of inertia at different scales. Note the corrugated surfaces, ribs, and honeycomb structures and the fractal character of the *Isthmia* shell. Parts **(A)** and **(C)** from C. Kages; **(B)** from F. Hinz.

analogous structures in the shells of unicellular organisms (see Figures 1, 7, and 8). The formula for the uniaxial moment of inertia is:

$$I_y = \int_A Z^2 dA$$

The units are thus m⁴, and increasing the thickness of a beam or a shell in the direction of the bending force results in an exponential improvement of its stiffness.

The axiom of equal stress describes another principle of efficient engineering: A structure is always only as strong as its weakest member, for example, a girder that is so thin that stress values within it exceed that of the ultimate stress will break (Mattheck 2004). On the other hand, "lazy" structural members in which stress values are low are not functional but costly because they have to be built and may be a disadvantage to an organism if it needs to be light to compete within its ecosystem. A homogeneous distribution of stress within a structure is, therefore, besides a high moment of inertia, a sign for an efficient use of the construction material.

C. Lightweight Constructions of Phytoplankton Armor

Examples of protist armor show that the principles of stable lightweight constructions are realized in many different ways. All

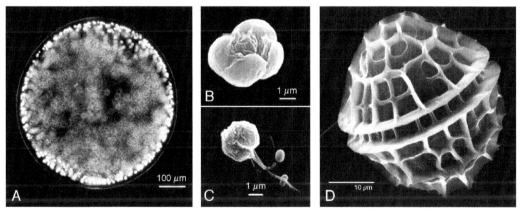

FIGURE 8. Organic covers. Left: *Phaeocystis* colony **(A)**, naked colonial cell **(B)**, and free-living flagellate **(C)**. Right: The stiff, lightweight cellulose armor of a dinoflagellate *Protoceratium spinulosum* **(D)**. Part **(D)** from Hallegraeff (1988). (See color plate.)

of them strictly comply with the previously mentioned laws of good engineering. Although very little is known on the material properties of protist armor, the available data indicate that natural selection has led to powerful solutions.

An example for tough, tensile armor was found in *Phaeocystis* colonies that are surrounded by a skin, which only ruptures after extreme deformations and contains only pores smaller than 4.4 nm—an effective barrier against viruses (Hamm *et al.* 1999; Figure 8). The blooms of *Phaeocystis* colonies quickly break down when the colony skins lose these properties, which usually happens in conjunction with nutrient depletion. This approach of a tough, tensile, and collective armor is successful but not often realized in autotrophic marine protists. We suggest that tensile covers do not enable structural differentiation; *Phaeocystis* colonies and other phytoplankton colonies with tensile covers, such as *Volvox* or *Coelosphaerium*, look very much alike—they resemble inflated objects. The rapid disintegration of such structures under unfavorable growth conditions suggests that they cannot be used for less productive stages.

More durable and more versatile in their forms are the thecae of dinoflagellates first described in detail in monographs published by Stein in 1878–1883. The thecae, or armor, of dinoflagellates is made of cellulose-like polymers, and their forms suggest compressive strength of this material (see Figure 8). It is known that many dinoflagellate cysts, which can be composed of the very refractory sporopollenin or even calcite, are resistant to bacterial attack and thus well preserved in the fossil record (Evitt 1985).

The highest diversity and the most obvious function as armor can be found in the silica shells, the frustules, of diatoms. Diatoms often combine several principles of lightweight constructions such as I-beams, honeycomb sandwich constructions, corrugated forms, or supporting ribs in a single shell at different size scales (see Figure 7).

The reason for this is, probably, that the attacking organisms likewise use tools of very different sizes to crack or puncture the shells. A fractal appearance of some diatom shells, that is, similar structures repeated at different size scales (see Figure 7), may also result from the fact that material cross-section and pressure on a shell both scale with the square of the length scale.

D. Spines and Large Size

Spines keep potential predators at a distance but, in addition, can injure the feeding/digestive system of organisms trying to feed on them. Diatoms often possess extensive, highly structured (i.e., barbed and latticed) siliceous spines, which are often hollow and harbor parts of the cytoplasm, including chloroplasts (e.g., in *Chaetoceros*), but there are also chitinous, threadlike processes such as those of *Thalassiosira*. The effect of spines as a deterring mechanism is difficult to quantify, but it is known that copepods bite off spines from diatoms individually before feeding on them. Spines, which are not integrated within an exoskeleton but connected to an endoskeleton, such as found in silicoflagellates (e.g., *Distephanus*), may be harmful to larger organisms (metazoans) but are inefficient against pathogens and small protists.

Spines have often been interpreted as structures that reduce sinking velocity of diatoms, as they increase the drag of the cells considerably without adding much weight to them. Although this function is plausible, it is only worthwhile on large, mechanically defended organisms, as reduction of size would have the same effect and could, in addition, increase the growth rate of a unicellular alga, as its smaller diffusive boundary layer would permit more efficient nutrient uptake.

Large size in combination with a certain mechanical resilience can, in addition to offering other beneficial effects (Finkel 2001), definitely help in making an armor more efficient. Thus, the costs for producing a given thickness of armor decreases, within limits, with the s/v ratio. If predators have difficulties swallowing or engulfing a cell, they cannot use their enzymes as efficiently. However, large cells usually grow slower than small cells, and larger size in combination with similar shell thickness leads to decreased stability. Recent results have shown that, although some copepods can exploit biomass of diatoms of approximately their own size by fracturing the cells at specific sites, others, which are unable to fracture the shells, are completely excluded from grazing (Jansen 2006).

E. Other Functional Explanations

Several other explanations for presence and form of the shells of eukaryotes have been proposed. However, in our opinion, they do not provide an equally plausible and straightforward explanation on both materials *and* complexity of forms present in phytoplankton shells, as does selection pressure to develop stable lightweight constructions against physical attack by grazers.

The diatom frustule is far beyond the requirements of rigidity to support a vacuole, as demonstrated by the auxospore, which first forms a huge vacuole supported by thin scales within which the silica frustules subsequently develop. The idea that the honeycomb pattern of many diatoms may act as photonic crystals that concentrate light on the chloroplasts is not consistent with the fact that the chloroplasts in large diatoms with loculate areolae (pores surrounded by honeycomb structures) are evenly distributed directly at the proximate side of the shell, thus not where a focus would be. Still, the analysis of the optical properties of the patterned silica or calcium carbonate is needed. Optical effects such as iridescence have been observed in diatom frustules but are also present in mother of pearl inside the shells of many bivalves and gastropods, where an optical significance is not plausible. Iridescence caused by multiple reflections from multilayered, semi-

transparent surfaces is thus more likely to result from mechanical necessities.

A proton buffering role of diatom shells to stabilize external carboanhydrase activity has been postulated (Milligan and Morel 2002), but it would, if efficient, explain the presence of silica but not its intricate forms. Also, a negative sorting effect of the silica pore structure of diatoms as an adaptation against attack by pathogens, as proposed by Hale and Mitchell (2001), is problematic, as the silica shells vary strongly in their types of external surfaces. Strong evidence for the hypothesis that phytoplankton covers serve as armor is given by the possibility to induce colony formation of *Phaeocystis* (Tang 2003) and a reinforcement of diatom shells by the presence of grazing copepods (Pondaven *et al.* 2007). Such induced responses are well described from freshwater and terrestrial systems.

IV. CONCLUSIONS

The potential of mechanical defense in unicellular organisms, especially those occurring in the plankton, is still underestimated. Most likely, this neglect is related to the small size of unicellular organisms, in spite of Feynman's remark that there is plenty of room (e.g., for sophisticated and complex structures) in the microscopic world and in spite of the current enthusiasm about nanotechnology. In order to link structural features of an organism to a function, knowledge of the physical and ecological context, and thus of the factors that cause significant selection pressure, is crucial. We have given an overview of the mechanical challenges faced by these organisms. The criteria typical for stable lightweight constructions are realized in many aspects of phytoplankton shells. Efficient implementation of lightweight principles are reflected in the fact that these biogenic constructions can be used to improve professionally engineered anthropogenic structures. The time is, therefore, ripe for reconsideration of the role of defense in the evolution of unicellular organisms, in particular the plankton. Given the panoply of attack techniques extending from viruses to fish to which protists are exposed, we have deemed it pointless to draw artificial boundaries and have accordingly defined armor with its German connotations as all forms of mechanical defense against pathogens, parasites, and ingestors. These in turn will range from tough cell walls that withstand puncturing to long spines that deter ingestion. Armor evolves in response to a specific form of attack. Indeed the co-evolution of attack and defense systems, encapsulated in the term evolutionary arms race, is an acknowledged driving force in natural selection of terrestrial plants, leading to speciation and ultimately shaping the structure and function of ecosystems. Autotrophic marine protists will be shaped by the same processes.

Acknowledgments

We thank Stephanie Kühn, Friedel Hinz, Christina Kage, and Jan Michels for generously providing us with photographs to use as illustrations for this chapter.

References

Alcaraz, M., Paffenhofer, G.H., and Strickler, J.R. (1980). Catching the algae: a first account of visual observations on filter feeding calanoids. *Evolution and Ecology of Zooplankton Communities*. W.C. Kerfoot, ed. Univ. Press of New England, pp. 241–248.

Azam, F., Fenchel, T., Field, J.G., Gray, J.S., Meyer-Reil, L.A., and Thingstad, F. (1983). The ecological role of water-column microbes in the sea. *Mar. Ecol. Prog. Ser.* **10:** 257–263.

Beklemishev, K.V. (1954). The discovery of silicious formations in the epidermis of lower crustacea. *Dokl. Akad. Nauk SSSR* **97:** 543–545.

Bratbak, G., Egge, J.K., and Heldal, M. (1993). Viral mortality of the marine alga *Emiliania huxleyi* (Haptophyceae) and termination of algal blooms. *Mar. Ecol. Prog. Ser.* **93:** 39–48.

Cembella, A.D. (2003). Chemical ecology of eukaryotic microalgae in marine ecosystems. *Phycologia* **42:** 420–447.

Córdova, J.L., Cárdenas, L., Cárdenas, L., and Yudelevich, A. (2002). Multiple bacterial infection of *Alexandrium catenella* (Dinophyceae) *J. Plankton Res.* **24:** 1–8.

Darwin, C. (1859). *On the Origin of the Species by Means of Natural Selection*. London, John Murray.

Decho, A.W. (1990). Microbial exopolymer secretions in ocean environments: their role(s) in food webs and marine processes: *Oceanogr. Mar. Biol. Ann. Rev.* **28:** 75–153.

Drebes, G. (1974). *Marines Phytoplankton. Eine Auswahl der Helgoländer Planktonalgen (Diatomeen, Peridineen)*. Stuttgart, G. Thieme.

Ehrenberg, C.G. (1838). *Die Infusionsthierchen als volkommene Organismen. Ein Blick in das tiefere organische Leben der Natur*. Leipzig, Leopold Voss.

Ehrlich, P.R., and Raven, P.H. (1964). Butterflies and plants: a study in coevolution. *Evolution* **18:** 586–608.

Evitt, W.R. (1985). *Sporopollenin Dinoflagellate Cysts: Their Morphology and Interpretation*. Dallas, American Association of Stratigraphic Palynologists Foundation, pp. 333.

Finkel, Z.V. (2001). Light absorption and the size scaling of light-limited growth and photosynthesis in marine diatoms. *Limnol. Oceanogr.* **46:** 86–94.

Friedman, M.M., and Strickler, J.R. (1975). Chemoreceptors and feeding in calanoid copepods (Arthropoda: Crustacea) *PNAS* **72:** 4185–4188.

Fowler, S.W., and Fisher, N.S. (1983). Viability of marine phytoplankton in zooplankton fecal pellets. *Deep Sea Res.* **30:** 963–969.

Gersonde, R., and Harwood, D.M. (1990). Lower Cretaceous diatoms from ODP Leg 113 Site 693 Weddell Sea. Part 1: vegetative cells. *Proceedings of the Ocean Drilling Program, Scientific Results*, Volume 113. P.F. Barker, D.M. Kennett, A. Masterson, and N.J. Stewart, eds. College Station, TX, Ocean Drilling Program, pp. 403–425.

Gordon, J.E. (1978). *Structures or Why Things Don't Fall Down*. Gretna, LA, Pelican.

Gould, S.J., and Lewontin, R. (1979). The spandrels of San Marco and the Panglossion paradigm: a critique of the adaptationist programme. *Proc. R. Soc. Lond.* **205:** 581–598.

Haeckel, E. (1862). *Die Radiolarien (Rhizopoda Radiaria)*. Berlin, G. Reimer.

Haeckel, E. (1866). *Generelle Morphologie der Organismen*. (2nd Vol.) Berlin, G. Reimer.

Haeckel, E. (1904). *Kunstformen der Natur*. Leipzig und Wien, Verlag des Bibliographischen Instituts.

Haellegraeff, G.M. (1988). *Plankton: A Microscopic World*. Leiden, NY, Brill.

Hale, M.S., and Mitchell, J.G. (2001). Functional morphology of diatom frustule microstructures: hydrodynamic control of Brownian particle diffusion and advection. *Aquat. Microb. Ecol.* **24:** 287–295.

Hamm, C., Merkel, R., Springer, O., Jurkojc, P., Maier, C., Prechtel, K., and Smetacek, V. (2003). Architecture and material properties of diatom shells provide effective mechanical protection. *Nature* **421:** 841–843.

Hamm, C.E., Simson, D., Merkel, R., and Smetacek, V. (1999). Colonies of *Phaeocystis globosa* are protected by a thin but tough skin. *Mar. Ecol. Prog. Ser.* **187:** 101–111.

Hardy, A.C. (1956). *The Open Sea, Its Natural History: The World of Plankton*. London, Collins, pp. 355.

Hensen, V. (1887). Über die Bestimmung des Planktons oder des im Meere treibenden Materials an Pflanzen und Thieren, nebst Anhang. *V. Bericht d. Kommission zur Wiss. Untersuch. d. Deutschen Meere zu Kiel* 1882–1886 **12–14:** 1–107.

Jackson, A.P., Vincent, J.F.V., and Turner, R.M. (1990). Comparison of nacre with other ceramic composites. *J. Mater. Sci.* **25:** 3173–3185.

Jacobsen, A., Bratbak, G., and Heldal, M. (1996). Isolation and characterization of a virus infecting *Phaeocystis pouchetii* (Prymnesiophyceae). *J. Phycol.* **32:** 923–927.

Jacobson, M. (1999). A brief history of dinoflagellate feeding research. *J. Eukar. Microbiol.* **46:** 376–381.

Jansen, S. (2006). *Feeding Behaviour of Calanoid Copepods and Analyses of Their Faecal Pellets*. PhD Thesis, University of Bremen, 154 pp.

John, D.M., Hawkins, S.J., and Price, J.H. (1992). *Plant-Animal Interactions in the Marine Benthos*. New York, Oxford University Press.

Knoll, A.H. (2003). The geological consequences of evolution. *Geobiology* **1:** 3–14.

Koehl, M.A.R. (1984). Mechanisms of particle capture by copepods at low Reynolds number: Possible modes of selective feeding. *Trophic Interactions within Aquatic Ecosystems*. D.G. Meyers and J.R. Strickler, eds. Boulder, CO, Westview Press, pp. 135–166.

Kühn, S. (1995). *Infection of Marine Diatoms by Parasitoid Protists*. Ph.D. Thesis, University of Bremen, 150 pp.

Kühn, S.F. (1998). Infection of *Coscinodiscus* spp. by the parasitoid nanoflagellate *Pirsonia diadema*: II. Selective infection behaviour for host species and individual host cells. *J. Plankton Res.* **20:** 443–454.

Lange, M., Chen, Y., and Medlin, L.K. (2002). Molecular genetic delineation of *Phaeocystis* species (Prymnesiophyceae) using coding and non-coding regions of nuclear and plastid genomes, *Eur. J. Phycol.* **37:** 77–92.

Maloj, A., and Harris, R.P. (1993). Inhibition of copepod grazing by diatom exudates—a factor in the development in the mucus aggregates. *Mar. Ecol. Prog. Ser.* **96:** 33–42.

Mattheck, C. (2004). *The Face of Failure in Nature and Engineering*. Karlsruhe, Verlag Forschungszentrum Karlsruhe GmbH.

Mayr, E. (2001). *What Evolution Is*. New York, Basic Books.

Meyer, F.S., ed. (1888). *Handbuch der Ornamentik: zum Gebrauche für Musterzeichner, Architekten, Schulen und Gewerbetreibende sowie zum Studium im Allgemeinen*. Leipzig, Seemann.

Michels, J. (2003). *Signifikanz von Morphologie und Stabilität der Mandibel-Gnathobasen für die Ernährungsweise antarktischer Copepoden*. Diplomarbeit, University of Kiel, 136 pp.

Milligan, A.J., and Morel, F.M. (2002). A proton buffering role for silica in diatoms. *Science* **297:** 1848–1850.

Nagai, S., and Imai, I. (1998). Killing of a giant diatom *Coscinodiscus wailesii* Gran by a marine bacterium *Alteromonas* sp. Isolated from the Seto Inland Sea of Japan. *Harmful Algae*. B. Reguera, J. Blanco, M.L. Fernández, and T. Wyatt, eds. Vigo, Xunta de Galicia and Intergovernmental Oceanographic Commission of UNESCO, pp. 402–405.

Okumura, K., and de Gennes, P.-G. (2001). Why is nacre strong? Elastic theory and fracture mechanics for biocomposites with stratified structures, *Eur. Phys. J.* **4:** 121–127.

Parsons, T.R., Takahashi, M., and Hargrave, B. (1977). *Biological Oceanographic Processes*, 3rd ed. Oxford, Pergamon Press.

Paffenhöfer, G.-A., and Lewis, K.D. (1990). Perceptive performance and feeding behavior of calanoid copepods. *J. Plankton Res.* **12:** 933–946.

Passow, U. (2002). Transparent exopolymer particles (TEP) in the marine environment. *Prog. Oceanogr.* **55:** 287–333.

Pernthaler, J. (2005). Predation on prokaryotes in the water column and its ecological implications. *Nature Rev. Microbiol.* **3:** 537–546.

Pfannkuche, O., and Lochte, K. (1993). Open ocean pelago-benthic coupling: Cyanobacteria as tracers of sedimenting salp faeces. *Deep Sea Res.* **40:** 727–737.

Pomeroy, L.R. (1974). The ocean's food web, a changing paradigm. *BioScience* **24:** 499–504.

Pondaven, P., Gallinari, M., Chollet, S., Bucciarelli, E., Sarthou, G., Schultes, S., and Jean, F. (2007). Grazing-induced changes in cell wall silicification in a marine diatom. *Protist* **158:** 21–28.

Rausher, M.D. (2001). Co-evolution and plant resistance to natural enemies. *Nature* **411:** 857–862.

Raymont, J.E.G. (1963). *Plankton and Productivity in the Ocean*. Oxford, Pergamon Press Ltd.

Schultes, S. (2004). *The Role of Mesozooplankton Grazing in the Biogeochemical Cycle of Silicon in the Southern Ocean*. Ph.D. Thesis, University of Bremen, 123 pp.

Smetacek, V. (2001). A watery arms race. *Nature* **411:** 745.

Smetacek, V., Assmy, P., and Henjes, J. (2004). The role of grazing in structuring Southern Ocean pelagic ecosystems and biogeochemical cycles. *Antarct. Sci.* **16:** 541–558.

Smetacek, V., Montresor, M., and Verity, P. (2002). Marine productivity: Footprints of the past and steps into the future. *Phytoplankton Productivity*. P. Williams, D.N. Thomas, and C.S. Reynolds, eds. Blackwell Sciences, pp. 350–369.

Smith, D.E., Tans, S.J., Smith, S.B., Grimes, S., Anderson, D.L., and Bustamante, C. (2001). The bacteriophage straight phi29 portal motor can package DNA against a large internal force. *Nature* **413:** 748–752.

Sommer, U. (1994). Planktologie. Berlin, Springer-Verlag.

Sournia, A. (1982). Form and function in marine phytoplankton. *Biol. Rev.* **57:** 347–394.

Stein, F. (1878). *Der Organismus der Infusionsthiere. III. Flagellaten I.* Leipzig, W. Engelmann.

Stein, F. (1883). *Der Organismus der Infusionstiere. III. Abt. Der Organismus der Arthrodelen Flagellaten. Einleitung und Erklärung der Abbildungen.* Leipzig, W. Engelmann.

Stewart, J.R., and Brown, R.M. (1969). *Cytophaga* that kills or lyses algae. *Science* **164:** 1523–1524.

Suttle, C.A. (2005). Viruses in the sea. *Nature* **437:** 356–361.

Tang, K.W. (2003). Grazing and colony size development in *Phaeocystis globosa* (Prymnesiophyceae): the role of a chemical signal. *J. Plankton Res.* **25:**831–842.

Tillmann, U. (2004). Interactions between planktonic microalgae and protozoan grazers. *J. Eukar. Microbiol.* **51:** 156–168.

Tillmann, U., and Reckermann, M. (2002). Dinoflagellate grazing on the raphidophyte *Fibrocapsa japonica. Aquat. Microbial Ecol.* **26:** 247–257.

Wexels Riser, C., Jansen, S., Bathmann, U., and Wassmann, P. (2003). Grazing of *Calanus helgolandicus* on *Dinophysis norvegica* during bloom conditions in the North Sea: evidence from investigations of faecal pellets. *Mar. Ecol. Prog. Ser.* **256:** 301–304.

Young, J.R., and Geisen, M. (2002) Xenospheres—associations of coccoliths resembling coccosphere. *J. Nannoplankton Res.* **24:** 27–35.

15

Does Phytoplankton Cell Size Matter? The Evolution of Modern Marine Food Webs

Z. V. FINKEL

Many physiological rates and ecological and evolutionary patterns are affected by the size of organisms involved; for example, metabolic rate, elemental requirements, sinking rates, abundance, biomass, diversity, home range, and longevity (Smayda 1970; Shuter 1978; Peters 1983a; Agusti *et al.* 1987; Bonner 1988; Kiorboe 1993; Brown 1995; Stemmann *et al.* 2004). As the foundation of aquatic food webs, the size of the cells that compose the phytoplankton community has a fundamental influence on the structure and function of aquatic ecosystem (Figure 1). In this chapter, I review the relationship between organism size and metabolic rate and the resulting consequences for ecological patterns of abundance, diversity, and food web structure and function; summarize the current state of knowledge of how phytoplankton community size structure has changed over geological time; and explore the potential consequences of macroevolutionary shifts in phytoplankton community size structure to the evolution of food web assembly, structure, and function over geological time.

I. SIZE MATTERS: FROM PHYSIOLOGICAL RATES TO ECOLOGICAL AND EVOLUTIONARY PATTERNS

A. Size Scaling of Physiological Rates

"Almost all aspects of the life of a phytoplankton cell are influenced, more or less, by its size…The mechanisms underlying the size-dependent patterns have undoubtedly steered the general course of phytoplankton evolution, but the organisms that do not abide by the rules reveal the wonderful diversity of ways in which cells have managed to disobey the 'laws' scripted for them"—Chisholm 1992

Size scaling laws are a remarkably general and widely observed phenomenon in biology (Kleiber 1947; Peters 1983a; Bonner 1988; Brown 1995; Kerr and Dickie 2001). The most fundamental is the relationship between an organism's size and its metabolic rate because this governs the rate of the individual's interaction with the environment (Peters 1983a; Brown 1995). From bacteria to large mammals, body size can be used to predict metabolic rate:

$$B = B_0 M^b \qquad (1)$$

where b is the size scaling exponent of the relationship between the metabolic rate (B) and the organism's size (M) (see Figure 1). The size scaling exponent tends to be ¾ under standard temperature and optimal growth conditions due to the geometric constraints of transportation networks within organisms (Banavar *et al.* 2002). Phytoplankton cell volumes span 10 orders of magnitude; assuming b = ¾, then, phytoplankton size can account for ~6 orders of magnitude in metabolic rate. A large variety of size scaling exponents have been reported from field and experimental observations of the metabolic rates of phytoplankton (Banse 1976; Taguchi 1976; Schlesinger *et al.* 1981; Sommer 1989; Finkel 2001). When resource or energy supply does not match the demands of growth, and the acquisition of resources is size dependent (such as light-harvesting and nutrient diffusion in phytoplankton), then the size scaling exponent b changes with resource availability (Finkel and Irwin 2000; Finkel 2001; Finkel *et al.* 2004). Regardless of the explanation, or the exact exponent, the size dependence of metabolic rates appears to influence many fundamental macroecological and

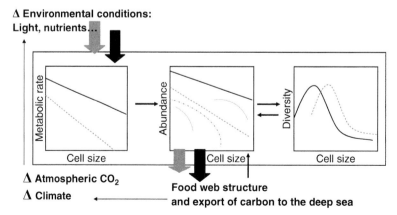

FIGURE 1.   The interrelationships between climate, the size scaling of metabolic rate, abundance, and diversity of phytoplankton. The dark lines represent before a change in climate and the lighter grey arrows and dashed lines represent after a change in climate.

evolutionary patterns (Gould 1966; Peters 1983a; Bonner 1988; Brown 1995; Kerr and Dickie 2001; Niklas and Enquist 2001; Trammer 2002; Irwin *et al.* 2006).

B. Size–Abundance Relationship

Classic studies on marine plankton indicated that the abundance (A) of organisms per unit volume or area is often inversely related to organism mass:

$$A = cM \qquad (2)$$

where the size scaling exponent (ξ) is often approximately −1 (see Figure 1) (Sheldon and Parsons 1967; Sheldon *et al.* 1972), such that biomass per unit volume or area is the same for all logarithmically separated size classes. Subsequent field and lab studies have documented considerable variability in the size scaling exponent associated with marine phytoplankton communities, ξ, often ranging from − $2/3$ to − $5/3$ (Agusti 1987; Agusti and Kalff 1989; Boss *et al.* 2001). Small phytoplankton cells often dominate in stable, oligotrophic environments such as the open ocean, whereas larger cells increasingly dominate in eutrophic environments that are characterized by environmental variability, such as coastal areas (Malone 1971; Peters 1983b; Sprules and Munawar 1986; Chisholm 1992; Li 2002). This suggests that both ξ and the intercept c increase with limiting resource availability (or chlorophyll concentration) (Peters 1983b; Sprules and Munawar 1986; Chisholm 1992; Duarte *et al.* 2000; Li 2002). The changes in size structure with resource availability and total biomass have often been interpreted as a release from grazer control (Malone 1971; Armstrong 2003; Irigoien *et al.* 2004; Morin and Fox 2004). Competition (Grover 1989, 1991), deviations from steady-state (Sprules and Munawar 1986), and the scale-free self-organization of complex adaptive systems (Rinaldo *et al.* 2002) have also been proposed as mechanisms to explain the size scaling of

abundance, but none of these hypotheses explains the increasing dominance of large cells under high nutrient conditions. Alternatively, it has been demonstrated that the size scaling of cellular nutrient requirements and growth can explain the power-law relationship between cell size and abundance, the dominance of small phytoplankton cells under oligotrophic conditions, and the relative increase in abundance of larger phytoplankton cells under eutrophic conditions (Irwin *et al.* 2006).

C. Size–Diversity Relationship

Species diversity (S, number of species) is often a skewed log-normal function of organism size, where maximum species diversity for the log-normal distribution occurs at an intermediate organism size (see Figures 1 and 2) (Van Valen 1973; May 1978; Fenchel 1993; Brown 1995; Gaston and Blackburn 2000). There are known deviations from this relationship; some taxonomic groups have several peaks in diversity at different sizes, or diversity can be independent of organism size (Gaston and Blackburn 2000). Although the majority of the pioneering work on the size–diversity relationship has focused on mammals and birds, Fenchel's compilation of 18,500 aquatic free-living species, including bacteria and protozoa, indicates intermediate-sized species are most diverse in marine and freshwater habitats (Fenchel 1993). An analysis of the size distribution of phytoplankton species from the Provosili-Guillard national culture collection indicates that phytoplankton species diversity may also be a skewed log-normal function of cell size (Figure 2). The data available may be biased due to sampling techniques and sampling effort; for example, documentation of the diversity of picoeukaryotes has increased dramatically with the recent application of molecular techniques to identify species from field assemblages (Moon-Van Der Staay *et al.* 2001; Not *et al.*

FIGURE 2. An analysis of the size distribution of all phytoplankton species with size information from the Provasoli-Guillard national culture collection of marine phytoplankton indicating phytoplankton species diversity may be a skewed log-normal function of cell size (see Figure 1).

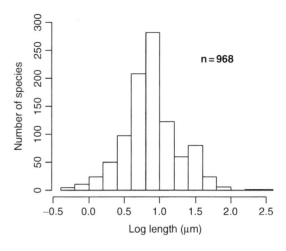

2002; Vaulot *et al.* 2002). Similar discoveries of cryptic diversity across all size classes of phytoplankton may follow as these techniques are applied equally to the whole phytoplankton community. The recent demonstration that phytoplankton species richness scales with habitat area (Smith *et al.* 2005) suggests there may also be a relationship between habitat size and size structure of phytoplankton communities. Ideally, a global database of current phytoplankton diversity and associated cell size should be compiled. Many hypotheses have been proposed to explain the pattern in the size scaling of diversity; the most likely factors include the size scaling of origination (Gillooly *et al.* 2005) and extinction rates (Norris 1991) as a function of different environmental and ecological conditions over the history of the taxonomic group (Stanley 1973b; Dial and Marzluff 1988; McShea 1994).

D. Size Matters: Food Web Structure and Function

1. Functional Composition of the Marine Food Web

Early descriptions of marine feeding relationships focused on the relationship among nutrient concentrations, phytoplankton and zooplankton, and the transfer of primary production to organisms of human interest such as fish (Ryther 1969).

Improvements in the measurement of bacterial abundance have led to the increasing recognition of the importance of bacteria in the marine food web. Bacteria consume dissolved organic matter, stimulate recycling of organic matter, and provide fuel for a variety of small heterotrophs (protozoan ciliates and flagellates) that can be consumed by larger zooplankton and transferred to higher trophic levels (Pomeroy 1974; Azam *et al.* 1983). The grazer food chain (phytoplankton, zooplankton, fish) and the microbial food web have often been studied independently, but their integral coupling and resulting influence on the partitioning of matter and energy is widely recognized (Azam *et al.* 1983, 1993; Sherr and Sherr 1994; Thingstad and Hagström, 1997). Accumulating evidence indicates that marine viruses (and perhaps archaea and pathogenic bacteria) also appear to play an important role in the planktonic food web, altering the turnover time of different organisms and affecting the supply of dissolved organic material to bacteria (DeLong and Karl 2005; Suttle 2005).

2. Big Things (Often) Eat Smaller Things

Although there are many individual exceptions, the hierarchical size differences between consumers and their prey is the basis of the strong relationship between organism size and trophic level within the food web (Dussart 1965; Parsons and Takahashi 1973; Sieburth *et al.* 1978; Cohen *et al.* 2003; Jennings and Mackinson 2003) (Figure 3). Meta-analyses of aquatic and terrestrial systems suggest that for a majority of animal species in natural food webs, a larger predator consumes a smaller prey (Cohen *et al.* 1993). The size difference between predator and prey may be smallest in the microbial web where bacteria (~1 μm) are consumed by ciliates and flagellates (many of which are between 5 and 20 μm) (Sherr and Sherr 1994). Small heterotroph grazers can be consumed by larger zooplankton and can contribute to food webs with 5 or more trophic levels (Sherr and Sherr 1994; Link 2002). In comparison, a grazing food web dominated by phytoplankton (10–100 μm) should result in food webs with relatively few trophic levels because the phytoplankton are too big for most flagellate and ciliate grazers and instead are prey for larger zooplankton such as copepods (0.2–28 mm) (Huys and Boxshall 1991), which are often a direct food source for fish (Ryther 1969). Each trophic transfer is associated with large losses (> 80%) of matter and energy (Parsons and deLange Bloom 1974; Cohen *et al.* 1993, 2003; Jennings and Mackinson 2003), so food webs dominated by the microbial food web will theoretically result in a decrease in matter and energy transfer to higher trophic levels.

Larger individuals often consume prey with a larger range of masses than smaller predators (increased diet breadth) (Cohen *et al.* 1993, 2003; Hansen *et al.* 1994) and as a result the predator to prey mass ratio tends to change with increasing body mass of the predator (Cohen *et al.* 1993; Jonnson and Ebenman 1998). This increase in the breadth of available prey with increasing size of the consumer has profound consequences for food web structure. Models indicate that changes in the predator–prey body size ratio with predator size can affect the resilience and probability of stability of Lotka–Volterra food chains of more than three trophic levels (Jonnson and Ebenman 1998). An increase in the breadth of prey masses consumed by larger predators will tend to increase the connectance (proportion of total possible binary trophic links) in the food web. Increases in connectance and species diversity will tend to decrease

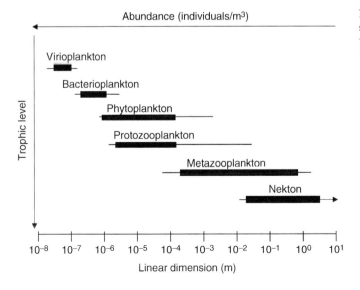

FIGURE 3. Size, abundance, and trophic structure of the planktonic marine food web. Adapted from Sieburth *et al.* 1978.

the strength of interaction between predator–prey pairs and generally increase the complexity of the food web (Cohen *et al.* 2003; Woodward *et al.* 2005).

There is considerable evidence that many phytoplankton grazers discriminate between prey based on differences in size as well as nutritional quality or toxicity (Mullin 1963; Parsons *et al.* 1967; Irigoien *et al.* 2000, 2003; Teegarden *et al.* 2001). For example, *Calanus finmarchicus,* a main food source for many commercially important fish, has a general preference for diatoms relative to ciliates and smaller phytoplankton species (Irigoien *et al.*, 2003). In general, it is assumed that a predator has a Gaussian-type preference for prey of a certain size (Hansen *et al.* 1994; Loeuille and Loreau 2005). Experimental measurements of the size selectivity of protozoan predators on bacterial prey do appear to be unimodal, but the acceptable prey size range varies considerably among predators (Chrzanowski and Simek 1990; Hansen *et al.* 1994). It seems credible that most predators will have sharp upper and lower size boundaries beyond which they cannot physically consume prey, depending on the feeding apparatus, creating size refugia that can cause complex patterns in food web structure (Chase 1999). For example, the adult females of the marine copepod *Calanus pacificus* cannot effectively consume diatoms with diameters less than 3.8 µm, and below a certain biomass concentration of prey, ingestion rate (mg C/copepod/time) is greatest for larger diatoms (Frost 1972).

3. Organism Size and Trophic Cascades

Field observations and theoretical approaches suggest that it is possible that the size selectivity of predators for prey can cause a coincident cascading change in the size distribution of all organisms in the food web. Brooks and Dodson (1965) discovered that small- to medium-sized

lakes with populations of the zooplankton predator *Alosa* (commonly known as the alewife) were characterized by much smaller-bodied zooplankton populations than lakes that lacked functionally similar fish populations. They hypothesized that predators prefer the largest possible prey that can be consumed and that larger zooplankton species are more metabolically efficient than smaller zooplankton. There is some evidence that marine bacteriovores selectively feed on larger, faster growing bacteria, altering the size structure of the standing stock bacteria (Gonzalez *et al.* 1990). Although this size efficiency hypothesis provides a consistent explanation for the patterns in zooplankton populations in the lakes of southern New England, the proposal that larger zooplankton are metabolically more efficient has been questioned (Peters 1991, 1992; Hopcroft *et al.* 2001), and the theory has generally failed to predict the abundance and size structure of phytoplankton and zooplankton in larger lakes and the ocean (Brooks and Dodson 1965; Frank *et al.* 2005; Scheffer *et al.* 2005). The increase in the diet breadth of larger zooplankton provides a potentially simpler alternative explanation for the relatively high abundance of larger versus smaller zooplankton under low predation pressure (Brooks and Dodson 1965; Cohen *et al.* 2003). The prevalence of omnivory and high connectance (Link 2002) and low interaction strength between individual predator–prey pairs may contribute to the relative rarity of trophic cascades in marine food webs (Scheffer *et al.* 2005). These differences may lessen as we continue to "fish down the food web" (Pauly *et al.* 1998).

The once cod-dominated northwest Atlantic food web now shows evidence of a trophic and size cascade where the removal and subsequent lack of recovery of the cod, *Gadus morhua,* as well as a large number of other predatory fish, is associated with an increase in the abundance of smaller pelagic fish and benthic macroinvertebrates, most notably the northern snow crab and shrimp (Worm and Myers 2003), a decrease in abundance of small-bodied invertebrates, a slight increase in the abundance of phytoplankton, and a corresponding decrease in surface nitrate concentrations (Frank *et al.* 2005). This is in contrast with the majority of observations in the surface ocean that show a strong correlation between low nitrate (or light, phosphate or iron) concentrations and low chlorophyll concentrations. It seems probable that there is similar potential for a dynamic change in the standing stock and size of phytoplankton in response to the change in limiting nutrient concentration that could then transfer up the food web where the size of the organisms within each trophic level is the outcome of the dynamic interaction of environmental conditions that affect net photosynthetic production, the size structure of the phytoplankton community, and size-selective predation of organisms in the upper trophic levels. Mesocosm enrichment experiments in the oligotrophic coastal Mediterranean support this hypothesis: increasing nutrient inputs resulted in an increase in the relative abundance of large phytoplankton and the ratio of phytoplankton to heterotrophic biomass (Duarte *et al.* 2000). Meta-analyses of food webs strongly support the assertion that both bottom-up and top-down processes control real food webs. Primary production determines the biomass in all trophic levels and top-down processes generally weaken with each trophic level and are often not observable at the base of the food web (Brett and Goldman 1996). The generally weak effect of trophic cascades at the base of the food web may be due to nutrient contributions from consumers (excretion, sloppy feeding, etc.) from all trophic levels (Attayde and Hannsson 2001). Much more work is required to determine the size selectivity of predators for their prey, especially with and without the presence of multiple prey, predators and their higher-level consumers.

II. RESOURCE AVAILABILITY, PRIMARY PRODUCTION, AND SIZE STRUCTURE OF PLANKTONIC AND BENTHIC FOOD WEBS

In modern marine food webs, there is often a strong coincident association among resource availability, primary productivity, total chlorophyll, export production, and the size structure of the phytoplankton community. When light and inorganic nutrients are abundant, as in coastal and upwelling zones and eddies, total chlorophyll tends to be high and large phytoplankton cells are abundant. In contrast, when light and nutrient availability are low, such as in the oceanic gyres or the iron-limited high-nutrient low-chlorophyll regions, phytoplankton communities are dominated by extremely small picoplankton (Malone 1971; Falkowski *et al.* 1991; Chisholm 1992; Le Bouteiller *et al.* 1992; McGillicuddy and Robinson 1997; Le Borgne *et al.* 2002; Li 2002; Sweeney *et al.* 2003). This general trend is supported by vast amounts of field data; for example, more than a decade of data collected predominantly from the North Atlantic documents clear increases in the abundance of large phytoplankton and decreases in the abundance of small picoplankton with increasing bulk chlorophyll concentration in the surface ocean (Li 2002). Environmental conditions such as temperature can have different effects on the grazing versus the microbial food web and the transfer of carbon through the food web and export of particles to the deep sea (Pomeroy and Deibel 1986; Pomeroy *et al.* 1991).

Field evidence strongly supports a strong correlation between the increase in primary production and chlorophyll standing stock with corresponding changes in planktonic and benthic primary and secondary consumers. Large phytoplankton cells tend to be grazed by large zooplankton (Steele and Baird 1961; Mullin 1963; Sherr and Sherr 1994; Savenkoff *et al.* 2000). In contrast, when smaller phytoplankton species dominate the autotroph community, heterotrophic dinoflagellates, ciliates, and smaller zooplankton can come to dominate the zooplankton community (Sherr and Sherr 1994; Savenkoff *et al.* 2000). In the equatorial Pacific, the long-term monitoring at the Hawaiian Ocean Time Series documents long-term oscillations in the taxonomic structure of the phytoplankton community from small cyanobacteria to larger eukaryotes (Karl 1999). Cyanobacterial dominance of the phytoplankton community is associated with the low availability of fixed nitrogen, low export flux, and famine in the benthic community on the underlying abyssal plane. In contrast, the dominance of the phytoplankton community by larger eukaryotic phytoplankton species is associated with higher population abundance and increases in the size of the organisms in the benthos (Smith *et al.* 1997; Karl 1999; Smith *et al.* 2002).

There is a well-established global association between increases in resource supply (water-column depth, distance from coast) and the size of benthic communities (Thiel 1975; Brown *et al.* 2001). Specific positive associations have been established between the magnitude of phytoplankton export and the abundance and size of benthic organisms such as nematodes in the central equatorial Pacific (Brown *et al.* 2001). The extremely tight correlation between equatorial upwelling, primary productivity, export production, and macrofaunal abundance led to the hypothesis that changes in primary productivity on decadal or greater time scales could yield profound changes in deep-sea benthic communities that may be evident in the fossil record (Smith *et al.* 1997).

The interaction between organisms and their environment have resulted in large-scale changes in the size of phytoplankton and marine invertebrates over time (Hallam 1975; Bambach 1993; Schmidt *et al.* 2004; Finkel *et al.* 2005; Huntley *et al.* 2006). The size of organisms in the food web affects

a number of the primary descriptors of food web structure and function, including species abundance and diversity (S), the number of trophic links, the minimum and maximum number of trophic levels or trophic height, connectance (C, proportion of total possible binary trophic links present in the food web), and complexity (S*C) (Brown and Gillooly 2003; Cohen *et al.* 2003; Woodward *et al.* 2005). Limited paleodata do not permit detailed or complete analyses on all these descriptors; nonetheless, application of the main physiological and ecological principles of the effects of organism size on food web structure and function may provide additional insight into the evolutionary history of marine food webs from observations of changes in the size structure of fossil phytoplankton and zooplankton assemblages.

III. SIZE AND THE EVOLUTION OF MARINE FOOD WEBS

A. Increase in the Maximum Size of Living Organisms Through Time

The maximum size of living organisms has increased over geological time (Bonner 1988). Life began as prokaryotic microbes. Eukaryotic phytoplankton, which become increasingly dominant in the deep Proterozoic, are often larger and morphologically more complex than the fossils of cyanobacteria, which dominated the phytoplankton in the Paleoproterozoic and Mesoproterozoic (Knoll *et al.* 2006; Knoll *et al.*, Chapter 8, this volume). The origination of new faunal groups is frequently associated with an increase in the maximal body size (Gould 1966). For example, the transition to the Ediacaran fauna (575–543 million years ago [Ma]) introduced a number of organisms that ranged from centimeters to ~1 m in the case of *Dickinsonia* (Carroll 2001). The largest organisms to date are the flowering plants, with their origins in the Cretaceous. There is some suggestion that this trend does not extend indefinitely but saturates once the disadvantages of getting larger can no longer be overcome by the advantages of an increase in complexity (Bonner 1988).

B. Organism Size Within Lineages Through Time (Cope's Rule)

It has been suggested that the fossil record documents an evolutionary tendency for taxonomic groups to evolve toward larger physical size, commonly termed Cope's rule (Newell 1949; Nicol 1964; Stanley 1973b; Pearson 1998). Although there are numerous counter-examples (Stanley 1973b), secular increase in organism size has been documented in a wide variety of organisms including foraminifera (Norris 1991; Kaiho 1999), ammonites and bivalves (Hallam 1975), and many vertebrate groups (Cope 1885; Nicol 1964; Alroy 1998). Several different hypotheses have been postulated to explain the evolutionary tendency toward larger organism size within lineages, including: competitive advantage over siblings (Castle 1932), improved ability to capture food or avoid predation, increased intelligence, greater reproductive success, extended individual longevity (Stanley 1973b), and increased complexity (Bonner 1988). The most parsimonious explanation for Cope's rule is the tendency for higher extinction rates for larger versus smaller species after a perturbation, such as a mass extinction, followed by speciation. Species radiation in conjunction with passive evolutionary mechanisms will tend to result in increases in both the maximum and minimum size, with no change in the mean body size of the group (Stanley 1973b; McShea 1994; Gould 1997; Pearson 1998). In other words, species radiation within a group will often result in an increase in the diversity and size range within a lineage, with or without specific size-dependent selection pressures (Stanley 1973b; McShea 1994; Gould 1997). If external environmental or biological factors select for species of a specific size, there will be a corresponding change in the average size of the species within the lineage through time. A combination of size bias in origination or extinction, physiologically imposed boundaries on minimum and maximum size, and active selection pressures can result in complex temporal patterns in the evolution of body size (McShea 1994). Furthermore, different selection pressures may act on individuals of different size, resulting in a large variety of size distributions within a lineage (Carroll 2001).

C. Climatically Driven Macroevolutionary Change in Organism Size

Macroevolutionary change in organism size can be a dynamic response of size-dependent selection in response to temporal changes in environmental forcing (Stanley 1973b; McShea 1994; Finkel et al. 2005). For example, a recent examination of marine planktonic diatoms found that the extreme minimum and maximum size of the diatom frustule has expanded in concert with species diversity through the Cenozoic, but the average size of the diatom frustules within the communities is highly correlated with climate change (Figure 4) (Finkel et al. 2005). The average size of the diatoms has followed changes in the vertical temperature gradient in the tropical ocean; an indicator of both the stability of the water column and the average temperature gradient between the equator and poles (Wright 2001; Zachos et al. 2001). The thermal gradient affects the availability of light and nutrients in the surface ocean, altering the types of niches available to plankton and other marine organisms (Rea 1994; Rutherford et al. 1999; Zachos et al. 2001; Schmidt et al. 2004; Finkel et al. 2005). The macroevolutionary changes in the size of diatom frustule over the Cenozoic are consistent with size changes observed for single species of marine diatoms in response to temperature and upwelling zones over hundreds of thousands to several millions of years (Wimpenny 1936; Burckle and McLaughlin 1977; Burckle et al. 1981; Sorhannus et al. 1988).

FIGURE 4. The average size of the frustule of the dominant marine diatom species (± 1 standard error, dotted line) through the Cenozoic. Modified from Finkel *et al.* 2005.

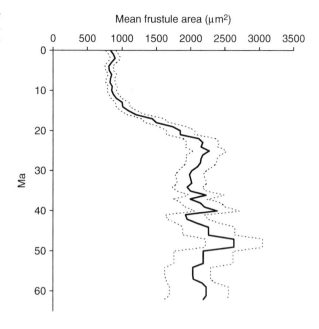

Zooplankton groups also exhibit size changes that correspond to changes in water-column structure. Schmidt *et al.* (2004) demonstrated that the greater than 150-μm foram community, as determined from the individuals within specific assemblages, increases with upper water-column stratification in low latitudes in the Neogene. This shift in the size of the foram community reflects an indistinguishable combination of macroecological and macroevolutionary factors because the measure reported is affected by the relative abundance of species of different sizes as well as shifts in the extinction and/or origination of species of different sizes. Regardless, it appears that there is a shift in the size structure of marine zooplankton that follows the largest decrease in the mean size of the diatoms in the early to mid-Miocene. Currently, there is insufficient temporal resolution to determine if the trends in increasing size between the planktonic foraminifera and the diatoms are causally related. In the modern ocean, there is often an association between the size structure of the phytoplankton community, the abundance and size of organisms in higher trophic levels, and the export of photosynthetically produced carbon into the deep sea, which increases the ocean's capacity to act as a sink for atmospheric carbon dioxide (Ryther 1969; Azam *et al.* 1983; Laws *et al.* 2000; Finkel *et al.* 2005). This suggests that if climate-induced environmental change in nutrient and light availability alters the size structure of marine phytoplankton and the associated marine food web, there is the potential for a climatic feedback (see Figure 1). Observation of these macroecological and macroevolutionary shifts in the size structure of the plankton lead to the question: what effect did these changes have on the evolution of marine food web structure and function over the Cenozoic?

D. The Evolution of the Modern Marine Food Web

Size-structured food webs can emerge when selection acts from the bottom-up (Loeuille and Loreau 2005). Increases in inorganic nutrient input result in evolutionary increases in organism size, total community biomass, and maximum trophic height, consistent with patterns observed in marine systems (Bambach 1993). Furthermore, Loeuille and Loreau (2005) identify that prey size range and exploitation competition are key factors that can drive the evolution of many food web characteristics. As the variance in body sizes that can be consumed decreases, only species that are the correct size to eat specific prey evolve, resulting in simple food chains. Exploitation competition acts to decrease the fitness of individuals of similar body size and is critical for the development of diversity (Loeuille and Loreau 2005). In contrast, Stanley's cropping hypothesis advocates that the introduction of consumers to "crop" abundant phytoplankton assemblages is required to catalyze the diversification of predators and prey, and likely stimulated the explosion in metazoan diversity in the Cambrian (Stanley 1973a). Further extending the cropping hypothesis, Butterfield proposed the evolution of herbivorous meso-zooplankton, which could transfer primary production from small photoautotrophs to larger animals, was key to the diversification of the metazoans in the Early Cambrian (Butterfield 1997, 2001a, b).

The end of the Cretaceous marks one of the largest mass extinction events of the Phanerozoic, providing an opportunity to examine the coincident re-establishment of the marine food web and size structure of the plankton. In the marine environment, an estimated ~50% of genera went extinct (Sepkoski 1984). The calcareous plankton were seriously affected; only 9 of the 131 late Maastrichtian species survived (Bown et al. 2004). The extinction event corresponded with the collapse of the export flux of organic carbon to the deep sea, which took several million years to recover (D'Hondt et al. 1998). D'Hondt proposed that primary production likely recovered quickly, but without established assemblages of zooplankton consumers most of the primary production was recycled through the microbial food web. The recovery of primary and secondary consumers was likely required to re-establish the carbon flux to

the deep sea (D'Hondt et al. 1998; D'Hondt 2005). This suggests that food supply to the benthos would have been seriously limited for several million years, a conclusion that is hard to reconcile with the timing and level of diversification of the deep-sea benthic foraminifera in the Early Cenozoic (Miller et al. 1992).

The fossil record of the dominant extant herbaceous zooplankton such as the crustaceans, the copepods, euphausiids, amphipods, and decapods is extremely poor, and there is no fossil record of the planktonic urochordates, salps, and appendicularians (Rigby and Milsom 2000), making it extremely difficult to reconstruct the detailed evolution of the planktonic marine food web over the Cenozoic. Many of the groups associated with the microbial food web (heterotrophic bacteria, flagellates, and ciliates) would likely not have been seriously affected by the end-Cretaceous mass extinction or would have recovered quickly due to their inherently high rates of reproduction (Sherr and Sherr 1994). Some moderately large species of phytoplankton were present in the early Cenozoic (Finkel et al. 2005), but many groups, including the planktonic forams and many metazoans, were much reduced in diversity, and many of the surviving species were small (Rigby and Milsom 2000; Gallagher 2003; D'Hondt 2005). It is probable that the larger organisms, due to their inherently slower reproductive rates and higher minimum resource requirements, took much longer to recover from the mass extinction event. The coincident recovery of the deep-sea carbon flux, the size of the planktonic forams, and the increase in fish teeth accumulation suggest a partial recovery of the planktonic marine food web that likely included the establishment of large phytoplankton or large grazing zooplankton, approximately 62 Ma (D'Hondt 2005).

Based on the diversity of the fossil records of the major eukaryotic phytoplankton groups, there was a significant change in the taxonomic structure of the marine phytoplankton community in response to changes in climate (Falkowski *et al.* 2004; Katz *et al.* 2004; de Vargas *et al.*, Chapter 12; Kooistra *et al.*, Chapter 11, this volume). The calcareous nannoplankton and organic-walled cyst-forming dinoflagellates diversified through much of the Mesozoic but then declined through the Cenozoic as the diversity of the diatoms rose (Harwood and Nikolaev 1995; MacRae *et al.* 1996; Bown *et al.* 2004). The diatoms have a number of unique eco-physiological abilities relative to the other major phytoplankton groups; they have an outer wall of opaline silica, they can take up large amounts of nutrients very quickly, and they store inorganic nutrients in vacuoles (Raven 1987, 1997; Stolte and Riegman 1995; Lomas and Gilbert 2000) that they can use to maintain high rates of growth when ambient concentrations of nutrients are temporarily low (Grover 1991). In addition, at a given size, diatoms have the fastest maximum growth rates of all the phytoplankton groups (Langdon 1988; Tang 1995; Raven *et al.* 2006). The unique eco-physiology of the diatoms represents a new innovation for unicellular photoautotrophs to take advantage of temporal nutrient variability at certain frequencies that may have been increasingly important as the equator-to-pole temperature gradient increased through the Cenozoic (Finkel *et al.* 2005). Assuming that the changes in fossil diversity reflect a change in abundance, the diatoms become an increasingly dominant part of the phytoplankton community through the Cenozoic.

The establishment of the diatoms altered the food resources available to the planktonic primary consumers and likely affected the evolution of the copepods, isopods, and appendicularians, which in turn likely affected the evolutionary establishment of particular species in the upper trophic levels. As prey, the diatoms are particularly suited to zooplankton species that are large enough to ingest them whole or break their opaline frustule (Hamm *et al.* 2003; Hamm and Smetacek, Chapter 14, this volume). It is interesting to note that many copepods have mandibles lined with teeth reinforced with silica, indicating co-evolution with diatoms (Hamm *et al.* 2003; Hamm and Smetacek, Chapter 14, this volume). An increase in biomass and the size of secondary and higher consumers would also have contributed to an increase in the transfer of resources to the benthos, because larger predators tend to have larger fecal pellets with a higher sinking flux (Uye and Kaname 1994; Taguchi and Saino 1998). In addition, because diatoms are often large and heavy and tend to aggregate, the dominance of phytoplankton community by diatoms may have led to an increase in the export of primary production to the deep sea, providing increased resources to the benthic community and likely contributing to their documented increases in abundance, biomass, and size through the Cenozoic (Bambach 1993).

In general, the diversity of prey is linearly related to the diversity of predators (Rosenzweig 1995), and larger prey are often associated with larger predators (Cohen *et al.* 2003). A phytoplankton species that can decrease its size below the threshold of its main consumers would have both an increase in its metabolic rate and a decrease in mortality through grazing. A limitation to this diversification may be imposed by nonscalable cellular components associated with photoautotrophy that set the theoretical minimum cell size to ~1 μm (Raven 1994; Raven *et al.* 2006). Once consumer species evolved to consume these smallest phytoplankton species, this downward selection pressure should somewhat dissipate. As an alternative strategy, a phytoplankter can increase its size above the threshold of its main consumer, but this has a trade-off in growth rate, efficiency of light absorption (per mass of pigment) (Finkel 2001; Finkel *et al.* 2004), and rates of nutrient diffusion and sinking (Pasciak and Gavis 1974; Gavis 1976). Larger zooplankton will also have a decreased metabolic rate; if the increase in size is more costly to the prey than the predator, then the upward size trend will cease.

Some taxonomic groups have evolved strategies to effectively increase their physical size (as experienced by their consumers) with a minimum decrease in metabolic rate. For example, many diatom species have large silica spines or other appendages and can form large chains. Diatom species are known to have nutrient storage vacuoles (Sicko-Goad *et al.* 1984; Raven 1987, 1997), silica spines and other appendages, and the ability to form chains, all of which allow them to increase their effective size to avoid predation but mitigate the negative effects of increased cell size on growth and sinking rates. From this perspective, the evolutionary trajectory in the size of phytoplankton species may be a consequence of an evolutionary arms race with its predators (Dawkins and Krebs 1979; Smetacek, 2001; Smetacek *et al.* 2004). For example, some armored heterotrophic dinoflagellates use a large pseudopodial feeding veil that permits them to extracellularly digest their prey and therefore feed on diatoms chains that can be considerably larger than themselves (Gaines and Taylor 1984; Sherr and Sherr 1994; Hamm and Smetacek this volume). The evolution of some larger diatom species through the Cenozoic (even though median frustule size decreased) does correspond with a change in the size of the larger planktonic forams (Schmidt *et al.* 2004); these trends in increasing size between predators and prey may be causally related or due to response to the same environmental variables (Bengtson 2002). Interpretation of macroevolutionary and microevolutionary trajectories in the size of organisms is further complicated by factors such as the existence of multiple extant predators at different trophic levels with the ability to consume prey of a variety of sizes, omnivory and mixotrophy, competition for (and recycling of) inorganic and organic nutrients by heterotrophic and photosynthetic organisms, and mortality caused by parasitoids and viruses (Litchman, Chapter 16, this volume).

Phytoplankton interact and affect climate through the uptake of inorganic materials, the output of oxygen, and the production of reduced carbon compounds that form the base of the food web. The fate of photosynthetically reduced carbon is determined largely by the consumers in the food web. The size structure of the phytoplankton likely contributes to the size of the zooplankton predators, and in turn they affect the size structure of the higher trophic levels on both ecological and evolutionary time scales. The fossil record documents biotic responses to a large range of climate conditions, including the size of the organisms present. Further application of the general relationships that exist among environmental conditions, organism size and physiology, ecology, and evolution may provide additional information for the potential for feedbacks between climate and biota (see Figure 1). Continued investigation through experiments, field observation, analysis and simulation, and the development of theory will likely yield further insights into the generality and basis of the relationships between organism size and metabolic rate, abundance, diversity, and food web structure.

Acknowledgments

I thank P.G.F. and A.K. for inviting me to be a part of this volume and for their insightful editorial comments. This work was supported by NSERC Discovery and New Brunswick Innovation Fund grants.

References

Agusti, S. (1987). Algal cell size and the maximum density and biomass of phytoplankton. *Limnol. Oceanogr.* **32:** 983–986.

Agusti, S., Duarte, C.M., and Kalff, J. (1987). Algal cell size and the maximum density and biomass of phytoplankton. *Limnol. Oceanogr.* **32:** 983–986.

Agusti, S., and Kalff, J. (1989). The influence of growth conditions on the size dependence of maximal algal density and biomass. *Limnol. Oceanogr.* **34:** 1104–1108.

Alroy, J. (1998). Cope's rule and the dynamics of body mass evolution in North American fossil mammals. *Science* **280:** 731–734.

Armstrong, R.A. (2003). A hybrid spectral representation of phytoplankton growth and zooplankton response: the "control rod" model of plankton inter-action. *Deep Sea Res. II* **50:** 2895–2916.

Attayde, J.L., and Hannsson, L.A. (2001). Fish-mediated nutrient recycling and the trophic cascade in lakes. *Can. J. Fish. Aquat. Sci.* **58:** 1924–1931.

Azam, F., Fenchel, T., Field, J.G. Gray, J. S., Meyer-Reil, L.A., and Thingstad, F. (1983). The ecological role of water-column microbes in the sea. *Mar. Ecol. Prog. Ser.* **10:** 257–263.

Azam, F., Smith, D.C., Steward, G.F., and Hagström, Å. (1993). Bacteria-organic matter coupling and its significance for oceanic carbon cycling. *Microbial Ecol.* **28:** 167–179.

Bambach, R.K. (1993). Seafood through time: changes in biomass, energetics, and productivity in the marine ecosystem. *Paleobiology* **19:** 372–397.

Banavar, J.R., Damuth, J., Maritan, A., and Rinaldo, A. (2002). Supply-demand balance and metabolic scaling. *Proc. Natl. Acad. Sci. U S A* **99:** 10506–10509.

Banse, K. (1976). Rates of growth, respiration and photosynthesis of unicellular algae as related to cell size—a review. *J. Phycol.* **12:** 135–140.

Bengtson, S. (2002). Origins and early evolution of predation. *Paleontol. Soc. Papers* **8:** 289–317.

Bonner, J.T. (1988). *The Evolution of Complexity by Means of Natural Selection.* Princeton University Press.

Boss, E., Twardowski, M.S., and Herring, S. (2001). The shape of the particulate beam attenuation spectrum and its relation to the size distribution of oceanic particles. *Appl. Optics* **40:** 4885–4893.

Bown, P.R., Lees, J.A., and Young, J.R. (2004). Calcareous nannoplankton evolution and diversity. *Coccolithophores: From Molecular Processes to Global Impact.* H. Thierstein and J.R. Young, eds. Heidelberg, Germany, Springer-Verlag, pp. 481–508.

Brett, M.T., and Goldman, C.R. (1996). A meta-analysis of the freshwater trophic cascade. *Proc. Natl. Acad. Sci. U S A* **93:** 7723–7726.

Brooks, J.L., and Dodson, S.I. (1965). Predation, body size, and composition of plankton. *Science* **150:** 28–35.

Brown, C.J., Lambshead, P.J.D., Smith, C.R., Hawkins, L.E., and Farley, R. (2001). Phytodetritus and the abundance and biomass of abyssal nematodes in the central, equatorial Pacific. *Deep Sea Res. Part I Oceanogr. Res. Papers* **48:** 555–565.

Brown, J.H. (1995). *Macroecology.* Chicago, University of Chicago Press.

Brown, J.H., and Gillooly, J.F. (2003). Ecological food webs: high-quality data facilitate theoretical unification. *Proc. Natl. Acad. Sci. U S A* **100:** 1467–1468.

Burckle, L., and McLaughlin, R.B. (1977). Size changes in the marine diatom *Coscinodiscus nodulifer* A. Schmidt in the equatorial Pacific. *Micropaleontology* **23:** 216–222.

Burckle, L., Shackleton, N.J., and Bromble, S.L. (1981). Late Quaternary stratigraphy for the equatorial Pacific based upon the diatom *Coscinodiscus nodulifer. Micropaleontology* **27:** 352–355.

Butterfield, N.J. (1997). Plankton ecology and the Proterozoic-Phanerozoic transition. *Paleobiology* **23:** 247–262.

Butterfield, N.J. (2001a). Cambrian food webs. *Palaeobiology II: A Synthesis.* Oxford, Blackwell Scientific, pp. 40–43.

Butterfield, N.J. (2001b). Ecology and evolution of the Cambrian plankton. *Ecology of the Cambrian Radiation.* New York, Columbia University Press, pp. 200–216.

Carroll, S.B. (2001). Chance and necessity: the evolution of morphological complexity and diversity. *Nature* **409:** 1102–1109.

Castle, W.E. (1932). Body size and body proportions in relation to growth rates and natural selection. *Science* **76:** 365–366.

Chase, J.M. (1999). Food web effects of prey size refugia: variable interactions and alternative stable equilibria. *Am. Naturalist* **154:** 559–570.

Chisholm, S.W. (1992). Phytoplankton size. *Primary Productivity and Biogeochemical Cycles in the Sea.* P.G. Falkowski and A.D. Woodhead, eds. New York, Plenum Press, pp. 213–237.

Chrzanowski, T.H., and Simek, K. (1990). Prey-size selection by freshwater flagellated protozoa. *Limnol. Oceanogr.* **35:** 1429–1436.

Cohen, J.E., Jonsson, T., and Carpenter, S.R. (2003). Ecological community description using the food web, species abundance, and body size. *Proc. Natl. Acad. Sci. U S A* **100:** 1781–1786.

Cohen, J.E., Pimm, S.L., Yodzis, P., and Saldana, J. (1993). Body sizes of animal predators and animal prey in food webs. *J. Anim. Ecol.* **62:** 67–78.

Cope, E.D. (1885). On the evolution of the Vertebrata, progressive and retrogressive. II. The vertebrate line. *Am. Naturalist* **19:** 144–148.

Dawkins, R., and Krebs, J. R. (1979). Arms races between and within species. *Proceedings of the Royal Society of London, series B.* **205:**489–511.

DeLong, E.F., and Karl, D.M. (2005). Genomic perspectives in microbial oceanography. *Nature* **437:** 336–342.

D'Hondt, S. (2005). Consequences of the Cretaceous/Paleogene mass extinction for marine ecosystems. *Annu. Rev. Ecol. Evol. Syst.* **36:** 295–317.

D'Hondt, S., Donaghay, P., Zachos, J., Luttenberg, D., and Lindinger, M. (1998). Organic carbon fluxes and ecological recovery from the Cretaceous-Tertiary mass extinction. *Science* **282:** 276–279.

deVargas, C., Aubry, M-P., Probert, I., and Young, J. (2007). Origin and evolution of coccolithophores: From coastal hunters to oceanic farmers. *Evolution of Primary Producers in the Sea.* P.G. Falkowski and A.H. Knoll, eds. Boston, Elsevier, pp. 251–285.

Dial, K.P., and Marzluff, J. M. (1988). Are the smallest organisms the most diverse? *Ecology* **69:** 1620–1624.

Duarte, C.M., Agusti, S., Gasol, J.M., Vaque, D., and Vazquez-Dominguez, E. (2000). Effect of nutrient supply on the biomass structure of planktonic communities: an experimental test on a Mediterranean coastal community. *Mar. Ecol. Prog. Ser.* **206:** 87–95.

Dussart, B.M. (1965). Les differentes categories de plankton. *Hydrobiologia* **26:** 72–74.

Falkowski, P.G., Katz, M.E., Knoll, A.H., Quigg, A.S., Raven, J.A., Schofield, O., and Taylor, F.J.R. (2004). The evolution of modern eukaryotic phytoplankton. *Science* **305:** 354–360.

Falkowski, P.G., Ziemann, D., Kolber, Z., and Bienfang, P.K. (1991). Nutrient pumping and phytoplankton response in a subtropical mesoscale eddy. *Nature* **352:** 544–551.

Fenchel, T. (1993). There are more small than large species? *Oikos* **68:** 375–378.

Finkel, Z.V. (2001). Light absorption and size scaling of light-limited metabolism in marine diatoms. *Limnol. Oceanogr.* **46:** 86–94.

Finkel, Z.V., and Irwin, A.J. (2000). Modeling size-dependent photosynthesis: light absorption and the allometric rule. *J. Theor. Biol.* **204:** 361–369.

Finkel, Z.V., Irwin, A.J., and Schofield, O. (2004). Resource limitation alters the 3/4 size scaling of metabolic rates in phytoplankton. *Mar. Ecol. Prog. Ser.* **273:** 269–279.

Finkel, Z.V., Katz, M., Wright, J., Schofield, O., Falkowski, P. (2005). Climatically driven macroevolutionary patterns in the size of marine diatoms over the Cenozoic. *Proc. Natl. Acad. Sci. U S A* **102:** 8927–8932.

Frank, K.T., Petrie, B., Choi, J.S., and Leggett, W.C. (2005). Trophic cascades in a formerly cod-dominated ecosystem. *Science* **308:** 1621–1623.

Frost, B.W. (1972). Effects of size and concentration of food particles on the feeding behavior of the marine planktonic copepod *Calanus pacificus*. *Limnol. Oceanogr.* **17:** 805–815.

Gaines, G., and Taylor, F.J.R. (1984). Extracellular digestion in marine dinoflagellates. *J. Plankton Res.* **6:** 1057–1061.

Gallagher, W.B. (2003). Oligotrophic oceans and minimalist organisms: collapse of the Maastrichtian marine ecosystem and Paleocene recovery in the Cretaceous-Tertiary sequence in New Jersey. *Netherlands J. Geosci. Geol. Mijnbouw* **82:** 225–231.

Gaston, K.J., and Blackburn, T.M. (2000). *Pattern and Process in Macroecology*. London, Blackwell Science.

Gavis, J. (1976). Munk and Riley revisited: nutrient diffusion transport and rates of phytoplankton growth. *J. Mar. Res.* **34:** 161–179.

Gillooly, J.F., Allen, A.P., West, G.B., and Brown, J.H. (2005). The rate of DNA evolution: effects of body size and temperature on the molecular clock. *Proc. Natl. Acad. Sci. U S A* **102:** 140–145.

Gonzalez, J.M., Sherr, E.B., and Sherr, B.F. (1990). Size-selective grazing on bacteria by natural assemblages of estuarine flagellates and ciliates. *Appl. Environ. Microbiol.* **56:** 583–589.

Gould, S.J. (1966). Allometry and size in ontogeny and phylogeny. *Biol. Rev.* **41:** 587–640.

Gould, S.J. (1997). Cope's rule as psychological artefact. *Nature* **385:** 199–200.

Grover, J.P. (1989). Influence of cell shape and size on algal competitive ability. *J. Phycol.* **25:** 402–405.

Grover, J.P. (1991). Resource competition in a variable environment: phytoplankton growing according to the variable-internal-stores model. *Am. Naturalist* **138:** 811–835.

Hallam, A. (1975). Evolutionary size increase and longevity in Jurassic bivalves and ammonites. *Nature* **258:** 493–496.

Hamm, C.E., Merkel, R., Springer, O. Jurkojc, P., Maier C., Prechtel, K., and Smetacek, V. (2003). Architecture and material properties of diatom shells provide effective mechanical protection. *Nature* **421:** 841–843.

Hamm, C., and Smetacek, V. (2007). Armor: why, when, and how. *Evolution of Primary Producers in the Sea*. P.G. Falkowski and A.H. Knoll, eds. Boston, Elsevier, pp. 311–332.

Hansen, B., Bjornsen, P.K., and Hansen, P.J. (1994). The size ratio between planktonic predators and their prey. *Limnol. Oceanogr.* **39:** 395–403.

Harwood, D.M., and Nikolaev, V.A. (1995). Cretaceous diatoms: morphology, taxonomy, biostratigraphy. *Siliceous Microfossils*. Short courses in paleontology. C.D. Blome, P.M. Whalen, and R. Katherine, eds. New Orleans, Paleontological Society, pp. 81–106.

Hopcroft, R.R., Roff, J.C., and Chavez, F.P. (2001). Size paradigms in copepod communities: a re-examination. *Hydrobiologia* **453/454:** 133–141.

Huntley, J.W., Xiao, S., and Kowalewski, M. (2006). 1.3 Billion years of acritarch history: an empirical morphospace approach. *Precambrian Res.* **144:** 52–68.

Huys, R., and Boxshall, G. A. (1991). *Copepod Evolution*, v. 159. London, The Ray Society.

Irigoien, X., Head, R.N., Harris, R.P., Cummings, D., and Harbour, D. (2000). Feeding selectivity and egg production of *Calanus helgolandicus* in the English Channel. *Limnol. Oceanogr.* **45:** 44–54.

Irigoien, X., Huisman, J., and Harris, R.P. (2004). Global biodiversity patterns of marine phytoplankton and zooplankton. *Nature* **429:** 863–867.

Irigoien, X., Titelman, J., Harris, R.P., Harbour, D., and Castellani, C. (2003). Feeding of *Calanus finmarchicus* nauplii in the Irminger Sea. *Mar. Ecol. Prog. Ser.* **262:** 193–200.

Irwin, A.J., Finkel, Z.V., Schofield, O.M.E., and Falkowski, P.G. (2006). Scaling-up from nutrient physiology to the size-structure of phytoplankton communities. *J. Plankton Res.* **28:** 459–471.

Jennings, S., and Mackinson, S. (2003). Abundance-body mass relationships in size-structured food webs. *Ecol. Lett.* **6:** 971–974.

Jonnson, T., and Ebenman, B. (1998). Effects of predator-prey body size ratios on the stability of food chains. *J. Theor. Biol.* **193:** 407–417.

Kaiho, K. (1999). Evolution in the test size of deep-sea benthic foraminifera during the past 120 m.y. *Mar. Micropaleontol.* **37**: 53–65.

Karl, D.M. (1999). A sea of change: biogeochemical variability in the North Pacific subtropical gyre. *Ecosystems* **2**: 181–214.

Katz, M.E., Finkel, Z.V., Gryzebek, D., Knoll, A.H., and Falkowski, P.G. (2004). Eucaryotic phytoplankton: evolutionary trajectories and global biogeochemical cycles. *Annu. Rev. Ecol. Evol. Syst.* **35**: 1–320.

Kerr, S.R., and Dickie, L.M. (2001). *The Biomass Spectrum: A Predator-Prey Theory of Aquatic Production.* Complexity in ecological systems series. New York, Columbia University Press.

Kiorboe, T. (1993). Turbulence, phytoplankton cell size, and the structure of pelagic food webs. *Adv. Mar. Biol.* **29**: 1–72.

Kleiber, M. (1947). Body size and metabolic rate. *Physiol. Rev.* **27**: 511–541.

Knoll, A.H., Javaux, E.J., Hewitt, D., and Cohen, P. (2006). Eukaryotic organisms in Proterozoic oceans. *Phil. Trans. R. Soc. Lond. B* **361**: 1023–1038.

Knoll, A.H., Summons, R.E., Waldbauer, J.R., and Zumberge, J.E. (2007). The geological succession of primary producers in the oceans. *Evolution of Primary Producers in the Sea.* P.G. Falkowski and A.H. Knoll, eds. Boston, Elsevier, pp. 133–163.

Kooistra, W.H.C.F., Gersonde, R., Medlin, L.K., and Mann, D.G. (2007). The origin and Evolution of the diatoms: Their adaptation. *Evolution of Primary Producers in the Sea.* P.G. Falkowski and A.H. Knoll, eds. Boston, Elsevier, pp. 207–249.

Langdon, C. (1988). On the cause of interspecific differences in the growth-irradiance relationship for phytoplankton. II. A general review. *J. Plankton Res.* **10**: 1291–1312.

Laws, E.A., Falkowski, P.G., Smith, W.O.J., Ducklow, H., and McCarthy, J. J. (2000). Temperature effects on export production in the open ocean. *Global Biogeochemical Cycles* **14**: 1231–1246.

Le Borgne, R., Feely, R.A., and Mackey, D.J. (2002). Carbon fluxes in the equatorial Pacific: a synthesis of the JGOFS programme. *Deep Sea Res. II* **49**: 2425–2442.

Le Bouteiller, A., Blanchot, J., and Rodier, M. (1992). Size distribution patterns of phytoplankton in the western Pacific: towards a generalization for the tropical open ocean. *Deep Sea Research 1* **39**: 805–823.

Li, W.K.W. (2002). Macroecological patterns of phytoplankton in the northwestern North Atlantic Ocean. *Nature* **419**: 154–157.

Link, J. (2002). Does food web theory work for marine ecosystems? *Mar. Ecol. Prog. Ser.* **230**: 1–9.

Litchman, E. (2007). Resource competition and the ecological success of phytoplankton. *Evolution of Primary Producers in the Sea.* P.G. Falkowski and A.H. Knoll, eds. Boston, Elsevier, pp. 351–375.

Loeuille, N., and Loreau, M. (2005). Evolutionary emergence of size-structured food webs. *PNAS* **102**: 5761–5766.

Lomas, M.W., and Gilbert, P.M. (2000). Comparisons of nitrate uptake, storage, and reduction in marine diatoms and flagellates. *J. Phycol.* **36**: 903–913.

MacRae, R.A., Fensome, R.A., and Williams, G.L. (1996). Fossil dinoflagellate diversity, originations, and extinctions and their significance. *Can. J. Bot.* **74**: 1687–1694.

Malone, T.C. (1971). The relative importance of nannoplankton and netplankton as primary producers in tropical oceanic and neritic phytoplankton communities. *Limnol. Oceanogr.* **16**: 633–639.

May, R.M. (1978). The dynamics and diversity of insect faunas. *Diversity of Insect Faunas.* L.A. Mound, N. Waloff, eds. Oxford, Blackwell, pp. 188–204.

McGillicuddy, D.J.J., and Robinson, A.R. (1997). Eddy-induced nutrient supply and new production in the Sargasso Sea. *Deep Sea Res.* **44**: 1427–1450.

McShea, D.W. (1994). Mechanisms of large-scale evolutionary trends. *Evolution* **48**: 1747–1763.

Miller, K.G., Katz, M.E., and Berggren, W.A. (1992). Cenozoic deep-sea benthic foraminifera: a tale of three turnovers. *Studies in Benthic Foraminifera.* Sendai, Tokai University Press, pp. 67–75.

Moon-Van Der Staay, S.Y., De Wachter, R., and Vaulot, D. (2001). Oceanic 18S rDNA sequences from picoplankton reveal unsuspected eukaryotic diversity. *Nature* **409**: 607–610.

Morin, P.J., and Fox, J.W. (2004). Diversity in the deep blue sea. *Nature* **429**: 814.

Mullin, M.M. (1963). Some factors affecting the feeding of marine copepods of the genus *Calanus*. *Limnol. Oceanogr.* **8**: 239–250.

Newell, N.D. (1949). Phyletic size increase, an important trend illustrated by fossil invertebrates. *Evolution* **3**: 103–124.

Nicol, D. (1964). Cope's rule and Precambrian and Cambrian invertebrates. *J. Paleontol.* **38**: 968–974.

Niklas, K.J., and Enquist, B.J. (2001). Invariant scaling relationships for interspecific plant biomass production rates and body size. *Proc. Natl. Acad. Sci. U S A* **98**: 2922–2927.

Norris, R.D. (1991). Biased extinction and evolutionary trends. *Paleobiology* **17**: 388–399.

Not, F., Simon, N., Biegala, I.C., and Vaulot, D. (2002). Application of fluorescent *in situ* hybridization coupled with tyramide signal amplification (FISH-TSA) to assess eukaryotic picoplankton composition. *Aquat. Microbiol Ecol.* **28**: 157–166.

Parsons, T.R., and deLange Bloom, B.R. (1974). The control of ecosystem processes in the sea. *Biol. Oceanic Pacific.* Vancouver, British Columbia University, Institute of Oceanography, pp. 29–58.

Parsons, T.R., LeBrasseur, R.J., and Fulton, J.D. (1967). Some observations on the dependance of Zooplankton grazing on the cell size and concentration of phytoplankton blooms. *J. Oceanogr. Soc. Japan* **23:** 10–17.

Parsons, T.R., and Takahashi, M. (1973). *Biological Oceanographic Processes*. New York, Pergamon Press.

Pasciak, W.J., and Gavis, J. (1974). Transport limitation of nutrient uptake in phytoplankton. *Limnol. Oceanogr.* **19:** 881–888.

Pauly, D., Christensen, V., Dalsgaard, J., Froese, R., and Torres, Jr., F.T. (1998). Fishing down marine food webs. *Science* **279:**860–863.

Pearson, P. (1998). Cricket and Cope's rule. *Palaeontol. Newslett.* **38:** 9–11.

Peters, R.H. (1983a). *The Ecological Implications of Body Size*. Cambridge University Press.

Peters, R.H. (1983b). Size structure of the plankton community along the trophic gradient of Lake Memphremagog. *Can. J. Fish. Aquat. Sci.* **40:** 1770–1778.

Peters, R.H. (1991). Lessons from the size efficiency hypothesis I. The general refuge concept. *Form and Function in Zoology*. G. Lanzavecchia and R. Valvassori, eds. Mucchi, Modena, Selected Symposia and Monographs U.Z.I, pp. 335–361.

Peters, R.H. (1992). Lessons from the size efficiency hypothesis II. The mire of complexity. *Hydrobiologia* **235/236:** 435–455.

Pomeroy, L.R. (1974). The ocean's food web, a changing paradigm. *Bioscience* **24:** 499–504.

Pomeroy, L.R., and Deibel, D. (1986). Temperature regulation of bacterial activity during the spring bloom in Newfoundland coastal waters. *Science* **233:** 359–361.

Pomeroy, L.R., Wiebe, W.J., Deibel, D., Thompson, R.J., Rowe, G.T., and Pakulski, J.D. (1991). Bacterial responses to temperature and substrate concentration during the Newfoundland spring bloom. *Mar. Ecol. Prog. Ser.* **75:** 143–159.

Raven, J.A. (1987). The role of vacuoles. *New Phytol.* **106:** 357–422.

Raven, J.A. (1994). Why are there no picoplanktonic O_2 evolvers with volumes less than 10^{-19} m^3 ? *J. Plankton Res.* **16:** 565–580.

Raven, J.A. (1997). The vacuole: a cost-benefit analysis. *The Plant Vacuole. Advances in Botanical Research Incorporating Advances in Plant Pathology*. R.A. Leigh and D. Sanders, eds. San Diego, Academic Press, pp. 59–82.

Raven, J.A., Finkel, Z.V., and Irwin, A.J. (2006). Picophytoplankton: bottom-up and top-down controls on ecology and evolution. *Vie et Milieu* **55:** 209–215.

Rea, D.K. (1994). The paleoclimatic record provided by eolian deposition in the deep sea: the geologic history of wind. *Rev. Geophys.* **32:** 159–195.

Rigby, S., and Milsom, C. V. (2000). Origins, evolution, and diversification of zooplankton. *Annu. Rev. Ecol. Syst.* **31:** 293–313.

Rinaldo, A., Maritan, A., Cavender-Bares, K.K., and Chisholm, S.W. (2002). Cross-scale ecological dynamics and microbial size spectra in marine ecosystems. *Proc. R. Soc. Lond. Ser. B* **269:** 2051–2059.

Rosenzweig, M.L. (1995). *Species Diversity in Space and Time*. Cambridge, Cambridge University Press.

Rutherford, S., D'Hondt, S., and Prell, W. (1999). Environmental controls on the geographic distribution of zooplankton diversity. *Nature* **400:** 749–752.

Ryther, J.H. (1969). Photosynthesis and fish production in the sea. *Science* **166:** 72–76.

Savenkoff, C., Vezina, A.F., Roy, S., Klein, B. Lovejoy, C., Therriault, J.-C., Legendre, L., Rivkin, R., Berube, C., Tremblay, J.-E., and Silverberg, N. (2000). Export of biogenic carbon and structure and dynamics of the pelagic food web in the Gulf of St. Lawrence Part 1. Seasonal variations. *Deep Sea Res. Part II Top. Studies Oceanogr.* **47:** 585–607.

Scheffer, M., Carpenter, S., and de Young, B. (2005). Cascading effects of overfishing marine systems. *Trends Ecol. Evol.* **20:** 579–581.

Schlesinger, D.A., Molot, L.A., and Shuter, B.J. (1981). Specific growth rate of freshwater algae in relation to cell size and light intensity. *Can. J. Fish. Aquat. Sci.* **38:** 1052–1058.

Schmidt, D.N., Thierstein, H.R., Bollmann, J., and Schiebel, R. (2004). Abiotic forcing of plankton evolution in the Cenozoic. *Science* **207:** 207–210.

Sepkoski, J.J. (1984). A kinetic model of phanerozoic taxonomic diversity. III. Post-Paleozoic families and mass extinctions. *Paleobiology* **10:** 246–267.

Sheldon, R.W., and Parsons, T.R. (1967). A continuous size spectrum for particulate matter in the sea. *J. Fish. Res. Board Can.* **24:** 909–915.

Sheldon, R.W., Prakash, A., and Sutcliffe, W.H., Jr. (1972). The size distribution of particles in the ocean. *Limnol. Oceanogr.* **17:** 327–340.

Sherr, E.B., and Sherr, B.F. (1994). Bactivory and herbivory: key roles of phagotrophic protists in pelagic food webs. *Microbial Ecol.* **28:** 223–235.

Shuter, B.G. (1978). Size dependence of phosphorus and nitrogen subsistence quotas in unicellular microorganisms. *Limnol. Oceanogr.* **23:** 1248–1255.

Sicko-Goad, L.M., Schekske, C.L., and Stoermer, E.F. (1984). Estimation of intracellular carbon and silica content of diatoms from natural assemblages using morphometric techniques. *Limnol. Oceanogr.* **29:** 1170–1178.

Sieburth, J.M., Smetacek, V., and Lenz, J. (1978). Pelagic ecosystem structure: heterotrophic compartments of the plankton and their relationship to plankton size fractions. *Limnol. Oceanogr.* **23:** 1256–1263.

Smayda, T.J. (1970). The suspension and sinking of phytoplankton in the sea. *Oceanogr. Mar. Biol. Annu. Rev.* **8:** 353–414.

Smetacek, V. (2001). A watery arms race. *Nature* **411:** 745.

Smetacek, V., Assmy, P., and Henjes, J. (2004). The role of grazing in structuring Southern Ocean pelagic ecosystems and biogeochemical cycles. *Antarctic Sci.* **16:** 541–558.

Smith, C.R., Berelson, W., Demaster, D.J., Dobbs, F.C., Hammond, D., Hoover, D.J., Pope, R.H., and Stephen, M. (1997). Latitudinal variations in benthic processes in the abyssal equatorial Pacific: control by biogenic particle flux. *Deep Sea Res. Part II Top. Studies Oceanogr.* **44**: 2295–2317.

Smith, Jr., K.L., Baldwin, R.J., Karl, D.M., and Boetius, A. (2002). Benthic community responses to pulses in pelagic food supply: North Pacific Subtropical Gyre. *Deep Sea Res. Part I Oceanogr. Res. Papers* **49**: 971–990.

Smith, V.H., Foster, B.L., Grover, J.P., Holt, R.D., Liebold, M.A., and deNoyelles, Jr., F. (2005). Phytoplankton species richness scales consistently from laboratory microcosms to the world's oceans. *Proc. Natl. Acad. Sci. U S A* **102**: 4393–4396.

Sommer, U. (1989). Maximal growth rates of Antarctic phytoplankton: only weak dependence on cell size. *Limnol. Oceanogr.* **34**: 1109–1112.

Sorhannus, U., Fenster, E.J., Burckle, L.H., and Hoffman, A. (1988). Cladogenetic and anagenetic changes in the morphology of *Rhizosolenia praebergonii* Mukhina. *Historical Biol.* **1**: 185–205.

Sprules, W.G., and Munawar, M. (1986). Plankton size spectra in relation to ecosystem productivity, size and perturbation. *Can. J. Fish. Aquat. Sci.* **43**: 1789–1794.

Stanley, S.M. (1973a). An ecological theory for the sudden origin of multicellular life in the Late Precambrian. *Proc. Natl. Acad. Sci. U S A* **70**: 1486–1489.

Stanley, S.M. (1973b). An explanation for Cope's rule. *Evolution* **27**: 1–25.

Steele, J.H., and Baird, I.E. (1961). Relations between primary production, chlorophyll and particulate carbon. *Limnol. Oceanogr.* **6**: 68–78.

Stemmann, L., Jackson, G.A., and Ianson, D. (2004). A vertical model of particle size distributions and fluxes in the midwater column that includes biological and physical processes-Part I: model formulation. *Deep Sea Res. I* **51**: 865–884.

Stolte, W., and Riegman, R. (1995). Effect of phytoplankton cell size on transient state nitrate and ammonium uptake kinetics. *Microbiology* **141**: 1221–1229.

Suttle, C.A. (2005). Viruses in the sea. *Nature* **437**: 356–361.

Sweeney, E.N., McGillicuddy, D.J., and Buesseler, K.O. (2003). Biogeochemical impacts due to mesoscale eddy activity in the Sargasso Sea as measured at the Bermuda Atlantic Time Series (BATS) site. *Deep Sea Res. II* **50**: 3017–3039.

Taguchi, S. (1976). Relationship between photosynthesis and cell size of marine diatoms. *J. Phycol.* **12**: 185–189.

Taguchi, S., and Saino, T. (1998). Net zooplankton and the biological pump off Sanriku, Japan. *J. Oceanogr.* **54**: 573–582.

Tang, E.P.Y. (1995). The allometry of algal growth rates. *J. Plankton Res.* **17**: 1325–1335.

Teegarden, G.J., Campbell, R.G., and Durbin, E.G. (2001). Zooplankton feeding behaviour and particle selection in natural plankton assemblages containing toxic Alexandrium spp. *Mar. Ecol. Prog. Ser.* **218**: 213–226.

Thiel, H. (1975). The size structure of the deep-sea benthos. *Int. Rev. Gesamten Hydrobiol.* **60**: 575–606.

Thingstad, F., and Hagström, Å. (1997). Accumulation of degradable DOC in surface waters: is it caused by a malfunctioning microbial loop? *Limnol. Oceanogr.* **42**: 398–404.

Trammer, J. (2002). Power formula for Cope's rule. *Evol. Ecol. Res.* **4**: 147–153.

Uye, S.-I., and Kaname, K. (1994). Relations between fecal pellet volume and body size form major zooplankters of the inland Sea of Japan. *J. Oceanogr.* **50**:43–49.

Van Valen, L. (1973). Body size and numbers of plants and animals. *Evolution* **27**: 27–35.

Vaulot, D., Romari, K., and Not, F. (2002). Are autotrophs less diverse than heterotrophs in marine picoplankton. *Trends Microbiol.* **10**: 266–267.

Wimpenny, R.S. (1936). The size of diatoms. I. The diameter variation of *Rhizosolenia styliformis* Brightw. and *R. alata* Brightw. in particular and of pelagic marine diatoms in general. *Mar. Biol. Assoc. J.* **21**: 29–60.

Woodward, G., Ebenman, B., Emmerson, M.C. Montoya, J.M., Olesen, J.M., Valid, A., and Warren, P.H. (2005). Body size in ecological networks. *Trends Ecol. Evol.* **20**: 1–8.

Worm, B., and Myers, R.A. (2003). Meta-analysis of cod-shrimp interactions reveals top-down control in oceanic food webs. *Ecology* **84**: 162–173.

Wright, J.D. (2001). Cenozoic climate—oxygen isotope evidence. *Encyclopedia of Ocean Sciences*. J. Steele, S. Thorpe, and K. Turekian, eds. London, Academic Press, pp. 415–426.

Zachos, J., Pagani, M., Sloan, L., Thomas, E., and Billups, K. (2001). Trends, rhythms, and aberrations in global climate 65 Ma to present. *Science* **292**: 686–693.

16

Resource Competition and the Ecological Success of Phytoplankton

ELENA LITCHMAN

Phytoplankton are major primary producers in the aquatic realm, responsible for almost half of global net primary production (Field *et al.* 1998). Their abundance and community structure directly impact higher trophic levels and key biogeochemical cycles. Phytoplankton are an extremely diverse, polyphyletic group that includes both prokaryotic and eukaryotic forms. What makes phytoplankton so successful? Several fundamental processes, such as photosynthesis, growth, resource acquisition, and grazer avoidance, to a large degree define ecological niche of phytoplankton. The success of phytoplankton depends on how efficiently they acquire resources, transform them into growth, and avoid being eaten or infected. Diverse selective pressures on phytoplankton increase the efficiency of these processes and allow species to persist under changing conditions. Resource competition is one of the key ecological processes that controls species composition, diversity, and succession of phytoplankton communities. In this chapter I discuss the role of resource competition in structuring past, present, and future phytoplankton communities.

I. RESOURCE ACQUISITION AND MEASURES OF COMPETITIVE ABILITY

The resources that phytoplankton need include carbon, light, macro- and micronutrients such as nitrogen, phosphorus, and iron, and certain vitamins. The ability to utilize these resources effectively will give a phytoplankton species a competitive advantage over other members of the community and thus contribute to its overall fitness. I briefly review how the utilization of resources such as nutrients and light is described and how competitive ability for these resources is defined.

A. Nutrients

Nutrients are dissolved in the aquatic medium and taken up through the cellu-

lar surface. The rate of nutrient uptake, V, is often described as a saturating function of nutrient, R (Michaelis–Menten kinetics). The key parameters that characterize the efficiency of nutrient uptake are the maximum uptake rate, V_{max}, and half-saturation constant for uptake, K.

$$V = V_{max} \frac{R}{K + R} \qquad (1)$$

Acquired resource is then used for growth and, consequently, growth, μ, depends on the internal nutrient concentration or cellular quota, Q, as follows:

$$\mu = \mu_\infty \left(1 - \frac{Q_{min}}{Q} \right) \qquad (2)$$

where μ_∞ is the growth rate of species at an infinite quota, Q is the internal nutrient concentration (nutrient quota), and Q_{min} is the minimum quota (when growth rate equals 0). Equation 2 is known as the Droop model (Droop 1968, 1973). Equations 1 and 2 are widely used to describe phytoplankton nutrient uptake and growth. A similar model of phytoplankton growth based on internal nutrient concentration was proposed by Caperon (1968). Droop and other internal stores models have been shown to be more accurate than models of growth based on the external nutrient concentration (e.g., Monod model), especially under fluctuating nutrient supply (Turpin and Harrison 1979; Turpin 1988; Grover 1991a). However, even these internal stores models with a separate Michaelis–Menten equation for uptake still oversimplify nutrient uptake and growth processes. A more detailed overview of nutrient uptake in phytoplankton is given in Raven (1980) and Riebesell and Wolf-Gladrow (2002). Examples of more mechanistic models of phytoplankton growth can be found in Kooijman and others (Kooijman *et al.* 2004; Vrede *et al.* 2004).

How do these different traits that affect nutrient uptake and requirements combine to determine the outcome of nutrient competition?

Resource competition theory shows that the species that reduces the limiting resource to the lowest concentration in monoculture is the best competitor at equilibrium in a well-mixed system (Tilman 1982). This critical concentration is termed R^*. It is, therefore, possible to predict the outcome of competition by measuring R^*s in monocultures for different species for a given nutrient. It is also possible to derive R^* from a particular model of nutrient uptake and growth. Given Droop model (Equations 1 and 2), the break-even nutrient concentration, R^*, can be expressed as follows:

$$R^* = \frac{K\mu_\infty Q_{min} m}{V_{max}(\mu_\infty - m) - \mu_\infty Q_{min} m},\qquad (3)$$

where symbols are as in Equations 1 and 2 and m is mortality. Thus R^* integrates many physiological parameters into a synthetic metric of competitive ability. Figure 1 shows how R^* depends on the individual parameters of nutrient uptake and growth.

Resource competition theory has been successfully used to determine the outcome of laboratory experiments (Tilman 1977) and to explain the distribution of species along resource supply gradients (Tilman et al. 1982) and patterns of species diversity in different lakes (Interlandi and Kilham 2001). A review by Tilman and colleagues provides a more detailed introduction into the theory of resource competition, as applied to phytoplankton (Tilman et al. 1982).

It must be stressed that this theory is based on well-mixed and equilibrium conditions. Deviations from spatial and temporal homogeneity may alter the selective value of physiological traits and can open up new mechanisms for species coexistence (see Section II).

B. Light

Light, another essential resource, has a complex pattern of spatial and temporal variability (Falkowski 1984; Ferris and Christian 1991). The unavoidable vertical light gradient makes the dynamics of light competition more complex than those of nutrient competition. Recent work of Huisman and colleagues showed that light-dependent growth of monocultures can be used to predict the outcome of competition for light, similar to nutrient competition (Huisman and Weissing 1994; Weissing and Huisman 1994; Huisman et al. 1999a). Phytoplankton growth is modeled as a function of light availability at a given depth (we use the original notation of Huisman et al.):

$$\frac{d\omega_i}{dt} = \frac{1}{z}\int_0^z P_i[I(s)]\omega_i ds - D\omega_i$$

where ω_i is the biomass and $p_i(I)$ is the specific production rate of species i as an increasing function of light intensity, $I(s)$ is the light intensity as a decreasing function of depth s, z is the total depth of the water column, and D is the loss rate.

Light is attenuated exponentially by phytoplankton biomass and background turbidity:

$$I(s) = I_{in}\exp\left(-\left[\sum_{j-1}^n k_j\omega_j s + K_{bg}s\right]\right)$$

where I_{in} is the incident light intensity, k_j is the specific light attenuation coefficient of species j, K_{bg} is the total background turbidity due to nonphytoplankton components, and n is the total number of phytoplankton species. The model predicts that each species has a critical light intensity I_{out} at which its biomass remains constant. A species with the lowest critical light intensity is the best light competitor, as it is able to reduce I_{out} below the critical intensities of other species (Huisman and Weissing 1994; Huisman et al. 1999a). The I_{out} as the measure of light competitive ability is similar to the R^* measure of the nutrient competitive ability; the key difference is, however, that I_{out} depends on the incoming light (Huisman and Weissing 1994), while R^* is independent of nutrient inflow (Tilman 1982).

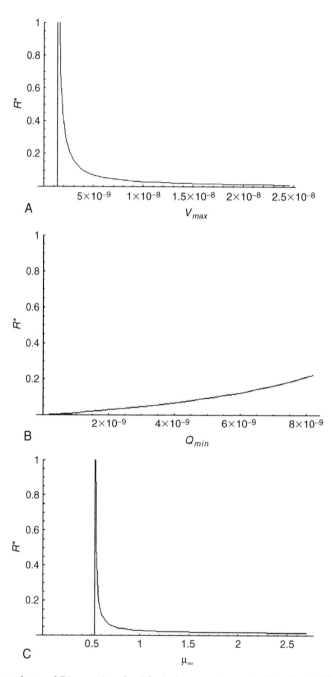

FIGURE 1.   The dependence of R^* on major physiological parameters, as in Equation 3. R^* as a function of **(A)** V_{max}, **(B)** Q_{min}, **(C)** μ_∞,

(Continued)

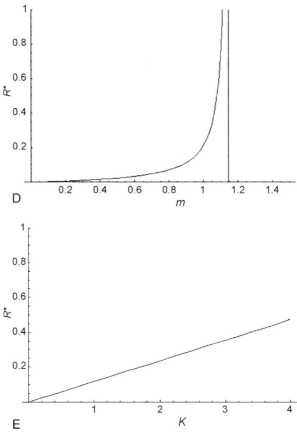

FIGURE 1. Cont'd   **(D)** m, and **(E)** K. Parameter values are as follows: V_{max} = $1.23^{*}10^{-8}$ µmol nutrient cell^{-1} day^{-1}, Q_{min} = $1.64^{*}10^{-9}$ µmol nutrient cell^{-1}, μ_{∞} = 1.35 day^{-1}, m = 0.5 day^{-1}, and K = 0.2 µmol nutrient L^{-1}.

II. THE ROLE OF SPATIAL AND TEMPORAL HETEROGENEITY IN RESOURCE COMPETITION IN PHYTOPLANKTON

In aquatic ecosystems, the effects of spatial and temporal heterogeneity are especially strong, as physical forcing and biological processes are tightly coupled due to similar characteristic time scales of physical forcing and biological responses (Steele 1985). Major resources for phytoplankton are heterogeneously distributed in space, both along the vertical and horizontal dimensions and in time. Here I discuss the implications of the vertical heterogeneity in resource distribution: a vertical gradient in light, with irradi-

ances decreasing exponentially with depth and an opposing gradient in nutrient concentration (increase with depth) (Figure 2). Phytoplankton also often exhibit vertically heterogeneous distributions that can be both the cause and the consequence of the vertical heterogeneity in light and nutrients.

A. Heterogeneity in Nutrient Distribution

Low nutrient concentrations in the euphotic zone are due to biological uptake and often lead to severe nutrient limitation of phytoplankton during the periods of high water-column stability. However, a number of physical processes episodically supply

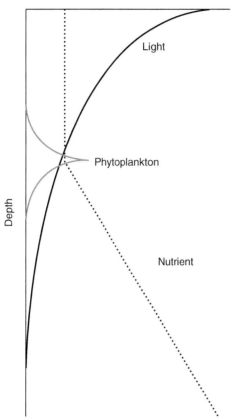

FIGURE 2. Typical heterogeneous distributions of light, a limiting nutrient, and a subsurface chlorophyll maximum in the water column.

nutrients to the upper water column, thus causing a temporal heterogeneity in resource supply. Turbulent eddies may entrain nutrient-rich waters from the pycnocline, leading to higher rates of primary productivity in the upper mixed layer (Falkowski *et al.* 1991; Garcon *et al.* 2001). If nutrient concentrations are elevated in the deeper water, increased nutrient fluxes will occur to the upper mixed layer due to mixing created by internal waves (Holligan *et al.* 1985). Upwelling brings nutrients from the nutrient-rich deep waters into the euphotic zone, where uptake by phytoplankton can lead to enhanced growth. Similarly, Ekman transport and pumping near the coast creates upwelling conditions that distribute more nutrients to the euphotic zone (Bauer *et al.* 1991). Such nutrient pulses occur at multiple temporal

scales; for example, extremely slow Rossby waves with periods of ca. 3 months were shown to inject nitrate into the upper water column, increasing productivity and shifting phytoplankton community structure to fast-growing species of haptophytes and pelago-phytes (Sakamoto *et al.* 2004). Long-lived baroclinic eddies (life span of ca. 150 days) were shown to increase nitrate concentrations up to eightfold and promote growth of large phytoplankton (Vaillancourt *et al.* 2003). Understanding how episodic mesoscale physical processes in the ocean such as eddies (Falkowski *et al.* 1991; McGillicuddy *et al.* 1998), localized upwelling (Martin *et al.* 2002), and others influence phytoplankton production, and community structure, and thus the rest of the food web, is an active area of research. There are also much faster processes that can lead to nutrient injections into the upper part of the water column, such as Langmuir circulation and small-scale turbulent processes (Thorpe 2004a, b).

Theoretical and experimental studies of phytoplankton nutrient competition show that dynamics and outcome of competition strongly depend on frequency of nutrient pulses (Turpin and Harrison 1979; Grover 1991c; Stolte and Riegman 1996; Tozzi *et al.* 2004). Under high-frequency nutrient pulses (e.g., daily), the conditions are similar to the continuous nutrient supply, and species that win in a constant nutrient environment (species with the lowest R^*) dominate in such a regime as well, as fast fluctuations are physiologically averaged out (Grover 1991c). In contrast, under low-frequency nutrient pulses (with a several day period) species that may be inferior competitors under constant nutrient supply may gain competitive advantage (Grover 1991c; Stolte and Riegman 1996). Such species may have superior storage abilities through high maximum nutrient uptake rate and/or a large maximum nutrient quota, and thus be non-equilibrium strategists (Grover 1991c; Stolte and Riegman 1996). Assuming that diatoms are better adapted to store nitrate, a limiting nutrient, compared to coccolithophores,

Tozzi *et al.* (2004) showed with a simulation model that pulsed nutrient regimes favor diatoms over coccolithophores.

Hutchinson (1961) proposed that nonequilibrium conditions could explain the paradox of the plankton in which many species of phytoplankton limited by few resources coexist in a seemingly homogeneous environment, whereas according to the competitive exclusion principle, the number of coexisting species cannot exceed the number of limiting resources (Gause 1934; Hardin 1960). Theoretical studies show that fluctuations in limiting resources (e.g., nutrients) may allow coexistence of more than one species on a single limiting resource, while under constant resource supply a species capable of reducing the resource concentration to the lowest level would competitively displace other species (Armstrong and McGehee 1976, 1980; Levins 1979; Hsu 1980). The dynamics and outcome of competition may also depend on how predictable fluctuation regimes are: Generally, under more randomly fluctuating conditions, the advantage of non-equilibrium specialists decreases (Grover 1991c).

B. Heterogeneity in Light Distribution

Light, another essential resource for phytoplankton, also fluctuates on a wide range of temporal scales with different amplitudes. Due to a vertical gradient in light distribution and different mixing patterns phytoplankton experience temporal fluctuations in light supply on a wide range of time scales. Many of the hydrodynamic processes that alter nutrient regimes (see previous section) also cause light fluctuations for phytoplankton. Turbulent eddies in the upper mixed layer change the light exposure of phytoplankton on time scales of minutes to hours (Denman and Gargett 1983). Contracting and expanding internal waves can move phytoplankton tens of meters up and down through the vertical light gradient (Denman and Gargett 1983).

Light fluctuations may significantly change phytoplankton photosynthesis and primary productivity (e.g., Marra 1978; Falkowski and Wirick 1981; Mallin and Paerl 1992). Moreover, our own studies show that fluctuating light may also alter growth rates in a species-specific way, delay competitive exclusion, promote coexistence, and change community composition (Litchman 1998, 2000; Litchman and Klausmeier 2001). We found that fast light fluctuations may change competitive outcome and lead to a dominance by nonequilibrium specialists, species that are inferior competitors under constant light but have high maximum growth rates (Litchman and Klausmeier 2001). In addition, slow light fluctuations could lead to stable coexistence of multiple species and thus increase diversity (Litchman and Klausmeier 2001). In general, fluctuating conditions select for fast-growing species, whereas constant environments select for best equilibrium competitors (Grover 1991c; Litchman and Klausmeier 2001).

Light fluctuations may also mediate competition for a limiting nutrient due to the light-dependent nutrient uptake (Macisaac and Dugdale 1972; Litchman *et al.* 2004). Depending on the costs and benefits of maintaining nutrient uptake in the dark, different strategies of nutrient use are optimal under different light regimes: If the cost of maintaining nutrient uptake in the dark is high and algae experience short periods of darkness or low light (shallow mixing and/or long day length), the optimal strategy is not to take up nutrient in the dark. In contrast, if the periods of low light or darkness are long (deep mixing or short day length), the optimal strategy would be to maintain some uptake in the dark, despite the costs (Litchman *et al.* 2004). It may be that different functional groups of phytoplankton are more adapted to a particular strategy, depending on the physical conditions they are associated with and other physiological adaptations. Diatoms often dominate under low water-column stability (Li 2002), that is, deep mixing conditions, and thus may experience prolonged periods of low light or darkness. Some studies suggest that diatoms are more efficient at maintaining nitrate uptake in the dark compared to flagellates (Clark *et al.* 2002).

The composite measure of competitive abilities, R^*, depends nonlinearly on resource fluctuations, and it is possible in some cases to derive such a relationship analytically to predict the outcome of resource competition under different fluctuation regimes (Litchman and Klausmeier 2001; Litchman *et al.* 2004).

C. Vertical Heterogeneity in Phytoplankton Distribution

Vertical gradients in the distribution of limiting nutrients may lead to the diversification of nutrient acquisition strategies (Troost *et al.* 2005a). Similarly, vertical gradient in irradiance allows for niche separation and coexistence of different strategies of light utilization under stratified conditions: High-light–adapted species grow in the upper part of the water column, whereas low-light–adapted species inhabit greater depths, often forming subsurface chlorophyll maxima (Reynolds 1997).

There is an emerging literature on modeling the vertical distribution of phytoplankton, begun by Riley *et al.* in 1949. Lande and Lewis (1989) did not consider population dynamics but rather the effect of mixing on the vertical distribution of a physiological state, photoadaptation. Early models of population dynamics (Shigesada and Okubo 1981; Britton and Timm 1993) focused on mathematical proofs and treated competition phenomenologically. Huisman and Weissing (1994, 1995; Weissing and Huisman 1994) developed a mechanistic model of phytoplankton growth and competition in perfectly well-mixed water columns. The assumption of total mixing allowed them to integrate over the water column to arrive at a simpler, nonspatial model. They found that although there is an inevitable light gradient, the competitive exclusion principle still holds, allowing one species when competing for light and two when competing for light and nutrients. Later, Huisman *et al.* (1999a, b) relaxed the assumption of

perfect mixing by considering a reaction-diffusion model of growth and competition in incompletely mixed water columns. Considering light limitation only (no nutrients), they showed that incomplete mixing can allow two species to coexist on one limiting resource but only over a narrow range of parameter values (Huisman *et al.* 1999b). Phytoplankton sinking has been considered by Huisman *et al.* (2002), Diehl (2002), and Yoshiyama and Nakajima (2002).

Yoshiyama and Nakajima (2002) made a step toward physical realism by considering vertical heterogeneity in mixing rates, placing a well-mixed layer above a poorly mixed one. They considered a single species, limited by light and nutrients. They found that under low nutrients, a heterogeneous distribution of phytoplankton arose in the deep layer; under high nutrients, the biomass was uniformly distributed throughout the surface mixed layer; and in between, multiple stable states were possible (although this may depend on the assumption that phytoplankton and nutrients have different diffusion coefficients across the thermocline).

Klausmeier and Litchman (2001) investigated the vertical distribution of a single species of phytoplankton in poorly mixed water columns, with light- and nutrient-dependent growth. Light is supplied at the top of the water column and nutrients are supplied at the bottom, diffusing across the sediment-water interface. They included a novel feature not found in the other models reviewed so far: phytoplankton behavior in the form of active movement up gradients in potential growth rate. Direct numerical solution of the reaction-diffusion-advection model showed that phytoplankton could form thin layers of biomass when mixing is weak and swimming is strong. The width of the layer is determined by the ratio of the turbulent diffusion coefficient and the swimming speed.

A novel analytical approach was developed by Klausmeier and Litchman (2001) to determine the location and biomass of the phytoplankton layer under poor mixing and rapid swimming. By treating the distribution of phytoplankton as an infinitely thin layer, they could solve for the optimal depth of the layer as an evolutionarily stable strategy, using methods from evolutionary game theory. Competition for nutrients and light in a poorly mixed water column could result in three different kinds of vertically heterogeneous distributions: at low nutrient and background attenuation levels, a benthic layer forms; at high nutrient levels, a surface scum forms; and at intermediate nutrient levels, a deep chlorophyll maximum (DCM) occurs (Klausmeier and Litchman 2001). The opposing gradients of light and a limiting nutrient may lead to stable coexistence of multiple species at different depths if those species exhibit a trade-off in nutrient versus light competitive ability (Klausmeier et al. unpublished).

Phytoplankton in the DCM may exhibit population fluctuations and chaotic behavior because sinking of phytoplankton and the upward diffusion of nutrients occur on different time scales: Phytoplankton sinking out of the optimal irradiance decline and do not completely utilize upward diffusing nutrients. When the upward diffusing nutrients reach depth with adequate irradiance, phytoplankton growth is stimulated (Huisman et al. 2006). Seasonal changes in irradiance may also facilitate chaotic oscillations in phytoplankton density; these oscillations, in turn, may promote coexistence of multiple species in the DCM (Huisman et al. 2006).

III. PHYSIOLOGICAL TRADE-OFFS

The ability of phytoplankton to acquire and utilize resources can be characterized by key physiological parameters that differ among species. These differences in physiological traits arise from physical and chemical constraints on physiological functions. These constraints define various trade-offs in resource utilization in all organisms, and these trade-offs lie at the heart of niche-based nonneutral theories explaining species diversity (Tilman 2000; Bohannan et al. 2002).

A trade-off occurs when a trait that confers an advantage for performing one function simultaneously confers a disadvantage for performing another function (Bohannan et al. 2002). I discuss some relevant trade-offs in phytoplankton and illustrate how such trade-offs can promote species coexistence and diversity.

A lower R^* can be achieved by increasing the maximum nutrient uptake rate, V_{max}, growth rate at infinite nutrient quota, μ_∞, by decreasing half-saturation constant for uptake, K, minimum nutrient quota, Q_{min}, and/or mortality, m. An evolutionary pressure to increase nutrient competitive ability, that is, to lower the R^*, may act on all these parameters simultaneously. Interestingly, the dependence of R^* on individual physiological parameters is nonlinear for all parameters except for the half-saturation constant for uptake, K (see Figure 1). This means that for certain ranges of these parameters a change in parameter value could lead to a greater change in competitive ability than for other parameter ranges. In addition, the influence of these parameters on the R^* may be unequal, so that a relatively similar change in one parameter compared to another may lead to a much greater change in R^* (see Figure 1).

If these parameters were independent, it would be possible to select for a "superspecies" with all the parameters at values that would lead to the lowest R^* possible. However, as our recent work indicates, key parameters of nutrient uptake and growth are correlated with each other, and these correlations may represent real physiological trade-offs (Litchman et al. submitted). For example, we found significant positive correlations between V_{max} and K and between Q_{min} and V_{max} (Litchman et al. submitted). Such trade-offs are likely to preclude the evolution of a superspecies and, instead, can promote diversity of eco-physiological strategies of resource utilization. A consideration of physiological trade-offs in an ecological context that defines selective pressures should improve our understanding of the mechanisms structuring past, present, and future phytoplankton communities.

A. Nutrient Utilization Trade-Offs

Various physiological trade-offs that lead to distinct ecological strategies of nutrient utilization in phytoplankton can be explained mechanistically. For example, a positive correlation between the maximum uptake rate and half-saturation constant for nitrate uptake can be explained based on the model of nutrient uptake by Aksnes and Egge (1991). Briefly, a limited surface area used for uptake of a particular nutrient could be partitioned into uptake sites in at least two contrasting ways: large number of uptake sites with a smaller active area of individual sites or fewer uptake sites with a larger area of individual uptake sites (Litchman et al. submitted). According to the formulas of Aksnes and Egge (1991), the former strategy would result in high $Vmax$ and high K, whereas the latter strategy would lead to low $Vmax$ and low K (Litchman et al. submitted).

Data suggest that there may also be trade-offs in the utilization of and competitive abilities for different nutrients. Oceanic centric diatoms have lower iron requirements than coastal species (Brand et al. 1983; Maldonado and Price 1996). However, there may be a trade-off in requirements for iron and another micronutrient, copper. Peers et al. (2005) showed that oceanic diatoms have significantly higher requirements for copper (necessary for iron utilization and growth) compared to coastal forms, demonstrating a trade-off between iron and copper requirements (Peers et al. 2005). Classic work by Tilman and colleagues revealed trade-offs in competitive ability for silicon versus phosphorus among freshwater diatoms (Tilman 1977; Tilman et al. 1982). It is unknown, however, how universal such trade-offs are.

B. Light Utilization Trade-Offs

There are also various physiological trade-offs in light utilization. Often, species that can grow at very low irradiances have their growth saturated or inhibited at lower irradiances, compared to species that cannot grow at low irradiances (Ryther 1956; Falkowski 1980; Richardson et al. 1983). For example, green algae have significantly higher irradiances at which minimum and maximum growth is achieved, compared to other major taxonomic groups (Richardson et al. 1983). Experimental work by Stomp and colleagues (Stomp et al. 2004) demonstrated that a trade-off in utilization of different parts of the light spectrum in algae can lead to species coexistence. Two isolates of unicellular cyanobacteria of the genus *Synechococcus* differ in their accessory pigment composition: an isolate BS4 contains phycocyanin and hence is capable of absorbing light in the orange-red part of the spectrum, whereas the isolate BS5 contains phycoerythrin and is able to absorb photons in the green-yellow part of the spectrum. When the two isolates were competing for light, the phycocyanin-containing isolate won under red light. Conversely, the phycoerythrin-containing cyanobacterium won under green light. Competition under white light did not result in competitive exclusion, as the two isolates were utilizing different parts of the spectrum and were able to coexist (Stomp et al. 2004).

C. Trade-Offs in Nutrient Competitive Ability Versus Light Competitive Ability

Another important trade-off is that between nutrient and light competitive ability (Leibold 1997). An elegant study by Strzepek and Harrison (2004) demonstrated a trade-off between iron requirements and light utilization in marine diatoms: the open ocean species have significantly lower iron requirements compared to coastal species due to changes in photosynthetic apparatus. By lowering their cellular iron content, open ocean diatoms compromise their ability to utilize rapidly fluctuating light; this ability is, however, less relevant in the open ocean than in coastal environment where fast light fluctuations occur due to tidal mixing (Strzepek and Harrison 2004). A higher minimum quota for nitrogen at low irradiance (Rhee and Gotham 1981) represents another example of a physiological trade-off between light and nutrient competitive abilities and is probably due to higher chlorophyll concentration to increase light utilization at low light.

D. Trade-Offs in Growth Rate Versus Competitive Ability

Phytoplankton often exhibit a trade-off between the maximum growth rate and equilibrium competitive ability, a so-called gleaner-opportunist trade-off (Grover 1997) that is essentially a trade-off between the r and K strategies (Kilham and Kilham 1980; Sommer 1981). Under nonequilibrium or high resource conditions, species with high maximum growth rates, but not necessarily the lowest R^*, may win competition (Grover 1997). Due to trade-offs in biochemical allocation, high maximum growth rates are often associated with high cellular concentration of phosphorus-rich ribosomes enabling high growth rates (the growth rate hypothesis, Sterner and Elser 2002; Klausmeier et al. 2004). Species that are good competitors for nutrients or light invest heavily in nitrogen-rich uptake proteins or light-harvesting structures (chlorophyll complexes). These differences in investment (growth machinery, i.e., ribosomes, versus resource acquisition machinery) will lead to differences in cellular N:P stoichiometry (Klausmeier et al. 2004). If environmental conditions select for fast-growing species, the overall N:P stoichiometry of phytoplankton may be shifted to low N:P ratios and, conversely, if the conditions select for good competitors, phytoplankton N:P ratios will be higher (Klausmeier et al. 2004).

E. Trade-Offs in Grazing Resistance Versus Competitive Ability

Aside from acquiring resources and growing, the other important thing in the life of phytoplankton is to avoid being eaten. Good competitors in phytoplankton are often highly susceptible to grazing, whereas poor competitors are grazer-resistant, thus exhibiting the competitive ability–grazer resistance trade-off (Grover 1995). Given this trade-off, both generalist and specialist grazers can preclude competitive exclusion and maintain species diversity in phytoplankton (Armstrong 1994; Grover 1995; Leibold 1996; Grover and Holt 1998).

IV. ECOLOGICAL STRATEGIES OF RESOURCE UTILIZATION IN MAJOR FUNCTIONAL GROUPS

Major divisions of phytoplankton (taxonomic groups) such as diatoms, coccolithophores, dinoflagellates, chlorophytes (including prasinophytes), and cyanobacteria are often separated into distinct functional groups, as these taxonomic groups have unique biogeochemical signatures. Functional groups in phytoplankton are defined as groups of "organisms related through common biogeochemical processes" and are not necessarily phylogenetically related (Iglesias-Rodriguez et al. 2002a). Do these major functional groups have distinct strategies of nutrient utilization?

In his seminal article, Margalef (1978) classified marine phytoplankton according to their responses to the nutrient and turbulence fields. The so-called Margalef mandala includes diatoms, coccolithophorids, and dinoflagellates. Diatoms occupy the high-nutrient–high-turbulence corner of the nutrient–turbulence space, coccolithophorids are associated with intermediate nutrient concentrations and turbulence, and dinoflagellates cluster at low turbulence and high (red tide forms) or low nutrients. Later, Sommer (1984) proposed three major strategies of resource (nutrient) utilization in phytoplankton: "velocity-adapted" species with high maximum nutrient uptake rates (V_{max}) and high maximum growth rates (μ_{max}) that are able to utilize nutrient pulses and grow fast, "storage-adapted" species with high V_{max} but lower μ_{max} that are able to acquire the limiting nutrient fast when it is available and store it for later use, and the "affinity-adapted" species with a low half-saturation constant for nutrient uptake (K) that is advantageous under severe nutrient limitation.

Early work on both marine and freshwater phytoplankton by Hutchinson, Eppley, Dugdale, and others indicated that different species exhibit contrasting strategies in utilizing major nutrients (Dugdale and Goering 1967; Hutchinson 1967; Eppley et al. 1969). Eppley et al. (1969) noted that species with high maximum growth rates tend to have low half-saturation constants (K) for uptake of nitrate and that oceanic species have lower K than coastal species. Dugdale (1967) hypothesized a positive association between the maximum nutrient uptake rate $Vmax$ and half-saturation constant for uptake K and that species from oligotrophic environments would exhibit low values of both parameters and, conversely, species from eutrophic environments would have high values of both parameters. Dugdale's early ideas were fully supported by later studies.

Litchman et al. (2006, submitted) have recently compiled a database of key parameters of nutrient (nitrate, ammonium, and phosphate) uptake and growth for major groups of marine eukaryotic phytoplankton, diatoms, dinoflagellates, coccolithophores, and chlorophytes. Our data compilation revealed that major phytoplankton functional groups appear to differ in their parameters of nutrient uptake and growth.

A. Diatoms

Diatoms tend to have significantly higher maximum uptake rates of nutrients than any other group (Litchman et al. 2006). This, together with their relatively high maximum growth rates, makes diatoms good nutrient competitors in general (Figure 3) and the "velocity" specialists that are able to effectively utilize nutrient pulses. Large diatoms tend to have lower growth rates, as there is a negative correlation between the maximum growth rate and cell size (Banse 1976; Chisholm, 1992) but high maximum nutrient uptake rates, which would make large diatoms more "storage-adapted" (Sommer 1984). Coincidentally, large diatoms have disproportionately large vacuoles compared to smaller diatoms (Sicko-Goad et al. 1984) and because nitrate can be stored in vacuoles (Raven 1987), large diatom size leads to a "storage-adapted" strategy.

FIGURE 3.   Competitive abilities (R's) of major taxonomic groups of marine phytoplankton for nitrate and phosphate, calculated according to Equation 3, using data from Litchman et al. (2006) and assuming mortality (m) of 0.1 day^{-1}. The lower R^* values indicate good competitive abilities at equilibrium.

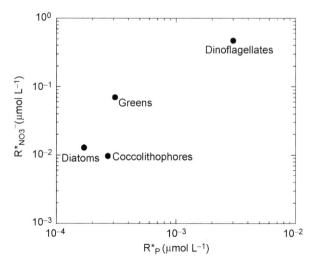

B. Coccolithophores

Coccolithophores also appear to be good nitrate, phosphorus, and iron competitors (see Figure 3), given their parameters for uptake of and growth on these nutrients (Sunda and Huntsman 1995; Riegman *et al.* 2000; Maldonado *et al.* 2001). In contrast to diatoms, they do not have particularly high maximum uptake rates but tend to have significantly lower half-saturation constants for nitrate uptake compared to other functional groups (Litchman *et al.* 2006, submitted) and are, thus, the affinity-adapted strategists. We should note, however, that this analysis is based on a single species of coccolithophorids, *Emiliania huxleyi,* as there are insufficient data on relevant parameters for other coccolithophores. Although *E. huxleyi* is not necessarily a typical coccolithophore (see de Vargas *et al.*, Chapter 12, this volume), it is an ecologically important species widely distributed in the world ocean (Iglesias-Rodriguez *et al.* 2002a, b).

The distribution patterns of diatoms and coccolithophores correspond broadly to their strategies of nutrient utilization. Contrasting nutrient regimes are also associated with distinct patterns of the physical structure of the water column and irradiance regimes: low nutrient conditions arise under stratified conditions characterized by high water-column stability and high irradiance, whereas high nutrient concentrations are often associated with deep mixing, low average irradiance, and large amplitude fluctuations in irradiance. Both groups appear to be adapted to the whole suite of physicochemical conditions that are associated with contrasting nutrient regimes. Coccolithophores are well adapted not only to oligotrophic conditions but to high irradiance that is often associated with stratified, low nutrient conditions: they tend to have higher half-saturation constants of light-dependent growth and are resistant to photoinhibition (Nanninga and Tyrrell 1996). In contrast, diatoms have low half-saturation constants for irradiance-dependent growth

(Falkowski 1980; Richardson *et al.* 1983, Geider and Osborne 1989) and are, therefore, adapted to low light characteristic of the high mixing conditions. Thus, phytoplankton nutrient utilization strategies in conjunction with their responses to physical environment, such as turbulence and light, to a large extent define ecological niches of the two groups. Diatom relative abundance is positively correlated with nitrogen (and phosphorus) concentrations (Schiebel *et al.* 2004) and negatively correlated with stability of the water column (Li 2002). In contrast, the coccolithophorid abundance is greater at low nitrate and phosphate and high water-column stability and irradiance (Cavender-Bares *et al.* 2001; Haidar and Thierstein 2001).

C. Green Algae

Green algae, including prasinophytes, can be characterized by intermediate values of nutrient uptake and growth parameters, except for ammonium and, thus, have intermediate competitive abilities (see Figure 2). Our data compilation as well as the data by Reay *et al.* (1999) indicate that chlorophytes may have a disproportionately high affinity for ammonium over nitrate. Nutrient uptake affinity (Healey 1980), which is the ratio of the maximum uptake rate (V_{max}) and the half-saturation constant for uptake (K), is an important characteristic of nutrient acquisition. A high affinity indicates a superior ability to acquire the nutrient (Healey 1980). All microalgae tend to prefer ammonium over nitrate, as nitrate is more energetically costly to assimilate due to its oxidized state (Syrett 1981). In prasinophytes, however, this preference is around 10-fold greater than in other groups (Litchman *et al.* submitted). We hypothesized that this may be related to the conditions at the time of origin of prasinophytes (Litchman *et al.* submitted). Prasinophyte algae likely appeared around 1.5 billion years ago (Ga) (Hedges 2002; Hedges *et al.* 2004; Yoon *et al.* 2004) in mid-Proterozoic when lower pO_2

(Anbar and Knoll 2002) could have caused the reduced form of nitrogen (i.e., ammonium) to be prevalent (Stumm and Morgan 1981). Interestingly, cyanobacteria, the earliest oxygenic photoautotrophs, evolved under anoxic conditions (Anbar and Knoll 2002; Hedges 2002) and also appear to have a very strong preference for ammonium over nitrate (Herrero *et al.* 2001).

The suboxic conditions at the time of prasinophyte origin may have led to a whole suite of adaptations in this group enabling their success under such conditions. Several studies indicate that a prasinophyte *Tasmanites* may have been the dominant phytoplankter during the ocean anoxic events (OAEs) in the Mesozoic, for example, early Toarcian OAE in the lower Jurassic (Palliani and Riding 1999; Palliani *et al.* 2002; van de Schootbrugge *et al.* 2005). Low oxygen conditions likely increased ammonium availability directly due to a reducing potential and indirectly by stimulating nitrogen fixation by cyanobacteria (Figure 4). Nitrogen fixation may occur more readily under low oxygen condition because enzyme nitrogenase is sensitive to oxygen (Pienkos *et al.* 1983) and because low oxygen conditions increase availability of iron (Stumm and Morgan 1981), which is required for nitrogenase synthesis (Berman-Frank *et al.* 2001; Kustka *et al.* 2002). Increased iron availability may have also stimulated prasinophytes directly, as they may have high iron requirements (Quigg *et al.* 2003). In addition, low oxygen conditions may have reduced the abundance of grazers and that would increase *Tasmanites* densities, as prasinophytes (at least modern species) are effectively controlled by grazers (Boyd and Harrison 1999). Increased *Tasmanites* densities may lead to accumulation of organic matter that would promote anoxic conditions thus creating a positive feedback in the ecosystem (see Figure 4).

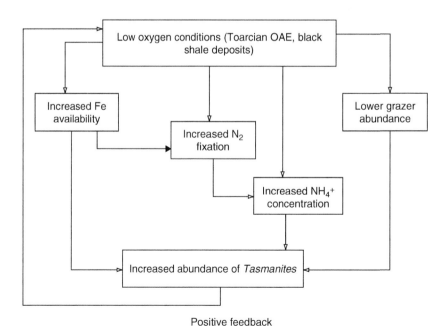

Positive feedback

FIGURE 4.   The hypothesized effects of anoxic conditions on *Tasmanites* dominance, including a possible positive feedback between *Tasmanites* dominance and anoxia. OAE, ocean anoxic event.

D. Dinoflagellates and the Role of Mixotrophy

Dinoflagellates have relatively unimpressive median values of key parameters of nutrient-dependent uptake and growth that result in poor competitive abilities for inorganic macronutrients (high R^*) (see Figure 3). For example, they have significantly higher half-saturation constants for nitrate uptake and significantly lower maximum uptake rates as well as lower maximum growth rates compared to diatoms and other functional groups of eukaryotic phytoplankton (Litchman *et al.* 2006 submitted). Dinoflagellates' ability to regulate their position in the water column by active swimming and a mixotrophic mode of nutrition likely contribute to their success in the ocean (Taylor 1987; Smayda 1997). Theoretical studies demonstrate that mixotrophy is advantageous in oligotrophic environments, whereas in more eutrophic environments an evolutionary specialization into autotrophs and heterotrophs occurs (Troost *et al.* 2005a, b). Phagotrophy in mixotrophs is more often a significant source of inorganic nutrients such as nitrogen, phosphorus, and iron rather than carbon, although at low light, prey ingestion may aid carbon acquisition as well (Raven 1997). A preliminary model study suggests that mixotrophy may also be advantageous in fluctuating environments (Litchman and Klausmeier unpublished).

These findings suggest an interesting hypothesis regarding the origin of eukaryotic phytoplankton. Plastid acquisition in eukaryotic phytoplankton occurred through primary (chlorophytes), secondary (cryptophytes, haptophytes, and stramenopiles), or tertiary (dinoflagellates) endosymbioses (Yoon *et al.* 2004), when an ingested autotroph was retained and not digested. We may hypothesize that this plastid enslavement was possible in an environment that selected for mixotrophy rather than for pure autotrophy and heterotrophy: the environment characterized by oligotrophy and/or resource fluctuations. Brasier and Lindsay (1998) made a similar suggestion on nutrient limitation potentially triggering endosymbiotic event(s). Some estimates for the occurrence of primary and secondary endosymbioses put these respective events before 1558 million years ago (Ma) and around 1200 Ma, in the late Paleoproterozoic and early Mesoproterozoic, respectively (Yoon *et al.* 2004). Interestingly, at least at the time of secondary endosymbiosis, the conditions in the ocean may have been severely nutrient (nitrogen)-limited (Anbar and Knoll 2002), which would select for a mixotrophic strategy via secondary endosymbiosis. By late Mesoproterozoic, nitrogen availability may have increased (Anbar and Knoll 2002; Yoon *et al.* 2004), and this may have stimulated autotrophic specialization and algal diversification (Figure 5).

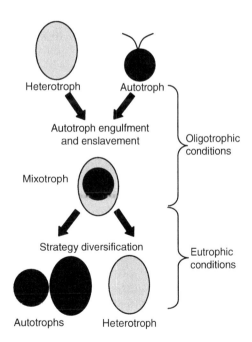

FIGURE 5. Hypothesis on the occurrence of endosymbiotic event under oligotrophic conditions that favor mixotrophy and the diversification of purely autotrophic and heterotrophic strategies under more eutrophic conditions.

E. The Role of Size

Key parameters of resource utilization and growth are correlated with cell size in phytoplankton (Eppley *et al*. 1969; Banse 1976; Shuter 1978; Smith and Kalff 1982; Finkel 2001 and Chapter 15, this volume; Litchman *et al*. submitted). Diffusion limitation of nutrient uptake increases with size (Munk and Riley 1952; Pasciak and Gavis 1974), and thus smaller cells should have a competitive advantage under nutrient-limited conditions. However, even in oligotrophic parts of the ocean where nutrient competition is severe, phytoplankton size spectra include not only extremely small-celled picoplankton but large diatoms as well (Hulburt 1967; Kilham and Kilham 1980; Gin *et al*. 1999; Venrick 1999). Why aren't large-celled species competitively displaced by small-celled species that are more effective nutrient competitors? First, a large cell size likely offsets the advantage of small-celled phytoplankton by affording an increased protection from grazing (Smetacek *et al*. 2004). Second, large size may be directly advantageous under fluctuating nutrient regimes: large-celled species, especially diatoms, have disproportionately large vacuoles that allow nutrient, nitrate in particular, to be stored until the next nutrient pulse while making it unavailable for smaller-sized species (Grover 1991c; Stolte and Riegman 1996). Third, large diatoms in oligotrophic gyres can migrate vertically: downward to reach nutrients at depths and upward to photosynthesize (Villareal *et al*. 1993, 1999), and the migration rate is positively correlated with cell size (Moore and Villareal 1996). This vertical migration can contribute significantly to new primary production in oligotrophic ocean (Villareal *et al*. 1999).

Cell size in different phytoplankton groups may evolve on long time scales due to selective pressures from changing physical environment such as sea level and the degree of turbulent mixing (Finkel *et al*. 2005 and Finkel this volume). Such long-term size changes may also significantly affect competitive abilities of phytoplankton on a geologic time scale.

F. Clonal Differences in Resource Utilization

There is an increasing body of evidence showing significant intraspecific genetic variation in phytoplankton, with different clones of the same species often having significantly different physiological parameters of growth, nutrient utilization, and, in case of toxic algae, toxin production (Medlin *et al*. 2000; Rynearson and Armbrust 2000, 2005; Carrillo *et al*. 2003). This genetic and physiological diversity within species allows adaptation to changing environments and, thus, species persistence. Genetic diversity in populations of phytoplankton is the basis for selection by biotic or abiotic selective forces and consequent rapid evolution. Grazing pressure on green alga *Chlorella vulgaris* selected for significantly more grazer-resistant cells (smaller cells) in just tens of generations (Yoshida *et al*. 2003). More grazer-resistant populations were also poorer competitors for nitrate, thus demonstrating a grazer-resistance–nutrient competitive ability trade-off (Yoshida *et al*. 2004). There is an intriguing indication that clonal diversity in phytoplankton may be maintained by algal viruses (Tarutani *et al*. 2000), and thus viruses may increase the ability of phytoplankton to adapt to changing environments.

V. FUTURE PHYTOPLANKTON COMMUNITIES

Phytoplankton clearly differ in their resource requirements, utilization strategies, and competitive abilities (e.g., Dugdale 1967; Margalef 1978; Sommer 1981, 1989; Reynolds 1984; Smayda 1997; Litchman *et al*. 2006). These differences underlie the immense diversity and both the short-term and long-term compositional changes of phytoplankton communities in nature. Predicting reorganization of phytoplankton communities under different global change scenarios is critical, as phytoplankton significantly impact higher trophic levels and elemental cycles, yet the task remains challenging. Recent models of phytoplankton communities have started incorporating multiple functional groups such as diatoms, coccolithophores, dinoflagellates, cyanobacteria, and green algae, as well as multiple nutrients (e.g., Bissett *et al.* 1999; Moore *et al.* 2002; Gregg *et al.* 2003; Litchman *et al.* 2006). Parameterization of such parameter-rich models is usually based on just a few laboratory studies of individual species that increases the uncertainty of the predictions (Anderson 2005). A better way of parameterizing multiple functional groups of phytoplankton in such models may be to compile numerous data that would account for variation among species within taxa, physiological status, and measurement methods. Knowing physiological requirements of major groups is essential in predicting future dynamics of these groups.

In a multigroup multinutrient model of phytoplankton that was parameterized based on extensive literature compilation, Litchman *et al.* (2006) attempted to model community reorganizations under different global change scenarios such as altered mixed layer depth dynamics and changes in nitrogen, phosphorus, and iron concentrations. The predicted communities differed in total biomass and relative abundances of major functional groups, diatoms, coccolithophores, and prasinophytes. For example, an increased iron concentration is predicted to stimulate growth of diatoms and prasinophytes at the expense of coccolithophores because higher iron concentrations alleviate iron competition and diminish coccolithophores' competitive advantage (coccolithophores appear to be good iron competitors; Sunda and Huntsman 1995; Maldonado *et al.* 2001). Prolonged stratification and shallower mixing depths are predicted to result in more severe nutrient limitation and higher irradiance and would consequently increase densities of coccolithophores, as they appear to be adapted to low nutrient concentrations and high irradiances (Eppley *et al.* 1969; Nanninga and Tyrrell 1996; Litchman *et al.* 2006). However, other factors not included in the model may also be important: increased CO_2 will likely lower the pH of the ocean, and this may have a drastic effect on calcifying organisms, including coccolithophores (Orr *et al.* 2005). The rates of calcification and growth of the coccolithophore *E. huxleyi* decrease under high pCO_2 and may shift community dominance to noncalcifying phytoplankton (Engel *et al.* 2005).

The attempts to predict composition and dynamics of future phytoplankton communities are in their infancy and the consideration of phytoplankton resource competition along with other ecological interactions is crucial for the success of such attempts.

VI. CHALLENGES AND FUTURE DIRECTIONS

Resource competition of phytoplankton is an area of ecology that has made significant advances over the last few decades and posed as well as answered many fundamental questions of ecology, such as the problem of competitive exclusion and species coexistence formulated in the Hutchinson's paradox of the plankton (Hutchinson 1961) and the concept of R^* as a measure of resource competitive ability (Tilman 1977, 1982; Tilman *et al.* 1982). Here, I briefly outline some challenges and promising directions in the area of phytoplankton resource competition that could lead to future advances in phytoplankton ecology and perhaps ecology in general.

A. Dynamic Regulation of Resource Utilization and Competitive Ability

Models of resource competition in phytoplankton generally assume constant parameter values for each species (e.g., Grover 1991c; Huisman and Weissing 1994; Ducobu *et al.* 1998). However, numerous experimental studies show that various parameters that determine competitive ability, such as nutrient-dependent uptake and growth parameters (e.g., see Equation 3) depend on the growth conditions and hence are not simply a number but a distribution around a mean value. For example, the maximum rate of nutrient uptake, V_{max}, can change several-fold depending on the degree of nutrient limitation or irradiance (see Harrison *et al.* 1989); minimum nutrient quota, Q_{min}, for nitrogen also depends on irradiance levels (Rhee and Gotham 1981). Often, these dependencies not included in the model do not affect the dynamics, and outcome of competition and models with constant coefficients are sufficient for predicting competitive interactions (Tilman 1977; Huisman *et al.* 1999a). There are, however, cases in which simple models seem inadequate, fluctuating conditions may bring up physiological responses that are significant enough to cause a disagreement between models and experiments (e.g., Grover 1991a, b, c, 1992; Litchman 2003). The mechanisms of dynamic regulation of resource acquisition and utilization in phytoplankton are beginning to be addressed through genomic and metabolomic approaches (Armbrust *et al.* 2004; Allen 2005; Parker and Armbrust 2005). The up- or down-regulation of nutrient uptake transporters and photosynthetic machinery occurs rapidly under varying degree of resource limitation (e.g., Hildebrand and Dahlin 2000; Song and Ward 2004; Allen *et al.* 2005; Hildebrand 2005) and may alter species' competitive ability (Klausmeier *et al.* 2007). Understanding the dynamics of acclimation of phytoplankton to changing resource fields and determining how such acclimation may influence resource competitive ability and outcome of competition poses a new challenge and would improve our ability to predict phytoplankton dynamics in general.

B. Resource Interaction

Most models of resource competition often assume no interaction among resources and describe growth rate of phytoplankton as a minimum function of all potentially limiting resources (Liebig's law of the minimum or, more precisely referred to as Blackman limitation, if rates are considered, see Cullen 1991). The parameters of growth and uptake for each resource are assumed independent of other resources in the models. In reality, the physiology of relevant processes is more complex. Phytoplankton ecologists have long known that uptake of a nutrient depends on light and other nutrients, often in a species-specific way (e.g., Wheeler *et al.* 1983; Kudela and Cochlan 2000). The uptake of nitrate is inhibited by ammonium, and such inhibition may significantly alter nitrate utilization (Syrett 1981; Dortch 1990). The uptake rate of nitrate, phosphate, and other nutrients also depends on irradiance, and, because this dependence is species-specific, variations in light regime may mediate nutrient competition (Ahn *et al.* 2002; Litchman *et al.* 2004). Do we have to take into account such relationships among resources to adequately model competitive interactions? A model that included light-ammonium-nitrate interaction was used to successfully predict nitrate dynamics in a high-nutrient low-chlorophyll (HNLC) ocean, which suggests that such models may be more powerful in cases of resource colimitation (Armstrong 1999).

C. Evolution of Competitive Ability

Along with the short-term flexibility and plasticity of phytoplankton's ability to acquire and compete for resources, phytoplankton competitive ability likely evolves in response to changes in the environment, due to both biotic and abiotic selective forces. Because of their large population numbers, short generation times and clonal reproduction, phytoplankton can evolve on relatively fast time scales (Lynch *et al.* 1991; Yoshida *et al.* 2003, 2004). Mutants or clones with better competitive ability due to variation in ecophysiological parameters will be selected under given environmental conditions.

However, because of the trade-offs among various parameters as well as among competitive abilities for different resources and other fitness components such as grazer resistance, the evolution is likely to be constrained so that the emergence of a superspecies is improbable. Rapid evolution in competitive ability can alter the dynamics and outcome of resource competition and food web interactions (Yoshida *et al.* 2003, 2004). Very little is known of how fast competitive ability can evolve, and whether the rate of evolutionary change in competitive ability differs among taxonomic groups or for different resources. Laboratory experiments with green alga *Chlamydomonas* growing under elevated CO_2 showed lack of evolutionary adaptation to increased CO_2 over 1000 generations (Collins and Bell 2004), whereas over comparable number of generations in *Chlorella* zooplankton grazing pressure led to a selection of significantly more grazer-resistant clones with poor nutrient competitive ability (Yoshida *et al.* 2004). The experimental investigation of evolution of competitive ability in phytoplankton is a promising venue of research, especially as environmental conditions change rapidly due to anthropogenic perturbations.

Existing models of resource competition in phytoplankton do not take into account possible evolution of competitive ability and the appearance of new phenotypes and, therefore, may not be capable of realistically describing the dynamics of competitive communities. A promising approach to modeling evolution in an ecological context that accounts for complex species interactions is the adaptive dynamics approach, where fitness is a function of frequencies of interacting genotypes (Geritz *et al.* 1998; Waxman and Gavrilets 2005). This approach extends the methods of evolutionary game theory and can be used to investigate what new competitive strategies and evolutionary lineages may emerge under different environmental conditions (Doebeli and Dieckmann 2005; Kisdi and Gyllenberg 2005; Waxman and Gavrilets 2005).

D. Phylogenetic Relationships

Experimental data indicate that phytoplankton from major taxonomic groups differ in their competitive abilities for various resources (nutrients and light). Differences in competitive abilities and strategies among taxonomic groups may, in part, be explained by the environmental conditions at the time of their origin (Falkowski *et al.* 2004; Katz *et al.* 2004; Tozzi *et al.* 2004; Litchman *et al.* submitted). At the same time, molecular and physiological data indicate that within major taxa, species from marine versus freshwater environments differ in their strategies of nutrient utilization (Song and Ward 2004; Litchman *et al.* submitted). For example, nitrogen reductase transcription in marine green algae is regulated differently compared to freshwater green algae, which may be due to differences in the degree of limitation by nitrate in marine versus freshwater systems (Song and Ward 2004). Even genetically close clones inhabiting different depths may differ in their nutrient or light physiologies and competitive abilities (Moore *et al.* 1998; Rocap *et al.* 2003; Stomp *et al.* 2004). Understanding the influence of phylogenetic relationships versus ecological factors on phytoplankton competitive abilities poses exciting new challenges that require a synergism of genetic, physiological, and ecological approaches.

E. Concluding Remarks

Resource competition is a fundamental ecological process that structures phytoplankton communities. Significant theoretical and experimental advances have been made in the field of phytoplankton resource competition in the past several decades. However, we are just beginning to understand how differences in competitive strategies among major functional groups, as well as individual species, contribute to phytoplankton diversity and community shifts in nature. Diverse aspects of resource competition, such as mechanistic trade-offs in competitive abilities, as well as their flexibility and evolution, and competition under variable conditions needs to be considered if we are to understand and/or predict the structure and function of the past, present, and future of phytoplankton communities.

Acknowledgments

The work on this chapter was in part supported by the NSF grants OCE-0084032, DEB0445265 (0610531), and DEB-0610532. Comments by C.A. Klausmeier, P.G. Falkowski, and A.H. Knoll are appreciated. This is Kellogg Biological Station contribution no. 1424.

References

Ahn, C.-Y., Chung, A.-S., and Oh, H.-M. (2002). Diel rhythm of algal phosphate uptake rates in P-limited cyclostats and simulation of its effect on growth and competition. *J. Phycol.* **38:** 695–704.

Allen, A.E. (2005). Beyond sequence homology: redundant ammonium transporters in a marine diatom are not functionally equivalent. *J. Phycol.* **41:** 4–6.

Allen, A.E., Ward, B.B., and Song, B.K. (2005). Characterization of diatom (Bacillariophyceae) nitrate reductase genes and their detection in marine phytoplankton communities. *J. Phycol.* **41:** 95–104.

Aksnes, D.L., and Egge, J.K. (1991). A theoretical model for nutrient uptake in phytoplankton. *Marine Ecology Progress Series* **70:** 65–72.

Anbar, A.D., and Knoll, A.H. (2002). Proterozoic ocean chemistry and evolution: a bioinorganic bridge? *Science* **297:** 1137–1142.

Anderson, T.R. (2005). Plankton functional type modeling: running before we can walk? *J. Plankton Res.* **27:** 1073–1081.

Armbrust, E.V., Berges, J.A., Bowler, C., and 42 others. (2004). The genome of the diatom *Thalassiosira pseudonana*: ecology, evolution, and metabolism. *Science* **306:** 79–86.

Armstrong, R.A. (1994). Grazing limitation and nutrient limitation in marine ecosystems—steady-state solutions of an ecosystem model with multiple food-chains. *Limnol. Oceanogr.* **39:** 597–608.

Armstrong, R.A. (1999). An optimization-based model of iron-light-ammonium colimitation of nitrate uptake and phytoplankton growth. *Limnol. Oceanogr.* **44:** 1436–1446.

Armstrong, R.A., and McGehee, R. (1976). Coexistence of two competitors on one resource. *J. Theor. Biol.* **56:** 499–502.

Armstrong, R.A., and McGehee, R. (1980). Competitive exclusion. *Am. Naturalist* **115:** 151–169.

Banse, K. (1976). Rates of growth, respiration and photosynthesis of unicellular algae as related to cell-size—review. *J. Phycol.* **12:** 135–140.

Bauer, S., Hitchcock, G.L., and Olson, D.B. (1991). Influence of monsoonally-forced Ekman dynamics upon surface-layer depth and plankton biomass distribution in the Arabian Sea. *Deep Sea Res. Part a Oceanogr. Res. Papers* **38:** 531–553.

Berman-Frank, I., Cullen, J.T., Shaked, Y., Sherrell, R.M., and Falkowski, P.G. (2001). Iron availability, cellular iron quotas, and nitrogen fixation in *Trichodesmium*. *Limnol. Oceanogr.* **46:** 1249–1260.

Bissett, W.P., Walsh, J.J., Dieterle, D.A., and Carder, K.L. (1999). Carbon cycling in the upper waters of the Sargasso Sea: I. Numerical simulation of differential carbon and nitrogen fluxes. *Deep Sea Res. Part I Oceanogr. Res. Papers* **46:** 205–269.

Bohannan, B.J., Jessup, M.C., Kerr, B., and Sandvik, G. (2002). Trade-offs and coexistence in microbial microcosms. *Antonie van Leeuwenhoek* **81:** 107–115.

Boyd, P., and Harrison, P.J. (1999). Phytoplankton dynamics in the NE subarctic Pacific. *Deep Sea Res. II* **46:** 2405–2432.

Brand, L.E., Sunda, W.G., and Guillard, R.R.L. (1983). Limitation of marine phytoplankton reproduction rates by zinc, manganese and iron. *Limnol. Oceanogr.* **28:** 1182–1198.

Brasier, M.D., and Lindsay, J.F. (1998). A billion years of environmental stability and the emergence of eukaryotes: new data from northern Australia. *Geology* **26:** 555–558.

Britton, N.F., and Timm, U. (1993). Effects of competition and shading in planktonic communities. *J. Math. Biol.* **31:** 655–673.

Caperon, J. (1968). Population growth response of *Isochrysis galbana* to nitrate variation at limiting concentrations. *Ecology* **49:** 866–872.

Carrillo, E., Ferrero, L.M., Alonso-Andicoberry, C., Basanta, A., Martin, A., Lopez-Rodas, V., and Costas, E. (2003). Interstrain variability in toxin production in populations of the cyanobacterium *Microcystis aeruginosa* from water-supply reservoirs of Andalusia and lagoons of Donana National Park (southern Spain). *Phycologia* **42:** 269–274.

Cavender-Bares, K.K., Karl, D.M., and Chisholm, S.W. (2001). Nutrient gradients in the western North Atlantic Ocean: Relationship to microbial community structure and comparison to patterns in the Pacific Ocean. *Deep Sea Res. I* **48:** 2373–2395.

Chisholm, S.W. (1992). Phytoplankton size. In *Primary Productivity and Biogeochemical Cycles in the Sea*. P.G. Falkowski and A.D. Woodhead, eds. New York, Plenum Press, pp. 213–237.

Clark, D.R., Flynn, K.J., and Owens, N.J.P. (2002). The large capacity for dark nitrate-assimilation in diatoms may overcome nitrate limitation of growth. *New Phytol.* **155:** 101–108.

Collins, S., and Bell, G. (2004). Phenotypic consequences of 1,000 generations of selection at elevated CO_2 in a green alga. *Nature* **431:** 566–569.

Cullen, J.J. (1991). Hypotheses to explain high-nutrient conditions in the sea. *Limnol. Oceanogr.* **36:** 1578–1599.

Denman, K.L., and Gargett, A.E. (1983). Time and space scales of vertical mixing and advection of phytoplankton in the upper ocean. *Limnol. Oceanogr.* **28:** 801–815.

deVargas, C., Aubry, M-P., Probert, I., and Young, J. (2007). Origin and evolution of coccolithophores: From coastal hunters to oceanic farmers. *Evolution of Primary Producers in the Sea.* P.G. Falkowski and A.H. Knoll, eds. Boston, Elsevier, pp. 251–285.

Diehl, S. (2002). Phytoplankton, light, and nutrients in a gradient of mixing depths: theory. *Ecology* **83:** 386–398.

Doebeli, M., and Dieckmann, U. (2005). Adaptive dynamics as a mathematical tool for studying the ecology of speciation processes. *J. Evol. Biol.* **18:** 1194–1200.

Dortch, Q. (1990). Review of the interaction between ammonium and nitrate uptake in phytoplankton. *Mar. Ecol. Prog. Ser.* **61:** 183–201.

Droop, M.R. (1968). Vitamin B12 and marine ecology, IV. The kinetics of uptake, growth and inhibition in *Monochrysis lutheri. J. Mar. Biol. Assoc. U K* **48:** 689–733.

Droop, M.R. (1973). Some thoughts on nutrient limitation in algae. *J. Phycol.* **9:** 264–272.

Ducobu, H., Huisman, J., Jonker, R.R., and Mur, L.R. (1998). Competition between a prochlorophyte and a cyanobacterium under various phosphorus regimes: comparison with the Droop model. *J. Phycol.* **34:** 467–476.

Dugdale, R.C. (1967). Nutrient limitation in sea—dynamics, identification and significance. *Limnol. Oceanogr.* **12:** 685–695.

Dugdale, R.C., and Goering, J.J. (1967). Uptake of new and regenerated forms of nitrogen in primary productivity. *Limnol. Oceanogr.* **12:** 196–206.

Engel, A., Zondervan, I., Aerts, K., and 15 others. (2005). Testing the direct effect of CO_2 concentration on a bloom of the coccolithophorid *Emiliania huxleyi* in mesocosm experiments. *Limnol. Oceanogr.* **50:** 493–507.

Eppley, R.W., Rogers, J.N., and McCarthy, J.J. (1969). Half-saturation constants for uptake of nitrate and ammonium by marine phytoplankton. *Limnol. Oceanogr.* **14:** 912–920.

Falkowski, P.G. (1980). Light-shade adaptation in marine phytoplankton. *Primary Productivity in the Sea.* P.G. Falkowski, ed. New York, Plenum Press, pp. 99–119.

Falkowski, P.G. (1984). Physiological-responses of phytoplankton to natural light regimes. *J. Plankton Res.* **6:** 295–307.

Falkowski, P.G., Katz, M.E., Knoll, A.H., Quigg, A., Raven, J.A., Schofield, and Taylor, F.J.R. (2004). The evolution of modern eukaryotic phytoplankton. *Science* **305:** 354–360.

Falkowski, P.G., and Wirick, C.D. (1981). A simulation-model of the effects of vertical mixing on primary productivity. *Mar. Biol.* **65:** 69–75.

Falkowski, P.G., Ziemann, D., Kolber, Z., and Bienfang, P.K. (1991). Role of eddy pumping in enhancing primary production in the ocean. *Nature* **352:** 55–58.

Ferris, M., and Christian, R. (1991). Aquatic primary production in relation to microalgal responses to changing light: a review. *Aquat. Sci.* **53:** 187–217.

Field, C., Behrenfeld, M., Randerson, J., and Falkowski, P. (1998). Primary production of the biosphere: integrating terrestrial and oceanic components. *Science* **281:** 237–240.

Finkel, Z.V. (2001). Light absorption and size scaling of light-limited metabolism in marine diatoms. *Limnol. Oceanogr.* **46:** 86–94.

Finkel, Z.V. (2007). Does phytoplankton cell size matter? The evolution of modern marine food webs. *Evolution of Primary Producers in the Sea.* P.G. Falkowski and A.H. Knoll, eds. Boston, Elsevier, pp. 333–350.

Finkel, Z.V., Katz, M.E., Wright, J.D., Schofield, O.M.E., and Falkowski, P.G. (2005). Climatically driven macroevolutionary patterns in the size of marine diatoms over the Cenozoic. *Proc. Natl. Acad. Sci. U S A* **102:** 8927–8932.

Garcon, V.C., Oschlies, A., Doney, S.C., McGillicuddy, D., and Waniek, J. (2001). The role of mesoscale variability on plankton dynamics in the North Atlantic. *Deep Sea Res. Part II Top. Studies Oceanogr.* **48:** 2199–2226.

Gause, G.F. (1934). *The Struggle for Existence.* Baltimore, Williams and Williams.

Geider, R.J., and Osborne, B.A. (1989). Respiration and microalgal growth—a review of the quantitative relationship between dark respiration and growth. *New Phytol.* **112:** 327–341.

Geritz, S.A.H., Kisdi, E., Meszena, G., and Metz, J.A.J. (1998). Evolutionarily singular strategies and the adaptive growth and branching of the evolutionary tree. *Evol. Ecol.* **12:** 35–57.

Gin, K.Y.H., Chisholm, S.W., and Olson, R.J. (1999). Seasonal and depth variation in microbial size spectra at the Bermuda Atlantic time series station. *Deep Sea Res Part I Oceanogr. Res. Papers* **46:** 1221–1245.

Gregg, W.W., Ginoux, P., Schopf, P.S., and Casey, N.W. (2003). Phytoplankton and iron: validation of a global three-dimensional ocean biogeochemical model. *Deep Sea Res. Part II Top. Studies Oceanogr.* **50:** 3143–3169.

Grover, J.P. (1991a). Dynamics of competition among microalgae in variable environments: experimental tests of alternative models. *Oikos* **62:** 231–243.

Grover, J.P. (1991b). Non-steady state dynamics of algal population growth: experiments with two chlorophytes. *J. Phycol.* **27:** 70–79.

Grover, J.P. (1991c). Resource competition in a variable environment: phytoplankton growing according to the variable-internal-stores model. *Am. Naturalist* **138:** 811–835.

Grover, J.P. (1992). Constant-and variable yield models of population growth: responses to environmental variability and implications for competition *J. Theor. Biol.* **158:** 409–428.

Grover, J.P. (1995). Competition, herbivory, and enrichment-nutrient-based models for edible and inedible plants. *Am. Naturalist* **145:** 746–774.

Grover, J.P. (1997). *Resource Competition.* London, Chapman and Hall.

Grover, J.P., and Holt, R.D. (1998). Disentangling resource and apparent competition: realistic models for plant-herbivore communities. *J. Theor. Biol.* **191:** 353–376.

Haidar, A.T., and Thierstein, H.R. (2001). Coccolithophore dynamics off Bermuda (N. Atlantic). *Deep-Sea Res. II* **48:** 1925–1956.

Hardin, G. (1960). The competitive exclusion principle. *Science* **131:** 1292–1298.

Harrison, P.J., Parslow, J.S., and Conway, H.L. (1989). Determination of nutrient uptake kinetic parameters: a comparison of methods. *Mar. Ecol. Prog. Ser.* **52:** 301–312.

Healey, F.P. (1980). Physiological indicators of nutrient deficiency in lake phytoplankton. *Can. J. Fish. Aquat. Sci.* **37:** 442–453.

Hedges, S.B. (2002). The origin and evolution of model organisms. *Nat. Genet.* **3:** 838–849.

Hedges, S.B., Blair, J.E., Venturi, M.L., and Shoe, J.L. (2004). A molecular timescale of eukaryote evolution and the rise of complex multicellular life. *BMC Evol. Biol.* **4:** 2.

Herrero, A., Muro-Pastor, A.M., and Flores, E. (2001). Nitrogen control in cyanobacteria. *J. Bacteriol.* **183:** 411–425.

Hildebrand, M. (2005). Cloning and functional characterization of ammonium transporters from the marine diatom *Cylindrotheca fusiformis* (Bacillariophyceae). *J. Phycol.* **41:** 105–113.

Hildebrand, M., and Dahlin, K. (2000). Nitrate transporter genes from the diatom *Cylindrotheca fusiformis* (Bacillariophyceae): mRNA levels controlled by nitrogen source and by the cell cycle. *J. Phycol.* **36:** 702–713.

Holligan, P.M., Pingree, R.D., and Mardell, G.T. (1985). Oceanic solitons, nutrient pulses and phytoplankton growth. *Nature* **314:** 348–350.

Hsu, S.B. (1980). A competition model for a seasonally fluctuating nutrient. *J. Math. Biol.* **9:** 115–132.

Huisman, J., Arrayas, M., Ebert, U., and Sommeijer, B. (2002). How do sinking phytoplankton species manage to persist? *Am. Nat.* **159:** 245–254.

Huisman, J., Jonker, R.R., Zonneveld, C., and Weissing, F.J. (1999a). Competition for light between phytoplankton species: experimental tests of mechanistic theory. *Ecology* **80:** 211–222.

Huisman, J., Oostveen, P.V., and Weissing, F.J. (1999b). Species dynamics in phytoplankton blooms: incomplete mixing and competition for light. *Am. Nat.* **154:** 46–68.

Huisman, J., Thi, N.N.P., Karl, D.M., and Sommeijer, B. (2006). Reduced mixing generates oscillations and chaos in the oceanic deep chlorophyll maximum. *Nature* **439:** 322–325.

Huisman, J., and Weissing, F.J. (1994). Light-limited growth and competition for light in well-mixed aquatic environments—an elementary model. *Ecology* **75:** 507–520.

Huisman, J., and Weissing, F.J. (1995). Competition for nutrients and light in a mixed water column—a theoretical-analysis. *Am. Naturalist* **146:** 536–564.

Hulburt, E.M. (1967). A note on regional differences in phytoplankton during a crossing of the southern North Atlantic Ocean in January, 1967. *Deep Sea Res.* **14:** 685–690.

Hutchinson, G.E. (1961). The paradox of the plankton. *Am. Naturalist* **95:** 137–145.

Hutchinson, G.E. (1967). *A Treatise on Limnology: Introduction to Lake Biology and the Limnoplankton.* New York, Wiley.

Iglesias-Rodriguez, M.D., Brown, C.W., Doney, S.C., Kleypas, J.A., Kolber, D., Kolber, Z., Hayes, P.K., and Falkowski, P.G. (2002a). Representing key phytoplankton functional groups in ocean carbon cycle models: coccolithophorids. *Global Biogeochemical Cycles* **16:** 47(1)–47(20).

Iglesias-Rodriguez, M.D., Saez, A.G., Groben, R., Edwards, K.J., Batley, J., Medlin, L.K., and Hayes, P.K. (2002b). Polymorphic microsatellite loci in global populations of the marine coccolithophorid *Emiliania huxleyi. Mol. Ecol. Notes* **2:** 495–497.

Interlandi, S.J., and Kilham, S.S. (2001). Limiting resources and the regulation of diversity in phytoplankton communities. *Ecology* **82:** 1270–1282.

Katz, M.E., Finkel, Z.V., Grzebyk, D., Knoll, A.H., and Falkowski, P.G. (2004). Evolutionary trajectories and biogeochemical impacts of marine eukaryotic phytoplankton. *Annu. Rev. Ecol. Evol. Syst.* **35:** 523–556.

Kilham, P., and Kilham, S.S. (1980). The evolutionary ecology of phytoplankton. *The Physiological Ecology of Phytoplankton.* I. Morris, ed. Berkeley, The University of California Press, pp. 571–597.

Kisdi, E., and Gyllenberg, M. (2005). Adaptive dynamics and the paradigm of diversity. *J. Evol. Biol.* **18:** 1170–1173.

Klausmeier, C.A., and Litchman, E. (2001). Algal games: the vertical distribution of phytoplankton in poorly mixed water columns. *Limnol. Oceanogr.* **46:** 1998–2007.

Klausmeier, C.A., Litchman, E., Daufresne, T., and Levin, S.A. (2004). Optimal nitrogen-to-phosphorus stoichiometry of phytoplankton. *Nature* **429:** 171–174.

Klausmeier, C.A., Litchman, E., and Levin, S.A. (2007). A model of flexible uptake of two essential resources. *J. Theor. Biol.* **246:** 278–289.

Kooijman, S., Andersen, T., and Kooi, B.W. (2004). Dynamic energy budget representations of stoichiometric constraints on population dynamics. *Ecology* **85:** 1230–1243.

Kudela, R.M., and Cochlan, W.P. (2000). Nitrogen and carbon uptake kinetics and the influence of irradiance for a red tide bloom off southern California. *Aquat. Microbial Ecol.* **21**: 31–47.

Kustka, A., Carpenter, E.J., and Sanudo-Wilhelmy, S.A. (2002). Iron and marine nitrogen fixation: progress and future directions. *Res. Microbiol.* **153**: 255–262.

Lande, R., and Lewis, M.R. (1989). Models of photoadaptation and photosynthesis by algal cells in a turbulent mixed layer. *Deep Sea Res.* **36**: 1161–1175.

Leibold, M.A. (1996). A graphical model of keystone predators in food webs: trophic regulation of abundance, incidence, and diversity patterns in communities. *Am. Naturalist* **147**: 784–812.

Leibold, M.A. (1997). Do nutrient-competition models predict nutrient availabilities in limnetic ecosystems? *Oecologia* **110**: 132–142.

Levins, R. (1979). Coexistence in a variable environment. *Am. Naturalist* **114**: 765–783.

Li, W.K.W. (2002). Macroecological patterns of phytoplankton in the northwestern North Atlantic Ocean. *Nature* **419**: 154–157.

Litchman, E. (1998). Population and community responses of phytoplankton to fluctuating light. *Oecologia* **117**: 247–257.

Litchman, E. (2000). Growth rates of phytoplankton under fluctuating light. *Freshwater Biol.* **44**: 223–235.

Litchman, E. (2003). Competition and coexistence of phytoplankton under fluctuating light: experiments with two cyanobacteria. *Aquat. Microbial Ecol.* **31**: 241–248.

Litchman, E., and Klausmeier, C.A. (2001). Competition of phytoplankton under fluctuating light. *Am. Naturalist* **157**: 170–187.

Litchman, E., Klausmeier, C.A., and Bossard, P. (2004). Phytoplankton nutrient competition under dynamic light regimes. *Limnol. Oceanogr.* **49**: 1457–1462.

Litchman, E., Klausmeier, C.A., Miller, J.R., Schofield, O.M., and Falkowski, P.G. (2006). Multi-nutrient, multi-group model of present and future oceanic phytoplankton communities. *Biogeosci. Discuss.* **3**: 585–606.

Litchman, E., Klausmeier, C.A., Miller, J.R., Schofield, O.M., and Falkowski, P.G. (In press). Multi-nutrient, multi-group model of present and future oceanic phytoplankton communities. *Biogeosciences.*

Litchman, E., Schofield, O.M., and Falkowski, P.G., Submitted. The role of functional traits and trade-offs in structuring phytoplankton communities: Scaling from cellular to ecosystem level.

Lynch, M., Gabriel, W., and Wood, A.M. (1991). Adaptive and demographic responses of plankton populations to environmental change. *Limnol. Oceanogr.* **36**: 1301–1312.

Macisaac, J.J., and Dugdale, R.C. (1972). Interactions of light and inorganic nitrogen in controlling nitrogen uptake in sea. *Deep Sea Res.* **19**: 209–232.

Maldonado, M.T., Boyd, P.W., LaRoche, J., Strzepek, R., Waite, A., Bawie, A.R., Croot, P.L., Frew, R.D., and Price, N.M. (2001). Iron uptake and physiological response of phytoplankton during a mesoscale Southern Ocean iron enrichment. *Limnology and Oceanography,* **46**, 1802–1808.

Maldonado, M.T., and Price, N.M. (1996). Influence of N substrate on Fe requirements of marine centric diatoms. *Mar. Ecol. Prog. Ser.* **141**: 161–172.

Mallin, M.A., and Paerl, H.W. (1992). Effects of variable irradiance on phytoplankton productivity in shallow estuaries. *Limnol. Oceanogr.* **37**: 54–62.

Margalef, R. (1978). Life forms of phytoplankton as survival alternatives in an unstable environment. *Oceanol. Acta* **1**: 493–509.

Marra, J. (1978). Effect of short-term variations intensity on photosynthesis of a marine phytoplankter: a laboratory simulation study. *Mar. Biol.* **46**: 191–202.

Martin, A.P., Richards, K.J., Bracco, A., and Provenzale, A. (2002). Patchy productivity in the open ocean. *Global Biogeochemical Cycles,* **16**. Art. No. 1025.

McGillicuddy, D.J.J., Robinson, A.R., Siegel, D.A., Jannasch, H.W., Johnson, R., Dickey, T.D., McNeil, J., Michaels, A.F., and Knap, A.H. (1998). Influence of mesoscale eddies on new production in the Sargasso Sea. *Nature* **394**: 263–266.

Medlin, L.K., Lange, M., and Nothig, E.M. (2000). Genetic diversity in the marine phytoplankton: a review and a consideration of Antarctic phytoplankton. *Antarctic Sci.* **12**: 325–333.

Moore, J.K., Doney, S.C., Kleypas, J.A., Glover, D.M., and Fung, I.Y. (2002). An intermediate complexity marine ecosystem model for the global domain. *Deep Sea Res. II* **49**: 403–462.

Moore, J.K., and Villareal, T.A. (1996). Size-ascent rate relationships in positively buoyant marine diatoms. *Limnol. Oceanogr.* **41**: 1514–1520.

Moore, L.R., Rocap, G., and Chisholm, S.W. (1998). Physiology and molecular phylogeny of coexisting Prochlorococcus ecotypes. *Nature* **393**: 464–467.

Munk, W.H., and Riley, G.A. (1952). Absorption of nutrients by aquatic plants. *J. Mar. Res.* **11**: 215–240.

Nanninga, H.J., and Tyrrell, T. (1996). Importance of light for the formation of algal blooms by *Emiliania huxleyi. Mar. Ecol. Prog. Ser.* **136**: 195–203.

Orr, J.C., Fabry, V.J., Aumont, O., and 24 others. (2005). Anthropogenic ocean acidification over the twenty-first century and its impact on calcifying organisms. *Nature* **437**: 681–686.

Palliani, R.B., Mattioli, E., and Riding, J.B. (2002). The response of marine phytoplankton and sedimentary organic matter to the early Toarcian (Lower Jurassic) oceanic anoxic event in northern England. *Mar Micropaleontol.* **46**: 223–245.

Palliani, R.B., and Riding, J.B. (1999). Relationship between the early Toarcian anoxic event and organic-walled phytoplankton in central Italy. *Mar. Micropaleontol.* **37**: 101–116.

Parker, M.S., and Armbrust, E.V. (2005). Synergistic effects of light, temperature, and nitrogen source on transcription of genes for carbon and nitrogen metabolism in the centric diatom *Thalassiosira pseudonana* (Bacillariophyceae). *J. Phycol.* **41**: 1142–1153.

Pasciak, W.J., and Gavis, J. (1974). Transport limitation of nutrient uptake in phytoplankton. *Limnol. Oceanogr.* **19**: 881–889.

Peers, G., Quesnel, S.A., and Price, N.M. (2005). Copper requirements for iron acquisition and growth of coastal and oceanic diatoms. *Limnol. Oceanogr.* **50**: 1149–1158.

Pienkos, P.T., Bodmer, S., and Tabita, F.R. (1983). Oxygen inactivation and recovery of nitrogenase activity in cyanobacteria. *J. Bacteriol.* **153**: 182–190.

Quigg, A., Finkel, Z.V., Irwin, A.J., Rosenthal, Y., Ho, T.Y., Reinfelder, J.R., Schofield, O., Morel, F.M.M., and Falkowski, P.G. (2003). The evolutionary inheritance of elemental stoichiometry in marine phytoplankton. *Nature* **425**: 291–294.

Raven, J.A. (1980). Nutrient transport in microalgae. *Adv. Microbial Physiol.* **21**: 47226.

Raven, J.A. (1987). The role of vacuoles. *New Phytol.* **106**: 357–422.

Raven, J.A. (1997). Phagotrophy in autotrophs. *Limnol. Oceanogr.* **42**: 198–205.

Reay, D.S., Nedwell, D.B., Priddle, J., and Ellis-Evans, J.C. (1999). Temperature dependence of inorganic nitrogen uptake: reduced affinity for nitrate at suboptimal temperatures in both algae and bacteria. *Appl. Environ. Microbiol.* **65**: 2577–2584.

Reynolds, C.S. (1984). *The Ecology of Freshwater Phytoplankton*. Cambridge, Cambridge University Press.

Reynolds, C.S. (1997). *Vegetation Processes in the Pelagic: A Model for Ecosystem Theory*. Oldendorf, Inter-Research.

Rhee, G.Y., and Gotham, I.J. (1981). The effect of environmental-factors on phytoplankton growth—light and the interactions of light with nitrate limitation. *Limnol. Oceanogr.* **26**: 649–659.

Richardson, K., Beardall, J., and Raven, J.A. (1983). Adaptation of unicellular algae to irradiance: an analysis of strategies. *New Phytol.* **93**: 157–191.

Riebesell, U., and Wolf-Gladrow, D.A. (2002). Supply and uptake of inorganic nutrients. *Phytoplankton Productivity: Carbon Assimilation in Marine and Freshwater Ecosystems*. P.J.L.B. Williams, D.N. Thomas, and C.S. Reynolds, eds. Malden, Blackwell Science, pp. 109–140.

Riegman, R., Stolte, W., Noordeloos, A.A.M., and Slezak, D. (2000). Nutrient uptake and alkaline phosphatase (EC 3:1:3:1) activity of *Emiliania huxleyi* (Prymnesiophyceae) during growth under N and P limitation in continuous cultures. *J. Phycol.* **36**: 87–96.

Riley, G.A., Stommel, H., and Bumpus, D.F. (1949). Quantitative ecology of the plankton of the western North Atlantic. Bull. Bingham Oceanogr. Collection, Vol. 12, article 3, Yale University.

Rocap, G., Larimer, F.W., Lamerdin, J., and 21 others. (2003). Genome divergence in two *Prochlorococcus* ecotypes reflects oceanic niche differentiation. *Nature* **424**: 1042–1047.

Rynearson, T.A., and Armbrust, E.V. (2000). DNA fingerprinting reveals extensive genetic diversity in a field population of the centric diatom *Ditylum brightwellii*. *Limnol. Oceanogr.* **45**: 1329–1340.

Rynearson, T.A., and Armbrust, E.V. (2005). Maintenance of clonal diversity during a spring bloom of the centric diatom *Ditylum brightwellii*. *Mol. Ecol.* **14**: 1631–1640.

Ryther, J.H. (1956). Photosynthesis in the ocean as a function of light intensity. *Limnol. Oceanogr.* **1**: 61–70.

Sakamoto, C.M., Karl, D.M., Jannasch, H.W., Bidigare, R.R. Letelier, R.M., Walz, P.M., Ryan, J.P., Polito, P.S., and Johnson, K.S. (2004). Influence of Rossby waves on nutrient dynamics and the plankton community structure in the North Pacific subtropical gyre. *J. Geophys. Res. Oceans* **109**.

Schiebel, R., Zeltner, A., Treppke, U.F., Waniek, J.J., Bollmann, J., Rixen, T., and Hemleben, C. (2004). Distribution of diatoms, coccolithophores and planktic foraminifers along atrophic gradient during SW monsoon in the Arabian Sea. *Mar. Micropaleontol.* **51**: 345–371.

Shigesada, N., and Okubo, A. (1981). Analysis of the self-shading effect on algal vertical-distribution in natural-waters. *J. Math. Biol.* **12**: 311–326.

Shuter, B.J. (1978). Size dependence of phosphorus and nitrogen subsistence quotas in unicellular microorganisms. *Limnol. Oceanogr.* **23**: 1248–1255.

Sicko-Goad, L.M., Schelske, C.L., and Stoermer, E.F. (1984). Estimation of intracellular carbon and silica content of diatoms from natural assemblages using morphometric techniques. *Limnol. Oceanogr.* **29**: 1170–1178.

Smayda, T.J. (1997). Harmful algal blooms: their ecophysiology and general relevance to phytoplankton blooms in the sea. *Limnol. Oceanogr.* **42**: 1137–1153.

Smetacek, V., Assmy, P., and Henjes, J. (2004). The role of grazing in structuring Southern Ocean pelagic ecosystems and biogeochemical cycles. *Antarctic Sci.* **16**: 541–558.

Smith, R.E.H., and Kalff, J. (1982). Size-dependent phosphorus uptake kinetics and cell quota in phytoplankton. *J. Phycol.* **18**: 275–284.

Sommer, U. (1981). The role of r-and K-selection in the succession of phytoplankton in Lake Constance. *Acta Oecologica* **2**: 327–342.

Sommer, U. (1984). The paradox of the plankton: Fluctuations of phosphorus availability maintain diversity of phytoplankton in flow-through cultures. *Limnol. Oceanogr.* **29**: 633–636.

Sommer, U. (1989). The role of competition for resources in phytoplankton succession. *Plankton Ecology: Succession in Plankton Communities*. U. Sommer, ed. New York, Springer, pp. 57–106.

Song, B.K., and Ward, B.B. (2004). Molecular characterization of the assimilatory nitrate reductase gene and its expression in the marine green alga *Dunaliella tertiolecta* (Chlorophyceae). *J. Phycol.* **40**: 721–731.

Steele, J.H. (1985). A comparison of terrestrial and marine ecological systems. *Nature* **313**: 355–358.

Sterner, R.W., and Elser, J.J. (2002). *Ecological Stoichiometry: The Biology of Elements from Molecules to the Biosphere*. Princeton, Princeton University Press.

Stolte, W., and Riegman, R. (1996). A model approach for size-selective competition of marine phytoplankton for fluctuating nitrate and ammonium. *J. Phycol.* **32:** 732–740.

Stomp, M., Huisman, J., de Jongh, F., Veraart, A.J., Gerla, D., Rijkeboer, M., Ibelings, B.W., Wollenzien, U.I.A., and Stal, L.J. (2004). Adaptive divergence in pigment composition promotes phytoplankton biodiversity. *Nature* **432:** 104–107.

Strzepek, R.F., and Harrison, P.J. (2004). Photosynthetic architecture differs in coastal and oceanic diatoms. *Nature* **431:** 689–692.

Stumm, W., and Morgan, J.J. (1981). *Aquatic Chemistry: An Introduction Emphasizing Chemical Equilibria in Natural Waters.* New York, Wiley and Sons.

Sunda, W.G., and Huntsman, S.A. (1995). Iron uptake and growth limitation in oceanic and coastal phytoplankton. *Mar. Chem.* **50:** 189–206.

Syrett, P.J. (1981). Nitrogen metabolism of microalgae. In *Physiological Bases of Phytoplankton Ecology.* T. Platt, ed. *Can. Bull. Fish Aquat. Sci.* Vol. 210, pp. 182–210.

Tarutani, K., Nagasaki, K., and Yamaguchi, M. (2000). Viral impacts on total abundance and clonal composition of the harmful bloom-forming phytoplankton *Heterosigma akashiwo. Appl. Environ. Microbiol.* **66:** 4916–4920.

Taylor, F.J.R., ed. (1987). *The Biology of Dinoflagellates.* Vol. 21, p. 785. Oxford, Blackwell Scientific Publications.

Thorpe, S.A. (2004a). Langmuir circulation. *Annu. Rev. Fluid Mechanics* **36:** 55–79.

Thorpe, S.A. (2004b). Recent developments in the study of ocean turbulence. *Annu. Rev. Earth Planetary Sci.* **32:** 91–109.

Tilman, D. (1977). Resource competition between planktonic algae: an experimental and theoretical approach. *Ecology* **58:** 338–348.

Tilman, D. (1982). *Resource Competition and Community Structure.* Princeton, NJ, Princeton University Press.

Tilman, D. (2000). Causes, consequences and ethics of biodiversity. *Nature* **405:** 208–211.

Tilman, D., Kilham, S.S., and Kilham, P. (1982). Phytoplankton community ecology. The role of limiting nutrients. *Ann. Rev. Ecol. Syst.* **13:** 349–372.

Tozzi, S., Schofield, O., and Falkowski, P. (2004). Historical climate change and ocean turbulence as selective agents for two key phytoplankton functional groups. *Mar. Ecol. Prog. Ser.* **274:** 123–132.

Troost, T.A., Kooi, B.W., and Kooijman, S. (2005a). Ecological specialization of mixotrophic plankton in a mixed water column. *Am. Naturalist* **166:** E45–E61.

Troost, T.A., Kooi, B.W., and Kooijman, S. (2005b). When do mixotrophs specialize? Adaptive dynamics theory applied to a dynamic energy budget model. *Math. Biosci.* **193:** 159–182.

Turpin, D.H. (1988). Physiological mechanisms in phytoplankton resource competition. *Growth and*

Reproductive Strategies of Freshwater Phytoplankton. C.D. Sandgren, ed. Cambridge, Cambridge University Press, pp. 316–368.

Turpin, D.H., and Harrison, P.J. (1979). Limiting nutrient patchiness and its role in phytoplankton ecology. *J. Exp. Mar. Biol. Ecol.* **39:** 151–166.

Vaillancourt, R.D., Marra, J., Seki, M.P., Parsons, M.L., and Bidigare, R.R. (2003). Impact of a cyclonic eddy on phytoplankton community structure and photosynthetic competency in the subtropical North Pacific Ocean. *Deep Sea Res. Part I Oceanogr. Res. Papers* **50:** 829–847.

van de Schootbrugge, B., Bailey, T.R., Rosenthal, Y., Katz, M.E., Wright, J.D., Miller, K.G., Feist-Burkhardt, S., and Falkowski, P.G. (2005). Early Jurassic climate change and the radiation of organic-walled phytoplankton in the Tethys Ocean. *Paleobiology* **31:** 73–97.

Venrick, E. (1999). Phytoplankton species structure in the central North Pacific 1973–1996: variability and persistence. *J. Plankton Res.* **21:** 1029–1042.

Villareal, T., Pilskaln, C., Brzezinski, M., Lipschultz, F., Dennett, M., and Gardner, G.B. (1999). Upward transport of oceanic nitrate by migrating diatom mats. *Nature* **397:** 423–425.

Villareal, T.A., Altabet, M.A., and Culver-Rymsza, K. (1993). Nitrogen transport by vertically migrating diatom mats in the North Pacific Ocean. *Nature* **363:** 709–712.

Vrede, T., Dobberfuhl, D.R., Kooijman, S., and Elser, J.J. (2004). Fundamental connections among organism C : N : P stoichiometry, macromolecular composition, and growth. *Ecology* **85:** 1217–1229.

Waxman, D., and Gavrilets, S. (2005). 20 questions on adaptive dynamics. *J. Evol. Biol.* **18:** 1139–1154.

Weissing, F.J., and Huisman, J. (1994). Growth and competition in light gradient. *J. Theor. Biol.* **168:** 323–326.

Wheeler, P.A., Olson, R.J., and Chisholm, S.W. (1983). Effects of photocycles and periodic ammonium supply on three marine phytoplankton species. II. Ammonium uptake and assimilation. *J. Phycol.* **19:** 528–533.

Yoon, H.S., Hackett, Y.D., Ciniglia, C., Pinto, G., and Bhattacharya, D. (2004). A molecular timeline for the origin of photosynthetic eukaryotes. *Mol. Biol. Evol.* **21:** 809–818.

Yoshida, T., Hairston, N.G., and Ellner, S.P. (2004). Evolutionary trade-off between defence against grazing and competitive ability in a simple unicellular alga, *Chlorella vulgaris. Proc. R. Soc. Lond. Ser. B Biol. Sci.* **271:**, 1947–1953.

Yoshida, T., Jones, L.E., Ellner, S.P., Fussmann, G.F., and Hairston, N.G. (2003). Rapid evolution drives ecological dynamics in a predator-prey system. *Nature* **424:** 303–306.

Yoshiyama, K., and Nakajima, H. (2002). Catastrophic transition in vertical distributions of phytoplankton: Alternative equilibria in a water column. *J. Theor. Biol.* **216:** 397–408.

17

Biological and Geochemical Forcings to Phanerozoic Change in Seawater, Atmosphere, and Carbonate Precipitate Composition

MICHAEL W. GUIDRY, ROLF S. ARVIDSON, AND FRED T. MACKENZIE

Although considerable work has gone into obtaining proxy information (chemical, mineralogical, and isotopic data obtained mainly from sediments and sedimentary rocks and used to decipher past environmental change) and in the development of models for the compositional history of Phanerozoic seawater (e.g., Berner *et al.* 1983; Lasaga *et al.* 1985; Hardie 1996; Wallmann 2001; Hansen and Wallmann 2003; Holland 2004), there has been little attempt to assess this history as a function of all the processes and mechanisms involved in controlling seawater composition. Changes in seawater composition, which are partly tied to changes in atmospheric composition and vice versa, play a strong role in biological evolution that, in turn, leads to feedbacks on the compositional changes in these reservoir systems (see Katz *et al.*, Chapter 18, this volume). Arvidson *et al.* (2006a) provided a new Earth system model termed *MAGic* (Mackenzie, Arvidson, Guidry *interactive cycles*) that describes fully the exogenic system of ocean, atmosphere, biota, and sediments and its coupling to the endogenic system (shallow mantle) in terms of the processes, fluxes, and reservoirs of 10 important elements involved in the behavior of the coupled system through the past 500 million years of the Phanerozoic Eon. In this chapter, we use this model to examine how the weathering fluxes of Ca, Mg, C, S (as sulfate), and P have influenced biogeochemical cycles (Berner *et al.* 1983; Lasaga *et al.* 1985) in the ocean over the past 500 million years. In addition, we calculate the fluxes of these five components in relation to their sink reservoirs and the effect of the changes in these fluxes and those from basalt–seawater reactions on the chemistry of seawater. We then go on to show that the age distributions of inorganic and biogenic carbonate phases in the Phanerozoic carbonate rock record are related to the kinetics of the precipitation rates of these phases that in turn are controlled by changing atmosphere–seawater composition.

As an example of the coupling between changes in atmosphere–seawater chemistry and biological evolution, one may look at the emergence of the coral organisms that formed massive reefs. Some paleontologists have noted that massive coral reefs are less common in the Paleogene than in the Neogene (Budd 2000; Perrin 2002). This change in abundance has been attributed to the decline in atmospheric CO_2, thus leading to less acidic ocean water and/or to the rising Mg:Ca ratio of seawater from the Paleogene to the Neogene that led to a transition from ocean waters whose chemical composition favored calcite precipitation (calcite seas; Sandberg 1983) to waters that favored aragonite precipitation (aragonite seas; Sandberg 1983). This change, in turn, gave rise to the expansion of coral reefs dominated

by corals that secrete carbonate skeletons of aragonite (Stanley 2006).

Figure 1 is a schematic diagram of the dynamical global biogeochemical model *MAGic* illustrating the general features of this model. The Earth's surface and near-surface global environment is divided into a series of entities termed reservoirs (boxes) and material transports among the boxes are described by physical, chemical, and biological processes that move materials from one box to the other. Associated with each process is a flux, that is, a rate of transfer usually given in terms of mass per unit time. Such an Earth system model is a global biogeochemical model, and in *MAGic*, each elemental cycle is explicitly coupled to corresponding cycles of other elements via a reaction network. This network incorporates the basic reactions controlling atmospheric carbon dioxide and oxygen concentrations, continental and seafloor weathering of silicate and carbonate rocks, net ecosystem productivity, basalt–seawater exchange reactions, precipitation and diagenesis of chemical sediments and authigenic silicates, oxidation–reduction reactions involving C, S, and Fe, and subduction–decarbonation reactions. Coupled reservoirs include shallow and deep cratonic silicate and carbonate rocks and sediments, seawater, atmosphere, oceanic sediments and basalts, and the shallow mantle (Arvidson *et al.* 2006a).

I. CONTINENTAL WEATHERING FLUXES AND CO$_2$

Weathering fluxes of Ca, Mg, C, SO$_4$, Si, and P are computed using equations similar to those in GEOCARB III (Berner and Kothavala 2001), that is, by evaluating a feedback function sensitive to CO_2, and hence temperature, and additional parameters reflecting the role of vascular land plant fertilization, runoff, paleogeography, and differential susceptibility of carbonate and silicate rock types to weathering. In addition, weathering rate constants were

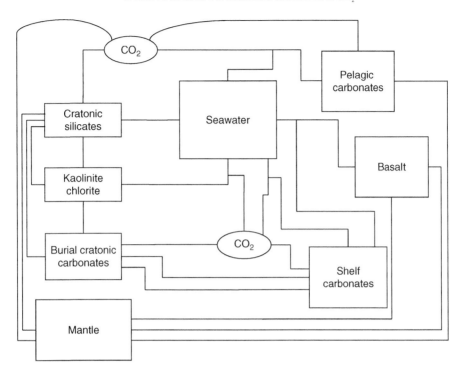

FIGURE 1. *MAGic* topology (Ca-Mg-SiO$_2$-CO$_2$ system only). Large rectangles and ovals denote reservoirs and lines with arrows denote fluxes.

also calibrated by additional time-dependent terms for land area, runoff, relief, and land plants (Figure 2). These parameters are adapted directly from GEOCARB III. Discussion of these terms is available in the various GEOCARB articles (e.g., Berner *et al.* 1983; Berner and Kothavala 2001) and in Arvidson *et al.* (2006a).

The weathering fluxes of Ca, Mg, dissolved inorganic carbon (DIC, mostly bicarbonate), sulfate, dissolved silica, and reactive P from land to the world's oceans during the past 500 million years are shown in Figures 3 and 4. In the modern world, Ca, Mg, HCO$_3$, and SO$_4$ form approximately 75% of the unpolluted total river flux of major dissolved elements to the ocean (Berner and Berner 1996). Dissolved silica contributes another 10% of the total flux. P is found in very low concentrations, at the micromolar level, but the weathering flux of P to the ocean is important to oceanic productivity, especially in coastal oceans, and on the

longer term geologic time scale. The Ca in river water is mainly derived from the weathering of calcite and dolomite and the dissolution of silicate minerals, particularly felsic plagioclase (Ca, Na aluminosilicate). Magnesium, (CaMg[CO$_3$]$_2$), is derived from dolomite weathering and the weathering of mafic silicate minerals such as amphibole, pyroxene, and biotite (Fe- and Mg-rich silicate minerals). The bicarbonate is derived from the weathering of carbonate and silicate rocks in the presence of CO_2-charged soil and ground waters, whereas the sulfate comes from the oxidative weathering of pyrite (FeS$_2$) and dissolution of CaSO$_4$ dispersed in rocks and in bedded form. For Mg, Ca, and SO$_4$, rainwater additions of salt recycled from the air–sea interface of the ocean are also important. The most recent estimates of the sources of these dissolved constituents in river waters derived from the present continental source consisting of 63 wt % average sediment and 37 wt %

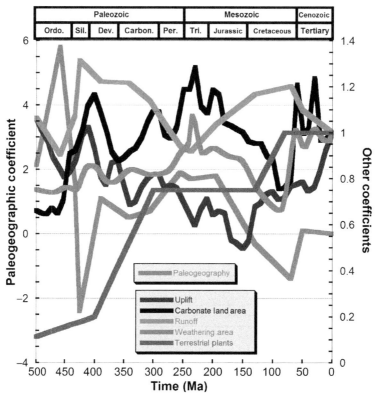

FIGURE 2.   Model weathering coefficients. (See color plate.)

FIGURE 3.   Ca, Mg, SO$_4$, and dissolved inorganic carbon (DIC) weathering fluxes. (See color plate.)

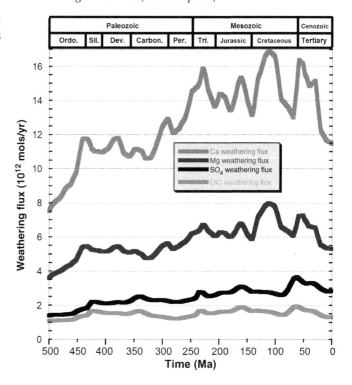

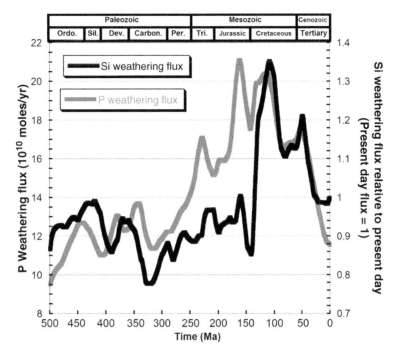

FIGURE 4. P and Si weathering fluxes.

crystalline crust (Mackenzie and Lerman 2006) show that the proportions are for Ca: 18% from dissolution of silicate minerals, 59.8% from dissolution of carbonate minerals, 19.4% from dissolution of evaporite minerals, and 2.9% from weathering reactions involving the acid H$_2$SO$_4$, rather than H$_2$CO$_3$, as a reactant; for Mg: 71.3, 24.4, and 1.6% from silicate, carbonate, and evaporite mineral dissolution, respectively, and 2.7% from reactions involving H$_2$SO$_4$; for HCO$_3^-$: 72% from atmospheric and soil CO$_2$ and 28% from weathering of carbonate minerals; and for SO$_4$: 75.6% from dissolution of evaporite minerals and 24.4% from reactions involving the H$_2$SO$_4$ weathering of rock minerals and the oxidative weathering of pyrite.

In Figure 3, notice in particular the irregular but continuous increase in the fluxes of Ca, Mg, and DIC (mostly HCO$_3^-$) from the start of the model simulation at 500 million years ago (Ma) up until about 100 Ma, the time of maximum invasion of the Cretaceous seas onto the continents. This trend certainly

reflects changes in the land area available for weathering and that occupied by carbonate rocks (see Figure 2), as modified by the location of the continents and runoff. After 100 Ma, the fluxes of these three elements all decline irregularly toward the present due to the combination of the factors mentioned previously. The flux of DIC varies by about 100% over the course of the 500 million years of model simulation, whereas those of Ca, Mg, and SO$_4$ vary by about 127%, 74%, and 171%, respectively, over this time frame. These fluxes from land to sea either accumulate in the ocean or are removed by processes and mechanisms whose nature and magnitude are discussed in Section III.

Dissolved silica is an *implicit* component of the *MAGic* model, although it has no specific source or sink reservoir. Dissolved Si river flux and that of reactive P are directly related to the equations that govern the stoichiometry of silicate and apatite weathering reactions on land (Arvidson *et al.* 2006a) by CO$_2$-charged soil and ground waters. Figure 4 shows the fluxes of dissolved silica and reactive P to the ocean due mainly to the weathering of silicates and apatite, $Ca_5(PO_4)_3F$, on land by CO_2-charged soil and ground waters and, less importantly, by sulfuric acid weathering of these phases. The silica flux is normalized to the present-day river flux of roughly 6.5 × 10^{12} mol Si/year, and the reactive P flux is in units of 10^{12} mol P/year. The latter flux calculated for the present day is very close to the observational estimates, on the order of 10^{11} mol/year of reactive P (Mackenzie *et al.* 1993).

The relative flux of dissolved Si in rivers for the 500 million years shows generally depressed Si fluxes for the first 350 Ma ramping up at 150 Ma and then declining erratically toward the present. The apatite weathering flux shows a somewhat similar pattern but with the increase in flux starting about 250 Ma. These flux-age distributions appear to reflect a combination of forcings, most importantly the increase in weathering rates 400–300 Ma due to the evolution and spread of terrestrial plants, the generally cyclical pattern of runoff from the continents, and the amount of crystalline crust available for weathering. The relatively high rate during the Cretaceous is indicative of a confluence of factors involving high runoff, warm climate, and enhanced crystalline rock area (rich in P) relative to carbonate rock area.

II. THE GLOBAL BIOGEOCHEMICAL CYCLES OF CALCIUM, MAGNESIUM, CARBON, SULFUR, SILICA, AND PHOSPHORUS

In this section, we *briefly* describe the processes that control the global biogeochemical cycles of Ca, Mg, C, S, Si, and P in the context of the way in which they are represented in the *MAGic* model. The processes in these cycles and the magnitudes of their fluxes have exerted primary control over the composition of the ocean

and atmosphere through Phanerozoic time. A more detailed description is available in Arvidson *et al.* (2006a).

A. Calcium-Magnesium-Silicate-Carbonate-CO_2 Cycle

The biogeochemical cycle of Ca and Mg is the most important in terms of its interactions with C and the consequent long-term controls on seawater alkaline earth and C chemistry as well as atmospheric CO_2. In turn, changes in seawater chemistry and atmospheric CO_2 have influenced pelagic and benthic carbonate mineral biomineralization and production over geologic time (see later sections). Ca and Mg are initially weathered from continental silicates and carbonates by carbonic and sulfuric acids. Ca and Mg are also produced during weathering from dissolution of evaporite minerals. The resultant dissolved constituents are transported to the ocean by rivers and less importantly by ground waters. Once present in seawater, these components are redeposited in shelf environments as primary or secondary calcite ($CaCO_3$), dolomite, and $CaSO_4$, and $MgSO_4$ salts in evaporite deposits. Subsequent to deposition, they either re-enter the weathering cycle through uplift and exposure or can be removed to deeper burial regimes. In pelagic environments (slope, rise, and deep-sea), only calcium carbonate is deposited and no dolomite; thus, the total $CaCO_3$ flux is partitioned between shelf and pelagic regimes.

The precipitation rate of shelf dolomite in *MAGic* is controlled by seawater saturation state (Arvidson and Mackenzie 1999). In addition, the accumulation of dolomite is a function of the shelf area available for deposition of carbonate sediment. This function is estimated from Quaternary depositional fluxes in shallow water environments and the submerged shelf area as a function of time and is normalized to its Quaternary value. Dolomite is also formed in the shelf environment during the oxidation of organic matter (so-called organogenic dolomitization) tied to the rate of microbial sulfate reduction and production of DIC.

After deposition, pelagic calcite is either subducted to the mantle or removed to a metamorphic regime. Both of these pathways lead to decarbonation reactions that serve to return material to the weathering cycle. We should note that the mantle does not constitute a reservoir *per se* in *MAGic,* but simply a collection point for return fluxes to either the silicate (basaltic or continental crust) or atmospheric reservoirs. Basalt–seawater reactions result in the uptake of Mg and the release of Ca. Mg uptake is assumed to be first order with respect to the Mg^{2+} concentration in seawater and uptake is modified by a time-dependent parameter reflecting seafloor spreading rate.

The complementary release of Ca^{2+} from basalt during seawater–basalt interaction, although identical to the Mg uptake flux in the steady state (Quaternary) condition, is allowed to otherwise vary to balance with the uptake fluxes. This variation is related to the maintenance of a proton (i.e., CO_2) balance between seawater and hydrothermal fluids within the basalt (see detailed explanation in Arvidson *et al.* 2006a). In addition, diagenetic alteration of shelf and pelagic sediments also results in release of Ca to seawater that is first order with respect to sediment mass. Mg uptake by the sediment to form chlorite, a composition representative of a neoformed magnesium silicate clay mineral, is allowed to vary according to seawater Mg^{2+} mass. The chlorite thus formed reacts with buried calcite to form burial dolomite. The rate of burial dolomite formation is consequently limited by the size of the chlorite reservoir. This dolomite is eventually returned by uplift to the continental weathering regime, and burial calcite not consumed by dolomitization or uplift eventually returns CO_2 to the atmosphere and Ca to the silicate reservoir by metamorphic decarbonation.

Silica is released to river and ground waters during the chemical weathering of silicate minerals in rocks and in more recent geologic time and, less importantly, due to the dissolution of silica phytoliths in plants (see Katz *et al.*, Chapter 18, this volume). Since the appearance of organisms such as diatoms, radiolarians, dinoflagellates, and sponges that secrete siliceous opal-A skeletons, the dissolved silica entering the ocean from this weathering source, and from both low- and high-temperature basalt–seawater reactions, has been removed from the ocean primarily by the precipitation of biogenic silica. During early diagenesis on the seafloor some of this Si, upon dissolution in the pore waters of sediments, may precipitate in reverse weathering reactions, leading to the neoformation of silicate phases in the sediments (Mackenzie and Garrels 1966a, b; Mackenzie *et al.* 1981; Michalopoulas and Aller 1995). The net amount of opal-A buried in the sediments goes through a series of diagenetic reactions that convert it to quartz (SiO_2) on the longer time scale. Some of this silica may then react with calcium carbonate at the higher temperatures and pressures of subduction zones or in deep sedimentary basins to form calcium silicate ($CaSiO_3$) and release CO_2 to the atmosphere (Berner *et al.* 1983). Most of the silica is uplifted to the surface environment where it can be weathered or recycled (Wollast and Mackenzie 1983).

B. Organic Carbon and Phosphorus Subcycles

Organic matter cycling exerts a critical control on both long-term atmospheric CO_2 and O_2. An explicit assumption of the *MAGic* model is that net ecosystem production is ultimately limited by P availability (Smith and Mackenzie 1987; Tyrrell 1999). The control of the rate of marine organic matter production closely follows the model of Van Cappellen and Ingall (1996), in which the rate is proportional to the rate of organic P uptake following the Redfield ratio of C:P = 106:1. The organic P flux is in turn a first-order function of the mass of reactive inorganic P in the ocean and the ocean's ventilation rate. In *MAGic's* standard run, this ventilation rate is fixed, and organic P fluxes are computed by dividing Quaternary organic C fluxes by 250. This approach adopts Ingall and Van Cappellen's (1990) observations that the C:P ratios of organic matter vary from 200 for fully oxic to 4000 for completely anoxic sediments. The net burial of organic material on the shelf reflects the difference between the flux of organic C oxidized in the formation of pyrite, in turn tied to reactive Fe availability, and the pelagic organic C flux. Because of the very low rate of SO_4 reduction in pelagic sediments compared with shelf environments (Bender and Heggie 1984), significant pyrite formation does not occur in this reservoir. The partitioning of organic matter burial between shelf and pelagic regimes is fixed so that the shelf receives 95% of the total. Pelagic sediment organic C is eventually subducted and returned to the atmosphere as CO_2.

The mass of organic C residing in shelf sediments may be rapidly recycled and returned as CO_2. In addition, the *MAGic* model provides for the development of burial of terrestrial organic matter on land in coal basins, using an approach similar to that of Berner and Canfield (1989). Data derived from the organic C content of the rock record (Budyko *et al.* 1987) are used to calculate changes in the size of the reservoir of shelf organic C, and the assumption of a first-order uplift-weathering flux yields the required burial flux of organic C.

C. Sulfur Subcycle

S is delivered to the ocean as SO_4 derived from the weathering of pyrite and the first-order dissolution of the evaporite minerals $CaSO_4$ and $MgSO_4$. Although the sulfuric acid released from pyrite oxidation on the continent participates in silicate, carbonate, and phosphate mineral weathering reactions, the fluxes involved are minor compared to carbonic acid weathering of minerals and have no associated feedback functions in *MAGic*. SO_4 reduction occurs both biogenically within shallow marine sediment and abiogenically during hydrothermal reaction between seawater and basalt. The bicarbonate released in biogenic sulfate reduction is used to form calcite and dolomite. Reduced S formed by these pathways is returned through uplift and exposure of marine sediments and through release of volcanic H_2S and SO_2. Oxidation of this reduced S thus completes the cycle, consuming the oxygen initially liberated by organic matter burial and FeS_2 formation in sediments and yielding sulfuric acid for weathering.

III. OCEANIC SINKS

A. The Major Sink Processes

The dissolved constituents delivered to the ocean by rivers and ground water discharges and those from other sources during Phanerozoic time are removed from the oceans by various mechanisms and processes. These processes are relatively well known for the present-day ocean, although magnitudes are still somewhat controversial, but have not been calculated rigorously for any extended period of geologic time. As enunciated previously, variations in the input and output fluxes of dissolved constituents involving the ocean determine seawater composition during geologic time. The *MAGic* model permits us to make calculations of the magnitude of these fluxes and, in addition, the changes in seawater and atmospheric CO_2 (and O_2, not discussed here) composition deriving from changes in these fluxes. These variations in composition exert an influence on the evolution of life in the Phanerozoic that in turn through feedback mechanisms influences the composition of the ocean and atmosphere. In this section, we present the first full description and quantification of the sinks of Ca, Mg, HCO_3, SO_4, and P, five major components related to $CaCO_3$ biomineralization and production and to organic productivity through Phanerozoic time. Silica is treated in less detail because its immediate oceanic sink is mainly only biological precipitation in the opal-A ("amorphous silica") skeletons of siliceous organisms.

The flux of dissolved Ca to the oceans from weathering on land and that released through reactions at the seafloor involving basalt and seawater at low and high temperatures (Thompson 1983; Wollast and Mackenzie 1983; Mottl and Wheat 1994; de Villiers and Nelson 1999; Arvidson *et al.* 2006a) is removed from the ocean in five sinks (Figure 5): accumulation of $CaCO_3$ on the shelf and in deep sea sediments; accumulation of dolomite on the shelf; accumulation of $CaCO_3$ in the pores, cracks, and fractures in submarine basalts; and accumulation as evaporite $CaSO_4$. Mg mainly derived from weathering reactions on land is primarily removed from the ocean during dolomite formation on the shelves, in low-temperature, diagenetic reverse weathering reactions in seafloor sediments (e.g., Mackenzie and Garrels 1966a; Mackenzie *et al.* 1981;

Michalopoulos and Aller 1995; Holland 2004); during reactions involving basalts and seawater (Thompson 1983; Mottl and Wheat 1994); and during formation of evaporite $MgSO_4$ (Figure 6). The sinks for dissolved inorganic C released via weathering and seafloor basalt–seawater reactions are the accumulation of inorganic C in shelf $CaCO_3$ and dolomite, the accumulation of $CaCO_3$ in seafloor basalts, and the accumulation of calcite and aragonite in the deep sea (Figure 7). Organic C sinks include shelf and pelagic storage of organic matter and the burial of organic matter in coal basins (see Figure 7). The latter is especially important to the history of atmosphericoxygen (Berner and Canfield 1989; Berner 2004). Finally, the sinks of SO_4 derived from weathering reactions on land involve microbial reduction of SO_4 to FeS_2 during anaerobic diagenetic processes in sediments, reduction of seawater SO_4 circulating through ridge systems at high temperatures leading to formation of FeS_2, and the

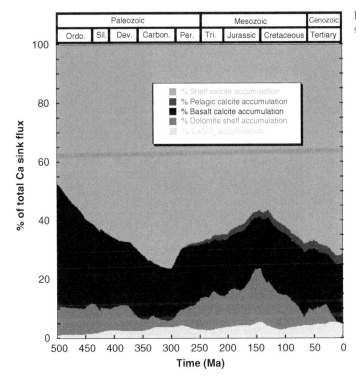

FIGURE 5. Ocean Ca relative sink strength. (See color plate.)

FIGURE 6. Ocean Mg relative sink strength. (See color plate.)

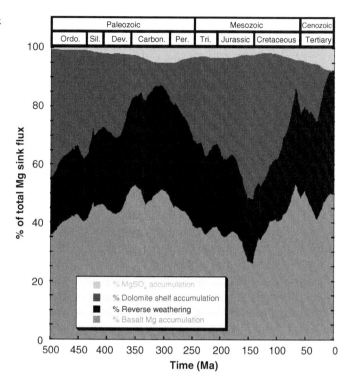

FIGURE 7. Ocean carbon relative sink strength. (See color plate.)

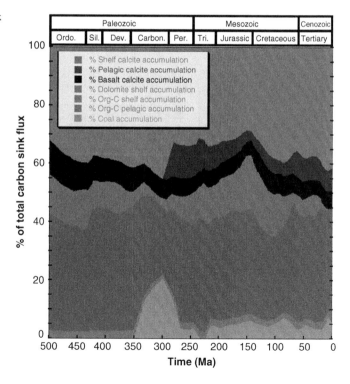

accumulation of evaporite $CaSO_4$ and $MgSO_4$ (Figure 8). The major sink of dissolved silica brought to the ocean via rivers and derived from basalt–seawater reactions is opal-A in the skeletons of siliceous organisms accumulated on the seafloor. The major reactive P sinks are in burial of organic matter, iron oxyhydroxide-absorbed P, and sedimentary carbonate fluorapatite (Figure 9).

B. Sink Trends Through Time

We now look at the trends in the sinks of Ca, Mg, HCO_3, SO_4, and P through much of Phanerozoic time and their relationship to processes occurring at Earth's surface and seawater and atmospheric CO_2 history. First, we shall consider the pattern of Ca and Mg flux variations as shown in Figures 5 and 6. It can be seen from the figures that an important sink of Mg, and less so for Ca, through the Phanerozoic Eon is in the accumulation of dolomite on the shelf. Storage of these elements in dolomite is relatively high for much of the Mesozoic and mid-Paleozoic and low in the Cenozoic and Permo-Carboniferous. The downward trend from the high dolomite accumulation rate of the Cretaceous toward the present is well established by observational records of the distribution of dolomite through Phanerozoic time (Wilkinson and Walker 1989; Wilkinson and Algeo 1989; Morse and Mackenzie 1990; Holland and Zimmermann 2000). The high of the mid-Paleozoic and the subsequent low of the Permo-Carboniferous have been debated in the literature based on interpretations of observations of dolomite mass-age distribution (Wilkinson and Algeo 1989; Mackenzie and Morse 1992; Holland and Zimmermann 2000; Veizer and Mackenzie 2004), but nevertheless, the results of the numerical simulations from *MAGic* suggest the possibility of a cyclic pattern in the dolomite accumulation flux-age distribution. This cyclic pattern mimics to some extent the first-order sea level curves of Vail *et al.* (1977) and Hallam (1984), with

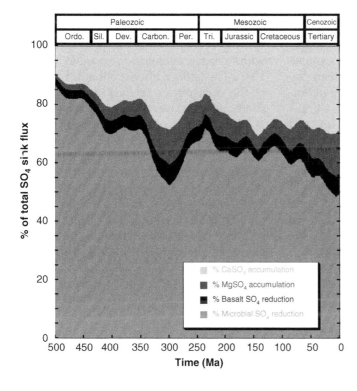

FIGURE 8. Relative SO_4 sink flux strengths (percentage of the total SO_4 flux) for the past 500 million years. (See color plate.)

FIGURE 9. Ocean P relative sink strength. (See color plate.)

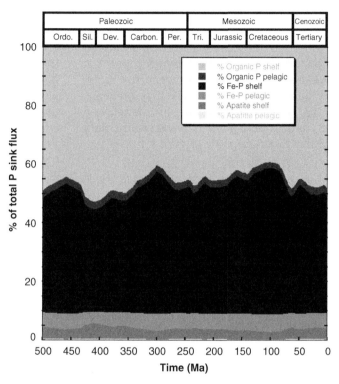

high sea level stands being times of high dolomite accumulation rates. The pattern also mimics that of the trend in the carbonate saturation state of seawater with respect to carbonate minerals (Figure 10; see also Locklair and Lerman, 2005). One can see from Figure 6 that as the sink for Mg in dolomite accumulation becomes more significant, the sinks in reverse weathering and submarine basalt uptake weaken.

For Ca, the most important sink through Phanerozoic time is in the accumulation of $CaCO_3$ in shelf carbonate deposits (see Figure 5). Until the advent of open ocean pelagic calcifying organisms, there probably was scant accumulation of biogenic $CaCO_3$ in the deep sea (Boss and Wilkinson 1991), and that accumulation only became significant in the early Mesozoic, resulting in the transfer of $CaCO_3$ from the shallow-water continental regime to the deep sea. Most, if not all, extant Paleozoic carbonates were deposited in cratonic, continental environments as mainly organo-detrital lime-

stones and not in the deep sea, as evidenced by the sedimentologic characteristics of the carbonate rock record and the near-absence of deep-sea carbonate deposits associated with Paleozoic ophiolites (Boss and Wilkinson 1991). The latter authors contend that some of the occurrences of dark limestone and rhythmically layered marble associated with the scant record of Paleozoic ophiolites may represent inorganic $CaCO_3$ deposition on the deep seafloor. The accumulation of calcite in submarine basalts is of further importance to the record of Ca accumulation in the deep sea. In this case, the Ca is derived from the reaction of primary basaltic minerals with seawater at hydrothermal temperatures and accumulates during the precipitation of $CaCO_3$ in the pores, cracks, and fractures of submarine basalt. Subduction of this $CaCO_3$, along with pelagic calcareous oozes, results in decarbonation at higher temperatures and pressures and ultimate release of CO_2 by volcanism to the Earth's atmosphere.

FIGURE 10. Ocean saturation state for calcite, aragonite, 15% MgCO$_3$, and dolomite for the past 500 million years. (See color plate.)

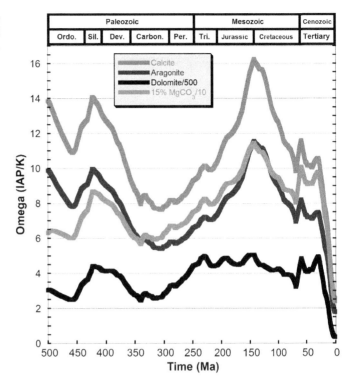

Evaporite minerals form an additional, albeit minor, sink for both Ca and Mg. As shown in Figures 5 and 6, there is a roughly continuous growth in the accumulation rates of CaSO$_4$ and MgSO$_4$ during the past 500 Ma. However, the age distribution of extant evaporites in the sedimentary rock record is irregular, reflecting the infrequent coincidence of requisite tectonic, paleogeographic, and paleoclimatic conditions necessary for evaporite deposition. For example, vast quantities of evaporite salts accumulated in the Upper Permian that may actually have resulted in a lowering of the mean salinity of seawater at this time by 1 to 4% (Holser 1984). The current version of the *MAGic* model does not incorporate these episodic evaporite depositional events and thus the modeled pattern of Ca and Mg accumulation reflects more the integrated strength of these evaporate sinks over time and not specific variation in their positional history.

The Phanerozoic oceanic sinks of organic and inorganic C derived from land, DIC derived from reactions at mid-ocean ridges, and organic C produced *in situ* by productivity in the ocean are shown in Figure 7. Shelf accumulation of both organic C and inorganic C in calcite and dolomite are very important sinks of C for the past 500 million years. The trend in the rate of accumulation of inorganic C in dolomite, as might be expected, mimics the trend in accumulation of Mg in this phase. Pelagic calcite accumulation of inorganic C becomes important in the Mesozoic and inorganic accumulation of CaCO$_3$ in submarine basalts occurs with only minor variation through the past 500 million years. Accumulation of terrestrially derived organic C is particularly important in the Permo-Carboniferous, a time of vast coal deposits, whereas pelagic accumulation of organic C varies only slightly throughout Phanerozoic time.

Figure 11 shows the changing ratio of the accumulation rates of reduced organic C to oxidized inorganic C during the past 500 million years (blue curve). The trend of

FIGURE 11. Reduced/oxidized C and S flux ratio for the past 500 million years. (See color plate.)

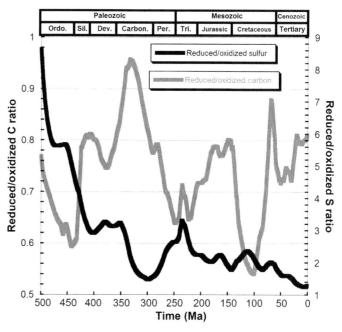

increasing reduced/oxidized C ratio from the Ordovician toward the Permo-Carboniferous is consistent with the $\delta^{13}C$ trend of enrichment of ^{13}C in low magnesian calcite shell materials (Veizer *et al.* 1999) during this interval of time. The model ratio dips at about 400Ma, coinciding with an observed depletion in ^{13}C from the $\delta^{13}C$-age curve. These relationships are what would be anticipated if the organic C burial rate is the main driver of the $\delta^{13}C$ record of carbonates during this time interval (Garrels and Lerman 1984). As the organic C burial rate increases, seawater, and the carbonates derived by precipitation, become enriched in ^{13}C. The Permo-Carboniferous increased burial of organic matter reflects in part enhanced burial of organic C at this time in coal basins. The general decline from the Permo-Carboniferous in the reduced/oxidized C burial ratio also accords with the erratic post-Permian decline in $\delta^{13}C$ and its trend toward the present. The low value of the reduced/oxidized C burial ratio during Cretaceous time is also possibly evidenced in the $\delta^{13}C$ record. Even though this is a time of slightly enhanced organic C burial

in coal basins, other factors appear to be playing a role in controlling the $\delta^{13}C$ of carbonates.

The major two repositories for SO_4 during the Phanerozoic are the microbial reduction of SO_4 during early diagenesis and the consequent formation of pyrite (FeS_2) and accumulation of $CaSO_4$ and $MgSO_4$ in evaporite basins (see Figure 8). The burial of sedimentary pyrite, as well as the more minor hydrothermal formation of pyrite in basalt, both represent a source of atmospheric O_2. The integrated amount of SO_4 sulfur that has accumulated in sediments as reduced S during the past 500 million years is about 30% of the total S in the extant exogenic reservoirs of ocean, continental and oceanic sediments, and evaporites. Estimates of the value of the fraction of S in the reduced state vary from 0.29 to 0.69 (Holser *et al.* 1988) for all of geologic time.

Figure 11 also shows the changing ratio of the accumulation rates of reduced S to oxidized S for the past 500 million years (black curve). The ratio drops precipitously for the first approximately 200Ma into the Permian. This drop strongly follows the erratic drop in the $\delta^{34}S$ of S structurally bound in calcitic shells as well as in evaporitic SO_4 deposits for this interval of time (Kampschulte 2001; Veizer and Mackenzie 2004). This is what would be anticipated because as the accumulation rate of reduced S relative to that of oxidized S decreases, the ocean should become depleted in ^{34}S, and the $\delta^{34}S$ of oxidized S in calcitic shells and in evaporite SO_4 minerals should fall. The trend of reduced / oxidized S accumulation rates generally continues to mimic the $\delta^{34}S$ curve after the Permian, even displaying a relatively sharp increase at about 245 Ma, the Permian–Triassic boundary, a time of apparently rapidly increasing $\delta^{34}S$.

Figure 12 shows total mass (and hence concentration because the volume of the ocean is taken as today's volume of 1.37×10^{21} L throughout the 500 million years of *MAGic* calculations) of dissolved reactive P through the past 500 million years. Because P is the ultimate limiting nutrient in the *MAGic* model and phytoplankton produc-

tivity is modeled as a first-order relationship with respect to P concentration, these changes in the concentration of P in the ocean are what drive the changes in organic productivity in the ocean through time (see Figure 7). Nitrogen and its role in ocean productivity are not as yet a feature of the *MAGic* model but will be in later versions.

To some extent, as modified by the various sink strengths of P through time, the time course of the apatite-weathering flux from the land to the sea is reflected in the time course of the mass (concentration) of P in the ocean. Because P drives ocean productivity, the higher the concentration of P in the ocean, the greater the organic productivity, and in addition, the greater the rate of accumulation of organic C in sediments (see Figure 7). These latter two rates are relatively low around 500, 300, and 100 Ma, and presently. At present and 500 Ma, the climate was relatively cold (icehouses), and the weathering rate of apatite on land was reduced, hence the concentration

FIGURE 12. Ocean P mass (moles P) for the past 500 million years.

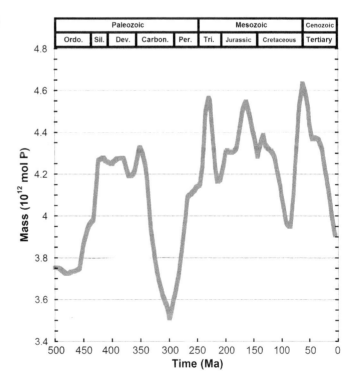

of P in the ocean was low as was organic matter production and accumulation. This was also true around 300 Ma (icehouse). However, at this time and around 100 Ma, the lower P concentration and hence productivity in the open ocean also appears to some extent to be due to the abundance of coal forests that led to land-derived P being sequestered in the organic matter produced in these forests and buried in coastal swamp sediments, later to become coals.

IV. SOME TRENDS IN CARBONATE ROCK FEATURES

In this section we discuss some age trends in the features of carbonate rocks that are particularly relevant to ocean–atmosphere evolution. Here we concentrate on the distribution with carbonate rock age of dolomite, oolite mineralogy, and skeletal mineralogy and its strontium content through the Phanerozoic Eon. Much of the discussion concerning the observational data is based on the recent synthesis articles of Veizer and Mackenzie (2004) and Holland (2004) and the book by Mackenzie and Lerman (2006).

For several decades, it was assumed that the Mg:Ca ratio of carbonate rocks increases with increasing rock age (see Daly 1909; Chilingar 1956; Vinogradov and Ronov 1956a, b). In these summaries, the Mg content of North American and Russian Platform carbonates is relatively constant for the last 100 million years and then increases gradually with age, at a time very close to, if not the same as, the commencement of the general increase in Mg content of pelagic limestones (Renard 1986). The dolomite content in deep-sea sediments also increases erratically with increasing age back to ~125 million years before present (Lumsden 1985). Thus, the increase in Mg content of carbonate rocks with increasing age into at least the early Cretaceous appears to be a global phenomenon. In the 1980s, the accepted truism that dolomite abundance increases relative to limestone with increasing age was

challenged by Given and Wilkinson (1987). They re-evaluated all the existing data and concluded that dolomite abundances do vary significantly but cyclically throughout the Phanerozoic and did not simply increase systematically with age.

If we accept the Given and Wilkinson cyclic trend for the dolomite mass-age distribution in the Phanerozoic, then the period-averaged mass ratio of calcite to dolomite is anomalously high for the Cambrian and Permian System rocks. For the remainder of the Phanerozoic, the ratio appears to oscillate within the 1.1 ± 0.6 range, except for the limestone-rich Tertiary. In addition, comparison with the generalized Phanerozoic sea-level curves of Vail et al. (1977) and Hallam (1984) further illustrates that dolomite accumulation rates and extant dolomite mass per unit time are more abundant at times of higher sea levels. These cycles in dolomite accumulation and extant mass-age distribution correspond to the two Phanerozoic super cycles of Fischer (1984) and to the oscillatory and submergent tectonic modes of Mackenzie and Pigott (1981).

Although still somewhat controversial (e.g., Bates and Brand 1990), the textures of oolites (sand-sized usually internally concentrically banded grains of $CaCO_3$) appear to vary during the Phanerozoic Eon. Sorby (1879) first pointed out the petrographic differences between ancient and modern oolites: Ancient oolites commonly exhibit relict textures of a calcitic origin, whereas modern oolites are dominantly made of aragonite. Sandberg (1975) reinforced these observations by study of the textures of some Phanerozoic oolites and a survey of the literature.

Following this classical work, Sandberg (1983, 1985) and several other investigators (Mackenzie and Pigott 1981; Wilkinson et al. 1985; Bates and Brand 1990) attempted to quantify further this relationship. It appears that, although originally aragonitic oolites are found throughout the Phanerozoic, an oscillatory trend in the relative percentage of calcite versus aragonite oolites may be superimposed on a long-term evolutionary decrease in oolites with an inferred original calcitic mineralogy. The maxima in calcitic oolite abundance appear to coincide with the two major maxima in the sea level curves of Vail *et al.* (1977) and Hallam (1984). Wilkinson *et al.* (1985) found that the best correlation between various datasets representing global eustasy and oolite mineralogy is that of inferred mineralogy with percentage of continental freeboard (extent of flooding of the continents by the sea). Sandberg (1983) further concluded that the cyclic trend in oolite mineralogy correlates with cyclic trends observed for the inferred mineralogy of marine carbonate cements in sediments.

In some similarity to the trends observed for the inorganic precipitates of oolites, the mineralogy of some calcareous fossils during the Phanerozoic also exhibits a cyclic pattern, with calcitic organisms that make massive skeletons being particularly abundant during the high sea levels of the early to mid-Paleozoic and the Cretaceous (Stanley and Hardie 1998; Kiessling 2002). The strontium

content of biologically precipitated calcite with little Mg content (Steuber and Veizer 2002; Lear *et al.* 2003) also follows the trend in skeletal mineralogy, being higher during times of high dolomite accumulation rates, an abundance of calcitic oolites and cements, and high sea level. It is noteworthy that the major groups of pelagic and benthic organisms contributing to carbonate sediments in today's ocean first appeared in the fossil record during the middle Mesozoic and progressively became more abundant.

V. ATMOSPHERE AND SEAWATER COMPOSITION

In terms of Phanerozoic atmospheric CO_2 changes, both the various renditions of GEOCARB (Berner 1991, 1994; Berner and Kothavala 2001) and the *MAGic* model (Arvidson *et al.* 2006a) exhibit similar results (Figure 13). During the Phanerozoic, there is a distinct pattern of oscillation of

FIGURE 13. Atmospheric CO_2 as calculated from the GEOCARB and *MAGic* models and compared with the proxy paleosol data of Yapp and Potts (1992) and the stomatal index data of Retallack (2001). (See color plate.)

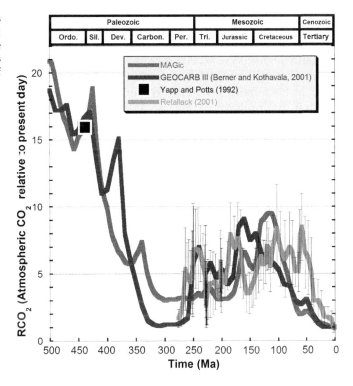

atmospheric CO_2. This pattern was recognized initially in model calculations of the global C cycle by Berner and coworkers (Berner *et al.* 1983; Berner and Kothavala 2001) and then later supported by proxy records of atmospheric CO_2 found in paleosol (fossilized soils) iron oxyhydroxides and $CaCO_3$, marine sedimentary C, the fossil stomatal record of plants, and the boron isotopic composition of carbonate fossils. Atmospheric CO_2 concentrations rose steeply to relatively high levels, perhaps reaching 7500 ppmv, approximately 550 Ma and then fell to low concentrations 350–250 Ma. Concentrations then rose again into the Middle Mesozoic, about 150–100 Ma, and perhaps reached levels of 3000 ppmv and then fell again somewhat erratically to the levels of the Pleistocene and Holocene Epochs of 180–280 ppmv. The modeled Phanerozoic long-term oscillations are a result of the interplay between the out-gassing of volcanic and metamorphic CO_2 and weathering changes consuming CO_2, for example, during uplift of mountain ranges and the fall of sea level. The large downward trend in atmospheric CO_2 during the Paleozoic probably reflects to some extent the appearance of vascular land plants that accelerated the rates of weathering and introduced a new sink of C in bacterially resistant organic matter that accumulated in marine and nonmarine sediments (Berner 2004). Later, in the Mesozoic Era, the overall downward trend from the Cretaceous on probably reflects the appearance of the angiosperms with their numerous roots and root hairs, and later on the C_4 grasses, and their role in accelerating weathering rates at a time when global sea level was falling and the land area for weathering was increasing (Berner 2004; Katz *et al.*, Chapter 18, this volume). Although there are model disagreements regarding actual paleo-P_{CO2} estimates, there is generally broad agreement among models (e.g., *MAGic* and GEOCARB) in the long-term general trends in pCO_2 over the Phanerozoic.

Geologists have long recognized that relatively long periods of time spanning tens of millions of years existed in the Phanerozoic when the climate was warm and that these extended periods of time were interspersed with shorter periods of cold and continental-scale glaciations. For example, during the warmth of the Cretaceous, 65–145 Ma, dinosaur fossils show that these organisms lived in what is now Alaska, and the $\delta^{18}O$ record in planktonic and benthic calcareous organisms show that ocean bottom temperatures, which are now about 2°C, were a warm 15°C at this time. In the Phanerozoic, there are at least two hothouse (greenhouse) periods (545–360 Ma and 240–30 Ma) and two icehouse periods (360–240 Ma and the last one beginning in the Cenozoic 30–35 Ma). Geological evidence shows that major continental-scale ice sheets were present during late Neoproterozoic time and during the two icehouses of the Phanerozoic. Thus, it appears that there is a first-order agreement between the continental glaciation record and atmospheric CO_2 levels for the Phanerozoic, supporting the conclusion that atmospheric CO_2 plays an important role in long-term climatic change.

This conclusion has been recently challenged by Shaviv and Veizer (2003) and Veizer (2005) who believe that changes in the sun's energy field and the generation of cosmic rays modulated climate through the Phanerozoic and may even be responsible to some extent for modern climatic change. Thus, some of the cooler periods of Phanerozoic time as deduced from the $\delta^{18}O$ record that do not follow the trend between atmospheric CO_2 and climate may be a result of changing cosmic ray flux. However, it should be kept in mind that the $\delta^{18}O$ record might be biased by both diagenesis and changes in seawater and atmospheric CO_2 composition over time.

As atmospheric CO_2 concentrations varied through the Phanerozoic so, too, did seawater composition. Marine evaporites provide important clues to the history of seawater composition, including dissolved C species, because they can be used to place compositional bounds on the three major seawater components that are part of the global carbonate–silicate geochemical cycle: Ca, Mg, and SO_4. Researchers have been able to compile values of Phanerozoic seawater concentrations of Mg^{2+}, Ca^{2+}, and SO_4^{2-} based on analyses of fluid inclusions in marine halite (e.g., Hardie 1996; Holland 2004). These analyses show that seawater concentrations of Mg^{2+}, Ca^{2+}, and SO_4^{2-} have varied by a factor of two or slightly more over the Phanerozoic. The variation of Ca^{2+} and Mg^{2+} in seawater implies that the C chemistry of seawater has changed over time, including its saturation state with respect to carbonate minerals.

The relationship between Ca-Mg carbonate mineral composition, solubility, precipitation kinetics, and seawater chemistry is complex; even this statement understates many longstanding problems, adequate discussion of which is well beyond the scope of this chapter. In the case of dolomite, key relationships still elude definitive understanding. Arvidson and Mackenzie (1999) did quantify robust relationships between dolomite's precipitation rate, temperature, and saturation state and also demonstrated that formation of the mineral at Earth surface conditions can occur at rates that are geologically rapid. They showed that the rate is particularly sensitive to carbonate ion activity (and thus carbonate alkalinity), a relationship that is key to the correspondence between continental weathering fluxes and dolomite accumulation rates. The relationship between dolomite abundance and sulfate is more problematic. Early work by Baker and Kastner (1981) demonstrated that sulfate inhibits dolomite growth at high temperature (200°C), challenging the importance of Mg:Ca ratio and leading to a longstanding notion that current levels of seawater SO_4 were at least part of the explanation for the paucity of dolomite in modern sediments. However, the idea that dolomite growth is promoted in low-SO_4 or

SO_4-free systems may be strictly true only at high temperature, given the occurrence of dolomite in SO_4-rich hypersaline environments (although SO_4 concentrations would be buffered by $CaSO_4$ formation in these settings; Hardie 1987). In addition to providing enhancements in dissolved Mg:Ca ionic ratio, evaporative hypersaline environments may also promote dolomite formation through a decrease in the activity of water, thus promoting dehydration of aqueous Mg ion (Folk and Land 1975; Lippman 1973), a key reaction step that must limit the crystal growth of ordered dolomite. More recent sorption experiments also suggest that magnesium sulfate aqueous (or surface) complexes may facilitate a reaction path involving carbonation–dehydration and lattice attachment of magnesium ion; in this case, SO_4 may actually catalyze dolomite nucleation on carbonate surfaces (Brady et al. 1996). Given these uncertainties, we shall avoid the assertion of a putative link between dolomite and SO_4; however, a positive relationship between variations in model saturation state (driven primarily by carbonate alkalinity) and dolomite accumulation fluxes is entirely consistent with experimental and field observations of dolomite reactions and its mass-age distribution.

Several authors have tried to document variations in the CO_2–carbonic acid system and carbonate saturation of seawater for portions of the Phanerozoic (e.g., Lasaga et al. 1985; Tyrrell and Zeebe 2004; Locklair and Lerman 2005; Arvidson et al. 2006a). Figure 14 shows the calculated variations in the various seawater CO_2–carbonic acid parameters. As atmospheric CO_2 and the concentrations of Ca, Mg, and sulfate varied through the Phanerozoic, so did the C chemistry of seawater. Generally, during times of high atmospheric CO_2 (hothouses or greenhouses), the carbonic acid content and the total alkalinity of seawater were relatively high and the pH low. DIC, dominated by HCO_3^-, followed the same pattern as total alkalinity. At intervening times of

FIGURE 14. CO$_2$–carbonic acid system parameters of seawater. (See color plate.)

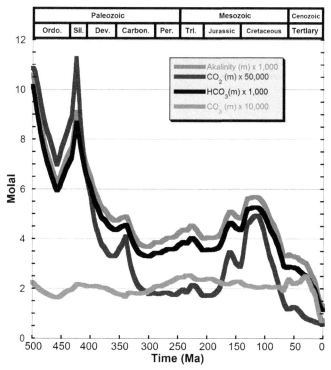

low atmospheric CO$_2$ (icehouses), the converse was true. Despite the lower pH, the saturation state of seawater with respect to calcite was generally high during hothouse climatic conditions (see Figure 10), reflecting high total alkalinity and Ca concentrations of seawater at this time. Conversely, saturation states were depressed during icehouse conditions. The saturation state of seawater with respect to dolomite reflects concentration changes in carbonate alkalinity, Ca, and Mg, dependencies that create a more complex pattern in terms of saturation state. Although evaporite fluid inclusion data and other modeling results support the conclusion that seawater composition varied significantly during the Phanerozoic, the *MAGic* calculations reinforce and extend this understanding to the carbonic acid system and seawater saturation state with respect to carbonate minerals.

Figures 5, 6, and 7 show the fluxes of carbonate-related elements in accumulation of calcite and dolomite on the shelf and pelagic calcite in the deep sea during the Phanerozoic. Unless abiotic deposition of calcite occurred in the deep sea during pre-Mesozoic time, there was scant deep-sea carbonate deposition during the Paleozoic. Both the calcite and dolomite shelf carbonate flux records during the Phanerozoic show peaks and troughs that represent changing seawater C chemistry and global first-order changes in sea level. The dolomite shelf flux is particularly interesting because it exhibits a strong cyclicity that accords with the rise and fall of global atmospheric CO$_2$ and sea level, as well as seawater total alkalinity and Mg:Ca and SO$_4$:Ca ratios, during the Phanerozoic. This supports the idea that cyclical variations in dolomite mass in the rock record are related to changes in seawater chemistry and the availability of appropriate shallow-water environments.

The general pattern of cyclicity in atmospheric CO$_2$ levels and ocean chemistry during Phanerozoic time was interrupted abruptly at certain times, resulting in major reorganizations of the ocean–atmosphere system and effects on biological evolution. An example of one type of reorganization involves events that take place at times of oceanic anoxia in shelf and abyssal marine environments, such as during the mid-Cretaceous and Late Devonian. For those times, Kump *et al.* (2005) envision a situation similar to that at the beginning of the development of modern anoxic basins that exist today, such as the Black Sea. In these basins, mildly acidic and sulfidic deep waters are separated from the atmosphere by an oxygenated surface layer at the base of which is a zone of dissolved reduced S concentrations increasing with increasing depth, the sulfide chemocline. The flux of H_2S across the chemocline is regulated by the photosynthetic and chemosynthetic green, purple, and colorless bacteria that exist at the chemocline. If in the past H_2S concentrations below the chemocline increased beyond a critical threshold during anoxic events, then the sulfide chemocline could have risen to the surface, leading to high fluxes of H_2S to the atmosphere and possibly toxic levels of H_2S in the atmosphere. The surface ocean would then be populated by the bacteria mentioned previously. In addition, if this occurred, the abundance of hydroxyl radical (OH·) consumed in scavenging the enhanced flux of H_2S to the atmosphere from the ocean would be reduced to very low levels. The depletion of OH· would also lead to higher concentrations of CH_4 in the atmosphere, perhaps exceeding 100 ppmv, because the OH· radical is also responsible for the destruction of atmospheric CH_4. Furthermore, the ozone shield may have been partially or totally destroyed because of the reaction of H_2S in the stratosphere with atomic O, leading to depletion of atomic O in the stratosphere, which is the most important reactant in the formation of stratospheric ozone. Such a series of catastrophic events could act as a "kill mechanism" (Kump *et al.* 2005), leading to the mass extinctions and

reorganization of the global biogeochemical cycles at the end of the Permian, Late Devonian, and in the Cenomanian-Turonian of the Cretaceous. However, as the Earth's history shows, such events did not lead to permanent changes.

VI. DISCUSSION AND CONCLUSIONS

In the prevailing view, the first-order changes in sea level are driven by the accretion of ridges: High accretion rate results in high sea level and low accretion rate results in low sea level. These sea level changes are related to changes in atmospheric CO_2 levels, seawater composition, and features of the extant carbonate rock record (e.g., Morse and Mackenzie 1990; Veizer and Mackenzie 2004). Extended times of global high sea level were times of enhanced atmospheric pCO_2 levels, higher temperatures (hothouse), lower seawater Mg:Ca and SO_4:Ca ratios (Zimmermann 2000; Lowenstein *et al.* 2001; Dickson 2002; Horita *et al.* 2002), higher seawater strontium concentrations (Steuber and Veizer 2002; Lear *et al.* 2003), and higher alkalinity and carbonate saturation state and lower pH of seawater (Arvidson *et al.* 2000; Arvidson *et al.* 2006a). In addition, it appears that the environmental conditions necessary for enhanced accumulation of dolomite and early dolomitization, formation of calcitic ooids and cements, and preponderance of calcitic reef-building organisms are best met during extended times of global high sea levels, with ubiquitous shallow-water and sabkha-like environments. These are the times of the calcite seas of Sandberg (1983) that probably should be termed the calcite/dolomite seas. The converse of all of these is true for first-order global sea level low stands and icehouse conditions. These are times of aragonite seas with reduced dolomite accumulation rates, abundant aragonitic oolites, and a greater propensity for skeletal carbonate mineralogies and marine carbonate cements of aragonite and calcite containing significant Mg.

Stanley and Hardie (1998, 1999, 2002) concluded that the Phanerozoic long-term oscillations in the carbonate mineralogy of calcifying organisms that produced massive skeletons, large reefs, or voluminous bodies of carbonate sediment (termed hyper-calcifying organisms such as tabulate, rugose and scleractinian corals, coralline algae, and calcareous nannoplankton) are related to the Mg:Ca ratio of seawater. For example, during the time of the generally low Mg:Ca ratio and high Ca concentration of seawater approximately coinciding with the last hothouse (greenhouse) and climaxing in the Cretaceous, chalks dominated by coccoliths formed massive carbonate deposits in the warm shallow seas of the Late Cretaceous. Stanley and Hardie (1998) further concluded that it was variable seafloor spreading rates that led to the changing Mg:Ca ratios of seawater and hence to effects on biocalcification and the long-term cyclical distribution patterns of the chemistry and mineralogy of calcareous taxa, marine carbonate sedimentation, and reef growth during the Phanerozoic Eon.

The basis for the conclusion that the Mg:Ca ratio is the governing factor in the biomineralization of the hypercalcifying organisms is the fact that the Mg:Ca ratio of a solution can control the Mg:Ca ratio and the skeletal mineralogy of both inorganic and biogenic carbonate precipitates (e.g., Mackenzie *et al.* 1983, see Figure 15; Lowenstam and Weiner 1989; Morse and Mackenzie 1990; Stanley *et al.* 2002). This is true to some extent, but other factors can influence both the mineralogy and chemistry of marine carbonate precipitates. In the case of both inorganic and biogenic precipitates, temperature is an important factor: At lower temperature the same group of magnesian calcite calcifying organisms will contain less Mg in the skeleton than at high temperature (Chave 1954); and magnesian calcite precipitated inorganically at

low temperature will contain less Mg than that at high temperature (Morse and Mackenzie 1990). In addition, the saturation state of the seawater, particularly in terms of carbonate ion activity, will play a role in the nature of the carbonate precipitate: Obviously it is difficult for an organism to precipitate a certain carbonate composition if the bulk solution is undersaturated with respect to that phase, and this will not happen if the precipitate is inorganic in origin. Furthermore and importantly, the rate of precipitation of an inorganic carbonate precipitate and the rate of calcification of many modern calcifying organisms are related to several factors, including temperature, carbonate saturation state, and the DIC content of the water. The rate can influence the composition of the precipitate. Stanley *et al.* (2002) grew the coralline algal *Amphiroa* sp. in artificial seawater of differing Mg:Ca mol ratios and observed that the lower the ratio in the seawater, the less Mg incorporated in the coralline algal skeleton. They used the results of these experiments to conclude that the Mg:Ca ratio of seawater controls the composition of the biogenic magnesian calcite precipitate and that many taxa producing calcite high in Mg content in the modern oceans may have produced calcite low in Mg content in the Late Cretaceous hothouse seas. However, Mackenzie and Agegian (1989; Agegian 1985) grew another coralline algal, *Porolithon gardineri,* under varying conditions, but constant seawater Mg:Ca ratio, and showed that the Mg content of the organism varied with carbonate saturation state, temperature, and growth rate: the higher the value of any of these three variables, the higher the Mg content of the *Porolithon* skeleton (Andersson *et al.* 2005). Thus, the controls on the mineralogy and chemistry of both inorganic and biogenic precipitates are complex and may not be simply related to a single environmental variable but perhaps to the confluence of variables.

As mentioned previously, the model calculations of *MAGic* for the hothouse (greenhouse) and icehouse intervals of Phanerozoic time show that the former were times of high atmospheric CO_2 concentrations and temperatures, low seawater Mg:Ca and SO_4:Ca ratios, high DIC (mainly total alkalinity) concentrations, low pH, and high carbonate mineral saturation states (Calcite I and II seas of Stanley and Hardie 1999). In addition, the strontium content of seawater was relatively high. The opposite is true for the icehouses (Aragonite I, II, and III seas of Stanley and Hardie 1999).

For all inorganic carbonate precipitates, their overall rates of precipitation have been shown to be a function of saturation state and temperature (Morse and Mackenzie 1990; Arvidson and Mackenzie 1999) and follow the parabolic rate law

$$r = k(\Omega-1)^n \qquad (1)$$

where Ω is the carbonate saturation index, n is the order of reaction, and k is the rate constant. Thus, during hothouse intervals of time with higher temperatures and seawater carbonate saturation states (see Figure 10), one would anticipate that the rates of precipitation of all inorganic carbonate phases would be higher than those during icehouse conditions. For the biogenic phases, it is also likely that at these times, their rates of precipitation (biomineralization) would be higher, if for no other reason than the fact that seawater has a higher DIC content and carbonate saturation state (Andersson et al. 2005). However, it is possible that precipitation rates of calcite with low Mg content could outcompete those of calcites with high Mg content and aragonite under specific conditions, in which (1) less Mg in the seawater was available to act as an inhibitor of calcite precipitation (Berner 1975; Morse 1983) and (2) rates of calcite precipitation were enhanced by higher sea surface temperature and seawater saturation state. The effects of saturation state and temperature are coupled and complex: At constant composition in a closed (isochemical) system, an increase in temperature will shift carbonate mineral equilibria in the direction of increasing oversaturation (or decreasing undersaturation) because of simple changes in the equilibrium constants of the relevant species. In addition, temperature also has a direct kinetic effect, whose leverage is reflected by the magnitude of the activation energy for the overall precipitation reaction. Thus, increasing temperature, even absent changes in bulk chemistry, tends to favor phases having more negative enthalpies (retrograde solubility) and larger energies of activation.

Higher rates of calcite precipitation with low Mg content could have led to mineralogical shifts favoring oolites and skeletal and nonskeletal carbonates composed of calcite deficient in Mg. The reduction in the accumulation of aragonite in sediments would raise the concentration of strontium in seawater during hothouse conditions and lead to skeletal calcites with increased strontium concentrations, as is observed (Steuber and Veizer 2002; Lear et al. 2003; Veizer and MacKenzie 2004). Thus, we conclude that the changing cyclic character of inorganic and biogenic aragonite and calcite precipitates through the Phanerozoic Eon strongly reflects the relationship between precipitation kinetics of these phases as mediated by sea surface temperature and seawater chemistry.

Dolomite is a difficult precipitate to form experimentally at low temperatures and is not found in abundance in modern marine environments, being most common in unique environmental settings and typically forming from compositionally modified seawater solutions at warm temperatures. This has led to the paradigm of a "dolomite problem" of why occurrences of dolomite are so scarce in the modern marine environment, and yet large bedded dolomite deposits are characteristic of much of the pre-Tertiary rock record. As Land (1985) has pointed out, "there are dolomites and dolomites"; that is, there are a number of environmental settings and solution compositions under which dolomite may form. In addition, bacteria are known to be involved in reactions involving dolomite formation (e.g., Vasconcelos and McKenzie 1997; Roberts *et al.* 2004), and our model results are consistent with the importance of reactions in which alkalinity may be supplied by the microbially mediated oxidation of organic C. The dependencies of (inorganic) dolomite precipitation rate on saturation state and temperature are complex and have not been resolved in detail. In addition, the mechanistic role of Mg:Ca ratio in dolomite formation is also not well understood, and recent work on the interaction of Mg with carbonate mineral surfaces has confirmed complex relationships (Davis *et al.* 2000; Arvidson *et al.* 2006b). Despite these uncertainties, our model results in terms of the relationship between fluxes and distribution (see Figures 5, 6, and 7) are consistent with dolomite being favored during periods of elevated alkalinity, temperature, and perhaps even most importantly, expanded shallow-water depositional environments of interior seas, conditions characteristic of hothouse regimes.

Geological and biological processes have acted in concert to alter atmospheric and seawater chemistry over the Phanerozoic. The evolution and rise of various planktonic siliceous and calcareous organisms over this period are direct evidence to this interplay of these processes. Katz *et al.* (Chapter 18, this volume) use carbon isotopic records from carbonates and organic matter in conjunction with sulfur isotopes from evaporites to reconstruct C burial, pCO_2, pO_2, and pH for the past 250 million years. Their results indicate that organic C burial, and consequently atmospheric oxygen concentrations, have increased since the early Jurassic. In addition, they indicate that pCO_2 and pH have decreased since the early Cretaceous. These results are used as evidence to argue for the impact of the evolution and expansion of larger red-celled eukaryotic phytoplankton since the early Jurassic, which has led to the modern-day dominance of diatoms, coccolithophores, and dinoflagellates.

MAGic results indicate a sharp increase in dissolved Si weathering input from the late Jurassic to the mid-Cretaceous (0.85–1.35 relative to the present-day flux; see Figure 4). This increased Si availability may have supported the emergence of the diatoms in the late Jurassic to early Cretaceous. *MAGic* results also indicate that during the late Jurassic to mid-Cretaceous, reactive P input to the oceans reached a Phanerozoic maximum (see Figure 4). The biological evolution and radiation of the eukaryotic phytoplankton also occurred during a hothouse period. These geochemical results support the arguments laid out in Katz *et al.* (Chapter 18, this volume) for the interplay of geochemical and biological processes that influenced long-term ocean C burial, pCO_2, pO_2, and pH for the past 250 million years.

Acknowledgments

This research was supported by NSF ATM (ATM04–394051). We thank Paul Falkowski of Rutgers University and Andrew Knoll of Harvard University for their thoughtful review and comments on the initial manuscript. School of Ocean and Earth Science and Technology Contribution No. 7127.

References

Agegian, C.R. (1985). The biogeochemical ecology of *Porolithon gardineri* (foslie), Ph.D. thesis. University of Hawaii, Honolulu, 178 pp.

Andersson, A.J., Mackenzie, F.T., and Lerman, A. (2005). Coastal ocean and carbonate systems in the high CO2 world of the anthropocene. *Am. J. Sci.* **305:** 875–918.

Arvidson, R.S., Collier, M., Davis, K.J., Vinson, M.D., and Luttge, A. (2006b). Magnesium inhibition of calcite dissolution kinetics. *Geochim. Cosmochim. Acta* **70:** 583–594.

Arvidson, R.S., and Mackenzie, F.T. (1999). The dolomite problem: control of precipitation kinetics by temperature and saturation state. *Am. J. Sci.* **299:** 257–288.

Arvidson, R.S., Mackenzie, F.T., and Guidry, M.W. (2000). Ocean/atmosphere history and carbonate precipitation rates: a solution to the "dolomite problem"? *Marine Authigenesis: From Global to Microbial.* C.R. Glenn, L. Prévôt-Lucas, and J. Lucas, eds.

S.E.P.M. Special Publication No. 65. Tulsa, SEPM, Tulsa, pp. 1–5.

Arvidson, R.S., Mackenzie, F.T., and Guidry, M.W. (2006a). MAGic: a Phanerozoic model for the geochemical cycling of major rock-forming components. *Am. J. Sci.* **306**: 135–190.

Baker, P.A., and Kastner, M. (1981). Constraints on the formation of sedimentary dolomite. *Science* **213**: 214–216.

Bates, N.R., and Brand, U. (1990). Secular variation of calcium carbonate mineralogy: an evaluation of oöid and micrite chemistries. *Geol Rundsch* **79**: 27–46.

Bender, M.L., and Heggie, D.T. (1984). Fate of organic carbon reaching the deep sea floor: a status report. *Geochim. Cosmochi. Acta* **48**: 977–986.

Berner, R.A. (1975). The role of magnesium in the crystal growth of calcite and aragonite from sea water. *Geochim. Cosmochim. Acta* **39**: 489–504.

Berner, R.A. (1991). A model for atmospheric CO_2 over Phanerozoic time. *Am. J. Sci.* **291**: 339–376.

Berner, R.A. (1994). GEOCARB II: a revised model of atmospheric CO_2 over Phanerozoic time. *Am. J. Sci.* **294**: 56–91.

Berner, R.A. (2004). *The Phanerozoic Carbon Cycle: CO_2 and O_2.* New York, Oxford University Press.

Berner, E.K., and Berner, R.A. (1996). *Global Environment: Water, Air, and Geochemical Cycles.* Upper Saddle River, NJ, Prentice Hall.

Berner, R.A., and Canfield, D.E. (1989). A new model for atmospheric oxygen over Phanerozoic time. *Am. J. Sci.* **289**: 333–361.

Berner, R.A., and Kothavala, Z. (2001). GEOCARB III: a revised model of atmospheric CO_2 over Phanerozoic time. *Am. J. Sci.* **301**: 182–204.

Berner, R.A., Lasaga, A.C., and Garrels, R.M. (1983). The carbonate-silicate geochemical cycle and its effect on atmospheric carbon dioxide over the past 100 million years. *Am. J. Sci.* **283**: 641–683.

Boss, S.K., and Wilkinson, B.H. (1991). Planktogenic/eustatic control of cratonic/oceanic carbonate accumulation. *J. Geol.* **99**: 497–513.

Budyko, O.M.I., Ronov, A.B., and Yanshin, A.L. (1987). *History of the Earth's Atmosphere.* Translated from Russian by S.F. Lemeshko and V.G. Yanuta. Berlin, Springer-Verlag.

Brady, P.V., Krumhansl, J.L., and Papenguth, H.W. (1996). Surface complexation clues to dolomite growth. *Geochim. Cosmochim. Acta* **60**: 727–731.

Budd, A.F. (2002). Diversity and extinction in the Cenozoic history of Caribbean reefs. *Coral Reefs* **19**: 25–35.

Chave, K.E. (1954). Aspects of the biogeochemistry of magnesium 1. calcareous marine organisms. *J. Geol.* **62**: 266–283.

Chilingar, G.V. (1956). Relationship between Ca/Mg ratio and geologic age. *Am. Assoc. Petrol. Geol. Bull.* **40**: 2256–2266.

Daly, R.A. (1909). First calcareous fossils and evolution of limestones. *Geol. Soc. Am. Bull.* **20**: 153–170.

de Villiers, S., and Nelson, B.K. (1999). Detection of low-temperature hydrothermal fluxes by seawater Mg and Ca anomalies. *Science* **295**: 721–723.

Davis, K.J., Dove, P.M., and De Yoreo, J.J. (2000). The role of Mg2+ as an impurity in calcite growth. *Science* **290**: 1134–1137.

Dickson, J.A.D. (2002). Fossil echinoderms as monitor of the Mg/Ca Ratio of Phanerozoic oceans. *Science* **298**: 1222–1224.

Fisher, A.G. (1984). The two Phanerozoic super cycles. *Catastrophes in Earth history.* W.A. Berggren and J.A. Vancouvering, eds. Princeton, Princeton University Press, pp. 129–148.

Folk, R.L., and Land, L.S. (1975). Mg/Ca ratio and salinity: two controls over crystallization of dolomite. *Amer. Assoc. Petrol. Geol. Bull.* **59**: 60–68.

Garrels, R.M., and Lerman, A. (1984). Couple of the sedimentary sulfur and carbon cycles—an improved model. *Am. J. Sci.* **284**: 989–1007.

Given, R.K., and Wilkinson, B.H. (1987). Dolomite abundance and stratigraphic age-constraints on rates and mechanisms of dolomite formation. *J. Sediment. Petrol.* **57**: 1068–1078.

Hallam, A. (1984). Pre-Quaternary sea-level changes. *Ann. Rev. Earth Planet. Sci. Let.* **12**: 205–243.

Hansen, K.W., and Wallmann, K. (2003). Cretaceous and Cenozoic evolution of seawater composition, atmospheric O_2 and CO_2: a model perspective. *Am. J. Sci.* **303**: 94–148.

Hardie, L.A. (1987). Dolomitization: a critical review of some current views. *J. Sediment. Petrol.* **57**: 166–183.

Hardie, L.A. (1996). Secular variations in seawater chemistry: An explanation for the coupled secular variation in the mineralogies of marine limestones and potash evaporties over the past 600 m.y. *Geology* **24**: 279–283.

Holland, H.D. (2004). The geologic history of seawater. *Treatise on Geochemistry: The Oceans and Marine Geochemistry.* H.D. Holland and K.K. Turekian, eds., Vol. 6. Oxford, Elsevier-Pergamon, pp. 583–625.

Holland, H.D., and Zimmermann, H. (2000). The dolomite problem revisited. *Int. Geol. Rev.* **42**: 481–490.

Holser, W.T. (1984). Gradual and abrupt shifts in ocean chemistry during Phanerozoic time. *Patterns of Change in Earth Evolution.* H.D. Holland and A.F. Trendall, eds. Heidelberg, Springer, pp. 123–143.

Holser, W.T., Schidlowski, M., Mackenzie, F.T., and Maynard, J.B. (1988). Geochemical cycles of carbon and sulfur. *Chemical Cycles in the Evolution of the Earth.* C.B. Gregor, R.M. Garrels, R.T. Mackenzie, and J.B. Maynard, eds. New York, Wiley, pp. 105–173.

Horita, J., Zimmermann, H., and Holland, H.D. (2002). Chemical evolution of seawater during the Phanerozoic—implications from the record of marine evaporates. *Geochim. Cosmochim. Acta* **66**: 3733–3756.

Ingall, E.D., and Van Cappellen, P. (1990). Relation between sedimentation rate and burial of organic phosphorus and organic carbon in marine sediments. *Geochim. Cosmochim. Acta* **54**: 373–386.

Kampschulte, A. (2001). Schwefelisotopenuntersuchungen an strukturell substituierten Sulfaten in marinen Karbonaten des Phanerozoikums-Implikationen für die geochemische Evolution des Meerwassers und Korrelation verschiedener Stoffkreisläufe. Ph.D. Thesis. Ruhr Unviersität, Bochum.

Katz, M.E., Fennel, K., and Falkowski, P.G. (2007). Geochemical and biological consequences of phytoplankton. *Evolution of Primary Producers in the Sea*. P.G. Falkowski and A.H. Knoll, eds. Boston, Elsevier, pp. 405–429.

Kiessling, W. (2002). Secular variations in the Phanerozoic reef ecosystem. *Phanerozoic Reef Patterns*. W. Kiessling, E. Flügel, and J. Golonka, eds. SEPM Special Publication No. 72. Tulsa, SEPM, pp. 625–690.

Kump, L.R., Pavlov, A., and Arthur, M.A. (2005). Massive release of hydrogen sulfide to the surface ocean and atmosphere during intervals of oceanic anoxia. *Geology* 33: 397–400.

Land, L.S. (1985). The origin of massive dolomite. *J. Geol. Ed.* 33: 112–125.

Lasaga, A.C., Berner, R.A., and Garrels, R.M. (1985). An improved geochemical model of atmospheric CO_2 fluctuations over the past 100 million years. *The Carbon Cycle and Atmospheric CO_2: Natural variations Archean to Present*. E.T. Sundquist and W.S. Broecker, eds. Geophysical Monograph Vol. 32. Washington, DC, American Geophysical Union, pp. 397–411.

Lear, C.H., Elderfield, H., and Wilson, P. (2003). A Cenozoic seawater Sr/Ca record from benthic foraminiferal calcite and its application in determining global weathering fluxes. *Earth Planet. Sci. Lett.* 208: 69–84.

Lippmann, F. (1973). *Sedimentary Carbonate Minerals, Minerals, Rocks and Inorganic Materials*. Monograph Series of Theoretical and Experimental Studies, vol. 4. Berlin, Springer-Verlag.

Locklair, R.E., and Lerman, A. (2005). A model of Phanerozoic cycles of carbon and calcium in the global ocean: evaluation and constraints on ocean chemistry and input fluxes. *Chem. Geol.* 217: 113–126.

Lowenstam, H.A., and Weiner, S. (1989). *On Biomineralization*. New York, Oxford University Press.

Lowenstein, T.K., Timofeeff, M.N., Brennan, S.T., Hardie, L.A., and Demicco, R.V. (2001). Oscillations in Phanerozoic sweater chemistry: evidence from fluid inclusions. *Science* 294: 1086–1088.

Lumsden, D.N. (1985). Secular variations in dolomite abundance in deep marine sediments. *Geology* 13: 766–769.

Mackenzie, F.T., and Agegian, C.R. (1989). Biomineralization and tentative links to plate tectonics. *Origin, Evolution, and Modern Aspects of Biomineralization in Plants and Animals*. R.E. Crick, ed. New York, Plenum Press, pp. 11–27.

Mackenzie, F.T., Bischoff, W.D., Bishop, F.C., Loijens, M., Schoonmaker, J., and Wollast, R. (1983). Magnesian calcites: Low-temperature occurrence, solubility

and solid-solution behavior. *Carbonates: Mineralogy and Chemistry*. R.J. Reeder, ed. Reviews in Mineralogy Vol. 11. Washington, DC, Mineralogical Society of America, pp. 97–144.

Mackenzie, F.T., and Garrels, R.M. (1966a). Chemical mass balance between rivers and oceans. *Am. J. Sci.* 264: 507–525.

Mackenzie, F.T., and Garrels, R.M. (1966b). Silica-bicarbonate balance in the ocean and early diagenesis. *J. Sediment. Petrol.* 36: 1075–1084.

Mackenzie, F.T., and Lerman, A. (2006). *Carbon in the Outer Shell of the Earth*. Berlin, Springer.

Mackenzie, F.T., and Morse, J.W. (1992). Sedimentary carbonates through Phanerozoic time. *Geochim. Cosmochim. Acta* 56: 3281–3295.

Mackenzie, F.T., and Pigott, J.D. (1981). Tectonic controls of Phanerozoic sedimentary rock cycling. *J. Geol. Soc. Lond.* 138: 183–196.

Mackenzie, F.T., Ristvet, B.L., Thorsetenson, D.C., Lerman, A., and Leeper, R.H. (1981). Reverse weathering and chemical mass balance in a coastal environment. *River Inputs to Ocean Systems*. J.M. Marten, J.D. Burton, and D. Eisma, eds. Switzerland, UNEP and UNESCO, pp. 152–187.

Mackenzie, F.T., Ver, L.M., Sabine, C., Lane, M., and Lerman, A. (1993). C, N, P, S global biogeochemical cycles and modeling of global change. *Interactions of C, N, P, and S Biogeochemical Cycles and Global Change*. R. Wollast, F.T. Mackenzie, and L. Chou, eds. New York, Springer-Verlag, pp. 1–62.

Michalopoulos, P., and Aller, R.C. (1995). Rapid clay mineral formation in Amazon delta sediments: reverse weathering and oceanic elemental cycles. *Science* 270: 614–617.

Morse, J.W. (1983). The Kinetics of calcium carbonate dissolution and precipitation. *Carbonates: Mineralogy and Chemistry*. R.J. Reeder, ed. Reviews in Mineralogy Vol. 11. Washington, DC, Mineralogical Society of America, pp. 227–264.

Morse, J.W., and Mackenzie, F.T. (1990). *Geochemistry of Sedimentary Carbonates*. New York, Elsevier.

Mottl, M., and Wheat. C.G. (1994). Hydrothermal circulation through mid-ocean ridge flanks: fluxes of heat and magnesium. *Geochim. Cosmochim. Acta* 58: 2225–2237.

Perrin, C. (2002). Tertiary: the emergence of modern reef ecosystems. *Phanerozoic Reef Patterns*. W. Kiessling, E. Flügel, and J. Golonka, eds. SEPM Special Publication No. 72, Tulsa, SEPM, pp. 587–621.

Renard, M. (1986). Pelagic carbonate chemostratigraphy (Sr, Mg, ^{18}O, ^{13}C). *Mar. Micropaleontol.* 10: 117–164.

Retallack, G. J. (2001). A 300-million-year record of atmospheric carbon dioxide from fossil plant cuticles. *Nature* 411: 287–290.

Roberts, J.A., Bennett, P.C., Macpherson, G.L., González, L.A., and Milliken, K.L. (2004). Microbial precipitation of dolomite in groundwater: Field and laboratory experiments. *Geology* 32: 277–280.

Sandberg, P.A. (1975). New interpretation of Great Salt Lake ooïds and of ancient non-skeletal carbonate mineralogy. *Sedimentology* 22: 497–538.

Sandberg, P.A. (1983). An oscillating trend in non-skeletal carbonate mineralogy. *Nature* 305: 19–22.

Sandberg, P.A. (1985). Nonskeletal aragonite and pCO2 in the Phanerozoic and Proterozoic. *The Carbon Cycle and Atmospheric CO2: Natural Variations Archean to Present*. E.T. Sundquist and W.S. Broecker, eds. Geophysical Monographs Series Vol. 32. Washington, DC, American Geophysical Union, pp. 585–594.

Shaviv, N.J., and Veizer, J. (2003). Celestial driver of Phanerozoic climate? *GSA Today* 13: 4–10.

Smith, S.V., and Mackenzie, F.T. (1987). The ocean as a net heterotrophic system: implications from the carbon biogeochemical cycle. *Global Biogeochem. Cycles* 1: 187–198.

Sorby, H.C. (1879). The structure and origin of limestones. *Proc. Geol. Soc. Lond.* 35: 56–95.

Stanley, S.M. (2006). Influence of seawater chemistry on biomineralization through Phanerozoic time: paleontological and experimental evidence. *Palaeogeogr. Palaeoclimat. Palaeoecol.* 232: 214–236.

Stanley, S.M., and Hardie, L.A. (1998). Secular oscillations in the carbonate mineralogy of reef-building and sediment-producing organisms driven by tectonically forced shifts in seawater chemistry. *Palaeogeogr. Palaeoclimat. Palaeoecol.* 144: 3–19.

Stanley, S.M., and Hardie, L.A. (1999). Hypercalcification: paleontology links plate tectonics and geochemistry to sedimentology. *GSA Today* 9: 1–7.

Stanley, S.M., Ries, J.B., and Hardie, L.A. (2002). Low-magnesium calcite produced by coralline algae in seawater of Late Cretaceous composition. *Proc. Natl. Acad. Sci. USA* 99: 15323–15326.

Steuber, T., and Veizer, J. (2002). A Phanerozoic record of plate tectonic control of seawater chemistry and carbonate sedimentation. *Geology* 30: 1123–1126.

Thompson, G. (1983). Basalt-seawater interaction. *Hydrothermal Processes at Seafloor Spreading Centers*. P.A. Rona, K. Boström, L. Laubier, and K.L. Smith Jr., eds. Marine Sciences. NATO Conference Series Vol. 12. New York, Plenum Press, pp. 255–278.

Tyrrell, T. (1999). The relative influence of nitrogen and phosphorus on oceanic primary production. *Nature* 400: 525–531.

Tyrrell, T., and Zeebe, R. E. (2004). History of carbonate ion concentration over the last 100 million years. *Geochim. Cosmochim. Acta* 68: 3521–3530.

Vail, P.R., Mitchum, R.W., and Thompson, S. (1977). Seismic stratigraphy and global changes of sea level. 4. Global cycle of relative changes of sea level. *AAPG Mem.* 26: 83–97.

Van Cappellen, P., and Ingall, E.D. (1996). Redox stabilization of the atmosphere and oceans by phosphorous-limited marine productivity. *Science* 271: 493–496.

Vasconcelos, C., and McKenzie, J.A. (1997). Microbial mediation of modern dolomite precipitation and diagenesis under anoxic conditions (Lagoa Vermelha, Rio De Janeiro, Brazil). *J. Sed. Petr.* 67: 378–390.

Veizer, J. (2005). Celestial climate driver: a perspective from four billion years of the carbon cycle. *Geosci. Canada* 32: 13–28.

Veizer, J., Ala, D., Azmy, K., Bruckschen, P., Buhl, D., Bruhn, F., Carden, G.A.F., Diener, A., Ebneth, S., Godderis, Y., Korte, C., Pawellek, F., Podlaha, O.G., and Strauss, H. (1999). $^{87}Sr/^{86}Sr$, $\delta^{13}C$ and $\delta^{18}O$ evolution of Phanerozoic seawater. *Chem. Geol.* 161: 59–88.

Veizer, J., and Mackenzie, F.T. (2004). Evolution of sedimentary rocks. *Treatise on Geochemistry: Sediments, Diagenesis, and Sedimentary Rocks*. F.T. Mackenzie, ed. Vol. 7. Oxford, Elsevier-Pergamon, pp. 369–407.

Vinogradov, A.P., and Ronov, A.B. (1956a). Composition of the sedimentary rocks of the Russian platform in relation to the history of its tectonic movements. *Geochemistry* 6: 533–559.

Vinogradov, A.P., and Ronov, A.B. (1956b). Evolution of the chemical composition of clays of the Russian platform. *Geochemistry* 2: 123–129.

Wallmann, K. (2001). Controls on the Cretaceous and Cenozoic evolution of seawater composition, atmospheric CO_2 and climate. *Geochim. Cosmochim. Acta* 65: 3005–3025.

Wilkinson, B.H., and Algeo, T.J. (1989). Sedimentary carbonate record of calcium magnesium cycling. *Am. J. Sci.* 289: 1158–1194.

Wilkinson, B.H., Owen, R.M., and Carroll, A.R. (1985). Submarine hydrothermal weathering, global eustasy and carbonate polymorphism in Phanerozoic marine oolites. *J. Sedim. Petrol.* 55: 171–183.

Wilkinson, B.H., and Walker, J.C.G. (1989). Phanerozoic cycling of sedimentary carbonate. *Am. J. Sci.* 289: 525–548.

Wollast, R., and Mackenzie, F.T. (1983). The global cycle of silica. *Silicon Geochemistry and Biogeochemistry*. S.R. Ashton, ed. New York, Academic Press, pp. 39–76.

Yapp, C.J., and Poths, H. (1992). Ancient atmospheric CO_2 pressures inferred from natural goethites. *Nature* 355: 342–344.

Zimmermann, H. (2000). Tertiary seawater chemistry—implications from fluid inclusions in primary marine halite. *Am. J. Sci.* 300: 723–767.

18

Geochemical and Biological Consequences of Phytoplankton Evolution

MIRIAM E. KATZ, KATJA FENNEL, AND
PAUL G. FALKOWSKI

I. INTRODUCTION

The early evolution of Earth's atmosphere, oceans, and lithosphere strongly influenced, and in turn was influenced by, the evolution of life. Reconstructing the interactions and feedbacks between life and abiotic systems is one of the most challenging problems in understanding the history of the planet.

Analyses of the stable isotopes of five of the six major light elements that comprise all life on Earth (H, C, N, O, and S; there is only one stable isotope of P) have provided clues about the co-evolution of life and geophysical processes on Earth. Of these, the isotopic records of carbon and sulfur are perhaps the most useful in reconstructing redox chemistry resulting from biological evolution.

Because these two elements are extremely abundant on the surface of the planet and are capable of accepting and donating four and eight electron equivalents, respectively, on geological time scales, their global biological chemistry exerts a major control on the redox conditions of the atmosphere and oceans. In this chapter, we examine the role of marine photoautotrophs on Earth's carbon cycle, with an emphasis on the impact of these organisms on the redox changes inferred from the isotopic signals preserved in the geological record.

A. The Two Carbon Cycles

There are two major carbon cycles operating on Earth. The "geological" (or "slow") carbon cycle consists of a linked set of wholly abiotic, acid-base reactions in which CO_2 is outgassed primarily via volcanism, becomes hydrated to form carbonic acid, and is subsequently consumed, primarily by reactions with magnesium- and calcium-bearing silicate rocks, to form magnesium and calcium carbonates. The cycle is completed when the carbonates are subducted along active continental margins, heated in the mantle, and outgassed again. In this cycle, no electron transfers occur and there are no significant changes in the isotopic compositions of the various carbon pools (Berner 2004). This geological carbon cycle, which is strictly dependent upon tectonic processes, operates on time scales of hundreds of millions of years; on such time scales, it is generally assumed that the balance between volcanism and weathering largely determines the CO_2 concentration in Earth's atmosphere (e.g., Berner, 1991).

The "biological" (or "fast") carbon cycle is biologically driven and is based on redox reactions, which are at the core the fundamental chemistry of life (Falkowski 2001). Redox reactions are characterized by transfers of electrons with or without protons, and like acid-base reactions, they always occur in pairs. CO_2 is a potential electron sink; in its most reduced form, it can accept

up to four electrons and protons. Such reduction reactions are endergonic and do not occur spontaneously on Earth's surface without biological catalysis. The biologically catalyzed reduction of CO_2 invariably leads to the production of an oxidant. If oxygenic photosynthesis is the catalytic reaction, then H_2O is the source of the electrons and protons, and the oxidized water is converted to O_2. In the steady state, the oxidant produced by the photoautotrophs is rapidly consumed (reduced) by heterotrophs to reoxidize the organic matter. However, if some small fraction of the organic matter is buried in sediments, it can escape biological reoxidation; the buried organic matter is, in effect, a sink of reductant. Electron balance requires that if reductants are buried in the lithosphere, oxidants must be produced somewhere else; in this case, this occurs in the atmosphere and oceans (Holland 1984). Hence, burial of organic matter implies oxidation of the surface of Earth (Hayes et al. 1999; Hayes and Waldbauer 2006).

The biological reduction of CO_2 to form organic matter is invariably associated with a large kinetic isotopic fractionation of up to $\sim 25\%$. Hence, long-term burial of organic matter depletes the atmosphere and ocean of ^{12}C, resulting in ^{13}C-enriched carbonate pools (e.g., Broecker and Peng 1982; Kump and Arthur 1999). Similarly, as long-term burial of organic matter depletes the atmosphere and ocean of ^{12}C, organic matter becomes continuously enriched in ^{13}C over time. These two fractionations are the basis of the isotopic mass balance quantified as:

$$f_w * \delta^{13}C_w + f_v * \delta^{13}C_v = f_{carb} * \delta^{13}C_{carb} + f_{org} * \delta^{13}C_{org} \quad (1)$$

where f = fraction (i.e., a dimensionless ratio in which the sum of all "fs" in the equation equals 1.0), w = weathering processes, v = volcanic/hydrothermal outgassing, carb = carbonate, and org = organic carbon.

This equation is a biogeochemical analogue of the physical notion that, in the steady state, "what goes up, must come down." The left-hand side of the equation represents the input of carbon to the atmospheric/oceanic/terrestrial ecosystems from the lithosphere, and the right-hand side represents the output of organic matter and carbonates from the ocean and land to the lithosphere. Clearly, net oxidation of Earth's atmosphere means that the equation cannot be in steady state; rather, the burial of organic carbon must exceed its oxidation in this scenario (Berner and Canfield 1989).

The average isotopic value of mantle CO_2 is ca. $-5‰$, whereas that of organic matter is ca. $-25‰$. Hence, to balance the input with the output, the ratio of buried organic carbon relative to total carbon is about 0.20 to 0.25. That is, for every one atom of carbon buried as organic matter, between four and five atoms are buried as carbonate (see Guidry *et al.*, Chapter 17, this volume). Hence, changes in marine $\delta^{13}C_{carb}$ and $\delta^{13}C_{org}$ through time serve as sedimentary archives of changes in carbon sources and sinks, thereby providing the best monitor available to reconstruct the geological carbon cycle and its role in the oxidation state of the planet's surface.

B. The "Great Oxidation Event" and the Wilson Cycle

Although the evolution of oxygenic photosynthesis provided a mechanism to produce large amounts of O_2 from the ubiquitous and abundant sources of reductant (water) and energy (solar radiation), this process, in and of itself, was not sufficient to oxidize Earth's atmosphere and oceans. Net oxidation requires net long-term burial of organic carbon in the lithosphere, where it is protected from reoxidation. Let us briefly examine how this process evolved.

The two basic physical processes involved in continental evolution are the Archimedes Principle and heat conduction. The former holds that less dense bodies will float on more dense bodies; the latter suggests that thicker, more insulated bodies do not conduct heat as efficiently as thinner, less insulated bodies. Let us put these two concepts to work in understanding how tectonic processes alter continental configuration and how that, in turn, influences the carbon cycle and the oxidation state of Earth.

It is generally believed there were no continents in the early Archean Eon (Drake and Righter 2002). The growth of continents required repeated subduction of dense, basaltic mantle rocks, where heating and recrystallization in the presence of water vapor produced silicate-rich (so-called felsic) rocks such as granites. The granitic rock aggregated over time into larger units to form cratons (the large, heterogenous lithopheric bases of continents), which are lighter than the underlying mantle, and therefore float on that semifluid structure.

Cratons are thick, about three to four times thicker than the denser basalts (called mafic rocks) that form the oceanic crusts. Near Earth's surface, the heat that emanates from within the core is conducted about three times more efficiently through the thin oceanic crust than it is through the thicker continental crust. This differential heat dissipation leads to the build-up of thermal energy below large cratons, causing massive pressures beneath them. As the heat and pressure build, the craton (continent) eventually thins and fractures, and pieces of continents slowly rift apart, creating a new oceanic ridge spreading center between the continental fragments. Heat-driven convection in the mantle below the lithosphere drives the tectonic plates apart with the fragmented continents attached, and new oceanic crust forms at the intervening spreading center. The oceanic crust becomes cooler and denser as it ages and eventually subsides as it moves away from the spreading center. When this old basaltic crust becomes so dense that it subsides below the less dense adjacent continental crust, a new subduction zone is created, the whole process reverses itself, and the ocean basin is consumed as the continents ultimately reassemble. The episodic break-up, dispersal, and subsequent reassembly of supercontinents occurs over ~300–500 million year (myr) intervals (e.g., Valentine and Moores 1974; Fischer, 1984; Rich *et al.* 1986; Worsley *et al.* 1986) and is known as the "Wilson cycle," named after its conceptual discoverer, J. Tuzo Wilson (Wilson 1966).

A critical point in the discussion of the Wilson cycle is the exchange between oceanic crust and cratons. As oceanic crust subducts along continental margins, a fraction of the oceanic material can be "shaved off" during orogenic (mountain-building) processes along the active margin. In this way, some oceanic sediments and crust can accrete onto the cratons, effectively removing these sediments from the Wilson cycle. Hence, organic matter buried on the ocean margins can be displaced onto the cratons, where it may be sequestered for hundreds of millions (and even billions) of years, even though portions of the cratons themselves are returned to the oceans via weathering of continental rocks. This process of continental accretion is critical to allowing oxidants to accumulate in the atmosphere; without cratons, the oxygen concentration on Earth would be much lower than it is at present. Hence, the net oxidation of Earth 2.3 billion years ago (Ga), known as the "Great Oxidation Event" (Holland 1997), implies that the rates of burial and sequestration of organic carbon exceeded the rates of weathering and oxidation. This process almost certainly was enabled by the formation of large cratons during this period in Earth's history. However, although the tectonic processes on Earth are absolutely essential to planetary geochemistry, like photosynthesis, tectonics alone is also insufficient to lead to a net oxidation of Earth's atmosphere and oceans. Form and function of the photoautotrophs play an integral role.

II. THE ROLE OF PHYTOPLANKTON IN THE GEOLOGICAL CARBON CYCLE

Marine phytoplankton constitute less than 1% of Earth's photosynthetic biomass today, yet they are responsible for more than 45% of our planet's annual net primary production (Field *et al.* 1998). How has primary and export production affected the carbon cycle in the past? To answer this question, we begin with a review of the macroevolutionary trends of marine phytoplankton.

A. Early Phytoplankton Evolution

The evolutionary succession of marine photoautotrophs began in the Archean Eon with the origin of photosynthesis, perhaps as early as 3.8 Ga (see Knoll *et al.*, Chapter 8, this volume). Eukaryotes first appeared in the fossil record ~1800 million years ago (Ma) (Han and Runnegar 1992; Knoll 1994; Javaux *et al.* 2001), yet molecular biomarkers show that both prokaryotic cyanobacteria and eukaryotic algae evolved by 2700 Ma (Brocks *et al.* 1999; Summons *et al.* 1999), if not earlier (Knoll 2003; Knoll *et al.*, Chapter 8, this volume). A schism occurred early in the evolution of eukaryotic photoautotrophs and gave rise to the two major plastid superfamilies, the "green" (chlorophyll *b*-containing) and "red" (chlorophyll *c*-containing) plastid groups (Figure 1). By 1200 Ma, the red algae appear to have been among the first group of organisms to differentiate into multicellular forms. Cyanobacteria dominated primary production during most of the Paleoproterozoic; eukaryotic green algae became increasingly important toward the end of the Proterozoic (543 Ma) (Tappan 1980; Knoll 1989, 1992; Lipps 1993; Knoll *et al.*, Chapter 8, this volume).

The now-extinct eukaryotic acritarchs were unicellular, noncolonial, organic-walled microfossils of unknown polyphyletic affinity that include the phycomata and vegetative cells of the green Prasinophyceae algae (Stover *et al.* 1996). They appeared in the fossil record by ~1700–1900 Ma (Zhang 1986; Summons *et al.* 1992) and were clearly an important component in the fossil record in the mid- to late Proterozoic. Acritarchs diversified beginning ~800–900 Ma (Knoll 1994; Knoll *et al.* 2006) during the early stages of rifting

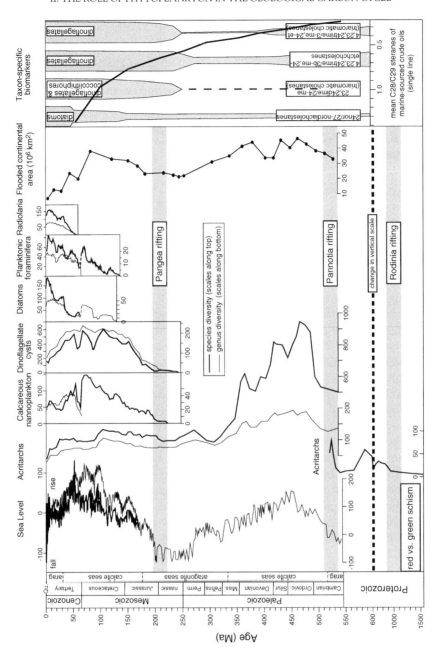

FIGURE 1.   Comparison of eukaryotic phytoplankton diversity curves with zooplankton diversity curves (this study), sea-level change (Mesozoic-Cenozoic: Miller *et al.* 2005, dark line, scale at top; Haq *et al.* 1987, light line, scale at bottom; Paleozoic: Vail *et al.* 1977, light line, scale at bottom), flooded continental areas (Ronov 1994) (after Katz *et al.* 2004). Phytoplankton and zooplankton species (dark line, scale at top) and genus (light line, scale at bottom) diversities are from published studies or were compiled for this study from publicly available databases: calcareous nannofossils (species—Bown *et al.* 2004; genus—Spencer-Cervato *et al.* 1999), dinoflagellates (Stover *et al.* 1996), diatoms (Spencer-Cervato 1999), acritarchs (Proterozoic: Knoll 1994; Phanerozoic: R.A. MacRae, unpublished data), planktonic foraminifera (Spencer-Cervato 1999), and radiolaria (Spencer-Cervato 1999). All records are adjusted to the Berggren *et al.* (1995) (Cenozoic), Gradstein (Gradstein *et al.* 1995) (Mesozoic), and GSA (http://rock.geosociety. org/science/timescale/timescl.htm) (Paleozoic) time scales. Taxon-specific biomarkers (Moldowan and Jacobson 2000) and C_{28}:C_{29} sterane ratios (Grantham and Wakefield 1988) provide a record of increased biomass preservation of eukaryotic phytoplankton in the Mesozoic-Cenozoic. Episodes of supercontinent rifting are shaded.

of the supercontinent Rodinia, perhaps in response to the expansion of physical space as well as ecological niches along growing continental margins (see Figure 1). Acritarchs underwent a second expansion in concert with the early Paleozoic radiations of marine invertebrates (Knoll 1994); their diversity peaked in the mid-Paleozoic and declined rapidly in the late Devonian to early Mississippian.

B. The Rise of the Red Lineage

In the early Mesozoic, the red eukaryotic phytoplankton began to displace green eukaryotic algae in the marine realm (e.g., Falkowski *et al.* 2004b) (see Figure 1). Marine prasinophytes declined in the Jurassic, although this group of green algae is extant as a minor constituent. A period of transition to red-line–dominated primary production began in the Triassic to Early Jurassic, ultimately resulting in the dominance of coccolithophores, dinoflagellates, and diatoms in the contemporary ocean.

The first unequivocal appearance of dinoflagellates in the fossil record occurs as organic-walled relict cysts in Middle Triassic continental margin sediments (Stover *et al.* 1996). Although molecular biomarker studies indicate that dinoflagellates may have existed as far back as the Neoproterozoic (Summons and Walter 1990; Moldowan and Talyzina 1998; Knoll *et al.*, Chapter 8, this volume), these biomarkers did not become prominent constituents of marine bitumens until the Triassic (Moldowan *et al.* 1996; Moldowan and Jacobson 2000), when microfossils more clearly document their expansion and radiation (Fensome *et al.* 1996; Stover *et al.* 1996) (see Figure 1).

The calcareous nannoplankton (dominated by coccolithophorids) originated in the Late Triassic and were the second group in the red lineage to radiate in the fossil record (Bown *et al.* 2004) (see Figure 1), at about the same time that molecular biomarkers of coccolithophorids became common (Moldowan and Jacobson 2000). The earliest nannoplankton have been identified in Carnian sediments

from the southern Alps (Janofske 1992; Bown 1998; Bown *et al.* 2004) and Nevada (F. Tremolada, personal communication).

The diatoms were the last of the three major groups of the red lineage to emerge in the Mesozoic. The highly soluble siliceous diatom frustules may impart a preservational bias to the fossil record of diatoms. Reports of diatom frustules in Jurassic sediments (Rothpletz 1896) have proven difficult to replicate, yet molecular biological clock estimates (Medlin *et al.* 2000) and molecular biomarkers (Moldowan and Jacobson 2000) indicate that diatoms may have evolved earlier but remained minor components in the marine realm until the Cretaceous. The first unequivocal fossil record of diatoms document Early Cretaceous radiations (Harwood and Nikolaev 1995; Kooistra *et al.*, Chapter 11, this volume). Early diatom morphologies were dominated by cylindrical and long-cylindrical forms that show very little variation (Gersonde and Harwood 1990; Harwood and Gersonde 1990); the morphological similarity among early diatoms suggests a common ancestor of early Mesozoic origin (Harwood and Nikolaev 1995). Late Cretaceous diatom morphologies were dominated by discoidal and biddulphioid frustules (Harwood and Nikolaev 1995). Diatoms appeared in nonmarine environments by 70 Ma (Chacon-Baca *et al.* 2002).

Regardless of the exact timing of the evolutionary origins of coccolithophores, dinoflagellates, and diatoms, fossil and biomarker data document the major expansion of all three groups in the Mesozoic (Grantham and Wakefield 1988; Harwood and Nikolaev 1995; Stover *et al.* 1996; Moldowan and Jacobson 2000; Bown *et al.* 2004) (see Figure 1). They began their evolutionary trajectories to ecological prominence when the supercontinent Pangea began to break apart and the Atlantic Ocean basin opened in the Late Triassic–Early Jurassic (~200 Ma), marking the opening phase of the current Wilson cycle (Wilson 1966; Worsley *et al.* 1986).

When Pangea was fully assembled, the larger-celled phytoplankton with high nutrient requirements most likely were concentrated along the supercontinent margins where nutrient supply was greatest, whereas the small-celled plankton (e.g., cyanobacteria) likely better competed in the large, oligotrophic Panthalassic Ocean (Finkel, Chapter 15, this volume). The balance between large- and small-celled phytoplankton began to shift as Pangea started to rift in the Jurassic. Sea level rose as Pangea fragmented and the Atlantic Ocean basin widened, flooding broad continental shelves and low-lying inland areas (see Figure 1). The fragmentation of the continents and creation of a new ocean basin produced an increase in the total length of coastline where many plankton lived. Nutrients (such as phosphate) that were previously locked up in the large continental interior of Pangea were transported to newly formed shallow seas and distributed over wider shelf areas and longer continental margins; in addition, models predict that the hydrological cycle accelerated when Pangea rifted, delivering more nutrients to the oceans (Wallman 2001). These changes were profound: Greater nutrient availability coupled with expanded ecospace and ecological niches appears to have selected for the large-celled phytoplankton that lived along continental margins and contributed to their rapid radiation and evolution. Accordingly, the diversities of eukaryotic phytoplankton of the red lineage parallel sea-level rise through the Mesozoic (see Figure 1).

The Cretaceous–Tertiary boundary bolide impact caused mass extinctions (Alvarez et al. 1980) that are recorded in the fossil records of the calcareous nannoplankton and, to a lesser extent, the diatoms and dinoflagellates (see Figure 1). Dinoflagellates and calcareous nannoplankton recovered to near pre-extinction diversity levels by the earliest Eocene (~55 Ma), only to decline through the rest of the Cenozoic as long-term sea level fell.

In contrast to the other phytoplankton, diatom diversity has increased through the Cenozoic despite shrinking niche availability due to falling sea level. Increased bioavailability of silica may have contributed to this radiation of diatoms. Diatoms require orthosilicic acid to form extremely strong shells called frustules (Falciatore and Bowler 2000). Silica is supplied to the oceans primarily from continental weathering, a process that was accelerated by the mobilization of silica from soils by grasses because grasses contain up to 15% silica in phytoliths (micromineral deposits in cell walls). As grasses evolved, radiated, and expanded, increased transfer of silica to the oceans (primarily via fluvial erosion and, secondarily, via aeolian transport) increased the bioavailability of silica for diatom growth (Falkowski et al. 2004a, b). This mechanism may account for the close correlation between the evolutionary histories of grasses, terrestrial grazing animals, and diatoms.

The oldest grass phytoliths (70 Ma, Late Cretaceous) are preserved in coprolites of sauropod dinosaurs (Prasad et al. 2005); diatoms also became fairly common in the Late Cretaceous (Harwood and Nikolaev 1995). Although grasslands remained sparse in the Paleocene and Eocene (65–33.7 Ma), the increased abundance of phytoliths in the marine record beginning in the mid-Eocene (Retallack 2001) indicates increased delivery of bioavailable silica to the oceans (Falkowski et al. 2004a, b). Near the Eocene–Oligocene boundary, grasslands expanded (Retallack 2001), grazers displaced browsers (Janis and Damuth 2000), and diatoms diversified (Spencer-Cervato 1999) (see Figure 1). A second pulse of diversification occurred in the middle to late Miocene, when the further expansion of grasslands (including a shift from C_3 to C_4 grasses) (Retallack 1997, 2001) was accompanied by a second pulse of diatom diversification at the species level (Falkowski et al. 2004a, b). Through the Cenozoic, diatoms have been increasingly successful in outcompeting radiolaria (siliceous zooplankton) for silica (Harper and Knoll 1975; Kidder and Erwin 2001) (see Section IV.B for discussion of Phanerozoic siliceous sedimentation).

The long-term success of diatoms in the Cenozoic may also be attributed in part to increasing latitudinal thermal gradients and decreasing deep-ocean temperatures that have contributed to greater vertical thermal stratification in the latter half of the Cenozoic (Falkowski *et al.* 2004a). These changes increased the importance of wind-driven upwelling and mesoscale eddy turbulence, which provide nutrients to the upper ocean. Sporadic nutrient influx to the euphotic zone may favor diatoms over coccolithophores and dinoflagellates, and a change in concentration and pulsing of nutrients may favor small diatoms over large diatoms. As a result, there has been a significant decrease in the average size of diatoms in the Cenozoic, with periods of change concentrated in the mid- to late Eocene and early to mid-Miocene (Finkel *et al.* 2005).

C. Biological Overprint of the Geological Carbon Cycle

There is a clear link between the history of phytoplankton evolution and the carbon cycle. In the contemporary ocean, marine phytoplankton are responsible for almost half of Earth's annual net primary production (Field *et al.* 1998) and are major contributors to export production (e.g., Falkowski and Raven 1997; Dugdale *et al.* 1998; Smetacek 1999; Laws *et al.* 2000). The efficiency of export of organic matter from the surface to the ocean interior depends on specific phytoplankton groups (Falkowski 1998; Armstrong *et al.* 2002). Although zooplankton fecal pellets (Urrere and Knauer 1981) and marine snow (Shanks and Trent 1980) can accelerate the sinking of organic matter (Smayda 1969, 1970), the direct flux of large eukaryotic phytoplankton accounts for a large fraction of the export flux. In addition, many eukaryotic phytoplankton have armor that is denser than seawater (Hamm and Smetacek, Chapter 14, this volume), predisposing them to sink. Although cyanobacteria numerically dominate the oceanic phytoplankton community today, their extremely small size (ca. 10^{-2}–$10^2 \, \mu m^3$) greatly reduces their sinking rate. In contrast, eukaryotic phytoplankton, which range in size from ca. 10^1 to more than $10^6 \, \mu m^3$, have much greater potential sinking rates and also are much more likely to be incorporated into fecal pellets of macrozooplankton, which also contribute significantly to the overall export of organic matter (Roman *et al.* 2002).

On time scales of centuries, the export of organic matter from the surface to the ocean interior helps to maintain a lower partial pressure of CO_2 in the atmosphere (Volk and Hoffert 1985). However, on geological time scales, some small fraction of the exported carbon becomes incorporated into marine sediments, where it is effectively removed from the mobile carbon pools of the atmosphere and oceans (Falkowski *et al.* 2003). In addition to the size and composition of particles, the fraction of sinking material that reaches the sediments also depends on depth of the water column (Betzer *et al.* 1984; Martin *et al.* 1987). Carbon burial is most efficient on shallow continental margins (Premuzic 1980; Hedges and Keil 1995), where both export production and sedimentation rates are high (Falkowski *et al.* 1994). Hence, the area of shallow seas and length of coastlines along tectonically passive margins potentially influences the net burial of the export flux. Therefore, the taxonomic composition and cell size of phytoplankton communities that have the potential to control export flux, combined with the geographic extent of shallow seas and coastlines determined by tectonic processes, has the potential to determine the fraction of organic matter buried in marine sediments. These processes, in turn, impact the carbon cycle, the redox state of the oceans, and the long-term concentration of CO_2 in the atmosphere (Hayes and Waldbauer 2006).

III. THE PHANEROZOIC CARBON ISOTOPE RECORD

The sequestration of organic matter in marine sediments leads to changes in the isotopic composition of the mobile pools of inorganic carbon in the ocean and atmosphere. Hence, knowledge of isotopic variations in both carbonates and organic matter can be used to constrain the numerical solution to the isotopic mass balance equation (Equation 1). Let us now consider the carbon isotopic records in carbonates and organic matter throughout the Phanerozoic.

There are two major Phanerozoic carbon isotope compilations. Veizer *et al.* (1999) compiled $\delta^{13}C_{carb}$ of many fossil groups including brachiopods, belemnites, oysters, planktonic and benthic foraminiferas, and corals. Although there are some issues regarding the preservation of some of the samples in the compilation, selected data have been used extensively to infer oxygen and CO_2 concentrations over the past 550 million years (Berner 2001, 2004, 2006). Hayes *et al.* (1999) compiled published and unpublished $\delta^{13}C_{org}$ and then modified the data based on several criteria to produce a smoothed $\delta^{13}C_{org}$ curve for the past 800 million years; unfortunately, the authors have neither electronic nor paper files of their unmodified datasets (Hayes, personal communication). Hence, it is not possible to construct a high-resolution record from the low-resolution, time-averaged, modified data reported in Hayes *et al.* (1999).

In order to constrain the carbon isotopic mass balance, we developed high-resolution datasets from the Jurassic to present (Falkowski *et al.* 2005; Katz *et al.* 2005a). Because this interval spans the record of extant marine sediments, it is more accessible to isotopic analyses than older periods in Earth's history. We produced new isotopic curves from this interval for two major reasons: (1) to utilize uniform, unaltered, and suitable sample material that provide the highest quality isotopic analyses and (2) to achieve tight stratigraphic control of high-resolution datasets. Bulk sediment samples were analyzed because they best characterize the carbon outflow from the ocean/atmosphere/biosphere, and provide the average $\delta^{13}C$ of the total carbonate and organic carbon sequestered in marine sediments (Shackleton 1987; Katz *et al.* 2005a), allowing us to monitor long-term changes in the global carbon cycle through time. We specifically elected not to use carbon isotope records generated from specific organisms, which may reflect the different environments where each of those organisms lived (e.g., nearshore surface ocean versus deep-ocean bottom water), rather than the average $\delta^{13}C$ output from the system.

We analyzed samples from open ocean Atlantic Deep Sea Drilling Project (DSDP) and Ocean Drilling Program (ODP) boreholes with well-documented magnetobiostratigraphies that provide excellent age control and minimize the risk of undetected unconformities. Using open ocean locations circumvents problems that may be encountered in analyzing epicontinental sections, such as unconformities associated with sea-level changes and local overprint of geochemical signals (e.g., Smith *et al.* 2001). It was necessary, however, to use an epicontinental section for the older record because there is little to no pre-Middle Jurassic ocean floor left. We chose a single location (Mochras Borehole, Wales) that spans the entire Lower Jurassic, with well-documented lithology and biostratigraphy (Ivimey-Cook 1971; Woodland 1971).

We determined the isotopic composition of both carbonates (Shackleton and Hall 1984; Katz *et al.* 2005a) and organic matter (Falkowski *et al.* 2005) from a series of Atlantic marine sediment cores from the Lower Jurassic through the Cenozoic (Figure 2). These carbon isotope data are coeval, high-resolution (225,000-year average sampling interval) records from both carbonates and organic matter that cover the past 205 million years, providing full constraint on the carbon sinks. Correlations to shorter duration records of transient

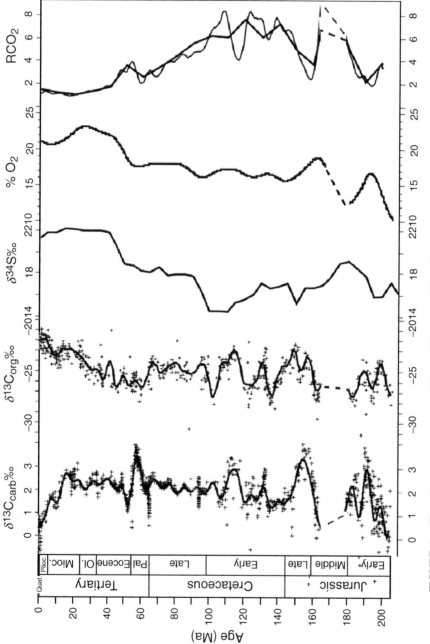

FIGURE 2.   Three isotopic signatures ($\delta^{13}C_{carb}$, $\delta^{13}C_{org}$, and $\delta^{34}S_{sulf}$; Kampschulte and Strauss 2004; Falkowski et al. 2005; Katz et al. 2005a) were used to reconstruct atmospheric O_2 (Falkowski et al. 2005) and atmospheric CO_2 (reported as levels relative to today, RCO_2; the low-resolution record is from Katz et al. 2005b; the high-resolution record was produced for this study) for the Jurassic-Cenozoic. We generated RCO_2 curves using two generations of Berner's GEOCARB/GEOCARBSULF modeling series (Berner and Kothavala 2001; Berner 2006).

$\delta^{13}C$ excursions (thousand-year scale) and $\delta^{13}C$ events (3–20 million years) establish the global nature of the $\delta^{13}C_{carb}$ dataset (Katz *et al.* 2005a). The $\delta^{13}C_{carb}$ record reveals a 190 million-year increase of ~1.1‰ from the early Jurassic through the mid-Miocene and a subsequent ~2.5‰ decrease (see Figure 2). Statistical analysis of the regression of the ~1.1‰ long-term increase indicates that the slope is significantly different from zero ($P < 0.05$). The long-term increase is supported by combining Figures 2 and 3 from Hayes *et al.* (1999); when spliced together, the resulting $\delta^{13}C_{carb}$ record shows a long-term increase of ~1‰ from 200–20 Ma that was not identified. Over the same time period, statistical analysis of the $\delta^{13}C_{org}$ record also reveals a long-term secular increase of ~5‰ (Falkowski *et al.* 2005), indicating that organic matter became depleted in the light carbon isotope.

A. Jurassic to Mid-Miocene 1.1‰ $\delta^{13}C_{carb}$ Increase

The simultaneous increases in $\delta^{13}C_{carb}$ and $\delta^{13}C_{org}$ highlight a long-term increase in $\delta^{13}C$ of the mobile carbon reservoir (see Figure 2), which could have been driven by a combination of two processes: (1) an increase in the $\delta^{13}C$ of input carbon ($\delta^{13}C_{input}$) and/or (2) an increase in the fraction of organic carbon buried (f_{org}). To investigate the extent to which each of these two processes contributed to the long-term $\delta^{13}C$ increases, we ran model simulations based on a derivation of Equation 1:

$$f_{org} = \left[\left(\delta^{13}C_{input} \right) - \left(f_{carb} * \delta^{13}C_{carb} \right) \right] / \delta^{13}C_{org} \quad (2)$$

Four model runs are shown (Figure 3) that use the $\delta^{13}C_{carb}$ (Katz *et al.* 2005a) and $\delta^{13}C_{org}$ (Hayes *et al.* 1999; Falkowski *et al.* 2005) datasets to calculate the burial fractions of carbonate versus organic carbon. In the first set of model runs, $\delta^{13}C_{input}$ was allowed to vary according to parameters described in GEOCARB III (Berner and Kothavala 2001), which include various feedbacks and variables such as the influence of land plants on weathering, global temperature, paleogeography, and continental water discharge (see Berner and Kothavala 2001 for details of the model). In the second set of simulations, $\delta^{13}C_{input}$ was held constant at −5‰, based on the assumption that carbonate and organic carbon weathering from the continents averages out to this contemporary mantle carbon value (−5‰) over long time periods (e.g., Kump and Arthur 1999). The results from both of these analyses indicate that increases in $\delta^{13}C_{carb}$ and $\delta^{13}C_{org}$ require an f_{org} increase of ~0.05–0.1, regardless of whether $\delta^{13}C_{input}$ is allowed to vary (Berner and Kothavala 2001) or is held constant (Kump and Arthur 1999), and regardless of whether the high- (Falkowski *et al.* 2005) or low- (Hayes *et al.* 1999) resolution $\delta^{13}C_{org}$ dataset was used (see Figure 3). Hence, the long-term increase in the burial of organic carbon implied in the isotopic records requires greater burial efficiency (i.e., long-term sequestration) of organic matter in marine and/or terrestrial environments.

Sensitivity tests establish that neither excess burial of organic matter (marine and/or terrestrial) nor increase in $\delta^{13}C_{input}$ alone can account for the measured $\delta^{13}C$ changes (Berner and Kothavala 2001; Katz *et al.* 2005a). Rather, increases in both the extraction of ^{12}C from, and supply of ^{13}C to, the mobile carbon reservoir are required to account for the measured changes in $\delta^{13}C_{carb}$ and $\delta^{13}C_{org}$ (Falkowski *et al.* 2005; Katz *et al.* 2005a). Following this logic, we conclude that the concurrent changes in the isotopic composition of both organic and inorganic carbon pools must have occurred through (1) an increase in net burial of organic carbon in the lithosphere, which implies a net increase in the oxidation state of the atmosphere (Falkowski *et al.* 2005; Katz *et al.* 2005a), and (2) an increase in $\delta^{13}C$ of carbon introduced to the mobile carbon reservoir from volcanic outgassing and weathering of continental rocks (Caldeira 1992; Schrag 2002; Katz *et al.* 2005a).

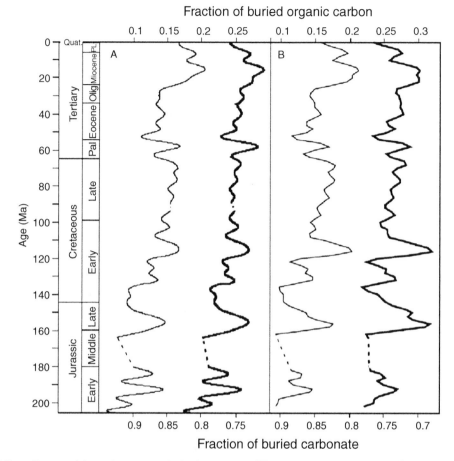

FIGURE 3. Four model simulations use the high-resolution $\delta^{13}C_{carb}$ (Katz *et al.* 2005a) and $\delta^{13}C_{org}$ (**Panel A** uses Hayes *et al.* 1999; **Panel B** uses Falkowski *et al.* 2005) datasets with two input (source) carbon scenarios (dark line: $\delta^{13}C_{input} = -5‰$; light line: variable $\delta^{13}C_{input}$ according to Berner and Kothavala [2001] parameters) to predict variations in the burial of carbonate (f_{carb}) relative to organic carbon (f_{org}) through time. Each model run shows that the measured $\delta^{13}C_{carb}$ and $\delta^{13}C_{org}$ datasets require an f_{org} increase of ~0.05–0.1, despite different datasets and input carbon parameters. This requires greater burial efficiency (i.e., long-term sequestration) of organic matter in marine and/or terrestrial environments through time.

Near the Eocene–Oligocene boundary (~33 Ma), $\delta^{13}C_{org}$ began to increase more rapidly (Hayes *et al.* 1999), whereas the rate of increase in $\delta^{13}C_{carb}$ (Shackleton and Hall 1984) remained relatively constant until ~15 Ma (see Figure 2). These records indicate that f_{org} increased during this interval to the highest level of the past 205 million years (see Figure 3), culminating in the "Monterey Carbon Excursion" in which large amounts of organic-rich, diatomaceous sediments were deposited in marginal basins (Vincent and Berger 1985).

B. 2.5‰ $\delta^{13}C_{carb}$ Decrease Since the Mid-Miocene

$\delta^{13}C_{carb}$ values have decreased by ~2.5‰ since ~15 Ma (Shackleton and Hall 1984), whereas $\delta^{13}C_{org}$ values have continued to increase (Hayes *et al.* 1999) (Figure 2). This requires an increase in ^{12}C in the mobile carbon reservoir through greater ^{12}C supply and/or less ^{12}C burial. Because Sr and Os isotopic composition varies with rock type, $^{87}Sr/^{86}Sr$ and $^{187}Os/^{186}Os$ measured in marine sediments can be used to interpret continental weathering rates and continental source rock types. These isotopic records indicate a shift in the Neogene to continental source rocks rich in organic carbon, which may have increased the supply of ^{12}C to the oceans even though continental weathering rates may have decreased (Ravizza 1993; Derry and France-Lanord 1996; Turekian and Pegram 1997). Although an additional ~194,000 Gt of carbon from organic carbon weathering (at the expense of carbonate weathering) can account for the entire 2.5‰ $\delta^{13}C_{carb}$ decrease, the higher $\delta^{13}C_{org}$ values can in part be better attributed to the increasing importance of β-carboxylation photosynthetic pathways in marine phytoplankton and C_4 pathways in terrestrial plants in the latter part of the Cenozoic (Katz et al. 2005a). These pathways produce ^{13}C-enriched organic matter (i.e., higher $\delta^{13}C$ values) relative to organic matter produced through the C_3 photosynthetic pathway.

For most of the Phanerozoic, marine and terrestrial photoautotrophs used a C_3 photosynthetic pathway to fix carbon (Falkowski and Raven 1997). The long-term drawdown of CO_2 associated with greater organic carbon burial since the break-up of Pangea was a key factor that selected for new photosynthetic pathways in marine and terrestrial ecosystems. Diatoms have β-carboxylation pathways that use HCO_3^- as a substrate (Morris 1987; Reinfelder et al. 2000) and do not discriminate as strongly against ^{13}C (Falkowski 1997). These organisms are responsible for a disproportionate fraction of carbon export in the modern ocean (Dugdale et al. 1998; Smetacek 1999), especially along continental margins (Sancetta et al. 1991; Bienfang 1992). The rapid radiation of diatoms in the mid-Cenozoic may have resulted in marine export production that was enriched in ^{13}C (Katz et al. 2005a). In the mid- to late Miocene oceans, an expansion of bloom-forming diatoms (L. Burckle and R. Sombratto, personal communication) further contributed to export production of ^{13}C- enriched organic matter. In the terrestrial realm, grasslands

expanded throughout most of the world in the late Miocene (6–8 Ma), accompanied by a shift in dominance from C_3 to C_4 grasses, resulting in ^{13}C-enriched terrestrial biomass (Cerling et al. 1997; Still et al. 2003). Consequently, ^{13}C-enriched terrestrial organic matter was ultimately transferred to and sequestered in the oceans (France-Lanord and Derry 1994; Hodell 1994) at the same time that ^{13}C-enriched diatoms continued to expand and export ^{13}C-enriched organic matter to the seafloor. The rise of β-carboxylation and C_4 photosynthetic pathways can account for a 1.1‰ decrease in $\delta^{13}C_{carb}$ (based on the measured $\delta^{13}C_{org}$, and assuming that $\delta^{13}C_{input}$ remained the same). The remaining ~1.4‰ decrease requires an additional ~110,000 Gt of organic carbon weathered from organic-rich shales (Katz et al. 2005a), as discussed previously.

IV. FEEDBACKS IN BIOGEOCHEMICAL CYCLES

Various components of global paleobiogeochemical systems can undergo complex interactions and feedbacks through time. Evidence of these interactions is recorded in geological records. Fossil evidence for phytoplankton evolutionary pulses appear to have been associated with the past three Wilson cycles. In turn, changes in continental configurations and evolutionary shifts have influenced marine sedimentation patterns and global climates. In this chapter, we have suggested there may be a causal relationship between the large-scale tectonics of the current Wilson cycle and biogeochemical cycles driven, at least in part, by changing phytoplankton community structure. Let us now take a closer look at past Wilson cycles, macroevolutionary changes, and biogeochemical cycles.

A. Phytoplankton Community Structure and the Wilson Cycle

Earlier studies suggested a correlation between evolutionary pulses in the marine realm and the Wilson cycle (e.g., Valentine and Moores 1974; Fischer 1984; Rich *et al.* 1986; Worsley *et al.* 1986) and provided a basis for developing hypotheses about causal rather than casual linkages between these two processes (Katz *et al.* 2004). Biological diversity is maintained by isolation or fluctuations in habitat areas (MacArthur and Wilson 1967; Rosenweig 1995). The apparent correlation between phytoplankton species diversity and the Wilson cycle suggests that the break-up of supercontinents created new, unstable physical habitats for the eukaryotic phytoplankton groups that lived along continental perimeters (Katz *et al.* 2004). As the supercontinent Rodinia rifted in the Late Proterozoic, early eukaryotic acritarchs began to expand (see Figure 1). At this time, surface waters were somewhat oxic, whereas subsurface water masses were either anoxic or sulfidic (Anbar and Knoll 2002). Higher nutrient levels and better oxygenated waters in coastal regions may have favored early phytoplankton of the red lineage, whereas the phytoplankton of the green lineage outcompeted in oligotrophic open ocean surface waters that were underlain by anoxic or sulfidic deepwaters that delivered a different suite of trace elements to the surface (Whitfield 2001; Anbar and Knoll 2002; Katz *et al.* 2004).

Acritarchs diversified again in the Early Paleozoic when the supercontinent Pannotia rifted. Sea-level rise and flooded continental area associated with rifting are highly correlated with increasing diversity of Paleozoic acritarch genera and Mesozoic calcareous nannoplankton (e.g., coccolithophorids) and dinoflagellates (see Figure 1). Although flooded continental area is a small percentage of the total oceanic area suitable for phytoplankton habitat, the shallow seas appear to have contributed proportionally more to niche space because of high nutrient input, high rates of primary production, and habitat heterogeneity (Katz *et al.* 2004).

Although there is a conceptual connection between the Wilson cycle and phytoplankton diversification, more evidence is needed to establish a conclusive causal link between the tectonic cycles and evolutionary pulses. Regardless of the causal link, there also appears to be a strong connection among the current Wilson cycle, evolutionary changes in phytoplankton community structure, and long-term changes in the carbon cycle.

When Pangea was fully assembled in the Triassic, the larger-celled plankton with high nutrient requirements most likely were concentrated along the supercontinent perimeter where nutrient supplies were highest, whereas the small-celled plankton (e.g., cyanobacteria) probably better competed in the large, oligotrophic Panthalassic Ocean. As Pangea started to rift and the Atlantic Ocean basin widened in the Jurassic, sea level rose and flooded broad continental shelves and low-lying inland areas (see Figure 1). The fragmentation of the continents and creation of a new ocean basin increased the total length of coastline where many plankton lived. Nutrients (such as phosphate) that were previously locked up in the large continental interior of Pangea were transported to newly formed shallow seas and distributed over wider shelf areas and longer continental margins. Furthermore, models predict that the accelerated hydrological cycle associated with Pangea fragmentation delivered more nutrients to the oceans (Wallman 2001). These changes were profound: Greater nutrient availability coupled with expanded ecospace and ecological niches appears to have selected for the large-celled phytoplankton that lived along continental margins and contributed to their rapid radiation and evolution. This is supported by both diversity curves and biomarker studies that show increases in biomass of coccolithophores, dinoflagellates, and diatoms in the Mesozoic (see Figure 1).

In the Mesozoic, the radiation of large-celled eukaryotic marine phytoplankton (Grantham and Wakefield 1988; Moldowan and Jacobson 2000) that were efficient export producers contributed to an overall increase in export production through time (Bambach 1993; Falkowski *et al.* 2003; Katz *et al.* 2005a). Much of the export production is concentrated along continental margins today (Walsh 1988; Laws *et al.* 2000), where up to 90% of organic carbon burial occurs (Hedges and Keil 1995). In the Mesozoic, substantial amounts of organic carbon were sequestered on the newly formed passive continental margins of the Atlantic and on flooded continental interiors (Claypool *et al.* 1977; Arthur *et al.* 1984; Jenkyns and Clayton 1997; Bralower 1999) as Pangea broke apart. This in effect has provided a long-term storehouse of ^{12}C-enriched organic matter during the current Wilson cycle (Katz *et al.* 2005a). Global sediment budgets indicate that an order of magnitude more sediment is deposited in ocean basins than is subducted (Rea and Ruff 1996) and that the long-term marine sedimentary system can be at steady state only over a complete Wilson cycle (Mackenzie and Pigott 1981; Worsley *et al.* 1986; Rea and Ruff 1996); however, sedimentary accretion on cratons has the potential to keep the system out of balance even over several Wilson cycles (e.g., Katz *et al.* 2005a; Hayes and Waldbauer 2006), as discussed previously.

These results support the hypothesis that the Phanerozoic Wilson cycles drove the greenhouse–icehouse cycles (Fischer 1984), in what is essentially a carbon redox-mediated climate system (Worsley and Nance 1989). In this scenario, volcanic CO_2 outgassing during continental fragmentation created greenhouse climates, and atmospheric CO_2 drawdown due to weathering processes eventually switched the planet to an icehouse mode. Icehouse intervals with major glaciations tend to be associated with times of either supercontinent assembly or maximum continent dispersal (Worsley and Nance 1989). There may well be a significant biological component that contributes to the climate switch (Katz *et al.* 2005a). This important additional biological loop connects changes in phytoplankton community structure that contributed to greater efficiency of organic carbon burial beginning in the Early Jurassic to the excess

carbon burial that drove the net oxidation of Earth's surface reservoirs and atmospheric CO_2 drawdown during the opening phase of the current Wilson cycle. This ultimately contributed to the climate change from the greenhouse conditions of the Mesozoic to the icehouse conditions that characterize the latter half of the Cenozoic (Katz *et al.* 2005a).

B. Biological Impact on Global Sedimentation Patterns

The distribution of carbonate deposition in the oceans over hundreds of millions of years has been determined by a complex set of parameters. Continental distributions coupled with global sea level determine the area of flooded tropical shelves, which in turn factors into the amount of shallow-water carbonate accumulation (Wilkinson and Algeo 1989; Walker *et al.* 2002). The poleward migration of continents over the past 540 million years has decreased carbonate accumulation, as has falling sea level through the Cenozoic, by decreasing the areal extent of shallow tropical seas. As tropical carbonate deposition declined, the carbonate saturation of the oceans rose (Wilkinson and Algeo 1989; Walker *et al.* 2002). A positive feedback loop may ultimately have favored the expansion of open ocean calcifying phytoplankton and zooplankton: as discussed previously, the expansion of the larger-celled eukaryotic phytoplankton of the red-plastid lineage, coupled with the opening of the Atlantic Ocean basin and global sea-level rise, resulted in greater organic carbon burial beginning in the Early Jurassic as the supercontinent Pangea rifted. Excess organic carbon burial drove the drawdown of atmospheric CO_2. Declining pCO_2 and pH, combined with rising carbonate saturation, may have facilitated the rise of coccolithophores and planktonic foraminifera; expansion of these two groups close the loop by further contributing to export production and declining pCO_2.

The rise of coccolithophores beginning in the early Mesozoic may also have been favored by ocean chemistry associated with the Wilson cycle. As seawater cycles through mid-ocean ridges, calcium is added and magnesium is removed. Total ridge length and seafloor spreading rates can alter Mg:Ca ratios in seawater. High Mg:Ca ratios tend to occur during times of supercontinent assembly when ridge length is shortest, characterized by deposition of aragonite and high-magnesium calcite (called "aragonite seas") (see Figure 1). In contrast, low magnesium-calcite deposition tends to characterize times of continental break-up (called "calcite seas") (Sandberg 1975; Hardie 1996). Low Mg:Ca and high Ca^{+2} concentration in seawater favors calcification in several groups of marine organisms, including coccolithophores; hence, there is a correspondence between these organisms and "calcite sea" intervals (Stanley and Hardie 1998), including the expansion of coccolithophores in the Mesozoic calcite seas (see Figure 1). This correlation further links the expansion of coccolithophores to the opening phase of the current Wilson cycle, organic carbon burial increase, and declining pCO_2 and pH.

The expansion of open-ocean calcifiers (see Figure 1) altered global carbonate depositional patterns. Prior to the Mesozoic, most marine calcifying organisms lived in shallow coastal and shelf regions; as a result, carbonate deposition was concentrated in these areas, whereas deposition of pelagic carbonates was minimal. As two groups of carbonate-secreting plankton expanded in the Mesozoic oceans—coccolithophores with their calcitic-plated armor (see de Vargas et al., Chapter 12, this volume), and planktonic foraminifera with their carbonate tests (e.g., Tappan and Loeblich 1988)—the loci of marine carbonate deposition gradually expanded from shallow shelf areas to the deeper ocean (Sibley and Vogel 1976; Southam and Hay 1981) and the carbonate compensation depth (CCD) deepened (e.g., van Andel 1975; Wilkinson and Algeo 1989).

Pelagic carbonate sedimentation has come to dominate as sea level and shelf area have declined since the Late Cretaceous, tropical shelf area has decreased as continents moved poleward, and the pelagic carbonate reservoir has increased at the expense of the shallow-water carbonate reservoir since the Mesozoic (Wilkinson and Algeo 1989).

Similarly, evolutionary trends in biosiliceous marine organisms have affected silica deposition through time (Maliva et al. 1989; Kidder and Erwin 2001). Inorganic precipitation and microbial activity in shallow seas were responsible for silica deposition in the Late Proterozoic and Cambrian. In the Paleozoic and much of the Mesozoic, siliceous sponges contributed most to shelf siliceous deposits (Racki 1999), whereas radiolaria were responsible for deep-ocean silica deposition (Casey 1993; Kidder and Erwin 2001). With the rise of diatoms in the Cretaceous, siliceous sponges largely vacated continental shelves and populated the deeper slope regions (Maldonado et al. 1999). Diatoms have outcompeted radiolaria for silica through the Cenozoic (Harper and Knoll 1975) and control silica removal from the oceans today (Treguer et al. 1995). As diatoms expanded in the Cenozoic oceans, radiolarian test weight declined and tests underwent structural changes to adapt to declining silica availability (Harper and Knoll 1975).

C. Effects of Carbon Burial on Atmospheric Gases

During the current Wilson cycle, the long-term increase in $\delta^{13}C$ of the mobile carbon reservoir resulted from biological and tectonic processes that acted in concert to increase the efficiency of organic carbon burial, driving the 190 million-year–long depletion of ^{12}C from the ocean-atmosphere system since the Early Jurassic. As discussed previously, this was most likely the result of increases in both f_{org} and $\delta^{13}C_{input}$, which supplied more ^{13}C to and extracted more ^{12}C from the mobile carbon reservoir during the opening phase of the current Wilson cycle (Falkowski *et al.* 2005; Katz *et al.* 2005a). Because the ultimate source of the buried organic matter was oxygenic photosynthesis, the increase in organic carbon burial should have acted to draw down atmospheric CO_2 levels, while enriching the atmosphere with O_2.

1. Atmospheric CO_2

In the contemporary ocean, diatoms are responsible for ~60% of the sinking flux of particulate organic carbon (POC) (Smetacek 1999; Kooistra *et al.*, Chapter 11, this volume), whereas coccolithophores are responsible for up to 80% of the particulate inorganic carbon (PIC) flux, primarily in the form of calcite (de Vargas *et al.*, Chapter 12, this volume). Precipitation of calcite produces CO_2 that can be released to the atmosphere (Berner *et al.* 1983). Hence, diatoms and coccolithophores potentially can biologically modify Earth's carbon cycle, and specifically, atmospheric and surface ocean CO_2 inventories. This suggests that biological processes can modify the slow, abiotic carbon cycle, an idea that can be explored by modeling CO_2 inventories through time.

The history of Phanerozoic atmospheric CO_2, reported as ratios (R) relative to contemporary levels (RCO_2), has been reconstructed using models based on carbon isotope records and from proxy CO_2 records, including plant stomatal index, boron isotopes, planktonic foraminiferal carbon isotope, and paleosol carbon isotopes (see Royer *et al.* 2004) for summary). In general, reconstructions based on modeled RCO_2 are consistent with proxy CO_2 records. Early Paleozoic high RCO_2 fell during the Devonian land plant explosion and continued to decline to low levels during the Permo-Carboniferous, which was characterized by the longest glacial period during the Phanerozoic. RCO_2 increased beginning in the Triassic, reaching relatively high levels in the Jurassic and Early Cretaceous, and then declined through the Late Cretaceous and Cenozoic.

Published RCO_2 reconstructions that use models based on carbon isotope records have been done in 10 million year time averages (Berner and Kothavala 2001; Berner 2006), providing long-term trends in RCO_2 that do not capture the detailed changes in the global carbon cycle that are indicated by our higher resolution isotopic records. Using our high-resolution $\delta^{13}C_{carb}$ and $\delta^{13}C_{org}$ records, we modeled the short-term (1 million year) variations superimposed on the long-term RCO_2 trends in the Jurassic-Cenozoic. Our model uses the $\delta^{13}C_{carb}$ and $\delta^{13}C_{org}$ data (Falkowski *et al.* 2005; Katz *et al.* 2005a) to parameterize the GEOCARB III (Berner and Kothavala 2001) and GEOCARBSULF (Berner 2006) models of the global carbon cycle (see Figure 2). The GEOCARB models use isotope mass balance constraints to describe the long-term geochemical carbon cycle by describing the transfer of carbon between sediment/rock reservoirs and the ocean/atmosphere system. The fundamental processes are drawdown of CO_2 from the atmosphere during weathering of silicate rocks and subsequent carbonate precipitation in the ocean, breakdown of carbonate minerals via metamorphism, and the burial, weathering, and thermal breakdown of organic carbon (Berner 1991). The GEOCARB models take into account many feedbacks in the carbon system, such as the dependence of temperature and runoff on atmospheric CO_2 levels, the effects of large vascular land plants on calcium–magnesium silicate weathering, the effects of changing paleogeography on land temperatures and weathering, and the enhancement of weathering by gymnosperms versus angiosperms (Berner and Kothavala 2001).

We show two RCO_2 curves using our data in two generations of Berner's GEOCARB modeling series (Berner and Kothavala 2001; Berner 2006) (see Figure 2); both curves are roughly within the high and low RCO_2 ranges given by GEOCARB III. The lower resolution curve (Katz *et al.* 2005a) uses GEOCARBSULF (Berner 2006) in

10 million-year averages, with peak RCO_2 in the Jurassic-Early Cretaceous and decline through the Late Cretaceous-Cenozoic. Our higher resolution RCO_2 curve shows changes of up to ~50% over several ~10 million-year intervals in the Late Jurassic and Early Cretaceous; these CO_2 fluctuations are consistent with lower resolution proxy records. Short-term CO_2 fluctuations of the same magnitude have been recorded during glacial–interglacial cycles, when CO_2 ranged from 180 to 280 ppm (~50% variability) on the 100 thousand-year scale. Hence, our high-resolution RCO_2 reconstruction is consistent with published model and proxy records but provides much more temporal detail. The ~10 million-year fluctuations can be attributed to changes in the biological processes that are responsible for export production and/or the geological processes that are responsible for sediment preservation. Our results suggest that the long-term decline in RCO_2 is in part the result of the long-term increase in organic carbon sequestration, that is, CO_2 drawdown has been mediated by biological processes.

2. Atmospheric O_2

The carbon, sulfur, and iron cycles together are the dominant controls on the redox conditions that determine atmospheric oxygen levels through the biological processes of oxygenic photosynthesis and bacterial sulfate reduction, coupled with the geological processes that are responsible for organic carbon and pyrite sedimentation and iron oxidation (Hayes and Waldbauer 2006). The iron cycle is a small, yet important, component that contributes to average redox conditions and pO_2, but its geologic history cannot be readily reconstructed through direct measurements (Hayes and Waldbauer 2006). Therefore, the carbon and sulfur cycles are perhaps the most important of the biogeochemical cycles that can be directly measured and used to reconstruct the history of O_2 using the stable isotopic records of these elements. Three isotopic signatures ($\delta^{13}C_{carb}$,

$\delta^{13}C_{org}$, and $\delta^{34}S_{sulf}$; Kampschulte and Strauss 2004; Falkowski et al. 2005; Katz et al. 2005a) were used to reconstruct atmospheric O_2 over the past 205 million years (Falkowski et al. 2005) by hindcasting from the present value of 21% using the isotope mass balance model of Berner and colleagues (Berner et al. 2000) (see Figure 2). These results indicate that atmospheric O_2 increased throughout the Mesozoic from a low in the Triassic to ~18% of the total atmospheric volume by the Late Cretaceous. Levels may have reached as high as 23% O_2 in the Eocene, followed by a small decline over the last ~10 million years. Contemporary O_2 levels were reached by 50 Ma.

Except for a few points near 200 Ma and 180 Ma, the results suggest there was sufficient oxygen throughout the Mesozoic to allow fires to burn (Chaloner 1989). Although several previous models (Berner and Canfield 1989; Berner et al. 2000; Hansen and Wallman 2003; Bergman et al. 2004) predict O_2 greater than 21% in the Cretaceous, they are based on very different sets of assumptions with predicted temporal evolution of atmospheric O_2 that are qualitatively quite different. Of these models, only one (Berner et al. 2000) uses isotopic mass balance and agrees with the predicted long-term increase of O_2 during the past 205 million years (Falkowski et al. 2005). The abundance of different rock types that reflect redox-sensitive mineral distributions (Tappan 1974; Berner and Canfield 1989) supports an increase in oxygen over this time. In contrast, the models by Hansen and Wallmann (2003) and Bergman et al. (2004) do not show this rise, do not use isotopic records to numerically constrain their predictions, and rely on many more assumptions than models based on isotopic mass balance.

3. The Evolution of Placental Mammals

The fairly rapid decline in oxygen at the end-Permian through Early Triassic may have been a major factor in the extinction of terrestrial animals (mostly reptiles) (Huey and Ward 2005). In contrast, the rise of oxygen over the ensuing 150 million years almost certainly contributed to evolution of large animals (Figure 4). Animals with relatively high oxygen demands evolved by the Late Triassic, such as small mammals and theropod dinosaurs (the group that includes living birds) (Carroll 1988; Asher *et al.* 2005). The metabolic demands of birds and mammals are three- to sixfold higher per unit biomass than those of reptiles (Else and Hulbert 1981). In addition, placental reproduction requires relatively high ambient oxygen concentrations because of the inefficiency of gas transport through the placenta (Mortola 2001; Andrews 2002). We note that placental evolution is not unique to mammals; it is found in 54 species of extant reptiles (Shine 2005). However, the overall metabolic demands of these placental reptiles are four- to sixfold lower per unit biomass than that of mammals. Moreover, mammal fetuses have relatively high oxygen demands, which scale directly with body size (Schmidt-Nielson 1970). Few extant mammals reproduce above elevations of ca. 4500 m, corresponding to atmospheric oxygen levels in the Early Jurassic (Falkowski *et al.* 2005; Huey and Ward 2005). Ultimately, there must be sufficient atmospheric O_2 to meet both the mammalian mother's and fetus' requirements.

Although the reproductive strategies of the earliest mammals are not known with certainty, both fossil records and molecular clock models suggest that the evolution and superordinal diversification of placental mammals occurred between 65 and 100 Ma (Murphy *et al.* 2001; Springer *et al.* 2003; Asher *et al.* 2005). This radiation corresponds to a period of relatively high and stable oxygen levels in the atmosphere (see Figure 4). All modern placental mammal orders appear in the fossil record by the Eocene, corresponding to a relatively sharp increase in atmospheric oxygen levels (Falkowski *et al.* 2005).

Atmospheric oxygen may also have been a factor in determining evolutionary trends in mammal size (see Figure 4). Being homeotherms, mammals have extremely high metabolic demands. These demands require not only high caloric intake relative to poikilotherms but also high rates of oxygen supply. Oxygen is delivered to tissue via arterial networks, culminating in capillaries that serve as the actual point of diffusion of gases to tissues. Muscles are among the tissues with the highest oxygen demand. In mammals, the density of capillaries per unit muscle scales to the power of 0.87 (Weibel and Hoppeler 2005). Hence, the larger the animal, the lower the capillary density per unit muscle tissue; large mammals require high ambient O_2 levels to obtain maximal metabolic rates. The Cretaceous/Tertiary mass extinction at 65 Ma provided the ecological opportunity for an increase in small- to medium-sized mammals in the first few million years following the event (Alroy 1999). However, the increase in average mammalian body size was significantly smaller than that of Eocene placental mammals (see Figure 4). In fact, it has long been recognized that the lack of a herbivorous mammalian megafauna for the entire duration of the Paleocene is one of the more interesting and peculiar lags in the postextinction recovery phase (Wing *et al.* 1992). A second increase in mean body mass occurred in the early through middle Eocene (50–40 Ma) (Alroy 1999), followed by additional (but less dramatic) size increases through the Miocene.

Hence, a secular increase in atmospheric oxygen over the past 205 million years broadly corresponds with three main aspects of vertebrate evolution—endothermy, placentation, and size (Falkowski *et al.* 2005). Particularly notable are high stable O_2 levels during the time of placental mammal origins and diversification, along with a close correspondence between marked increases in both atmospheric oxygen levels and mammalian body size during the early to middle Eocene. Although increases in mammalian body size, morphological disparity, and inferred ecological heterogeneity during this interval may have been influenced as well by

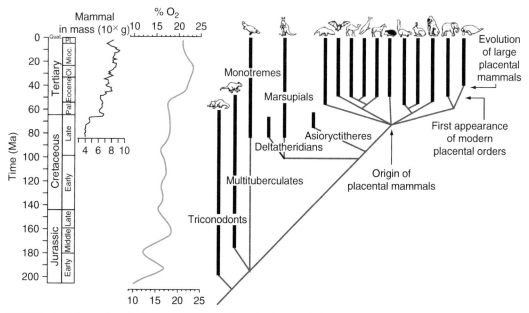

FIGURE 4.   Mammal evolutionary events based on fossil, morphological (Murphy *et al.* 2001), and molecular (Murphy *et al.* 2001; Springer *et al.* 2003) evidence compared with oxygen concentrations in Earth's atmosphere modeled over the past 205 million years using carbon and sulfur isotope datasets; O_2 levels approximately doubled over this time from 10% to 21%, punctuated by rapid increases in the Early Jurassic and from the Eocene to the present. Changes in average mammalian body mass is taken from Alroy (1999). Vertical black bars represent known fossil ranges; dashed lines represent inferred phylogenetic branching. Only some of the ordinal-level placental mammals are shown. Reproduced from Falkowski *et al.* (2005). (See color plate.)

other environmental factors, the correlation between evolutionary changes in mammalian body size and increased atmospheric O_2 has a physiological basis related to placental mammal reproduction (Wing *et al.* 1992).

V. CONCLUDING REMARKS

Geological and biological processes acted in concert to modify atmospheric and seawater chemistry through time: Tectonic outgassing and erosional processes are the primary suppliers of most major elements in geochemical cycles; biologically mediated redox processes alter mobile elemental reservoirs before geological processes sequester (remove) elements from these mobile reservoirs. This biological overprint of geological processes forms complex feedback loops in biogeochemical cycles that are recorded in geochemical proxy records;

these proxies, studied in conjunction with numerical models, provide a window on paleobiogeochemical interactions that form the network of Earth's system.

Carbon isotope records provide critical information that can be used to reconstruct changes in redox conditions and biological processes that affected past atmospheric and seawater chemistry. We use carbon isotope records of carbonates and organic matter, in conjunction with sulfur isotopes of sulfates, in model simulations to reconstruct carbon burial, pCO_2, and pO_2. These model results indicate that organic carbon burial and pO_2 have increased since the Early Jurassic, whereas pCO_2 has decreased since the Early Cretaceous.

We attribute these secular changes in the carbon cycle to the interplay between geological and biological processes: The evolution and expansion of the larger-celled eukaryotic phytoplankton of the red-plastid lineage acted to increase export production, whereas continental margin ecospace and sediment storage capacity increased with the opening of the Atlantic Ocean basin and global sea-level rise as the supercontinent Pangea rifted. The resulting excess organic carbon burial increased the oxidation state of Earth's surface reservoirs while drawing down atmospheric CO_2, which in turn acted as a strong selective agent in both marine and terrestrial primary producers, resulting in the rise in C_4 and β-carboxylation photosynthetic pathways in the latter part of the Cenozoic. At the same time, atmospheric O_2 levels approximately doubled. Our analysis suggests that the rise of oxygen may have been a key factor in the evolution, radiation, and subsequent increase in the average size of placental mammals during the Cenozoic.

Acknowledgments

This study was supported by NSF OCE 0084032 Biocomplexity: The Evolution and the Radiation of Eucarytoic Phytoplankton Taxa (EREUPT). This research used samples provided by the Ocean Drilling Program (ODP), which is sponsored by the U.S. National Science Foundation (NSF) and participating countries under management of Joint Oceanographic Institutions (JOI), Inc. We thank A.H. Knoll and R.A. Berner for discussions.

References

Alroy, J. (1999). The fossil record of North American mammals: evidence for a Paleocene evolutionary radiation. *Syst. Biol.* **48:** 107–118.

Alvarez, L.W., Alvarez, W., Asaro, F., and Michel, H.V. (1980). Extraterrestrial cause for the Cretaceous-Tertiary extinction. *Science* **208:** 1095–1108.

Anbar, A.D., and Knoll, A.H. (2002). Proterozoic ocean chemistry and evolution: a bioinorganic bridge? *Science* **297:** 1137–1142.

Andrews, R. (2002). Low oxygen: a constraint on the evolution of viviparity in reptiles. *Physiol. Biochem. Zool.* **75:** 145–154.

Armstrong, R.A., Lee, C., Hedges, J.I., Honjo, S., and Wakeham, S.G. (2002). A new, mechanistic model for organic carbon fluxes in the ocean, based on the quantitative association of POC with ballast minerals. *Deep Sea Res. II* **49:** 219–236.

Arthur, M.A., Dean, W.E., and Stow, D.A.V. (1984). Models for the deposition of Mesozoic-Cenozoic fine-grained organic-C rich sediment in the deep sea. *Fine-Grained Sediments: Deep-Water Processes and Facies.* D.A.V. Stow and D.J.W. Piper, eds. Great Britain, Geological Society of London Special Publication, pp. 527–559.

Asher, R.J., Meng, J., Wible, J.R., McKenna, M.C., Rougier, G.W., Dashzeveg, D., and Novacek, M.J. (2005). Stem lagomorpha and the antiquity of glires. *Science* **307:** 1091–1094.

Bambach, R.K. (1993). Seafood through time: changes in biomass, energetics, and productivity in the marine ecosystem. *Paleobiology* **19:** 372–397.

Berggren, W.A., Kent, D.V., Swisher, C.C., and Aubry, M.-P. (1995). A revised Cenozoic geochronology and chronostratigraphy. *Geochronology, Time Scales and Global Stratigraphic Correlations: A Unified Temporal Framework for an Historical Geology.* W. A. Berggren, D. V. Kent, and J. Hardenbol, eds. Tulsa OK, SEPM (Society for Sedimentary Geology), pp. 129–212.

Bergman, N.M., Lenton, T.M., and Watson, A.J. (2004). COPSE: A new model of biogeochemical cycling over Phanerozoic time. *Am. J. Sci.* **304:** 397–437.

Berner, R.A. (1991). A model for atmospheric CO_2 over Phanerozoic time. *Am. J. Sci.* **291:** 339–376.

Berner, R.A. (2001). Modeling atmospheric O_2 over Phanerozoic time. *Geochim. Cosmochim. Acta* **65:** 685–694.

Berner, R.A. (2004). *The Phanerozoic Carbon Cycle: CO_2 and O2.* Oxford University Press.

Berner, R.A. (2006). GEOCARBSULF: a combined model for Phanerozoic atmospheric O_2 and CO_2. *Geochim. Cosmochim. Acta* **70:** 5653–5664.

Berner, R., and Canfield, D. (1989). A model for atmospheric oxygen over Phanerozoic time. *Am. J. Sci.* **289:** 333–361.

Berner, R.A., and Kothavala, Z. (2001). GEOCARB III: a revised model of atmospheric CO_2 over Phanerozoic time. *Am. J. Sci.* **301:** 182–204.

Berner, R.A., Lasaga, A., and Garrels, R. (1983). The carbonate-silicate geochemical cycle and its effect of atmospheric carbon dioxide over the past 100 million years. *Am. J. Sci.* **283:** 641–683.

Berner, R.A., Petsch, S.T., Lake, J.A., Beerling, D.J., Popp, B.N., Lane, R.S., Laws, E.A., Westley, M.B., Cassar, N., Woodard, F.I., and Quick, W.P. (2000). Isotope fractionation and atmospheric oxygen: implications for Phanerozoic O_2 evolution. *Science* **287:** 1630–1633.

Betzer, P.R., Showers, W.J., Laws, E.A., Ditullio, G.R., and Kroopnick, P.M. (1984). Primary productivity and particle fluxes on a transect of the equator at 153 degrees West in the Pacific Ocean. *Deep Sea Res.* **31:** 1–12.

Bienfang, P.K. (1992). The role of coastal high latitude ecosystems in global export production. *Primary Productivity and Biogeochemical Cycles in the Sea.* P.G. Falkowski and A. Woodhead, eds. New York, Plenum Press, pp. 285–297.

Bown, P.R. (1998). *Calcareous Nannofossil Biostratigraphy.* London, Kluwer Academic Publishers.

Bown, P.R., Lees, J.A., and Young, J. R. (2004). Calcareous nannoplankton diversity and evolution through time. *Coccolithophores: From Molecular Processes to Global Impact*. H. Thierstein, and J. R. Young, eds. Berlin, Springer-Verlag, pp. 481–507.

Bralower, T.J. (1999). The record of global change in mid-Cretaceous (Barremian-Albian) sections from the Sierra Madre, northeast Mexico. *J. Foraminiferal Res.* **29**: 418–437.

Brocks, J.J., Logan, G.A., Buick, R., and Summons, R.E. (1999). Archean molecular fossils and the early rise of eukaryotes. *Science* **285**: 1033–1036.

Broecker, W.S., and Peng, T.-H. (1982). *Tracers in the Sea*. Palisades NY, LDEO of Columbia University.

Caldeira, K. (1992). Enhanced Cenozoic chemical weathering and the subduction of pelagic carbonate. *Nature* **357**: 578–581.

Carroll, R.L. (1988). *Vertebrate Paleontology and Evolution*. New York, Freeman.

Casey, R.E. (1993). Radiolaria. *Fossil Prokaryotes and Protists*. J.H. Lipps, ed. Boston, Blackwell Scientific, pp. 249–284.

Cerling, T.E., Harris, J.M., MacFadden, B.J., Leakey, M.G., Quade, J., Eisenmann, V., and Ehleringer, J.R. (1997). Global vegetation change through the Miocene/Pliocene boundary. *Nature* **389**: 153–158.

Chacon-Baca, E., Beraldi-Campesi, H., Cevallos-Ferriz, S. R. S., Knoll, A. H., and Golubic, S. (2002). 70 Ma nonmarine diatoms from northern Mexico. *Geology* **30**:279–281.

Chaloner, W. (1989). Fossil charcoal as an indicator of palaeoatmospheric oxygen level. *J. Geol. Soc.* **146**: 171–174.

Claypool, G.E., Lubeck, C.M., Bayeinger, J.P., and Ging, T.G. (1977). Organic geochemistry. *Geological Studies on the COST No. B-2 Well, U.S. Mid-Atlantic Outer Continental Shelf Area*. P.A. Scholle, ed. U.S.G.S. Circular, pp. 46–59.

Derry, L.A., and France-Lanord, C. (1996). Neogene Himalayan weathering history and river $^{87}Sr/^{86}Sr$: impact on the marine Sr record. *Ear. Planet. Sci. Lett.* **142**: 59–74.

de Vargas, C. Probert, I., Aubry, M.P., and Young, J. (2007). Origin and evolution of coccolithophores: from coastal hunters to oceanic farmers. *Evolution of Primary Producers in the Sea*. P.G. Falkowski and A.H. Knoll, eds. Boston, Elsevier, pp. 251–285.

Drake, M.J., and Righter, K. (2002). Determining the composition of the Earth. *Nature* **416**: 39–44.

Dugdale, R., and Wilkerson, F. (1998). Silicate regulation of new production in the equatorial Pacific upwelling. *Nature* **391**: 270–273.

Else, P., and Hulbert, A. (1981). Comparison of the "mammal machine" and the "reptile machine": energy production. *Am. J. Physiol. Regul. Integr. Comp. Physiol.* **240**: 3–9.

Falciatore, A., and Bowler, C. (2000). Revealing the molecular secrets of marine diatoms. *Ann. Rev. Plant Biol.* **53**: 109–130.

Falkowski, P.G. (1997). Photosynthesis: the paradox of carbon dioxide efflux. *Curr. Biol.* **7**: R637–R639.

Falkowski, P.G. (1998). Biogeochemical controls and feedbacks on ocean primary productivity. *Science* **281**: 200–206.

Falkowski, P.G. (2001). Biogeochemical cycles. *Encyclopedia of Biodiversity*. S.A. Levin, ed. San Diego, Academic Press, pp. 437–453.

Falkowski, P.G., Biscaye, P.E., and Sancetta, C. (1994). The lateral flux of biogenetic particles from the eastern North American continental margin to the North Atlantic Ocean. *Cont. Shelf Res.* **41**: 583–601.

Falkowski, P.G., Katz, M.E., Knoll, A.H., Quigg, A., Raven, J.A., Schofield, O., and Taylor, M. (2004b). The evolutionary history of eukaryotic phytoplankton. *Science* **305**: 354–360.

Falkowski, P.G., Katz, M.E., Milligan, A., Fennel, K., Cramer, B.S., Aubry M.P., Berner, K.A., Novacek, M., and Zupole, W.M. (2005). The rise of oxygen over the past 205 million years and the evolution of large placental mammals. *Science* **309**: 2202–2204.

Falkowski, P., Laws, E.A., Barbar, R.T., and Murray, J.W. (2003). Phytoplankton and their role in primary, new and export production. M.J.R. Fasham, ed. *Ocean Biogeochemistry. The Role of the Ocean Carbon Cycle in Global Change*. Berlin, Springer-Verlag, pp. 99–121.

Falkowski, P.G., and Raven, J. A. (1997). *Aquatic Photosynthesis*. Malden, MA, Blackwell Science.

Falkowski, P.G., Schofield, O., Katz, M.E., van de Schootbrugge, B., and Knoll, A. (2004a). Color wars: why is the land green and the ocean red? *Coccolithophores: From Molecular Processes to Global Impact*. Berlin, H. Thierstein, and J.R. Young, eds. Springer-Verlag, pp. 429–453.

Fensome, R.A., MacRae, R.A., Moldowan, J.M., Taylor, F.J.R., and Williams, G.L. (1996). The early Mesozoic radiation of dinoflagellates. *Paleobiology* **22**: 329–338.

Field, C., Behrenfeld, M., Randerson, J., and Falkowski, P. (1998). Primary production of the biosphere: integrating terrestrial and oceanic components. *Science* **281**: 237–240.

Finkel, Z.V. (2007). Does phytoplankton cell size matter? The evolution of modern marine food webs. *Evolution of Primary Producers in the Sea*. P.G. Falkowski and A.H. Knolls, eds. Boston, Elsevier, pp. 333–350.

Finkel, Z.V., Katz, M.E., Wright, J.D., Schofield, O.M.E., and Falkowski, P.G. (2005). Climatically-driven evolutionary change in the size of diatoms over the Cenozoic. *PNAS* **102**: 8927–8932.

Fischer, A.G. (1984). The two Phanerozoic supercycles. *Catastrophes and Earth History*. Princeton University Press, pp. 129–150.

France-Lanord, C., and Derry, L.A. (1994). $d^{13}C$ of organic carbon in the Bengal Fan: source evolution and transport of C_3 and C_4 plant carbon to marine sediments. *Geochim. Cosmochim. Acta* **58**: 4809–4814.

Gersonde, R., and Harwood, D. M. (1990). Lower Cretaceous diatoms from ODP Leg 113 Site 693 (Weddell

Sea). Part 1: vegetative cells. *Proceedings of the Ocean Drilling Program, Scientific Results*. P.F. Barker, J.P. Kennett, and S.B. O'Connell, eds. College Station TX, Ocean Drilling Program, pp. 365–402.

Gradstein, F.M., Agterberg, F.P., Ogg, J.G., Hardenbol, H., vanVeen, P., Thierry, J., and Huang, Z. (1995). A Triassic, Jurassic, and Cretaceous time scale. *Geochronology, Time Scales and Global Stratigraphic Correlations: A Unified Temporal Framework for an Historical Geology*. W.A. Berggren, D.V. Kent, and J. Hardenbol, eds. Tulsa OK, SEPM (Society for Sedimentary Geology), pp. 95–126.

Grantham, P.J., and Wakefield, L.L. (1988). Variations in the sterane carbon number distributions of marine source rock derived crude oils through geological time. *Org. Geochem.* **12:** 61–73.

Guidry, M., Arvidson, R.S., and MacKenzie, F.T. (2007). Biological and geochemical forcings to phanerozoic change in seawater, atmosphere, and carbonate precipitate composition. *Evolution of Primary Producers in the Sea*. P.G. Falkowski and A.H. Knoll, eds. Boston, Elsevier, pp. 377–403.

Hamm, C., and Smetacek, V. (2007). Armor: why, when, and how? *Evolution of Primary Producers in the Sea*. P.G. Falkowski and A.H. Knoll, eds. Boston, Elsevier, pp. 311–332.

Han, T.M., and Runnegar, B. (1992). Megascopic eukaryotic algae from the 2.1-billion-year-old Negaunee iron-formation, Michigan. *Science* **257:** 232–235.

Hansen, K.W., and Wallman, K. (2003). Cretaceous and Cenozoic evolution of seawater composition, atmospheric O_2 and CO_2: a model perspective. *Am. J. Sci.* **303:** 94–148.

Haq, B.U., Hardenbol, J., Vail, P.R. (1987). Chronology of fluctuating sea levels since the Triassic (250 million years ago) to present. *Science* **235:** 1156–1167.

Hardie, L.A. (1996). Secular variation in seawater chemistry: an explanation for the coupled secular variation in the mineralogies of marine limestones and potash evaporites over the past 600 m.y. *Geology* **24:** 279–283.

Harper, H.E., Jr., and Knoll, A.H. (1975). Silica, diatoms, and Cenozoic radiolarian evolution. *Geology* **3:** 175–177.

Harwood, D.M.M., and Gersonde, R. (1990). Lower Cretaceous diatoms from ODP Leg 113 Site 693 (Weddell Sea). Part 2: Resting spores, Chrysophycean cysts, and endoskeletal dinoflagellate, and notes on the origin of diatoms. *Proceedings of the Ocean Drilling Program*. Ocean Drilling Program, College Station, TX, pp. 403–426.

Harwood, D.M., and Nikolaev, V.A. (1995). Cretaceous diatoms: morphology, taxonomy, biostratigraphy. *Siliceous Microfossils*. C.D. Blome, P.M. Whalen, and K.M. Reed, eds. Lawrence, KS, Paleontological Society Short Courses in Paleontology, pp. 81–106.

Hayes, J., and Waldbauer, J. (2006). The carbon cycle and associated redox processes through time. *Phil Trans R. Soc. B* **361:** 931–950.

Hayes, J.M., Strauss, H., and Kaufman, A.J. (1999). The abundance of ^{13}C in marine organic matter and isotopic fractionation in the global biogeochemical cycle of carbon during the past 800 Ma. *Chem. Geol.* **161:** 103–125.

Hedges, J.I., and Keil, R.G. (1995). Sedimentary organic matter preservation: an assessment and speculative synthesis. *Mar. Chem.* **49:** 81–115.

Hodell, D.A. (1994). Magnetostratigraphic, biostratigraphic, and stable isotope stratigraphy of an Upper Miocene drill core from the Sale Briqueterie (northwestern Morocco): a high-resolution chronology for the Messinian stage. *Paleoceanography* **9:** 835–855.

Holland, H.D. (1984). *The Chemical Evolution of the Atmosphere and Oceans*. Princeton Series in Geochemistry. Princeton University Press.

Holland, H.D. (1997). Geochemistry—evidence for life on Earth more than 3850 million years ago. *Science* **275:** 38–39.

Huey, R.B., and Ward, P.D. (2005). Hypoxia, global warming, and terrestrial Late Permian extinctions. *Science* **308:** 398–401.

Ivimey-Cook, H.C. (1971). Stratigraphical palaeontology of the Lower Jurassic of the Llandbedr (Mochras Farm) Borehole. The Llandbedr (Mochras Farm) Borehole. A.W. Woodland, ed. Institute of Geological Sciences, Natural Environment Research Council, Great Britain, Report no. 71/81, p. 115.

Janis, C.M., and Damuth, J. (2000). Mammals. *Evolutionary Trends*. K.J. McNamara, ed. London, J. Belknap, pp. 301–345.

Janofske, D. (1992). Calcareous nannofossils of the Alpine Upper Triassic. *Nannoplankton Research*. B. Hamrsmid and J.R. Young, eds. Knihovnicka ZPZ, pp. 870–109.

Javaux, E.J., Knoll, A.H., and Walter, M.R. (2001). Morphological and ecological complexity in early eukaryotic ecosystems. *Nature* **412:** 66–69.

Jenkyns, H.C., and Clayton, C.J. (1997). Lower Jurassic epicontinental carbonates and mudstones from England and Wales: chemostratigraphic signals and the early Toarcian anoxic event. *Sedimentology* **44:** 687–706.

Kampschulte, A., and Strauss, H. (2004). The sulfur isotopic evolution of Phanerozoic seawater based on the analysis of structurally substituted sulfate in carbonates. *Chem. Geol.* **204:** 255–286.

Katz, M.E., Fennel, K., Berner, R.A., and Falkowski, P.G. (2005b). *Long-Term Trends in the Global Carbon Cycle: Biogeochemical Records of the Past ~200 myrs*. American Geophysical Union Ann. Mtg. abstract.

Katz, M.E., Finkel, Z.V., Grzebyk, D., Knoll, A.H., and Falkowski, P.G. (2004). Evolutionary trajectories and biogeochemical impacts of marine eukaryotic phytoplankton. *Annu. Rev. Ecol. Evol. Syst.* **35:** 523–556.

Katz, M.E., Wright, J.D., Miller, K.G., Cramer, B.S., Fennel, K., and Falkowski, P.G. (2005a). Biological overprint of the geological carbon cycle. *Mar. Geol.* **217:** 323–338.

Kidder, D.L., and Erwin, D.H. (2001). Secular distribution of biogenic silica through the Phanerozoic:

comparison of silica-replaced fossils and bedded cherts at the series level. *J. Geol.* **109**: 509–522.

Knoll, A.H. (1989). Evolution and extinction in the marine realm: some constraints imposed by phytoplankton. *Phil. Trans. R. Soc. Lond.* **325**: 279–290.

Knoll, A.H. (1992). The early evolution of eukaryotes: a geological perspective. *Science* **256**: 622–627.

Knoll, A.H. (1994). Proterozoic and Early Cambrian protists: evidence for accelerating evolutionary tempo. *Proc. Natl. Acad. Sci. USA* **91**: 6743–6750.

Knoll, A.H. (2003). *Life on a Young Planet: The First Three Billion Years of Evolution on Earth.* Princeton University Press.

Knoll, A.H., Javaux, E.J., Hewitt, D., and Cohen, P. (2006). Eukaryotic organisms in Proterozoic oceans. *Phil. Trans. R. Soc. B* **361**: 1023–1038.

Knoll, A.H., Summons, R.E., Waldbauer, J.R., and Zumberge, J.E. (2007). The geological succession of primary producers in the oceans. *Evolution of Primary Producers in the Sea.* P.G. Falkowski and A.H. Knoll, eds. Boston, Elsevier, pp. 133–163.

Kooistra, W.H.C.F., Gersonde, R., Medlin, L.K., and Mann, D.G. (2007). The origin and evolution of the diatoms: their adaptation to a planktonic existence. *Evolution of Primary Producers in the Sea.* P.G. Falkowski and A.H. Knoll, eds. Boston, Elsevier, pp. 207–249.

Kump, L.R., and Arthur, M.A. (1999). Interpreting carbon-isotope excursions: carbonates and organic matter. *Chem. Geol.* **161**: 181–198.

Laws, E.A., Falkowski, P.G., Smith, W.O., Jr., and McCarthy, J.J. (2000). Temperature effects on export production in the open ocean. *Global Biogeochem. Cycles* **14**: 1231–1246.

Lipps, J.H. (1993). *Fossil Prokaryotes and Protists.* Oxford, Blackwell, p. 342.

MacArthur, R.H. and Wilson, E.O. (1967). *The Theory of Island Biogeography.* Monographs in population biology, v. 1.

Mackenzie, F.T., and Pigott, J.D. (1981). Tectonic controls of Phanerozoic sedimentary rock cycling. *J. Geol. Soc. Lond.* **138**: 183–196.

Maldonado, M., Carmona, M.C., Uriz, M.J., and Cruzado, A. (1999). Decline in Mesozoic reef-building sponges explained by silicon limitation. *Nature* **401**: 785–788.

Maliva, R.G., Knoll, A.H., and Siever, R. (1989). Secular change in chert distribution; a reflection of evolving biological participation in the silica cycle. *Palaios* **4**: 519–532.

Martin, J., Knauer, G., Karl, D., and Broenkow, W. (1987). VERTEX: carbon cycling in the northeast Pacific. *Deep Sea Res.* **34**: 267–285.

Medlin, L.K., Kooistra, W.C.H.F., and Schmid, A.M.M. (2000). A review of the evolution of the diatoms—a total approach using molecules, morphology and geology. *The Origin and Early Evolution of the Diatoms: Fossil, Molecular and Biogeographical Approaches.* A. Witkowski and J. Sieminska, eds. Krakow, Poland, Polish Academy of Sciences, pp. 13–35.

Miller, K.G., Kominz, M.A., Browning, J.V., Wright, J.D., Mountain, G.S., Katz, M.E., Sugarman, P.J., Cramer, B.S., Christie-Blick, N., Pekar, S.F. (2005). The Phanerozoic record of global sea-level change. *Science* **310**: 1293–1298.

Moldowan, J.M., Dahl, J., Jacobson, S.R., Huizing, B.J., Fago, F.J., Shetty, R., Watts, D.S., and Peters, K.E. (1996). Chemostratigraphic reconstruction of biofacies: molecular evidence linking cyst-forming dinoflagellates with pre-Triassic ancestors. *Geology* **24**: 159–162.

Moldowan, J.M., and Jacobson, S.R. (2000). Chemical signals for early evolution of major taxa: biosignatures and taxon-specific biomarkers. *Int. Geol. Rev.* **42**: 805–812.

Moldowan, J.M., and Talyzina, N.M. (1998). Biogeochemical evidence for dinoflagellate ancestors in the Early Cambrian. *Science* **281**: 1168–1170.

Morris, I. (1987). Paths of carbon assimilation in marine phytoplankton. *Primary Productivity in the Sea.* P.G. Falkowski, ed. New York, Plenum, pp. 139–159.

Mortola, J. (2001). *Respiratory Physiology of Newborn Mammals: A Comparative Perspective.* Baltimore, Johns Hopkins University Press.

Murphy, W.J., Eizirik, E., O'Brien, J.O., Madsen, D., Scally, M., Douady, C.J., Teeling, E., Ryder, D.A., Stanhope, M.J., DeJong, W.W., and Springer, M.S. (2001). Resolution of the early placental mammal radiation using Bayesian phylogenetics. *Science* **294**: 2348–2351.

Prasad, V., Strömberg, C.A.E., Alimohammadian, H., and Sahni, A. (2005). Dinosaur coprolites and the early evolution of grasses and grazers. *Science* **310**: 1177–1180.

Premuzic, E.T. (1980). Organic carbon and nitrogen in the surface sediments of world oceans and seas: Distribution and relationship to bottom topography. *Brookhaven National Lab Formal Report* **51084**.

Racki, G. (1999). Silica-secreting biota and mass extinctions: survival patterns and processes. *Palaeogeog. Palaeoclim. Palaeoecol.* **154**: 107–132.

Ravizza, G. (1993). Variations of the ^{187}Os/^{186}Os ratio of seawater over the past 28 million years as inferred from metalliferous carbonates. *Ear. Planet. Sci. Lett.* **118**: 335–348.

Rea, D.K., and Ruff, L.J. (1996). Composition and mass flux of sediment entering the world's subduction zones: implications for global sediment budgets, great earthquakes, and volcanism. *Ear. Planet. Sci. Lett.* **140**: 1–12.

Reinfelder, J.R., Kraepeil, A.M.L., and Morel, F.M.M. (2000). Unicellular C_4 photosynthesis in a marine diatom. *Nature* **407**: 996–999.

Retallack, G.J. (1997). Cenozoic expansion of grasslands and climatic cooling. *J. Geol.* **109**: 407–426.

Retallack, G.J. (2001). Neogene expansion of the North American prairie. *Palaios* **12**: 380–390.

Rich, J.E., Johnson, G.L., Jones, J.E., and Campsie, J. (1986). A significant correlation between fluctuations in seafloor spreading rates and evolutionary pulsations. *Paleoceanography* **1**: 85–95.

Roman, M.R., Adolf, H.A., Landry, M.R., Madin, L.P., Steinberg, D.K., and Zhang, X. (2002). Estimates of oceanic mesozooplankton production: a comparison using the Bermuda and Hawaii time-series data. *Deep Sea Res. Part II Top. Studies Oceanogr.* **49**: 175–192.

Ronov, A.B. (1994). Phanerozoic transgressions and regressions on the continents: a quantitative approach based on areas flooded by the sea and areas of marine and continental deposition. *Am. J. Sci.* **294**: 777–801.

Rosenweig, M.L. (1995). *Species Diversity in Space and Time.* Cambridge, CUP.

Rothpletz, A. (1896). Über die flysch-fucoiden und einige andere fossile Algen, sowie über liasische, diatomeen fuhrende Hornschwamme. *Zeitschrift Deutsch Geolog. Gesellschaft* **48**: 854–914.

Royer, D., Berner, R. A., Montanez, I.P., Tabor, N.J., and Beerling, D.J. (2004). CO_2 as a primary driver of Phanerozoic climate. *GSA Today* **14**: 4–10.

Sancetta, C., Villareal, T., and Falkowski, P.G. (1991). Massive fluxes of rhizosolenoid diatoms: A common occurrence? *Limnol. Oceanogr.* **36**: 1452–1457.

Sandberg, P.A. (1975). New interpretations of Great Salt Lake ooids and of ancient nonskeletal carbonate mineralogy. *Sedimentoloty* **22**: 497–538.

Schmidt-Nielson, K. (1970). Energy metabolism, body size, and problems of scaling. *Fed. Proc.* **29**: 1524–1532.

Schrag, D.P. (2002). *Control of Atmospheric CO_2 and Climate over the Phanerozoic American Geophysical Union,* Fall Meeting abstract.

Shackleton, N.J. (1987). Oxygen isotopes, ice volume, and sea level. *Q. Sci. Rev.* **6**: 183–190.

Shackleton, N.J., and Hall, M.A. (1984). Carbon isotope data from Leg 74 sediments. *Init. Repts. DSDP.* T.C. Moore, Jr., P.D. Rabinowitz, eds. Washington DC, U.S. Government Printing Office, pp. 613–619.

Shanks, A., and Trent, J. (1980). Marine snow: sinking rates and potential role in vertical flux. *Deep Sea Res.* **27**: 137–143.

Shine, R. (2005). Life-history evolution in reptiles. *Annu. Rev. Ecol. Evol. Syst.* **36**: 23–46.

Sibley, D.F., and Vogel, T.A. (1976). Chemical Mass Balance of the Earth's crust: the calcium dilemma (?) and the role of pelagic sediments. *Science* **192**: 551–553.

Smayda, T.J. (1969). Some measurements of the sinking rate of fecal pellets. *Limnol. Oceanogr.* **14**: 621–625.

Smayda, T.J. (1970). The suspension and sinking of phytoplankton in the sea. *Oceanogr. Mar. Bio. Annu. Rev.* **8**: 353–414.

Smetacek, V. (1999). Diatoms and the ocean carbon cycle. *Protist* **150**: 25–32.

Smith, A.B., Gale, A.S., and Monks, N.E.A. (2001). Sea-level change and rock-record bias in the Cretaceous: a problem for extinction and biodiversity studies. *Paleobiology* **27**: 241–253.

Southam, J.R., and Hay, W.W. (1981). Global sedimentary mass balance and sea level changes. *The Sea.* C. Emiliani, ed. New York, Wiley-Interscience, pp. 1617–1684.

Spencer-Cervato, C. (1999). The Cenozoic deep sea microfossil record: explorations of the DSDP/ODP sample set using the Neptune database. *Palaeontol. Electronica* **2**: 1–270.

Springer, M.S., Murphy, W.J., Eizirik, E., and O'Brien, S.J. (2003a). Placental mammal diversification and the Cretaceous-Tertiary boundary. *Proc. Nat. Acad. Sci. USA* **100**: 1056–1061.

Stanley, G.D., Jr. and Hardie, L.A. (1998). Secular oscillations in the carbonate mineralogy of reef-building and sediment-producing organisms driven by tectonically forced shifts in seawater chemistry. *Palaeogeogr. Palaeoclimatol. Palaeoecol.* **144**: 3–19.

Still, C.J., Berry, J.A., Collatz, G.J., and DeFries, R.S. (2003). Global distribution of C_3 and C_4 vegetation: carbon cycle implications. *Glob. Biogeochem. Cycles* **17**:doi:1029/2001GB001807.

Stover, L.E., Brinkhuis, H., Damassa, S.P. *et al.* (1996). Mesozoic-Tertiary dinoflagellates, acritarchs & prasinophytes. *Palynology: Principles and Applications.* J. Jansonius and D.C. McGregor, eds. Amer. Assoc. Strat. Palynologists Foundation, pp. 641–750.

Summons, R., Jahnke, L., Hope, J., and Logan, G. (1999). 2-Methylhopanoids as biomarkers for cyanobacterial oxygenic photosynthesis. *Nature* **400**: 554–557.

Summons, R.E., Thomas, J., Maxwell, J.R., and Boreham, C.J. (1992). Secular and environmental constraints on the occurrence of dinosterane in sediments. *Geochim. Cosmochim. Acta* **56**: 2437–2444.

Summons, R.E., and Walter, M.R. (1990). Molecular fossils and microfossils of prokaryotes and protists from Proterozoic sediments. *Am. J. Sci.* **290–A**: 212–244.

Tappan, H. (1974). Molecular oxygen and evolution. *Molecular Oxygen in Biology: Topics in Molecular Oxygen Research.* O. Hayaishi, ed. Amsterdam, North-Holland Publishing Company, pp. 81–135.

Tappan, H. (1980). *The Palaeobiology of Plant Protists.* San Francisco, W.H. Freeman, pp. 1028.

Tappan, H., and Loeblich, J. (1988). Foraminiferal evolution, diversification, and extinction. *J. Paleontol.* **62**: 695–714.

Treguer, P., Nelson, D.M., van Bennekom, A.J., DeMaster, D.J., Leynaert, A., and Queguiner, B. (1995). The silica balance in the world ocean: a reestimate. *Science* **268**: 375–379.

Turekian, K.K., and Pegram, W.J. (1997). Os isotope record in a Cenozoic deep-sea core: its relation to global tectonics and climate. *Tectonic Uplift and Climate Change.* W.F. Ruddiman, ed. New York, Plenum Press, pp. 383–397.

Urrere, M., and Knauer, G. (1981). Zooplankton fecal pellet fluxes and vertical transport of particulate organic material in the pelagic environment. *J. Plankton Res.* **3**: 369–387.

Vail, P.R., Mitchum, R.M., Todd, R.G., Widmier, J.M., Thompson III, S. Sangree, J.B., Bubb, J.N., and Hatelid, W.G. (1977). Seismic Stratigraphy and global changes of sea level. *Seismic Stratigraphic—Applications of Hydrocarbon Exploration.* C.E. Payton, Memoirs of the American Association of Petroleum Geologists. **26**: 49–205.

Valentine, J.W., and Moores, E.M. (1974). Plate tectonics and the history of life in the oceans. *Sci. Am.* **230:** 80–89.

van Andel, T.H. (1975). Mesozoic/Cenozoic calcite compensation depth and global distribution of calcareous sediments. *Earth Planetary Sci. Lett.* **26:** 187–194.

Veizer, J., Ala, K., Azmy, K., Brunckschen, P., Buhl, D., Bruhn, F., Carden, G.A.F., Diener, A., Ebneth, S., Godderis, Y., Jasper, T., Korte, C. Pawallek, F., Podlaha, O.G., and Strauss, H. (1999). ^{87}Sr/^{86}Sr, δ^{13}C, and δ^{18}O evolution of Phanerozoic seawater. *Chem. Geol.* **161:** 59–88.

Vincent, E., and Berger, W.H. (1985). Carbon dioxide and polar cooling in the Miocene: the Monterey hypothesis. *The Carbon Cycle and Atmospheric CO2: Natural Variations Archean to Present.* Geophys. Monogr. E.T. Sundquist and W.S. Broecker, eds. Washington DC, AGU, pp. 455–468.

Volk, T., and Hoffert, M.I. (1985). Ocean carbon pumps: Analysis of relative strengths and efficiencies in ocean-driven atmospheric CO_2 changes. *Geophys. Monogr.* **32:** 99–110.

Walker, L.J., Wilkinson, B.H., and Ivany, L.C. (2002). Continental drift and Phanerozoic carbonate accumulation in shallow-shelf and deep-marine settings. *J. Geol.* **110:** 75–87.

Wallman, K. (2001). Controls on the Cretaceous and Cenozoic evolution of seawater composition, atmospheric CO_2 and climate. *Geochim. Cosmochim. Acta* **65:** 3005–3025.

Walsh, J.J. (1988). *On the Nature of Continental Shelves.* San Diego, Academic Press.

Weibel, E.R., and Hoppeler, H. (2005). Exercise-induced maximal metabolic rate scales with muscle aerobic capacity. *J Exp. Biol.* **208:** 1635–1644.

Whitfield, M. (2001). Interactions between phytoplankton and trace metals in the ocean. *Adv. Mar. Biol.* **41:** 3–128.

Wilkinson, B.H., and Algeo, T.J. (1989). Sedimentary carbonate record of calcium-magnesium cycling. *Am. J. Sci.* **289:** 1158–1194.

Wilson, J.T. (1966). Did the Atlantic close and then re-open? *Nature* **211:** 676–681.

Wing, S.L., Sues, H.D., Potts, R., DiMichele, W.A., and Behrensmeyer, A.K. (1992). Evolutionary paleoecology. *Terrestrial Ecosystems Through Time: Evolutionary Paleoecology of Terrestrial Plants and Animals.* A.K. Behrensmeyer, J. Damuth, W.A. DiMichele, R. Potts, H.D. Sues, and S.L. Wing, eds. Chicago, University of Chicago Press, pp. 1–14.

Woodland, A.W. (1971). The Llanbedr (Mochras Farm) Borehole. Institute of Geological Sciences, Natural Environment Research Council, Great Britain, Report no. 71/81, p. 115.

Worsley, T.R., and Nance, R.D. (1989). Carbon redox and climate control through Earth history: a speculative reconstruction. *Palaeogeogr. Palaeoclimatol. Palaeoecol.* **75:** 259–282.

Worsley, T.R., Nance, R.D., and Moody, J.B. (1986). Tectonic cycles and the history of the Earth's biogeochemical and paleoceanographic record. *Paleoceanography* **1:** 233–263.

Zhang, Z. (1986). Clastic facies microfossils from the Chuanlingguo Formation near Jixian, North China. *J. Micropaleontol.* **5:** 9–16.

Index

Hopanoids, 154
Horizontal gene transfer, 21, 24
Hunting Formation, 118
Hyalodiscus, 214
Hyalolithus neolepis, 220, 271
Hyalosphenia papilio, 93
Hydrocarbon skeletons, 135, 137, 138f
Hydrogen, 2
  hypothesis, 64–66
    defence of, 66–68
Hydrogenases, 15
Hydrogenosomes, 56, 70
5′ hydroxymethyluracil, 192
Hydroxylated quartz surfaces, 219
Hydrozoa, 80
Hymenomonadaceae, 267, 269

I
Ichaetoceros, 215
Ichnoreticulina elegans, 297
Ichthyosauria, 175
Incertae sedis, 88, 97–98
Independent asexual reproduction, 258
Infrared (IR) photons, 29
Ingestors, 321–323
Intrinsic biotic, 275
IR photons. *See* Infrared photons
Iron, 7, 31, 38
  evolution of, 10
Iron-sulfur complexes, 14–15
ISB. *See* Isua Supracrustal Belt
IsiA-Pcb Chl-proteins, 39, 42–44
  phylogenetic tree of, 43f
Isochrysidaceae, 257, 270
Isochrysis galbana, 254
Isomorphic, 294
Isorenieratene, 24
Isotope record, 413–417, 424–425
Isotopic signatures, 414f
Isthmia, 327f
Isua Supracrustal Belt (ISB), 22, 23

J
Jakobids, 87
Jurassic, 134, 173, 400, 411, 415–416

K
K. *See* Potassium
Kerkis bispinosa, 224
Knoblike structures, 258
K/T boundary. *See* Cretaceous–Tertiary boundary

L
Ladinian, 171, 173
Larapintine Petroleum System, 145
Lateral gene transfer (LGT), 60, 80
Lauderia, 215
Lazarus, 169
Lead (Pb), 11
Lepidodinium viride, 93, 96, 124
Leptocylindrus, 214
LGT. *See* Lateral gene transfer
LHC. *See* Light-harvesting complex
LI818 clade, 47
Licmophora, 215
Life, energy requirements for, 8
Light, 353
  heterogeneity in distribution of, 357–358
  utilization trade-offs, 360
Light micrographs, 298
Light-harvesting antennas, 37–38, 45–47
  evolution of, 39–40
Light-harvesting complex (LHC), 39–40, 45–49
  members of, 47f
  prokaryotic ancestry of, 48–49
Lioloma, 237
Liposomes, 12
Lithodesmiales, 215
Lithodesmium, 237
Lithosphere, 407
Lobose amoebae, 80
Loculate areolae, 329
Lophotrochozoa, 80
Lotka-Volterra food chains, 337
Lower Cambrian Lükati Formation, 146
LSU, 125, 195, 262, 267

M
M. vibrans, 79
Macroevolution, 275–279
MAGic, 378, 382, 384, 393, 400
Magnesium (Mg), 14, 31, 379, 380f, 395, 396, 398
  biogeochemical cycles of, 382–384
Manganese, 11, 30
Margulis' theory, 56, 57, 68
Marine reptiles, 177
Marinoan glaciation, 152
Marsupiomonas, 299
Maximum likelihood (ML), 115, 117
  chronograms of, 119f
  unrooted, 116f
Mayorella viridis, 93
McArthur Basin, 151
MCMC chain, 117
Medlin, 221
Megalodontid bivalves, 173
Megalodontoidea, 173
Melosira, 214
Mesodinium rubrum, 93, 97
Mesomycetozoa, 79
Mesoproterozoic, 121, 365
Mesostigma viride, 290, 296, 302
Mesotrophic masses, 276
Mesozoic, 4, 148, 156, 173, 324, 394
  coccolithophores in, 420
  phytoplankton in, 418–419
Mesozoic-Paleocene (MP) mode, 279
Metals, 10–12
  transition, 11
Metazoa, 79–80
2-MeHI. *See* 2-methylhopane
2-methylhopane (2-MeHI), 145, 151
Methylotrophs, 142
Mg. *See* Magnesium
$MgSO_4$, 389
Microbial mats, 23
Microfossils, 23, 134. *See also* Fossils
  fossils and, 137
  Paleozoic primary production, 146–147
Miller experiment, 8–9

438
INDEX

Paulinella chromatophora, 94, 111, 114, 121, 127–128
Pavlovophyceae, 257
Pb. *See* Lead
PBS. *See* Phycobilisomes
Pedinomonas, 299
Pelagial, protists and, 315–316
Pelagic parasitoids, 319–321
Pelagophyceae, 84
Pelomyxa, 80
Pennates, 215–216, 217f, 218
PEP. *See* Phosphoenolpyruvate
Peridinin, 199, 200
  structure of, 195f
Peridinium foliaceum, 124
Perispira ovum, 97
Perizonium, 218
Permian, 166, 170, 272
Permian–Triassic boundary, 145
Permo-Carboniferous, 389, 390
Permo–Triassic (P/Tr) boundary, 228, 265
Phaedarea, 82
Phaeocystales, 264
Phaeocystis, 258, 319, 320, 321, 328
Phaeodactylum tricornutum, 48, 232, 239, 240
Phaeodarea, 82
Phaeophila dendroides, 297
Phaeothamniophyceae, 84
Phagotrophy, 67, 121, 200, 258
Phanerozoic Eon, 141f, 151, 166, 181, 343
  atmosphere in, 393–397
  carbon in, 140f
  change in, 377–378
  dinosteranes in, 140f
  isotope record, 413–417
  sink trends in, 387–392
Phosphate, 8
Phosphoenolpyruvate (PEP), 232
Phospho-ribose-adenosine, 17
Phosphorus, 2, 381
  biogeochemical cycles of, 382–384
  ocean mass, 391f
  relative sink strength, 388f
  subcycles, 383–384

Photoautotrophy, 121
  drifting, 316–318
  sessile, 316–318
Photochemistry, 7
  oceanic, 8–10
  ions in, 10t
  prebiotic, 8–10
Photosynthesis, 1, 2, 8
  advent of, 21
  core structure of, 25–28
  in dinoflagellates, 193
  early evidence for, 22–25
  in eukaryotes, 78f, 98–99
    oxygenic, 89–98
  evolution of, 25f, 109–110
  on land, 3–4
  transition between types of, 29–33
Photosystem, chromist, 230–231
Photosystem I, 3, 27, 37, 42, 46
Photosystem II, 3, 26, 37, 42
  reaction centers of, 28–29
Phototrophs, 22–25
Phycobilins, 39
Phycobiliprotein family, evolutionary origin of, 41f
Phycobilisomes (PBS), 39f, 40–42, 49
  cyanobacterial, 41
  energy cascade, 40f
Phycocyanin, 40
Phycoerythrin, 40
Phycomata, 174
Phylogenetic analysis, of plastids, 123f
Phylogenetic relationships, 369
PhyML, 115, 123f
Phytobenthos, 317
Phytoliths, 229
Phytophthoran infestans, 84, 122
Phytoplankton, 176f, 304, 315, 333–334, 405–408
  armor, 327–328
  biomarkers, 143–146
  in carbon cycle, 408–412
  community structure, 417–419
  diversity curves, 409
  evolution, 408–410
  fossils, 142–143
  future communities, 366–367

in Late Triassic, 177–179
lineages, 227–229
lipid structures of, 136f
Mesozoic, 418–419
physiological trade-offs, 359–361
  competitive abilities, 360–361
    grazing resistance versus competitive ability, 361
    growth rate versus competitive ability, 361
    light utilization, 360
    nutrient utilization, 360
  resource competition in, 355–359
  rise of, 142–146
    modern, 146
  size of, 334–339
    food web structure and function, 336–339
    increase in, 340–341
    within lineages, 341
    macroevolutionary, 341–342
    scaling, 334–335
    size-abundance relationship, 335
    size-diversity relationship, 335–336
  vertical heterogeneity in, 358–359
Pi homeostasis, 111
Pigment, 30–31, 136–137, 198, 289
Pinguiochrysidales, 84
Placental mammals, 422–424
Placodontia, 175
Plankton, 174–179
  assistance from, 182
  control, 181
  cryptic diversity in, 237–239
  dinoflagellates in, 200–202
  success of diatoms in, 227–237
Plantae monophyly, 114–121, 288
  molecular clock analyses, 120–121
Plasmalemma, 211
Plasmodium, 122
Plastid endosymbiosis, 109–114
  secondary, 121–124
  tertiary, 124–127

.

Printed and bound by CPI Group (UK) Ltd, Croydon, CR0 4YY

03/10/2024

01040309-0019